Bioactive Polysaccharides

Bioactive Polysaccharides

Shaoping Nie
Nanchang University, Jiangxi, China

Steve W. Cui
Agriculture and Agri-Food Canada, Ontario, Canada

Mingyong Xie
Nanchang University, Jiangxi, China

ELSEVIER

ACADEMIC PRESS
An imprint of Elsevier

Academic Press is an imprint of Elsevier
125 London Wall, London EC2Y 5AS, United Kingdom
525 B Street, Suite 1800, San Diego, CA 92101-4495, United States
50 Hampshire Street, 5th Floor, Cambridge, MA 02139, United States
The Boulevard, Langford Lane, Kidlington, Oxford OX5 1GB, United Kingdom

British Library Cataloguing-in-Publication Data
A catalogue record for this book is available from the British Library

Library of Congress Cataloging-in-Publication Data
A catalog record for this book is available from the Library of Congress

ISBN: 978-0-12-809418-1

For Information on all Academic Press publications
visit our website at https://www.elsevier.com/books-and-journals

Publisher: Andre G. Wolff
Acquisition Editor: Nina Bandeira
Editorial Project Manager: Karen R. Miller
Production Project Manager: Punithavathy Govindaradjane
Cover Designer: Alan Studholme

Typeset by MPS Limited, Chennai, India

Contents

Preface

Bioactive Polysaccharides is intended to be a comprehensive reference book in the area of bioactive polysaccharides. After the exploration of protein and nucleic acid, research into polysaccharides could be considered as the most important cutting-edge topic to explore the profound mysteries of life. Polysaccharides are polymeric carbohydrate macromolecules consisting of long chains of sugar units bound together by glycosidic linkages and/or combined with different branch chains. They are rich in natural resources, with a complex structure and various functional activities. Bioactive polysaccharides are those polysaccharides which possess significant bioactivities like immunomodulatory effects, antitumor, antidiabetic, antioxidative effects, and beneficial effects on intestinal health, etc. To understand the bioactive polysaccharides, a systematic methodology has been established to provide a new insight into the structure – activity relationship of polysaccharides. To sum up, the bioactivities and structure – function relationship of the polysaccharides could provide much information into the applications of polysaccharides.

This book is designed to cover the most recent advances in polysaccharide research, with a focus on bioactivities intended for pharmaceutical, medicinal, and nutraceutical fields, as a research reference book for scientists and graduate students in chemistry, biology, pharmaceutical and food sciences. The aim of this book is not only to provide the basic knowledge and methodologies for studying bioactive polysaccharides, but also to emphasize understanding of the selected bioactive polysaccharides and provide detailed examples to show the knowledge and techniques in the preparation process, elucidation of structural features, chemical modifications, solubility, molecular weight and molecular weight distribution, conformational properties, bioactivities, and applications. There are eleven chapters in the book covering overviews on the structure of polysaccharides, structural diversity, and bioactivities (see Chapter 1: Introduction); methodologies for studying bioactive polysaccharides (see Chapter 2: Methodologies for Studying Bioactive Polysaccharides); beta-Glucans and their derivatives (from fungi and bioprocesses) (see Chapter 3: Beta-Glucans and Their Derivatives); *Cordyceps* polysaccharides (see Chapter 4: *Cordyceps* Polysaccharides), Complex glucomannan from *Ganoderma atrum* (see Chapter 5: Complex Glucomannan From *Ganoderma atrum*); Dendronan from *Dendrobium officinale* and aloe (see Chapter 6: Glucomannans From *Dendrobium officinale* and Aloe); Tea polysaccharides (see Chapter 7: Tea Polysaccharide); *Psyllium* polysaccharides (see Chapter 8: Psyllium Polysaccharide); Cereal beta-glucan (see Chapter 9: Cereal Beta-Glucan); other herbal polysaccharides (see Chapter 10: Other Herbal Polysaccharides); and practical applications of bioactive polysaccharides (see Chapter 11: Practical Applications of Bioactive Polysaccharides).

We hope this comprehensive book can serve well as an uncomplicated introduction to different disciplines of bioactive polysaccharides for researchers and students. Consulting the literature cited in the chapters is suggested for readers who need further information. Corrections, improvements, and suggestions from readers for all the chapters are welcome and appreciated.

We would like to acknowledge all our students' contributions to all work that is cited in this book. Also, our sincere thanks go to our students: Jielun Hu, Xiaojun Huang, Qiang Yu, Junqiao Wang, and Leming Jiang for collecting and organizing the references and preparing some figures and tables. We deeply appreciate the assistance from Elsevier in publishing this book.

Shaoping Nie, Steve W. Cui, and Mingyong Xie

Chapter 1

Introduction

Polysaccharides are widely found in plants, animals, microorganisms, and algae. Besides proteins and polynucleotides, they are also major and essential biomacromolecules in living organisms, and have important biological functions in the life activities of signaling pathways, cell–cell connections, and recognition of molecules in the immune system of the host. Applications of polysaccharides permeate nearly all facets of our lives, such as health care, food industry, and materials science. Not only have those natural polysaccharides been proved to have certain beneficial functions to human health, such as in anticancer, immuneenhancement, antiinflammatory, and hypoglycemic effects (Chang, 2002; Jiang et al., 2010; Li et al., 2011), their derivatives considered to be semisynthetic polysaccharides are also investigated (Ma et al., 2012; Tao et al., 2009; Wang et al., 2009a,b). Nowadays, systematic methodology has been established to rapidly analyze the structure of polysaccharide (Yang and Zhang, 2009), and has provided a new insight to explore the interaction between polysaccharide and host (Janeway and Medzhitov, 2002; Leung et al., 2006). It has become possible to conceive of strategies to understand the structure – activity relationship (SAR) of polysaccharides.

1.1 DEFINITION OF POLYSACCHARIDES AND THEIR RESEARCH HISTORY

Polysaccharides belong to one of the important biomacromolecules, which are rich in natural resources with a complex structure and various functional activities. Polysaccharides are polymeric carbohydrate macromolecules consisting of long chains of sugar units bound together by glycosidic linkages and/or combined with different branch chains. They vary in chemical structure from linear to branched polymers. Polysaccharides can be categorized as homopolysaccharides and heteropolysaccharides, depending on their monosaccharide building units, the polysaccharide is called a homopolysaccharide or homoglycan when all the monosaccharides are the same type, such as cellulose; but they are called heteropolysaccharides or heteroglycans when more than one type of monosaccharide is present, such as pectin, etc. Furthermore, many polysaccharides exist in nature by combining with

Bioactive Polysaccharides. DOI: http://dx.doi.org/10.1016/B978-0-12-809418-1.00001-0

protein or peptide or lipid, we can call them complex carbohydrates or glycoconjugates (Lee, 2012; Belitz et al., 2004; Nie, 2006).

Due to the complexity of their structure and the limitations of experimental methods, research into polysaccharides has always lagged behind that on protein and nucleic acid. As early as about 100 years ago, the famous Germany scientist, Dr. Fischer, opened the research area into polysaccharides. Around the 1950s–60s, the main research into polysaccharides was focused on their chemical composition and primary structures. Until the 1970s, the combination of two traditional research areas, carbohydrate chemistry and biochemistry, made it possible for researchers to explore the potential influences of polysaccharides on cell and molecular biology, and the research into polysaccharides was revived. In 1988, a professor at Oxford University, Dr. Dwek, first mentioned the concept of glycobiology, which opened a new research area which was based on the knowledge of carbohydrate chemistry, immunology, and molecular biology, using carbohydrate chains in the biomacromolecule or oligo-/polysaccharide itself as the research object, exploring the functional effect of polysaccharides or carbohydrate chains (as bioinformational factors) on multi cell and high level life entity (Rademacher et al., 1988). People came to realize that the polysaccharides were not only a kind of supporting matrix and energy source in life, but they also play an important role in several life phenomena and physiological processes, such as biosynthesis reactions in cells. In a manner of speaking, each important biogenic activity requires the presence of carbohydrate chains (Kiyohara et al., 2008). After the explorations into proteins and nucleic acids, research into polysaccharides has been another important cutting-edge topic to explore the profound mystery of life.

The history of polysaccharides as immunemodulators goes back to the middle of the last century. Though bacterial filtrate exhibited an effect on inducing hemorrhage and necrosis of mouse carcinoma in tumor-transplant mice, it was not until 1943 that polysaccharide was proved to be the potent antitumor ingredient, when a polysaccharide-rich fraction obtained from *Serratia marcescens* showed such an antitumor effect (Creech et al., 1949; Jones, 1979; Shear et al., 1943). In 1958, Brandner found zymosan could promote tumor loss in Sarcoma 180-bearing mice, and the mechanism might be explained via a host defense reaction (Bradner et al., 1958). Since then, a great deal of effort has been devoted to the development and utilization of polysaccharide.

Mushrooms are considered as macrofungi, and their polysaccharides were investigated first. They have a long history of medicinal applications, as well as a known nutritional value. Lentinan is a glucan isolated from *Lentinus edodes*, also known as *Cortinellus edodes* (Berk.) Sing, which is a common edible mushroom. In 1969, two fractions of lentinan were demonstrated that could obviously prevent tumor growth in Sarcoma 180-bearing mice (Bradner et al., 1958; Chihara et al., 1970). Further, the structural characterization showed lentinan was composed of mostly β-(1→3)-D-glucose linkages in the backbone and side chain of both β-(1→3)-glucose and β-(1→6)-glucose residues (Sasaki and Takasuka, 1976).

Zákány et al. (1980) investigated the influence of lentinan in the regression and retardation of transplanted tumors, and the results showed it dramatically inhibited the growth of tumors in a syngeneic system. A similar phenomenon was observed by Suga et al. (1984). The mechanism of this antitumor effect was more likely due to its activity on immune response (Dennert and Tucker, 1973). Ng and Yap (2002) extracted the lymphocytes from AKR mice prefed with lentinan and inoculated them into colon-carcinoma bearing nude mice, and found that tumors formed in the mice inoculated with lymphocytes were much smaller compared to nude mice without lymphocyte inoculation. Flow cytometric analysis of peripheral blood lymphocytes from cancer patients was performed, and it was found that patients after lentinan administration glycn had a higher ratio of $CD11^{-}$ $CD8^{+}$ cells/$CD11^{+}$ $CD8^{+}$ cells, meanwhile, lower levels of IL-6, GM-CSF, and PGE2 were observed (Matsuoka et al., 1997). Particularly, in the phase III study of lentinan, patients with recurrent or advanced gastrointestinal cancer were treated with lentinan in combination with chemotherapeutic agents, and were found to have a longer lifespan and higher survival rates, which is significant compared with the nonlentinan administration group (Taguchi et al., 1985). In addition, the same antitumor efficacy was observed in the treatment of stomach cancer (Taguchi, 1987). To date, lentinan was developed as an adjuvant to chemotherapy for cancer, and has been clinically used in China and Japan (Chen et al., 2013a,b,c). Besides lentinan, many bioactive polysaccharides are being investigating by different research groups worldwide.

1.2 OVERVIEW OF THE STRUCTURE OF POLYSACCHARIDE, AND STRUCTURAL DIVERSITY

Most biologically active polysaccharides are composed of glucose, fucose, mannose, galactose, arabinose, ribose, xylose, galacturonic acid, and glucuronic acid. According to the literature, fungal polysaccharides are mostly found to be β-glucans and hetero-β-glucans, heteroglycans, α-mannan, α-mannan-β-glucan complexes, glycoprotein or glycopeptides, and proteoglycan. Plant polysaccharides are found to be β-(1→3)-glucans, heteroglycans, sulfated polysaccharide, arabinans, acetylated glucomannans, arabinogalactan I and II, pectins, pectic acid, and rhamnogalacturonan I and II (Jiang et al., 2010). They are polymers comprised of monosaccharide linked through acetal bonds, i.e., glycosidic linkage (Fig. 1.1). The monosaccharide units generally occur in furanose or pyranose rings. Only taking the monosaccharide, ring size, location, linkage, and sterochemical variation of chiral carbon atoms for the oligosaccharide into consideration, there are theoretically about 84 billion

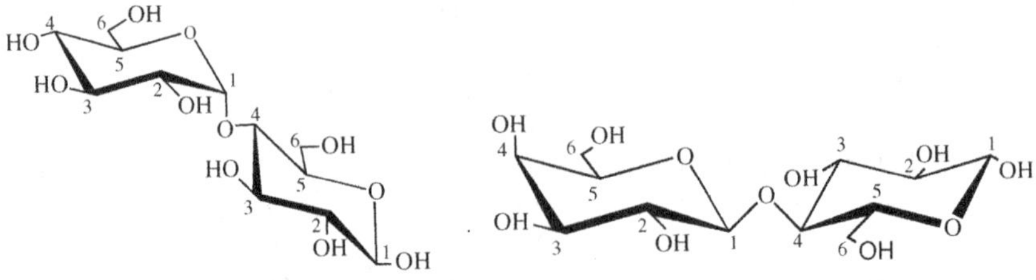

FIGURE 1.1 Schematic formation of a glycosidic bond.

possible tetrasaccharides; however, not all of these can be found in nature (Edgar, 2010).

Though it is hard to elucidate the structure of polysaccharides, a systematic methodology was established to achieve this (Yang and Zhang, 2009). The procedure covers the isolation and purification of polysaccharide, and determination of the molecular weight (Harding et al., 1991), together with FT-IR spectroscopy (Kačuráková et al., 1999; Synytsya et al., 2003), periodate oxidation, partial acid hydrolysis, methylation, Smith degradation (Harris et al., 1984; Kvernheim et al., 1987), and GC-MS-based methods (Ruiz-Matute et al., 2011) for the investigation of both monosaccharides and glycosidic linkages. Further, the anomeric configuration of each sugar residue, sequence of monosaccharides, and degree of branching could be evaluated by one-dimensional and two-dimensional NMR spectroscopy (Bubb, 2003). To date, the structures of hundreds polysaccharides have been characterized (some examples are listed in Table 1.1). The conformation of polysaccharides includes the conformation of each monosaccharide, orientation of monosaccharide, and flexibility of the spatial structure. Corresponding to its structurally diverse properties, polysaccharides show a wide spectrum of bioactivities, such as immuneregulation (Lee et al., 2004), antitumor (Zong et al., 2012), antiinflammatory (Liao and Lin, 2013; Pereira et al., 2012), and antiviral activity (Jiao et al., 2012), and so on.

The diversity of polysaccharides is mainly due to biological diversity. Many natural factors would closely associate with polysaccharide biosynthesis, such as synthesis of parts or organs, development stage, and growing environment. Polysaccharides from spores of *Ganoderma lucidum* were shown to have a backbone of (1→3)-β-linked glucans (Bao et al., 2001), while the backbone of two heteroglycans, PL-1 and PL-4, isolated from fruit bodies of *G. lucidum*, were comprised of 1,3-, 1,4-, and 1,6-linked-β-glucopyranosyl residues, 1,6-linked β-D-mannopyranosyl residues (PL-4), and 1,4-linked-α-glucopyranosyl residues, 1,6-linked-β-galactopyranosyl residues (PL-1), respectively (Bao et al., 2002a,b). Wang and Tan (2012) found that polysaccharide content of 3 year-old *Polygonatum odoratum* was higher than

TABLE 1.1 Structure Features of Polysaccharides From Various Sources

Source	Mw (Da)	Monosaccharide Composition	Backbone	References
Pleurotus florida	180,000	Glucose	(1→3,6)-linked-β-D-Glcp	Maji et al. (2012)
			(1→6)-linked-β-D-Glcp	
Auricularia polytricha	120,000	Glucose	1,3-β-glucan	Song and Du (2012)
			1,4-α-glucan	
			1,3-α-glucan	
Cistanche deserticola Y. C. Ma	10,000	Glucose	(1→4)-α-D-glucan	Dong et al. (2007)
Cordyceps sinensis	–	Glucose, Mannose, Galactose	(1→4)-linked-α-D-Glcp	Nie et al. (2011)
Lentinus squarrosulus (Mont.) Singer	196, 000	Galactose, Fucose, Glucose	(1→4)-linked-α-D-Glcp	Bhunia et al. (2010)
			(1→6)-linked-β-D-Glcp	
			(1→6)-linked-β-D-Galp	
			(1→4,6)-linked-D-Glcp	
			(1→3,6)-linked-D-Glcp	
Ganoderma lucidum	83,000	Rhamnose, Glactose, Gluose	1,4-linked α-D-Glcp	Bao et al. (2002a,b)
			1,6-linked β-D-Galp	
Ganoderma lucidum	200, 000	Glucose, Mannose	1,3-, 1,4-, 1,6-linked β-D-Glp	Bao et al. (2002a,b)
			1,6-linked β-D-Manp	
Ophiopogon japonicus	35,200	Arabinose, Glucose, Galatose	1,6-linked Glc	Chen et al. (2011)
			1,4-linked Glc	
			1,4,6-linked Glc	

(Continued)

TABLE 1.1 (Continued)

Source	Mw (Da)	Monosaccharide Composition	Backbone	References
Dendrobium huoshanense	73,000	Glucose, Xylose, Galactose	1,6-linked-β-D-Glcp	Pan et al. (2013b)
			1,4-linked-β-D-Glcp 1,4,6-linked-β-D-Glcp	
Saussurea involucrate	170,000	Arabinose, Rhamnose, Xylose	(1→4)-linked-β-Ara	Wang et al. (2004)
			(1→5)-linked-β-Ara	
			(1→4)-linked-β-Gla	
			(1→6)-linked-β-Gla	
			(1→4)-linked-β-GlaA	
Poria cocos	12,000	Glucose, Rhamnose, Xylose, Mannose	(1→3)-β-D-glucan	Wang et al. (2004)
Angelica sinensis (*Oliv.*) Diels	5100	Glucose, Arabinose	1,4-α-D-Glc*p*	Cao et al. (2006)
Cordyceps militaris	42,000	Mannose, Galactose, Glucose	(1→6)-linked-β-D-Man*p*	Yu et al. (2007)
Plantago asiatica L.	1849,000	Rhamnose, Arabinose, Xylose, Mannose, Glucose, Galatose	1,4-linked-β-Xyl*p*	Yin et al. (2012a)
Plantago asiatica L.	–	Rhamnose, Arabinose, Xylose, Glucose, Galatose	1,4-linked-β-Xyl*p*	Yin et al. (2012b,c)
Astragalus	36,000	Glucose	(1→4)-α-D-glucan	Li et al. (2009)
Radix Astragali	1334,000	Rhamnose, Arabinose, Glucose, Galactose, Galactose acid	1,2,4-linked-Rha*p*	Yin et al. (2012c)
			1,4-linked-α-Glc*p*	
			1,4-linked-α-GalA*p*6Me	
			1,3,6-linked-β-Gal*p*	

that of 2 year-old *P. odoratum*. Yao et al. (2011) reported polysaccharide contents in *Lycii* species were ranked in descending order as: *L. yunnanense* >*L. barbarum* >*L. cylindricum* >*L. chinense* >*L. dasystemum* >*L. truncatum* >*L. chinense* var. potaninii > *L. barbarum* var. auranticarpum >*L. ruthenicum* >*L. dasystemum* var. rubricaulium. Wu (2012a) isolated two water-soluble acidic polysaccharides fractions from the fruiting bodies of *Cordyceps militaris*, and found these two polysaccharides were composed of monosaccharides of dramatically different ratios. Procedures of extraction and purification of polysaccharides are also critical factors related to its properties. Generally, hot-water, dilute acidic or alkali aqueous solutions, and aqueous NaCl solutions are used to extract polysaccharide. Novel polysaccharides from *Pleurotus ostreatus* mushrooms are extracted by hot-water (PH), cold-water (PC), and hot-aqueous NaOH (PB). The PC fraction was formed by (1→3)-α and (1→6)-α-linked galactopyranosyl residues, whereas the PH was an α-glucan composed of (1→4)-α-linkages, and PB was a β-glucan having (1→3) and (1→6) glycosidic bonds (Palacios et al., 2012).

Besides the natural factors, polysaccharide derivatives also contribute to the structural diversity, which can also be classified as a semisynthetic polysaccharide. Several chemical modifications have been applied to modulate the physicochemical or biological properties of polysaccharides, such as carboxymethylation (Jing et al., 2009a,b; Silva et al., 2004), sulfation (Wang et al., 2009a,b), acetylation (Ma et al., 2012) and phosphorylation (Liu et al., 2011a,b), and oxidation (Bae et al., 2011) modification. In addition, some semisynthetic polysaccharides have been developed into various drug delivery systems (Shah et al., 2011).

1.3 OVERVIEW ON BIOACTIVITIES

Polysaccharides are not only energy and/or structural resources, but also play important biological roles in many life activities. Many polysaccharide bioactivities are confirmed in vivo and in vitro, such as their role as antidiabetics, in anticancer and immunemodulation, as antiinflammatories, and in hypolipidemic and hypoglycemic activity.

1.3.1 Anticancer

Cancer, especially lung, stomach, colon, and breast cancer, is a main cause of death all over the world. Surgery, chemotherapy, and radiotherapy are usually the main treatment for cancer, but it is still hard to achieve a satisfactory clinical effect. Many polysaccharides have been found to be potential anticancer agents (Zong et al., 2012). These include, especially, lentinan from *L. edodes*, protein-bound polysaccharide (PSK) from cultured mycelium of *Coriolus versicolor*, schizophyllan from *Schizophyllum commune*, polysaccharide from *G. lucidum* (PSG), and *Poria cocos*, which have all been used as adjuvants in clinical applications. Table 1.2 presents the anticancer effects of different polysaccharides.

In in vitro studies, a series of human carcinoma cell lines were established to investigate the gastric carcinoma cell line anticancer activity of polysaccharides, such as the cervical carcinoma cell of Hela cells, the lung cancer cell line of A549 cells, the gastric carcinoma cell line of BGC-823 cells, and the MCF-7 breast carcinoma cell line. Besides, mouse carcinoma cell lines, such as the colon carcinoma CT26 cell line, hepatic carcinoma cell line of H22, and sarcoma 180 cells have also been used. Much in vitro research shows that polysaccharide could not only inhibit cancer cell growth by cell cycle arrest directly, but could also induce the apoptosis of cancer cells, which are a possible mechanism of its antitumor activity.

Polysaccharides from traditional medicines, such as *Angelica sinensis* (Jin et al., 2012), *Cyclocarya paliurus* (Batal.) Iljinskaja (Xie et al., 2013), *C. militaris* (Zhang et al., 2010), *Oldenlandia diffusa* (Yang et al., 2010), and *Glycyrrhiza* (Chen et al., 2013a,b,c) are reported to possibly inhibit the proliferation of Hela cells. In addition, such antiproliferation effects may also be associated with an apoptosis-inducing effect via the mitochondrial pathway, involving changing expressions of Bcl-2 family protein and activating caspase protein, increasing intracellular Ca^{2+}concentration, and modulating PI3K/AKT signaling transduction (Cao et al., 2010; Kim et al., 2010a,b,c; Zhu and Zhang, 2013).

Polysaccharides from *Tricholoma matsutake* (You et al., 2013), *Codonopsis pilosula* (Yang et al., 2013a,b), and *Lentinus polychrous* Lev. (Thetsrimuang et al., 2011) showed notable antiproliferation on A549 (one type of nonsmall cell lung cancer cells) in vitro. Zhu and Song (2006) found polysaccharide from Ginseng could inhibit the growth of A549 by G0/G1 phase cell cycle arrest. Lu et al. (2013) found A549 cells treated with a polysaccharide fraction from *Coixlachryma-jobi* L. formed wrinkles on the cell surface and showed thinner microvilli, increased protrusions, and increased synthesis of caspase-3 and caspase-9. Wu et al. (2012b) reported polysaccharide from *Armillaria mellea* could induce disruption of the cell cycle in the phase of G0/G1, and could also activate the caspase-3 and caspase-9 protein of A549 cells.

HepG2 is a human live carcinoma cell line. Many mushroom polysaccharides, such as *Astragalus* (Zhu and Song, 2006), *Phellinus linteus* (Wang et al., 2012a,b), *Sarcodon aspratus* (Yan et al., 2013a,b), *Agaricus bisporus* and *Lactarius rufus* (Pires et al., 2013), and *Phellinus baumii mycelia* (Qin et al., 2011) are found to have a significant antitumor effect against HepG2. In addition, Wang et al. (2013a) reported polysaccharide from *Grifola frondosa* could stimulate S-phase arrest in HepG2 and cause apoptosis through a notch 1/NFκB-mediated caspase pathway. Li et al. (2013a,b) found polysaccharide from *P. linteus* induced S-phase arrest in HepG2 cells by reducing

TABLE 1.2 Anticancer Effects of Polysaccharides

Source	Mw (Da)	Cell Line/Model	References
Portulaca oleracea L.	253,688	HeLa cells	Zhao et al. (2013a,b)
Amomum villosum	61,400	HepG2 cells	Zhang et al. (2013a)
Sargassum plagiophyllum C. Agardh	35,000	HepG2, A549 cells	Suresh et al. (2013)
Punica granatum	–	HepG2, HCT116, A375 cells	Joseph et al. (2013)
Coix lachryma-jobi L.	12,000	A549 cells	Lu et al. (2013)
Codonopsis pilosula	145,000	A549 cells	Yang et al. (2013b)
Armillaria mellea	460,000	A549 cells	Wu et al. (2012b)
Wolfberry fruit	–	A549, MCF-7, LoVo cells	He et al. (2012)
Flammulina velutipes	28,000 268,000	A549, BGC-823 cells	Yang et al. (2012)
Camellia sinensis	–	BGC-823 cells	Xu et al. (2012a)
Cyclina sinensis	–	BGC-823 cells	Jiang et al. (2011)
Lycium barbarum	–	QGY7703 cells	Zhang et al. (2005b)
Taxus yunnanensis	3,694, 000	MCF-7, K562 cells	Yan et al. (2013a,b)
Lentinus polychrous Lev.	–	MCF-7	Thetsrimuang et al. (2011)
Menispermum dauricum	91,000 58,000	SKOV3 cells.	Thetsrimuang et al. (2011)
Artemisia capillaris	58,000	CNE-2 cells	Feng et al. (2013)
Stachys floridana Schuttl. ex Benth	168,000	HT29 cells	Ma et al. (2013)
Grifola frondosa	28,000	SGC-7901 cells	Shi et al. (2007)
Grifola umbellate	250,000 50,000	H22 cells	Bi et al. (2013)

(Continued)

TABLE 1.2 (Continued)

Source	Mw (Da)	Cell Line/Model	References
Scolopendra subspinipes mutilans L. Koch	–	S180 and H22 tumor-bearing mice	Zhao et al. (2012a)
Artemisia argyi	5169	S180 tumor-bearing mice	Bao et al. (2013)
Antrodia camphorata	–	S180 tumor-bearing mice	Liu et al. (2004)

the expression of calreticulin and inducing the pathway of P27kip1-cycil A/D1/E-CDK2. Ji et al. (2011) reported *Capparis spionosa* L polysaccharide induced HepG2 apoptosis by controlling Bax/Bcl-2 in the Ca^{2+} pathway. Some polysaccharides may show anticancer effects on a wide spectrum of carcinoma cells. For example, polysaccharide from *C. militaris* could also inhibit HepG2 cells' growth (Zhang et al., 2010), and polysaccharide from brown seaweed *Sargassum pallidum* significantly inhibits the growth of A549 cells, HepG2 cells, and MGC-803 cells (a human gastric cancer cell line) (Ye et al., 2008). Xin et al. (2012) reported polysaccharides from *Polygala tenuifolia* could stimulate high cytotoxicity in a variety of human cancer cells in the order of A549 >MCF-7 >HT-29 >HepG-2 >SK-N-AS cells.

Besides research in vitro, tumor transplanted mice are commonly used as models in vivo. Polysaccharide from *Salvia miltiorrhiza* (SMP-W1) showed a significantly toxic influence in hepatocellular carcinoma H22 cells, both in vivo and in vitro. Furthermore, the body weight, thymus index, and spleen index in tumor-bearing mice were greatly increased by treatment of SMP-W1 (Liu et al., 2013a,b). Interestingly, there are some polysaccharides that have certain antitumor effects in vivo, but not in vitro. Wang et al. (2013c) reported polysaccharide from *Mortierella hepiali* obviously inhibited the growth of implanted H22 tumor in vivo of mice, but not on H22, A549, and HepG2 cells in vitro. Zhao et al. (2014b) found polysaccharide from *Schisandra chinensis* (Turcz.) Baill (SCPP11) cannot prevent HepG2 cells' proliferation directly in vitro, but can notably reduce the tumor growth in Heps-tumor-bearing mice. Further study shown the immune organ indexes of Heps-tumor-bearing mice were significantly increased after administration of SCPP11. Meanwhile, SCPP11 could remarkably increase the activity of phagocytosis and NO production of RAW 264.7 cells. Similarly, polysaccharide from *Ganoderma atrum* could inhibit the growth of tumor in CT26 tumor-bearing mice, but could not directly kill CT26 cells in vitro. However, peritoneal macrophage treated with PSG showed an enhanced phagocytosis and higher cytolytic activity against CT26 cells (Zhang et al., 2013b). These results demonstrate that immuneenhancement is another important pathway through which the antitumor activity of polysaccharides is carried out.

Furthermore, some phenomena indicate there are still other mechanisms involved in antitumor activity. For example, Gao et al. (2011) found polysaccharide from *A. sinensis* only had a weak antiproliferation effect on A549 at high concentrations, but notably down-regulated the cell adhesion ability in a time- and dose-dependent manner, and reduced its invasive and migratory abilities.

1.3.2 Immunoregulation

The immune system is in a complex homeostasis, in which recognition of self is an important feature. Thus, the immune response can be considered as a heuristic process in which foreign antigens, in the context of particular self-antigens, perturbs equilibrium, producing a regulatory shift leading finally to positive responses (immunity) or negative ones (tolerance) (Green et al., 1983). Immune response is maintained by lymphocyte pools that are kept at fixed levels by homeostatic processes. The mechanisms that could affect these thresholds are unclear and various, and they are termed immunoregulation (Crispin et al., 2004).

Thought of as a nondigestible diet fiber, polysaccharide can be helpful in shaping the gut microbiota, which is closely related to host health (Marzorati et al., 2010; Maslowski and Mackay, 2011), and many studies showed that some polysaccharides could affect the immune system, which is pivotally responsible for immuneosurveillance (Table 1.3). It is also believed that the anticancer activity of polysaccharide is partly associated with its immunological enhancement effect.

Immune organs are composed of central immune organs (thymus, bonemarrow, and bursa fabricius) and peripheral immune organs (lymph node, spleen, and mucosa associated lymphoid tissue). The immune organ weight, spleen index, and thymus index are established to investigate the status of immune organs in vivo. Immunosuppression is always observed in both cancer patients and tumor-bearing animals. So far, many polysaccharides from traditional medicine, such as *Astragalus membranaceus* (Yang et al., 2013a,b), *Lycium barbarum* (Nan et al., 2012), *S. miltiorrhiza* (Liu et al., 2013a,b), and *C. militaris* (Mi et al., 2012) are found to have a notable effect on enhancing such indexes.

Immune cells like macrophages, dendritic cells (DCs), T lymphocytes and B lymphocytes, and monocytes are important in the immune system, they enable the host immune system to recognize and take up antigen, and then generate a series of immune responses (Banchereau and Steinman,

TABLE 1.3 Immunological Enhancement Activities of Polysaccharides

Source	Organ Weight	Cell Type	Effects on Immune Cells	References
Porphyra vietnamensis	Body weight↑, thymus weight↑	–	–	Bhatia et al. (2013)
Aconitum coreanum	Body weight↑, thymus index↑, spleen index↑, peripheral white blood cells↑	–	–	Li et al. (2013a)
Dendrobium nobile Lindl.	Thymus index↑	–	–	Luo and Fan (2011)
Potentilla anserina	Thymus Index↑, spleen index↑	PM	Phagocytosis↑	Chen et al. (2010a)
		SL	Proliferation↑	
Tremella aurantialba	–	SL	Proliferation↑	Du et al. (2009)
Schisandra chinensis (Turcz.) Baill	Thymus index↑, spleen index↑	PM	Phagocytosis↑	Chen et al. (2012c)
		SL	Proliferation ↑	
Scolopendra subspinipes mutilans L. Koch	Thymus index↑	SL	Proliferation ↑	Zhao et al. (2012a)
		NK	Activity↑	
		CT	Cytotoxicity↑	
Boschniakia rossica	–	PM	Phagocytosis↑, NO↑, TNF-α↑	Liu et al. (2011b)
Rhizoma Arisaematis	Body weight↑, spleen index↑	SL	Proliferation ↑	Chen et al. (2012a)
		CT	Cytotoxicity↑	
Combretum racemosum (P. Beauv.)	–	PMBC	IL-1β↑, IL-6↑, IL-10↑, TNF-α↑	Schepetkin et al. (2013)
Lentinus edodes	–	RAW 264.7	NO↑	Xu et al. (2012b)
Dendrobium officinale	–	RAW 264.7	Phagocytosis↑, NO↑, TNF-α↑, IL-1α↑, IL-6↑, IL-10↑	Meng et al. (2013)
Vernonia kotschyana	–	RAW 264.7	NO↑	Inngjerdingen and Meskini (2011)

(Continued)

TABLE 1.3 (Continued)

Source	Organ Weight	Cell Type	Effects on Immune Cells	References
Dendrobuim officinale	–	RAW264.7	Phagocytosis↑, NO↑, IL-2↑, IL-4↑, TNF-α↑, IL-1β↑	Xia et al. (2012)
		NK	Proliferation ↑	
		NK	Cytotoxicity↑	
Ganoderma lucidum	Thymus index↑, spleen index↑	NK	Cytotoxicity↑	Huang and Ning (2010)
Ganoderma lucidum	–	DC	Maturation↑	Meng et al. (2011b)
Rehmannia glutinose	–	DC	Maturation↑	Zhang et al. (2013c)
Polyporus umbellatus (Per) Fr	–	DC	Maturation↑	Li et al. (2010)
Cordycps militaris	–	DC	Maturation↑	Kim et al. (2010c)

PM, peritoneal macrophage; *SL*, spleen lymphocyte; *NK*, natural killer cell; *DC*, dendritic cell; *CT*, cytotoxic T lymphocyte; *PMBC*, peripheral blood mononuclear cell.

1998; Kumar et al., 2011). Macrophages initiate the innate immune response by recognizing, taking up, and then secreting inflammatory cytokines. However, activated macrophages are not only involved in both innate and adaptive immune response, but also are an important bridge between innate and adaptive immunity (Chen et al., 2010a,b). Xiao et al. (2011) found that LBPF4-OL, a polysaccharide from *L. barbarum* L., could stimulate spleen cell proliferation through macrophage activation. Phagocytosis, nitric oxide (NO), and reactive oxygen species (ROS) production and cytokine production of macrophages are key functions related to its activity. For example, *P. linteus* polysaccharide showed a macrophage-mediate antitumor effect through stimulating the phagocytosis of peritoneal macrophage and enhancing its production of NO and tumor necrosis factor-α (TNF-α) (Kim et al., 2004). Polysaccharide from *Ganoderma sinense* could stimulate the secretion of TNF-α, IL-12, and IL-1β of the murine macrophage RAW 264.7 cell line (He et al., 2015). Macrophage activation by polysaccharide is found to be primarily mediated by serious PRRs, i.e., Toll-like receptors (TLRs), complement receptors 3 (CR3), scavenger receptor (SR), mannose receptor (MR), and Dectin-1, and sometimes via an endocytosis-dependent pathway (Schepetkin and Quinn, 2006).

Lymphocytes are mainly mediators of adaptive immunity, which consists of cellular immunity and humoral immunity. Many polysaccharides are found that could stimulate the proliferation of lymphocytes and regulate their cytokine production, which is a crucial event in the activation cascade of adaptive immunity. Recently, DCs as professional antigen presenting cells aroused worldwide interest for their vital function in initiating the adaptive immune response. Though their number proportionally is relatively low, DCs located in most tissues are specialized in capturing and processing antigens, moreover, they could open tight junctions in epithelial cells of intestinal mucosa and send dendrites outside the epithelium to sample bacteria directly (Banchereau and Steinman, 1998; Schepetkin and Quinn, 2006). Generally, DCs remain in an immature state, however, once they take up a maturation signal from the microenvironment, they develop dramatically. During this maturation, DCs change from antigen take-up cells to antigen present cells, which have a profound influence on immunity. So far, it has been found that polysaccharides could regulate immunity by inducing DC maturation. For instance, *Rehmannia glutinosa* polysaccharide could enhance the ability of DCs to stimulate T lymphocyte proliferation and the ability for antigen presenting (Huang et al., 2013). Polysaccharide from *Angelica dahurica* not only stimulate the expression of the major histocompatibility complex II (MHC II) and CD 86, but also the synthesis of NO and IL-12 on DCs. Recently, TLRs have been found expressed on DCs, and triggering these would result in maturation development. As is well-known, lipopolysaccharide (LPS), a typical TLR-4 ligand, is a strong maturation inducer of DCs. Kim et al. (2010c) reported cordlan polysaccharide isolated from mushroom *C. militaris* could induce $TLR4^{+/+}$ DCs' maturation from C3H/HeN mice. Zhu and Zhang (2013) reported functional and phenotypic maturation of DC induced *L. barbarum* polysaccharide was notably down-regulated by pretreating DCs with antibody for anti-TLR2 or anti-TLR4.

1.3.3 Antidiabetics and Kidney Repair

As a metabolic disease, diabetes mellitus (DM) has become a worldwide concern, as it obviously influences humans' standards of life, and is related to several serious complications, such as nephropathy and cardiovascular diseases. According to the International Diabetes Federation (IDF) in 2013, over 382,000,000 individuals are living with DM, the number of DM cases was 13-fold that in 1980, and it will increase to 600,000,000 by 2030 (Kokil et al., 2015). For patients with DM, diabetic nephropathy and retinopathy can give rise to end-stage renal disease and blindness, while atherosclerosis always influences life expectancy (Rask-Madsen and King, 2013). In addition, enormous financial stress on overstretched health care budgets has been caused by DM in the developed and developing world, and the global expenditure was reached $548,000,000,000 in 2013 (Morales and Morris, 2014).

Kidney repair for the host is very important in diabetic nephropathy. Diabetic nephropathy is found in nearly 40% of patients who are diagnosed with type 2 diabetes, and it has become the main contributor to end-stage renal disease in the United States and Europe, and is related to 25%–42% of cases. The rate of development of diabetic nephropathy has been in the range of 5% and 10% of patients with diabetes (type 2) and microalbuminuria annually (Parving et al., 2001). The primary cause of patients' chronic kidney disease is diabetic nephropathy, which is connected with increased cardiovascular mortality (Gross et al., 2005).

Oral antidiabetic drugs and insulin are the main therapies for DM. However, insulin resistance and side-effects appeared after continuous use of these drugs. Therefore, seeking helpful, innoxious and inexpensive drugs for patients of DM are necessary. Non- noxious biological macromolecules, especially polysaccharides, exhibit remarkable efficacy on DM, and have been demonstrated in several previous studies. According to these inspiring findings, the demand for seeking antidiabetic polysaccharides for effective therapeutics for DM is necessary and important.

Herbal medicine is an abundant source for DM drug research because of its wide utilization in DM in Africa, Asia, and India. Because of the improvement in technologies and intensive research on DM pathological pathways, specific studies should be carried out to better understand the role of herbal medicines in DM. Polysaccharides, which were the leading active sections of many antidiabetic plants, have captured much attention from researchers (Tong et al., 2008). As an example, in China, *A. membranaceus*, called "Huang Qi," was widely utilized for diabetes as a crucial substrate of herb prescriptions in Chinese native medicine, while polysaccharide was observed to be the main bioactive component (Cheng et al., 2011). *Astragalus* polysaccharide (APS) is isolated from *A. membranaceus*, which is a water-soluble heteroglycan. There are two main structural components in APS, which are APS-I (Mw, 1.7×10^6 Da), and APS-II (Mw, 1.2×10^6 Da), the monosaccharide composition of APS-I is arabinose and glucose with a molar ratio of 1:3.45, while APS-II is composed of rhamnose, arabinose, and glucose with a molar ratio of 1: 6.25: 17.86 (Jian et al., 2013). Table 1.4 lists the antidiabetic effect of APS on levels of cells and animals.

In addition, mushrooms were regarded as an abundant source of various nutrients, including dietary fibers, vitamins, and minerals. Mushrooms were also considered as functional foods, due to their beneficial effects on human health. A lot of commercial pharmaceutical products resulting from this polysaccharide were showed proven results in clinical trials. For example, *G. lucidum*, known as "Ling-Zhi" in Chinese, and was considered as an agent for preserving health and providing therapy (Tie et al., 2012; Xiao

TABLE 1.4 The Antidiabetic Effects of *Astragalus* Polysaccharide

Model	Effects and Mechanisms	References
L6 myotubes	Stimulate glucose uptake through the AMP-AMPK-AS160 pathway	Zhao et al. (2012b)
C2C12 myotubes	Improve palmitate-induced insulin resistance by inhibiting PTP1B and NF-B	Zhou et al. (2011)
C57BL/6J Mice	Up-regulate the expression of galectin-1 in muscle of T1DM mice, resulting in the apoptosis of CD8 + T cells	Wei et al. (2010)
Hamsters	Reduce myocardial collagen (I and III) expression and lower cardiac MMP-2 activity, myocardial Ang II levels, myocardial chymase expression, and p-ERK1/2 kinase expression	Chen et al. (2012b)
C57BL/6J mice	Prevent the development of diabetic cardiomyopathy through cardiac PPAR-mediated regulatory pathway	Liu et al. (2009a)
KKAy mice	Restore insulin-induced protein kinase B Ser-473 phosphorylation and GLUT4 translocation in skeletal muscle	Liu et al. (2013b)
KKAy mice	Improve insulin sensitivity and decrease myostatin expression in skeletal muscle through down-regulating ROS-ERK-NF-B pathway	Cheng et al. (2011)
Hamsters	Up-regulated cholesterol-7a-hydroxylase and LDL-receptor gene expressions	Liu et al. (2014a)
3T3-L1 adipocytes	Increase adiponectin secretion and reduce IL-6 secretion	Ye et al. (2013)
KKAy mice	Down-regulate the ubiquitination levels of IRS-1 and the nuclear expression of ubiquity	Ma et al. (2015)

et al., 2012; Zhang et al., 2003; Zheng et al., 2012; Pan et al., 2015; Meng et al., 2011a,b). The antidiabetic influence of polysaccharides obtained from *G. lucidum* is listed in Table 1.5.

In addition, many therapeutic effects of *G. atrum* have been found (Yi et al., 2007) (Fig. 1.2). PSG-1 exhibited beneficial effects on insulin resistance and increased metabolism of serum lipid through decreasing Bax protein and enhancing expression of Bcl-2 protein in pancreatic tissues (Zhu et al., 2013a,b). In addition, PSG-1 showed potential protection of endothelial cells against diabetic arteriosclerosis by activating the pathway of PI3K/Akt/eNOS (Zhu et al., 2014).

The antidiabetic activities of polysaccharides from other plants and mushrooms are presented in Table 1.6.

TABLE 1.5 The Antidiabetic Effects of Polysaccharides From *Ganoderma lucidum*

	Effects and Mechanisms	References
Gl-PS	Protect against alloxan-induced pancreatic islets damage in vitro and in vivo due to scavenging free radicals and inhibition of NF-B activity	Zheng et al. (2012)
	Elevate wound angiogenesis by decreasing mitochondria oxidative stress and inhibiting activity of MnSOD and GPx activities, as well as decreasing redox enzyme p66Shc expression and phosphorylation	Zhang et al. (2003)
	Prevent apoptosis of pancreatic cells and enhance cell regeneration through up-regulating Bcl-2 and PDX-1, and down-regulating Bax, iNOS, and Casp-3 mRNA expression	Xiao et al. (2012)
GLPs	Decrease mRNA expression levels of several key enzymes (GP, FBPase, PEPCK, and G6Pase) involved in gluconeogenesis and/or glycogenolysis	Pan et al. (2015)
GLP-II	Attenuate myocardial collagen cross-linking in diabetic rats through decreasing AGE level and augmenting antioxidant enzymes activities	Pan et al. (2013a)
FYGL-*n*	Enhance insulin sensitivity and decrease hepatic glucose output along with increased levels of adipocyte and skeletal muscle glucose disposa in the late stage of diabetes	Yi et al. (2007)
	Inhibit the PTP1B activity on a competitive mechanism in vitro	Meng et al. (2011a)

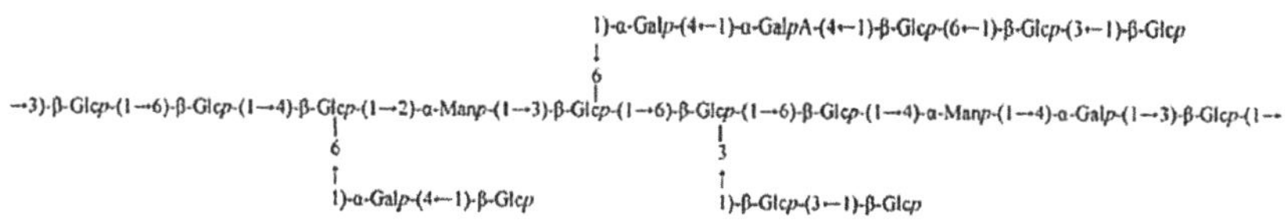

FIGURE 1.2 The chemical structure of PSG-1 from *Ganoderma atrum*.

The alarming statistics for DM suggest that it is necessary to consider it as a threat to the world population's health, because it is related to an enhanced emergence of complications, including leg amputation, myocardial infarction, retinopathy, and nephropathy. A multitude of interests have been focused on the development of antidiabetic polysaccharide drugs.

There has been great interest in producing polysaccharide drugs with antidiabetic effects over the past 10 years. Some in vivo study results are promising, but it is necessary that more in-depth research and clinic trials are

TABLE 1.6 The Antidiabetic Effects of Other Polysaccharides From Plants and Mushrooms

	Origin	Part	Effects	References
Plants VBTLP	Vaccinium bracteatum		BG, ↓; TC, ↓; LDL-C, ↓	Wang et al. (2013b)
POCS	Zea mays		BG, ↓; TC, ↓; TG, ↓	Zhao et al. (2012c)
ZSP	Zizyphus jujube		BG, ↓; insulin, ↓; TC, ↓; TG, ↓; HDL-C, ↑; LDL-C, ↓; VLDL-C, ↓; HOMA-IR, ↑; HOMA-β, ↑; AI, ↓	Zhao et al. (2014b)
ATP	Acacia tortilis		α-D-glucosidase activities, ↓; FBG, ↑; BG, ↓; TC, ↓; TG, ↓; HDL, ↑; LDL, ↓; VLDL, ↓	Bisht et al. (2013)
LLP	Lilium lancifolium		SOD, ↑; GPx, ↑; CAT, ↑; MDA, ↓; BG, ↓; structure integrity of pancreatic islet, ↑	Zhang et al. (2014)
PF40	Pseudostellaria heterophylla		BG, ↓; INS, ↓; CHOl, ↑; TG, ↓; HDL-C, ↑; LDL-C, ↓; TNF-α, ↓; IL-10, ↑; adiponectin, ↑; leptin, ↓	Hu et al. (2013)
SMPW1	Salvia miltiorrhiza		BG, ↓; FBG, ↑; FINS, ↑; SOD, ↑; TBARS, ↓;	Wei et al. (2012)
SGP-2	Sarcandra glabra		α-D-glucosidase activities, ↓; HOMA-IR, ↑; BG, ↓; insulin, ↓; TC, ↓; TG, ↓; HDL/LDL, ↑; FFA, ↓; T-AOC, ↑; SOD, ↑; CAT, ↑; GSH-Px, ↑; MDA, ↓; PK, ↑; HK, ↑	Liu et al. (2014b)
TTP	Talinum triangulare		FBG, ↑; TC, ↓; TG, ↓; HDL, ↑; HDL/TC, ↑; LDL, ↓; FINS, ↑; HOMA-IR, ↑	Xu et al. (2014)

(Continued)

TABLE 1.6 (Continued)

	Origin	Part	Effects	References
HPS3	Hedysarum polybotrys		FBG, ↓; TC, ↓; TG, ↓; HDL-C, ↑; LDL-C, ↓; BG, ↓; SOD, ↑; NOS, ↑; NO, ↑; MDA, ↓; T-AOC, ↑; INS, ↑; ISI, ↑; C-P, ↑; IL-6, ↓; leptin, ↓; TNF-, ↓; FFA, ↓; HG, ↑	Hu et al. (2010)
CPP	Portulaca oleracea		BW, ↓; BG, ↓; TC, ↓; TG, ↓; HDL-C, ↑; insulin, ↑	Gong et al. (2009)
Mushroom EPS	Agaricus bisporus		BW, ↓; BG, ↓	Zhang et al. (2009)
TMP	Tremella aurantialba	Mycelia	BW, ↓; TC, ↓; phospholipid, ↓; TG, ↓; SOD, ↑; GPx, ↑; CAT, ↑; GR, ↑; MDA, ↓	Huang et al. (2014)
	Pleurotus tuber-regium	Culture media	BW, ↓; BG, ↓; insulin, ↓; adiponectin, ↑; TC, ↓; TG, ↓; LDL, ↓; HDL, ↑; PPAR- _, ↑	Zhao et al. (2014a)
PLP	Phellinus linteus	Mycelial culture	BW, ↓; BG, ↓	Kim et al. (2010b)
PLP	Phellinus linteus	Mycelial culture	BW, ↓; BG, ↓; Insulin, ↓; IL-2, ↓; IFN- _, ↓; IL-4, ↑; TNF-α, ↓	Kim et al. (2005)
β-Glucan	Agaricus blazei	Fruiting body	BW, ↓; BG, ↓; TC, ↓; TG, ↓; HDL-C, ↑; insulin, ↑	Kim et al. (2005)
EPS	Laetiporus sulphureus	Mycelial culture	BW, ↓; BG, ↓; TC, ↓; TG, ↓; iNOS, ↓; SOD2, ↑; GPx, ↑; CAT, ↑; insulin, ↑	Hwang and Yun (2010)

carried out to verify the data from animal trials. However, there are many challenges to developing efficient methods for the isolation and modification of polysaccharides, and to illustrate the relationship of structure – activity. Thus, a positive area of study is yet to come. Some natural polysaccharides have been introduced to the market according to their hypoglycemic effects. The possible mechanisms of polysaccharides against DM have been summarized in six aspects: (1) plasma insulin enhancement, and pancreatic glucagon decrease; (2) improvement in insulin sensitivity and insulin resistance control; (3) α-glycosidase enzyme inhibition in the bowel, and down-regulation of absorption of carbohydrates; (4) hepatic glycogen improvement, and sugar dysplasia suppression; (5) up-regulated use of glucose for the peripheral tissue; (6) scavenging of free radicals and peroxidation of lipids.

Along with the antidiabetic effects of polysaccharides, diabetes-associated kidney tissue damage can also often be repaired by polysaccharide to some extent. For example, the polysaccharide from *Liriope spicata* (LSP) enhanced renal function, and attenuated the histo-pathological damage to kidney tissue (Xiao et al., 2013). It exhibited strong antioxidant activity and prevented kidney and liver damage by diabetes in vivo (Chen et al., 2013a,b,c).

1.3.4 Antioxidant

The word "antioxidant" has become more and more popular in modern society as it has obtained publicity through mass media (Clancey, 2010). A biologically relevant definition of antioxidants is "synthetic or natural substances preventing or delaying deterioration through oxygen in air when added to products." Antioxidants are usually enzymes or other organic substances that can counteract the injurious influence of oxidation in animal tissues in biochemistry and medicine.

Many polysaccharide extracts from plants and mushrooms have been reported to be excellent natural antioxidants. However, the antioxidant effect of pure polysaccharides was considered to be weak compared with free radical scavenging agents, such as Trolox or pyrrolidine dithiocarbamate. For naturally produced polysaccharides, only polyelectrolytes, e.g., phosphorylated or sulfated glycans and LPS, showed considerably higher scavenging activities (Tsiapali et al., 2001). Most reported antioxidant polysaccharides were crude polysaccharides or polysaccharide conjugates which contained protein, uronic acid, and/or other undefined compounds.

The antioxidant activities of polysaccharides are always controversial, due to its unclear scavenging mechanism on free radicals and ROS. The most probable pathway for the antioxidant activity of polysaccharide has been discussed as being via the hydroxyl groups, which could donate hydrogen to stabilize the free radicals (Gülçin, 2006; Yi et al., 2008). However, many pure polysaccharides, e.g.: starch, dextran, and isolated tea polysaccharide (TPS) fraction, were reported to possess very low scavenging effects (Chen et al., 2009a,b; Wang et al., 2013a,b,c,d; Xiao and Jiang, 2015). Most of the reported antioxidative polysaccharides were polysaccharide conjugates containing protein, peptide, and phenols, etc. Nie et al. (2008) reported that the TPS – protein conjugate antioxidant abilities depended on the content of protein. Chen et al. (2004) identified that uronic acids were the key factor for the antioxidant activities of tea polysaccharide.

It is very interesting that Zhang et al. (2016) found the active factors of polysaccharide from *G. atrum* (PSG) for antioxidant activity to be the phenolic compounds/proteins. Chen et al. (2013a,b,c) revealed that the antioxidant activities assayed by FRAP and ORAC showed a significant correlation with the phenolic content, which corresponded with another report (Siu et al., 2014). The correlation between DPPH scavenging ability and phenolic content was moderately significant. For the correlation between antioxidant activities and protein content, the linear regressions were less significant than for phenolic content. On the contrary, negative R^2 was obtained from the regression analysis between antioxidant activity and neutral sugar content in all three assays, and no linear correlation was found for uronic acid content. It could be concluded that the phenolic compounds and proteins bonded/cross-linked with PSG were the pivotal components responsible for its reported in vitro antioxidant activities. It has been reported that the phenolic compounds from mushrooms possess excellent antioxidant activities, with a strong dose – response correlation (Cheung et al., 2003; Ferreira et al., 2009). The phenolic acids, such as coromaric acid, and phenolic derivatives, such as rutin, which are commonly found in mushroom extracts, are important phenolic compounds which exert antioxidant activities. The hydrogens existing in the phenolics were considered to provide H-atom or electron donation to terminate the oxidative reactions (Jovanovic et al., 1999). On the other hand, proteins attaching with polysaccharides could strengthen the negative charge of molecules, and provide a greater chance to stabilize the radicals and stop the initiation and propagation of further oxidative reactions. The typical amino acids responsible for the significant antioxidant activities of peptides and/or proteins include nucleophilic sulfur-containing aromatic amino acid, amino acids, and imidazole-containing amino acid (Davies and Dean, 1997). These amino acids have been found in PSG (Yi et al., 2008), and might attribute to its strong antioxidant activities. This provides more information for us to understand the mechanism of antioxidant activity exerted by polysaccharides, and to reassess the antioxidant activities claimed for polysaccharides from natural resources.

1.3.5 Antiinflammatory

Inflammation is a process where the human body tries to counteract potential injurious agents, including invading viruses, bacteria, and other pathogens

(Henderson et al., 1996; Hersh et al., 1998; Ulevitch and Tobias, 1995). It is a complicated physiological response and is involved in the pathogenesis of various diseases, such as atherosclerosis, cancer, and coronary artery disease (Coussens and Werb, 2002; García, 2005; Ridker, 2002). Macrophages are important in inflammation's initiation, maintenance, and resolution, and they have three major functions in inflammation: immunomodulation. phagocytosis, and antigen presentation (Fujiwara and Kobayashi, 2005). The antiinflammatory activity of polysaccharides is usually investigated using an LPS-induced macrophage model. A polysaccharide from *Bupleurum smithii var. parvifolium* could inhibit LPS-induced production of NO and IL-6, IL-1β, and TNF-α (Cheng et al., 2010). A sulfated polysaccharide extracted from *Sargassum hemiphyllum* not only inhibited the LPS-induced secretion of inflammatory cytokines, but also reduced mRNA expressions of iNOS, IL-β, and COX-2 in a dose-dependent manner (Hwang et al., 2011). Some special animal models are also used. Wu et al. (2010) reported polysaccharide isolated from Golden needle mushroom could significantly decrease CD4+ and CD8+ cell levels and antiintercellular adhesion molecule-1 and myeloperoxidase levels in the serum of burned rats, while it enhanced antiinflammation cytokine IL-10 production. Paiva et al. (2011) found heterofucan from the *Lobophora variegate* not only decreased serum TNF-α, but also decreased cells infiltrated in the synovial membrane in rats with zymosan-induced arthritis, which indicated an antiinflammatory activity in vivo.

Although inflammation plays an important role in humans, inflammation also exhibited effects on the host through various aspects of pharmacological, biochemical, and molecular controls (Boraschi et al., 1998; Dinarello, 2000; Nicod, 1993; Rouveix, 1997; Turcanu and Williams, 2001). It could lead to inflammation, tissue destruction, fever, and death after proinflammatory mediators or cytokines like NO, TNF-α, IL-6, and PGE2 were administered to humans (Dinarello, 2000). One of these mediators is NO, which exerts effects on the inflammatory response, and is formed by nitric oxide synthases (NOSs) or iNOS in the cells of macrophages and other cells. Large quantities of NO could induce a multitude of enzymes and proteins which are essential for reactions of inflammation (Nijkamp and Parnham, 2011; Aga et al., 2004; Defranco et al., 1995; Mestre et al., 2001; Chan and Riches, 2001). In addition, expression of iNOS and production of NO are mediated by LPS, and these were reported to induce septic shock (Jacobs and Ignarro, 2002). Prostaglandins are found to control the aggregation of platelet and formation of thrombus in the inflammatory process. One of the prostaglandins is E2, which results from the pathway of cyclooxygenase. The sepsis can be ameliorated by the preventative of activity of COX-2 (Knöferl et al., 2001). These cytokines are as important as targets for the treatment of inflammatory disease, and it is crucial to properly understand the inflammatory basis for a complete understanding of cancer, atherosclerosis, and other diseases. NO is related to the modulation of bioactivities in immune, neural, and vascular systems (Kilbourn et al., 2000).

A multitude of studies have shown that polysaccharides from plants are considered as energy sources and important bioactive substances in numerous life processes. This has been largely studied to analyze the structure and bioactive mechanisms of polysaccharides on diseases. The various curative influences of increasing natural polysaccharides have been investigated, and even used in clinical therapies (Wang and Fang, 2005).

1.3.6 Other

Reactive nitrogen species and ROS are free radicals generated during the metabolism of cells, while antioxidative defenses, such as superoxide dismutase and glutathione peroxidase, serve as scavengers of ROS and RNS. Under physiological conditions, the balance between free radical generation and the defense system is maintained, which is involved in the process of cell development and growth. Once this balance is broken, either by overproduction of ROS, or by deficits in the antioxidative system, the surplus free radicals would lead to cellular damage, which is the cause of many diseases (Fang et al., 2009; Klaus et al., 2011; Sun et al., 2010; Wang et al., 2012a, b). In addition, many polysaccharides can up-regulate the activities of SOD antioxidant enzyme, catalase, and the activities of nonenzymic antioxidant (vitamin C, vitamin E, and reduced glutathione) in vivo, which indicates notable antioxidant activity. *L. barbarum* polysaccharides enhanced the SOD activity in the liver, kidney, and blood of streptozotocin (STZ)-induced diabetic rats (Li, 2007). *G. lucidum* polysaccharide significantly enhanced the activities of antioxidant enzymes in rats with cervical cancer (Chen et al., 2009a,b). *G. lucidum* polysaccharides could also restore the decreased nonenzymic antioxidant levels in STZ-induced diabetic rats (Jia et al., 2009).

Polysaccharides also show hypolipidemic and hypoglycemic effects. For instance, Mao et al. (2009) found treatment of polysaccharide from *A. membranaceus* down-regulated body weight, postprandial hyperglycemia, and triglycerides in insulin resistant mice induced by a high-fat diet. Yu et al. (2013) reported oral administration of *Rosae Laevigatae Fructus* polysaccharide could significantly decrease levels of serum total cholesterol, low-density lipoprotein cholesterol, and triglycerides in rats with hyperlipidemia. These indicate a potential use for polysaccharide in the therapy for diabetes and it complications.

1.4 THE STRUCTURE – BIOACTIVITY RELATIONSHIP OF POLYSACCHARIDES

Compounds that could interact with the immune system to improve or inhibit a host response's specific aspects are considered as biologic response modifiers (BRM). So far, many polysaccharides, including natural or synthetic polysaccharides, have been shown to act as potent BRM (Leung et al., 2006). Their up-regulating or down-regulating effects on immune responses depend on many factors, such as route, dose, and administration timing. Polysaccharides have at least two factors that influence their function, one is their mechanical characteristics, and the other is the multiple regulatory processes or interactions between cells, molecules, or extracellular space (Jedrzejas, 2000).

1.4.1 Viscosity

Viscosity is the internal friction of a fluid or its trend to resist flow (Bourne, 2002). For liquids, it is the informal concept of "thickness"; for instance, honey has a higher viscosity than water (Bourne, 2002).

Viscosity is a very important rheological property of polysaccharides. Some viscous dietary fiber, especially, can be used to restructure foods. Some human studies suggested that the inclusion of guar gum (Fairchild et al., 1996; Sierra et al., 2001; Torsdottir et al., 1989), alginates (Ikegami et al., 1990; Torsdottir et al., 1991), and β-glucan (Behall et al., 2006; Jenkins et al., 2002; Yokoyama et al., 1997) into meals can attenuate postprandial insulin responses and glycemia. This special property attracts much attention in the treatment of DM (Aro et al., 1981; Czubayko, 2000; Jenkins et al., 1977, 1978). So far, it has been demonstrated that dietary fiber from various sources may exhibit different behaviors on physiological responses associated with viscosity in the gastrointestinal tract, and viscous dietary fibers are most likely to be therapeutically useful in modifying postprandial hyperglycemia (Dikeman et al., 2006; Edwards et al., 1987; Gallaher et al., 1993; Jenkins et al., 1978; Vuksan et al., 2009). Besides, a fiber preload with high viscosity may result in moderating a decrease in subsequent food intake (Kristensen and Jensen, 2011; Vuksan et al., 2009).

Guar gum obtained from *Cyamopsis tetragonolobus* or *Cyamopsis psoraloides* is an outstanding plant gum (Chudzikowski, 1971). Guar gum ingestion will up-regulate the viscosity of the contents in the stomach and small intestine, which would impact glucose absorption (Blackburn and Johnson, 1981; Cherbut et al., 1990). Like other gums, guar gum's viscosity is related to concentration, ionic strength, time, pH, and agitation type (Mudgil et al., 2015). However, intrinsic viscosity variation with molecular weight had the relationship: $[\eta] = 3.8 \times 10^{-4}\,Mw^{0.723}$ (Robinson et al., 1982).

Alginate is an anionic polysaccharide primarily existing in alga's intercellular space o. The alginate polymer contains three block types: homopolymeric blocks for mannuronic acid (MM) and for guluronic acid (GG), and blocks with a changing sequence (MG). In particular, blocks of GG are bound most strongly to ions of calcium in the formation of gel, and further,

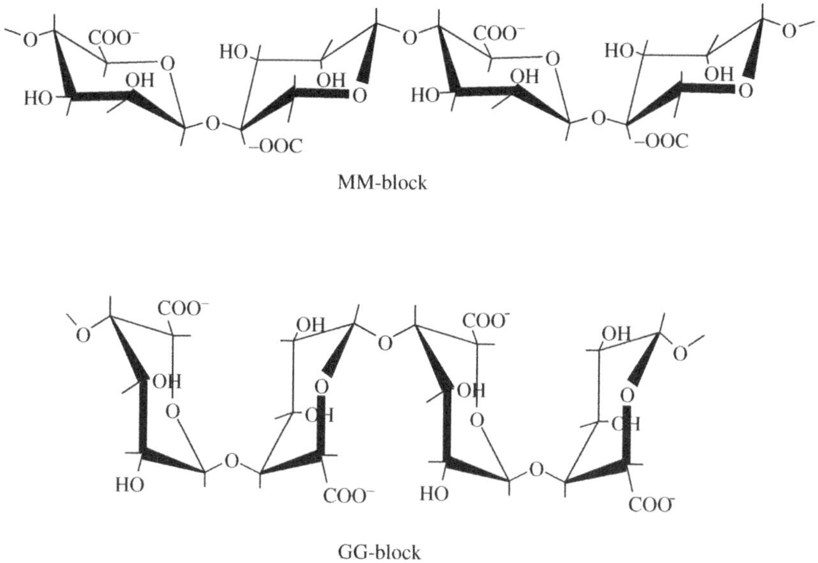

FIGURE 1.3 Schematics of M-block and G-block of alginate.

an egg-box model has been proposed (Morris, 1986). The relationship between the intrinsic viscosity and molecular weight of alginate samples can be: $[\eta] = 2.0 \times 10^{-5}$ Mw (Smidsrød et al., 1968). However, the intrinsic viscosity is affected not only by the molecular weight, but also by the flexibility of the polymer chains of alginate, partly because chains containing predominantly GG blocks are less flexible than those containing predominantly MM blocks (Smidsrød et al., 1973) (Fig. 1.3).

1.4.2 Molecular Weight

Molecular weight is the mass of a molecule. It is calculated as the sum of the atomic mass of each component atom multiplied by the number of atoms of that element in the molecular formula. Molecular weight is a basic property, and its measurement is involved in the physical characteristics of polysaccharides. Polysaccharides are polydisperse in molecular weight, meaning that each polysaccharide contains chains of various monosaccharide units leading to a distribution of molecular weight (Kennedy and Panesar, 2005).

Though Mw may affect the viscosity of polysaccharide, which could modify the digestion and absorption in the gastrointestinal tract, it also could directly influence bioactivity. For instance, heparin is usually used for treating venous thrombosis, but clinical results have found low-molecular weight heparins to be more effective and safe (Levine et al., 1996; Koopman et al., 1996; Prandoni et al., 1992; Rasmussen et al., 2009).

To investigate the effect of Mw on polysaccharides, batches of polysaccharides of gradient Mw were obtained, and then their bioactivities were compared in vitro and in vivo. Calazans classified the polysaccharide from *Zymomonas mobilis* (Levan) by their viscosity average Mw and tested their antitumor effect on S180-bearing mice. The results showed that maximum inhibition was found on levan of viscosity average Mw around 210,897 (Calazans et al., 2000). Chen et al. (2008) obtained three water-soluble polysaccharide conjugate fractions, i.e., TPC-1, TPC-2, and TPC-3, from *Camellia sinensis* using chromatography and ion-exchange chromatography. Their Mws were 26.8×10^4, 11.8×10^4, and 4.2×10^4, respectively, and TPC-3 showed the best level of antioxidant activity. Not only the natural polysaccharides were purified, many degraded polysaccharides were also prepared to study such Mw − bioactivity relationships. Polysaccharide from *Porphyridium cruentum* (EPS, 2918.7 kDa) with a high Mw has a high apparent viscosity and poor water-solubility. After hermetical microwave treatment, three low-Mw derivatives of 256.2, 60.6, and 6.55 kDa were obtained, and they were found to have notable antioxidant activity (Sun et al., 2009). Polysaccharide from *Ulva pertusa* Kjellm (ulvan) was degraded using H_2O_2, and the antioxidant activity of the degraded polysaccharide also increased with decreasing Mw, especially the polysaccharide of the lowest Mw (28.2 kDa), which had most preventative effects from superoxide and hydroxyl radicals (Qi et al., 2005). While in vivo experiment showed the high Mw ulvan was effective in lowering the serum total and LDL-cholesterol, low-molecular weight ulvan was effect in lowering TG and HDL-cholesterol (Yu et al., 2003).

The relationship between Mw and the activities of polysaccharide is more likely due to Mw partly affecting the polymerization and conformation of polysaccharide. Schizophyllan is a highly potent antitumor polysaccharide, and it is a kind of β-(1→3)-glucan. It has been found that schizophyllan of Mw above 90,000 in aqueous solution exists as a rigid triple helix, schizophyllan of Mw between 5000 and 90,000 exists both as triple helices and a single chain, whereas a Mw below 5000 only exists as a single chain. Simultaneously, it is considered the immunopharmacological activities of schizophyllan were dependent on its conformation, for schizophyllan of Mw higher than 10,000 exhibited antitumor activity and the activity was enhanced with increasing Mw up to 90,000 (Norisuye, 1985; Yanaki et al., 1983), while the single helix exhibited a higher priming effect on LPS-triggered synthesis of TNF-α, and NO of peritoneal macrophage in vivo (Ohno et al., 1995).

1.4.3 Conformation

The conformation of carbohydrate was a three-dimensional structure for through-bond and through-space physical forces resulting from its molecular structure. It is necessary to understand the properties of carbohydrate conformation to explain their mechanisms, which may be helpful in designing carbohydrate-based vaccines and other therapeutic agents (Kirschner and Woods, 2001). The chain of a polysaccharide is dependent on the conformation of the monosaccharide structural unit and the types and positions of linkages in the chain. Regular conformations are considered to reflect the presence of a partial periodic sequence in the chain, rather than irregular conformations. There are several typical conformation properties of glucan and other polysaccharides will be discussed later (Belitz et al., 2004).

The conformation properties of polysaccharides consist of the monosaccharide conformation, monosaccharide orientation with respect to each other, and the spatial structural flexibility. Zhang et al. (2007) reported the order of chain rigidity of glucan was (1→4)-β- >(1→3)-α- >(1→4)-α- >(1→3)-β- >(1→6)-α- >(1→6)-β-, while β-glucans containing primarily 1→6 linkages show less activity, which may have resulted from their inherent flexibility.

Polysaccharides in aqueous solution show different conformations of chain, such as conformations of single helix, random coil, rod-like, double helix, and sphere-like. For instance, polysaccharide purified from *C. militaris* is a high Mw polysaccharide with a random coil conformation of β-1,4-branched -β-1,6-galactoglucomannan, which could stimulate the secretion of NO, IL-1β, and TNF-α on RAW 264.7 (Lee et al., 2010a,b). Two α-D-glucans isolated from hot-water and alkaline extracts of the mycelial biomass of *Cordyceps sinensis* (Cs-Hk1) were named WIPS and AIPS, respectively. Both WIPS and AIPS exhibited a random coil structure, but AIPS was more homogenous with a polydispersity index of 0.370 AIPS, whereas that of WIPS was 0.187. Besides, AIPS exhibited higher significant antitumor and immunestimulating effects (Yan et al., 2011). α-(1→4)-Glucan heteropolysaccharide, an antitumor polysaccharide from *Rhizoma Panacis Japonici*, presented as a spherical conformation in aqueous solution (Huang et al., 2010). Many (1→3)-β-glucans, such as lentinan, PSG, and schizophyllan, have been considered as effective agents in antitumor action. Moreover, the triple helical conformation of (1→3)-β-glucan is considered as an important feature of the structure, because of various bioactivities. Zhang et al. (2001) found that lentinan was a triple helix in an aqueous solution of 0.9% NaCl, but was a single flexible chain in dimethylsulfoxide (DMSO), while the antitumor effect of lentinan of the triple-helix sample is high, but that of a single flexible chain sample almost disappeared (Zhang et al., 2005a,b), similar to a result reported by Surenjav et al. (2006). Norisuye et al. (1980) found schizophyllan dissolves in water as a triple helix, which is not recoverable broken to a single chain in DMSO. However, α-(1→3)-β-glucan isolated from the fruit body of *Auricularia auricula-judae*, which dissolves in water as single-stranded helical chains, also possesses an anticancer effect (Zhang and Yang, 1995). Elucidating the conformation of polysaccharides will be helpful in understanding their bioactivities.

1.4.4 Substituent Group

1.4.4.1 Sulfate Group

Sulfated polysaccharides are widely spread in marine algae, and exhibit many beneficial human properties (Costa et al., 2010; Jiao et al., 2011; Wijesekara et al., 2011). Sulfated polysaccharides from *Monostroma nitidum* had a direct cytotoxic influence in AGS and Hela cells (Karnjanapratum and You, 2011). Sulfated polysaccharide from *Corallina officinalis* possessed considerable antioxidant properties (Yang et al., 2011). Sulfated galactan from *Codium fragile* could activate RAW 264.7 by stimulating the production of NO and the synthesis of IL-6, IL-1β, TNF-α, IL-12p40, and IL-10mRNA (Lee et al., 2010a,b). Besides, sulfated polysaccharides also possess notable antiviral and anticoagulant effects (Mestechkina and Shcherbukhin, 2010). Hayashi et al. (1996) found sulfated polysaccharide isolated from *Spirulina platensis* (Ca-PS) was a potential anti-HIV-1 and HSV-1 agent, for it not only interfered with viral replication, but also inhibited the formation of syncytium. Wijesinghe et al. (2011) reported the administration of sulfated polysaccharide purified from *Ecklonia cava* could prolong bleeding time in the tail bleeding assay in rats. In addition, research proved that their activities were partly attributed to the sulfate content, as weaker effects were observed on the desulfated derivative. The mechanism of sulfated polysaccharides inhibition of virus replication is generally through the prevention of virus binding to the cells of host (Baba et al., 1988; Mitsuya et al., 1988).

So far, a chemical grafting modification of sulfation is largely used for improving polysaccharides' biological properties. For instance, Wang et al. (2010a) reported sulfation could enhance the antiviral effect of lentinan. There are a great number of methods for polysaccharide sulfated modification, such as the sulfuric acid, chlorosulfuric acid (CAS)-pyridine (Pyr) method, sulfurtrioxide-pyridine, and sulfurtrioxide-dimethylacetamid methods. The degree of substitution (DS) of sulfate derivatives is an important parameter associated to its activity (Ghosh et al., 2008). Wang et al. (2010c) found sulfated *L. barbarum* polysaccharides of $DS_{1.5}$ and $DS_{1.9}$ can notably enhance the proliferation of lymphocytes and promote serum antibody titer in chickens. Liu et al. (2009b) reported polysaccharides of *Phellinus ribis* (PRP) had no obvious influence on the growth of HepG2 cells. However, sulfated PRPs (PRP-S) of DS between 1.52 and 2.20 can prevent HepG2 cells' growth, but PRP-S of $DS_{0.62}$ do not. It is likely that sulfated polysaccharides of DS between 1.5 and 2.0 exhibit higher enhanced activities. In some case, low-Mw sulfated polysaccharide can also possess a dramatical antiviral activity, particularly when its sulfate content is high. Besides, the substitution positions are also important for the activity of derivatives. The hydroxyl group of C-6 and C-2 are most reactive, while that of C-4, and C-3 would also be partially substituted. Yang et al. (2005) reported the

6-O-SO_3^- was the key for the anticoagulant activity of sulfated lacquer polysaccharides. The sulfated modification could not only increase the solubility of water, but also change the charge density and the conformation of polysaccharide. In addition, this modification may sharply impact the conformation of polysaccharide, partly due to a notable degradation which may occur during preparation, which is also involved in explaining the activity-enhancement modification. For example, the surface conformation of *laminarin* was dramatically change from cloud-like and spongy to block-like and flaky (Ji et al., 2013). Wang et al. (2010b) prepared sulfated polysaccharide of *Artemisia sphaerocephala* (ASP), and found the sulfate groups enhanced the steric hindrance of ASP, which resulted in a relatively expanded conformation of the sulfated derivatives. However, this modification will not always work; a T cell stimulating polysaccharide from spores of *G. lucidum* was sulfated using piperidine-chlorosulfonic agent, then it was found that the activity disappeared (Bao et al., 2002a,b).

1.4.4.2 Carboxymethyl Group

Carboxymethylation is also a grafting chemical modification for polysaccharide, which the production of has been used in a wide variety of fields. A well-known relevant carboxymethyl derivative is carboxymethyl cellulose, which is applied in food, pharmaceutical, paper-making, textile, and construction industries. Commonly, a method of carboxymethylation uses monochloroacetic acid as an etherified reagent in a sodium hydroxide aqueous solution, which is based on the Williamson synthesis. The advantages of carboxymethylation are technical simplicity, low cost of the chemicals, environmental protection, and nontoxicity of the products (Verraest et al., 1995).

Now much research indicates that carboxymethylation modification is an effective method to modify the bioactivity of polysaccharides. For example, carboxymethylated *G. lucidum* polysaccharide showed enhanced hydroxyl radical, hydrogen peroxide, and superoxide anion radical scavenging activities (Jing et al., 2009a,b), and carboxymethylated κ-carrageenan exhibited good antimicrobial properties against *Escherichia coli* and *Staphylococcus aureus*, especially when the antimicrobial activity depended on the DS of the carboxymethyl group (Fan et al., 2011). One of the mechanisms of carboxymethylation in modifying the bioactivities of polysaccharides is by increasing its water-solubility. Yang et al. (2011) found the solubility of a poorly water-soluble crude polysaccharide from *Auricularia auricular* (AAP) was enhanced to 0.6 mg/mL after carboxymethylation, and its antioxidant activity was significantly increased twofold. Alkali-extract linear α-(1→3)-D-glucan isolated from the fruit bodies of *Amanita muscaria* is a water-insoluble polysaccharide without any antitumor effects, however, its water-soluble carboxymethylated derivative showed potent antitumor activity against Sarcoma 180 in tumor-bearing mice (Kiho et al., 1994).

1.4.4.3 Other Substitution Groups

Plants' cells are in a wall including various, complex, substituted polymers (polysaccharide). One of the substituents found in many polysaccharides is *O*-linked acetyl (Gille and Pauly, 2011). Acetylated polysaccharides are also found in fungi. Ukai et al. (1983) reported partially *O*-acetylated α-D-mannan from *Dictyophora indusiata* showed marked inhibitory effects on scald hyperalgesia, indicating a potent antiinflammatory activity. Recently, acetylation modification has been used to introduce acetyl group into polysaccharides to modify their activities. Acetylated *U. pertusa* showed enhanced scavenging activity against hydroxyl radical and chelating ability (Qi et al., 2006). Polysaccharide from mushroom *Inonotus obliquus* showed a hyperbranched conformation with a lower Mw and intrinsic viscosity after being acetylated, while its antioxidant abilities on ferric-reducing power and lipid peroxidation inhibition activity were enhanced (Ma et al., 2012).

Sulfonation is another common modification for polysaccharides, derivatives of which are likely found to be potent antiviral agents. Chinese lacquer polysaccharide showed coagulation-promoting and antitumor activities. However, sulfonated lacquer polysaccharide exhibited potent anti-HIV activity, while its antitumor activity disappeared (Lu et al., 2000). Sulfonated derivative of polysaccharide from *Polygonatum cyrtonema* Hua was found to possess significant inhibitory activity against HSV (Liu et al., 2011a,b). In addition, sulfonated polysaccharide may help to improve other properties of polysaccharide, e.g., Unursaikhan reported *O*-sulfonation enhanced the anticancer effect of (1→3)-α-D-glucan derived from *L. edodes*, both in vivo and in vitro (Unursaikhan et al., 2006).

Furthermore, phosphorylation is also used to introduce phosphate groups to polysaccharides, which may notably affect the anticoagulant and antioxidant properties of them. Dace et al. (1997) found phosphorylated derivatives of oat spelt xylan K-carrageenan showed enhanced anticoagulation in an Mw-depended manner, while the Mw increase may be due to formation of the phosphodiester and diphosphodiester bonds between the chains of the phosphorylated polysaccharides. Jing et al. (2009a) reported phosphorylated fucoidan exhibited excellent ability on scavenging hydroxyl and superoxide radical, the mechanism of which may be because phosphate is a polyelectrolytes group and could activate the hydrogen atom of the anomeric carbon.

1.5 PERSPECTIVES

Over the past decades, numerous studies have focused on elucidating the structures and bioactivities of polysaccharides. Not only natural polysaccharides, but also semisynthetic polysaccharides, have been proved to possess various bioactivities. Recently, research on the relationship between their structure and activity has been increasing.

Similar to proteomics and genomics, the conception of "glycol-informatics" was proposed as a new branch of bioinformatics, which refers to informatics tools available for assessing the covalent and three-dimensional structure of glycans and glycol-conjugates. Polysaccharides are dramatically different from protein and nucleic acids, not only due to their extremely complex structure, but also to their nontemplate-driven biosynthesis; classical bioinformatics algorithms developed for proteomics or genomics are not suitable for glycol-informatics. Though the development of glycol-informatics is lagging far behind proteomics and genomics, high-throughput and automated techniques which allow increasingly rapid structural analysis of polysaccharides, combined with utilization of carbohydrate databases and computer molecular simulated technology would strongly promote a vigorous development of glycol-informatics (Pérez and Mulloy, 2005).

Besides investigation into the SAR of polysaccharides, to explore the medicinal value of polysaccharide sources still need to deal with other problems, one of which is the stability of polysaccharide preparation, and another the patent protection of polysaccharide-derived products (Novak and Vetvicka, 2009). Recent advances in chemical and enzymatic synthesis of oligosaccharides make it possible to prepare designed oligosaccharides reliably, which is now used as an effective method in biological research (Boltje et al., 2009). In particular, some synthetic oligosaccharides are found to possess comparable activity to natural polysaccharides. For example, glucan-like synthetic oligosaccharides prepared by Jamois et al. (2005) showed a similar immune stimulatory effect as natural (1→3)-β-glucan, and synthetic oligosaccharides prepared by Li et al. (2007) were potent radical scavenging agents. Moreover, synthetic oligosaccharides would also provide a new insight to investigate the SAR of polysaccharides.

REFERENCES

Aga, M., Watters, J.J., Pfeiffer, Z.A., Wiepz, G.J., Sommer, J.A., Bertics, P.J., 2004. Evidence for nucleotide receptor modulation of cross talk between MAP kinase and NF-kappa B signaling pathways in murine RAW 264.7 macrophages. Am. J. Cell Phys. Cell. Ph. 286, 1165–1166.

Aro, A., Uusitupa, M., Voutilainen, E., Hersio, K., Korhonen, T., Siitonen, O., 1981. Improved diabetic control and hypocholesterolaemic effect induced by long-term dietary supplementation with guar gum in type 2 (insulin-independent) diabetes. Diabetologia 21, 29–33.

Baba, M., Pauwels, R., Balzarini, J., Arnout, J., Desmyter, J., Clercq, E.D., 1988. Mechanism of inhibitory effect of dextran sulfate and heparin on replication of human immunodeficiency virus in vitro. P. Natl. Acad. Sci. 85, 6132–6136.

Bae, I.Y., Kim, H.Y., Lee, S., Lee, H.G., 2011. Effect of the degree of oxidation on the physicochemical and biological properties of *Grifola frondosa* polysaccharides. Carbohyd. Polym. 83, 1298–1302.

Banchereau, J., Steinman, R.M., 1998. Dendritic cells and the control of immunity. Nature 392, 245–252.

Bao, X., Liu, C., Fang, J., Li, X., 2001. Structural and immunological studies of a major polysaccharide from spores of *Ganoderma lucidum* (Fr.) Karst. Carbohyd. Res. 332, 67–74.

Bao, X., Yuan, H., Wang, C., Liu, J., Lan, M., 2013. Antitumor and immunomodulatory activities of a polysaccharide from Artemisia argyi. Carbohyd. Polym. 98, 1236–1243.

Bao, X.F., Wang, X.S., Dong, Q., Fang, J.N., Li, X.Y., 2002a. Structural features of immunologically active polysaccharides from *Ganoderma lucidum*. Phytochemistry 59, 175–181.

Bao, X.F., Zhen, Y., Ruan, L., Fang, J.N., 2002b. Purification, characterization, and modification of T lymphocyte-stimulating polysaccharide from spores of *Ganoderma lucidum*. Chem. Pharm. Bull. 50, 623–629.

Behall, K.M., Scholfield, D.J., Hallfrisch, J.G., 2006. Barley β-glucan reduces plasma glucose and insulin responses compared with resistant starch in men. Nutr. Res. 26, 644–650.

Belitz, H.D., Grosch, W., Schieberle, P., 2004. Translation from the fifth German edition by M. Burghagen. Food Chemistry, third revised ed. Springer, Germany.

Bhatia, S., Rathee, P., Sharma, K., Chaugule, B.B., Kar, N., Bera, T., 2013. Immuno-modulation effect of sulphated polysaccharide (Porphyran) from Porphyra vietnamensis. Int. J. Biol. Macromol. 57, 50–56.

Bhunia, S.K., Dey, B., Maity, K.K., Patra, S., Mandal, S., Maiti, S., et al., 2010. Structural characterization of an immunoenhancing heteroglycan isolated from an aqueous extract of an edible mushroom, Lentinus squarrosulus (Mont.) Singer. Carbohyd. Res. 345, 2542–2549.

Bi, Y., Ye, M., Yan, H., Jian, X., Wang, Q., 2013. Biological and physicochemical properties of two polysaccharides from the mycelia of Grifola umbellate. Carbohyd. Polym. 95, 740–745.

Bisht, S., Kant, R., Kumar, V., 2013. α-d-Glucosidase inhibitory activity of polysaccharide isolated from Acacia tortilis gum exudate. Int. J. Biol. Macromol. 59, 214–220.

Blackburn, N.A., Johnson, I.T., 1981. The effect of guar gum on the viscosity of the gastrointestinal contents and on glucose uptake from the perfused jejunum in the rat. Brit. J. Nutr. 46, 239–246.

Boltje, T.J., Buskas, T., Boons, G.J., 2009. ChemInform abstract: opportunities and challenges in synthetic oligosaccharide and glycoconjugate research. Nat. Chem. 1, 611–622.

Boraschi, D., Cifone, M.G., Falk, W., Flad, H.D., Tagliabue, A., Martin, M.U., 1998. Cytokines in inflammation. Joint Workshop of the Deutsche Gesellschaft fur Immunologie (DGfI) and the Gruppo di Cooperazione in Immunologia (GCI) Assergi (L'Aquila, Italy), February 8–11, 1998. Eur. Cytokine Netw. 9, 205–212.

Bourne, M.C., 2002. Food Texture and Viscosity: Concept and Measurement. Academic Press, San Diego.

Bradner, W.T., Clarke, D.A., Stock, C.C., 1958. Stimulation of host defense against experimental cancer I. Zymosan and Sarcoma 180 in Mice. Cancer Res. 18, 347–351.

Bubb, W.A., 2003. NMR spectroscopy in the study of carbohydrates: characterizing the structural complexity. Concepts Magn. Reson. A 19A, 1–19.

Calazans, Gc.M.T., Lima, R.C., França, F.P.D., Lopes, C.E., 2000. Molecular weight and antitumour activity of *Zymomonas mobilis* levans. Int. J. Biol. Macromol. 27, 245–247.

Cao, W., Li, X.Q., Liu, L., Yang, T.H., Li, C., Fan, H.T., et al., 2006. Structure of an anti-tumor polysaccharide from *Angelica sinensis* (Oliv.) Diels. Carbohyd. Polym. 66, 149–159.

Cao, W., Li, X.Q., Wang, X., Fan, H.T., Zhang, X.N., Hou, Y., et al., 2010. A novel polysaccharide, isolated from *Angelica sinensis* (Oliv.) Diels induces the apoptosis of cervical cancer HeLa cells through an intrinsic apoptotic pathway. Phytomedicine 17, 598–605.

Chan, E.D., Riches, D.W., 2001. IFN-gamma + LPS induction of iNOS is modulated by ERK, JNK/SAPK, and p38(mapk) in a mouse macrophage cell line. Am. J. Cell Phys. Cell. Ph. 280, C441–450.

Chang, R., 2002. Bioactive polysaccharides from traditional Chinese medicine herbs as anticancer adjuvants. J. Altern. Complem. Med. 8, 559–565.

Chen, H., Zhang, M., Xie, B., 2004. Quantification of uronic acids in tea polysaccharide conjugates and their antioxidant properties. J. Agr. Food Chem. 52, 3333–3336.

Chen, H., Zhang, M., Qu, Z., Xie, B., 2008. Antioxidant activities of different fractions of polysaccharide conjugates from green tea (*Camellia sinensis*). Food Chem. 106, 559–563.

Chen, S.K., Minlang, T., Huang, J.R., Chen, R.H., 2009a. In vitro antioxidant activities of low-molecular-weight polysaccharides with various functional groups. J. Agr. Food Chem. 57, 2699–2704.

Chen, X.P., Chen, Y., Li, S.B., Chen, Y.G., Lan, J.Y., Liu, L.P., 2009b. Free radical scavenging of *Ganoderma lucidum* polysaccharides and its effect on antioxidant enzymes and immunity activities in cervical carcinoma rats. Carbohyd. Polym. 77, 389–393.

Chen, J.R., Yang, Z.Q., Hu, T.J., Yan, Z.T., Niu, T.X., Wang, L., et al., 2010a. Immunomodulatory activity in vitro and in vivo of polysaccharide from Potentilla anserina. Fitoterapia 81, 1117–1124.

Chen, W., Zhang, W., Shen, W., Wang, K., 2010b. Effects of the acid polysaccharide fraction isolated from a cultivated *Cordyceps sinensis* on macrophages in vitro. Cell. Immunol. 262, 69–74.

Chen, X., Jin, J., Tang, J., Wang, Z., Wang, J., Jin, L., et al., 2011. Extraction, purification, characterization and hypoglycemic activity of a polysaccharide isolated from the root of Ophiopogon japonicus. Carbohyd. Polym. 83, 749–754.

Chen, G., Xu, J., Miao, X., Yi, H., Liu, X., Ju, Y., et al., 2012a. Characterization and antitumor activities of the water-soluble polysaccharide from Rhizoma Arisaematis. Carbohyd. Polym. 90, 67–72.

Chen, W., Xia, Y., Zhao, X., Wang, H., Yu, M., Li, Y., et al., 2012b. The critical role of Astragalus polysaccharides for the improvement of PPARα-mediated lipotoxicity in diabetic cardiomyopathy. Plos One 7, 8441–8449.

Chen, Y., Tang, J., Wang, X., Sun, F., Liang, S., 2012c. An immunostimulatory polysaccharide (SCP-IIa) from the fruit of *Schisandra chinensis* (Turcz.) Baill. Int. J. Biol. Macromol. 50, 844–848.

Chen, J., Jin, X., Chen, J., Liu, C., 2013a. Glycyrrhiza polysaccharide induces apoptosis and inhibits proliferation of human hepatocellular carcinoma cells by blocking PI3K/AKT signal pathway. Tumour Biol. 34, 1381–1389.

Chen, X., Tang, J., Xie, W., Wang, J., Jin, J., Ren, J., et al., 2013b. Protective effect of the polysaccharide from Ophiopogon japonicus on streptozotocin-induced diabetic rats. Carbohyd. Polym. 94, 378–385.

Chen, Y.W., Hu, D.J., Cheong, K.L., Li, J., Xie, J., Zhao, J., et al., 2013c. Quality evaluation of lentinan injection produced in China. J. Pharmaceut. Biomed. 78–79, 176–182.

Cheng, X.Q., Li, H., Yue, X.L., Xie, J.Y., Zhang, Y.Y., Di, H.Y., et al., 2010. Macrophage immunomodulatory activity of the polysaccharides from the roots of Bupleurum smithii var. parvifolium. J. Ethnopharmacol. 130, 363–368.

Cheng, Y., Tang, K., Wu, S., Liu, L., Qiang, C., Lin, X., et al., 2011. Astragalus polysaccharides lowers plasma cholesterol through mechanisms distinct from statins. Plos One 6, e27437.

Cherbut, C., Albina, E., Champ, M., Doublier, J.L., Lecannu, G., 1990. Action of Guar Gums on the viscosity of digestive contents and on the gastrointestinal motor function in pigs. Digestion 46, 205–213.

Cheung, L.M., Cheung, P.C.K., Ooi, V.E.C., 2003. Antioxidant activity and total phenolics of edible mushroom extracts. Food Chem. 81, 249–255.

Chihara, G., Hamuro, J., Maeda, Y., Arai, Y., Fukuoka, F., 1970. Fractionation and purification of the polysaccharides with marked antitumor activity, especially lentinan, from *Lentinus edodes* (Berk.) Sing. (an edible mushroom). Cancer Res. 30, 2776–2781.

Chudzikowski, R.J., 1971. Guar gum and its application. J. Soc. Cosmet. Chem. 22, 43–60.

Clancey, W.J., 2010. A Transactional Perspective on the Practice-Based Science of Teaching and Learning. Theor. Learn. Stud. Instruct. Practice 1, 247–278.

Costa, L.S., Fidelis, G.P., Cordeiro, S.L., Oliveira, R.M., Sabry, D.A., Câmara, R.B.G., et al., 2010. Biological activities of sulfated polysaccharides from tropical seaweeds. Biomed. Pharmacother. 64, 21–28.

Coussens, L.M., Werb, Z., 2002. Inflammation and cancer. Nature 420, 860–867.

Creech Jr, H.J., Hankwitz, R., Wharton, D.R., 1949. Further studies of the immunological properties of polysaccharides from *Serratia marcescens* (Bacillus prodigiosus) the effects of passive and active immunization on the lethal activity of the polysaccharides. Cancer Res. 9, 150–157.

Crispin, J.C., Vargas, M.I., Alcocer-Varela, J., 2004. Immunoregulatory T cells in autoimmunity. Autoimmun. Rev. 3, 45–51.

Czubayko, F., 2000. Beneficial effects of high dietary fiber intake in patients with type 2 diabetes mellitus. New Engl. J. Med. 342, 1392–1398.

Dace, R., Mcbride, E., Brooks, K., Gander, J., Buszko, M., Doctor, V.M., 1997. Comparison of the anticoagulant action of sulfated and phosphorylated polysaccharides. Thromb. Res. 87, 113–121.

Davies, M., Dean, R.T., 1997. Radical-mediated Protein Oxidation: From Chemistry to Medicine.

Defranco, A.L., Hambleton, J., Mcmahon, M., Weinstein, S.L., 1995. Examination of the role of MAP kinase in the response of macrophages to lipopolysaccharide. Prog. Clin. Biol. Res. 392, 407–420.

Dennert, G., Tucker, D., 1973. Antitumor polysaccharide lentinan. A T cell adjuvant. J. Natl. Cancer I 51, 1727–1729.

Dikeman, C.L., Murphy Jr, M.R., Fahey, G., 2006. Dietary fibers affect viscosity of solutions and simulated human gastric and small intestinal digesta. J. Nutr. 136, 913–919.

Dinarello, C.A., 2000. Proinflammatory cytokines. Chest 118, 503–508.

Dong, Q., Yao, J., Fang, J.N., Ding, K., 2007. Structural characterization and immunological activity of two cold-water extractable polysaccharides from Cistanche deserticola Y. C. Ma. Carbohyd. Res. 342, 1343–1349.

Du, X.J., Zhang, J.S., Yang, Y., Ye, L.B., Tang, Q.J., Jia, W., et al., 2009. Structural elucidation and immuno-stimulating activity of an acidic heteropolysaccharide (TAPA1) from Tremella aurantialba. Carbohyd. Res. 344, 672–678.

Edgar, K.J., 2010. Polysaccharide chemistry: frontiers and challenges, ACS Symp. Series, 1017. Oxford University Press, pp. 1–12.

Edwards, C.A., Blackburn, N.A., Craigen, L., Davison, P., Tomlin, J., Sugden, K., et al., 1987. Viscosity of food gums determined in vitro related to their hypoglycemic actions. Am. J. Clin. Nutr. 46, 72–77.

Fairchild, R.M., Ellis, P.R., Byrne, A.J., Luzio, S.D., Mir, M.A., 1996. A new breakfast cereal containing guar gum reduces postprandial plasma glucose and insulin concentrations in normal-weight human subjects. Brit. J. Nutr. 76, 63–73.

Fan, L., Wang, L., Gao, S., Wu, P., Li, M., Xie, W., et al., 2011. Synthesis, characterization and properties of carboxymethyl kappa carrageenan. Carbohyd. Polym. 86, 1167–1174.

Fang, J., Seki, T., Maeda, H., 2009. Therapeutic strategies by modulating oxygen stress in cancer and inflammation. Adv. Drug Deliver. Rev. 61, 290–302.

Feng, G., Wang, X., You, C., Cheng, X., Han, Z., Liang, Z., et al., 2013. Antiproliferative potential of Artemisia capillaris polysaccharide against human nasopharyngeal carcinoma cells. Carbohyd. Polym. 92, 1040–1045.

Ferreira, I.C., Barros, L., Abreu, R.M., 2009. Antioxidants in wild mushrooms. Curr. Med. Chem. 16, 1543–1560.

Fujiwara, N., Kobayashi, K., 2005. Macrophages in inflammation. Current Drug Targets Inflammation & Allergy 4, 281–286.

Gallaher, D.D., Hassel, C.A., Lee, K.J., Gallaher, C.M., 1993. Viscosity and fermentability as attributes of dietary fiber responsible for the hypocholesterolemic effect in hamsters. J. Nutr. 123, 244–252.

Gao, M., Zhang, J.H., Zhou, F.X., Xie, C.H., Han, G., Fang, S.Q., et al., 2011. *Angelica sinensis* suppresses human lung adenocarcinoma A549 cell metastasis by regulating MMPs/TIMPs and TGF-β1. Oncol. Rep. 27, 585–593.

García, dT.J., 2005. Inflammation, atherosclerosis, and coronary artery disease. New Engl. J. Med. 353, 1685–1695.

Ghosh, T., Chattopadhyay, K., Marschall, M., Karmakar, P., Mandal, P., Ray, B., 2008. Focus on antivirally active sulfated polysaccharides: from structure-activity analysis to clinical evaluation. Glycobiology 19, 2–15.

Gille, S., Pauly, M., 2011. O-Acetylation of Plant Cell Wall Polysaccharides. Front. Plant Sci. 3, 12.

Gong, F., Li, F., Zhang, L., Li, J., Zhang, Z., Wang, G., 2009. Hypoglycemic effects of crude polysaccharide from Purslane. Int. J. Mol. Sci. 10, 880–888.

Green, D.R., Flood, P.M., Gershon, R.K., 1983. Immunoregulatory T-cell pathways. Annu. Rev. Immunol. 1, 439–463.

Gross, J.L., de Azevedo, M.J., Silveiro, S.P., Canani, L.H., Caramori, M.L., Zelmanovitz, T., 2005. Diabetic nephropathy: diagnosis, prevention, and treatment. Diabetes Care 28, 164–176.

Gülçin, İ., 2006. Antioxidant activity of caffeic acid (3,4-dihydroxycinnamic acid). Toxicology 217, 213–220.

Harding, S.E., Vårum, K.M., Stokke, B.T., Smidsrød, O., 1991. Molecular weight determination of polysaccharides. Adv. Carbohyd. Analy. 1, 63–144.

Harris, P.J., Henry, R.J., Blakeney, A.B., Stone, B.A., 1984. An improved procedure for the methylation analysis of oligosaccharides and polysaccharides. Carbohyd. Res. 127, 59–73.

Hayashi, K., Hayashi, T., Kojima, I., 1996. A natural sulfated polysaccharide, calcium spirulan, isolated from Spirulina platensis: in vitro and ex vivo evaluation of anti-herpes simplex virus and anti-human immunodeficiency virus activities. AIDS Res. Hum. Retrov. 12, 1463–1471.

He, N., Yang, X., Jiao, Y., Tian, L., Zhao, Y., 2012. Characterisation of antioxidant and antiproliferative acidic polysaccharides from Chinese wolfberry fruits. Food Chem. 133, 978–989.

He, T.B., Huang, Y.P., Yang, L., Liu, T.T., Gong, W.Y., Wang, X.J., et al., 2015. Structural characterization and immunomodulating activity of polysaccharide from Dendrobium officinale. Int. J. Biol. Macromol. 83, 34–41.

Henderson, B., Poole, S., Wilson, M., 1996. Bacterial modulins: a novel class of virulence factors which cause host tissue pathology by inducing cytokine synthesis. Microbiol. Rev. 60, 316–341.

Hersh, D., Weiss, J., Zychlinsky, A., 1998. How bacteria initiate inflammation: aspects of the emerging story. Curr. Opin. Microbiol. 1, 43–48.

Hu, F.D., Li, X.D., Zhao, L.G., Feng, S.L., Wang, C.M., 2010. Antidiabetic properties of purified polysaccharide from Hedysarum polybotrys. Can. J. Physiol. Pharm. 88, 64–72.

Hu, J., Pang, W., Chen, J., Bai, S., Zheng, Z., Wu, X., 2013. Hypoglycemic effect of polysaccharides with different molecular weight of Pseudostellaria heterophylla. BMC Complem. Altern. M 13, 1–9.

Huang, S.Q., Ning, Z.X., 2010. Extraction of polysaccharide from *Ganoderma lucidum* and its immune enhancement activity. Int. J. Biol. Macromol. 47, 336–341.

Huang, Z., Ren, H., Duan, X., Zhang, L., 2010. Chain conformation and bioactivity of water-soluble polysaccharide extracted from Rhizoma Panacis Japonici. Biopolymers 93, 383–390.

Huang, Y., Jiang, C., Hu, Y., Zhao, X., Shi, C., Yu, Y., et al., 2013. Immunoenhancement effect of *Rehmannia glutinosa* polysaccharide on lymphocyte proliferation and dendritic cell. Carbohyd. Polym. 96, 516–521.

Huang, H.Y., Korivi, M., Yang, H.T., Huang, C.C., Chaing, Y.Y., Tsai, Y.C., 2014. Effect of pleurotus tuber-regium polysaccharides supplementation on the progression of diabetes complications in obese-diabetic rats. Chinese J. Physiol. 57, 198–208.

Hwang, H.S., Yun, J.W., 2010. Hypoglycemic effect of polysaccharides produced by submerged mycelial culture of Laetiporus sulphureus on streptozotocininduced diabetic rats. Biotechnol. Bioproc. E 15, 173–181.

Hwang, P.A., Chien, S.Y., Chan, Y.L., Lu, M.K., Wu, C.H., Kong, Z.L., et al., 2011. Inhibition of Lipopolysaccharide (LPS)-induced inflammatory responses by *Sargassum hemiphyllum* sulfated polysaccharide extract in RAW 264.7 macrophage cells. J. Agr. Food Chem. 59, 2062–2068.

Ikegami, S., Tsuchihashi, F., Harada, H., Tsuchihashi, N., Nishide, E., Innami, S., 1990. Effect of viscous indigestible polysaccharides on pancreatic-biliary secretion and digestive organs in rats. J. Nutr. 120, 353–360.

Inngjerdingen, K.T., Meskini, S., 2011. Chemical and biological characterization of polysaccharides from wild and cultivated roots of Vernonia kotschyana. J. Ethnopharmacol. 139, 350–358.

Jacobs, A., Ignarro, L., 2002. Lipopolysaccharide-induced expression of interferon-beta mediates the timing of inducible nitric-oxide synthase induction in RAW 264.7 macrophages. J. Biol. Chem. 276, 47950–47957.

Jamois, F., Ferrières, V., Guégan, J.P., Yvin, J.C., Plusquellec, D., Vetvicka, V., 2005. Glucan-like synthetic oligosaccharides: iterative synthesis of linear oligo-beta-(1,3)-glucans and immunostimulatory effects. Glycobiology 15, 393–407.

Janeway, C.A., Medzhitov, R., 2002. Innate immune recognition. Annu. Rev. Immunol. 20, 197–216.

Jedrzejas, M.J., 2000. Structural and functional comparison of polysaccharide-Degrading enzymes. Crit. Rev. Biochem. Mol. 35, 221–251.

Jenkins, D.A., Derek, T., Hockaday, R., Howarth, R., Apling, E.C., Wolever, T.S., et al., 1977. Treatment of diabetes with guar gum: reduction of urinary glucose loss in diabetics. Lancet 2, 779–780.

Jenkins, D.J., Wolever, T.M., Leeds, A.R., Gassull, M.A., Haisman, P., Dilawari, J., et al., 1978. Dietary fibres, fibre analogues, and glucose tolerance: importance of viscosity. Brit. Med. J. 1, 1392–1394.

Jenkins, A.L., Jenkins, D.J., Zdravkovic, U., Würsch, P., Vuksan, V., 2002. Depression of the glycemic index by high levels of beta-glucan fiber in two functional foods tested in type 2 diabetes. Eur. J. Clin. Nutr. 56, 622–628.

Ji, Y.B., Dong, F., Gao, S.Y., Yu, M., 2011. Study on *Capparis spionosa L.* Polysaccharide (CSPS) Induced HepG2 Apoptosis by Controlling Ca^{2+} Path. Adv. Mater. Res. 282, 203–208.

Ji, C.F., Ji, Y.B., Meng, D.Y., 2013. Sulfated modification and anti-tumor activity of laminarin. Exp. Ther. Med. 6, 1259–1264.

Jia, J., Zhang, X., Hu, Y.S., Wu, Y., Wang, Q.Z., Li, N.N., et al., 2009. Evaluation of in vivo antioxidant activities of *Ganoderma lucidum* polysaccharides in STZ-diabetic rats. Food Chem. 115, 32–36.

Jian, L., Zhang, J.F., Lu, J.Z., Zhang, D.L., Li, K., Su, K., et al., 2013. Astragalus polysaccharide stimulates glucose uptake in L6 myotubes through AMPK activation and AS160/TBCID4 phosphorylation. Acta Pharmacol. Sin. 34, 137–145.

Jiang, M.H., Zhu, L., Jiang, J.G., 2010. Immunoregulatory actions of polysaccharides from Chinese herbal medicine. Expert Opin. Ther. Tar. 14, 1367–1402.

Jiang, C., Wang, M., Liu, J., Gan, D., Zeng, X., 2011. Extraction, preliminary characterization, antioxidant and anticancer activities in vitro of polysaccharides from *Cyclina sinensis*. Carbohyd. Polym. 84, 851–857.

Jiao, G., Yu, G., Zhang, J., Ewart, H., 2011. Chemical structures and bioactivities of sulfated polysaccharides from marine algae. Mar. Drugs 9, 196–223.

Jiao, G., Yu, G., Wang, W., Zhao, X., Zhang, J., Ewart, S.H., 2012. Properties of polysaccharides in several seaweeds from Atlantic Canada and their potential anti-influenza viral activities. J. Ocean U. China 11, 205–212.

Jin, M., Zhao, K., Huang, Q., Xu, C., Shang, P., 2012. Isolation, structure and bioactivities of the polysaccharides from *Angelica sinensis* (Oliv.) diels: a review. Carbohyd. Polym. 89, 713–722.

Jing, W., Zhang, Q.B., Zhang, Z.S., Zhang, J.J., Li, P.C., 2009a. Synthesized phosphorylated and aminated derivatives of fucoidan and their potential antioxidant activity in vitro. Int. J. Biol. Macromol. 44, 170–174.

Jing, X., Wei, L., Yao, W., Pang, X., Yin, D., Gao, X., 2009b. Carboxymethylation of a polysaccharide extracted from *Ganoderma lucidum* enhances its antioxidant activities in vitro. Carbohyd. Polym. 78, 227–234.

Jones, G.R.N., 1979. Early mitochondrial damage in the induction of haemorrhagic necrosis in the Crocker sarcoma (S 180) by endotoxin. J. Cancer Res. Clin. 93, 245–254.

Joseph, M.M., Aravind, S.R., George, S.K., Varghese, S., Sreelekha, T.T., 2013. A galactomannan polysaccharide from *Punica granatum* imparts in vitro and in vivo anticancer activity. Carbohyd. Polym. 98, 1466–1475.

Jovanovic, S.V., Steenken, S., Boone, C.W., Simic, M.G., 1999. H-Atom Transfer is a preferred antioxidant mechanism of curcumin. J. Am. Chem. Soc. 121, 9677–9681.

Karnjanapratum, S., You, S.G., 2011. Molecular characteristics of sulfated polysaccharides from *Monostroma nitidum* and their in vitro anticancer and immunomodulatory activities. Int. J. Biol. Macromol. 48, 311–318.

Kačuráková, M., Wellner, N., Ebringerová, A., Hromádková, Z., Wilson, R.H., Belton, P.S., 1999. Characterisation of xylan-type polysaccharides and associated cell wall components by FT-IR and FT-Raman spectroscopies. Food Hydrocolloids 13, 35–41.

Kennedy, J.F., Panesar, P.S., 2005. Food carbohydrates: chemistry, physical properties, and applications. In: Cui, S.W. (Ed.), Carbohyd. Polym., 62. CRC Press/Taylor and Francis Group, Boca Raton, FL, USA.

Kiho, T., Yoshida, I., Katsuragawa, M., Sakushima, M., Usui, S., Ukai, S., 1994. Polysaccharides in Fungi. XXXIV. A polysaccharide from the fruiting bodies of *Amanita muscaria* and the antitumor activity of its carboxymethylated product. Biol. Pharm. Bull. 17, 1460–1462.

Kilbourn, R.G., De Angelo, J., Bonaventura, J., 2000. Prevention or inhibition of pathogenic syndromes caused by nitric oxide overproduction using iron-containing hemoprotein. US. Patent.

Kim, G.Y., Choi, G.S., Lee, S.H., Park, Y.M., 2004. Acidic polysaccharide isolated from *Phellinus linteus* enhances through the up-regulation of nitric oxide and tumor necrosis factor-alpha from peritoneal macrophages. J. Ethnopharmacol. 95, 69–76.

Kim, Y.W., Kim, K.H., Choi, H.J., Lee, D.S., 2005. Anti-diabetic activity of beta-glucans and their enzymatically hydrolyzed oligosaccharides from *Agaricus blazei*. Biotechnol. Lett. 27, 483–487.

Kim, H.G., Song, H., Yoon, D.H., Song, B.W., Sang, M.P., Sung, G.H., et al., 2010a. Cordyceps pruinosa extracts induce apoptosis of HeLa cells by a caspase dependent pathway. J. Ethnopharmacol. 128, 342–351.

Kim, H.M., Kang, J.S., Kim, J.Y., Park, S.K., Kim, H.S., Lee, Y.J., et al., 2010b. Evaluation of antidiabetic activity of polysaccharide isolated from *Phellinus linteus* in non-obese diabetic mouse. Int. Immunopharmacol. 10, 72–78.

Kim, H.S., Kim, J.Y., Kang, J.S., Kim, H.M., Yong, O.K., Hong, I.P., et al., 2010c. Cordlan polysaccharide isolated from mushroom *Cordyceps militaris* induces dendritic cell maturation through toll-like receptor 4 signalings. Food Chem. Toxicol. 48, 1926–1933.

Kirschner, K.N., Woods, R.J., 2001. Solvent interactions determine carbohydrate conformation. P. Natl Acad. Sci. USA 98, 10541–10545.

Kiyohara, H., Matsuzaki, T., Matsumoto, T., Nagai, T., Yamada, H., 2008. Elucidation of structures and functions through Peyer's patches of responsible carbohydrate chains in intestinal immune system modulating polysaccharides from Japanese medicinal herbs. Yakugaku Zasshi 128, 709–716.

Klaus, A., Kozarski, M., Niksic, M., Jakovljevic, D., Todorovic, N., Griensven, L.J.L.D.V., 2011. Antioxidative activities and chemical characterization of polysaccharides extracted from the basidiomycete *Schizophyllum commune*. LWT-Food Sci. Technol. 44, 2005–2011.

Knöferl, M.W., Diodato, M.D., Schwacha, M.G., Cioffi, W.G., Bland, K.I., Chaudry, I.H., 2001. Cyclooxygenase-2-mediated regulation of Kupffer cell interleukin-6 production following trauma-hemorrhage and subsequent sepsis. Shock 16, 479–483.

Kokil, G.R., Veedu, R.N., Ramm, G.A., Prins, J.B., Parekh, H.S., 2015. Type 2 diabetes mellitus: limitations of conventional therapies and intervention with nucleic acid-based therapeutics. Chem. Rev. 115, 4719–4743.

Koopman, M.M., Prandoni, P., Piovella, F., Ockelford, P.A., Brandjes, D.P., van der Meer, J., et al., 1996. Treatment of venous thrombosis with intravenous unfractionated heparin administered in the hospital as compared with subcutaneous low-molecular-weight heparin administered at home. New Engl. J. Med. 334, 682–687.

Kristensen, M., Jensen, M.G., 2011. Dietary fibres in the regulation of appetite and food intake. Importance of viscosity. Appetite 56, 65–70.

Kumar, H., Kawai, T., Akira, S., 2011. Pathogen recognition by the innate immune system. Int. Rev. Immunol. 30, 16–34.

Kvernheim, A.L., Hazell, R.G., Lund, H., Berg, J.E., Eberson, L., 1987. Methylation analysis of polysaccharides with butyllithium in dimethyl sulfoxide. Acta Chem. Scand. 41, 150–152.

Lee, F., 2012. Basic food chemistry. Springer Science & Business Media, New York.

Lee, C.J., Huo, Y.S., Lee, L.H., Lu, C.H., Koizumi, K., 2004. Immunomodulating fungal and plant polysaccharides; Biochemistry, immunologic activity and clinical application. J. Tradit. Med. 21, 67–80.

Lee, J.B., Ohta, Y., Hayashi, K., Hayashi, T., 2010a. Immunostimulating effects of a sulfated galactan from *Codium fragile*. Carbohyd. Res. 345, 1452–1454.

Lee, J.S., Kwon, J.S., Yun, J.S., Pahk, J.W., Shin, W.C., Lee, S.Y., et al., 2010b. Structural characterization of immunostimulating polysaccharide from cultured mycelia of *Cordyceps militaris*. Carbohyd. Polym. 80, 1011–1017.

Leung, M.Y.K., Liu, C., Koon, J.C.M., Fung, K.P., 2006. Polysaccharide biological response modifiers. Immunol Lett. 105, 101–114.

Levine, M., Gent, M., Hirsh, J., Leclerc, J., Anderson, D., Weitz, J., et al., 1996. A comparison of low-molecular-weight heparin administered primarily at home with unfractionated heparin administered in the hospital for proximal deep-vein thrombosis. New Engl. J. Med. 334, 677–681.

Li, X.M., 2007. Protective effect of *Lycium barbarum* polysaccharides on streptozotocin-induced oxidative stress in rats. Int. J. Biol. Macromol. 40, 461–465.

Li, X.M., Shi, Y.H., Wang, F., Wang, H.S., Le, G.W., 2007. In vitro free radical scavenging activities and effect of synthetic oligosaccharides on antioxidant enzymes and lipid peroxidation in aged mice. J. Pharmaceut. Biomed. 43, 364–370.

Li, R., Chen, W.C., Wang, W.P., Tian, W.Y., Zhang, X.G., 2009. Extraction, characterization of Astragalus polysaccharides and its immune modulating activities in rats with gastric cancer. Carbohyd. Polym. 78, 738–742.

Li, X., Xu, W., Chen, J., 2010. Polysaccharide purified from *Polyporus umbellatus* (Per) Fr induces the activation and maturation of murine bone-derived dendritic cells via toll-like receptor 4. Cell. Immunol. 265, 50–56.

Li, Q., Dong, J., Hu, Z., Li, S., 2011. Research progress on hypoglycemic active ingredients in natural medicines. Guangdong Chem. Ind. 8, 38.

Li, H., Sun, M., Xu, J., Li, H., Zang, M., Cui, Y., 2013a. Immunological response in H22 transplanted mice undergoing *Aconitum coreanum* polysaccharide treatment. Int. J. Biol. Macromol. 55, 295–300.

Li, Y.G., Ji, D.F., Zhong, S., Liu, P.G., Lv, Z.Q., Zhu, J.X., et al., 2013b. Polysaccharide from *Phellinus linteus* induces S-phase arrest in HepG2 cells by decreasing calreticulin expression and activating the P27kip1-cyclin A/D1/E-CDK2 pathway. J. Ethnopharmacol. 150, 187–195.

Liao, C.H., Lin, J.Y., 2013. Purified active lotus plumule (*Nelumbo nucifera* Gaertn) polysaccharides exert anti-inflammatory activity through decreasing Toll-like receptor-2 and -4 expressions using mouse primary splenocytes. J. Ethnopharmacol. 147, 164–173.

Liu, J.J., Huang, T.S., Hsu, M.L., Chen, C.C., Lin, W.S., Lu, F.J., et al., 2004. Antitumor effects of the partially purified polysaccharides from Antrodia camphorata and the mechanism of its action. Toxicol. Appl. Pharm. 201, 186–193.

Liu, M., Wu, K., Mao, X., Wu, Y., Ouyang, J., 2009a. Astragalus polysaccharide improves insulin sensitivity in KKAy mice: regulation of PKB/GLUT4 signaling in skeletal muscle. J. Ethnopharmacol. 127, 32–37.

Liu, Y., Liu, C., Tan, H., Zhao, T., Cao, J., Wang, F., 2009b. Sulfation of a polysaccharide obtained from *Phellinus ribis* and potential biological activities of the sulfated derivatives. Carbohyd. Polym. 77, 370–375.

Liu, X.X., Wan, Z.J., Shi, L., Lu, X.X., 2011a. Preparation and antiherpetic activities of chemically modified polysaccharides from *Polygonatum cyrtonema* Hua. Carbohyd. Polym. 83, 737–742.

Liu, Y., Sheng, Y., Yuan, G., Wang, Y., Wei, H., Guan, M., et al., 2011b. Purification and physicochemical properties of different polysaccharide fractions from the water extract of *Boschniakia rossica* and their effect on macrophages activation. Int. J. Biol. Macromol. 49, 1007–1011.

Liu, L., Jia, J., Zeng, G., Zhao, Y., Qi, X., He, C., et al., 2013a. Studies on immunoregulatory and anti-tumor activities of a polysaccharide from *Salvia miltiorrhiza* Bunge. Carbohyd. Polym. 92, 479–483.

Liu, M., Qin, J., Hao, Y., Luo, J., Luo, T., Wei, L., 2013b. Astragalus polysaccharide suppresses skeletal muscle myostatin expression in diabetes: involvement of ROS-ERK and NF-κB pathways. Oxid. Med. Cell. Longev. 2013, 782497.

Liu, H., Bai, J., Weng, X., Wang, T., Li, M., 2014a. Amelioration of insulin resistance in rat cells by Astragalus polysaccharides and associated mechanisms. Exp. Ther. Med. 7, 1599–1604.

Liu, W., Zheng, Y., Zhang, Z., Yao, W., Gao, X., 2014b. Hypoglycemic, hypolipidemic and antioxidant effects of *Sarcandra glabra* polysaccharide in type 2 diabetic mice. Food Funct. 5, 2850–2860.

Lu, R., Yoshida, T., Nakashima, H., Premanathan, M., Aragaki, R., Mimura, T., et al., 2000. Specific biological activities of Chinese lacquer polysaccharides. Carbohyd. Polym. 43, 47–54.

Lu, X., Liu, W., Wu, J., Li, M., Wang, J., Wu, J., et al., 2013. A polysaccharide fraction of adlay seed (*Coix lachryma* - jobi L.) induces apoptosis in human non-small cell lung cancer A549 cells. Biochem. Bioph. Res. Co. 430, 846–851.

Luo, A., Fan, Y., 2011. Immune stimulating activity of water-soluble polysaccharide fractions from *Dendrobium nobile* Lindl. Afr. J. Pharm. Pharmaco. 5, 625–631.

Ma, H.T., Hsieh, J.F., Chen, S.T., 2015. Anti-diabetic effects of *Ganoderma lucidum*. Phytochemistry 114, 109–113.

Ma, L., Chen, H., Zhang, Y., Zhang, N., Fu, L., 2012. Chemical modification and antioxidant activities of polysaccharide from mushroom *Inonotus obliquus*. Carbohyd. Polym. 89, 371–378.

Ma, L., Qin, C., Wang, M., Gan, D., Lin, C., Ye, H., et al., 2013. Preparation, preliminary characterization and inhibitory effect on human colon cancer HT-29 cells of an acidic polysaccharide fraction from *Stachys floridana* Schuttl. ex Benth. Food Chem. Toxicol. 60, 269–276.

Maji, P.K., Sen, I.K., Behera, B., Maiti, T.K., Mallick, P., Sikdar, S.R., et al., 2012. Structural characterization and study of immunoenhancing properties of a glucan isolated from a hybrid mushroom of *Pleurotus florida* and *Lentinula edodes*. Carbohyd. Res. 358, 110–115.

Mao, X.Q., Feng, Y., Wang, N., Yong, W., Feng, Z., Ke, W., et al., 2009. Hypoglycemic effect of polysaccharide enriched extract of *Astragalus membranaceus* in diet induced insulin resistant C57BL/6J mice and its potential mechanism. Phytomedicine 16, 416–425.

Marzorati, M., Verhelst, A., Luta, G., Sinnott, R., Verstraete, W., Wiele, T.V.D., et al., 2010. In vitro modulation of the human gastrointestinal microbial community by plant-derived polysaccharide-rich dietary supplements. Int. J. Food Microbiol. 139, 168–176.

Maslowski, K.M., Mackay, C.R., 2011. Diet, gut microbiota and immune responses. Nat. Immunol. 12, 5–9.

Matsuoka, H., Seo, Y., Wakasugi, H., Saito, T., Tomoda, H., 1997. Lentinan potentiates immunity and prolongs the survival time of some patients. Anticancer Res. 17, 2751–2755.

Meng, G., Zhu, H., Yang, S., Wu, F., Zheng, H., Chen, E., et al., 2011a. Attenuating effects of *Ganoderma lucidum* polysaccharides on myocardial collagen cross-linking relates to advanced glycation end product and antioxidant enzymes in high-fat-diet and streptozotocin-induced diabetic rats. Carbohyd. Polym. 84, 180–185.

Meng, J., Hu, X., Shan, F., Hua, H., Lu, C., Wang, E., et al., 2011b. Analysis of maturation of murine dendritic cells (DCs) induced by purified *Ganoderma lucidum* polysaccharides (GLPs). Int. J. Biol. Macromol. 49, 693–699.

Meng, L.Z., Lv, G.P., Hu, D.J., Cheong, K.L., Xie, J., Zhao, J., et al., 2013. Effects of polysaccharides from different species of Dendrobium (Shihu) on macrophage function. Molecules 18, 5779–5791.

Mestechkina, N.M., Shcherbukhin, V.D., 2010. Sulfated polysaccharides and their anticoagulant activity (review). Appl. Biochem. Micro. 46, 267–273.

Mestre, J., Mackrell, P., De, Stapleton, P., Tanabe, T., Daly, J., 2001. Redundancy in the signaling pathways and promoter elements regulating cyclooxygenase-2 gene expression in endotoxin-treated macrophage/monocytic cells. J. Biol. Chem. 276, 3977–3982.

Mi, W., Xin, Y.M., Rui, L.Y., Tao, Q., Xiao, Y.W., Ke, Y.Z., et al., 2012. *Cordyceps militaris* polysaccharides can enhance the immunity and antioxidation activity in immunosuppressed mice. Carbohyd. Polym. 89, 461–466.

Mitsuya, H., Looney, D.J., Kuno, S., Ueno, R., Wongstaal, F., Broder, S., 1988. Dextran sulfate suppression of viruses in the HIV family: inhibition of virion binding to CD4 + cells. Science 240, 646–649.

Morales, D.R., Morris, A.D., 2014. Metformin in cancer treatment and prevention. Annu. Rev. Med. 66, 17–29.

Morris, E.R., 1986. Molecular interactions in polysaccharide gelation. Brit. Polym. J. 18, 14–21.

Mudgil, D., Barak, S., Khatkar, B.S., 2015. Guar gum: processing, properties and food applications-a review. J. Food Sci. Tech. Mys. 51, 409–418.

Nan, Y., Wang, R., Yuan, L., Niu, Y., 2012. Effects of *Lycium barbarum* polysaccharides (LBP) on immune function of mice. Afr. J. Microbiol. Res. 6, 4757–4760.

Ng, M.L., Yap, A.T., 2002. Inhibition of human colon carcinoma development by lentinan from shiitake mushrooms (*Lentinus edodes*). J. Altern. Complem. Med. 8, 581–589.

Nicod, L.P., 1993. Cytokines. 1. Overview. Thorax 48, 660–667.

Nie, S.P., 2006. Study on the Primary Structure and Conformation of Tea Glycoprotein and Its Bio-Diversity. Nanchang University, Nanchang, China.

Nie, S.P., Xie, M.Y., Fu, Z.H.H., Wan, Y.Q., Yan, A.I.P., 2008. Study on the purification and chemical compositions of tea glycoprotein. Carbohyd. Polym. 71, 626–633.

Nie, S.P., Cui, S.W., Phillips, A.O., Xie, M.Y., Phillips, G.O., Al-Assaf, S., et al., 2011. Elucidation of the structure of a bioactive hydrophilic polysaccharide from *Cordyceps sinensis* by methylation analysis and NMR spectroscopy. Carbohyd. Polym. 84, 894–899.

Nijkamp, F.P., Parnham, M.J., 2011. Principles of Immunopharmacology. Springer, Dordrecht.

Norisuye, T., 1985. Triple-stranded helical structure of schizophyllan and its antitumor activity. Die Makromolekulare Chemie 14, 105–118.

Norisuye, T., Yanaki, T., Fujita, H., 1980. Triple helix of a *Schizophyllum commune* polysaccharide in aqueous solution. J. Polym. Sci. Pol. Phys. E 18, 547–558.

Novak, M., Vetvicka, V., 2009. Glucans as biological response modifiers. Endocr. Metab. Immune 9, 67–75.

Ohno, N., Miura, N.N., Chiba, N., Adachi, Y., Yadomae, T., 1995. Comparison of the immunopharmacological activities of triple and single-helical schizophyllan in mice. Biol. Pharm. Bull. 18, 1242–1247.

Paiva, A.A.D.O., Castro, A.J.G., Nascimento, M.S., Will, L.S.E.P., Santos, N.D., Araújo, R.M., et al., 2011. Antioxidant and anti-inflammatory effect of polysaccharides from Lobophora variegata on zymosan-induced arthritis in rats. Int. Immunopharmacol. 11, 1241–1250.

Palacios, I., García-Lafuente, A., Guillamón, E., Villares, A., 2012. Novel isolation of water-soluble polysaccharides from the fruiting bodies of *Pleurotus ostreatus* mushrooms. Carbohyd. Res. 358, 72–77.

Pan, D., Zhang, D., Wu, J., Chen, C., Xu, Z., Yang, H., et al., 2013a. Antidiabetic, antihyperlipidemic and antioxidant activities of a novel proteoglycan from *Ganoderma lucidum* fruiting bodies on db/db mice and the possible mechanism. PLos One 8, 191–195.

Pan, L.H., Feng, B.J., Wang, J.H., Zha, X.Q., Luo, J.P., 2013b. Structual characterization and anti-glycation activity in vitro of a water-soluble polysaccharide from *Dendrobium huoshanense*. J. Food Biochem 37, 313–321.

Pan, D., Wang, L., Chen, C., Hu, B., Zhou, P., 2015. Isolation and characterization of a hyperbranched proteoglycan from *Ganoderma lucidum* for anti-diabetes. Carbohyd. Polym. 117, 106–114.

Parving, H.H., Lehnert, H., Brøchner-Mortensen, J., Gomis, R., Andersen, S., Arner, P., 2001. Effect of irbesartan on the development of diabetic nephropathy in patients with type 2 diabetes. New Engl. J. Med. 345, 870–878.

Pereira, L.D.P., Silva, R.O.D., Silva, K.E.S.D., Assreuy, A.M.S., Pereira, M.G., 2012. Polysaccharide fractions of Caesalpinia ferrea pods: potential anti-inflammatory usage. J. Ethnopharmacol. 139, 642–648.

Pérez, S., Mulloy, B., 2005. Prospects for glycoinformatics. Curr. Opin. Struc. Biol. 15, 517–524.

Pires, A.D.R.A., Ruthes, A.C., Cadena, S.M.S.C., Acco, A., Gorin, P.A.J., Iacomini, M., 2013. Cytotoxic effect of *Agaricus bisporus* and *Lactarius rufus* β- d -glucans on HepG2 cells. Int. J. Biol. Macromol. 58, 95–103.

Prandoni, P., Carta, M., Cogo, A., Ruol, A., Vigo, M., Casara, D., et al., 1992. Comparison of subcutaneous low-molecular-weight heparin with intravenous standard heparin in proximal deep-vein thrombosis. Lancet 339, 441–445.

Qi, H., Zhao, T., Zhang, Q., Li, Z., Zhao, Z., Xing, R., 2005. Antioxidant activity of different molecular weight sulfated polysaccharides from *Ulva pertusa* Kjellm (Chlorophyta). J. Appl. Phycol. 17, 527–534.

Qi, H., Zhang, Q., Zhao, T., Hu, R., Zhang, K., Li, Z., 2006. In vitro antioxidant activity of acetylated and benzoylated derivatives of polysaccharide extracted from *Ulva pertusa* (Chlorophyta). Bioorg. Med. Chem. Lett. 16, 2441–2445.

Qin, X., Jie, S., Zhao, M., Zhang, K., Ren, L., 2011. Immunostimulatory and anti-tumor activity of a water-soluble polysaccharide from *Phellinus baumii mycelia*. World J. Microb. Biot. 27, 1017–1023.

Rademacher, T.W., Parekh, R.B., Dwek, R.A., 1988. Glycobiology. Annu. Rev. Biochem. 57, 785–838.

Rask-Madsen, C., King, G., 2013. Vascular complications of diabetes: mechanisms of injury and protective factors. Cell Metab. 17, 20–33.

Rasmussen, M.S., Jørgensen, L.N., Wille-Jørgensen, P., 2009. Prolonged thromboprophylaxis with low molecular weight heparin for abdominal or pelvic surgery. Cochrane Db. Syst. Rev. 1, CD004318.

Ridker, P.M., 2002. On evolutionary biology, inflammation, infection, and the causes of atherosclerosis. Circulation 105, 2–4.

Robinson, G., Ross-Murphy, S.B., Morris, E.R., 1982. Viscosity-molecular weight relationships, intrinsic chain flexibility, and dynamic solution properties of guar galactomannan. Carbohyd. Res. 107, 17–32.

Rouveix, B., 1997. Clinical pharmacology of cytokines. Eur. Cytokine Netw. 8, 291–293.

Ruiz-Matute, A.I., Hernández-Hernández, O., Rodríguez-Sánchez, S., Sanz, M.L., Martínez-Castro, I., 2011. Derivatization of carbohydrates for GC and GC-MS analyses. J. Chromatogr. B. 879, 1226–1240.

Sasaki, T., Takasuka, N., 1976. Further study of the structure of lentinan, an anti-tumor polysaccharide from *Lentinus edodes*. Carbohyd. Res. 47, 99–104.

Schepetkin, I.A., Quinn, M.T., 2006. Botanical polysaccharides: macrophage immunomodulation and therapeutic potential. Int. Immunopharmacol. 6, 317–333.

Schepetkin, I.A., Kouakou, K., Yapi, A., Kirpotina, L.N., Jutila, M.A., Quinn, M.T., 2013. Immunomodulatory and hemagglutinating activities of acidic polysaccharides isolated from *Combretum racemosum*. Int. Immunopharmacol. 15, 628–637.

Shah, N., Shah, T., Amin, A., 2011. Polysaccharides: a targeting strategy for colonic drug delivery. Expert Opin. Drug Del. 8, 779–796.

Shear, M.J., Turner, F.C., Perrault, A., Shovelton, T., 1943. Chemical treatment of tumors. V. Isolation of the hemorrhage-producing fraction from *Serratia marcescens* (*Bacillus prodigiosus*) Culture Filtrate. J. Natl Cancer I. 4, 81–97.

Shi, B.J., Nie, X.H., Chen, L.Z., Liu, Y.L., Tao, W.Y., 2007. Anticancer activities of a chemically sulfated polysaccharide obtained from *Grifola frondosa* and its combination with 5-Fluorouracil against human gastric carcinoma cells. Carbohyd. Polym. 68, 687–692.

Sierra, M., Garcia, J.J., Fernández, N., Diez, M.J., Calle, A.P., Sahagún, A.M., et al., 2001. Effects of ispaghula husk and guar gum on postprandial glucose and insulin concentrations in healthy subjects. Eur. J. Clin. Nutr. 55, 235–243.

Silva, D.A., Paula, R.C.M.D., Feitosa, J.P.A., Brito, A.C.F.D., Maciel, J.S., Paula, H.C.B., 2004. Carboxymethylation of cashew tree exudate polysaccharide. Carbohyd. Polym. 58, 163–171.

Siu, K.C., Chen, X., Wu, J.Y., 2014. Constituents actually responsible for the antioxidant activities of crude polysaccharides isolated from mushrooms. J. Funct. Foods 11, 548–556.

Smidsrød, O., Haug, A., Westman, S., Gronowitz, S., Christiansen, H., Rosén, U., 1968. A light scattering study of alginate. Acta Chem. Scand. 22, 797–810.

Smidsrød, O., Glover, R.M., Whittington, S.G., 1973. The relative extension of alginates having different chemical composition. Carbohyd. Res. 27, 107–118.

Song, G., Du, Q., 2012. Structure characterization and antitumor activity of an α β-glucan polysaccharide from *Auricularia polytricha*. Food Res. Int. 45, 381–387.

Suga, T., Shiio, T., Maeda, Y.Y., Chihara, G., 1984. Antitumor activity of lentinan in murine syngeneic and autochthonous hosts and its suppressive effect on 3-methylcholanthrene-induced carcinogenesis. Cancer Res. 44, 5132–5137.

Sun, L., Wang, C., Shi, Q., Ma, C., 2009. Preparation of different molecular weight polysaccharides from *Porphyridium cruentum* and their antioxidant activities. Int. J. Biol. Macromol. 45, 42–47.

Sun, Y., Li, X., Yang, J., Liu, J., Kennedy, J.F., 2010. Water-soluble polysaccharide from the fruiting bodies of *Chroogomphis rutilus* (Schaeff.: Fr.) O. K. Miller: isolation, structural features and its scavenging effect on hydroxyl radical. Carbohyd. Polym. 80, 720–724.

Surenjav, U., Zhang, L., Xu, X., Zhang, X., Zeng, F., 2006. Effects of molecular structure on antitumor activities of (1→3)-β-d-glucans from different *Lentinus edodes*. Carbohyd. Polym. 63, 97–104.

Suresh, V., Senthilkumar, N., Thangam, R., Rajkumar, M., Anbazhagan, C., Rengasamy, R., et al., 2013. Separation, purification and preliminary characterization of sulfated polysaccharides from Sargassum plagiophyllum and its in vitro anticancer and antioxidant activity. Process Biochem. 48, 364–373.

Synytsya, A., ČopıKová, J., Matějka, P., Machovič, V., 2003. Fourier transform Raman and infrared spectroscopy of pectins. Carbohyd. Polym. 54, 97–106.

Taguchi, T., 1987. Clinical efficacy of lentinan on patients with stomach cancer: end point results of a four-year follow-up survey. Cancer Detect. Prev. 1, 333–349.

Taguchi, T., Furue, H., Kimura, T., Kondo, T., Hattori, T., Itoh, I., et al., 1985. Results of phase III study of lentinan. Gan to Kagaku Ryoho Cancer Chemother. 12, 366–378.

Tao, Y., Zhang, Y., Zhang, L., 2009. Chemical modification and antitumor activities of two polysaccharide–protein complexes from *Pleurotus tuber-regium*. Int. J. Biol. Macromol. 45, 109–115.

Thetsrimuang, C., Khammuang, S., Chiablaem, K., Srisomsap, C., Sarnthima, R., 2011. Antioxidant properties and cytotoxicity of crude polysaccharides from *Lentinus polychrous* Lév. Food Chem. 128, 634–639.

Tie, L., Yang, H.Q., An, Y., Liu, S.Q., Han, J., Xu, Y., et al., 2012. *Ganoderma lucidum* polysaccharide accelerates refractory wound healing by inhibition of mitochondrial oxidative stress in type 1 diabetes. Cell. Physiol. Biochem. 29, 583–594.

Tong, H., Liang, Z., Wang, G., 2008. Structural characterization and hypoglycemic activity of a polysaccharide isolated from the fruit of *Physalis alkekengi L.* Carbohyd. Polym. 71, 316–323.

Torsdottir, I., Alpsten, M., Andersson, H., Einarsson, S., 1989. Dietary guar gum effects on postprandial blood glucose, insulin and hydroxyproline in humans. J. Nutr. 119, 1925–1931.

Torsdottir, I., Alpsten, M., Holm, G., Sandberg, A.S., Tölli, J., 1991. A small dose of soluble alginate-fiber affects postprandial glycemia and gastric emptying in humans with diabetes. J. Nutr. 121, 795–799.

Tsiapali, E., Whaley, S., Kalbfleisch, J., Ensley, H.E., Browder, I.W., Williams, D.L., 2001. Glucans exhibit weak antioxidant activity, but stimulate macrophage free radical activity. Free Radical Bio. Med. 30, 393–402.

Turcanu, V., Williams, N.A., 2001. Cell identification and isolation on the basis of cytokine secretion: a novel tool for investigating immune responses. Nat. Med. 7, 373–376.

Ukai, S., Kiho, T., Hara, C., Kuruma, I., Tanaka, Y., 1983. Polysaccharides in fungi. XIV. Anti-inflammatory effect of the polysaccharides from the fruit bodies of several fungi. Journal of pharmacobio-dynamics 6, 983–990.

Ulevitch, R.J., Tobias, P.S., 1995. Receptor-dependent mechanisms of cell stimulation by bacterial endotoxin. Annu. Rev. Immunol. 13, 437–457.

Unursaikhan, S., Xiaojuan, X.U., Zeng, F., Zhang, L., 2006. Antitumor activities of -sulfonated derivatives of (1→3)-α--Glucan from different. Biosci. Biotech. Bioch. 70, 38–46.

Verraest, D.L., Peters, J.A., Batelaan, J.G., Bekkum, H.V., 1995. Carboxymethylation of inulin. Carbohyd. Res. 271, 101–112.

Vuksan, V., Panahi, S., Lyon, M., Rogovik, A.L., Jenkins, A.L., Leiter, L.A., 2009. Viscosity of fiber preloads affects food intake in adolescents. Nutr. Metab. Cardiovas. 19, 498–503.

Wang, Q., Fang, Y., 2005. Analysis of sugars in traditional Chinese drugs. J. Chromatogr. B 812, 309–324.

Wang, Y.L., Tan, Q.J., 2012. Comparison of polygonatum odoratum quality of different strains and different growth years. Guizhou Agri. Sci. 39, 50–52.

Wang, Y., Zhang, M., Dong, R., Shashkov, A.S., Kilcoyne, M., Savage, A.V., et al., 2004. Chemical components and molecular mass of six polysaccharides isolated from the sclerotium of *Poria cocos*. Carbohyd. Res. 339, 327–334.

Wang, J., Zhang, L., Yu, Y., Cheung, P.C., 2009a. Enhancement of antitumor activities in sulfated and carboxymethylated polysaccharides of *Ganoderma lucidum*. J. Agr. Food Chem. 57, 10565–10572.

Wang, Y.J., Yu, Y.Z., Mao, J.W., 2009b. Carboxymethylated β-glucan derived from *Poria cocos* with biological activities. J. Agr. Food Chem. 57, 10913–10915.

Wang, D., Guo, Z., Ma, X., Hu, Y., Huang, X., Fan, Y., et al., 2010a. Effects of sulfated lentinan on cellular infectivity of avian infectious bronchitis virus. Carbohyd. Polym. 79, 461–465.

Wang, J., Guo, H., Zhang, J., Wang, X., Zhao, B., Yao, J., et al., 2010b. Sulfated modification, characterization and structure-antioxidant relationships of *Artemisia sphaerocephala* polysaccharides. Carbohyd. Polym. 81, 897–905.

Wang, J., Hu, Y., Wang, D., Liu, J., Zhang, J., Abula, S., et al., 2010c. Sulfated modification can enhance the immune-enhancing activity of *Lycium barbarum* polysaccharides. Cell. Immunol. 263, 219–223.

Wang, G., Dong, L., Zhang, Y., Ji, Y., Xiang, W., Zhao, M., 2012a. Polysaccharides from *Phellinus linteus* inhibit cell growth and invasion and induce apoptosis in HepG2 human hepatocellular carcinoma cells. Biologia. 67, 247–254.

Wang, R., Chen, P., Jia, F., Tang, J., Ma, F., 2012b. Optimization of polysaccharides from *Panax japonicus* C.A. Meyer by RSM and its anti-oxidant activity. Int. J. Biol. Macromol. 50, 331–336.

Wang, C.L., Meng, M., Liu, S.B., Wang, L.R., Hou, L.H., Cao, X.H., 2013a. A chemically sulfated polysaccharide from Grifola frondos induces HepG2 cell apoptosis by notch1–NF-κB pathway. Carbohyd. Polym. 95, 282–287.

Wang, L., Zhang, Y., Xu, M., Wang, Y., Cheng, S., Liebrecht, A., et al., 2013b. Anti-diabetic activity of Vaccinium bracteatum Thunb. leaves' polysaccharide in STZ-induced diabetic mice. Int. J. Biol. Macromol. 61, 317–321.

Wang, Y., Zhao, H., Miao, X., Liu, D., Jiang, H., Liu, P., et al., 2013c. Structural determination and antitumor activities of a water-soluble polysaccharide from *Mortierella hepiali*. Fitoterapia 86, 13–18.

Wang, Y., Zhao, Y., Andrae-Marobela, K., Okatch, H., Xiao, J., 2013d. Tea polysaccharides as food antioxidants: an old woman's tale? Food Chem. 138, 1923–1927.

Wei, C., Yu, M.H., Li, Y.M., Chen, W.J., Xia, Y.P., 2010. Beneficial effects of astragalus polysaccharides treatment on cardiac chymase activities and cardiomyopathy in diabetic hamsters. Acta Diabetol. 47 (Suppl 1), 35–46.

Wei, Z., Zheng, L., Zhang, Z., Hai, C.X., 2012. Protective effect of a water-soluble polysaccharide from *Salvia miltiorrhiza* Bunge on insulin resistance in rats. Carbohyd. Polym. 89, 890–898.

Wijesekara, I., Pangestuti, R., Kim, S.K., 2011. Biological activities and potential health benefits of sulfated polysaccharides derived from marine algae. Carbohyd. Polym. 84, 14–21.

Wijesinghe, W.A.J.P., Athukorala, Y., Jeon, Y.J., 2011. Effect of anticoagulative sulfated polysaccharide purified from enzyme-assistant extract of a brown seaweed *Ecklonia cava* on Wistar rats. Carbohyd. Polym. 86, 917–921.

Wu, D.M., Duan, W.Q., Yong, L., Ying, C., 2010. Anti-inflammatory effect of the polysaccharides of Golden needle mushroom in burned rats. Int. J. Biol. Macromol. 46, 100–103.

Wu, F., Yan, H., Ma, X., Jia, J., Zhang, G., Guo, X., et al., 2012a. Comparison of the structural characterization and biological activity of acidic polysaccharides from *Cordyceps militaris* cultured with different media. World J. Microb. Biot. 28, 2029–2038.

Wu, J., Zhou, J., Lang, Y., Yao, L., Xu, H., Shi, H., et al., 2012b. A polysaccharide from *Armillaria mellea* exhibits strong in vitro anticancer activity via apoptosis-involved mechanisms. Int. J. Biol. Macromol. 51, 663–667.

Xia, L., Liu, X., Guo, H., Zhang, H., Zhu, J., Ren, F., 2012. Partial characterization and immunomodulatory activity of polysaccharides from the stem of *Dendrobium officinale* (Tiepishihu) in vitro. J. Funct. Foods 4, 294–301.

Xiao, J.B., Jiang, H., 2015. A review on the structure-function relationship aspect of polysaccharides from tea materials. Crit. Rev. Food Sci. 55, 930–938.

Xiao, R.Z., Wen, X.Z., Yong, X.Z., Qi, C.H., Yan, H., Zhong, F.W., et al., 2011. Macrophages, rather than T and B cells are principal immunostimulatory target cells of *Lycium barbarum L.* polysaccharide LBPF4-OL. J. Ethnopharmacol. 136, 465–472.

Xiao, C., Wu, Q.P., Cai, W., Tan, J.B., Yang, X.B., Zhang, J.M., 2012. Hypoglycemic effects of *Ganoderma lucidum* polysaccharides in type 2 diabetic mice. Arch. Pharm. Res. 35, 1793–1801.

Xiao, Z.Q., Wang, Y.L., Yue, Y.D., Zhang, Y.T., Chen, C.P., Wan, L.S., et al., 2013. Preventive effects of polysaccharides from *Liriope spicata* var. prolifera on diabetic nephropathy in rats. Int. J. Biol. Macromol. 61, 114–120.

Xie, J.H., Liu, X., Shen, M.Y., Nie, S.P., Zhang, H., Li, C., et al., 2013. Purification, physicochemical characterisation and anticancer activity of a polysaccharide from *Cyclocarya paliurus* leaves. Food Chem. 136, 1453–1460.

Xin, T., Zhang, F., Jiang, Q., Chen, C., Huang, D., Lv, Y., et al., 2012. Purification and antitumor activity of two acidic polysaccharides from the roots of *Polygala tenuifolia*. Carbohyd. Polym. 90, 1671–1676.

Xu, R., Hong, Y., Yi, S., Tu, Y., Zeng, X., 2012a. Preparation, preliminary characterization, antioxidant, hepatoprotective and antitumor activities of polysaccharides from the flower of tea plant (*Camellia sinensis*). Food Chem. Toxicol. 50, 2473–2480.

Xu, X., Chen, P., Zhang, L., Ashida, H., 2012b. Chain structures of glucans from *Lentinus edodes* and their effects on NO production from RAW 264.7 macrophages. Carbohyd. Polym. 87, 1855–1862.

Xu, W., Zhou, Q., Yin, J.J., Yao, Y., Zhang, J.L., 2014. Anti-diabetic effects of polysaccharides from Talinum triangulare in streptozotocin (STZ)-induced type 2 diabetic male mice. Int. J. Biol. Macromol. 72, 575–579.

Yan, J.K., Wang, W.Q., Li, L., Wu, J.Y., 2011. Physiochemical properties and antitumor activities of two α-glucans isolated from hot water and alkaline extracts of Cordyceps (Cs-HK1) fungal mycelia. Carbohyd. Polym. 85, 753–758.

Yan, C., Hu, M., Cui, W., Yang, Y., Chen, J., Ding, J., et al., 2013a. Characterization and in vitro antitumor activity of polysaccharides from the mycelium of *Sarcodon aspratus*. Int. J. Biol. Macromol. 52, 52–58.

Yan, C., Yin, Y., Zhang, D., Yang, W., Yu, R., 2013b. Structural characterization and in vitro antitumor activity of a novel polysaccharide from *Taxus yunnanensis*. Carbohyd. Polym. 96, 389–395.

Yanaki, T., Ito, W., Tabata, K., Kojima, T., Norisuye, T., Takano, N., et al., 1983. Correlation between the antitumor activity of a polysaccharide schizophyllan and its triple-helical conformation in dilute aqueous solution. Biophys. Chem. 17, 337–342.

Yang, L., Zhang, L.M., 2009. ChemInform abstract: chemical structural and chain conformational characterization of some bioactive polysaccharides isolated from natural sources. Cheminform. 76, 349–361.

Yang, J., Du, Y., Huang, R., Wan, Y., Yan, W., 2005. The structure–anticoagulant activity relationships of sulfated lacquer polysaccharide: effect of carboxyl group and position of sulfation. Int. J. Biol. Macromol. 36, 9–15.

Yang, P.M., Dai, L., Qin-Qing, L.I., Cao, G.S., Wei, Y.L., 2010. Effect of polysaccharides of *Oldenlandia diffusa* on growth of Bel-7402 and hela cells. Res. Practice Chinese Med. 3, 015.

Yang, L., Zhao, T., Wei, H., Zhang, M., Zou, Y., Mao, G., et al., 2011. Carboxymethylation of polysaccharides from *Auricularia auricula* and their antioxidant activities in vitro. Int. J. Biol. Macromol. 49, 1124–1130.

Yang, Y., Liu, D., Wu, J., Chen, Y., Wang, S., 2011. In vitro antioxidant activities of sulfated polysaccharide fractions extracted from *Corallina officinalis*. Int. J. Biol. Macromol. 49, 1031–1037.

Yang, W., Pei, F., Shi, Y., Zhao, L., Fang, Y., Hu, Q., 2012. Purification, characterization and anti-proliferation activity of polysaccharides from *Flammulina velutipes*. Carbohyd. Polym. 88, 474–480.

Yang, B., Bing, X., Sun, T., 2013a. Antitumor and immunomodulatory activity of *Astragalus membranaceus* polysaccharides in H22 tumor-bearing mice. Int. J. Biol. Macromol. 62, 287–290.

Yang, C., Gou, Y., Chen, J., An, J., Chen, W., Hu, F., 2013b. Structural characterization and antitumor activity of a pectic polysaccharide from *Codonopsis pilosula*. Carbohyd. Polym. 98, 886–895.

Yao, X., Li-Jia, X.U., Xiao, W., Peng, Y., Xiao, P.G., 2011. Analysis of *Lycium barbarum* polysaccharide from different lycii fructus. Herald of Medicine 4, 004.

Ye, H., Wang, K., Zhou, C., Liu, J., Zeng, X., 2008. Purification, antitumor and antioxidant activities in vitro of polysaccharides from the brown seaweed *Sargassum pallidum*. Food Chem. 111, 428–432.

Ye, Y., Deng, T., Wan, X.Y., Ouyang, J.P., Liu, M., Mao, X.Q., 2013. The role of quantitative changes in the epxression of insulin receptor substrate-1 and nuclear ubiquitin in abnormal glycometabolism in the livers of KKay mice and the relative therapeutic mechanisms of Astragalus polysaccharide. Int. J. Mol. Med. 33, 341–350.

Yi, C., Xie, M.Y., Gong, X.F., 2007. Microwave-assisted extraction used for the isolation of total triterpenoid saponins from *Ganoderma atrum*. J. Food Eng. 81, 162–170.

Yi, C., Xie, M.Y., Nie, S.P., Chang, L., Wang, Y.X., 2008. Purification, composition analysis and antioxidant activity of a polysaccharide from the fruiting bodies of *Ganoderma atrum*. Food Chem. 107, 231–241.

Yin, J., Lin, H., Li, J., Wang, Y., Cui, S.W., Nie, S., et al., 2012a. Structural characterization of a highly branched polysaccharide from the seeds of *Plantago asiatica L.* Carbohyd. Polym. 87, 2416–2424.

Yin, J.Y., Chan, C.L., Yu, H., Lau, Y.K., Han, X.Q., Cheng, S.W., et al., 2012b. Separation, structure characterization, conformation and immunomodulating effect of a hyperbranched heteroglycan from Radix Astragali. Carbohyd. Polym. 87, 667–675.

Yin, J.Y., Lin, H.X., Nie, S.P., Cui, S.W., Xie, M.Y., 2012c. Methylation and 2D NMR analysis of arabinoxylan from the seeds of *Plantago asiatica L.* Carbohyd. Polym. 88, 1395–1401.

Yokoyama, W.H., Hudson, C.A., Knuckles, B.E., Chiu, M.C.M., Sayre, R.N., Turnlund, J.R., et al., 1997. Effect of barley β-Glucan in durum wheat pasta on human glycemic response. Cereal Chem. 74, 293–296.

You, L., Gao, Q., Feng, M., Yang, B., Ren, J., Gu, L., et al., 2013. Structural characterisation of polysaccharides from *Tricholoma matsutake* and their antioxidant and antitumour activities. Food Chem. 138, 2242–2249.

Yu, P., Li, N., Liu, X., Zhou, G., Zhang, Q., Li, P., 2003. Antihyperlipidemic effects of different molecular weight sulfated polysaccharides from *Ulva pertusa* (Chlorophyta). Pharmacol. Res. 48, 543–549.

Yu, R., Yang, W., Song, L., Yan, C., Zhang, Z., Zhao, Y., 2007. Structural characterization and antioxidant activity of a polysaccharide from the fruiting bodies of cultured *Cordyceps militaris*. Carbohyd. Polym. 70, 430–436.

Yu, C.H., Dai, X.Y., Chen, Q., Zang, J.N., Deng, L.L., Liu, Y.H., et al., 2013. Hypolipidemic and antioxidant activities of polysaccharides from Rosae Laevigatae Fructus in rats. Carbohyd. Polym. 94, 56–62.

Zákány, J., Chihara, G., Fachet, J., 1980. Effect of lentinan on tumor growth in murine allogeneic and syngeneic hosts. Int. J. Cancer. 25, 371–376.

Zhang, L., Yang, L., 1995. Properties of *Auricularia auricula-judae* β-D-glucan in dilute solution. Biopolymers. 36, 695–700.

Zhang, L., Zhang, X., Zhou, Q., Zhang, P., Zhang, M., Li, X., 2001. Triple helix of l[beta]l-D-Glucan from *Lentinus edodes* in 0.5 M NaCl aqueous solution characterized by light scattering. Polym. J. 33, 317–321.

Zhang, H.N., He, J.H., Yuan, L., Lin, Z.B., 2003. In vitro and in vivo protective effect of *Ganoderma lucidum* polysaccharides on alloxan-induced pancreatic islets damage. Life Sci. 73, 2307–2319.

Zhang, L., Li, X., Xu, X., Zeng, F., 2005a. Correlation between antitumor activity, molecular weight, and conformation of lentinan. Carbohyd. Res. 340, 1515–1521.

Zhang, M., Chen, H., Huang, J., Li, Z., Zhu, C., Zhang, S., 2005b. Effect of *Lycium barbarum* polysaccharide on human hepatoma QGY7703 cells: inhibition of proliferation and induction of apoptosis. Life Sci. 76, 2115–2124.

Zhang, M., Cui, S.W., Cheung, P.C.K., Wang, Q., 2007. Antitumor polysaccharides from mushrooms: a review on their isolation process, structural characteristics and antitumor activity. Trends Food Sci. Tech. 18, 4–19.

Zhang, Z.C., Lian, B., Huang, D.M., Cui, F.J., 2009. Compare activities on regulating lipid-metabolism and reducing oxidative stress of diabetic rats of tremella Aurantialba Broth's Extract (TBE) with Its Mycelia Polysaccharides (TMP). J. Food Sci 74, H15–H21.

Zhang, A.L., Jia-Hui, L.U., Zhang, N., Zheng, D., Zhang, G.R., Teng, L.R., 2010. Extraction, purification and anti-tumor activity of polysaccharide from mycelium of mutant *Cordyceps militaris*. Chem. Res. Chinese U 26, 798–802.

Zhang, D., Li, S., Xiong, Q., Jiang, C., Lai, X., 2013a. Extraction, characterization and biological activities of polysaccharides from Amomum villosum. Carbohyd. Polym. 95, 114–122.

Zhang, S., Nie, S., Huang, D., Li, W., Xie, M., 2013b. Immunomodulatory effect of *Ganoderma atrum* polysaccharide on CT26 tumor-bearing mice. Food Chem. 136, 1213–1219.

Zhang, Z., Meng, Y., Guo, Y., Xin, H., Liu, Q., Wang, X., et al., 2013c. *Rehmannia glutinosa* polysaccharide induces maturation of murine bone marrow derived Dendritic cells (BMDCs). Int. J. Biol. Macromol. 54, 136–143.

Zhang, T., Gao, J., Jin, Z.Y., Xu, X.M., Chen, H.Q., 2014. Protective effects of polysaccharides from *Lilium lancifolium* on streptozotocin-induced diabetic mice. Int. J. Biol. Macromol. 65, 436–440.

Zhang, H., Cui, S.W., Nie, S.P., Chen, Y., Wang, Y.X., Xie, M.Y., 2016. Identification of pivotal components on the antioxidant activity of polysaccharide extract from *Ganoderma atrum*. Bioact. Carbohyd. Dietary Fibre 7, 9–18.

Zhao, H., Li, Y., Wang, Y., Zhang, J., Ouyang, X., Peng, R., et al., 2012a. Antitumor and immunostimulatory activity of a polysaccharide-protein complex from Scolopendra subspinipes mutilans L. Koch in tumor-bearing mice. Food Chem. Toxicol. 50, 2648–2655.

Zhao, M., Zhang, Z.F., Ding, Y., Wang, J.B., Li, Y., 2012b. Astragalus polysaccharide improves palmitate-induced insulin resistance by inhibiting PTP1B and NF-κB in C2C12 myotubes. Molecules 17, 7083–7092.

Zhao, W., Yin, Y., Yu, Z., Liu, J., Feng, C., 2012c. Comparison of anti-diabetic effects of polysaccharides from corn silk on normal and hyperglycemia rats. Int. J. Biol. Macromol. 50, 1133–1137.

Zhao, R., Gao, X., Cai, Y., Shao, X., Jia, G., Huang, Y., et al., 2013a. Antitumor activity of *Portulaca oleracea L.* polysaccharides against cervical carcinoma in vitro and in vivo. Carbohyd. Polym. 96, 376–383.

Zhao, T., Mao, G., Mao, R., Ye, Z., Zheng, D., Feng, W., et al., 2013b. Antitumor and immunomodulatory activity of a water-soluble low molecular weight polysaccharide from *Schisandra chinensis* (Turcz.) Baill. Food Chem. Toxicol. 55, 609–616.

Zhao, X.H., Qian, L., Yin, D.L., Zhou, Y., 2014a. Hypolipidemic effect of the polysaccharides extracted from pumpkin by cellulase-assisted method on mice. Int. J. Biol. Macromol. 64, 137–138.

Zhao, Y., Yang, X., Ren, D., Wang, D., Xuan, Y., 2014b. Preventive effects of jujube polysaccharides on fructose-induced insulin resistance and dyslipidemia in mice. Food Funct. 5, 1771–1778.

Zheng, J., Yang, B., Yu, Y., Chen, Q., Huang, T., Li, D., 2012. *Ganoderma lucidum* polysaccharides exert anti-hyperglycemic effect on streptozotocin-induced diabetic rats through affecting β-cells. Comb. Chem. High T. Scr. 15, 542–550.

Zhou, X., Xu, Y., Yang, G., Fang, L., 2011. Increased galectin-1 expression in muscle of Astragalus polysaccharide-treated Type 1 diabetic mice. J. Nat. Med. 65, 500–507.

Zhu, C.P., Zhang, S.H., 2013. *Lycium barbarum* polysaccharide inhibits the proliferation of HeLa cells by inducing apoptosis. J. Sci. Food Agr. 93, 149–156.

Zhu, J., Song, Y., 2006. Experimental study of the A549 cell apoptosis of human non-small cell lung cancer induced by ginseng polysaccharide in vitro. Zhongguo Zhong xi yi jie he za zhi 26 (Suppl), 95–97.

Zhu, J., Zhang, Y., Shen, Y., Zhou, H., Yu, X., 2013a. *Lycium barbarum* polysaccharides induce Toll-like receptor 2- and 4-mediated phenotypic and functional maturation of murine dendritic cells via activation of NF-κB. Mol. Med. Rep. 8, 1216–1220.

Zhu, K., Nie, S., Li, C., Lin, S., Xing, M., Li, W., et al., 2013b. A newly identified polysaccharide from *Ganoderma atrum* attenuates hyperglycemia and hyperlipidemia. Int. J. Biol. Macromol. 57, 142–150.

Zhu, K.X., Nie, S.P., Li, C., Gong, D., Xie, M.Y., 2014. Ganoderma atrum polysaccharide improves aortic relaxation in diabetic rats via PI3K/Akt pathway. Carbohyd. Polym. 103, 520–527.

Zong, A., Cao, H., Wang, F., 2012. Anticancer polysaccharides from natural resources: a review of recent research. Carbohyd. Polym. 90, 1395–1410.

Chapter 2

Methodologies for Studying Bioactive Polysaccharides

The bioactivities of polysaccharides are highly dependent on their structural conformation, such as molar mass, extent of side chains/groups or substitution, and their distribution on the backbone. How to isolate the polysaccharides from complex matrix networks, while minimizing any loss of the desired bioactivity, is one of the most important tasks to deal with.

Polysaccharides extracted from natural sources show various important bioactivities, including anti-tumor, immunomodulatory, and anti-inflammatory effects, which are strongly influenced by their structures and conformations. Thus, the relationship between chemical structures, chain conformations of polysaccharides, and their biological activities is important and significant. However, polysaccharides are constituted of different monosaccharides, which are linked by various glucosidic bonds. Some polysaccharides have hyper-branched structures. Moreover, polysaccharides often have high molecular weights, and tend to form aggregates in solution that can mask the behavior of individual macromolecules. As a consequence, to characterize the chemical structures and chain conformations of polysaccharides is not an easy task.

The bioactive activity of the polysaccharides can be evaluated by cell culture experiments, animal studies, and clinical trials.

2.1 ISOLATION, PURIFICATION, AND CHARACTERIZATION

Generally, bioactive polysaccharides are present together with various other components, such as proteins, polynucleotides, lipids, extractives, lignin, and some inorganic mineral substances. However, the desired biological activities of the naturally bioactive polysaccharides may be undermined by other compounds, which may even cause antagonistic effects or undesirable toxicity. In other words, pure bioactive polysaccharides could enable the safe, reproducible, and accurate dosage for experimental or therapeutic applications, and also enable the investigation of the structure – activity relationship, facilitating the development of new compounds with similar or more highly desirable bioactivities (Colegate and Molyneux, 2007). Thus, the isolation of natural polysaccharides that have biological activities toward organisms from various sources plays an important role in the investigation and application of bioactive polysaccharides.

Bioactive Polysaccharides. DOI: http://dx.doi.org/10.1016/B978-0-12-809418-1.00002-2

Fungi, algae, and plants are often powdered, hydrolyzed using cellulase, or experience ultrasonication for removal of the fat and substances which are oil soluble (Yang et al., 2008). Polysaccharides are extracted using water, aqueous NaCl solution, and dilute acidic or alkaline aqueous solutions. Hydrogen peroxide is usually used to decolor polysaccharides. The Sevag method is normally used to remove the free protein in polysaccharides (Schappacher et al., 2005). The Sevag reagent contains chloroform and *n*-butanol (4/1, v/v), which can cause proteins to become gels and precipitate.

In recent years, novel and effective extraction methods, such as supercritical fluid extraction, microwave-assisted extraction, and, the most promising, hot-water extraction, have received increasing attention due to their environmentally friendly process, higher extraction efficiency, cost-effectiveness, and structure-preservation ability (Normand et al., 2014; Tao et al., 2013; Zhao et al., 2013). For example, Cheng et al. (2013) compared hot-water, ultrasonic assisted, enzyme, and microwave-assisted extractions to isolate bioactive polysaccharides, and discovered that the four polysaccharides yielded had similar physicochemical properties; however, some activities of the polysaccharides obtained by hot-water extraction were stronger than those isolated with other methods. Hot-water extraction combined with some novel supporting methods, such as microwave, ultrasonic, and enzymatic pretreatment, increases the extraction efficiency and yield of products. For example, enzymatic pretreatment of the raw material before extraction normally results in a reduction in extraction time, lowers energy consumption, minimizes the usage of solvents, increases the yield, and maximally preserves biological activities of the product when compared to non-enzymatic methods (Chen et al., 2014; Dong et al., 2011; Jia et al., 2013; Puri et al., 2011). Recently, ionic liquids have also been developed to extract polysaccharides at low temperatures and over a short time (Abe et al., 2010).

Further purification of bioactive polysaccharides from crude extracts is of high importance, to ensure understanding of the relationship between structures and the safety of future biomedical, pharmaceutical, and food applications. This can be done with a combination of several techniques, such as ethanol precipitation, fractional precipitation, ion-exchange chromatography, gel filtration, and affinity chromatography, by taking advantage of particular properties of the desired compound such as acidity, polarity, and molecular size (Jin et al., 2013). They can be separated into two kinds:

1. Polysaccharides with different M_w and sizes are separated using methanol, ethanol, and acetone, or gel chromatography.
2. Acidic and neutral polysaccharides can be separated by cetyltrimethyl ammonium bromide or cetylpyridinium chloride. They can also be separated by ion-exchange cellulose chromatography (Dong et al., 2007; Mazumder et al., 2004), DEAE-Toyopearl column (Urai et al., 2007), or DEAE-Sepharose fast-flow column (Yang et al., 2007).

2.2 STRUCTURAL ANALYSIS

The primary structure of a complex carbohydrate is known only when all of the following characteristics have been elucidated: (1) composition of monosaccharide; (2) patterns of linkage; (3) size of ring; (4) anomeric configuration; (5) monosaccharide residue sequences and repeating units; (6) substitutions; (7) molecular weight distribution (Mcneil et al., 1982).

Polysaccharide structure analysis requires different techniques. Various methods have been used extensively and successfully to sequence complex carbohydrates (Wang and Cui, 2005). Related methods can be classified as chemical methods, instrumental methods, and biological method.

2.2.1 Chemical Methods

The hydrolysis method has been set up to cleave glycosidic bonds. One of the earliest chemical degradation methods used involves the acid-catalyzed partial hydrolysis of a complex carbohydrate into oligosaccharide fragments, which are then purified (Mcneil et al., 1982). The hydrolytic agents often used are trifluoroacetic acid, sulfuric acid, or hydrofluoric acid, and others (Biermann, 1988; Ruiter et al., 1992). Hydrolysis with trifluoroacetic acid, which could be removed after hydrolysis, is found to be difficult for β-(1−4)-linked cellulose hydrolysis (Selvendran et al., 1979), hydrolysis of mucins (Griggs et al., 1971), hydrolysis of *Haemophilus influenzae* polysaccharides (Ip et al., 1992), and hydrolysis of glycosaminoglycans (Höök et al., 1974).

2.2.1.1 Methanolysis

Another method for polysaccharide analysis is to analyze their sugar residues using methanolysis. Methanolysis with acid is better than hydrolysis with acid for hemicelluloses analysis (Sundberg et al., 1996), because it can be done in only one step. Methanolysis with acid by hydrochloric acid (in anhydrous methanol) could result in neutral and acid sugar units in pectin (Quemener et al., 1993; Bertaud et al., 2002).

2.2.1.2 Acetolysis

Partial acetolysis was applied to 16 dextrans and obtained some novel information on the relationship of the kojibiose yields with pneumococcus antiserum (type-12) reactivity (Suzuki and Hehre, 1964). This reaction could result in peracetylation and selected glycosidic bond cleavage during acetolysis treatment of the dextran, which contained 1,2-α, 1,3-α, and 1,6-α glucosidic linkages; 1,6-α linkages were cleaved in the first place, then 1,2-α and 1,3-α linkages were left as gluco-bioses (kojibiose and nigerose) containing these linkages (Fujimoto et al., 1962).

2.2.1.3 Periodate Oxidation and Smith Degradation

Periodate oxidation carried out by reduction with either hydrogen or a Raney nickel catalyst under pressure, or with sodium borohydride in aqueous solution between 60 and 100 shows a common method which can be set up for the polysaccharides as well. Results give information on the nature and amount of glycosidic linkages in the polysaccharide (Abdelakher et al., 1952). The previous investigations showed that access by periodate ions (IO_4^-) to oxidizable glycol groups of acid polysaccharides, such as heparin and polygalacturonate, was reduced by the intense negative electrostatic field of the polyanion (Scott and Dorling, 1969). The utility of the method of periodate oxidation was extended by applying structural analysis of water-insoluble dextrans (Rankin and Jeanes, 1954).

A combination of periodate oxidation, reduction, and mild acid hydrolysis is named the Smith Degradation. The Smith degradation produces glycerol from terminal, non-reducing groups of glycogen glucans (Hizukuri and Osaki, 1978). Then, the consequent characterization of the surviving monosaccharides and/or oligosaccharides could give information on the original polysaccharides fine structure (Wang and Cui, 2005).

2.2.1.4 Methylation Analysis

Methylation analysis is a useful method for polysaccharide structure chemistry. It involves methylation of the polysaccharide and hydrolysis to monomeric methylated sugars, which are then identified and estimated quantitatively. However, the method provides little information on the sugar residues relative order, or their anomeric nature (Björndal et al., 1970). Thus, methylation analysis is commonly carried out to evaluate the position of linkages between monosaccharide residues in oligo- and polysaccharides (Harris et al., 1984).

There are a series of improved methods for methylation analysis, such as the Haworth method (Haworth, 1915), the Menzies method (Fear and Menzies, 1926), the Hakomori method (Hakomori, 1964), the Harris method (Harris et al., 1984), Ciucanu method (Ciucanu and Kerek, 1984), and the Needs method (Needs and Selvendran, 1993). Particularly, an improved method by Harris et al. (1984), which gives high production for the acetylated alditols in methylated sugars, and allows investigation for glycosyl linkages (oligo- or poly-saccharides) to be done in only one day, is convenient.

2.2.2 Instrumental Methods

2.2.2.1 UV-Vis Spectrophotometry and FTIR Spectroscopy

The contents of uronic acids in polysaccharides are determined by spectrophotometric methods, for example, carbazole/sulfuric acid and m-hydroxydiphenyl/sulfamate (Radhakrishnamurthy and Berenson, 2002; Blumenkrantz

TABLE 2.1 **The Stretching Vibration (ν) of Some Pyranoses (Stockdale et al., 2009)**

Type	Stretching Vibration (cm^{-1})
α-D-Glucose	855–833
β-D-Glucose	905–876
α-D-Galactose	839–810
β-D-Galactose	914–866
α-D- or α-D-Mannose	843–818
β-D-Mannose	898–888
α-D-Xylose	760–740
β-D- or β-L-Arabinose	855–830

and Asboe-Hansen, 1973). The uronic acids in polysaccharides can also be analyzed quantitatively by HPLC (Chen et al., 2004a).

The spectroscopy by FTIR is carried out to find the molecules vibrations, and the polar bonds among various atoms. Types of monosaccharide, and bonds of glucosidic and function groups in structures of polysaccharides can be evaluated by FTIR spectroscopy (Mathlouthi and Koenig, 1986; Stockdale et al., 2009). Among the 1100–1010 cm^{-1} range, three peaks of absorption were obtained for pyranoside, while there were another two for furanoside. The representative pyranoside absorptions are shown in Table 2.1. Raman spectroscopy is also used to analyze the vibrations of molecules and nonpolar bonds in the same atom (Zhu et al., 2006). In addition, the helical conformation and the biomacromolecules plane fold can be investigated. The Raman bands of several saccharides are given in Table 2.2 (Gierlinger et al., 2006).

2.2.2.2 Chromatography and Mass Spectroscopy

2.2.2.2.1 GC and HPLC

Methods of GC and HPLC are used for the hydrolysis of polysaccharides using TFA, or after derivation through methylation (Yang et al., 2007; Ciucanu and Kerek, 1984), oxidation with periodic acid (Abdelakher et al., 1952), and Smith degradation (Datta et al., 1999). For example, the oligosaccharides from heparin were separated by SAX-HPLC (Datta et al., 1999). The monosaccharide linkages of three pectic arabinogalactan type polysaccharides were evaluated by the GC-MS method after reduction of uronic acid groups (Nergard et al., 2006), methyl esterification, and methylation.

TABLE 2.2 The Raman Bands of Some Saccharides

Type	Band (cm^{-1})		References
	Skeletal Modes of Pyranose Rings	Modes of Anomeric Carbons	
α-D-Glucose	440	842	Stockdale et al. (2009)
β-D-Glucose	440	900	Gelder et al. (2007); Stockdale et al. (2009)
Trehalose	449	839	De Gussem et al. (2005); Gelder et al. (2007)
α-D-Maltose		848	Bell et al. (1994)
β-D-Isomaltose		905	Bell et al. (1994); Stockdale et al. (2009)
Amylose	481	865	De Gussem et al. (2005); Fechner et al. (2005); Kizil et al. (2002); Schuster et al. (2000); Stockdale et al. (2009); Zhbankov et al. (2000)
Amylopectin	477	850	Fechner et al. (2005); Gelder et al. (2007); Schuster et al. (2000); Stockdale et al. (2009); Zhbankov et al. (2000)
Dextran	543	844	Stockdale et al. (2009); Zhbankov et al. (2000)

2.2.2.2.2 Mass Spectroscopy

Mass spectroscopy is used for analyzing the masses of electrically charged molecules. The composition of neutral glycosyl-residues for polysaccharides are obtained by GC-MS evaluation of the alditol acetate derivatives. The methodologies first require the production of volatile derivatives. The chosen derivatization was permethylation, and it served the multiple purpose of conferring volatility (Dell and Morris, 2001).

2.2.2.2.3 FAB-MS

The hardware consists of: (1) an atom gun (or ion gun, see later) which is either mounted on the source housing of the mass spectrometer or, if small enough, inside the housing on the source itself; (2) a sample probe to the end of which is attached a small metal target onto which the sample is loaded; and (3) suitable source-optics for the efficient extraction of ions into the analyzer of the mass spectrometer. FAB-MS is usually desirable to define molecular weights by analyzing underivatized samples, because chemically labile functional groups will then be observed. FAB-MS of underivatized samples is the procedure of choice for confirming that O-acyl groups are present in an oligosaccharide or glycoconjugate. Per-*O*-acetyl and per-*O*-methyl derivatives are used extensively for sequence analysis, and for providing molecular weight information at very high sensitivity. It is an ideal procedure for monitoring the progress of many of the chemical and enzymic reactions commonly used in carbohydrate chemistry (Dell, 1987).

2.2.2.2.4 ESI-MS

Conventional ESI MS (Fenn et al., 1990; Meng et al., 1988) includes pumping the solution into the ion source, and was found to result in signals of weak ion (Burlingame et al., 1996; Reinhold et al., 1995). The nano ESI, however, could produce signals of ion that were similar for peptides and classes of carbohydrate compounds (Wilm and Mann, 1994; Bahr et al., 1997).

2.2.2.2.5 MALDI-TOF

The use of MALDI for the investigation of neutral oligosaccharide is more advanced than ESI, especially for applications including the amount of glycoprotein mixtures. Mixture quantitation of permethylated carbohydrate (with the MALDI) was found to be reproduced (Viseux et al., 1999). Maybe one of the primary advantages of MALDI for analysis of glycan is that it could have excellent glycan profiles of neutral glycan mixtures, because the trend of the technique is to produce only $[M + Na]^+$ ions, unlike electrospray that gives the spectra containing ions in various charge states with ions from various adducts (Harvey, 2012).

The MALDI-TOF mass spectrometric and HPAEC evaluations showed that xylans (enzyme hydrolysis generated fragments) show their structure with a β-(1→4)-linked xylopyranosyl residue backbone with the three 4-*O*-methyl glucuronic acid per 97 units of xylopyranosyl substituted (Mazumder et al., 2005). MALDI-TOF-MS and ESI MS/MS were used to investigate the low-molecular weight fraction of a fucoidan extracted using ethanol (Anastyuk et al., 2009). The isolation and identification of the isomeric oligosaccharides resulted from the digestion of AX with endoxylanase which can be achieved by off-line coupling of the MALDI-CID technique (Bubb, 2003).

2.2.2.3 NMR Spectroscopy

2.2.2.3.1 Liquid-State NMR

The NMR spectroscopy, particularly NMR with liquid-state, has come to be regarded as a potential instrument to characterize polysaccharide structures (Maslen et al., 2007). Almost all polysaccharides are soluble in either water or DMSO, so liquid-state NMR experiments generally choose deuterated water and DMSO to dissolve polysaccharides.

Two-dimensional (2D) NMR technologies are applied to figure out the polysaccharides fine structures like monosaccharide composition, and sorts and configurations of glucosidic bond. Two-dimensional NMR (correlated) is so far the most easily utilized and common technology (Abraham et al., 1988; Dou et al., 2013; Duus et al., 2001). It can be used for heteronuclear and homonuclear spin systems. For methods of homonuclear correlation, the related technology covers total correlation spectroscopy (TOCSY), double quantum filtered-COSY (DQF-COSY), and correlation spectroscopy (COSY). The 2D experiment related to heteronuclear, in particular for coupled systems of ^{13}C-^{1}H, has demonstrated great value, like C-COSY, H, heteronuclear single-quantum coherence, and heteronuclear multiple-quantum coherence.

Nuclear Overhauser enhancement spectroscopy (NOESY) and rotating frame Overhauser enhancement spectroscopy (ROESY) could be productive 2D methods to acquire the polysaccharide structures and configurations (Adeyeye et al., 2003; Suárez et al., 2006). It is of interest to find that DOSY may be useful for determining M_w and its dispersion for polysaccharides.

The common methods to ascertain polysaccharide structures by NMR spectroscopy of 1D and 2D are summed up in five sections (Leeuwen et al., 2008; Duus et al., 2001; Castro et al., 2008): (1) residues of saccharide; (2) monosaccharide composition; (3) glycosidic bond and sequence; (4) anomeric configuration; and (5) substituted group positions. Table 2.3 shows some chemical variations of some substituted groups.

Latterly, experiments with 1D and 2D heteronuclear or homonuclear systems, such as HMBC, HSQC, DQF-COSY, and NOESY, are frequently applied for analyzing the structures and the configurations of polysaccharides with bioactivities, e.g., antimutagenic polysaccharide from *Sepiella maindroni* ink (Liu et al., 2008a), a Lentinus heterogalactan (Carbonero et al., 2008)

TABLE 2.3 Chemical Shifts of Substituted Groups of Polysaccharides

Substituted Groups	Chemical Shift			References
	^{1}H	^{13}C	^{31}P	
Methyl	1.28	17.8	–	Liu et al. (2008a)
Acetyl	2.02	24.8	–	Liu et al. (2008a)
Carboxyl	–	176.7, 177.3	–	Liu et al. (2008a)
Phosphate	–	–	−0.78, −1.67	Ganeshapillai et al. (2008); Liu et al. (2008a)

anti-inflammatory with activities, a heteroglycan acquired from *Volvariella bombycina*, a kind of edible mushroom (Das et al., 2008), an exopolysaccharide isolated from *Leuconostoc citreum* and *Weissella colysac* (Maina et al., 2008), a polysaccharide produced by the edible mushroom Pleurotus sajor-caju (Roy et al., 2008), a polysaccharide obtained from the mushroom *Volvariella diplasia* (Ghosh et al., 2008), a β-(1→3)-D-glucan (Tada et al., 2007) obtained from *Sparassis crispa*, a β-(1→6)-α-D-glucan (Cui et al., 2008) with antioxidant activities from *Pueraria lobata* (Willd.) Ohwi, a polysaccharide (Maciel et al., 2008) from *Gracilaria birdiae*, and Fucogel, a polysaccharide (Guetta et al., 2003) from bacteria. Three-dimensional NMR experiments, or even higher multiple-dimensional experiments, could be implemented by combining experiments in 2D with various times of evolution (Duus et al., 2001). They link the reactions of various nuclei along the x, y, and z dimensions. The x and y dimensions commonly consist of proton-proton coupling (e.g., NOESY and TOCSY) (Mopper et al., 2007), related with another nucleus in the z plane (e.g., HMBC and HSQC). Augmenting the isolation of resonances in the ^{13}C dimension is the primary strength for the application of high-dimension NMR trials, leading to better resolution of the spectra (Xu and Bush, 1996; Allerdings et al., 2005). The techniques of NOESY-TOCSY (1D, 2D, and 3D) and TOCSY-NOESY (3D) tests can be used to determine the structure of a lipopolysaccharide (LPS) from *Vibrio cholerae* O139 (Cox et al., 1996).

2.2.2.3.2 Solid-State NMR

The line widths of solid-state NMR are prevalent now, mainly because of the anisotropic properties (Xue, 2012). This is on the base of magic-angle spinning that is of importance to obtain high-resolution ^{13}C solid-state NMR spectra (Gladden, 1995). The strength of the solid ^{13}C signals could be up-regulated through the process of cross-polarization, where the polarization changes from ^{1}H to ^{13}C (Gladden, 1995).

2.2.3 Biological Methods

Using enzymes in experiments of polysaccharide structures has become so popular that it is now proper to figure out the progress made by the application of such techniques. For the enzymic analysis of polysaccharide structure, hydrolytic enzymes are the most effective, and other types of enzyme are rarely mentioned in this article (Marshall, 1974). Enzymes depolymerizing polysaccharides may have an endo or an em action pattern, and may hydrolyze, or cut by elimination. Endo-enzyme cleavage is a random type of depolymerization. A glycosidic linkage positioned internally in the chain of a glycan, resulted in a series of oligosaccharide fragments, of which some can be further split.

2.3 CONFORMATIONS

The properties of polysaccharide conformation cover monosaccharide composition, monosaccharide orientation, and flexibility of the spatial structure. The related methods are as follows.

2.3.1 Light Scattering

Polysaccharide size in solution is determined by the method of dynamic light scattering (DLS) (Goh et al., 2006a,b; Li et al., 2006; Ma et al., 2008). Hydrodynamic radius (R_h) is determined by the Einstein–Stokes equation:

$$R_h = \frac{k_B T}{6\pi\eta D} \tag{2.1}$$

where D is diffusion coefficient, and η, T, and k_B, are solvent viscosity, absolute temperature, and the Boltzmann constant, respectively. The next equations show the correlations among $[\eta]$, A_2, M_w, R_g, and R_h, and the polysaccharide conformation in solution can be obtained from that equation.

$$\rho = \frac{R_g}{R_h} \tag{2.2}$$

$\rho \geqslant 2$ when the compound extended chains, ρ is 1.5–1.8 when they are flexible random coils, and 1.3 in a θ solvent, and for compact spheres ρ is about 0.775 (Nichifor et al., 1999).

$$[\eta] = KM_w^{-b} \tag{2.3}$$

K and α are constant for a polymeric compound at a certain temperature in a certain solvent. The value of α is between 0.5 and 0.8 for flexible random coils in a suitable solvent (Huang and Zhang, 2005), and it is around 0.5 when coils are compact in the θ solvent, and the value is above 0.8 and more than 1.0 for inflexible chains of the polymer (Ma et al., 2008).

$$R_g = kM_w^{\alpha'} \tag{2.4}$$

In a given solvent, k and α' remain constants when a certain polymer is at a certain temperature. Generally, a α' of around 0.7 indicates the stiff rod-like conformation, α' levels are around 0.6 for free random coils in a given solvent, and the value is about 0.5 for flexible random coils in a theta solvent (Huang and Zhang, 2005; Goh et al., 2006a,b; Tao and Zhang, 2006). The α' level is approximately 0.33 when there are terse coils in the θ solvent (Tao and Zhang, 2008).

For instance, lentinan is the β-(1→3)-D-glucan, including residues of glycopyranosyl branched at the glucose unit in position 6 (Fig. 2.1). Five residues of β-(1→3)-glucose and two side branches of β-(1→6)-glucose constitute the repeating unit. At 130–145°C, the progress of conformation variations of lentinan was from linear chains to chains of winding (Fig. 2.2).

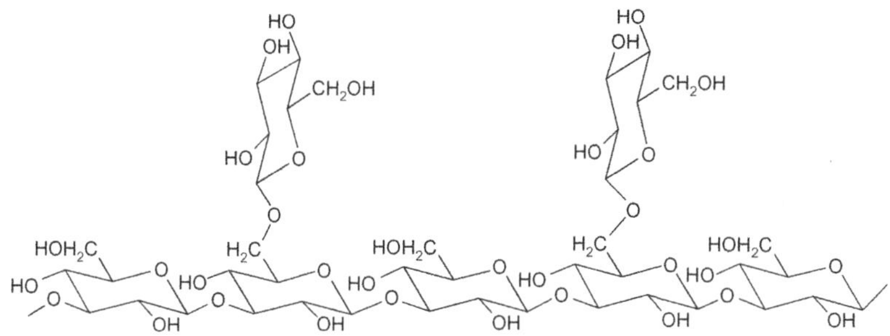

FIGURE 2.1 Chemical structure of lentinan.

FIGURE 2.2 Schematic representation of the thermally-induced conformation transition of lentinan solution at 137°C (Wang et al., 2008).

It was triggered by destroyed intra-molecular hydrogen and hydrogen linkages among molecules in lentinan. The thermally induced conformational variation was irreversible.

For polysaccharides isolated from *Ganoderma lucidum*, when water was chosen as the solvent, Huggins' constants were in the range of 0.59–0.93, indicating that aggregations of molecules occurred up to a point (Lai and Yang, 2007). However, dimethylsulfoxide was a prior choice of solvent because low Huggins' parameters were in the range of 0.39–0.48. The relationship of R_g and M_w of the (1→6)-α-D-glucan extracted from the root of the plant was shown by $R_g = 1.12 \times 10^{-2}\ M_w^{0.56 \pm 0.008}$, implying the polysaccharide existed in the aqueous solution as the random coil (Cui et al., 2008).

Conformations of several other polysaccharides were analyzed, with q and M_L values being shown in Table 2.4.

TABLE 2.4 The Chain Parameters of Some Polysaccharides

Polysaccharides	q (nm)	M_L (nm^{-1})	Flexibility	Solvent	References
Schizophyllan	170–230	2150	Triple helix	H_2O	Yanaki et al. (1980)
Lentinan	100	2240	Triple helix	0.9% aqueous NaCl solution	Zhang et al. (2005)
Xanthan	120	1940	Double helix	H_2O	Sato et al. (1984)
Succinoglycan	50	1500	Double helix or aggregate composed of paired single helices	0.1 mol/L aqueous NaCl solution	Kido et al. (2001); Nakanishi and Norisuye (2003)
β-(1→3) Glucan isolated from Auricularia auricular	90	1030	Single helix	H_2O	Zhang and Yang (1995)
Curdlan	6.8	890	Semi-flexible chains	0.3 mol/L NaOH	Nakata et al. (1998)
Water-insoluble glucan isolated from *Poria cocos* mycelia	6.3	760	Extended flexible chains	0.25 mol/L LiCl/Me_2SO	Ma et al. (2008)
Sulfated derivative of above glucan	13.1	1060	Semi-stiff chains	0.15 mol/L aqueous NaCl solution	Ma et al. (2008)
Polyguluronate	5.2	359	Semi-flexible chains	H_2O	Lee et al. (2002)
Wheat β-glucan	4.4	–	Extended random coils	0.5 mol/L NaOH	Mccleary and Matheson (1987)
Mannoxylan isolated from *Ganoderma tsugae*	4.2	832	Flexible random coils	0.25 mol/L LiCl/DMSO	Peng et al. (2005)
Water-insoluble glucan isolated from sclerotium of *Pleurotus tuber-regium*	3.1	408	Dense random-coil chains	DMSO	Zhang et al. (2001)

2.3.2 Molecular Modeling

The method of molecular simulation has a key role in conformational property evaluation, taking the structure visualization into consideration, and polymer conformation and could be necessary to trial data like 2D NMR, light scattering, X-ray, and HPSEC. Geometry optimization can be carried out by Insight II of the MSI molecular modeling package (Version 4.0.0) on a Silicon Graphics O2 workstation (Van Gunsteren and Berendsen, 1990).

2.3.2.1 *Force Fields*

The force field of AMBER-H has a revision for the effects of exo-anomeric and anomeric effects, and was particularly used for research into carbohydrates by Homans (1990). The automatic power field was induced by adding novel data to the force field of AMBER, and then was used to find the oligosaccharides conformation characters. It could provide a basic improvement, and thus better parameterizations and force fields for carbohydrate modeling (Mazeau and Rinaudo, 2004). Furthermore, the force field of PCFF was used for RMMC simulation.

2.3.2.2 *Minimization of Molecules*

Minimization of the energy was determined by the first 10,000 steps for the method of steepest descent before the method of Polak–Ribiere conjugate gradient, and finally the Quasi-Newton method (Wu et al., 2012)., The non-bond cut-offs (Li et al., 2012) were fixed at 2.0, 9.5, and 1 Å for spline width, group size, and buffer width, respectively. The dielectric constant, $\varepsilon = 80$, was set up for mimicking the effects of the solvation (Petkowicz et al., 1998). Figs. 2.3 and 2.4 present possible snapshots of some examples of polymers.

It was still a major task to determine the relationship of "fine structure" conformation, because it is hard to get information about the fine structures. However, the computer modeling for molecules having fine structures with different orientations and patterns of distribution is a better technology used to distinguish the conformational characters with the experimental data. Modeling of molecules is now becoming an indispensable conformation which is supplementary to experimental data (Chandrasekaran et al., 1998). For instance, computer simulation for chains of galactomannan was evaluated by many scientists, for example, Bergamini et al. (1995) obtained conformation properties of the oligo-galactomannans by force field of MM3 and NMR results.

2.4 ANIMAL STUDIES AND CLINICAL TRIALS

2.4.1 Animal Models

Zhou et al. (2015) evaluated hypoglycemic activity of *Misgurnus anguillicaudatus* polysaccharide in mice with STZ-induced diabetes. For one part, MAP induced an increase in levels of serum insulin and contents of

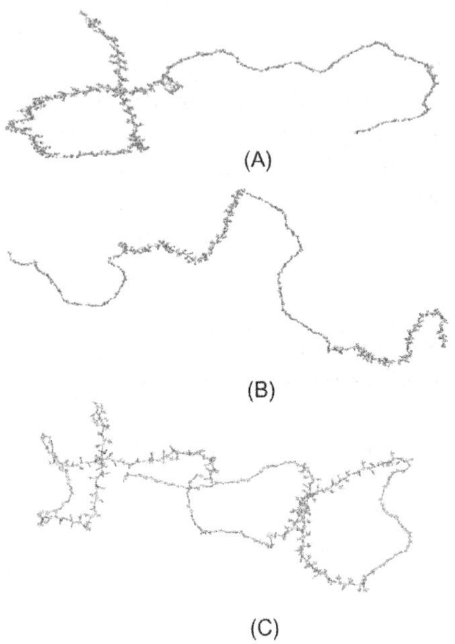

FIGURE 2.3 Snapshots of (GlcpA) Xylan chain models of various block distribution (A: 0; B: 1:2; and C: 1:1). GlcpA to Xylp ratios were kept as 1:2 for the calculation.

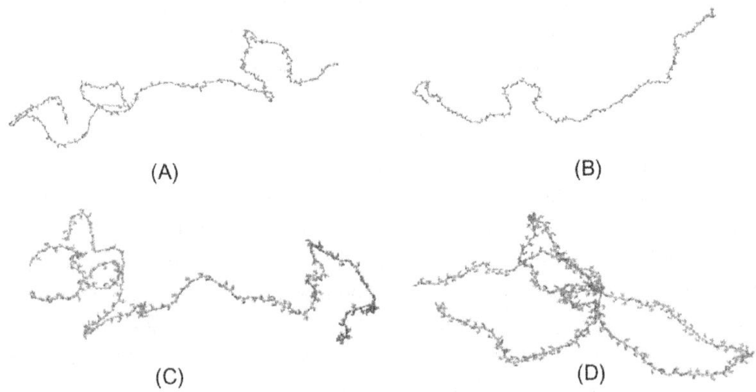

FIGURE 2.4 Snapshots of (GlcpA) Xylan chain models with different GlcpA to Xylp ratios (A: 1:3; B: 1:2; C: 2:3; and D: 1:1). Randomly distributed side chains were selected for the calculation.

pancreatic insulin. For another part, MAP decreased the improvement of expression of PEPCK mRNA and liver glycogen contents.

Hu et al. (2014) set up a study on the effects of fucosylated chondroitin sulfate on the metabolism of glucose and insulin signaling in the liver of C57BL/6 mice with insulin-resistance. Results found that CHS up-regulated the expression of mRNA of GS, IRS-2, IR, and PKB, and in insulin-resistant mice liver, thus stimulating glucose metabolism by enhancing the PI3K/PKB/GSK-3b signaling pathway.

2.4.2 Clinical Trials

Polysaccharides and their derivatives have advantages over synthetic polymers, because they are non-toxic, biodegradable, biocompatible, and less expensive compared to their synthetic counterparts. All these merits endow polysaccharides and their derivatives with a broad spectrum of applications in different areas, such as in biomedical or pharmaceutical, food, and cosmetic applications. Nowadays, polysaccharides play important roles in traditional disease control and health care; meanwhile, many new application areas are also being explored, such as in tissue engineering, drug delivery, wound treatment (both internal and external), cancer prevention, diagnosis, and therapy, and in the treatment of bacterial and viral diseases of polysaccharide and their derivatives following functionalization (Edgar et al., 2009; Lindblad et al., 2009; Symonette et al., 2014).

2.4.2.1 Tissue Engineering

Exploitation of polysaccharides and their derivatives for tissue engineering applications, such as biological signaling, cell adhesion, cell proliferation, cell differentiation, cell responsive degradation, and remodeling, is attracting a great deal of interest in medical research for guiding and promoting new tissue regeneration or in defining the shape and structure of cell growth (Symonette et al., 2014). A variety of polysaccharides, such as alginate, chitin, chitosan, hyaluronic acid, cellulose, chondroitin sulfate, starch, and their derivatives have been developed as biomaterials for tissue engineering applications (Oliveira and Reis, 2011). Application of polysaccharides as scaffolds in tissue engineering needs to fulfill requirements such as biocompatibility and non-toxicity, biodegradability with a controllable degradation rate, appropriate porosity, and structural integrity (Symonette et al., 2014).

Chitin and chitosan possess the requisite properties to act as scaffolds for tissue engineering, with respect to their degradability, immunogenicity, and mechanical strength, and therefore have been developed for tissue engineering applications in the form of 3D hydrogels or porous sponges, and fibrous scaffolds or free-standing films, within which the appropriate cell types are seeded for *in vitro* or *in vivo* culture and evaluation (Wan and Tai, 2013). The design of 3D chitin/chitosan hydrogel and sponge scaffolds, or 2D scaffolds and free-standing films to support cartilage regeneration, to chemically interact with apatite for bone and tendon regeneration, to encapsulate stem cells and support their growth and differentiation for different stem cell therapies, and their utility in regenerative medicine have been recently reported elsewhere (Lu et al., 2012a,b; Suh et al., 2013; Xu et al., 2014). Moreover, compo sites of chitosan with hydroxyl apatite and grafted chitosan with carbon nanotubes have been developed as potential materials for artificial bone and bone regeneration in tissue engineering (Venkatesan and Kim, 2010). Other polysaccharide-based materials, such as hyaluronic acid, starch, and cellulose, have also been explored for bone, cartilage, and/or skin tissue engineering applications (Rinaudo, 2008).

2.4.2.2 *Wound Healing and Wound Dressing*

Due to their inherent biocompatibility, low toxicity, and pharmaceutical biomedical activity, various polysaccharides, such as chitin, chitosan, cellulose, hyaluronan, and alginate, have been widely used to prepare wound healing materials (Barud et al., 2013; Czaja et al., 2006, 2007; Hrynyk et al., 2012). For example, hyaluronan, a major extracellular component with unique hygroscopic, rheological, and viscoelastic properties, has been extensively developed for tissue repair purposes, due to its physicochemical properties and specific interactions with cells and extracellular matrix. It is generally accepted that hyaluronan plays multifaceted roles in the mediation of the tissue repair process, and is involved in all stages of wound healing, i.e., inflammation, granulation tissue formation, reepithelialization, and remodeling. Derivatives of hyaluronan, such as cross-linked, esterified, or other chemically modified products, have also been developed for tissue repair or wound healing purposes (Anilkumar et al., 2011; Chen and Giovanni, 1999). Notably, the wound healing-promoting activity of the materials is also important in the designing of materials for tissue engineering.

2.4.2.3 *Drug Delivery and Controlled Release*

Polysaccharides hold promising potential for drug delivery and controlled release applications, due to their advantages such as biocompatibility, low immunogenicity, and minimal cytotoxicity. Numerous polysaccharide-based drug delivery systems have been developed for specific targeted delivery or controlled release, for protection of drugs from premature degradation, for improving intracellular penetration and transportation, for enhancing the stability and bioavailability of drugs, or for the delivery of biomolecules such as genes, antigens, and small interfering RNA (Csaba and Alonso, 2009; Mao et al., 2010; Mizrahy and Peer, 2012; Valo et al., 2013). These delivery systems are generally prepared in the form of a 3-D cross-linking network (covalently or ionically cross-linked), the polyelectrolyte structures, and the polysaccharides–drug conjugate (Mizrahy and Peer, 2012). The release of the entrapped drugs or certain molecules can be triggered by a change of pH, ions, electrical or magnetic field, light, temperature, redox potential, or certain molecules (Alvarez-Lorenzo et al., 2013). Cellulose, chitin, or chitosan, and alginate, represent the three most abundant polysaccharide types and are discussed more in detail below.

Several other polysaccharides such as pectin, chitin, chitosan, guar gum, xanthan gum, gellan gum, dextran, and chondroitin have also been developed for drug delivery or controlled drug release (Adams, 2010; Esposito et al., 1998; Luo and Qin, 2014; Pachuau and Mazumder, 2013; Reddy et al., 2011; Tønnesen and Karlsen, 2002; Zhang et al., 2013a,b,c).

2.4.2.4 Treatment of Cancer

Amounts of polysaccharides were believed to have anti-cancer functions following animal trials (Moertel et al., 1994). In addition, during clinical studies, they were found to be good for the function of immune systems and inflammatory factor reduction (De et al., 1995). Lentinan was found to be beneficial for treating patients with cancer in China and Japan (Pisi et al., 1995). The anti-tumor activities of lentinan and its inhibition effects on viral and chemical oncogenesis using animal trials were investigated by Wasser and Weis (1999). They showed that lentinan could notably prevent metastasis using animal models. It was also found that lentinan could be a good compensatory agent for treating cancer in the future. Lentinan may also be taken to treat patients who had advanced gastric cancer and other tumors.

2.5 CELL CULTURE: ACTION OF MECHANISM – POLYSACCHARIDE – RECEPTORS INTERACTIONS

Polysaccharides play a role as immunomodulators which can augment host defense responses. Recently, polysaccharides from various bioactive sources have received much attention in the biomedical area, due to their therapeutic functions and low toxicity (Tzianabos, 2000; Wasser, 2002). Researchers showed that polysaccharides have many functional effects, such as antibiotic, anti-mutant, and antioxidant activities (Wijesekara et al., 2011; Ali et al., 2009; Yu et al., 2009). However, the mechanism of operation is unclear; it showed that their main mechanisms involve nonspecific and specific induction of the immune system (Tzianabos, 2000). Much research has analyzed signaling molecule networks and cell surface receptors. Indeed, the mechanism of anti-tumor, immunostimulatory, and other polysaccharide effects is considered to occur *via* immune cells (B cells, macrophages, and DC) stimulation and complement system modulation (Wang et al., 1997). The specific receptors on the immune cells can recognize the polysaccharide polymers (Schepetkin and Quinn, 2006). These receptors are thought to be pattern recognition molecules (Gordon, 2003). At present, there are several major types of receptors, Toll-like receptors (TLRs), mannose receptors, Dectin-1, Dectin-2, scavenger receptors (SR), and Type 3 Complement Receptors (CR3). The polysaccharide binds to pattern recognition receptors (PRRs), and the activation of the receptors result in cascades of intracellular signaling which lead to the induction of immune genes, resulting in the production of pro-inflammatory cytokines. However, few polysaccharides have been identified where both receptor and action mechanism research has been done.

2.5.1 Natural Killer Group 2D

Natural killer group 2D (NKG2D) is a C-type lectin-like activating receptor on natural killer cells (NK), and could trigger cytotoxicity by glycan binding (Higai and Imaizumi, 2009). An NKG2D-mediated anti-tumor response could have great influence in innate immunity. Tsujitani et al. (2008) reported the influence of polysaccharide K in the expression of the immunoreceptor NKG2D. PBMCs from a healthy body were co-cultured with cell line MKN-45 from human gastric cancer, the expression of NKG2D on cells of $CD8^+$ T was reduced, but the expression of NKG2D inhibited by tumor cells was then recovered by the intake of non-specific immunopotentiator, polysaccharide K dose-dependently (Maehara et al., 2011). The researchers showed it was important for the inhibition of NKG2D expression that immediate contact between $CD8^+$ T cells and gastric cancer cells occurred. The NKG2D expression was considered to be a good biomarker for controlling the efficacy of PSK. Chang et al. (2014) found that *G. lucidum* polysaccharide could improve the cytotoxicity of NK cells through stimulating NKG2D/NCR receptors and MAPKs pathways. The concentration of the NKG2D cell surface and proteins of NCR were promoted after being treated with *G. lucidum*. In addition, according to previous reports on NKG2D and NCRs, they reported that polysaccharide from *G. lucidum* could directly bind to NKG2D/NCR receptors (Chang et al., 2014). Thus, these results show that NCR was the polysaccharide immune receptor for NK cells.

2.5.2 TGF-β Receptors

It is widely known that immunosuppression in tumor-bearing individuals can be induced by various humoral factors. Polysaccharides could attenuate this inhibition. Many reports showed that polysaccharide K could have an immunosuppressive influence in patients with cancer, which could inhibit the production of transforming growth factor (TGF-β) (Matsunaga et al., 1990, 1998). Yamaguchi et al. (2004) also found that PSK put an end to the TGF-β-induced inhibition of LAK activity. They proposed that PSK acted on receptors of TGF-β to restrict the relationship between TGF-β and its receptors. There is also another report on the mechanism of PSKs anti-metastatic action. Zhang et al. (2000) reported that PSK reduces the invasiveness of cell lines of human pancreatic cancer by many mechanisms, including inhibition of induction of the latent TGF-β, and indirect restriction of MMP production by inhibiting the production of TGF-β. Furthermore, the relationship between epithelial–mesenchymal transition (EMT) with invasion and metastasis of cancer cells has been proved, and TGF-β has been reported to have a crucial role in the EMT process (Thiery and Sleeman, 2006). Therefore, TGF-β has a great function in the anti-metastatic action of the polysaccharide.

2.5.3 Toll-Like Receptor 4

TLRs are considered to be sensors for triggering leukocyte, microbial molecules, and inflammatory response mobilization (Kawai and Akira, 2011). The TLRs identification showed a development in the concept of innate immunity to prevent pathogens in microorganisms. TLR4 is considered to be important for the innate recognition of bacteria or fungi (Chaudhary et al., 1998; Medzhitov et al., 1997). Besides binding with ligand, activated TLR4 guides a response mainly by the stimulation of MAPKs and NF-κB (Akashi et al., 2000; Chaudhary et al., 1998; Medzhitov et al., 1998; Ogata et al., 2000; Wesche et al., 2013; Yamamoto and Akira, 2010; Zhang et al., 1999).

It has been reported that polysaccharides isolated from natural plants can not only activate TLR signaling, but also B cells and/or macrophages by TLR4-mediated signaling (Beutler, 2000; Shao et al., 2004b; Yoon et al., 2003). Shao et al. (2004b) found that GLP could bind directly with TLR4 and activate macrophages. Han et al. (2003) reported that the activities of polysaccharide from *Acanthopanax senticosus* (ASP) on B cells and macrophages were greatly inhibited through treatment of the cells with TLR4 and TLR2 antibodies, more than ASP. The results showed ASP has a crucial role in the macrophages by reacting with the TLRs and generating the following stimulation of mitogen-activated protein kinases (MAPKs: p38, Erk1/2, and JNK) and NF-κB. Ando et al. (2002) demonstrated that the NF-κB signaling pathway *via* TLR4 can by activated by safflower polysaccharides. Polysaccharides isolated from *Platycodon grandiflorum* also activated B cells by CD79b and TLR4 (Sang et al., 2001). In our laboratory, a polysaccharide from *Ganoderma atrum*, named as PSG-1, evokes anti-tumor activity *via* the TLR4-mediated signaling pathways of MAPK and NF-κB (Zhang et al., 2013a,b,c). Taken together, TLR4 plays an important role as a membrance receptor. As a receptor, TLR4 recognizes ligands such as specific bioactive polysaccharides and active cells, transmits signals, and initiates various bioactivities.

2.5.4 Toll-Like Receptor 2

TLR2 is often considered as the receptor to bacterial lipoproteins, while the bacterial lipopolysaccharide receptor is the complex TLR4/MD2 (Figueiredo et al., 2012; Kawai and Akira, 2011). TLR2 was pointed out as one of the receptors involved in fungal polysaccharide recognition (Veerdonk et al., 2008). TLR2 is responsible for the activation of dendritic cells (DC) and T cells by the anti-tumor effect of polysaccharide Krestin (PSK) from

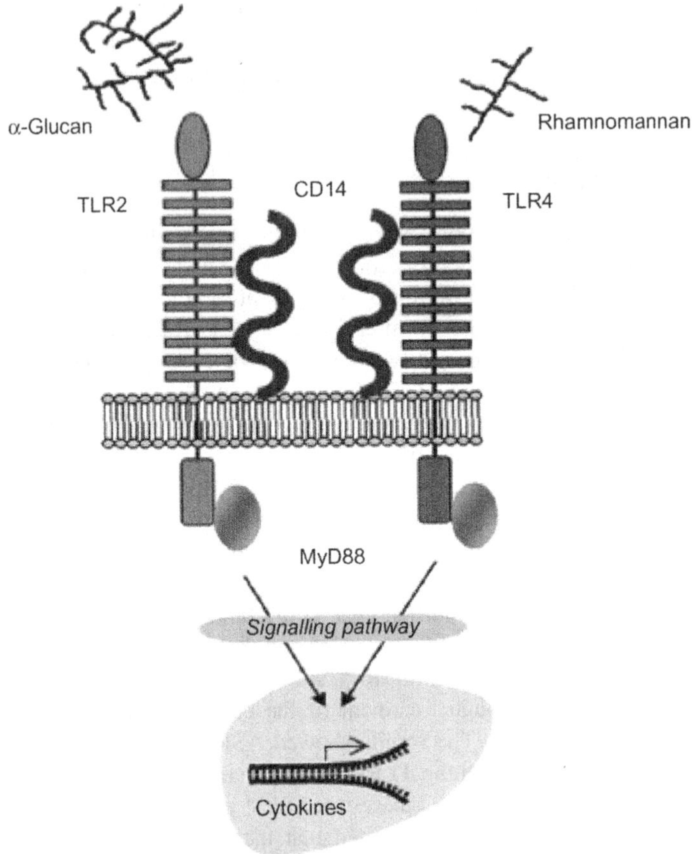

FIGURE 2.5 α-Glucans and rhamnomannans are recognized by TLR2 and TLR4, respectively.

mushroom, which can be explained because PSK does not inhibit tumor growth in TLR2$^{-/-}$ mice. So PSK is a selective agonist of TLR2 (Lu et al., 2010). Furthermore, Zhang et al. (2013a,b,c) found the polysaccharide isolated from the Marine fungus *Phoma herbarum* YS4108 resulted in the immunostimulating functions of YCP on B cells, which include TLR4 and TLR2, and the resultant stimulation of MAPK and signaling pathways of NF-κB. *Zymosan*, a particulate fragment of the *Saccharomyces cerevisiae* cell wall, is recognized by a complex formed by TLR2/TLR6, which could enhance macrophage responses (Gantner et al., 2003; Ozinsky et al., 2000; Brown et al., 2003). Polysaccharides from *G. lucidum* show immunomodulatory activities which are dependent on the activation of TLR2 and TLR4 (Lin et al., 2006) (Fig. 2.5). On the above basis, TLR2 is one of the key receptors for bioactive polysaccharides, mediates several signaling pathways, and elicits immunomodulatory and anti-inflammatory activities.

2.5.5 CD14

The pattern recognition receptor CD14 is a functional receptor for lipopolysaccharides (LPS). These LPS complexes are transferred to the soluble membrane CD14 present on cells, leading to cell activation (Matsushita et al., 2007). Sang et al. (2001) revealed that the specific anti-CD14 antibody could decrease macrophage NO production resulting from the *P. grandiflorum* polysaccharide, indicating the possible cellular binding sites of *P. grandiflorum*. Additionally, the expression of CD14 on murine peritoneal macrophages was enhanced after being treated with acidic polysaccharides extracted from *Panax ginseng* (Shin et al., 2002). Treatment of macrophages with type 2 capsular polysaccharide obtained from *Streptococcus pnemoniae* increased the expression of CD14; and the inhibiting antibody to CD14 partially changed TNF-alpha by polysaccharide. Therefore, macrophage-secretory and cellular activities can be induced by polysaccharide, possibly through a CD14-dependent pathway (Kim et al., 2000). Muroi and Tanamoto (2002) found that the polysaccharide has an important role in *Salmonella* lipopolysaccharide-induced stimulation of NF-κB *via* human CD14/TLR4/MD-2. Another study showed that *G. lucidum* polysaccharide enhances the surface expression of CD14, and further promotes LPS binding (Hua et al., 2007). In summary, CD14 on murine peritoneal macrophages is crucial for recognizing polysaccharides and for the endocytosis pathway of bacteria.

2.5.6 Dectin-1

Dectin-1 is the receptor of C-type lectin-like which is expressed on macrophages and dendritic cells (Brown et al., 2002). Some researchers have investigated the function of Dectin-1 in innate or adaptive immunity (Brown and Gordon, 2001; Drummond and Brown, 2011; Drummond et al., 2011). A water-soluble polysaccharide (GSG) from *G. lucidum* spores is able to induce macrophage activation through Dectin-1 and increase TNF-α production. The result showed that GSG exhibits immunomodulatory and anti-tumor activities, partially *via* the Dectin-1 receptor at least (Guo et al., 2009). Dectin-1 can also identify the polysaccharides from various fungal species, such as *S. cerevisiae*, and *Aspergillus fumigates* (Brown, 2006). Phagocytosis of heat-killed yeast, zymosan, by the macrophages is mediated by and large by the β-glucan receptor Dectin-1, while the mannose and complement receptors are not responsible for the process (Goldman, 1988). The β-glucan particles derived from fungus could lead to activation of leukocyte and inflammatory mediator production. Dectin-1 could mediate the production of TNF-α responding to zymosan and pathogens of live fungi (Brown et al., 2003). In summary, Dectin-1 plays an important role in natural polysaccharide (β-glucan) recognition, which could trigger the induction of pro-inflammatory and anti-inflammatory cytokines, inducing protective immune responses (Brown, 2006; Taylor et al., 2007).

2.5.7 Dectin-2

Dectin-2 is one of the members of receptors of C-type lectin especially for carbohydrates, with a "complex mannose-like" structure, and as such is able to recognize carbohydrates on pathogens including fungi and mycobacterium. It is mainly present on myeloid cells and maturing inflammatory monocytes (Sato et al., 2007; Feinberg et al., 2000; Mcgreal et al., 2006; Ariizumi et al., 2000; Philip et al., 2005). Some reports have revealed that Dectin-2 is the main receptor for α-mannans, because the production of cytokine was completely damaged in bone marrow-derived DCs from Dectin-2-deficient mice being treated with α-mannans, and it is of importance in the body's defense against *Candida albicans* (Saijo et al., 2010).

2.5.8 The Mannose Receptor

The mannose receptor (MR) is the prototype of multi-lectin receptor proteins, and establishes the relationship between innate and adaptive immunity (Ezekowitz et al., 1991; Lennartz et al., 1987; Taylor et al., 1990). If the surface and cell walls of carbohydrates were decorated by infectious agents, these patterns would be recognized by the mannose receptor. This pattern-recognition receptor functions as a molecular scavenger, which mediates endocytosis and phagocytosis (Weis et al., 1998).

The MR has five domains: a cytoplasmic carboxy-terminal domain; a transmembrane domain; an amino-terminal cysteine-rich region (Harris et al., 1994); eight tandem lectin-like carbohydrate recognition domains (CRDs); and a domain consisting of a fibronectin type II repeat. Each domain has a distinct function. The identification of fucose and mannose is limited to the CRDs (Mullin et al., 1994, 1997; Taylor et al., 1992), although some results (Fiete et al., 1998) show that a cysteine-rich domain can recognize the glycoproteins bearing sulfated sugars. East et al. (2002) studied the functions of the lectin-like domains Endo180, which appears to bind to mannose in a Ca^{2+}-dependent manner, *N*-acetylglucosamine, and fucose, except galactose. Furthermore, the macrophage MR binds to four serotypes of dengue virus (DV), and particularly to envelope glycoprotein. Antibodies specific for the MR significantly inhibited both macrophage infection, and infectious virus production in these cells (Miller et al., 2008). The MR has a major role in the clearance of serum glycoproteins and may handle the glycobiological code, and thus especially regulates serum concentrations of bioactive proteins, both in health and disease (Lee et al., 2002; Han et al., 2005). Zamze et al. (2002) found that the MR is able to bind to polysaccharides from *Streptococcus pneumoniae*, and to the lipopolysaccharides. Binding was $Ca(^{2+})$-dependent and inhibited with d-mannose. To sum up, the MR is important for specific polysaccharide uptake, and conduction of the polysaccharides stimulatory influence in macrophage.

2.5.9 Scavenger Receptors

Scavenger receptors (SR) could be expressed by the myeloid cells, including the macrophages, DCs, and the endothelial cells. They bind microorganisms and the products of them, including lipoteichoic acid lipopolysaccharide, and intracellular bacteria (Peiser et al., 2002). Some reports have shown that SR could serve as the PRRs for innate immune responses, because they could bind various pathogens. The interest in the function of SR in innate immunity is growing (Peiser et al., 2002; Gough and Gordon, 2000; Krieger, 1997; Pearson, 1996; Peiser and Gordon, 2001). Both Dushkin et al. (1996) and Vereschagin et al. (1998) have demonstrated that the soluble carboxymethylated glucan (CMG) ties to mouse peritoneal macrophages through scavenger receptors (SRs). Pearson et al. (1995) showed that the *Drosophila* CI, SR can bind laminarin and the other wall components of microbial cells. On the basis of this, Vereschagin et al. (1998) further reported that CMG interacted with macrophage SR and protects and increases the endotoxin clearance against endotoxic shock. Rice et al. (2002) also found that human monocyte SR could react with (1→3)-β-D-glucan polymers and SR, as human PRRs have a major role in the innate immune response. Nakamura et al. (2006) found that the macrophage SR ligand, fucoidan, enhanced activation for nitric oxide synthase (iNOS) promoter in RAW264.7 cells. In addition, *fucoidan* activated human monocyte and monocyte-derived macrophages, and stimulated much expression of IL-12 through scavenger receptors (Mytar et al., 2004). These results suggest that macrophage SRs might be involved in the recognition and signaling of glucan, playing key roles in innate immunity and host defense (Fig. 2.6).

2.5.10 CD19 and CD79

The CD19 has two Ig-like domains (Fujimoto et al., 1999). It appears on the B lymphocyte surface and follicular dendritic cells of the hematopoietic system as a single chain glycoprotein. The CD19 is an important regulator in the activation and differentiation of B cells (Fujimoto et al., 2000). Fujimoto et al. (1999) reported that B lymphocyte signal transduction was amplified by CD19 via controlling Src-family protein tyrosine kinase activation. The CD79 contains CD79a and CD79b components expressed specifically on B cells and B-cell neoplasms, which produces a complex with the B-cell receptor (BCR) (Chu and Arber, 2001). Han et al. (2003) pre-incubated B cells in the presence of anti-CD19 and anti-CD79b for 1 h, these antibodies significantly inhibited *Acanthopanax koreanum* (AK)-induced B-cell proliferation. The finding demonstrated that CD19 and CD79b could be used as AK polysaccharide receptors. The CD19 controls innate immunity through the signaling of the receptor RP105 in the B lymphocytes (Yazawa et al., 2003). We concluded that CD19 and CD79 both play important roles for polysaccharides in the regulation of signal transduction in B lymphocytes.

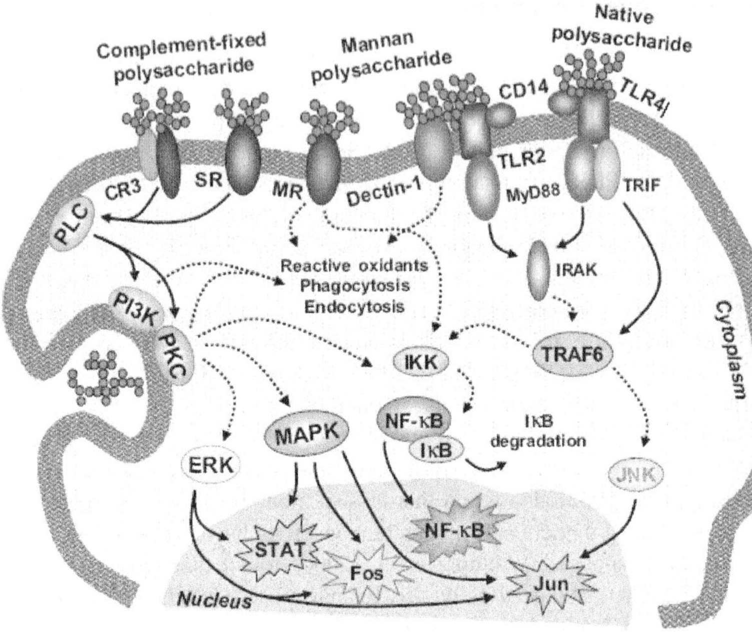

FIGURE 2.6 Schematic model illustrating potential signaling pathways involved in macrophage activation by bioactive polysaccharides.

2.5.11 Complement Receptor 3 (CR3)

The CR3 is the receptor responsible for β-glucans, complement (C3b)-opsonized particles, and microbial particles (Leung et al., 2006; Ross, 2000; Ross and Vĕtvicka, 1993). The CR3 was shown to take part in the response to many polysaccharides. Some studies reported that TNF-α production induced by polysaccharides produced by *Ganoderma formosanum* was blocked by an anti-CR3 antibody. These data suggested that receptor CR3 was necessary for the recognition of PS-F2 (Wang et al., 2012; Vetvicka et al., 1996; Mueller et al., 2000). Likewise, polysaccharide extracted from *Poria cocus* induced activation of NF-kappaB/Rel and expression of iNOS by the CR3, TLR4, CD14, and membrane receptor in macrophages (Lee et al., 2004). Yan et al. (1999) reported that β-glucan was able to bind to a lectin domain leading to this iC3b receptor's priming for the cytotoxicity of iC3b-opsonized target cells. The CR3 was found to have a direct attachment to unopsonized zymosan, while the attachment to zymosan could result in phagocytosis and a superoxide burst (Ross et al., 1985). On the other hand, Wright and Jong (1986) demonstrated that the LPS on *Escherichia coli* was found to bind to the CR3, which is a structurally homologous cell surface protein expressed on macrophages and stimulates CR3-dependent phagocytosis. So, macrophages expressing structurally homologous cell surface protein-CR3 could recognize plant polysaccharides, β-glucan, zymosan, LPS, etc. The CR3 is a necessary phagocyte and the NK cell receptor responsible for controlling cellular cytotoxic reactions, and thus is against the target cells bearing specific structures of carbohydrate.

2.5.12 Membrane Immunoglobulins

The B lymphocyte antigen receptors, membrane immunoglobulins (MIg), are integral proteins on B-cell surfaces binding foreign antigens, and take part in immune response regulation (Alhasan et al., 1994; Warner, 1974). Membrane immunoglobulins focus on the internalization of antigens following the appearance of T cells, and on the transmembrane transduction of signals resulting in cell activation or deletion (Cambier et al., 1994). The heterodimer of membrane Ig alpha and beta allows the cell to transduce the signal to the appearance of antigens on the cell surface. The signal enhanced the B cells growth and production of the antibody which was inside the plasma cell. Many studies on the leukemia lymphocytes of patients with chronic lymphocytic leukemia (CLL) suggest that IgM and IgD are the main receptor Igs on the surface of human lymphocytes (Fu and Kunkel, 1974; Fu et al., 1974; Rowe et al., 1973). Shao et al. (2004a) illustrated that anti-MIg Abs inhibited polysaccharides from *G. lucidum* (GLPS)-induced B-cell proliferation, suggesting that GLPS could directly bind with the MIg of B cells. So the MIg of B cells may be a potential cellular receptor for GLPS. In Shao's (2004b) report, results showed that polysaccharides from *Astragalus membranaceus* (APS) activate and bind to B cells via MIg in a TLR4-independent manner. Therefore, B-cell activation events initiated by receptor MIg transducing signals and mainly participate in immune response regulation.

Fig. 2.7 shows signal transduction pathways and the results of the Syk/CARD9-coupled receptor providing signals in antifungal immunity. The Syk signaling generates a mixture of MALT1, BCL10, and CARD9 that finally results in the stimulation of transcription elements, which include NF-κB. These factors of transcription affect key cytokines translation, thus helping to enhance Th1/Th17 differentiation, which then promotes the antifungal process from innate cells.

2.5.13 Receptor Protein 105/Myeloid Differentiation-1

Receptor protein 105 is monomeric, 105 kD in size, and one of the members of the leucine-rich repeat protein family that is expressed on mouse B cells and releases a growth-stimulating signal (Miyake et al., 1995, 1998). The molecule of RP105 transmits a stimulation signal that results in massive B-cell proliferation and resistance against apoptosis when cross-linked by an Ab. The MD-1 (an extracellular molecule) is related to RP105 and controls the ligand binding of RP105 by modulating its conformation (Ulevitch and

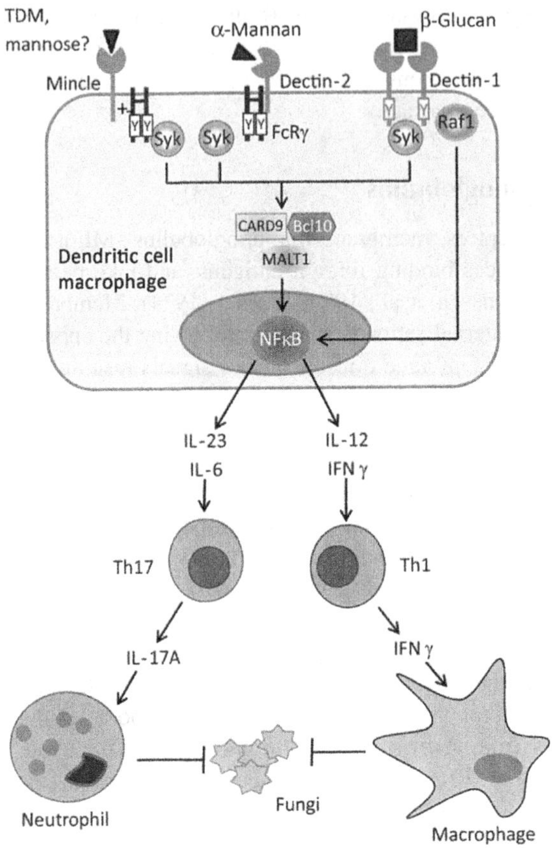

FIGURE 2.7 Signal transduction pathways and outcomes of several receptors signaling in antifungal immunity.

Tobias, 1995). The RP105/MD-1 is similar to TLR4/MD-2 in structure, which is directly associated with the expression of the B-cell surface and recognition of LPS. Lipopolysaccharide is a predominant ingredient of the Gram-negative bacteria that significantly stimulates the innate immune system (Ulevitch and Tobias, 1995). Some reports showed that RP105 has an important role in LPS responses (Nagai et al., 2002). Yamashita et al. (1996) also showed that activation mediated by RP105 allows normal B cells to be easily affected by anti-IgM-induced apoptosis. Ogata et al. (2000) found that the responses of B cells to LPS are controlled by another TLR protein, and can promote LPS signaling through TLR4. Nagia et al. (2005) also obtained similar results that RP105/MD-1 on the B cells, as well as TLR4/MD-2 and TLR2, is very important for innate Ab responding to LPS. The B cells could be beneficial to innate immunity by producing Abs, similar to natural immune receptors (Bendelac et al., 2001). On the above basis, RP105/MD-1-containing receptors on B cells are essential for the recognition of LPS and microbial lipoproteins, which is crucially responsible for activating B cells, contributes to immunity by producing Abs that directly enhance their phagocytosis, and promotes their immunogenicity.

Table 2.5 summarizes the immunomodulatory effects of different polysaccharides on macrophages. Table 2.6 shows the immune receptors on B cells.

In addition, treatment of immune cells with bioactive polysaccharides was shown to regulate different cell surface receptors' expression. For instance, Shin et al. (1997) found that the acidic polysaccharide extracted from *P. ginseng* could enhance the macrophage Fc-receptor's expression. The acidic polysaccharides from *P. ginseng* increased the CD14 expression, while it decreased the CR3 expression (Shin et al., 2002). The polysaccharide from *Bupleurum falcatum* L. could also increase expression of the Fc-receptor on the peritoneal macrophages through a process by an improvement in the intracellular Ca^{2+} followed by calmodulin activation (Matsumoto and Yamada, 1995).

In brief, the relationship between the PRRs and bioactive polysaccharides is very complicated. The bioactive polysaccharides can be specifically recognized by the PRRs, which also could mediate the receptors expression.

Bioactive polysaccharides from medicinal materials show a variety of immunopotentiating functions (Kim et al., 1996; Bao et al., 2001; Wasser, 2002). They could stimulate B cells, T cells, and macrophages (Kim et al., 1996). However, the polysaccharides from different sources have different functions which may be attributed to the polysaccharides origin, structure, and specific receptors. Most of bioactive polysaccharides act on/bind to specific receptors and exert biological effects. But the correlation between polysaccharide structure, specific receptors, and their immunomodulation activity still needs to be further investigated. It is reported that the receptor-binding protein responsible for the identification of bioactive polysaccharide isolated from bacteria and fungi were thought to be β-glucan receptor, SRs, TLRs, etc.; whether other kinds of polysaccharides share the same receptor with those polysaccharides remains unknown. The discovery of new receptors for bioactive polysaccharides is envisaged. The receptor – signaling pathway relationship still needs to be further understood. The data are limited only to NF-κB, MAPK signaling pathways, and mitochondria apoptosis pathways. The structural features of these polysaccharides may closely link to specific receptors. Because of various monosaccharide compositions and branchings of polysaccharides, the specificity of molecular recognition is still unclear. Furthermore, few results on the structure of polysaccharide in these studies have been reported. Therefore, there are still some difficulties in many cases in examining how to relate the specific structural properties of a bioactive polysaccharide with its bioactivity.

In the future work in this field, it is necessary to examine these structure–bioactivity correlations contributing to optimal immune cell-activating properties. We concluded that clearly understanding the polysaccharide receptor—function relationship provides a unique opportunity for investigators to understand the molecular mechanisms of polysaccharides. The use

TABLE 2.5 Immunomodulatory Effects of Polysaccharides on Macrophages

Source	Structural Feature	Macrophage Type	Dose	Protein	Endotoxin	Effect	Mediated Receptor/ Pathway	References
Acanthopanax senticosus		MPM	3–100	–	×	↑ NO, iNOS, IL-1β, IL-6, TNF-α	TLR4/TLR2 → MAPKs/ NF-κB	Sang et al. (2001)
Aloe vera	50–10,000 kDa β-(1,4)-D-Man	MPM	100–500	–	×	↑Migration, endocytosis, phagocytosis, cytotoxicity, MHC-II, NO, TNF-α	FcγR	Liu et al. (2006)
Angelica gigas		RAW264.7	1–100	–	×	↑ iNOS	CD14/CR3 → p38MAPK	Jeon and Kim (2001)
Astragalus membranaceus	3.5–1 580 kDa Rha, Xyl, Glc, Gal, Man, Fru	MPM, THP-1	100	×	×	↑ IL-1β, TNF-α	TLR4	Shao et al. (2004b)
Carthamus tinctorius	>100 kDa Glc, Gla, Ara, Xyl, Rha, Man	MPM	0.000 1 – 100	<3%	×	↑ INF-γ, IL-12, TNF-α	TLR4 → NF-κB, but not TLR2	Ando et al. (2002)
Cordyceps sinensis	82 kDa Glc, Man, Gal	NR8383	6.25–100	<1%	–	↑ Phagocytosis	Acid phosphatase	Cheung et al. (2009)
Fucoidan		MPM, RAW264.7	50	–	–	↑ NO, iNOS, iNOS promoter	SR → p38MAPK → AP-1, SR → NF-κB	Nakamura et al. (2006)

Ganoderma lucidum	585 kDa L-Fuc, D-Xyl D-Man, D-Gal, D-GlcNAc, D-Glc	MPM, THP-1	0.01–100	<6.5%	×	↑ IL-1β, IL-6, IL-12, IFN-γ, TNF-α	GM-CSF, G-CSF, M-CSF, TLR4→PTK(Src)→PLCγ1→PKC→MEK1→ERK, TLR4→PTK(Src)→Rac1→PAK→p38, TLR4→PTK→Rac1→PAK→JNK, ribosomal protein S7, transcriptional coactivator	Chen et al. (2004b); Shao et al. (2004a)
Opuntia polyacantha	168–733 kDa Gal, Xyl, Ara, Rha	THP-1, J774. A1	25 – 800	–	×	↑ NO, IL-6, TNF-α	NF-κB	Schepetkin et al. (2008)
Paenibacillus polymyxa	β-Glucans	RAW264.7	3 – 300	–	–	↑ NO	MAPKs, NF-κB, AP1	Chang and Jsgebru (2010)
Platycodon grandiflorum	Inulin-type polyfructose	RAW264.7	1 – 100	–	×	↑ NO, iNOS, TNF-α	TLR4→MAPKs→AP-1, TLR4→ NF-κB, CD14, CD11b	Sang et al. (2001); Yoon et al. (2004); Yoon et al. (2003)
Polyporus umbellatus	160 kDa D-Glc, D-Gal, D-Man	MPM	12.5 – 100	–	<0.4	↑ NO, IL-1β, TNF-α	TLR4→NF-κB, but not TLR2 and CR3	Li and Xu (2011)
Poria cocos	8 kD Man, Gal, Ara	RAW264.7	10 – 100	–	–	↑ NO, iNOS,	TLR4→NF-κB/Rel	Lee and Jeon (2003)

(*Continued*)

TABLE 2.5 (Continued)

Source	Structural Feature	Macrophage Type	Dose	Protein	Endotoxin	Effect	Mediated Receptor/ Pathway	References
Rheum tanguticum	60–80 kDa Man, Ara, Glc	RPM	200 mg/kg	–	–	↑ IFN-γ (*in vivo*)	MR	Liu et al. (2008b)
Silene vulgaris	Acidic arabinogalactan	RPM	15 – 150	–	–	↑ Phagocytosis	Ca^{2+}-dependent	Popov et al. (1999)
Streptococcus pnemoniae	D-GlcUA, D-Glc, L-Rha	MPM	5 – 50	×	–	↑ NO, TNF-α, cytotoxicity	CD14-dependent, TLR2	Sunghee et al. (2002)
Trametes versicolor	Glc, Man	J774.A, MPM	62.5 – 1000	–	×	↑ IL-6, TNF-α.	TLR4, but not Dectin1	Price et al. (2010)

×, not detected or negligible; –, not tested; ↑, increased; A, inhibition of receptor/kinase by specific antibody/inhibitor; AP-1, activator protein-1; B, competitive binding of receptor by reported saccharide ligand; C, resulting comparison between defined mutant and its wild-type; CR3, complement receptor-3; D, affinity adsorption and identification of polysaccharide receptor; MAPK, mitogen-activated protein kinases; MHC-II, major histocompability complex-II; MPM, mouse peritoneal macrophage; MR, mannose receptor; PKC, protein kinase C; PTK, protein tyrosine kinase; RPM, rat peritoneal macrophage; SAPK/JNK, stress-activated protein kinases/jun N-terminal kinase; SR, scavenger receptor; TLR2/4, toll-like receptor-2/4.

TABLE 2.6 Immune Receptor on B Cells

Source	Receptor	Cell	Effect	References
LPS	CD19	B lymphocytes cell	B-cell activation	Yazawa et al. (2003)
LPS	RP105/ MD-1	B lymphocytes cell	B-cell proliferation and activation	Nagai et al. (2002); Yamashita et al. (1996)
PSK	CD79	B-cell	B-cell development, activation, and differentiation	Chu and Arber (2001)
Ganoderma lucidum	mIg	B-cell	B-cell activation and signal transduction	Shao et al. (2004b); Taylor et al. (1992)
β-Glucans	CR3	B-cell	Mediate cellular cytotoxic reactions	Vetvicka et al. (1996); Yan et al. (1999)

of bioactive polysaccharides as alternative medicines for the treatment of cancer is popular in many countries. However, the limited use of polysaccharides is mainly resisted due to the lack of systematic mechanisms and clinical studies. A detailed deciphering of receptor usage by polysaccharides, and the structure–bioactivity relationship and mechanisms will be the main challenges for polysaccharide researchers.

REFERENCES

Abdelakher, M., Hamilton, J.K., Montgomery, R., Smith, F., 1952. A new procedure for the determination of the fine structure of polysaccharides. J. Am. Chem. Soc. 74, 4970–4971.

Abe, M., Fukaya, Y., Ohno, H., 2010. Extraction of polysaccharides from bran with phosphonate or phosphinate-derived ionic liquids under short mixing time and low temperature. Green Chem. 12, 1274–1280.

Abraham, R.J., Fisher, J., Loftus, P., 1988. Introduction to NMR Spectroscopy. Wiley, Chichester, 271p. Magn. Reson. Chem. 27, 1004–1004.

Adams, G.G., 2010. Polysaccharide drug delivery systems based on pectin and chitosan. Biotechnol. Genet. Eng. 27, 257–284.

Adeyeye, J., Azurmendi, H.F., Stroop, C.J., Sozhamannan, S., Williams, A.L., Adetumbi, A.M., et al., 2003. Conformation of the hexasaccharide repeating subunit from the *Vibrio cholerae* O139 capsular polysaccharide. Biochemistry 42, 3979–3988.

Akashi, S., Shimazu, R., Ogata, H., Nagai, Y., Takeda, K., Kimoto, M., et al., 2000. Cutting edge: cell surface expression and lipopolysaccharide signaling via the toll-like receptor 4-MD-2 complex on mouse peritoneal macrophages. J. Immunol. 164, 3471–3475.

Alhasan, S., Santini, R., Li, Q., Rosenspire, A., 1994. Membrane immunoglobulin is characterized by distinct structural subpopulations. Mol. Immunol. 30, 1593–1599.

Ali, B.H., Ziada, A., Blunden, G., 2009. Biological effects of gum arabic: a review of some recent research. Food Chem. Toxicol. 47, 1–8.

Allerdings, E., Ralph, J., Schatz, P.F., Gniechwitz, D., Steinhart, H., Bunzel, M., 2005. Isolation and structural identification of diarabinosyl 8-O-4-dehydrodiferulate from maize bran insoluble fibre. Phytochemistry 66, 113–124.

Alvarez-Lorenzo, C., Blanco-Fernandez, B., Puga, A.M., Concheiro, A., 2013. Crosslinked ionic polysaccharides for stimuli-sensitive drug delivery. Adv. Drug Deliv. Rev. 65, 1148–1171.

Anastyuk, S.D., Shevchenko, N.M., Nazarenko, E.L., Dmitrenok, P.S., Zvyagintseva, T.N., 2009. Structural analysis of a fucoidan from the brown alga Fucus evanescens by MALDI-TOF and tandem ESI mass spectrometry. Carbohyd. Res. 344, 779–787.

Ando, I., Tsukumo, Y., Wakabayashi, T., Akashi, S., Miyake, K., Kataoka, T., et al., 2002. Safflower polysaccharides activate the transcription factor NF-kappa B via Toll-like receptor 4 and induce cytokine production by macrophages. Int. Immunopharmac. 2, 1155–1162.

Anilkumar, T., Muhamed, J., Jose, A., Jyothi, A., Mohanan, P., Krishnan, L.K., 2011. Advantages of hyaluronic acid as a component of fibrin sheet for care of acute wound. Biologicals 39, 81–88.

Ariizumi, K., Shen, G.L., Shikano, S., Rd, R.R., Zukas, P., Edelbaum, D., et al., 2000. Cloning of a second dendritic cell-associated C-type lectin (dectin-2) and its alternatively spliced isoforms. J. Biol. Chem. 275, 11957–11963.

Bahr, U., Pfenninger, A., Karas, M., Stahl, B., 1997. High-sensitivity analysis of neutral underivatized oligosaccharides by nanoelectrospray mass spectrometry. Anal. Chem. 69, 4530–4535.

Bao, X., Liu, C., Fang, J., Li, X., 2001. Structural and immunological studies of a major polysaccharide from spores of *Ganoderma lucidum* (Fr.) Karst. Carbohyd. Res. 332, 67–74.

Barud, Hd.S., de Araújo Júnior, A.M., Saska, S., Mestieri, L.B., Campos, J.A.D.B., De Freitas, R.M., et al., 2013. Antimicrobial Brazilian propolis (EPP-AF) containing biocellulose membranes as promising biomaterial for skin wound healing. Evid. Based Complement. Alternat. Med. 2013, 703024.

Bell, A.F., Hecht, L., Barron, L.D., 1994. Disaccharide solution stereochemistry from vibrational Raman optical activity. J. Am. Chem. Soc. 116, 5155–5161.

Bendelac, A., Bonneville, M., Kearney, J.F., 2001. Autoreactivity by design: innate B and T lymphocytes. Nat. Rev. Immunol. 1, 177–186.

Bergamini, J.F., Boisset, C., Mazeau, K., Heyraud, A., Taravel, F.R., 1995. Conformational behavior of oligo-galactomannan chains inferred from NMR spectroscopy and molecular modeling. New J. Chem. 19, 115–127.

Bertaud, F., Sundberg, A., Holmbom, B., 2002. Evaluation of acid methanolysis for analysis of wood hemicelluloses and pectins. Carbohyd. Polym. 48, 319–324.

Beutler, B., 2000. Endotoxin, toll-like receptor 4, and the afferent limb of innate immunity. Curr. Opin. Microbiol. 3, 23–28.

Biermann, C.J., 1988. Hydrolysis and other cleavages of glycosidic linkages in polysaccharides. Adv. Carbohyd. Chem. Biochem. 46, 251–271.

Björndal, H., Hellerqvist, C.G., Lindberg, B., Svensson, S., 1970. Gas-liquid chromatography and mass spectrometry in methylation analysis of polysaccharides. Angew. Chem. Int. Ed. 9, 610–619.

Blumenkrantz, N., Asboe-Hansen, G., 1973. New method for quantitative determination of uronic acids. Anal. Biochem. 54, 484–489.

Brown, G.D., 2006. Dectin-1: a signalling non-TLR pattern-recognition receptor. Nat. Rev. Immunol. 6, 33–43.

Brown, G.D., Gordon, S., 2001. Immune recognition. A new receptor for beta-glucans. Nature 413, 36–37.

Brown, G.D., Taylor, P.R., Reid, D.M., Willment, J.A., Williams, D.L., Martinezpomares, L., et al., 2002. Dectin-1 is a major β-glucan receptor on macrophages. J. Exp. Med. 196, 407–412.

Brown, G.D., Herre, J., Williams, D.L., Willment, J.A., Marshall, A.S.J., Gordon, S., 2003. Dectin-1 mediates the biological effects of β-glucans. J. Exp. Med. 197, 1119–1124.

Bubb, W.A., 2003. NMR spectroscopy in the study of carbohydrates: characterizing the structural complexity. Concepts Magn. Reson. A 19A, 1–19.

Burlingame, A.L., Boyd, R.K., Gaskell, S.J., 1996. Mass spectrometry. Anal. Chem. 68, 599–652.

Cambier, J.C., Pleiman, C.M., Clark, M.R., 1994. Signal transduction by the B cell antigen receptor and its coreceptors. Annu. Rev. Immunol. 12, 457–486.

Carbonero, E.R., Gracher, A.H.P., Komura, D.L., Marcon, R., Freitas, C.S., Baggio, C.H., et al., 2008. *Lentinus edodes* heterogalactan: antinociceptive and anti-inflammatory effects. Food Chem. 111, 531–537.

Castro, C.D., Fregolino, E., Gargiulo, V., Lanzetta, R., Parrilli, M., 2008. A novel capsular polysaccharide from *Rhizobium rubi* strain DSM 30149. Carbohyd. Res. 343, 1482–1485.

Chandrasekaran, R., Balaji, P.V., Qasba, P.K., et al., 1998. Conformation of Carbohydrates. CRC Press.

Chang, C.J., Chen, Y.Y., Lu, C.C., Lin, C.S., Martel, J., Tsai, S.H., et al., 2014. *Ganoderma lucidum* stimulates NK cell cytotoxicity by inducing NKG2D/NCR activation and secretion of perforin and granulysin. Innate Immun. 20, 301–311.

Chang, Z.Q., Jsgebru, L., 2010. Mechanism of macrophage activation induced by beta-glucan produced from *Paenibacillus polymyxa* JB115. Biochem. Biophys. Res. Commun. 391, 1358–1362.

Chaudhary, P.M., Ferguson, C., Nguyen, V., Nguyen, O., Massa, H.F., Eby, M., et al., 1998. Cloning and characterization of two Toll/Interleukin-1 receptor-like genes TIL3 and TIL4: evidence for a multi-gene receptor family in humans. Blood 91, 4020–4027.

Chen, H., Zhang, M., Xie, B., 2004a. Quantification of uronic acids in tea polysaccharide conjugates and their antioxidant properties. J. Agric. Food Chem. 52, 3333–3336.

Chen, H.S., Tsai, Y.F., Lin, S., Lin, C.C., Khoo, K.H., Lin, C.H., et al., 2004b. Studies on the immuno-modulating and anti-tumor activities of *Ganoderma lucidum* (Reishi) polysaccharides. Bioorgan. Med. Chem. 12, 5595–5601.

Chen, S., Chen, H., Tian, J., Jia, W., Wang, Y., Xing, L., 2014. Enzymolysis-ultrasonic assisted extraction, chemical characteristics and bioactivities of polysaccharides from corn silk. Carbohyd. Polym. 101, 332–341.

Chen, W.Y.J., Giovanni, A., 1999. Fuctions of hyaluronan in wound repair. Hyaluronan 7, 147–156.

Cheng, H., Feng, S., Jia, X., Li, Q., Zhou, Y., Ding, C., 2013. Structural characterization and antioxidant activities of polysaccharides extracted from Epimedium acuminatum. Carbohyd. Polym. 92, 63–68.

Cheung, J.K.H., Li, J., Cheung, A.W.H., Zhu, Y., Zheng, K.Y.Z., Bi, C.W.C., et al., 2009. Cordysinocan, a polysaccharide isolated from cultured Cordyceps, activates immune responses in cultured T-lymphocytes and macrophages: signaling cascade and induction of cytokines. J. Ethnopharmacol. 124, 61–68.

Chu, P.G., Arber, D.A., 2001. CD79: a review. Appl. Immunohistochem. Mol. Morphol. 9, 97–106.

Ciucanu, I., Kerek, F., 1984. A simple and rapid method for the permethylation of carbohydrates. Carbohyd. Res. 131, 209–217.

Colegate, S.M., Molyneux, R.J., 2007. Bioactive Natural Products: Detection, Isolation, and Structural Determination, second ed. CRC Press.

Cox, A.D., Brisson, J.R., Varma, V., Perry, M.B., 1996. Structural analysis of the lipopolysaccharide from *Vibrio cholerae* O139. Carbohyd. Res. 290, 43–58.

Csaba, N., Alonso, M.J., 2009. Ionically crosslinked chitosan/tripolyphosphate nanoparticles for oligonucleotide and plasmid DNA delivery. Int. J. Pharm. 382, 205–214.

Cui, H., Liu, Q., Tao, Y., Zhang, H., Zhang, L., Ding, K., 2008. Structure and chain conformation of a (1→6)-α-D-glucan from the root of *Pueraria lobata* (Willd.) Ohwi and the antioxidant activity of its sulfated derivative. Carbohyd. Polym. 74, 771–778.

Czaja, W., Krystynowicz, A., Bielecki, S., Brown, R., 2006. Microbial cellulose--the natural power to heal wounds. Biomaterials 27, 145–151.

Czaja, W.K., Young, D.J., Kawecki, M., Brown Jr., R., 2007. The future prospects of microbial cellulose in biomedical applications. Biomacromolecules 8, 1–12.

Das, D., Maiti, D., Chandra, K., Mondal, S., Ojha, A.K., Roy, S.K., et al., 2008. NMR and MALDI-TOFMS analysis of a heteroglycan isolated from hot water extract of edible mushroom, *Volvariella bombycina*. Carbohyd. Res. 343, 2258–2265.

Datta, A.K., Basu, S., Roy, N., 1999. Chemical and immunochemical studies of the O-antigen from enteropathogenic *Escherichia coli* O158 lipopolysaccharide. Carbohyd. Res. 322, 219–227.

De, B.T., Dufaux, J., Roche, A., Counnord, J.L., Berthault, M.F., Denys, A., et al., 1995. Circulatory alterations induced by intra-arterial injection of iodized oil and emulsions of iodized oil and doxorubicin: experimental study. Radiology 194, 165–170.

De Gussem, K., Vandenabeele, P., Verbeken, A., Moens, L., 2005. Raman spectroscopic study of *Lactarius* spores (Russulales, Fungi). Spectrochim. Acta A: Mol. Biomol. Spectrosc. 61, 2896–2908.

Dell, A., 1987. F.A.B.-mass spectrometry of carbohydrates. Adv. Carbohyd. Chem. Biochem. 45, 19–72.

Dell, A., Morris, H.R., 2001. Glycoprotein structure determination by mass spectrometry. Science 291, 2351–2356.

Dong, J.Z., Wang, Z.C., Wang, Y., 2011. Rapid extraction of polysaccharides from fruits of *Lycium barbarum* L. J. Food Biochem. 35, 1047–1057.

Dong, Q., Yao, J., Fang, J.N., Ding, K., 2007. Structural characterization and immunological activity of two cold-water extractable polysaccharides from *Cistanche deserticola* Y. C. Ma. Carbohyd. Res. 342, 1343–1349.

Dou, S.Y., Yan, L.T., Liu, Y.C., Du, G., Zhou, P., 2013. Preparation and performance of organic–inorganic halide perovskites. J. Mater. Sci.: Mater. Electron. 24, 4862–4867.

Drummond, R.A., Brown, G.D., 2011. The role of Dectin-1 in the host defence against fungal infections. Curr. Opin. Microbiol. 14, 392–399.

Drummond, R.A., Saijo, S., Iwakura, Y., Brown, G.D., 2011. The role of Syk/CARD9 coupled C-type lectins in antifungal immunity. Eur. J. Immunol. 41, 276–281.

Dushkin, M.I., Safina, A.F., Vereschagin, E.I., Schwartz, Y.S., 1996. Carboxymethylated beta-1,3-glucan inhibits the binding and degradation of acetylated low density lipoproteins in macrophages in vitro and modulates their plasma clearance in vivo. Cell Biochem. Funct. 14, 209–217.

Duus, J., Gotfredsen, C.H., Bock, K., 2001. Carbohydrate structural determination by NMR spectroscopy: modern methods and limitations. Chem. Rev. 100, 4589–4614.

East, L., Rushton, S., Taylor, M.E., Isacke, C.M., 2002. Characterization of sugar binding by the mannose receptor family member, Endo180. J. Biol. Chem. 277, 50469–50475.

Edgar, K.J., Heinze, T., Buchanan, C.M., 2009. Polysaccharide Materials: Performance by Design. American Chemical Society.

Esposito, P., Colombo, I., Coceani, N., Curto, D.D., Chiellini, E., 1998. Lipid-based nanoparticles as drug delivery systems. J. Pharm. Pharmacol. 50, 14.

Ezekowitz, R.A., Sastry, K., Bailly, P., Warner, A., 1991. Molecular characterization of the human macrophage mannose receptor: demonstration of multiple carbohydrate recognition-like domains and phagocytosis of yeasts in Cos-1 cells. J. Exp. Med. 172, 1785–1794.

Fear, C.M., Menzies, R.C., 1926. CXXXI.—Applications of thallium compounds in organic chemistry. Part III. Alkylations. J. Chem. Soc. 129, 937–940.

Fechner, P.M., Wartewig, S., Kleinebudde, P., Neubert, R.H.H., 2005. Studies of the retrogradation process for various starch gels using Raman spectroscopy. Carbohyd. Res. 340, 2563–2568.

Feinberg, H., Park-Snyder, S., Kolatkar, A.R., Heise, C.T., Taylor, M.E., Weis, W.I., 2000. Structure of a C-type carbohydrate recognition domain from the macrophage mannose receptor. J. Biol. Chem. 275, 21539–21548.

Fenn, J.B., Mann, M., Meng, C.K., Wong, S.F., Whitehouse, C.M., 1990. Electrospray ionization–principles and practice. Mass Spectrom. Rev. 9, 37–70.

Fiete, D.J., Beranek, M.C., Baenziger, J.U., 1998. A cysteine-rich domain of the "mannose" receptor mediates GalNAc-4-SO4 binding. Proc. Natl. Acad. Sci. USA 95, 2089–2093.

Figueiredo, R.T., Bittencourt, V.C.B., Lopes, L.C.L., Sassaki, G., Barreto-Bergter, E., 2012. Toll-like receptors (TLR2 and TLR4) recognize polysaccharides of *Pseudallescheria boydii* cell wall. Carbohyd. Res. 356, 260–264.

Fu, S.M., Kunkel, H.G., 1974. Membrane immunoglobulins of B lymphocytes: inability to detect certain characteristic IgM and IgD antigens. J. Exp. Med. 140, 895–903.

Fu, S.M., Winchester, R.J., Kunkel, H.G., 1974. Occurrence of surface IgM, IgD, and free light chains of human lymphocytes. J. Exp. Med. 139, 451–456.

Fujimoto, K., Matsuda, K., Aso, K., 1962. Acetolysis of polysaccharides, part IV: acetolysis and acid hydrolysis of gluco-bioses. Tohoku J. Agric. Res. 13, 61–67.

Fujimoto, M., Poe, J.C., Jansen, P.J., Sato, S., Tedder, T.F., 1999. CD19 amplifies B lymphocyte signal transduction by regulating Src-family protein tyrosine kinase activation. J. Immunol. 162, 7088–7094.

Fujimoto, M., Fujimoto, Y., Poe, J.C., Jansen, P.J., Lowell, C.A., Defranco, A.L., et al., 2000. CD19 regulates Src family protein tyrosine kinase activation in B lymphocytes through processive amplification. Immunity 13, 47–57.

Ganeshapillai, J., Vinogradov, E., Rousseau, J., Weese, J.S., Monteiro, M.A., 2008. Clostridium difficile cell-surface polysaccharides composed of pentaglycosyl and hexaglycosyl phosphate repeating units. Carbohyd. Res. 343, 703–710.

Gantner, B.N., Simmons, R.M., Canavera, S.J., Akira, S., Underhill, D.M., 2003. Collaborative induction of inflammatory responses by Dectin-1 and Toll-like receptor 2. J. Exp. Med. 197, 1107–1117.

Gelder, J.D., Gussem, K.D., Vandenabeele, P., Moens, L., 2007. Reference database of Raman spectra of biological molecules. J. Raman Spectrosc. 38, 1133–1147.

Ghosh, K., Chandra, K., Roy, S.K., Mondal, S., Maiti, D., Das, D., et al., 2008. Structural investigation of a polysaccharide (Fr. I) isolated from the aqueous extract of an edible mushroom, *Volvariella diplasia*. Carbohyd. Res. 343, 1071–1078.

Gierlinger, N., Schwanninger, M., Reinecke, A., Burgert, I., 2006. Molecular changes during tensile deformation of single wood fibers followed by Raman microscopy. Biomacromolecules 7, 2077–2081.

Gladden, L.F., 1995. In: McBrierty, V.I., Packer, K.J. (Eds.), Nuclear Magnetic Resonance in Solid Polymers. Cambridge University Press, Cambridge, 1993. pp. xxii + 348, ISBN 0-521-30140-8. Polym. Int. 36, 213–213.

Goh, K.K., Pinder, D.N., Hall, C.E., Hemar, Y., 2006a. Rheological and light scattering properties of flaxseed polysaccharide aqueous solutions. Biomacromolecules 7, 3098–3103.

Goh, K.K., Haisman, D.R., Singh, H., 2006b. Characterisation of a high acyl gellan polysaccharide using light scattering and rheological techniques. Food Hydrocolloids 20, 176–183.

Goldman, R., 1988. Characteristics of the beta-glucan receptor of murine macrophages. Exp. Cell Res. 174, 481–490.

Gordon, S., 2003. Pattern recognition receptors: doubling up for the innate immune response. Cell 111, 927–930.

Gough, P.J., Gordon, S., 2000. The role of scavenger receptors in the innate immune system. Microbes Infect. 2, 305–311.

Griggs, L.J., Post, A., White, E.R., Finkelstein, J.A., Moeckel, W.E., Holden, K.G., et al., 1971. Identification and quantitation of alditol acetates of neutral and amino sugars from mucins by automated gas-liquid chromatography. Anal. Biochem. 43, 369–381.

Guetta, O., Mazeau, K., Auzely, R., Milas, M., Rinaudo, M., 2003. Structure and properties of a bacterial polysaccharide named Fucogel. Biomacromolecules 4, 1362–1371.

Guo, L., Xie, J., Ruan, Y., Zhou, L., Zhu, H., Yun, X., et al., 2009. Characterization and immunostimulatory activity of a polysaccharide from the spores of *Ganoderma lucidum*. Int. Immunopharmac. 9, 1175–1182.

Hakomori, S., 1964. A rapid permethylation of glycolipid, and polysaccharide catalyzed by methylsulfinyl carbanion in dimethyl sulfoxide. J. Biochem. 55, 205–208.

Han, S.B., Park, S.K., Ahn, H.J., Yoon, Y.D., Kim, Y.H., Lee, J.J., et al., 2003. Characterization of B cell membrane receptors of polysaccharide isolated from the root of *Acanthopanax koreanum*. Int. Immunopharmac. 3, 683–691.

Han, Y., Zhao, L., Yu, Z., Feng, J., Yu, Q., 2005. Role of mannose receptor in oligochitosan-mediated stimulation of macrophage function. Int. Immunopharmac. 5, 1533–1542.

Harris, N., Peters, L.L., Eicher, E.M., Rits, M., Raspberry, D., Eichbaum, Q.G., et al., 1994. The exon-intron structure and chromosomal localization of the mouse macrophage mannose receptor gene Mrcl: identification of a ricin-like domain at the N-terminus of the receptor. Biochem. Biophys. Res. Commun. 198, 682–692.

Harris, P.J., Henry, R.J., Blakeney, A.B., Stone, B.A., 1984. An improved procedure for the methylation analysis of oligosaccharides and polysaccharides. Carbohyd. Res. 127, 59–73.

Harvey, D.J., 2012. Analysis of carbohydrates and glycoconjugates by matrix-assisted laser desorption/ionization mass spectrometry: an update for 2007–2008. Mass Spectrom. Rev. 31, 183–311.

Haworth, W.N., 1915. III.--A new method of preparing alkylated sugars. J. Chem. Soc. 107, 8–16.

Higai, K., Imaizumi, Y.C., 2009. NKG2D and CD94 bind to heparin and sulfate-containing polysaccharides. Biochem. Biophys. Res. Commun. 386, 709–714.

Hizukuri, S., Osaki, S., 1978. A rapid Smith-degradation for the determination of non-reducing, terminal residues of (1→4)α-D-glucans. Carbohyd. Res. 63, 261–264.

Homans, S.W., 1990. A molecular mechanical force field for the conformational analysis of oligosaccharides: comparison of theoretical and crystal structures of Man alpha 1-3Man beta 1-4GlcNAc. Biochemistry 29, 9110–9118.

Höök, M., Lindahl, U., Iverius, P.H., 1974. Distribution of sulphate and iduronic acid residues in heparin and heparan sulphate. Biochem. J. 137, 33–43.

Hrynyk, M., Martins-Green, M., Barron, A.E., Neufeld, R.J., 2012. Alginate-PEG sponge architecture and role in the design of insulin release dressings. Biomacromolecules 13, 1478–1485.

Hu, S.W., Tian, Y.Y., Chang, Y.G., Li, Z.J., Xue, C.H., Wang, Y.M., 2014. Fucosylated chondroitin sulfate from sea cucumber improves glucose metabolism and activates insulin signaling in the liver of insulin-resistant mice. J. Med. Food. 17, 749–757.

Hua, K.F., Hsu, H.Y., Chao, L.K., Chen, S.T., Yang, W.B., Hsu, J., et al., 2007. *Ganoderma lucidum* polysaccharides enhance CD14 endocytosis of LPS and promote TLR4 signal transduction of cytokine expression. J. Cell. Physiol. 212, 537–550.

Huang, Q., Zhang, L., 2005. Solution properties of (1-- > 3)-alpha-D-glucan and its sulfated derivative from *Poria cocos* mycelia via fermentation tank. Biopolymers 79, 28–38.

Ip, C.C.Y., Manam, V., Hepler, R., Hennessey, J.P., 1992. Carbohydrate composition analysis of bacterial polysaccharides: optimized acid hydrolysis conditions for HPAEC-PAD analysis. Anal. Biochem. 201, 343–349.

Jeon, Y.J., Kim, H.M., 2001. Experimental evidences and signal transduction pathways involved in the activation of NF-kappa B/Rel by angelan in murine macrophages. Int. Immunopharmac. 1, 1331–1339.

Jia, S., Li, F., Liu, Y., Ren, H., Gong, G., Wang, Y., et al., 2013. Effects of extraction methods on the antioxidant activities of polysaccharides from *Agaricus blazei Murrill*. Int. J. Biol. Macromol. 62, 66–69.

Jin, M., Huang, Q., Ke, Z., Peng, S., 2013. Biological activities and potential health benefit effects of polysaccharides isolated from *Lycium barbarum* L. Int. J. Biol. Macromol. 54, 16–23.

Kawai, T., Akira, S., 2011. Toll-like receptors andtheir crosstalk with other innate receptors in infection and immunity. Immunity 34, 637–650.

Kido, S., Nakanishi, T., Norisuye, T., Kaneda, I., Yanaki, T., 2001. Ordered conformation of succinoglycan in aqueous sodium chloride. Biomacromolecules 2, 952–957.

Kim, H.M., Sang, B.H., Oh, G.T., Kim, Y.H., Dong, H.H., Hong, N.D., et al., 1996. Stimulation of humoral and cell mediated immunity by polysaccharide from mushroom phellinus linteus. Int. J. Immunopharmaco. 18, 295–303.

Kizil, R., Irudayaraj, J., Seetharaman, K., 2002. Characterization of irradiated starches by using FT-Raman and FTIR spectroscopy. J. Agric. Food Chem. 50, 3912–3918.

Krieger, M., 1997. The other side of scavenger receptors: pattern recognition for host defense. Curr. Opin. Lipidol. 8, 275–280.

Lai, L.S., Yang, D.H., 2007. Rheological properties of the hot-water extracted polysaccharides in Ling-Zhi (*Ganoderma lucidum*). Food Hydrocolloids 21, 739–746.

Lee, K.Y., Jeon, Y.J., 2003. Polysaccharide isolated from *Poria cocos* sclerotium induces NF-κB/Rel activation and iNOS expression in murine macrophages. Int. Immunopharmac. 3, 1353–1362.

Lee, K.Y., Bouhadir, K.H., Mooney, D.J., 2002. Evaluation of chain stiffness of partially oxidized polyguluronate. Biomacromolecules 3, 1129–1134.

Lee, K.Y., You, H.J., Jeong, H.G., Kang, J.S., Kim, H.M., Sang, D.R., et al., 2004. Polysaccharide isolated from *Poria cocos* sclerotium induces NF-κB/Rel activation and iNOS expression through the activation of p38 kinase in murine macrophages. Int. Immunopharmac. 4, 1029–1038.

Lee, S.J., Evers, S., Roeder, D., Parlow, A.F., Risteli, J., Risteli, L., et al., 2002. Mannose receptor-mediated regulation of serum glycoprotein homeostasis. Science 295, 1898–1901.

Leeuwen, S.S.V., Leeflang, B.R., Gerwig, G.J., Kamerling, J.P., 2008. Development of a 1 H NMR structural-reporter-group concept for the primary structural characterisation of α-D-glucans. Carbohyd. Res. 343, 1114–1119.

Lennartz, M.R., Cole, F.S., Shepherd, V.L., Wileman, T.E., Stahl, P.D., 1987. Isolation and characterization of a mannose-specific endocytosis receptor from human placenta. J. Biol. Chem. 262, 9942–9944.

Leung, M.Y.K., Liu, C., Koon, J.C.M., Fung, K.P., 2006. Polysaccharide biological response modifiers. Immunol. Lett. 105, 101–114.

Li, W., Cui, S.W., Wang, Q., 2006. Solution and conformational properties of wheat β-D-Glucans studied by light scattering and viscometry. Biomacromolecules 7, 446–452.

Li, W., Cui, S.W., Wang, Q., Yada, R.Y., 2012. Study of conformational properties of cereal β-glucans by computer modeling. Food Hydrocolloids 26, 377–382.

Li, X., Xu, W., 2011. TLR4-mediated activation of macrophages by the polysaccharide fraction from *Polyporus umbellatus* (pers.) Fries. J. Ethnopharmacol. 135, 1–6.

Lin, K.I., Kao, Y.Y., Kuo, H.K., Yang, W.B., Chou, A., Lin, H.H., et al., 2006. Reishi polysaccharides induce immunoglobulin production through the TLR4/TLR2-mediated induction of transcription factor Blimp-1. J. Biol. Chem. 281, 24111–24123.

Lindblad, M.S., Dahlman, O., Sjöberg, J., Albertsson, A.C., 2009. Modified galactoglucomannans from forestry waste-water for films and hydrogels. Polysaccharide materials: performance by design. ACS Symp. Ser. 1017, 185–198.

Liu, C., Leung, M.Y.K., Koon, J.C.M., Zhu, L.F., Hui, Y.Z., Yu, B., et al., 2006. Macrophage activation by polysaccharide biological response modifier isolated from *Aloe vera* L. var. chinensis (Haw.) Berg. Int. Immunopharmac. 6, 1634–1641.

Liu, C., Li, X., Li, Y., Feng, Y., Zhou, S., Wang, F., 2008a. Structural characterisation and antimutagenic activity of a novel polysaccharide isolated from *Sepiella maindroni* ink. Food Chem. 110, 807–813.

Liu, L., Guo, Z., Lv, Z., Sun, Y., Cao, W., Zhang, R., et al., 2008b. The beneficial effect of *Rheum tanguticum* polysaccharide on protecting against diarrhea, colonic inflammation and ulceration in rats with TNBS-induced colitis: the role of macrophage mannose receptor in inflammation and immune response. Int. Immunopharmac. 8, 1481–1492.

Lu, H., Yang, Y., Gad, E., Wenner, C.A., Chang, A., Larson, E.R., et al., 2010. Polysaccharide krestin is a novel TLR2 agonist that mediates inhibition of tumor growth via stimulation of CD8 T cells and NK cells. Clin. Cancer Res. 17, 67–76.

Lu, H.F., Lim, S.X., Meng, F.L., Narayanan, K., Toh, R.P.K., Gao, S., et al., 2012a. Efficient neuronal differentiation and maturation of human pluripotent stem cells encapsulated in 3D microfibrous scaffolds. Biomaterials 33, 9179–9187.

Lu, H.F., Narayanan, K., Lim, S.X., Gao, S., Leong, M.F., Wan, A.C., 2012b. A 3D microfibrous scaffold for long-term human pluripotent stem cell self-renewal under chemically defined conditions. Biomaterials 33, 2419–2430.

Luo, Y., Qin, W., 2014. Recent development of chitosan-based polyelectrolyte complexes with natural polysaccharides for drug delivery. Int. J. Biol. Macromol. 64, 353–367.

Ma, Z., Wang, J., Zhang, L., 2008. Structure and chain conformation of β-glucan isolated from *Auricularia auricula-judae*. Biopolymers 89, 614–622.

Maciel, J.S., Chaves, L.S., Souza, B.W.S., Teixeira, D.I.A., Freitas, A.L.P., Feitosa, J.P.A., et al., 2008. Structural characterization of cold extracted fraction of soluble sulfated polysaccharide from red seaweed *Gracilaria birdiae*. Carbohyd. Polym. 71, 559–565.

Maehara, Y., Tsujitani, S., Saeki, H., Oki, E., Yoshinaga, K., Emi, Y., et al., 2011. Biological mechanism and clinical effect of protein-bound polysaccharide K (KRESTIN ®): review of development and future perspectives. Surg. Today 42, 8–28.

Maina, N.H., Tenkanen, M., Maaheimo, H., Juvonen, R., Virkki, L., 2008. NMR spectroscopic analysis of exopolysaccharides produced by *Leuconostoc citreum* and *Weissella confusa*. Carbohyd. Res. 343, 1446–1455.

Mao, S., Sun, W., Kissel, T., 2010. Chitosan-based formulations for delivery of DNA and siRNA. Adv. Drug Deliv. Rev. 62, 12–27.

Marshall, J.J., 1974. Application of enzymic methods to the structural analysis of polysaccharides: part I. Adv. Carbohyd. Chem. Biochem. 30, 257–370.

Maslen, S.L., Goubet, F., Adam, A., Dupree, P., Stephens, E., 2007. Structure elucidation of arabinoxylan isomers by normal phase HPLC–MALDI-TOF/TOF-MS/MS. Carbohyd. Res. 342, 724–735.

Mathlouthi, M., Koenig, J.L., 1986. Vibrational spectra of carbohydrates. Adv. Carbohyd. Chem. Biochem. 44, 7–89.

Matsumoto, T., Yamada, H., 1995. Regulation of immune complexes binding of macrophages by pectic polysaccharide from *Bupleurum falcatum* L.: pharmacological evidence for the requirement of intracellular calcium/calmodulin on Fc receptor up-regulation by bupleuran 2IIb. J. Pharm. Pharmacol. 47, 152–156.

Matsunaga, K., Morita, I., Iijima, H., Endo, H., Oguchi, Y., Yoshimura, M., et al., 1990. Competitive action of a biological response modifier, PSK, on a humoral immunosuppressive factor produced in tumor-bearing hosts. J. Clin. Lab. Immunol. 31, 127–136.

Matsunaga, K., Hosokawa, A., Oohara, M., Sugita, N., Harada, M., Nomoto, K., 1998. Direct action of a protein-bound polysaccharide, PSK, on transforming growth factor-β. Immunopharmacology 40, 219–230.

Matsushita, A., Iwase, M., Kato, Y., Ichihara, S., Ichihara, G., Kimata, H., et al., 2007. Differential cardiovascular effects of endotoxin derived from *Escherichia coli* or *Pseudomonas aeruginosa*. Exp. Anim. 56, 339–348.

Mazeau, K., Rinaudo, M., 2004. The prediction of the characteristics of some polysaccharides from molecular modeling. Comparison with effective behavior. Food Hydrocolloids 18, 885–898.

Mazumder, S., Morvan, C., Thakur, S., Ray, B., 2004. Cell wall polysaccharides from chalkumra (*Benincasa hispida*) fruit. Part I. Isolation and characterization of pectins. J. Agric. Food Chem. 52, 3556–3562.

Mazumder, S., Lerouge, P., Loutelier-Bourhis, C., Driouich, A., Ray, B., 2005. Structural characterisation of hemicellulosic polysaccharides from *Benincasa hispida* using specific enzyme hydrolysis, ion exchange chromatography and MALDI-TOF mass spectroscopy. Carbohyd. Polym. 59, 231–238.

Mccleary, B.V., Matheson, N.K., 1987. Enzymic analysis of polysaccharide structure. Adv. Carbohyd. Chem. Biochem. 44, 147–276.

Mcgreal, E.P., Rosas MBrown, G.D., Zamze, S., Wong, S.Y., Gordon, S., Martinez, P.L., et al., 2006. The carbohydrate-recognition domain of Dectin-2 is a C-type lectin with specificity for high mannose. Glycobiology 16, 422–430.

Mcneil, M., Darvill, A.G., Åman, P., Franzén, L.E., Albersheim, P., 1982. Structural analysis of complex carbohydrates using high-performance liquid chromatography, gas chromatography, and mass spectrometry. Methods Enzymol. 83, 3–45.

Medzhitov, R., Preston-Hurlburt, P., Janeway Jr., C.A., 1997. A human homologue of the Drosophila Toll protein signals activation of adaptive immunity. Nature 388, 394–397.

Medzhitov, R., Preston-Hurlburt, P., Kopp, E., Stadlen, A., Chen, C., Ghosh, S., et al., 1998. MyD88 is an adaptor protein in the hToll/IL-1 receptor family signaling pathways. Mol. Cell. 2, 253–258.

Meng, C.K., Mann, M., Fenn, J.B., 1988. Of protons or proteins. Z. Phys. D: At., Mol. Clusters 10, 361–368.

Miller, J.L., Dewet, B.J.M., Martinez-Pomares, L., Radcliffe, C.M., Dwek, R.A., Rudd, P.M., et al., 2008. The mannose receptor mediates dengue virus infection of macrophages. PLoS Pathog. 4, e17.

Miyake, K., Yamashita, Y., Ogata, M., Sudo, T., Kimoto, M., 1995. RP105, a novel B cell surface molecule implicated in B cell activation, is a member of the leucine-rich repeat protein family. J. Immunol. 154, 3333–3340.

Miyake, K., Shimazu, R., Kondo, J., Niki, T., Akashi, S., Ogata, H., et al., 1998. Mouse MD-1, a molecule that is physically associated with RP105 and positively regulates its expression. J. Immunol. 161, 1348–1353.

Mizrahy, S., Peer, D., 2012. Polysaccharides as building blocks for nanotherapeutics. Chem. Soc. Rev. 41, 2623–2640.

Moertel, C.G., Johnson, C.M., Mckusick, M.A., Martin Jr., J.K., Nagorney, D.M., Kvols, L.K., et al., 1994. The management of patients with advanced carcinoid tumors and islet cell carcinomas. Ann. Intern. Med. 120, 302–309.

Mopper, K., Stubbins, A., Ritchie, J.D., Bialk, H.M., Hatcher, P.G., 2007. Advanced instrumental approaches for characterization of marine dissolved organic matter: extraction techniques, mass spectrometry, and nuclear magnetic resonance spectroscopy. Chem. Rev. 38, 419–442.

Mueller, A., Raptis, J., Rice, P.J., Kalbfleisch, J.H., Stout, R.D., Ensley, H.E., et al., 2000. The influence of glucan polymer structure and solution conformation on binding to (1→3)-β-D-glucan receptors in a human monocyte-like cell line. Glycobiology 10, 339–346.

Mullin, N.P., Hall, K.T., Taylor, M.E., 1994. Characterization of ligand binding to a carbohydrate-recognition domain of the macrophage mannose receptor. J. Biol. Chem. 269, 28405–28413.

Mullin, N.P., Hitchen, P.G., Taylor, M.E., 1997. Mechanism of Ca^{2+} and monosaccharide binding to a C-type carbohydrate-recognition domain of the macrophage mannose receptor. J. Biol. Chem. 272, 5668–5681.

Muroi, M., Tanamoto, K., 2002. The polysaccharide portion plays an indispensable role in Salmonella lipopolysaccharide-induced activation of NF-kappaB through human toll-like receptor 4. Infect. Immun. 70, 6043–6047.

Mytar, B., Gawlicka, M., Szatanek, R., Wołoszyn, M., Ruggiero, I., Piekarska, B., et al., 2004. Induction of intracellular cytokine production in human monocytes/macrophages stimulated with ligands of pattern recognition receptors. Inflamm. Res. 53, 100–106.

Nagai, Y., Shimazu, R., Ogata, H., Akashi, S., Sudo, K., Yamasaki, H., et al., 2002. Requirement for MD-1 in cell surface expression of RP105/CD180 and B-cell responsiveness to lipopolysaccharide. Blood 99, 1699–1705.

Nagai, Y., Kobayashi, T., Motoi, Y., Ishiguro, K., Akashi, S., Saitoh, S.I., et al., 2005. The radioprotective 105/MD-1 complex links TLR2 and TLR4/MD-2 in antibody response to microbial membranes. J. Immunol. 174, 7043–7049.

Nakamura, T., Suzuki, H., Wada, Y., Kodama, T., Doi, T., 2006. Fucoidan induces nitric oxide production via p38 mitogen-activated protein kinase and NF-κB-dependent signaling pathways through macrophage scavenger receptors. Biochem. Biophys. Res. Commun. 343, 286–294.

Nakanishi, T., Norisuye, T., 2003. Thermally induced conformation change of succinoglycan in aqueous sodium chloride. Biomacromolecules 4, 736–742.

Nakata, M., Kawaguchi, T., Kodama, Y., Konno, A., 1998. Characterization of curdlan in aqueous sodium hydroxide. Polymer 39, 1475–1481.

Needs, P.W., Selvendran, R.R., 1993. Avoiding oxidative degradation during sodium hydroxide/methyl iodide-mediated carbohydrate methylation in dimethyl sulfoxide. Carbohyd. Res. 245, 1–10.

Nergard, C.S., Kiyohara, H., Reynolds, J.C., Thomasoates, J.E., Matsumoto, T., Yamada, H., et al., 2006. Structures and structure-activity relationships of three mitogenic and complement fixing pectic arabinogalactans from the malian antiulcer plants *Cochlospermum tinctorium* A. Rich and *Vernonia kotschyana* Sch. Bip. ex Walp. Biomacromolecules 7, 71–79.

Nichifor, M., Lopes, A., Carpov, A., Melo, E., 1999. Aggregation in water of dextran hydrophobically modified with bile acids. Macromolecules 32, 7078–7085.

Normand, M.L., Mélida, H., Holmbom, B., Michaelsen, T.E., Inngjerdingen, M., Bulone, V., et al., 2014. Hot-water extracts from the inner bark of Norway spruce with immunomodulating activities. Carbohyd. Polym. 101, 699–704.

Ogata, H., Su, I., Miyake, K., Nagai, Y., Akashi, S., Mecklenbräuker, I., et al., 2000. The toll-like receptor protein RP105 regulates lipopolysaccharide signaling in B cells. J. Exp. Med. 192, 23–29.

Oliveira, J.T., Reis, R.L., 2011. Polysaccharide-based materials for cartilage tissue engineering applications. J. Tissue Eng. Regen. Med. 5, 421–436.

Ozinsky, A., Underhill, D.M., Fontenot, J.D., Haijar, A.M., Smith, K.D., Wilson, C.B., et al., 2000. The repertoire for pattern recognition of pathogens by the innate immune system is defined by cooperation between toll-like receptors. Proc. Natl. Acad. Sci. USA 97, 13766–13771.

Pachuau, L., Mazumder, B., 2013. Colonic drug delivery systems based on natural polysaccharides and their evaluation. Mini Rev. Med. Chem. 13, 1982–1991.

Pearson, A., Lux, A., Krieger, M., 1995. Expression cloning of dSR-CI, a class C macrophage-specific scavenger receptor from *Drosophila melanogaster*. Proc. Natl. Acad. Sci. USA 92, 4056–4060.

Pearson, A.M., 1996. Scavenger receptors in innate immunity. Curr. Opin. Immunol. 8, 20–28.

Peiser, L., Gordon, S., 2001. The function of scavenger receptorsexpressed by macrophages and their rolein the regulation of inflammation. Microbes Infect. 3, 149–159.

Peiser, L., Mukhopadhyay, S., Gordon, S., 2002. Scavenger receptors in innate immunity. Curr. Opin. Immunol. 14, 123–128.

Peng, Y., Zhang, L., Zhang, Y., Xu, X., Kennedy, J.F., 2005. Solution properties of water-insoluble polysaccharides from the mycelium of *Ganoderma tsugae*. Carbohyd. Polym. 59, 351–356.

Petkowicz, C.L.O., Reicher, F., Mazeau, K., 1998. Conformational analysis of galactomannans: from oligomeric segments to polymeric chains. Carbohyd. Polym. 37, 25–39.

Philip, R.T., Delyth, M.R., Sigrid, E.M.H., Gordon, D.B., Siamon, G., Simon, Y.C.W., 2005. Dectin-2 is predominantly myeloid restricted and exhibits unique activation-dependent expression on maturing inflammatory monocytes elicited in vivo. Eur. J. Immunol. 35, 2163–2174.

Pisi, E., Poussin, K., Paterlini, P., Bréchot, C., 1995. HCV-associated liver cancer without cirrhosis. Lancet 345, 413–415.

Popov, S.V., Popova, G.Y., Ovodova, R.G., Bushneva, O.A., Ovodov, Y.S., 1999. Effects of polysaccharides from *Silene vulgaris* on phagocytes. Int. J. Immunopharmaco. 21, 617–624.

Price, L.A., Wenner, C.A., Sloper, D.T., Slaton, J.W., Novack, J.P., 2010. Role for toll-like receptor 4 in TNF-alpha secretion by murine macrophages in response to polysaccharide Krestin, a *Trametes versicolor* mushroom extract. Fitoterapia 81, 914–919.

Puri, M., Sharma, D., Barrow, C.J., 2011. Enzyme-assisted extraction of bioactives from plants. Trends Biotechnol. 30, 37–44.

Quemener, B., Lahaye, M., Thibault, J.F., 1993. Studies on the simultaneous determination of acidic and neutral sugars of plant cell wall materials by HPLC of their methyl glycosides after combined methanolysis and enzymic prehydrolysis. Carbohyd. Polym. 20, 87–94.

Radhakrishnamurthy, B., Berenson, G.S., 2002. Effect of temperature and time of heating on the carbazole reaction of uronic acids and acid mucopolysaccharides. Anal. Chem. 35, 253–259.

Rankin, J.C., Jeanes, A., 1954. Evaluation of the periodate oxidation method for structural analysis of dextrans. J. Am. Chem. Soc. 76, 4435–4441.

Reddy, K., Mohan, G.K., Satla, S., Gaikwad, S., 2011. Natural polysaccharides: versatile excipients for controlled drug delivery systems. Asian J. Pharm. Sci. 6, 275–286.

Reinhold, V.N., Reinhold, B.B., Costello, C.E., 1995. Carbohydrate molecular weight profiling, sequence, linkage, and branching data: ES-MS and CID. Anal. Chem. 67, 1772–1784.

Rice, P.J., Kelley, J.L., Kogan, G., Ensley, H.E., Kalbfleisch, J.H., Browder, I.W., et al., 2002. Human monocyte scavenger receptors are pattern recognition receptors for (1--> 3)-beta-D-glucans. J. Leukoc. Biol. 72, 140–146.

Rinaudo, M., 2008. Main properties and current applications of some polysaccharides as biomaterials. Polym. Int. 57, 397–430.

Ross, G.D., 2000. Regulation of the adhesion versus cytotoxic functions of the Mac-1/CR3/alphaMbeta2-integrin glycoprotein. Crit. Rev. Immunol. 20, 197–222.

Ross, G.D., Větvicka, V.V., 0011. CR3 (CD11b, CD18): a phagocyte and NK cell membrane receptor with multiple ligand specificities and functions. Clin. Exp. Immunol. 92, 181–184.

Ross, G.D., Cain, J.A., Lachmann, P.J., 1985. Membrane complement receptor type three (CR3) has lectin-like properties analogous to bovine conglutinin as functions as a receptor for zymosan and rabbit erythrocytes as well as a receptor for iC3b. J. Immunol. 134, 3307–3315.

Rowe, D.S., Hug, K., Forni, L., Pernis, B., 1973. Immunoglobulin D as a lymphocyte receptor. J. Exp. Med. 138, 965–972.

Roy, S.K., Maiti, D., Mondal, S., Das, D., Islam, S.S., 2008. Structural analysis of a polysaccharide isolated from the aqueous extract of an edible mushroom, Pleurotus sajor-caju, cultivar Black Japan. Carbohyd. Res. 343, 1108–1113.

Ruiter, G.A.D., Josso, S.L., Colquhoun, I.J., Voragen, A.G.J., Rombouts, F.M., 1992. Isolation and characterization of β(1–4)-D-glucuronans from extracellular polysaccharides of moulds belonging to Mucorales. Carbohyd. Polym. 18, 1–7.

Saijo, S., Ikeda, S., Yamabe, K., Kakuta, S., Ishigame, H., Akitsu, A., et al., 2010. Dectin-2 recognition of α-mannans and induction of Th17 cell differentiation is essential for host defense against *Candida albicans*. Immunity 32, 681–691.

Sang, B.H., Park, S.H., Lee, K.H., Chang, W.L., Sang, H.L., Kim, H.C., et al., 2001. Polysaccharide isolated from the radix of *Platycodon grandiflorum* selectively activates B cells and macrophages but not T cells. Int. Immunopharmac. 1, 1969–1978.

Sato, K., Yang, X., Yudate, T., Chung, J.S., Wu, J., Lubyphelps, K., et al., 2007. Dectin-2 is a pattern recognition receptor for Fungi that couples with the Fc receptor γ chain to induce innate immune responses. J. Biol. Chem. 281, 38854–38866.

Sato, T., Norisuye, T., Fujita, H., 1984. Double-stranded helix of xanthan in dilute solution: evidence from light scattering. Polym. J. 16, 341–350.

Schappacher, M., Ricard, L., Fischer, J., Weiss, R., Bill, E., Montiel-Montoya, R., et al., 2005. Synthesis, structure and spectroscopic properties of two models for the active site of the oxygenated state of cytochrome P 540. Eur. J. Biochem. 168, 419–429.

Schepetkin, I.A., Quinn, M.T., 2006. Botanical polysaccharides: macrophage immunomodulation and therapeutic potential. Int. Immunopharmac. 6, 317–333.

Schepetkin, I.A., Xie, G., Kirpotina, L.N., Klein, R.A., Jutila, M.A., Quinn, M.T., 2008. Macrophage immunomodulatory activity of polysaccharides isolated from *Opuntia polyacantha*. Int. Immunopharmac. 8, 1455–1466.

Schuster, K.C., Ehmoser, H., Gapes, J.R., Lendl, B., 2000. On-line FT-Raman spectroscopic monitoring of starch gelatinisation and enzyme catalysed starch hydrolysis 1. Vib. Spectrosc. 22, 181–190.

Scott, J.E., Dorling, J., 1969. Periodate oxidation of acid polysaccharides. Histochemie 19, 295–301.

Selvendran, R.R., March, J.F., Ring, S.G., 1979. Determination of aldoses and uronic acid content of vegetable fiber. Anal. Biochem. 96, 282–292.

Shao, B.M., Dai, H., Xu, W., Lin, Z.B., Gao, X.M., 2004a. Immune receptors for polysaccharides from *Ganoderma lucidum*. Biochem. Biophys. Res. Commun. 323, 133–141.

Shao, B.M., Xu, W., Dai, H., Tu, P., Li, Z., Gao, X.M., 2004b. A study on the immune receptors for polysaccharides from the roots of *Astragalus membranaceus*, a Chinese medicinal herb. Biochem. Biophys. Res. Commun. 320, 1103–1111.

Shin, J.Y., Song, J.Y., Yun, Y.S., Yang, H.O., Rhee, D.K., Pyo, S., 2002. Immunostimulating effects of acidic polysaccharides extract of *Panax ginseng* on macrophage fuction. Immunopharm. Immunot. 24, 469–482.

Shin, K.S., Kiyohara, H., Matsumoto, T., Yamada, H., 1997. Rhamnogalacturonan II from the leaves of *Panax ginseng* C.A. Meyer as a macrophage Fc receptor expression-enhancing polysaccharide. Carbohyd. Res. 300, 239–249.

Stockdale, A., Davison, W., Zhang, H., 2009. Micro-scale biogeochemical heterogeneity in sediments: a review of available technology and observed evidence. Earth Sci. Rev. 92, 81–97.

Suárez, E.R., Syvitski, R., Kralovec, J.A., Noseda, M.D., Barrow, C.J., Ewart, H.S., et al., 2006. Immunostimulatory polysaccharides from *Chlorella pyrenoidosa*. A new galactofuranan. measurement of molecular weight and molecular weight dispersion by DOSY NMR. Biomacromolecules 7, 2368–2376.

Suh, H.J., Yang, H.S., Ra, K.S., Noh, D.O., Kwon, K.H., Hwang, J.H., et al., 2013. Peyer's patch-mediated intestinal immune system modulating activity of pectic-type polysaccharide from peel of *Citrus unshiu*. Food Chem. 138, 1079–1086.

Sundberg, A., Sundberg, K., Lillandt, C., Holmbom, B., 1996. Determination of hemicelluloses and pectins in wood and pulp fibres by acid methanolysis and gas chromatography. Nord. Pulp Paper Res. J. 11, 216–219.

Sunghee, U., Dongkwon, R., Suhkneung, P., 2002. Involvement of protein kinase C and tyrosin kinase in tumoricidal activation of macrophage induced by *Streptococcus pneumoniae* type II capsular polysaccharide. Int. Immunopharmac. 2, 129–137.

Suzuki, H., Hehre, E.J., 1964. The differentiation of serotype A and B dextrans by means of partial acetolysis 1. Arch. Biochem. Biophys. 104, 305–313.

Symonette, C.J., Kaur, M.A., Tan, X.C., Tolg, C., Ma, J., Perera, F., et al., 2014. Hyaluronan-phosphatidylethanolamine polymers form pericellular coats on keratinocytes and promote basal keratinocyte proliferation. Biomed. Res. Int. 2014, 727459.

Tada, R., Harada, T., Nagi-Miura, N., Adachi, Y., Nakajima, M., Yadomae, T., et al., 2007. NMR characterization of the structure of a β-(1$\rightarrow$3)-D-glucan isolate from cultured fruit bodies of *Sparassis crispa*. Carbohyd. Res. 342, 2611–2618.

Tao, S., Pranovich, A., Holmbom, B., 2013. Separation of polymeric galactoglucomannans from hot-water extract of spruce wood. Bioresour. Technol. 130, 198–203.

Tao, Y., Zhang, L., 2006. Determination of molecular size and shape of hyperbranched polysaccharide in solution. Biopolymers 83, 414–423.

Tao, Y., Zhang, L., 2008. Characterization of polysaccharide-protein complexes by size-exclusion chromatography combined with three detectors. Carbohyd. Res. 343, 2251–2257.

Taylor, M.E., Conary, J.T., Lennartz, M.R., Stahl, P.D., Drickamer, K., 1990. Primary structure of the mannose receptor contains multiple motifs resembling carbohydrate-recognition domains. J. Biol. Chem. 265, 12156–12162.

Taylor, M.E., Bezouska, K., Drickamer, K., 1992. Contribution to ligand binding by multiple carbohydrate-recognition domains in the macrophage mannose receptor. J. Biol. Chem. 267, 1719–1726.

Taylor, P.R., Tsoni, S.V., Willment, J.A., Dennehy, K.M., Rosas, M., Findon, H., et al., 2007. Dectin-1 is required for beta-glucan recognition and control of fungal infection. Nat. Immunol. 8, 31–38.

Thiery, J.P., Sleeman, J.P., 2006. Complex networks orchestrate epithelial-mesenchymal transitions. Nat. Rev. Mol. Cell Bio. 7, 131–142.

Tønnesen, H.H., Karlsen, J., 2002. Alginate in drug delivery systems. Drug Dev. Ind. Pharm. 28, 621–630.

Tsujitani, S., Osaki, T., Saito, H., Fukuda, K., Tatebe, S., Ikeguchi, M., 2008. The NKG2D expression on $CD8^+$T cells and the efficacy of polysaccharide K (PSK) in gastric cancer. J. Clin. Oncol. 26, 431–436.

Tzianabos, A.O., 2000. Polysaccharide immunomodulators as therapeutic agents: structural aspects and biologic function. Clin. Microbiol. Rev. 13, 523–533.

Ulevitch, R.J., Tobias, P.S., 1995. Receptor-dependent mechanisms of cell stimulation by bacterial endotoxin. Annu. Rev. Immunol. 13, 437–457.

Um, S.H., Son, E.W., Kim, B.O., Moon, E.Y., Rhee, D.K., Pyo, S., 2000. Activation of murine peritoneal macrophages by *Streptococcus pneumoniae* type II capsular polysaccharide: involvement of CD14-dependent pathway. Scand. J. Immunol. 52, 39–45.

Urai, M., Yoshizaki, H., Anzai, H., 2007. Structural analysis of an acidic,fatty acid ester-bonded extracellular polysaccharide produced by a pristane-assimilating marine bacterium, *Rhodococcus erythropolis* PR4. Carbohyd. Res. 342, 933–942.

Valo, H., Arola, S., Laaksonen, P., Torkkeli, M., Peltonen, L., Linder, M.B., et al., 2013. Drug release from nanoparticles embedded in four different nanofibrillar cellulose aerogels. Eur. J. Pharm. Sci. 50, 69–77.

Van Gunsteren, W.F., Berendsen, H.J.C., 1990. Computer simulation of molecular dynamics: methodology, applications, and perspectives in chemistry. Angew. Chem. Int. Ed. 29, 992–1023.

Veerdonk, F.L.V.D., Kullberg, B.J., Meer, J.W.V.D., Gow, N.A., Netea, M.G., 2008. Host–microbe interactions: innate pattern recognition of fungal pathogens. Curr. Opin. Microbiol. 11, 305–312.

Venkatesan, J., Kim, S.K., 2010. Chitosan composites for bone tissue engineering—an overview. Mar. Drugs 8, 2252–2266.

Vereschagin, E.I., van Lambalgen, A.A., Dushkin, M.I., Schwartz, Y.S., Polyakov, L., Heemskerk, A., et al., 1998. Soluble glucan protects against endotoxin shock in the rat: the role of the scavenger receptor. Shock 9, 193–198.

Vetvicka, V., Thornton, B.P., Ross, G.D., 1996. Soluble beta-glucan polysaccharide binding to the lectin site of neutrophil or natural killer cell complement receptor type 3 (CD11b/CD18) generates a primed state of the receptor capable of mediating cytotoxicity of iC3b-opsonized target cells. J. Clin. Invest. 98, 50–61.

Viseux, N., Costello, C.E., Domon, B., 1999. Post-source decay mass spectrometry: optimized calibration procedure and structural characterization of permethylated oligosaccharides. J. Mass Spectrom. 34, 364–376.

Wan, A.C., Tai, B.C., 2013. Chitin--a promising biomaterial for tissue engineering and stem cell technologies. Biotechnol. Adv. 31, 1776–1785.

Wang, C.L., Lu, C.Y., Pi, C.C., Zhuang, Y.J., Chu, C.L., Liu, W.H., et al., 2012. Extracellular polysaccharides produced by *Ganoderma formosanum* stimulate macrophage activation via multiple pattern-recognition receptors. BMC Complement. Altern. Med. 12, 1–10.

Wang, Q., Cui, S.W., 2005. Food Carbohydrates: Chemistry, Physical Properties, and Applications. CRC Press.

Wang, S.Y., Hsu, M.L., Hsu, H.C., Lee, S.S., Shiao, M.S., Ho, C.K., 1997. The anti-tumor effect of *Ganoderma lucidum* is mediated by cytokines released from activated macrophages and T lymphocytes. Int. J. Cancer 70, 699–705.

Wang, X., Xu, X., Zhang, L., 2008. Thermally induced conformation transition of triple-helical lentinan in NaCl aqueous solution. J. Phys. Chem. B. 112, 10343–10351.

Warner, N.L., 1974. Membrane immunoglobulins and antigen receptors on B and T lymphocytes. Adv. Immunol. 19, 67–216.

Wasser, S., 2002. Medicinal mushrooms as a source of antitumor and immunomodulating polysaccharides. Appl. Microbiol. Biotechnol. 60, 258–274.

Wasser, S.P., Weis, A.L., 1999. Therapeutic effects of substances occurring in higher *Basidiomycetes mushrooms*: a modern perspective. Crit. Rev. Immunol. 19, 65–96.

Weis, W.I., Taylor, M.E., Drickamer, K., 1998. The C-type lectin superfamily in the immune system. Immunol. Rev. 163, 19–34.

Wesche, H., Henzel, W.J., Shillinglaw, W., Li, S., Cao, Z., 2013. Pillars article: MyD88: an adapter that recruits IRAK to the IL-1 receptor complex. Immunity. 1997. 7: 837-847. J. Immunol. 190, 5–15.

Wijesekara, I., Pangestuti, R., Kim, S.K., 2011. Biological activities and potential health benefits of sulfated polysaccharides derived from marine algae. Carbohyd. Polym. 84, 14–21.

Wilm, M.S., Mann, M., 1994. Electrospray and Taylor-Cone theory, Dole's beam of macromolecules at last?. Int. J. Mass Spectrom. Ion Process. 136, 167–180.

Wright, S.D., Jong, M.T., 1986. Adhesion-promoting receptors on human macrophages recognize *Escherichia coli* by binding to lipopolysaccharide. J. Exp. Med. 164, 1876–1888.

Wu, Y., Li, W., Cui, W., Eskin, N.A.M., Goff, H.D., 2012. A molecular modeling approach to understand conformation–functionality relationships of galactomannans with different mannose/galactose ratios. Food Hydrocolloids 26, 359–364.

Xu, H., Yuan, X.D., Shen, B.D., Han, J., Lv, Q.Y., Dai, L., et al., 2014. Development of poly (N-isopropylacrylamide)/alginate copolymer hydrogel-grafted fabrics embedding of berberine nanosuspension for the infected wound treatment. J. Biomater. Appl. 28, 1376–1385.

Xu, Q., Bush, C.A., 1996. Molecular modeling of the flexible cell wall polysaccharide of Streptococcus mitis J22 on the basis of heteronuclear NMR coupling constants. Biochemistry 35, 14521–14529.

Xue, H.J., Yao, Y.F., 2012. Investigation of multiscale molecular dynamics in polymer solids: sold-state NMR approachs. Chin. J. Magn. Reson. 29, 143–162.

Yamamoto, M., Akira, S., 2010. Lipid a receptor TLR4-mediated signaling pathways. Adv. Exp. Med. Biol. 667, 59–68.

Yamaguchi, Y., Minami, K., Ohshita, A., Kawabuchi, Y., Noma, K., Toge, T., 2004. Enhancing effect of PS-K on IL-2-induced lymphocyte activation: possible involvement of antagonistic action against TGF-beta. Anticancer Res. 24, 639–648.

Yamashita, Y., Miyake, K., Miura, Y., Kaneko, Y., Yagita, H., Suda, T., et al., 1996. Activation mediated by RP105 but not CD40 makes normal B cells susceptible to anti-IgM-induced apoptosis: a role for Fc receptor coligation. J. Exp. Med. 184, 113–120.

Yan, J., Vetvicka, V., Xia, Y., Coxon, A., Carroll, M.C., Mayadas, T.N., et al., 1999. Beta-glucan, a "specific" biologic response modifier that uses antibodies to target tumors for cytotoxic recognition by leukocyte complement receptor type 3 (CD11b/CD18). J. Immunol. 163, 3045–3052.

Yanaki, T., Norisuye, T., Fujita, H., 1980. Triple helix of *Schizophyllum commune* polysaccharide in dilute solution. 3. Hydrodynamic properties in water. Macromolecules 13, 1462–1466.

Yang, B., Jiang, Y., Zhao, M., Shi, J., Wang, L., 2008. Effects of ultrasonic extraction on the physical and chemical properties of polysaccharides from longan fruit pericarp. Polym. Degrad. Stab. 93, 268–272.

Yang, Y., Zhang, J.S., Liu, Y.F., Tang, Q.J., Zhao, Z.G., Xia, W.S., 2007. Structural elucidation of a 3-O-methyl-D-galactose-containing neutral polysaccharide from the fruiting bodies of *Phellinus igniarius*. Carbohyd. Res. 342, 1063–1070.

Yazawa, N., Fujimoto, M., Sato, S., Miyake, K., Asano, N., Nagai, Y., et al., 2003. CD19 regulates innate immunity by the toll-like receptor RP105 signaling in B lymphocytes. Blood 102, 1374–1380.

Yoon, Y.D., Sang, B.H., Kang, J.S., Chang, W.L., Park, S.K., Lee, H.S., et al., 2003. Toll-like receptor 4-dependent activation of macrophages by polysaccharide isolated from the radix of *Platycodon grandiflorum*. Int. Immunopharmac. 3, 1873–1882.

Yoon, Y.D., Kang, J.S., Han, S.B., Park, S.K., Lee, H.S., Kang, J.S., et al., 2004. Activation of mitogen-activated protein kinases and AP-1 by polysaccharide isolated from the radix of *Platycodon grandiflorum* in RAW 264.7 cells. Int. Immunopharmac. 4, 1477–1487.

Yu, Z.H., Yin, L.H., Yang, Q., Liu, Y., 2009. Effect of *Lentinus edodes* polysaccharide on oxidative stress, immunity activity and oral ulceration of rats stimulated by phenol. Carbohyd. Polym. 75, 115–118.

Zamze, S., Martinez, P.L., Jones, H., Taylor, P.R., Stillion, R.J., Gordon, S., et al., 2002. Recognition of bacterial capsular polysaccharides and lipopolysaccharides by the macrophage mannose receptor. J. Biol. Chem. 277, 41613–41623.

Zhang, F.X., Kirschning, C.J., Mancinelli, R., Xu, X.P., Jin, Y., Faure, E., et al., 1999. Bacterial lipopolysaccharide activates nuclear factor-kappaB through interleukin-1 signaling mediators in cultured human dermal endothelial cells and mononuclear phagocytes. J. Biol. Chem. 274, 7611–7614.

Zhang, H., Morisaki, T., Matsunaga, H., Sato, N., Uchiyama, A., Hashizume, K., et al., 2000. Protein-bound polysaccharide PSK inhibits tumor invasiveness by down-regulation of TGF-β1 and MMPs. Clin. Exp. Metastasis 18, 345–351.

Zhang, L., Yang, L., 1995. Properties of *Auricularia auricula-judae* β-D-glucan in dilute solution. Biopolymers 36, 695–700.

Zhang, L., Zhang, M., Dong, J., Guo, J., Song, Y., Cheung, P.C.K., 2001. Chemical structure and chain conformation of the water-insoluble glucan isolated from *Pleurotus tuber-regium*. Biopolymers 59, 457–464.

Zhang, L., Li, X., Xu, X., Zeng, F., 2005. Correlation between antitumor activity, molecular weight, and conformation of lentinan. Carbohyd. Res. 340, 1515–1521.

Zhang, N., Wardwell, P.R., Bader, R.A., 2013a. Polysaccharide-based micelles for drug delivery. Pharmaceutics 5, 329–352.

Zhang, S.S., Nie, S.P., Huang, D.F., Huang, J.Q., Wang, Y.X., Xie, M.Y., 2013b. Polysaccharide from *Ganoderma atrum* evokes antitumor activity via Toll-like receptor 4-mediated NF-κB and mitogen-activated protein kinase signaling pathways. J. Agric. Food Chem. 61, 3676–3682.

Zhang, X., Ding, R., Zhou, Y., Zhu, R., Liu, W., Jin, L., et al., 2013c. Toll-like receptor 2 and Toll-like receptor 4-dependent activation of B cells by a polysaccharide from marine fungus *Phoma herbarum* YS4108. PLoS One 8, e60781.

Zhao, C., Yang, R.F., Qiu, T.Q., 2013. Ultrasound-enhanced subcritical water extraction of polysaccharides from *Lycium barbarum* L. Sep. Purif. Technol. 120, 141–147.

Zhbankov, R., Firsov, S., Korolik, E., Petrov, P., Lapkovski, M., Tsarenkov, V., et al., 2000. Vibrational spectra and the structure of medical biopolymers. J. Mol. Struct. 555, 85–96.

Zhou, J., Yan, J., Bai, Z., Li, K., Huang, K., 2015. Hypoglycemic activity and potential mechanism of a polysaccharide from the loach in streptozotocin-induced diabetic mice. Carbohyd. Polym. 121, 199–206.

Zhu, F., Isaacs, N.W., Hecht, L., Tranter, G.E., Barron, L.D., 2006. Raman optical activity of proteins, carbohydrates and glycoproteins. Chirality 18, R45–R50.

Chapter 3

Beta-Glucans and Their Derivatives

Polysaccharides can originate from diverse sources, they not only represent an essential component of cytoplasm, but also of the cell membrane and cell wall. As is known, the glucans are a heterogeneous group of glucose polymers that constitute important structural or skeletal components of the cell wall. According to the linkage patterns, glucans can be divided into two groups: α-glucan and β-glucan. Over the last half-century, fungal extracellular polymers have received a huge amount of attention due to their potential use in various industries, including food industries, cosmetics, pharmaceuticals, and so on. Particularly, fungi-derived β-glucans are now attracting great attention all over the world.

In this chapter, the source, preparation and purification method, as well as the structural properties of β-glucans are introduced. Further, the biological activities and the bioactivity–structure relationship, especially for the fungi β-1,3 or 1,6-glucans, are discussed.

3.1 SOURCES

3.1.1 Beta-Glucan From Fungi

Fungi are classified in the independent kingdom Fungi among organisms. They are heterotrophic, and obtain essential nutrients through breaking down and absorbing organic substances. Moreover, fungi are extremely abundant and diverse worldwide, and the generally accepted numbers of fungi species on earth is over 1,500,000. Many higher Basidiomycetes mushrooms contain bioactive polysaccharides in their fruit bodies, cultured mycelium, and cultured broth. So far, the data on mushroom polysaccharides are summarized for approximately 700 species of higher hetero- and homo-basidiomycetes. Mushrooms are currently evaluated for their nutritional value and acceptability, as well as for their pharmacological properties, and they provide polysaccharides and polysaccharide–protein complexes with various bioactivities. Beta-glucan is one of the key components of the fungal cell wall (Chan et al., 2009).

Bioactive Polysaccharides. DOI: http://dx.doi.org/10.1016/B978-0-12-809418-1.00003-4

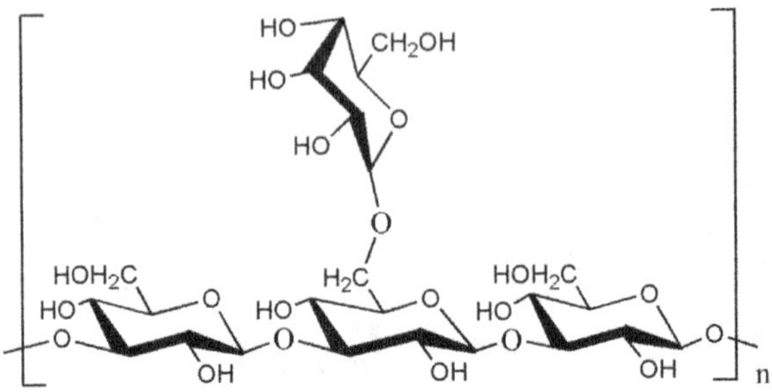

FIGURE 3.1 Structure of fungi and yeast β-(1,3)(1,6)-glucan.

Yeast (*Saccharomyces cerevisiae*) is a typical source of fungi-derived β-glucan. The cell wall in the yeast makes up 15%–30% of the dry weight of the cell, and 25%–50% of the volume based on calculations from electron micrographs. The cell wall of yeast is mainly composed of mannoprotein and fibrous β-(1,3)-glucan. The β-(1,3)-glucan–chitin complex is the major constituent of the inner wall, and β-(1,6)-glucan links the components of the inner and outer walls (Fig. 3.1). The average degree of polymerization of the yeast β-(1,3)-glucan is 1500, and its molecular mass is 240 kDa, while the average degree of the β-(1,6)-glucan is 150 degrees, and corresponds to a molecular mass of 24 kDa (Lipke and Ovalle, 1998).

3.1.2 Beta-Glucan From Bacteria

Polysaccharides from bacteria can be classified as capsular polysaccharides and exopolysaccharides. The cell wall of Gram-negative bacteria consists of an inner membrane and an outer membrane. The outer leaflet of the outer membrane is composed of lipopolysaccharide, which is a major determinant of the virulence of animal pathogens. A capsular layer may present as the outermost layer of the cell wall (Leung et al., 2006). The capsular polysaccharide is also identical to the polysaccharide chain of lipopolysaccharide in certain bacterial strains (Kogan et al., 1992). Exopolysaccharides are polysaccharides secreted to the culture medium by bacteria. Such exopolysaccharides may be either homopolymeric or heteropolymeric in composition, and their molecular weight may be from 10 to 1000 kDa. The homopolysaccharide is generally clustered into four groups, α-glucan, β-glucan, fructans, and polygalactan. The composition of heteropolysaccharides may include repeating units of glucose, galactose, rhamnose, and *N*-acetylglucosamine (GlcNAc), glucuronic acid (GlcA), or *N*-acetylgalactosamine (GalNAc) (Nwodo et al., 2012).

Curdlan is a well-known bacterial linear β-(1,3)-glucan (Fig. 3.2). For its unique rheological properties among natural and synthetic polymers, curdlan was used as a biothickening and gelling agent in the food industry. Curdlan

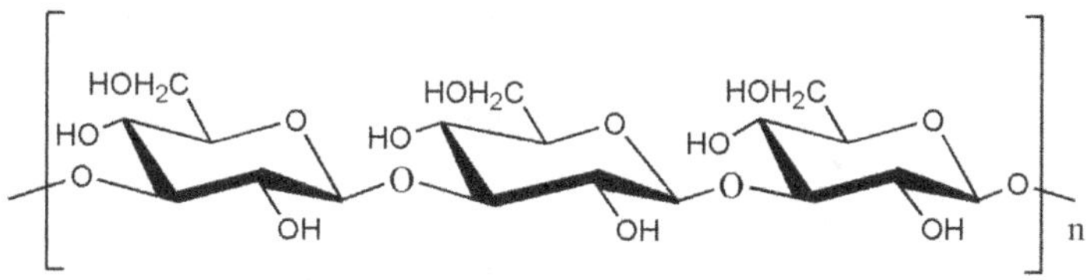

FIGURE 3.2 Structure of bacterial β-(1,3)-glucan curdlan.

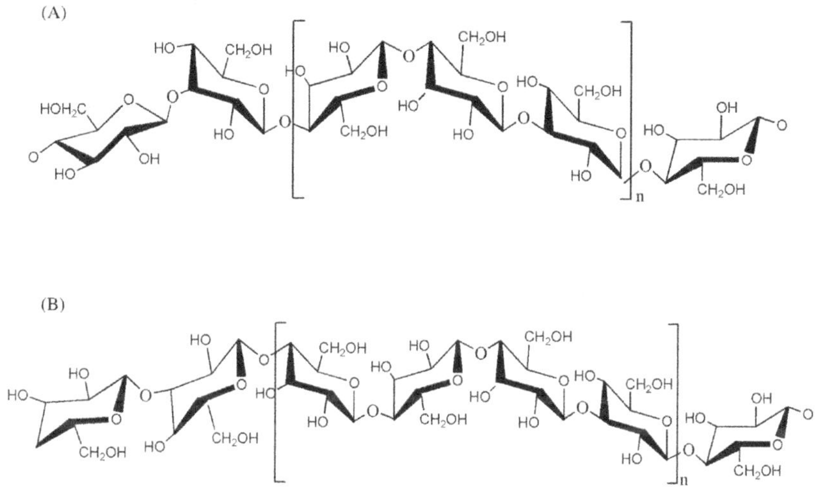

FIGURE 3.3 Structure of cereal β-glucan. (A) Trisaccharide unit; (B) Tetrasaccharide unit.

was firstly discovered in 1968, when Harada obtained it from *Alcaligenes faecalis* Var. *myuxogenes* 10C3 mutant K, and reported that the curdlan has an essentially unbranched structure, though it may contain a few internal (1,6)-glucosidic linkages (Nwodo et al., 2012). Curdlan consists of over 12,000 glucose units. It is insoluble in water, alcohols, and most organic solvents, but could dissolve in dilute bases, dimethylsulfoxide, and formic acid (Laroche and Michaud, 2007). Though it lacks solubility in water at ambient temperature, it can produce gels via the heating of an aqueous solution to around 55°C and subsequent cooling to ambient temperature. In contrast, curdlan is soluble in alkaline aqueous solutions, and gels can be formed simply by neutralization without heating.

3.1.3 Beta-Glucan From Plants and Algae

Beta-glucan is also isolated from plants and algae. Cereal β-(1,3)(1,4)-glucan occurs in the subaleurone and endosperm cell wall of grains (Fig. 3.3) (Cui and Wang, 2009). The content of β-glucan in wheat, rye, oats, and barley is generally in the range of <1%, 1%–2%, 3%–7%, and 3%–11%, respectively (Skendi et al., 2003). Significant structural differences in cereal β-glucan are characterized by the trisaccharide-to-tetrasaccharide ratios, which follow in the order of wheat (4.2–5.4), barley (2.8–3.3), and oat (2.0–2.4) (Cui and Wood, 2000).

Algal polysaccharides may be divided into three main groups on the basis of function: (1) fibrous glycans, including cellulose, chitin, mannans, glucomannans, and xylans; (2) matrix, or encrusting materials, corresponding to hemicellulose in higher plants and sometimes called mucilages; and (3) storage or food-reserve polysaccharides including starch, glycogen, inulin, and β-(1,3) glucans (Myklestad, 1989). We have summarized some sources of β-glucan include fungi, bacteria, plants, and algae in Table 3.1.

3.2 PREPARATION, EXTRACTION, AND PURIFICATION

Isolation of polysaccharides is an important process. There have been numerous researches on the isolation technology of polysaccharides from various plants or fungi in recent years. Except for curdlan and the scleroglucan produced by *Sclerotium rolfsiim*, the main drawback limiting the development of polysaccharides is lack of efficient isolation and purification methods, and their cost.

The isolation of polysaccharides is commonly performed in aqueous solvents. The extraction yield of polysaccharides from fungal material depends on the different natural samples, as well as different parameters, such as extraction temperature, extraction pressure, extraction time, and ratio of water volume to raw material. Recently, to improve the yield or reduce the time, various novel techniques including ultrasound-assisted extraction (UAE), microwave-assisted extraction (MAE), supercritical fluid extraction, and enzyme-assisted extraction (EAE) have been developed for isolating fungal polysaccharides.

3.2.1 Hot-Water Extraction

In general, water extraction is the most widely used traditional method for extracting polysaccharide, due to its efficiency in obtaining polysaccharides and its low cost. Extraction time, extraction temperature, and the ratio of water volume to raw material weight play an important role in the extraction process of polysaccharide. Normally, the lipid component and some other low-molecular weight substances in the raw material are firstly removed using 75%–95% (V/V) ethanol through a reflux divider (Li et al., 2013). Then the residue is subsequently dried and extracted with boiling water several times. The extracting liquids are separated and collected by centrifugation. To further purify the isolated polysaccharide, a combination of techniques are applied, such as fractional precipitation, ethanol precipitation,

TABLE 3.1 Some Sources of β-Glucan Include Fungi, Bacteria, and Plants

Source Species		Common Name	References
Fungi	*Lentinus edodes*	Lentinan	
	Saccharomyces cerevisiae	Yeast	
	Cordyceps sinensis		Yalin et al. (2005)
	Maitake		Fullerton et al. (1999)
	Entoloma lividoalbum		Maity et al. (2014)
	Lyophyllum decastes Sing		Ukawa et al. (2000)
	Pleurotus florida		Rout et al. (2005)
	Pleurotus eryngii		Carbonero et al. (2006)
	Pleurotus ostreatoroseus		Carbonero et al. (2006)
	Sparassis crispa		Ohno et al. (2003)
	Geastrum asccatum		Dore et al. (2007)
	Sparassis crispa		Nameda et al. (2003)
	Inonotus obliquus		Rhee et al. (2008)
	Agaricus blazei		Angeli et al. (2009)
Higher Plants	*Hordeum vulgare*	Barley	McCleary and Codd (1991)
	Avena sativa	Oats	Anttila et al. (2008)
	Triticum spp.	Wheat	Beresford and Stone (1983)
	Phoenix dactylifera Libyan dates		Ishurd and Kennedy (2005)
	Gossypium hirsutum Acala SJ-1	Cotton	Maltby et al. (1979)
Algae	*Ulva lactuca* Thuret		Lahaye et al. (1994)
	Laminaria digitata	Laminarin	Klarzynski et al. (2000)
	Rhodymenia pertusa		Whyte and Englar (1971)
	Chlorella pyrenoidosa		Reyes Suárez et al. (2008)
Bacteria	*Alcaligenes faecalis* var. myxogenes 10C3 mutant K	Curdlan	
	Alcaligenes faecalis var. myxogenes, IFO 13140		Sasaki et al. (1978)
	Pediococcus spp.		Llaubères et al. (1990)
	Acetobacter xylinum		Li et al. (2004)

acidic precipitation, ion-exchange chromatography, gel filtration, and affinity chromatography (Zhang et al., 2007).

However, extraction of polysaccharides from fungus remains a major problem, because they are heterogeneous with low water solubility, and the yield of water extraction is normally quite low. Pressurized water extraction (PWE) is an emerging technique that presents important advantages over traditional hot-water extractions. During the PWE process, the solvents are at high temperature and high pressure, and the high temperature makes possible the high solubility and high diffusion rate, while the high pressure keeps the solvent below its boiling point (Villares et al., 2012). For instance, Palanisamy et al. (2014) reported that an optimal extraction condition for cultivated mushrooms (*Agaricus bisporus* L., *Lentinus edodes* S., and *Pleurotus ostreatus*) is 200°C, and 5 cycles of 5 min each at 10.3 MPa. Tiffany found that a pressure of 10.1 MPa for 70 min at 28°C would result in a maximum recovery for extraction of α-(1,4)-glucans, and β-(1,6)-glucans (Lo et al., 2007).

Recently, an ultrahigh pressure extraction method has also been employed to isolate lentinan, in which the osmotic pressure difference across the cell membrane is increased to accelerate the speed of extraction, and can also kill some deleterious microorganisms. This is a short-time process without high temperature or alkali treatment that may degrade the polysaccharide and affect its bioactivity (Zhang et al., 2011).

3.2.2 Ultrasound-Assisted Extraction and Microwave-Assisted Extraction

Ultrasonic-assisted extraction is a fast and efficient technique that was developed for polysaccharide extraction. As is well-known, the application of UAE could offer many advantages including the reduction of solvents, temperature, and extraction time. It has been reported that the extraction rate of UAE is correlated to ultrasound power per unit volume of liquid (Cheung and Wu, 2013), and the kinetics of UAE strongly depend on the fungal materials, particularly the microstructure and morphology of the fungal cell and aggregation or dispersion of solid particles in the extracting liquid (Cheung et al., 2013). For instance, an optimum UAE condition of polysaccharide from *Ganoderma lucidum* (Fr.) Karst. was reported to be an ultrasonic frequency of 8 kHz, extraction temperature 95°C, extraction time 3 h, and the ratio of water to raw material was 12 (Chen et al., 2010a,b). Tian et al. (2012) reported an optimal extraction condition for *A. bisporus* polysaccharide. Under the conditions of ultrasonic power 230 W, extraction temperature 70°C, extraction time 62 min, and water–material ratio 30 (mL:g), the highest yield of *A. bisporus* polysaccharide was achieved, at 6.02%. The yield of polysaccharide from *L. edodes* under the condition of ultrasound power 340 W, water–material ratio 30:1 (mL:g), and ultrasound time 14 min was 14.39%, which is significantly higher than water heating extraction (Ke, 2014).

Microwave energy is a nonionizing type of electromagnetic radiation with wavelengths typically in the 1 mm to 1 m range, corresponding to 0.3–300 GHz. MAE was applied for the extraction of polysaccharide from the fruiting body of *Hericium erinaceum* mushroom. Comparison of the results obtained by MAE with those obtained by conventional hot-water extraction indicated that the extractability of the former for 5 min at 140°C was almost equivalent to that of the latter extraction for 6 h at 100°C (Ookushi et al., 2006). Besides, treatments by microwave are also used to improve the solubilization of polysaccharide. It is reported that microwave heating in a high-pressure vessel improves the dispersion of β-glucan solution in water without polymer degradation (Wang et al., 2002). A water-insoluble hyperbranched β-glucan obtained from sclerotia of *Pleurotus tuber-regium* was treat by microwave heating to produce an aqueous solution, and the soluble polysaccharide existed in a sphere-like conformation in 0.02% aqueous sodium azide solution (Tao and Xu, 2008).

In addition, a combination of both ultrasound- and MAE apparatus was employed by Huang and Ning (2010) to prepare *G. lucidum* polysaccharide. The optimal extraction conditions were ultrasonic power of 50 W, microwave power 284 W, extraction time of 701 s, and the ratio of water to raw material of 11.6 mL/g. You et al. (2014) reported that the ultrasound-microwave synergistic extraction condition of polysaccharide of *Tricholoma mongolicum lmai* mushroom was optimized to be an extraction time of 24.65 min, microwave power of 109.98 W, and water to raw material ratio of 21.62 mL/g.

3.2.3 Purification of Beta-Glucan

Polysaccharide enriched extracts obtained from the extraction procedure are further purified. Basically, the impurities are excluded from the polysaccharide through ethanol precipitation, neutral polysaccharides are separated from acidic ones using ion-exchange chromatography through DEAE-cellulose columns, and the neutral polysaccharides are then separated into α-glucan and β-glucan using gel filtration and affinity chromatography (Wasser, 2002). The acidic and neutral polysaccharide could also be separated using cetyl trimethyl ammonium bromide or cetyl pyridinum chloride, which may form a precipitated complex with the acidic polysaccharide (Yang and Zhang, 2009). Polysaccharide with a broad polydispersity can be fractionated by stepwise precipitation or preparative gel permeation chromatography, through which polysaccharides with different molecular weights and low polydispersity could be obtained (Zhang et al., 2007). And Sephadex, Sephacryl, and Sepherose are commonly used gels.

The solubilities of β-glucans and arabinoxylans in ammonium sulfate solution are different, and β-glucans can be precipitated at a much lower saturation level of ammonium sulfate (20%–55%) than arabinoxylans (55%–95%). Therefore, an ammonium sulfate precipitation was used to separate purified β-glucan from wheat β-glucan extract. Moreover, the purified wheat β-glucan was further fractionated into six fractions with molecular weights from 43,000 to 758,000 g/mol using an ammonium sulfate gradient precipitation method (Li et al., 2006).

3.3 STRUCTURAL FEATURES

It is important to elucidate the relationship between chemical structure, the chain conformations of polysaccharides, and their beneficial activities. However, polysaccharides derived from different sources are usually composed of various monosaccharides, and many of them have hyperbranched structures. To characterize the chemical structure and chain conformation of a polysaccharide is not easy work.

To determine the structure of glucans from different resources, a series of analytical methods have been developed. The chemical structure was analyzed by Fourier transform infrared spectroscopy (FITR), nuclear magnetic resonance spectroscopy (NMR), gas chromatography (GC), gas chromatography-mass spectrometer (GC-MS), and high-performance liquid chromatography (HPLC). The chain conformations of polysaccharides in solution were investigated using static and dynamic light scattering (Yang and Zhang, 2009).

The NMR techniques, including ^{1}H-^{1}H COSY, TOCSY, NOESY, ROESY and ^{1}H-^{13}C heteronuclear HSQC, HSQC-TOCSY, and HMBC, are the most effective nondestructive methods of structural analysis (Goffin et al., 2009). Chemical shifts and coupling constants of ^{1}H and ^{13}C identify anomeric forms of glucose residues and indicate the positions of glycosidic linkages or substituents, homonuclear 1H, and 1H correlation spectroscopy (COSY) detects interactions between neighbor protons in a glucose unit, while total correlation spectroscopy (TOCSY) may assign all the protons based on their interaction with one of them (commonly H-1). Heteronuclear single/multiple-quantum correlation spectroscopy (HSQC, HMQC) is used to assign C–H signals. Finally, both heteronuclear single- or multiple-bond correlation spectroscopy (HSBC, HMBC) and nuclear or rotating-frame overhauser effect spectroscopy (NOESY, ROESY) are able to determine interunit connections in the glucan macromolecule (Synytsya and Novák, 2013). According to the anomeric structure of a glucose unit, it is possible to distinguish three main groups of these polysaccharides: α-glucans, β-glucans, and mixed α,β-glucans. For example, extracellular glucan from the genus *Pleurotus* consist of 20% α-(1,4)-glucan and 80% β-(1,3)-glucan, and approximately 25% of the β-glucans were C-6 branched (Gutiérrez et al., 1996).

Glucans are high-molecular weight polymers of glucose. Two stereoisomers of glucose exist, with the naturally occurring D-glucose predominating over L-glucose, and the molecule can adopt a cyclic form as a glucopyranosyl. In general, six different types of β-glucans have been described: (1) linear β-(1,3)-glucans; (2) β-(1,3)-glucans with occasional β-(1,6) branching; (3) phosphorylated β-(1,3)-glucans with occasional β-(1,6) branching; (4) β-(1,3)-glucans with significant amounts of β-(1,6) branching; (5) glucans made mostly of β-(1,6)-linked glucose; and (6) glucans containing β-(1,3), β-(1,4) and β-(1,6) linkages (Ruiz-Herrera, 1991). Beta-(1,3)-glucan chains are the most abundant, and are composed of mainly linear molecules of about 1500 glucose units in length. However, β-(1,6)-glucan chains are less abundant and shorter, extending up to 140 glucose residues in length, with approximately 15% present in internally linked 3,6-disubstituted form, providing the polymer branch points. Most β-glucans from bacterial origin are unbranched 1,3-β-linked glycopyranosyl residues, whereas most β-glucans of yeast and fungi consists of 1,3-β-linked glycopyranosyl residues with small numbers of 1,6-β-linked branches (Volman et al., 2008). It is noteworthy that even if glucan is obtained from the same material, its structure is still closely related to its preparation process. An alkaline-soluble glucan obtained from the cell wall of yeast (*S. cerevisiae*) was obtained by serial enzymatic digestion, structural analysis showed that it contained 80%–85% β-(1,3)-linkages, 8%–12% β-(1,6)-linkages, and 3%–4% branched residues linked through C-1, C-3, and C-6 (Fleet and Manners, 1976). However, Manners reported a minor polysaccharide component isolated from the yeast glucan has a degree of polymerization of about 130–140, a highly branched structure, and a high proportion of β-(1,6)-glucosidic linkages (Manners et al., 1973).

3.4 MOLECULAR CHARACTERISTICS

Conformation of the β-(1,3)-glucan could be generally classified into three forms: triple helix, single helix, and random coil. The glucans having low-molecular weight or having a certain substituted group could exist as a random coiled conformer. A triple helical conformer is stable at neutral pH and ambient temperature. Both the intramolecular force which maintains the single helical conformation, and the intermolecular force which maintains the triple helical conformation are hydrogen bonds. An increase in temperature, hydroxide ion concentration, or chaotropic agent concentration will disrupt the hydrogen bonds, and cause transition from a triple helix to a single helix, and from helix to random coil (Leung et al., 2006). Beta-glucans prepared by hot-water extraction are more likely to have a triple helical conformation, whereas β-glucans prepared through alkaline extraction would have a single helical structure.

Dimethylsulfoxide is a commonly used powerful chaotropic agent. A linear water-insoluble β-(1,3)-glucan extracted from *G. lucidum* exists as a compact coil in dimethylsulfoxide, whereas its sulfation, carboxymethylation, hydroxyethylation, hydroxypropylation, and methylation derivatives are slightly expanded flexible chains in 0.9% aqueous NaCl solution (Wang and Zhang, 2009a,b). The triple helical conformation of sclerogulan was reported to be disentangled to a single chain by the oxidation of the β-glucosyl side residues, and a shielding charge on the side chains is helpful to recover the triple helical structure (Coviello et al., 1998).

The yeast glucan can be found as random coils or more organized conformations that constitute a network composed of a single helix chain associated in triple helices stabilized by inter- and intrahydrogen bonds (Zeković et al., 2005).

One of the well-studied medicinal fungal polysaccharides is lentinan, a glucan elaborated by the edible mushroom *L. edodes*. It is composed of a main chain of β-(1,3)-glucan residues to which β-(1,6)-glucose side groups are attached, and the molecular weight ranges from 400,000–600,000 Da. This water-soluble β-glucan exists mainly as triple-helical chains in 0.2M NaCl aqueous solution, and as single chain in dimethylsulfoxide. Especially, the irreversible helix-coil conformation transition occurred in a narrow range of dimethylsulfoxide concentration from the 80% to 85% (Zhang et al., 2002a,b). In addition, it was found that the conformation transition of lentinan from triple strand to single chain, and from extended chains to winding chains, were completed rapidly at 130–145°C (Wang et al., 2008). Zhang et al. (2004a,b,c) reported that lentinan exist as triple-helical chains at NaOH concentrations lower than 0.05 M, whereas it exists as single random coil chains at NaOH concentrations higher than 0.08 M.

Schizophyllan is a water-soluble β-glucan elaborated by *Schizophyllum commune* Fries. The molecular weight and the number of the repeating unit of schizophyllan were found to be 1.5×10^5 and 231, respectively (Sakurai and Shinkai, 2000). The triple helical structure of high-molecular weight β-(1,3)-glucan can be denatured either by increasing the temperature in aqueous solution above the thermal transition temperature of 135°C, or by dissolving the molecules in either dimethylsulfoxide, with a water weight fraction less than 0.13°C at 25°C, or in aqueous hydroxide. However, renaturation of the β-(1,3)-glucan by either cooling, dialysis to remove dimethylsulfoxide, or neutralization result in a yield of a labile single-stranded conformer that slowly reassociates to the triple helical structure or coexisting linear and cyclic topologies formed by at least part regeneration of the triple helical structure.

3.5 BIOACTIVITIES

3.5.1 Immune-Enhancement and Antitumor Effects

Cancer is one of the diseases that pose a serious threat to mankind's health and life nowadays. Though surgery, chemotherapy, and radiotherapy are used as the main treatments for cancer, cancer is still the leading cause of death. Modern medical practice mainly depends on the use of highly purified pharmaceutical compounds; researchers think the purity can be easily assessed and the pharmaceutical activity and toxicity can explain the clear relationship of structure–function; however, the curative effects are still not ideal. Although it is very difficult to translate into acceptable evidence-based western therapies, polysaccharides derived from traditional Eastern medicinal herbs are now considered as an important new source of anticancer agents (Sullivan et al., 2006). Particularly, mushroom polysaccharides have been used in clinical anticancer therapy in some countries.

3.5.1.1 Lentinan

It was in the 1960s when Chihara first reported that lentinan could significantly inhibit the growth of tumors in Sarcoma 180-bearing mice (Chihara et al., 1970, 1969), the antitumor activity of polysaccharides aroused worldwide attention, and numerous researches were focused on discovering this antitumor adjuvant resources. Numbers of researches demonstrated that lentinan possessed an inhibition effect on many types of cancer (Fujimoto et al., 2006; Ina et al., 2013). For instance, in in vitro bioassays, lentinan could inhibit the proliferation of cancer cells including the breast carcinoma cell line (MDA-MB-453 and MCF-7), the human nonmalignant breast epithelial cell line (MCF-10F), and myeloma cell lines (RPMI-8226 and IM-9) (Fang et al., 2006). Animal studies showed preadministration of lentinan could inhibit human colon carcinoma development in an AKR mouse model (Ng and Yap, 2002), and human clinical research proved that lentinan could prolong survival and improve quality of life in patients with gastric cancer (Nakano et al., 1999), colorectal cancer (Hazama et al., 2009), pancreatic cancer (Shimizu et al., 2009), or liver cancer (Isoda et al., 2008) (Fig. 3.4).

In most cases, β-glucans do not directly affect tumor cells, and their antitumor effect is closely related to its beneficial influence on the immune system. In other words, it helps the host to tolerate adverse biological stresses and exert an enhanced immunity against the development of cancer cells

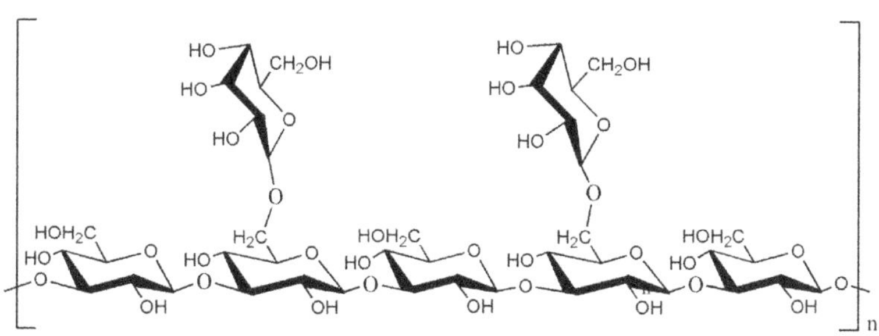

FIGURE 3.4 Structure of mushroom β-glucan lentinan.

by supporting some or all of the major biological systems. Since the antitumor effect of lentinan was blocked by an injection of antilymphocyte serum, one of the mechanisms responsible for the antitumor effect of lentinan may be the stimulation of cell-mediated response (Maeda et al., 1971). Similarly, Maruyama et al. (2006) found that tumor growth was suppressed in lentinan-treated tumor-bearing mice, and in further ex vivo experiments the tumor cytotoxicity was significantly reduced by treatment of splenocytes with anti-CD8 antibody. Before inducing the immunological effects, it is hypothesized that lentinan needs to first bind to host cells. A human leukocyte binding experiment showed that lentinan bound to monocytes and neutrophils, but not to lymphocytes in vitro, especially, the most avid lentinan binding was to monocytes. Meanwhile, in vivo experiments demonstrated that the percentage of lentinan binding monocytes in the peripheral blood of healthy volunteers increased significantly 3 and 4 h after lentinan was intravenously injected (Oka et al., 1996).

3.5.1.2 Ganoderma lucidum *Polysaccharide*

G. lucidum is another well-studied medicinal mushroom of the Basiodiomycetes family which has been used in traditional East Asian medicine for longevity and health promotion since ancient times. Nowadays, the anticancer activities of *G. lucidum*, both in vivo and in vitro, have certainly been demonstrated. For instance, Jiang et al. (2004a,b) reported *G. lucidum* suppresses the growth of human breast cancer cells and prostate cancer cells in a dose-dependent and time-dependent manner. Müller et al. (2006) reported *G. lucidum* extraction caused apoptosis in leukemia, lymphoma, and multiple myeloma cells. In addition, treatment of mice with oral *G. lucidum* polysaccharide was found to significantly reduce the tumor weight of Sarcoma-180 in a dose-dependent manner (Gao et al., 2005a,b).

The main two bioactive components of *G. lucidum* can be broadly grouped into triterpenes (Cheng et al., 2010; Lin et al., 2003; Mm et al., 2000) and polysaccharides. Particularly, the major component content of the polysaccharide is β-glucan, which contains a variety of linkages such as β-(1,3)-, β-(1,4)-, β-(1,6)-, β-(4,6)-, and β-(3,6) (Liu et al., 2010). In addition, the biologically active polysaccharides are mainly in the form of branched β-(1,3)-glucan.

Current data suggest that *G. lucidum* polysaccharides are potent immunomodulators with effects on both innate and adaptive immunity. Macrophages are important innate immune cells that are located near potential entry sites for microbial pathogens, and they are specialized for taking up particulate material by phagocytosis. Numbers of researches have demonstrated that *G. lucidum* polysaccharide stimulates the morphological change of macrophages, enhance its phagocytotic activity, and increase the cytokine release of macrophage, including TNF-α, NO, IL-12, and so on (Lin, 2005; Chaung et al., 2009; Ahmadi and Riazipour, 2007). Moreover, macrophage-mediated tumor cytotoxicity was increased through *G. lucidum* polysaccharide treatment as well (Reutter and Fan, 2010). Dectin-1 expressed on phagocytes, especially macrophage and dendritic cells, is a specific pattern recognition receptor for β-(1,3)-glucan. *G. lucidum* polysaccharide enhanced TNF-α secretion in RAW264.7 cells stably expressing wild-type Dectin-1, but not in Dectin-1 deficient RAW264.7 cells. Moreover, the secretion of TNF-α is dependent on the MAPK pathway triggering after the polysaccharide is recognized by Dectin-1 (Guo et al., 2009). Toll-like receptors also mediate the immune-stimulatory effect of *G. lucidum* polysaccharide. Shao found *G. lucidum* polysaccharide could induce IL-1β secretion on peritoneal macrophage from normal mice, but not from toll-like receptor-4 mutated C3H/HeJ mice, which suggested the polysaccharide activates macrophage through toll-like receptor 4 (Shao et al., 2004). Whereas Wang et al. (2014a,b) found the effect of *G. lucidum* polysaccharide on the secretion of NO was closely related to the interaction with toll-like receptor 2. Dendritic cells are professional antigen presenting cells that bridge the innate and adaptive immune systems. Cao and Lin (2002) reported that *G. lucidum* polysaccharide could increase the coexpression of CD11c and I-A.I-E molecules on the dendritic cell surface, promote mRNA expression of cytokine IL-12p40, and enhance the stimulatory capacity of dendritic cells to lymphocyte proliferation. Especially, Lin et al. (2005) found that the *G. lucidum* polysaccharide stimulated the expression of CD80, CD86, CD83, CD40, CD54, and human leukocyte antigen-DR on dendritic cells via the NF-κB and p38-MAPK pathway. Natural killer (NK) cells are large granular lymphocytes that do not belong to either T cell or B-cell lineage, and it is a part of the innate defense system and is able to kill certain tumor cells. It is reported that the *G. lucidum* polysaccharide could enhance NK cell activity in some advanced cancer patients, which are significantly decreased in the late stage of cancer (Gao et al., 2003, 2005a,b). An immune-suppressed mice model was established by injection of cyclophosphamide (CY), and *G. lucidum* polysaccharide treatment not only accelerated recovery of bone marrow cells, red blood cells, white blood cells, splenic NK cells, and NK T cell, but also enhanced the proliferation response of B cells and T cells (Zhu et al., 2007). For the adaptive immune cells, the *G. lucidum* polysaccharide could stimulate the proliferation and the immunoglobulin secretion of B cells directly. It could also enlarge B lymphocyte size, and the expression of CD71, and CD 25 molecules on the cell surface (Zhang et al., 2002a,b). Wang et al. (1997) reported that the *G. lucidum* polysaccharide promoted the release of the interferon-γ of T lymphocytes.

3.5.1.3 Yeast Beta-Glucans

One of the most exhaustively studied β-(1,3)-glucans is isolated from the baker's yeast *S. cerevisiae*. It is also an important potentiator for mucosal immunity in the digestive tract.

It is known that vertebrates do not possess β-glucanases, and cannot rapidly degrade β-glucan. When orally administered, β-glucans are metabolized slowly by oxidation, which allows transit from the stomach to the small intestine where immunostimulation of the gut-associated lymphoid tissue (GLAT) can occur (Thompson et al., 2010). The GALT, which protects the intestine, can be subdivided into several compartments including Peyer's patches, lamina propria, and intraepithelial leukocyte spaces (Cebra et al., 1998). Peyer's patches are aggregates of lymphoid follicles found throughout the mucosa and submucosa of the intestine. It contains plasma cells, macrophages, $CD4^+$ and $CD8^+$ T cells, and naïve B-cells, as well as dendritic cells (Schley and Field, 2002).

Orally administrated, the yeast β-(1,3)-glucans were taken up by macrophages that transported them to the spleen, lymph nodes, and bone marrow. Within the bone marrow, the macrophage degraded the large glucans into smaller soluble fragments that were taken by the CR3 of marginated granulocytes. These granulocytes were shown to kill inactivated C3b-opsonized tumor cells following their recruitment to a site of complement activation (Hong et al., 2004).

After being orally administered with yeast β-glucan for a week, there was a significant increase in the number of intraepithelial lymphocytes in the intestine, and especially the proportion of γδT cells expressing CD8 antigen was increased (Tsukada et al., 2003).

In radiotherapy for cancer, intraperitoneal injection of yeast β-glucans can greatly delay mortality in mice exposed to whole-body X-ray radiation, and inhibit tumor growth in tumor-bearing mice. In addition, the blood leukocyte and lymphocyte numbers were increased in the radiation-exposed mice, and the NK and lymphokine-activated activity of spleen cells were significantly increased, indicating that yeast glucan possesses radioprotection as well as the immune-enhancement activity (Gu et al., 2005).

Yeast β-glucans could also be used as a potent adjuvant for mAb therapy of cancer to elicit a novel granulocyte- and tissue macrophage-mediated tumor-killing mechanism. Beta-(1,3)-glucan was isolated from yeast through a series of alkaline and acid extractions, and was further dissolved in distilled water using sonication treatment.

3.5.1.4 Oat Beta-Glucan

The common oat (*Avena sativa*) is a species of cereal grain mainly grown for its utilization for human consumption as oatmeal, as well as for livestock feed, and it is one of the richest and most economical sources of soluble dietary fiber.

Oral gavage of oat β-glucan increased intestinal NF-κB transactivation in leukocytes and enterocytes, particularly in the proximal part of the small intestine, but not in the colon (Volman et al., 2010). Although human enterocyte cell lines do not express the β-glucan receptor in vitro, it can be explained that oat β-glucan firstly activates the intestinal leukocytes after consumption, and then increases cellular activation of enterocytes in turn.

It has been reported that oat glucan stimulated the production of IL-1 on macrophages in a time-dependent manner, and induced the production of IL-2, IFN-γ, and IL-4 secretion in a dose-dependent manner in vitro. Meanwhile, in vivo administration of oat glucan significantly enhanced the accumulation of macrophages in the peritoneal cavity, and pretreatment of mice with oat glucan confers protection against infection with *Staphylococcus aureus* (Estrada et al., 1997; Yun et al., 2003). By the way, Estrada et al. (1999) also found that the oat glucan did not influence immune responses of naïve cells in vitro, or of healthy steers in vivo, but it certainly restored the immune function of cells in steer when they were treated with dexamethasone. In addition, administration of oat β-glucan by a subcutaneous route resulted in a higher level of total serum immunoglobulins and antigen-specific immunoglobulins, and proliferative responses to *Eimeria vermiformis* sporozoite antigen of spleen lymphocyte was significantly increased, suggesting that oat β-glucan could up-regulate immunity to enhance resistance against bacterial infection (Yun et al., 1997; Yuri et al., 1998).

Macrophage activation effects are also involved in its antitumor activity. Oat β-glucan administration decreased lung tumor foci, increased macrophage cytotoxicity, and decreased the metastatic spread of injected melanoma cells (Murphy et al., 2004). Besides, it is also reported that oat β-glucan consumption could increase the number and function of neutrophils, which was impacted by fatiguing exercise (Murphy et al., 2007).

3.5.2 Antihypocholesterolemic and Antihpyerglycemic Effects

Heart disease is the leading cause of human death around the world. In addition to pharmacotherapy, lowering serum cholesterol levels through making dietary changes, such as reducing the intake of total fat and dietary cholesterol, is one way to reduce the risk of developing the disease. Interest in dietary fiber has increased due to its beneficial effects on blood cholesterol levels and glycemic response. Among the cereal grains, oats and barley have been reported to be the most effective in lowering serum total cholesterol and low density lipoprotein (LDL)-cholesterol in humans and animals (Pomeroy et al., 2001). Adding β-glucan-rich fiber in the daily diet is now considered as an effective measure to reduce serum lipid levels.

The first report on free-living volunteers to test the effect of rolled oats on serum cholesterol levels was published in 1963, which demonstrated that diets with rolled oat significantly decreased the cholesterol level in healthy volunteers (de Groot et al., 1963). Nowadays, the β-glucan component of oats is found to be able to attenuate blood postprandial glycemic and insulinemic responses, to lower blood total cholesterol and LDL cholesterol, and to improve high-density lipoprotein (HDL) cholesterol and blood lipid profiles, as well as to maintain body weight (Daou and Zhang, 2012). It is reported that muesli with 4 g of β-glucan reduces postprandial glucose and insulin levels compared to a breakfast based on high glycemic index, but that with 3 g of β-glucan gave no significant difference, which also indicates that a total of 4 g β-glucan from oats may be a critical level for a significant decrease in glucose and insulin responses in healthy people (Granfeldt et al., 2008). Interestingly, an oat β-glucan treatment could increase inducible nitric oxide synthase activity and phagocytosis, which were lower in hypercholesterolemic mice compared to normal mice (Dhillon and Bhatia, 2008).

Barley belongs to the genus *Hordeum*, and could be considered as one of the most ancient crop plants. Beta-glucan from barley should be regarded as an important functional ingredient for the cereal food industry (Brennan and Cleary, 2005). Flours of barley cultivars containing a gradient concentration of soluble β-glucan at 5.39% (Dolma group), 2.06% (DL-88 group), and 1.08% (BH-331 group), were used in the diets of rats to compare their hypocholesterolemic effect. After a 40-day-feeding trial, significantly lower levels of total serum cholesterol were observed for the test groups than the casein diet feeding group. The total cholesterol was lowest in the Dolma group (143 mg/100 mL), followed by the DL-88 group (180 mg/100 mL), which were significantly different from the BH-331 group (213 mg/100 mL). Significantly lower levels of LDL-cholesterol and higher levels of HDL-cholesterol were observed in the Dolma group, and HDL-cholesterol was found to increase with increasing levels of total and soluble β-glucan content in the diet (Kalra and Jood, 2000). Bryan reported that the cholesterol-lowering potency of barley was the same as oats (Delaney et al., 2003).

To date, fungal β-glucan is also found possess antihypocholesterolemic and antihyperglycemic effects (Giavasis, 2014; Lakhanpal and Rana, 2005). Rats which were given a *L. edodes* powder in a high-fat diet for 6 weeks showed a lower body weight gain, and their total fat deposition and plasma triacylglycerol level were significantly decreased (Handayani et al., 2011). Besides, a β-(1,3),(1,6)-glucan enriched fraction obtained from *L. edodes* mushrooms was found to be successfully used to produced cake (Kim et al., 2011). *G. lucidum* β-glucan is also reported to increase body weight and serum insulin levels, but decrease fasting blood glucose in streptozotocin-induced diabetic mice. Moreover, the total cholesterol, triglyceride, and LDL cholesterol level in the diabetic mice was decreased, whereas the HDL cholesterol level was increased (Li et al., 2011).

Pleurotus species contain a high β-glucan content (Giavasis, 2014; Refaie et al., 2009). Lots of researches demonstrate that feeding of *Pleurotus* mushroom showed a hypocholesterolemic effect (Bobek et al., 1996a,b; Gunde-Cimerman and Plemenitas, 2001). Consumption of *P. ostreatus* mushroom soup significantly decreased triacylglycerol concentration and oxidized LDL levels (Schneider et al., 2011), and feeding of 5% fruiting bodies of

P. ostreatus mushroom to hypercholesterolemic mice reduced the plasma total cholesterol by 28%, LDL-cholesterol by 55%, triglyceride by 34%, non-esterified fatty acid by 30%, and total live cholesterol levels by >34% (Hossain et al., 2003). Bobek reported that the amount of *P. ostreatus* mushroom in the diet was highly significantly negative correlated to the cholesterol levels in the serum (Bobek et al., 1996a,b). In addition, a 5% powder of *Pleurotus sajor-caju* and *Pleurotus florida* reduced the plasma total cholesterol level by 21% and 16%, respectively, and reduced the triglyceride level by 24% and 14%, respectively, it also decreased the LDL/HDL ratio by 64% and 41%, respectively (Alam et al., 2009).

Besides, β-glucan from other mushrooms are reported as potential hypocholesterolemic compounds. For instance, two edible mushrooms, namely *Auricularia auricular* and *Tremella fuciformis* have a high content of β-(1,3)-glucan. After a 4-week consumption of the mushroom diet, the serum total cholesterol concentration was significantly decreased in both the two mushroom groups. Meanwhile, the serum LDL-cholesterol level was also significantly decreased in the mushroom diet groups (Cheung, 1996a). Male Sprague-Dawley rats were fed two semisynthetic diets supplemented with 2% cholesterol and 1% extracellular β-glucan from submerged fermentation of *Volvariella volacea* straw mushroom. The experimental diets containing the two mycelial extracellular β-glucan induced a significant reduction in the levels of serum cholesterol, LDL-cholesterol, and liver total cholesterol in the rats (Cheung, 1996b). Jeong et al. (2010) reported that oral treatment of *A. bisporus* powder for 4 weeks resulted in a significant decrease in hepatic cholesterol and triglyceride concentration, as well as total cholesterol, LDL, and triglyceride concentration in plasma. In addition, the *A. bisporus* powder treatment could also reduce plasma glucose and triglyceride concentrations, liver enzyme activity, alanine aminotransferase and aspartate aminotransferase, and liver weight gain in the streptozotocin-induced diabetic male Sprague-Dawley rat model.

Four mechanisms have been proposed to explain how β-glucan lowers serum cholesterol: (1) β-glucan could bind to the bile acid in the intestinal lumen, so that less bile acid would circulate back to the liver, it will further stimulate production of more bile acids from cholesterol which were either made endogenously or captured from the circulation; (2) β-glucan could be fermented in the large bowel by colonic bacteria, which would result in the production of the short-chain fatty acids, these short-chain fatty acids are absorbed through the portal vein, and inhibit hepatic cholesterol synthesis by the action of HMG-CoA reductase (a rate-limiting enzyme required for cholesterol biosynthesis), or by increasing catabolism of LDL-cholesterol; (3) β-glucan could reduce postprandial serum insulin concentration by delaying gastric emptying, and further reduce hepatic cholesterol production through mediation of HMG-CoA reductase; and (4) β-glucan may interfere with the absorption of dietary fat by increasing intestinal viscosity, which causes the digesta to hold on to extra water, and slows its movement (Bell et al., 1999).

3.5.3 Antioxidant Activity

Increasing evidence highlights that oxidative stress may cause cell damage which would trigger both the physiological process of ageing (Harman, 2006) and many pathological progressions that eventually lead to serious health problems (Halliwell, 2005). There is a great deal of research showing that β-glucan from yeast and mushrooms are potential antioxidant agents.

Sepsis causes oxidative damage in organs by increasing malondialdehyde (MDA) levels and myeloperoxidase (MPO) activity, and decreasing glutathione (GSH) level; however, the yeast β-glucan treatment replenished the glutathione level in tissue and suppressed myeloperoxidase activity, indicating a protective effect against this oxidative injury in vivo (Şener et al., 2005).

Nicotine has long been considered to increase oxidative stress by stimulating the generation of reactive oxygen species in the periphery, and central nervous system. Nicotine-induced oxidative damage in the kidney is ameliorated by yeast β-glucan treatment. It not only increased glutathione levels, decreased both malondialdehyde level and myeloperoxidase activity, but also restored the morphological changes in the injured tissues, and improved the impairments in renal function (Sener et al., 2007).

Although the phenolic content in different subfractions of mushroom extracts is a potential antioxidant component (Choi et al., 2006; Cheung and Cheung, 2005; Cheung et al., 2003), the mushroom β-glucan also exhibits considerable antioxidant activities. For instance, studies of antioxidant activity with *Geastrum saccatum* mushroom showed that the β-glucan extract has an inhibitory action on the formation of hydroxyl radicals in a dose-dependent manner (Guerra Dore et al., 2007). *P. ostreatus* β-glucan increased markedly activities of antioxidant enzymes, as well as the value of ferric reducing activity and ascorbate concentration in human red blood cells hemolysates (Pietrzycka et al., 2006). The antioxidant properties of hot-water polysaccharide extracts of fruiting bodies of *A. bisporus*, *Agaricus brasiliensis*, *Phellinus linteus*, and *G. lucidum* containing β-glucan were compared, and the order was reported to be *G. lucidum* ≈ *P. linteus* > *A. brasiliensis* > *A. bisporus* (Kozarski et al., 2011).

Beyond the natural polysaccharides, a number of researches reported that chemically modified β-glucan also possesses powerful antioxidant activity. Sulfated lentinan exhibited a strong superoxide radical scavenging activity and hydroxyl radical scavenging activity, as well as inhibition of lipid peroxidation (Feng et al., 2010). Mice treated with a carboxymethyl chitin–glucan enriched diet were more resistant to oxidative stress induced by visible light-excited methylene blue (Lazarova et al., 2006), and the carboxymethyl chitin–glucan could also inhibit the genotoxicity and cytotoxicity in primary rat hepatocytes induced by hydrogen peroxide or 2,3-dimethoxy-1,4-naphthoquinone (Horvathova et al., 2008). Kogan et al. (2005) reported that the carboxymethylated yeast β-glucan not only exhibited high radical scavenging activity, but also decreased the plasma carbonyl content in the model of adjuvant arthritis in rats, which is an oxidative parameter associated with the progress of arthritic conditions. Carboxymethylated *G. lucidum* β-glucan was also reported to possess a high antioxidant activity, especially in scavenging of hydroxyl radicals and hydrogen peroxide (Jaehrig et al., 2008).

Though previous reports suggest that certain β-glucans may possess free radical scavenging activity, to which extent glucan is responsible for the antioxidative activity has always been controversial. Jaehrig et al. (2007, 2008) found that the antioxidative activity of the yeast cell wall protein exceeds that of β-glucan greatly. Particularly, aromatic side chains and free thiols from denatured proteins can be assumed as antioxidants. Similarly, Tsiapali reported that a series of β-(1,3)-glucans, including laminarin and schizophyllan, have been demonstrated to possess a weak free radical scavenging ability. Moreover, the antioxidant effect of the β-glucan did not correlate with the type of intrachain linkages, molecular weight, or degree of polymer branching, but it seems likely that the polymeric structure confers the additional free radical scavenging ability, because polymers showed significantly better free radical scavengers than monosaccharides (Tsiapali et al., 2001).

3.5.4 Antimicrobial and Antiviral Effects

In addition to the immune-enhancement and antitumor activity, antihypocholesterolemic and antihpyerglycemic activity, and antioxidant activity, some literature have demonstrated antimicrobial and antiviral effects of β-(1,3)-glucan.

A soluble branched β-(1,3)-glucan was obtained from a culture broth of the fungus *Sclerotinia sclerotiorum* IFO9395 (SSG), and a prechallenge SSG administration protected dose-dependently against both *Streptococcus pneumonia* sero type 4 and 6B infection in mice. Particularly, SSG injected postchallenge had a curative effect against type 6B pneumococcal infection (Hetland et al., 2000).

Peter reported that oral administration of a 1 mg/kg dose of β-glucan (glucan phosphate, laminarin, and scleroglucan) increased survival in mice challenged with *S. aureus* or *Candida albicans*. Especially, the data also demonstrated that the β-glucans were bound by gastrointestinal epithelial and GALT cells, and they modulated the expression of pattern recognition receptor in the GALT, increased IL-12 expression, and induced protection against infectious challenge (Rice et al., 2005).

A 0.1% dietary β-(1,3)(1,6)-glucan of Schizophyllum could enhance the host defense to prevent salmonella enterica *Serovar Enteritidis* infection in chicks, and its antimicrobial effect may depend on directly up-regulating the phagocytosis and bactericidal activity of abdominal macrophages in chicks (Chen et al., 2008).

Beta-glucan enriched polysaccharides from the fruiting body of *A. brasiliensis* mushrooms exhibit an antiviral activity against poliovirus typl in HEp-2 cells (Faccin et al., 2007). Since the polysaccharide had little effect on reducing viral adsorption and did not show any virucidal effect, it is suggested that the polysaccharide would influence the replication of poliovirus at the initial stage. In addition, Murphy et al. (2009) found oat β-glucan administration significantly reduced morbidity of herpes simplex virus 1 (HSV-1) respiratory tract infection in exercise-induced fatigue mice. However, Nieman et al. (2008) reported that an 18-day oat β-glucan supplementation shows no significant effect on the upper respiratory tract infection incidence in human endurance athletes.

In Table 3.2, we summarize some biological activities of β-glucans from various sources.

3.6 DISCUSSION: STRUCTURE–BIOACTIVITY RELATIONSHIP

Fungal β-glucans exhibit a variety of biological activity, there is no doubt that the significance of these bioactivities are closely correlated with their structural features, such as solubility, molecular weight, degree of branching (DB), polymer charge, and solution conformation. Although there is no clear conclusion to illuminate the exact structure–bioactivity relationship of β-glucans, the best conclusion at this time is that the immune-potentiation activity of β-(1,3)-glucans depends on a helical conformation, and on the presence of hydrophilic groups located on the outside surface of the helix (Bohn and BeMiller, 1995).

3.6.1 Conformation

In a biological system, the conformation or secondary structure often plays an important role in the identification process. Particularly, many reports suggest that specific physicochemical parameters may influence the interaction of glucan and its receptor. Moreover, the triple helical conformation of β-(1,3)-glucans is regarded as an important structural feature for their various bioactivities.

A lentinan sample with single chain conformation was prepared in dimethylsulfoxide, and the antitumor activity both in vivo and in vitro against sarcoma 180 solid tumor decreased compared to the natural triple helical lentinan (Surenjav et al., 2006). A single helix conformer-rich schizophyllan was prepared by alkaline treatment, and then its immune-pharmacological activities were compared with the normal triple-helix

TABLE 3.2 Biological Activity of β-Glucans

	Source	Activity	References
Immune-enhancement effects	*Phelinus linteus*	Stimulates proliferation of T lymphocyte induced by mixed lymphocyte reaction	Kim et al. (1996)
		Enhances cytotoxicity of cytotoxic T cells responding to alloantigen	
		Stimulates T-dependent and T-independent primary antibody response	
		Activate B cells	
	Grifola frondosa	Stimulates IL-10 production from $CD14^+$ and $CD13^+$ cell	Deng et al. (2009) Fang et al. (2012)
		Stimulates TNF-α production from $CD3^+$ cells	
		Stimulates IL-2 production from $CD56^+CD3^+$ cells	
		Stimulates phenotypical changes in $CD3^+CD56^+$ NK cell and $CD4^+CD25^+$ T cells	
		Induces TNF-α and IL-6 production in peritoneal macrophages	
		Induces TNF-α production in RAW 264.7 cells	
	Sparassis crispa Wulf.: Fr.	Modulates recovery rate of immune cells in CY-induced immunosuppressed mice	Ohno et al. (2002)
		Enhances IL-6 and IFN-γ production in CY-treat spleen cells	
	Ganoderma lucidum	Increases NK cell-mediated cytotoxicity	Chen et al. (2014)
		Inhibits tumor growth and rate of metastasis	
		Prolongs survival time in mice with radiation therapy	
	Russula albonigra	Increases NO production in RAW 264.7 cells	Nandi et al. (2014)
		Stimulates proliferation of splenocytes and thymocytes	

(Continued)

TABLE 3.2 (Continued)

	Source	Activity	References
	Sparassis crispa	Increases GM-CSF, IFN-γ, and IL-6 production of immune cells in Peyer's patches	Hida et al. (2013)
	Pleutratus florida	Increases oxidative radical production of neutrophils	Paul et al. (2012)
		Increases proliferation of lymphocytes	
		Increases IL-2 production of peripheral blood mononuclear cells	
		Increases NO production of macrophage	
	Pleurotus florida	Increases NO_2 production of intestinal intraepithelial leukocyte in vitro	Muthusamy et al. (2013)
		Stimulates proliferation of intestinal intraepithelial leukocyte in vitro	
		Increases phagocytic activity of intestinal intraepithelial leukocytes in vitro	
		Increases synthesis of IFN-γ mRNA of intraepithelial leukocytes in vitro	
	Aureobasidium pulluans ADK-34	Increases IgA production in Peyer's patch cells from normal mice	Tanioka et al. (2013)
		Increases IL-5 and IL-6 production in Peyer's patch cells from normal mice	
		Increases IL-6 production of Peyer's patch dendritic cells from normal mice	
		Increases IgA production in duodenum in normal mice and CY-treated mice	
		Increases IL-6 production in Peyer's patch cells from CY-treated mice	
		Increases IL-6 production of Peyer's patch dendritic cells from CY-treated mice	
	Entoloma lividoalbum	Increases NO production of macrophages	Maity et al. (2015)
		Stimulates proliferation of splenocytes	
		Stimulates proliferation of thymocytes	

Antitumor activity	*Sclerotinia sclerotiorum* IFO 9395	Inhibits solid form of sarcoma 180 in ICR mice	Ohno et al. (1986a,b)
	Agaricus blazei	Inhibits solid form of sarcoma 180 in ICR mice	Ohno et al. (2001)
	Grifola frondosa	Inhibits solid form of sarcoma 180 in ICR mice	Ohno et al. (1986a,b)
	Sparassis crispa	Inhibits solid form of sarcoma 180 in ICR mice	Ohno et al. (2000)
	Candida albicans IFO1385	Prolongs survival time of P815 mastocytoma-bearing mice	Tokunaka et al. (2002)
		Inhibits tumor growth in P815 mastocytoma-bearing mice	
		Stimulates antitumor effect of spleen cells	
	Agavicus blazei murill	Inhibits solid form of sarcoma 180 in ICR-JCL mice	Mizuno et al. (1990)
	Pleurotus tuber-regium	Inhibits solid form of sarcoma 180 in BALB/c mice	Zhang et al. (2001)
		Inhibits proliferation of HL-60 and HepG2 cells in vitro	
	Ganoderma tsugae	Inhibits solid form of sarcoma 180 in ICR mice	Wang et al. (1993)
	Flammulina velutipes	Inhibits solid form of sarcoma 180 in ICR mice	
		Shows no effect on the proliferation of sarcoma 180 cells in vitro	
	Auricularia polytricha	Inhibits solid form of sarcoma 180 in ICR mice	Song and Du (2012)
Antioxidant activity	*Dictyophora indusiata*	Reducing power was 0.31 at 1 mg/mL	
		Hydroxyl radical scavenging activity was 39.59% at 1 mg/mL	
		Superoxide radical scavenging activity was 47.18% at 1 mg/mL	
		DPPH scavenging ratio was 50.27% at 1 mg/mL	
	Russula albonigra	Reducing power was 0.5 at 500 μg/mL	Nandi et al. (2014)
		IC_{50} of hydroxyl radical scavenging activity was 265 μg/mL	
		IC_{50} of superoxide radical scavenging was 130 μg/mL	
		50% chelation of iron was found at a concentration of 300 μg/mL	

(*Continued*)

TABLE 3.2 (Continued)

	Source	Activity	References
	Polyporus dermoporus	Inhibition of hydroxyl radicals was 96% at 267 μg/mL	Dore et al. (2014)
		Inhibition of lipid peroxidation was 42.9% at 67 μg/mL	
		Inhibition of superoxide was 83.3% at 67 μg/mL	
	Jinqian mushroom	Hydroxyl radical scavenging activity was 24.30% at 5 mg/mL	Liu et al. (2014a,b)
		DPPH scavenging ratio was 89.84% at 5 mg/mL	
		ABT radical scavenging activity was 63.96% at 5 mg/mL	
		Iron chelating effect was 14.06% at 5 mg/mL	
	Huangshan floral mushroom	DPPH scavenging ratio was 79.46% at 5 mg/mL	Wang et al. (2015)
		IC_{50} of DPPH scavenging activity was 1.97 mg/mL	
		Hydroxyl radical scavenging activity was 74.18% at 5 mg/mL	
		IC_{50} of hydroxyl radical scavenging activity was 2.58 mg/mL	
		Iron chelating effect was 42.68% at 5 mg/mL	
		ABT radical scavenging activity was 22.51% at 5 mg/mL	
	Entoloma lividoalbum	Hydroxyl radical scavenging activity was 20.48% at 200 μg/mL	Maity et al. (2015)
		IC_{50} of hydroxyl radical scavenging activity was 400 μg/mL	
		Superoxide radical scavenging activity was 56.58% at 100 μg/mL	
		IC_{50} of superoxide radical scavenging was 75 μg/mL	
		Reducing power was 0.5 at 470 μg/mL	
	Termitomyces heimii	Decreases the ROS and NO in nicotine stimulated lymphocytes	Manna et al. (2015)
	Saccharomyces cerevisiae	Decreases RBARS formed in LPS stimulated human blood platelets	Saluk et al. (2013)
		Decreases the level of $O_2.^-$ formed in LPS stimulated human blood platelets	

schizophyllan. The results showed that both the two schizophyllans can significantly inhibit tumor growth in vivo, but the single helix conformer-rich schizophyllan possesses stronger immune-stimulating activity, whereas the triple-helix conformer is more significant for the recognition of the β-glucan receptor (Ohno et al., 1995).

Scleroglucan could stimulate TNF-α secretion of monocytes in a dose-dependent manner. However, once the triple helix conformation was denatured by NaOH treatment or heating to 135°C, an attenuated cytokine stimulating activity was observed (Falch et al., 2000). Mueller et al. (2000) reported that the rank order of competitive binding affinities for some glucan polymers were scleroglucan >>> schizophyllan >laminarin >glucan phosphate >glucan sulfate, and the scleroglucan exhibited a triple helical solution structure. Interestingly, the schizophyllan interacted nonselectively with two different non-CR3 binding sites on U937 cells, while scleroglucan, glucan sulfate, and laminarin interacted with one site preferentially.

Although β-glucans with a triple helical structure are reported more likely to possess ideal bioactivity according to most previous studies, a few researches also found that the beneficial bioactivity could also depend on its single helical structure. Saitô et al. (1991) found that schizophyllan with a single helical conformation could activate the coagulation factor G from limulus amebocyte lysate, and stimulate a host-mediated antitumor activity.

3.6.2 Molecular Weight and Solubility

Molecular weight may affect the polymerization and conformation of β-glucan. Yanaki et al. (1983) reported that schizophyllan of molecular weight above 9×10^4 exists as a rigid triple helix in aqueous solution, whereas that of molecular weight below 9×10^4 exists as both helices and single chains, and the fraction of triple helices decrease monotonically to zero as the molecular weight decreases to 5×10^3. However, the antitumor activity of aqueous schizophyllan against sarcoma 180 is related to the amount of triple helices relative to that of coexisting single coils in the solution, and an aqueous solution containing less than 50% triple helices showed no potent antitumor activity (Norisuye, 1985).

It is reported that *G. lucidum* β-glucans containing a higher branching ratio have higher activity, and higher molecular weight, and these glucans appear to be more effective than those with a lower molecular weight (Liu and Wang, 2005). Beta-(1,3)-glucan isolated from *Poria cocos* was phosphorylated, and the derivatives with molecular weight ranging from 2.6×10^4 to 26.8×10^4 and an extended chain conformation were potential antitumor adjuvants (Chen et al., 2009). A batch of β-glucan with molecular weights from 664 to 0.2 kDa were prepared, and the antioxidant analysis indicated that the antioxidant activity of β-glucan may be influenced more by its sources than molecules.

Solubility in water is one of the most important characteristic of β-glucan. The different solubility properties of apparently structurally similar polysaccharides has been explained on the basis of differences in polymer size, and in the DB which may interfere with chain packing (Ruiz-Herrera, 1991). The solubility of β-glucan increases as the degree of polymerization increases, and the β-glucans are completely insoluble in water when the degree of polymerization is greater than 100 (Mantovani et al., 2008).

Generally, soluble β-glucans appear to be stronger immunostimulators than insoluble β-glucans. Xiao et al. (2004) reported that soluble high-molecular weight yeast β-glucan could increase IFNγ-producing cells in a dose-dependent manner, whereas the insoluble β-glucan had no effect. A water-soluble, branched β-(1,3)-glucan of *Auricularia auricular-judae* exhibited potent inhibitory activity against implanted Sarcoma 180 solid tumors in mice, while the alkali-insoluble branched β-(1,3)-glucan showed essentially no inhibitory activity. Interestingly, the alkali-insoluble glucan was modified to be a water-soluble degraded glucan through periodate oxidation, borohydride reduction, and acid hydrolysis, then it exhibited potent antitumor activity, indicating that modification that introduces polyhydroxy groups to the β-(1,3)-glucan backbone may enhance the antitumor potency of the glucan (Misaki et al., 1981). However, Ishibashi et al. (2001) suggested that insoluble, as well as a high-molecular mass soluble form, is required for Grifola frondaso β-glucan to activate macrophages.

Particulate β-(1,3)-glucans have been found to be unsuitable for many human medical applications and responsible for significant adverse effects such as granuloma formation, microembolization, inflammation, and pain (Zeković et al., 2005). In order to improve the solubility of β-(1,3)-glucan, several derivation procedures including sulfation, phosphation, and carboxymethylation have been applied, and the solubility of the glucan derivative depends also on the anionic group introduced into the polysaccharide molecule. For instance, a water-insoluble β-(1,3)-glucan isolated from sclerotium of *P. cocos* did not show any antitumor activity; however, its water-soluble sulfation and carboxymethylation modified derivatives exhibit significant antitumor activity against sarcoma 180 cell and gastric carcinoma tumor cells (Wang et al., 2004). Besides, acid and alkaline hydrolysis, enzymatic digestion and ultrasound irradiation also have been applied to depolymerize the insoluble macromolecular structure to produce underivatized soluble β-(1,3)-glucan.

3.6.3 Carboxymethylation

Carboxymethylated glucan is one of the most studied soluble derivatives. It is commercially available in several countries, and has been proved to possess many human beneficial effects. Carboxymethylation modification of polysaccharide is usually carried out using sodium hydroxide and monochloroacetic acid. Carboxymethylation of polysaccharides is a well-studied conversion. According to the Williamson ether synthesis method, the polysaccharide is firstly activated with aqueous alkaline hydroxide, mostly sodium hydroxide, and then modified with monochloroacetic acid or its sodium salt to obtain the carboxymethyl polysaccharide derivative (Heinze and Koschella, 2005). Glucan becomes water-soluble due to the introduction of carboxymethyl substituents during the carboxymethylation modification. However, the chain conformation is also affected in this process. It is reported that the substituent needs space and there is interference with the helical conformation in some positions, particularly due to the substituents in the C-4 and C-2 position, while the substituents in the C-6 position stabilize the helical conformation because they protrude out of the helix (Kath et al., 1999), and the reactivity of the hydroxyl group of β-glucan is ordered as C-6 >C-4 >C-2 (Wang and Zhang, 2009a,b).

A carboxymethylated yeast glucan was prepared, briefly, the glucan was first suspended in a mixture of NaOH and isopropanol, and after stirring for 1 h, an appropriate amount of monochloroacetic acid was added to the reaction system to synthesize the carboxymethylated yeast β-glucan (Šandula et al., 1999). The modified glucan with a degree of substitution at 0.8 did not reveal any cytotoxic or genotoxic effects on cell viability (Magnani et al., 2009). More importantly, Darina reported that carboxymethylated yeast glucan promotes the efficiency of chemotherapy of Lewis lung carcinoma with CY. Not only was the growth of intramuscular tumor implants inhibited, but also the occurrence of lung metastases was increased in the carboxymethylated yeast β-glucan treated mice (Kogan et al., 2002). Similarly, a protective effect of carboxymethylated glucan against visible light-excited methylene blue-induced oxidative DNA lesion in V79 hamster lung cells was observed, and this beneficial effect was considered to be due to its scavenging activity of both hydroxyl radicals (·OH) and nonradical singlet oxygen ($^1O_2$) (Slameňová et al., 2003). Besides, Krizková et al. (2002) also found that the carboxymethylated yeast β-glucan possesses very high antioxidative activity, as well as expressive antimutagenic effects.

A water-soluble polysaccharide – protein complex was obtained from *P. tuber-regium* sclerotia, of which the sugar moiety is mainly β-(1,6)-, β-(1,4)-, and β-(1,3)-glucan (Tao and Zhang, 2008). The *P. tuber-regium* polysaccharide–protein complex was further chemically modified using carboxymethylation and sulfation to improve its antitumor activity (Tao et al., 2009). In the sulfation modification, the C-6 position of the glucan was fully substituted, the C-2, C-3, and C-4 positions were partially substituted by sulfate groups. The C-3, C-4, and C-6 positions were partially substituted during the carboxymethylation. The results showed that all of the native and modified polysaccharide inhibited the proliferation of Sarcoma 180 cells in vitro. However, much higher inhibition ratios were observed in the derivative-treated group. Zhang et al. (2004a,b,c, 2006) reported that carboxymethylated

P. tuber-regium β-glucan can suppress the growth of solid sarcoma 180 tumor in Balb/C mice in vivo, and inhibit the proliferation of leukemia HL-60 cell, hepatocarcinoma HepG2 cell, and human breast carcinoma MCH-7 cell in vitro as well.

Water-insoluble β-(1,3)-glucan from sclerotium of *Poria coco* was carboxymethylated to be a soluble polysaccharide. It is reported that the carboxymethylated β-glucan exists as a relatively extended flexible chain in 0.2M NaCl aqueous solution (Wang and Zhang, 2006). The derivatives showed an enhanced bile acid binding capacity in vitro, and the carboxymethylated β-glucan also exhibited antioxidant activity in vitro (Wang et al., 2009).

3.6.4 Sulfation

Sulfated polysaccharides have been shown to have potential antiviral activities. Most natural sulfated polysaccharides are complex polydisperse mixtures of macromolecules, which vary in their chemical structures and biological activities (Zhang et al., 2004a,b,c). Consequently, increasing researches have focused on producing water-soluble antiviral sulfated polysaccharides. As for the antiviral mechanism of sulfated polysaccharide, it is considered that SO_2^{4-} polyanions of sulfated polysaccharide can combine with virus or cells, and thus prohibit virus adsorption or inhibit some step of virus replication after entering the cell (Bagasra et al., 1991).

Polysaccharides are often sulfated using a chlorosulfonic acid–pyridine method. Briefly, a chlorosulfonic acid–pyridine complex was first prepared in an ice bath, and then polysaccharide would be added and stirred at a heating temperature to complete the phosphorylation modification reaction. After neutralizing with a NaOH solution, the sulfated polysaccharide was obtained through ethanol precipitation, dialysis, and lyophilization.

The antiviral effect of sulfated lentinan against avian infectious bronchitis virus (IBV) was investigated (Wang et al., 2010). All treatments, including preadding polysaccharide, postadding polysaccharide, and a mixture adding polysaccharide with IBV could significantly inhibit virus growth. Especially, the antivirus activity was higher than the natural lentinan. Wang reported that the C-6 position was preferentially substituted by the sulfate group, and the sulfated lentinan existed as an extended single flexible chain in 0.15 M NaCl solution. Although the sulfated glucan exhibited antitumor activity against sarcoma 180 cells in vitro, the inhibition ratios were lower than that of the triple-helix lentinan, but higher than that for the one with single random coil lentinan chains (Wang and Zhang, 2009a,b).

Liu et al. (2009, 2014a,b) reported that sulfated β-glucan of *Phellinus ribis* could block formation of new vessels in zebrafish, and inhibit the proliferation of human umbilical vein endothelial cells, suggesting that derivatives possess an antiangiogenic property. It also found that sulfated β-glucan showed a inhibitory effect against human ovary cancer SKOV-3 cells in vitro. Meanwhile, it could enhance the spleen and thymus index in H22-bearing mice at a dose of 20 mg/kg.

Beta-glucan of *Grifola frondosa* mycelia was sulfated. The molecular weight of the sulfated polysaccharide was determined to be 28 kDa, and the sulfur content was 16.4%. The sulfated β-glucan inhibited tumor growth and enhanced the peritoneal macrophage phagocytosis in sarcoma 180-bearing mice, and it improved the immunocompetence damaged by CY (Nie et al., 2006). Shi et al. (2007) also found that sulfated β-glucan distinctly inhibited human gastric carcinoma SGC-7901 cells grown in a dose-dependent manner, and induced apoptosis of the cancer cell line.

Multiple modifications of β-glucan have also been studied. A carboxymethylated-sulfated derivative β-(1,3)-glucan from *P. cocos* (CS-PCS3-II) was prepared by Chen et al. (2010a,b). The CS-PCS3 II exists as an extended flexible chain, the degrees of substitution of carboxymethyl and sulfate group in CS-PCS3-II were 1.05 and 0.36, respectively. The modified polysaccharide not only significantly inhibited Sarcoma 180 tumor in BALB/c mice by inducing necrosis and apoptosis, but also increased thymus indexes, spleen index, and hemolytic activity, as well as spleen antibody production in mice. It is postulated that the introduction of the carboxymethyl and sulfate group to PCS3-II increased its possible contact with the receptor of immune cells through hydrogen binding and electrostatic attraction. Wang reported that the carboxymethylated-sulfated *P. cocos* also possesses potential antioxidant activity.

3.6.5 Other Chemical Modifications

In addition to the common chemical modification methods above, some other modifications are also reported. *P. cocos* β-(1,3)-glucan was dissolved in dimethylsulfoxide and phosphorylated using phosphoric acid, and the phosphorylated polysaccharide showed strong inhibiting activities against the proliferation of sarcoma 180 cells (Chen et al., 2009).

Beta-glucan from *Inonotus obliquus* mushrooms was modified by acetylation, as well as sulfation and carboxymethylation (Ma et al., 2012). Among the three derivatives, acetylated polysaccharide resulted in lower molecular weight distribution, lower intrinsic viscosity, and hyperbranched conformation. Moreover, the acetylated polysaccharide showed higher antioxidant ability on ferric-reducing power and lipid peroxidation inhibition activity compared with the natural polysaccharide.

Smith degradation, which could be used to eliminate β-(1,6)-branching of β-glucan, and formalysis modification were applied to the β-(1,3)-glucan from *Tricholma giganteum* mushroom. Further analysis showed that four Smith degradation derivatives possessed higher antitumor activity compared to the original polysaccharide, but no influence was observed in the hydrolyzed derivatives (Mizuno et al., 1996). Beta-glucan from Chamsong-I mushrooms with a gradient DB from 19% to 50% were prepared to examine the influence of DB on their biological activity. It was showed that as the DB of β-glucan reduced, the inhibition effect on growth of cancer cells (MCF-7 breast cancer cells and Sarcoma 180 cells) and stimulation on nitric oxide production of RAW 264.7 cells was increased when DB was reduced to 32%. However, the activity further decreased as the DB further reduced (Bae et al., 2013).

Polysaccharide from *G. frondosa* was esterified using ethanol as the esterification reagent (Chan et al., 2011). Further antitumor analysis showed that chemically modified *G. frondosa* polysaccharides, including phosphorylated polysaccharide, esterified polysaccharide, and acetylated polysaccharide exhibited an adjuvant effect in rat glioma C-6 cells bearing mice, and the rank order observed among the derivatives of phosphorylation, esterification, acetylation, and the natural crude polysaccharide was phosphorylated polysaccharide>esterified polysaccharide>acetylated polysaccharide>crude polysaccharide.

REFERENCES

Ahmadi, K., Riazipour, M., 2007. Effect of *Ganoderma lucidum* on cytokine release by peritoneal macrophages. Iran. J. Immunol. 4, 220–226.

Alam, N., Amin, R., Khan, A., Ara, I., Shim, M.J., Lee, M.W., et al., 2009. Comparative effects of oyster mushrooms on lipid profile, liver and kidney function in hypercholesterolemic rats. Mycobiology 37, 37–42.

Angeli, J.P.F., Ribeiro, L.R., Bellini, M.F., Mantovani, M.S., 2009. β-Glucan extracted from the medicinal mushroom *Agaricus blazei* prevents the genotoxic effects of benzo [a] pyrene in the human hepatoma cell line HepG2. Arch. Toxicol. 83, 81–86.

Anttila, H., Sontag-Strohm, T., Salovaara, H., 2008. Viscosity of beta-glucan in oat products. Agr. Food Sci. 13, 80–87.

Bae, I.Y., Kim, H.W., Yoo, H.J., Kim, E.S., Lee, S., Park, D.Y., et al., 2013. Correlation of branching structure of mushroom β-glucan with its physiological activities. Food Res. Int. 51, 195–200.

Bagasra, O., Whittle, P., Heins, B., Pomerantz, R.J., 1991. Anti-human immunodeficiency virus type 1 activity of sulfated monosaccharides: comparison with sulfated polysaccharides and other polyions. J. Infect. Dis. 164, 1082.

Bell, S., Goldman, V.M., Bistrian, B.R., Arnold, A.H., Ostroff, G., Forse, R.A., 1999. Effect of β-glucan from oats and yeast on serum lipids. Crit. Rev. Food Sci. 39, 189–202.

Beresford, G., Stone, B.A., 1983. (1→3),(1→4)-β-d-Glucan content of *Triticum grains*. J. Cereal Sci. 1, 111–114.

Bobek, P., Ozdin, L., Kajaba, I., 1996a. Dose-dependent hypocholesterolaemic effect of oyster mushroom (*Pleurotus ostreatus*) in rats. Physiol. Res./Academia Scientiarum Bohemoslovaca 46, 327–329.

Bobek, P., Ozdín, L., Kuniak, L., 1996b. Effect of oyster mushroom (*Pleurotus ostreatus*) and its ethanolic extract in diet on absorption and turnover of cholesterol in hypercholesterolemic rat. Food/Nahrung 40, 222–224.

Bohn, J.A., Bemiller, J.N., 1995. (1→3)-β-d-Glucans as biological response modifiers: a review of structure-functional activity relationships. Carbohyd. Polym. 28, 3–14.

Brennan, C.S., Cleary, L.J., 2005. The potential use of cereal (1→3,1→4)-β-d-glucans as functional food ingredients. J. Cereal Sci. 42, 1–13.

Cao, L., Lin, Z., 2002. Regulation on maturation and function of dendritic cells by *Ganoderma lucidum* polysaccharides. Immunol. Lett. 83, 163–169.

Carbonero, E.R., Gracher, A.H.P., Smiderle, F.R., Rosado, F.R., Sassaki, G.L., Gorin, P.A.J., et al., 2006. A β-glucan from the fruit bodies of edible mushrooms *Pleurotus eryngii* and *Pleurotus ostreatoroseus*. Carbohyd. Polym. 66, 252–257.

Cebra, J.J., Periwal, S.B., Lee, G., Lee, F., Shroff, K.E., 1998. Development and maintenance of the gut-associated lymphoid tissue (GALT): the roles of enteric bacteria and viruses. Clin. Dev. Immunol. 6, 13–18.

Chan, G.C., Chan, W.K., Sze, D.M., 2009. The effects of beta-glucan on human immune and cancer cells. J. Hematol. Oncol. 2, 1–11.

Chan, J.Y., Chan, E., Chan, S., Sze, S., Chan, M., Tsui, S., et al., 2011. Enhancement of in vitro and in vivo anticancer activities of polysaccharide peptide from *Grifola frondosa* by chemical modifications. Pharm. Biol. 49, 1114–1120.

Chaung, H., Huang, T., Yu, J., Wu, M., Chung, W., 2009. Immunomodulatory effects of β-glucans on porcine alveolar macrophages and bone marrow haematopoietic cell-derived dendritic cells. Vet. Immunol. Immunop. 131, 147–157.

Chen, K.L., Weng, B.C., Chang, M.T., Liao, Y.H., Chen, T.T., Chu, C., 2008. Direct enhancement of the phagocytic and bactericidal capability of abdominal macrophage of chicks by beta-1, 3-1, 6-glucan. Poultry Sci. 87, 2242–2249.

Chen, S., Chang, C., Hung, M., Chen, S., Wang, W., Tai, C., et al., 2014. The effect of mushroom beta-glucans from solid culture of *Ganoderma lucidum* on inhibition of the primary tumor metastasis. Evid-Based. Compl. Alt. Med. 2014, 252171.

Chen, X., Xu, X., Zhang, L., Zeng, F., 2009. Chain conformation and anti-tumor activities of phosphorylated (1→3)-β-d-glucan from *Poria cocos*. Carbohyd. Polym. 78, 581–587.

Chen, X., Wang, W., Li, S., Xue, J., Fan, L., Sheng, Z., et al., 2010a. Optimization of ultrasound-assisted extraction of *Lingzhi polysaccharides* using response surface methodology and its inhibitory effect on cervical cancer cells. Carbohyd. Polym. 80, 944–948.

Chen, X., Zhang, L., Cheung, P.C.K., 2010b. Immunopotentiation and anti-tumor activity of carboxymethylated-sulfated β-(1→3)-d-glucan from *Poria cocos*. Int. Immunopharmacol. 10, 398–405.

Cheng, C., Yue, Q., Wu, Z., Song, X., Tao, S., Wu, X., et al., 2010. Cytotoxic triterpenoids from *Ganoderma lucidum*. Phytochemistry 71, 1579–1585.

Cheung, L.M., Cheung, P.C.K., 2005. Mushroom extracts with antioxidant activity against lipid peroxidation. Food Chem. 89, 403–409.

Cheung, L.M., Cheung, P.C., Ooi, V.E., 2003. Antioxidant activity and total phenolics of edible mushroom extracts. Food Chem. 81, 249–255.

Cheung, P.C.K., 1996a. The hypocholesterolemic effect of two edible mushrooms: *Auricularia auricula* (tree-ear) and *Tremella fuciformis* (white jelly-leaf) in hypercholesterolemic rats1. Nutr. Res. 16, 1721–1725.

Cheung, P.C.K., 1996b. The hypocholesterolemic effect of extracellular polysaccharide from the submerged fermentation of mushroom. Nutr. Res. 16, 1953–1957.

Cheung, Y., Wu, J., 2013. Kinetic models and process parameters for ultrasound-assisted extraction of water-soluble components and polysaccharides from a medicinal fungus. Biochem. Eng. J. 79, 214–220.

Cheung, Y., Siu, K., Wu, J., 2013. Kinetic models for ultrasound-assisted extraction of water-soluble components and polysaccharides from medicinal fungi. Food Bioprocess Tech. 6, 2659–2665.

Chihara, G., Maeda, Y., Hamuro, J., Sasaki, T., Fukuoka, F., 1969. Inhibition of mouse sarcoma 180 by polysaccharides from *Lentinus edodes* (Berk.) sing. Nature 222, 687–688.

Chihara, G., Hamuro, J., Maeda, Y.Y., Arai, Y., Fukuoka, F., 1970. Fractionation and purification of the polysaccharides with marked antitumor activity, especially lentinan, from *Lentinus edodes* (Berk.) Sing. (an edible mushroom). Cancer Res. 30, 2776–2781.

Choi, Y., Lee, S.M., Chun, J., Lee, H.B., Lee, J., 2006. Influence of heat treatment on the antioxidant activities and polyphenolic compounds of Shiitake (*Lentinus edodes*) mushroom. Food Chem. 99, 381–387.

Coviello, T., Maeda, H., Yuguchi, Y., Urakawa, H., Kajiwara, K., Dentini, M., et al., 1998. Conformational characteristics of oxidized scleroglucan. Macromolecules 31, 1602–1607.

Cui, W., Wang, Q., 2009. Cell wall polysaccharides in cereals: chemical structures and functional properties. Struct. Chem. 20, 291–297.

Cui, W., Wood, P.J., 2000. Relationships between structural features, molecular weight and rheological properties of cereal β-D-glucans. ChemInform 31. Available from: http://dx.doi.org/10.1002/chin.200041294.

Daou, C., Zhang, H., 2012. Oat beta-glucan: its role in health promotion and prevention of diseases. Compr. Rev. Food Sci. Food Safety 11, 355–365.

de Groot, A.P., Luyken, R., Pikaar, N.A., 1963. Cholesterol-lowering effect of rolled oats. Lancet 282, 303–304.

Delaney, B., Nicolosi, R.J., Wilson, T.A., Carlson, T., Frazer, S., Zheng, G., et al., 2003. β-Glucan fractions from barley and oats are similarly antiatherogenic in hypercholesterolemic Syrian golden hamsters. J. Nutr. 133, 468–475.

Deng, G., Lin, H., Seidman, A., Fornier, M., D Andrea, G., Wesa, K., et al., 2009. A phase I/II trial of a polysaccharide extract from *Grifola frondosa* (Maitake mushroom) in breast cancer patients: immunological effects. J. Cancer Res. Clin. 135, 1215–1221.

Dhillon, P., Bhatia, A., 2008. Hypercholesterolemic and immunomodulatory effects of oat extracts containing ß-glucan. Res. J. Immunol. 1, 29–35.

Dore, C.M.G., Azevedo, T.C., de Souza, M.C., Rego, L.A., de Dantas, J.C., Silva, F.R., et al., 2007. Antiinflammatory, antioxidant and cytotoxic actions of β-glucan-rich extract from *Geastrum saccatum* mushroom. Int. Immunopharmacol. 7, 1160–1169.

Dore, C.M.P.G., Alves, M.G.D.C., Santos, M.D.G.L., de Souza, L.A.R., Baseia, I.G., Leite, E.L., 2014. Antioxidant and anti-inflammatory properties of an extract rich in polysaccharides of the mushroom polyporus dermoporus. Antioxidants 3, 730–744.

Estrada, A., Yun, C., Kessel, A.V., Li, B., Hauta, S., Laarveld, B., 1997. Immunomodulatory activities of oat β-glucan in vitro and in vivo. Microbiol. Immunol. 41, 991–998.

Estrada, A., Van Kessel, A., Laarveld, B., 1999. Effect of administration of oat beta-glucan on immune parameters of healthy and immunosuppressed beef steers. Can. J. Vet. Res. 63, 261.

Faccin, L.C., Benati, F., Rincão, V.P., Mantovani, M.S., Soares, S.A., Gonzaga, M.L., et al., 2007. Antiviral activity of aqueous and ethanol extracts and of an isolated polysaccharide from *Agaricus brasiliensis* against poliovirus type 1. Lett. Appl. Microbiol. 45, 24–28.

Falch, B.H., Espevik, T., Ryan, L., Stokke, B.T., 2000. The cytokine stimulating activity of (1→3)-β-d-glucans is dependent on the triple helix conformation. Carbohyd. Res. 329, 587–596.

Fang, J., Wang, Y., Lv, X., Shen, X., Ni, X., Ding, K., 2012. Structure of a β-glucan from *Grifola frondosa* and its antitumor effect by activating Dectin-1/Syk/NF-κB signaling. Glycoconjugate J. 29, 365–377.

Fang, N., Li, Q., Yu, S., Zhang, J., He, L., Ronis, M.J., et al., 2006. Inhibition of growth and induction of apoptosis in human cancer cell lines by an ethyl acetate fraction from shiitake mushrooms. J. Altern. Complem. Med. 12, 125–132.

Feng, Y., Li, W., Wu, X., He, L., Ma, S., 2010. Rapid and efficient microwave-assisted sulfate modification of lentinan and its antioxidant and antiproliferative activities in vitro. Carbohyd. Polym. 82, 605–612.

Fleet, G.H., Manners, D.J., 1976. Isolation and composition of an alkali-soluble glucan from the cell walls of *Saccharomyces cerevisiae*. Microbiology 94, 180–192.

Fujimoto, K., Tomonaga, M., Goto, S., 2006. A case of recurrent ovarian cancer successfully treated with adoptive immunotherapy and lentinan. Anticancer Res. 26, 4015–4018.

Fullerton, S.A., Samadi, A.A., Tortorelis, D.G., Choudhury, M.S., Mallouh, C., Tazaki, H., et al., 1999. Induction of apoptosis in human prostatic cancer cells with beta-glucan (Maitake mushroom polysaccharide). Mol. Urol. 4, 7–13.

Gao, Y., Gao, H., Chan, E., Tang, W., Xu, A., Yang, H., et al., 2005a. Antitumor activity and underlying mechanisms of ganopoly, the refined polysaccharides extracted from *Ganoderma lucidum*, in mice. Immunol. Invest. 34, 171–198.

Gao, Y., Tang, W., Dai, X., Gao, H., Chen, G., Ye, J., et al., 2005b. Effects of water-soluble Ganoderma lucidum polysaccharides on the immune functions of patients with advanced lung cancer. J. Med. Food 8, 159–168.

Gao, Y., Zhou, S., Jiang, W., Huang, M., Dai, X., 2003. Effects of Ganopoly® (a *Ganoderma lucidum* polysaccharide extract) on the immune functions in advanced-stage cancer patients. Immunol. Invest. 32, 201–215.

Giavasis, I., 2014. Bioactive fungal polysaccharides as potential functional ingredients in food and nutraceuticals. Curr. Opin. Biotech. 26, 162–173.

Goffin, D., Bystricky, P., Shashkov, A.S., Lynch, M., Hanon, E., Paquot, M., et al., 2009. A systematic NMR determination of α-d-glucooligosaccharides, effect of linkage type, anomeric configuration and combination of different linkages type on 13 C chemical shifts for the determination of unknown isomaltooligosaccharides. Bull. Korean Chem. Soc. 30, 2535–2541.

Granfeldt, Y., Nyberg, L., Björck, I., 2008. Muesli with 4 g oat β-glucans lowers glucose and insulin responses after a bread meal in healthy subjects. Eur. J. Clin. Nutr. 62, 600–607.

Gu, Y., Takagi, Y., Nakamura, T., Hasegawa, T., Suzuki, I., Oshima, M., et al., 2005. Enhancement of radioprotection and anti-tumor immunity by yeast-derived β-glucan in mice. J. Med. Food 8, 154–158.

Guerra Dore, C.M.P., Azevedo, T.C.G., de Souza, M.C.R., Rego, L.A., de Dantas, J.C.M., Silva, F.R.F., et al., 2007. Antiinflammatory, antioxidant and cytotoxic actions of β-glucan-rich extract from *Geastrum saccatum* mushroom. Int. Immunopharmacol. 7, 1160–1169.

Gunde-Cimerman, N., Plemenitas, A., 2001. Hypocholesterolemic activity of the genus pleurotus (Jacq.: Fr.) P. Kumm. (Agaricales s. I., Basidiomycetes). Int. J. Med. Mushroom. 3, 395–397.

Guo, L., Xie, J., Ruan, Y., Zhou, L., Zhu, H., Yun, X., et al., 2009. Characterization and immunostimulatory activity of a polysaccharide from the spores of *Ganoderma lucidum*. Int. Immunopharmacol. 9, 1175–1182.

Gutiérrez, A., Prieto, A., Martínez, A.T., 1996. Structural characterization of extracellular polysaccharides produced by fungi from the genus Pleurotus. Carbohyd. Res. 281, 143–154.

Halliwell, B., 2005. Free radicals and other reactive species in disease. eLS. doi:10.1038/npg.els.0003913.

Handayani, D., Chen, J., Meyer, B.J., Huang, X., 2011. Dietary Shiitake mushroom (*Lentinus edodes*) prevents fat deposition and lowers triglyceride in rats fed a high-fat diet. J. Obesity 2011, Article ID 258051, 8 pages.

Harman, D., 2006. Free radical theory of aging: an update. Ann. Ny Acad. Sci. 1067, 10–21.

Hazama, S., Watanabe, S., Ohashi, M., Yagi, M., Suzuki, M., Matsuda, K., et al., 2009. Efficacy of orally administered superfine dispersed lentinan (β-1, 3-glucan) for the treatment of advanced colorectal cancer. Anticancer Res. 29, 2611–2617.

Heinze, T., Koschella, A., 2005. Carboxymethyl ethers of cellulose and starch–a review. Macromol. Symp. 223, 13–40.

Hetland, G., Ohno, N., Aaberge, I.S., Løvik, M., 2000. Protective effect of β-glucan against systemic *Streptococcus pneumoniae* infection in mice. FEMS Immunol. Med. Microbiol 27, 111–116.

Hida, T.H., Kawaminami, H., Ishibashi, K., Miura, N.N., Adachi, Y., Ohno, N., 2013. Oral administration of soluble β-glucan preparation from the cauliflower mushroom, sparassis crispa (higher basidiomycetes) modulated cytokine production in mice. Int. J. Med. Mushroom. 15, 525–538.

Hong, F., Yan, J., Baran, J.T., Allendorf, D.J., Hansen, R.D., Ostroff, G.R., et al., 2004. Mechanism by which orally administered β-1, 3-glucans enhance the tumoricidal activity of antitumor monoclonal antibodies in murine tumor models. J. Immunol. 173, 797–806.

Horvathova, E., Eckl, P.M., Bresgen, N., Slamenova, D., 2008. Evaluation of genotoxic and cytotoxic effects of H2O2 and DMNQ on freshly isolated rat hepatocytes; protective effects of carboxymethyl chitin-glucan. Neuroendocrinol. Lett. 29, 644.

Hossain, S., Hashimoto, M., Choudhury, E.K., Alam, N., Hussain, S., Hasan, M., et al., 2003. Dietary mushroom (*Pleurotus ostreatus*) ameliorates atherogenic lipid in hypercholesterolaemic rats. Clin. Exp. Pharmacol. P. 30, 470–475.

Huang, S., Ning, Z., 2010. Extraction of polysaccharide from *Ganoderma lucidum* and its immune enhancement activity. Int. J. Biol. Macromol. 47, 336–341.

Ina, K., Kataoka, T., Ando, T., 2013. The use of lentinan for treating gastric cancer. Anti-cancer agent. Med. Chem. 13, 681.

Ishibashi, K., Miura, N.N., Adachi, Y., Ohno, N., Yadomae, T., 2001. Relationship between solubility of grifolan, a fungal 1, 3-β-D-glucan, and production of tumor necrosis factor by macrophages in vitro. Biosci. Biotech. Biochem. 65, 1993–2000.

Ishurd, O., Kennedy, J.F., 2005. The anti-cancer activity of polysaccharide prepared from Libyan dates (*Phoenix dactylifera* L.). Carbohyd. Polym. 59, 531–535.

Isoda, N., Eguchi, Y., Nukaya, H., Hosho, K., Suga, Y., Suga, T., et al., 2008. Clinical efficacy of superfine dispersed lentinan (beta-1, 3-glucan) in patients with hepatocellular carcinoma. Hepato-Gastroenterol. 56, 437–441.

Jaehrig, S.C., Rohn, S., Kroh, L.W., Fleischer, L., Kurz, T., 2007. In vitro potential antioxidant activity of (1→3),(1→6)-β-d-glucan and protein fractions from *Saccharomyces cerevisiae* cell walls. J. Agr. Food Chem. 55, 4710–4716.

Jaehrig, S.C., Rohn, S., Kroh, L.W., Wildenauer, F.X., Lisdat, F., Fleischer, L., et al., 2008. Antioxidative activity of (1→3), (1→6)-β-d-glucan from *Saccharomyces cerevisiae* grown on different media. LWT - Food Sci. Technol. 41, 868–877.

Jeong, S.C., Jeong, Y.T., Yang, B.K., Islam, R., Koyyalamudi, S.R., Pang, G., et al., 2010. White button mushroom (*Agaricus bisporus*) lowers blood glucose and cholesterol levels in diabetic and hypercholesterolemic rats. Nutr. Res. 30, 49–56.

Jiang, J., Slivova, V., Harvey, K., Valachovicova, T., Sliva, D., 2004a. *Ganoderma lucidum* suppresses growth of breast cancer cells through the inhibition of Akt/NF-κB signaling. Nutr. Cancer 49, 209–216.

Jiang, J., Slivova, V., Valachovicova, T., Harvey, K., Sliva, D., 2004b. *Ganoderma lucidum* inhibits proliferation and induces apoptosis in human prostate cancer cells PC-3. Int. J. Oncol. 24, 1093–1099.

Kalra, S., Jood, S., 2000. Effect of dietary barley β-glucan on cholesterol and lipoprotein fractions in rat. J. Cereal Sci. 31, 141–145.

Kath, F., Lange, S., Kulicke, W.M., 1999. Influence of the glycosidic linkage on the solution conformation of glucans. Die Angewandte Makromolekulare Chemie 271, 28–36.

Ke, L.Q., 2014. Optimization of ultrasonic extraction of polysaccharides from *Lentinus edodes* based on enzymatic treatment. J. Food Process. Pres. 39, 254–259.

Kim, H.M., Han, S.B., Oh, G.T., Kim, Y.H., Hong, D.H., Hong, N.D., et al., 1996. Stimulation of humoral and cell mediated immunity by polysaccharide from mushroom *Phellinus linteus*. Int. J. Immunopharmacol. 18, 295–303.

Kim, J., Lee, S.M., Bae, I.Y., Park, H.G., Gyu Lee, H., Lee, S., 2011. (1–3)(1–6)-β-Glucan-enriched materials from *Lentinus edodes* mushroom as a high-fibre and low-calorie flour substitute for baked foods. J. Sci. Food Agr. 91, 1915–1919.

Klarzynski, O., Plesse, B., Joubert, J., Yvin, J., Kopp, M., Kloareg, B., et al., 2000. Linear β-1, 3 glucans are elicitors of defense responses in tobacco. Plant Physiol. 124, 1027–1038.

Kogan, G., Jann, B., Jann, K., 1992. Structure of the *Escherichia coli* 0104 polysaccharide and its identity with the capsular K9 polysaccharide. Fems Microbiol. Lett. 91, 135–140.

Kogan, G., šandula, J., Korolenko, T.A., Falameeva, O.V., Poteryaeva, O.N., Zhanaeva, S.Y., et al., 2002. Increased efficiency of *Lewis lung* carcinoma chemotherapy with a macrophage stimulator—yeast carboxymethyl glucan. Int. Immunopharmacol. 2, 775–781.

Kogan, G., Staško, A., Bauerová, K., Polovka, M., šoltés, L., Brezová, V., et al., 2005. Antioxidant properties of yeast (1→3)-β-d-glucan studied by electron paramagnetic resonance spectroscopy and its activity in the adjuvant arthritis. Carbohyd. Polym. 61, 18–28.

Kozarski, M., Klaus, A., Niksic, M., Jakovljevic, D., Helsper, J.P.F.G., Van Griensven, L.J.L.D., 2011. Antioxidative and immunomodulating activities of polysaccharide extracts of the medicinal mushrooms *Agaricus bisporus, Agaricus brasiliensis, Ganoderma lucidum* and *Phellinus linteus*. Food Chem. 129, 1667–1675.

Krizková, L., Duracková, Z., Sandula, J., Slamenová, D., Sasinková, V., Sivonová, M., et al., 2002. Fungal beta-(1-3)-D-glucan derivatives exhibit high antioxidative and antimutagenic activity in vitro. Anticancer Res. 23, 2751–2756.

Lahaye, M., Jegou, D., Buleon, A., 1994. Chemical characteristics of insoluble glucans from the cell wall of the marine green alga *Ulva lactuca* (L.) Thuret. Carbohyd. Res. 262, 115–125.

Lakhanpal, T.N., Rana, M., 2005. Medicinal and nutraceutical genetic resources of mushrooms. Plant Genet. Resour. 3, 288–303.

Laroche, C., Michaud, P., 2007. New developments and prospective applications for β (1, 3) glucans. Recent Patents Biotechnol. 1, 59–73.

Lazarova, M., Labaj, J., Kogan, G., Slamenova, D., 2006. Carboxymethyl chitin-glucan enriched diet exhibits protective effects against oxidative DNA damage induced in freshly isolated rat cells. Neoplasma 53, 434.

Leung, M.Y.K., Liu, C., Koon, J.C.M., Fung, K.P., 2006. Polysaccharide biological response modifiers. Immunol. Lett. 105, 101–114.

Li, B., Dobruchowska, J.M., Gerwig, G.J., Dijkhuizen, L., Kamerling, J.P., 2013. Structural investigation of water-soluble polysaccharides extracted from the fruit bodies of *Coprinus comatus*. Carbohyd. Polym. 91, 314–321.

Li, F., Zhang, Y., Zhong, Z., 2011. Antihyperglycemic effect of *Ganoderma lucidum* polysaccharides on streptozotocin-induced diabetic mice. Int. J. Mol. Sci. 12, 6135–6145.

Li, W., Yajima, T., Saito, K., Nishimura, H., Fushimi, T., Ohshima, Y., et al., 2004. Immunostimulating properties of intragastrically administered Acetobacter-derived soluble branched (1, 4)-β-d-glucans decrease murine susceptibility to *Listeria monocytogenes*. Infect. Immun. 72, 7005–7011.

Li, W., Cui, S.W., Kakuda, Y., 2006. Extraction, fractionation, structural and physical characterization of wheat β-d-glucans. Carbohyd. Polym. 63, 408–416.

Lin, Z., 2005. Cellular and molecular mechanisms of immuno-modulation by *Ganoderma lucidum*. J. Pharmacol. Sci. 99, 144–153.

Lin, S., Li, C., Lee, S., Kan, L., 2003. Triterpene-enriched extracts from *Ganoderma lucidum* inhibit growth of hepatoma cells via suppressing protein kinase C, activating mitogen-activated protein kinases and G2-phase cell cycle arrest. Life Sci. 72, 2381–2390.

Lin, Y., Liang, Y., Lee, S., Chiang, B., 2005. Polysaccharide purified from *Ganoderma lucidum* induced activation and maturation of human monocyte-derived dendritic cells by the NF-κB and p38 mitogen-activated protein kinase pathways. J. Leukocyte Biol. 78, 533–543.

Lipke, P.N., Ovalle, R., 1998. Cell wall architecture in yeast: new structure and new challenges. J. Bacteriol. 180, 3735–3740.

Liu, G., Wang, X., 2005. Structure--activity relationship and anticancer mechanisms of *Ganoderma lucidum* polysaccharides. Mycosystema 25, 430–438.

Liu, W., Wang, H., Pang, X., Yao, W., Gao, X., 2010. Characterization and antioxidant activity of two low-molecular-weight polysaccharides purified from the fruiting bodies of *Ganoderma lucidum*. Int. J. Biol. Macromol. 46, 451–457.

Liu, Y., Du, Y., Wang, J., Zha, X., Zhang, J., 2014a. Structural analysis and antioxidant activities of polysaccharide isolated from *Jinqian mushroom*. Int. J. Biol. Macromol. 64, 63–68.

Liu, Y., Liu, Y., Jiang, H., Xu, L., Cheng, Y., Wang, P.G., et al., 2014b. Preparation, antiangiogenic and antitumoral activities of the chemically sulfated glucan from *Phellinus ribis*. Carbohyd. Polym. 106, 42–48.

Liu, Y., Liu, C., Tan, H., Zhao, T., Cao, J., Wang, F., 2009. Sulfation of a polysaccharide obtained from *Phellinus ribis* and potential biological activities of the sulfated derivatives. Carbohyd. Polym. 77, 370–375.

Llaubères, R., Richard, B., Lonvaud, A., Dubourdieu, D., Fournet, B., 1990. Structure of an exocellular β-d-glucan from *Pediococcus* sp., a wine lactic bacteria. Carbohyd. Res. 203, 103–107.

Lo, T.C.T., Tsao, H.H., Wang, A.Y., Chang, C.A., 2007. Pressurized water extraction of polysaccharides as secondary metabolites from *Lentinula edodes*. J. Agr. Food Chem. 55, 4196–4201.

Ma, L., Chen, H., Zhang, Y., Zhang, N., Fu, L., 2012. Chemical modification and antioxidant activities of polysaccharide from mushroom *Inonotus obliquus*. Carbohyd. Polym. 89, 371–378.

Maeda, Y.Y., Hamuro, J., Chihara, G., 1971. The mechanisms of action of anti-tumour polysaccharides. I. The effects of antilymphocyte serum on the anti-tumour activity of lentinan. Int. J. Cancer 8, 41–46.

Magnani, M., Calliari, C.M., de Macedo Jr., F.C., Mori, M.P., de Syllos Cólus, I.M., Castro-Gomez, R.J.H., 2009. Optimized methodology for extraction of (1→3)(1→6)-β-d-glucan from *Saccharomyces cerevisiae* and in vitro evaluation of the cytotoxicity and genotoxicity of the corresponding carboxymethyl derivative. Carbohyd. Polym. 78, 658–665.

Maity, P., Samanta, S., Nandi, A.K., Sen, I.K., Paloi, S., Acharya, K., et al., 2014. Structure elucidation and antioxidant properties of a soluble β-d-glucan from mushroom *Entoloma lividoalbum*. Int. J. Biol. Macromol. 63, 140–149.

Maity, P., Sen, I.K., Maji, P.K., Paloi, S., Devi, K.S.P., Acharya, K., et al., 2015. Structural, immunological, and antioxidant studies of β-glucan from edible mushroom *Entoloma lividoalbum*. Carbohyd. Polym. 123, 350–358.

Maltby, D., Carpita, N.C., Montezinos, D., Kulow, C., Delmer, D.P., 1979. β-1, 3-Glucan in developing cotton fibers structure, localization, and relationship of synthesis to that of secondary wall cellulose. Plant Physiol. 63, 1158–1164.

Manna, D.K., Nandi, A.K., Pattanayak, M., Maity, P., Tripathy, S., Mandal, A.K., et al., 2015. A water soluble β-glucan of an edible mushroom *Termitomyces heimii*: structural and biological investigation. Carbohyd. Polym. 134, 375–384.

Manners, D.J., Masson, A.J., Patterson, J.C., Björndal, H., Lindberg, B., 1973. The structure of a beta-(1--6)-D-glucan from yeast cell walls. Biochem. J. 135, 31–36.

Mantovani, M.S., Bellini, M.F., Angeli, J.P.F., Oliveira, R.J., Silva, A.F., Ribeiro, L.R., 2008. β-Glucans in promoting health: prevention against mutation and cancer. Mutat. Res. 658, 154–161.

Maruyama, S., Sukekawa, Y., Kaneko, Y., Fujimoto, S., 2006. Anti tumor activities of lentinan and micellapist in tumor-bearing mice. Gan to kagaku ryoho. Cancer Chemoth. 33, 1726–1729.

Mccleary, B.V., Codd, R., 1991. Measurement of (1→ 3),(1→ 4)-β-D-glucan in barley and oats: a streamlined enzymic procedure. J. Sci. Food Agr. 55, 303–312.

Misaki, A., Kakuta, M., Sasaki, T., Tanaka, M., Miyaji, H., 1981. Studies on interrelation of structure and antitumor effects of polysaccharides: antitumor action of periodate-modified, branched (1→3)-β-D-glucan of *Auricularia auricula-judae*, and other polysaccharides containing (1→3)-glycosidic linkages. Carbohyd. Res. 92, 115–129.

Mizuno, T., Hagiwara, T., Nakamura, T., Ito, H., Shimura, K., Sumiya, T., et al., 1990. Antitumor activity and some properties of water-soluble polysaccharides from "himematsutake," the fruiting body of *Agavicus blazei* Murill. Agri. Biol. Chem. 54, 2889–2896.

Mizuno, T., Ykohlui, P., Kinoshita, T., Zhuang, C., Ito, H., Mayuzumi, Y., 1996. Antitumor activity and chemical modification of polysaccharides from niohshimeji mushroom, *Tricholma giganteum*. Biosci. Biotech. Biochem. 60, 30–33.

Mm, B., Gao, J., Nakamura, N., Hartor, M., 2000. Triterpenes from the spores of *Ganoderma lucidum* and their cytotoxicity against meth-A and LLC tumor cells. Chem. Pharm. Bull. 48, 1026–1033.

Mueller, A., Raptis, J., Rice, P.J., Kalbfleisch, J.H., Stout, R.D., Ensley, H.E., et al., 2000. The influence of glucan polymer structure and solution conformation on binding to (1→3)-β-D-glucan receptors in a human monocyte-like cell line. Glycobiology 10, 339–346.

Müller, C.I., Kumagai, T., O'Kelly, J., Seeram, N.P., Heber, D., Koeffler, H.P., 2006. *Ganoderma lucidum* causes apoptosis in leukemia, lymphoma and multiple myeloma cells. Leukemia Res. 30, 841–848.

Murphy, E.A., Davis, J.M., Brown, A.S., Carmichael, M.D., Mayer, E.P., Ghaffar, A., 2004. Effects of moderate exercise and oat β-glucan on lung tumor metastases and macrophage antitumor cytotoxicity. J. Appl. Physiol. 97, 955–959.

Murphy, E.A., Davis, J.M., Brown, A.S., Carmichael, M.D., Ghaffar, A., Mayer, E.P., 2007. Oat beta-glucan effects on neutrophil respiratory burst activity following exercise. Med. Sci. Sport. Exer. 39, 639–644.

Murphy, E.A., Davis, J.M., Carmichael, M.D., Mayer, E.P., Ghaffar, A., 2009. Benefits of oat β-glucan and sucrose feedings on infection and macrophage antiviral resistance following exercise stress. Am. J. Physiol. 297, R1188–R1194.

Muthusamy, G., Joardar, S.N., Samanta, I., Isore, D.P., Roy, B., Kumar, T., 2013. β–Glucan from edible mushroom (*Pleurotus florida*) enhances mucosal immunity in poultry. Adv. Animal Vet. Sci. 1, 116–119.

Myklestad, S.M., 1989. Production, chemical structure, metabolism, and biological function of the (1→3)-linked, β3-D-glucans in diatoms. Biol. Oceanogr. 6, 313–326.

Nakano, H., Namatame, K., Nemoto, H., Motohashi, H., Nishiyama, K., Kumada, K., 1999. A multi-institutional prospective study of lentinan in advanced gastric cancer patients with unresectable and recurrent diseases: effect on prolongation of survival and improvement of quality of life. Kanagawa Lentinan Research Group. Hepato-Gastroenterol. 46, 2662.

Nameda, S., Harada, T., Miura, N.N., Adachi, Y., Yadomae, T., Nakajima, M., et al., 2003. Enhanced cytokine synthesis of leukocytes by a β-glucan preparation, SCG, extracted from a medicinal mushroom, *Sparassis crispa*. Immunopharm. Immunot. 25, 321–335.

Nandi, A.K., Samanta, S., Maity, S., Sen, I.K., Khatua, S., Devi, K.S.P., et al., 2014. Antioxidant and immunostimulant β-glucan from edible mushroom *Russula albonigra* (Krombh.) Fr. Carbohyd. Polym. 99, 774–782.

Ng, M., Yap, A., 2002. Inhibition of human colon carcinoma development by lentinan from shii-take mushrooms (*Lentinus edodes*). J. Altern. Compleme. Med. 8, 581–589.

Nie, X., Shi, B., Ding, Y., Tao, W., 2006. Preparation of a chemically sulfated polysaccharide derived from *Grifola frondosa* and its potential biological activities. Int. J. Biol. Macromol. 39, 228–233.

Nieman, D.C., Henson, D.A., Mcmahon, M., Wrieden, J.L., Davis, J.M., Murphy, E.A., et al., 2008. Beta-glucan, immune function, and upper respiratory tract infections in athletes. Med. Sci. Sport. Exer. 40, 1463–1471.

Norisuye, T., 1985. Triple-stranded helical structure of schizophyllan and its antitumor activity in aqueous solution. Die Makromolekulare Chemie 14, 105–118.

Nwodo, U.U., Green, E., Okoh, A.I., 2012. Bacterial exopolysaccharides: functionality and prospects. Int. J. Mol. Sci. 13, 14002–14015.

Ohno, N., Adachi, Y., Suzuki, I., Sato, K., Oikawa, S., Yadomae, T., 1986a. Characterization of the antitumor glucan obtained from liquid-cultured *Grifola frondosa*. Chem. Pharm. Bull. 34, 1709–1715.

Ohno, N., Suzuki, I., Yadomae, T., 1986b. Structure and antitumor activity of a/3-1, 3-glucan isolated from the culture filtrate of *Sclerotinia sclerotiorum* IFO 9395. Chem. Pharm. Bull 34, 1362–1365.

Ohno, N., Miura, N.N., Chiba, N., Adachi, Y., Yadomae, T., 1995. Comparison of the immunopharmacological activities of triple and single-helical schizophyllan in mice. Biol. Pharm. Bull. 18, 1242–1247.

Ohno, N., Miura, N.N., Nakajima, M., Yadomae, T., 2000. Antitumor 1, 3-. BETA.-glucan from cultured fruit body of *Sparassis crispa*. Biol. Pharm. Bull. 23, 866–872.

Ohno, N., Furukawa, M., Miura, N.N., Adachi, Y., Motoi, M., Yadomae, T., 2001. Antitumor. BETA.-glucan from the cultured fruit body of *Agaricus blazei*. Biol. Pharm. Bull. 24, 820–828.

Ohno, N., Harada, T., Masuzawa, S., Miura, N.N., Adachi, Y., Nakajima, M., et al., 2002. Antitumor activity and hematopoietic response of a b-glucan extracted from an edible and medicinal mushroom *Sparassis crispa* Wulf.: Fr. (Aphyllophoromycetideae). Int. J. Med. Mushroom. 4, 14.

Ohno, N., Nameda, S., Harada, T., Miura, N.N., Adachi, Y., Nakajima, M., et al., 2003. Immunomodulating activity of a b-glucan preparation, SCG, extracted from a culinary−medicinal mushroom, *Sparassis crispa* Wulf.: Fr. (Aphyllophoromycetideae), and application to cancer patients. Int. J. Med. Mushroom. 5, 359−368.

Oka, M., Hazama, S., Suzuki, M., Wang, F., Wadamori, K., Iizuka, N., et al., 1996. In vitro and in vivo analysis of human leukocyte binding by the antitumor polysaccharide, lentinan. Int. J. Immunopharmacol. 18, 211−216.

Ookushi, Y., Sakamoto, M., Azuma, J., 2006. Optimization of microwave-assisted extraction of polysaccharides from the fruiting body of mushrooms. J. Appl. Glycosci. (Japan) 53, 267−272.

Palanisamy, M., Aldars García, L., Gil Ramírez, A., Ruiz Rodríguez, A., Marín, F.R., Reglero, G., et al., 2014. Pressurized water extraction of β-glucan enriched fractions with bile acids-binding capacities obtained from edible mushrooms. Biotechnol. Progr. 30, 391−400.

Paul, I., Isore, D.P., Joardar, S.N., Samanta, I., Biswas, U., Maiti, T.K., et al., 2012. Orally administered β-glucan of edible mushroom (*Pleuratus florida*) origin upregulates innate immune response in broiler. Indian J. Anim. Sci. 82, 745−748.

Pietrzycka, A., St Pniewski, M., Waszkielewicz, A.M., Solarz, O., 2006. Effect of vita glucan on some antioxidant parameters of the human blood. In vitro study. Acta Pol. Pharm. 63, 547−551.

Pomeroy, S., Tupper, R., Cehun-Aders, M., Nestel, P., 2001. Oat b-glucan lowers total and LDL-cholesterol. Aust. J. Nutr. Diet. 58, 51−55.

Refaie, F.M., Esmat, A.Y., Daba, A.S., Taha, S.M., 2009. Characterization of polysaccharopeptides from *Pleurotus ostreatus* mycelium: assessment of toxicity and immunomodulation in vivo. Micol. Apl. Int. 21, 67−75.

Reutter, W., Fan, H., 2010. GLIS, a bioactive proteoglycan fraction from *Ganoderma lucidum*, displays anti-tumour activity by increasing both humoral and cellular immune response. Life Sci. 87, 628−637.

Reyes Suárez, E., Bugden, S.M., Kai, F.B., Kralovec, J.A., Noseda, M.D., Barrow, C.J., et al., 2008. First isolation and structural determination of cyclic β-(1→2)-glucans from an alga, *Chlorella pyrenoidosa*. Carbohyd. Res. 343, 2623−2633.

Rhee, S.J., Cho, S.Y., Kim, K.M., Cha, D., Park, H., 2008. A comparative study of analytical methods for alkali-soluble β-glucan in medicinal mushroom, Chaga (*Inonotus obliquus*). LWT-Food Sci. Technol. 41, 545−549.

Rice, P.J., Adams, E.L., Ozment-Skelton, T., Gonzalez, A.J., Goldman, M.P., Lockhart, B.E., et al., 2005. Oral delivery and gastrointestinal absorption of soluble glucans stimulate increased resistance to infectious challenge. J. Pharmacol. Exp. Ther. 314, 1079−1086.

Rout, D., Mondal, S., Chakraborty, I., Pramanik, M., Islam, S.S., 2005. Chemical analysis of a new (1→3)-,(1→6)-branched glucan from an edible mushroom, *Pleurotus florida*. Carbohyd. Res. 340, 2533−2539.

Ruiz-Herrera, J., 1991. Biosynthesis of β-glucans in fungi. Antonie Van Leeuwenhoek 60, 73−81.

Saitô, H., Yoshioka, Y., Uehara, N., Aketagawa, J., Tanaka, S., Shibata, Y., 1991. Relationship between conformation and biological response for (1→3)-β-d-glucans in the activation of coagulation Factor G from limulus amebocyte lysate and host-mediated antitumor activity. Demonstration of single-helix conformation as a stimulant. Carbohyd. Res. 217, 181–190.

Sakurai, K., Shinkai, S., 2000. Molecular recognition of adenine, cytosine, and uracil in a single-stranded RNA by a natural polysaccharide: schizophyllan. J. Am. Chem. Soc. 122, 4520–4521.

Saluk, J., Bijak, M., Ponczek, M.B., Nowak, P., Wachowicz, B., 2013. (1→3)-β-d-Glucan reduces the damages caused by reactive oxygen species induced in human platelets by lipopolysaccharides. Carbohyd. Polym. 97, 716–724.

Šandula, J., Kogan, G., Kačuráková, M., Machová, E., 1999. Microbial (1→3)-β-D-glucans, their preparation, physico-chemical characterization and immunomodulatory activity. Carbohyd. Polym. 38, 247–253.

Sasaki, T., Abiko, N., Sugino, Y., Nitta, K., 1978. Dependence on chain length of antitumor activity of (1→3)-β-D-glucan from *Alcaligenes faecalis* var. myxogenes, IFO 13140, and its acid-degraded products. Cancer Res. 38, 379–383.

Schley, P.D., Field, C.J., 2002. The immune-enhancing effects of dietary fibres and prebiotics. Brit. J. Nutr. 87, S221–S230.

Schneider, I., Kressel, G., Meyer, A., Krings, U., Berger, R.G., Hahn, A., 2011. Lipid lowering effects of oyster mushroom (*Pleurotus ostreatus*) in humans. J. Funct. Foods 3, 17–24.

Şener, G., Toklu, H., Ercan, F., Erkanlı, G., 2005. Protective effect of β-glucan against oxidative organ injury in a rat model of sepsis. Int. Immunopharmacol. 5, 1387–1396.

Sener, G., Toklu, H.Z., Cetinel, S., 2007. β-Glucan protects against chronic nicotine-induced oxidative damage in rat kidney and bladder. Environ. Toxicol. Phar. 23, 25–32.

Shao, B., Dai, H., Xu, W., Lin, Z., Gao, X., 2004. Immune receptors for polysaccharides from *Ganoderma lucidum*. Biochem. Bioph. Res. Co. 323, 133–141.

Shi, B.J., Nie, X., Chen, L., Liu, Y., Tao, W., 2007. Anticancer activities of a chemically sulfated polysaccharide obtained from *Grifola frondosa* and its combination with 5-Fluorouracil against human gastric carcinoma cells. Carbohyd. Polym. 68, 687–692.

Shimizu, K., Watanabe, S., Watanabe, S., Matsuda, K., Suga, T., Nakazawa, S., et al., 2009. Efficacy of oral administered superfine dispersed lentinan for advanced pancreatic cancer. Hepatogastroenterology 56, 240–244.

Skendi, A., Biliaderis, C.G., Lazaridou, A., Izydorczyk, M.S., 2003. Structure and rheological properties of water soluble β-glucans from oat cultivars of *Avena sativa* and *Avena bysantina*. J. Cereal Sci. 38, 15–31.

Slameňová, D., Lábaj, J., Križková, L.A., Via, X., Kogan, G., Šandula, J., et al., 2003. Protective effects of fungal (1→3)-β-d-glucan derivatives against oxidative DNA lesions in V79 hamster lung cells. Cancer Lett. 198, 153–160.

Song, G., Du, Q., 2012. Structure characterization and antitumor activity of an α β-glucan polysaccharide from *Auricularia polytricha*. Food Res. Int. 45, 381–387.

Sullivan, R., Smith, J.E., Rowan, N.J., 2006. Medicinal mushrooms and cancer therapy: translating a traditional practice into Western medicine. Perspect. Biol. Med. 49, 159–170.

Surenjav, U., Zhang, L., Xu, X., Zhang, X., Zeng, F., 2006. Effects of molecular structure on antitumor activities of (1→3)-β-d-glucans from different *Lentinus edodes*. Carbohyd. Polym. 63, 97–104.

Synytsya, A., Novák, M., 2013. Structural diversity of fungal glucans. Carbohyd. Polym. 92, 792–809.

Tanioka, A., Tanabe, K., Hosono, A., Kawakami, H., Kaminogawa, S., Tsubaki, K., et al., 2013. Enhancement of intestinal immune function in mice by β-D-Glucan from Aureobasidium Pullulans ADK-34. Scand. J. Immunol. 78, 61–68.

Tao, Y., Xu, W., 2008. Microwave-assisted solubilization and solution properties of hyperbranched polysaccharide. Carbohyd. Res. 343, 3071–3078.

Tao, Y., Zhang, L., 2008. Characterization of polysaccharide–protein complexes by size-exclusion chromatography combined with three detectors. Carbohyd. Res. 343, 2251–2257.

Tao, Y., Zhang, Y., Zhang, L., 2009. Chemical modification and antitumor activities of two polysaccharide–protein complexes from *Pleurotus tuber-regium*. Int. J. Biol. Macromol. 45, 109–115.

Thompson, I.J., Oyston, P.C., Williamson, D.E., 2010. Potential of the β-glucans to enhance innate resistance to biological agents. Expert Rev. Anti-infe. 2010 8 (3), 339–352.

Tian, Y., Zeng, H., Xu, Z., Zheng, B., Lin, Y., Gan, C., et al., 2012. Ultrasonic-assisted extraction and antioxidant activity of polysaccharides recovered from white button mushroom (*Agaricus bisporus*). Carbohyd. Polym. 88, 522–529.

Tokunaka, K., Ohno, N., Adachi, Y., Miura, N.N., Yadomae, T., 2002. Application of Candida solubilized cell wall β-glucan in antitumor immunotherapy against P815 mastocytoma in mice. Int. Immunopharmacol. 2, 59–67.

Tsiapali, E., Whaley, S., Kalbfleisch, J., Ensley, H.E., Browder, I.W., Williams, D.L., 2001. Glucans exhibit weak antioxidant activity, but stimulate macrophage free radical activity. Free Radical Bio. Med. 30, 393–402.

Tsukada, C., Yokoyama, H., Miyaji, C., Ishimoto, Y., Kawamura, H., Abo, T., 2003. Immunopotentiation of intraepithelial lymphocytes in the intestine by oral administrations of β-glucan. Cell. Immunol. 221, 1–5.

Ukawa, Y., Ito, H., Hisamatsu, M., 2000. Antitumor effects of (1→3)-β-D-glucan and (1→6)-β-D-glucan purified from newly cultivated mushroom, Hatakeshimeji (*Lyophyllum decastes* Sing.). J. Biosci. Bioeng. 90, 98–104.

Villares, A., Mateo-Vivaracho, L., Guillamón, E., 2012. Structural features and healthy properties of polysaccharides occurring in mushrooms. Agriculture 2, 452–471.

Volman, J.J., Ramakers, J.D., Plat, J., 2008. Dietary modulation of immune function by β-glucans. Physiol. Behav. 94, 276–284.

Volman, J.J., Mensink, R.P., Ramakers, J.D., de Winther, M.P., Carlsen, H., Blomhoff, R., et al., 2010. Dietary (1→3), (1→4)-β-d-glucans from oat activate nuclear factor-κB in intestinal leukocytes and enterocytes from mice. Nutr. Res. 30, 40–48.

Wang, J., Zhang, L., 2009a. Structure and chain conformation of five water-soluble derivatives of a β-d-glucan isolated from *Ganoderma lucidum*. Carbohyd. Res. 344, 105–112.

Wang, X., Zhang, L., 2009b. Physicochemical properties and antitumor activities for sulfated derivatives of lentinan. Carbohyd. Res. 344, 2209–2216.

Wang, Y., Zhang, L., 2006. Chain conformation of carboxymethylated derivatives of (1→3)-β-d-glucan from *Poria cocos* sclerotium. Carbohyd. Polym. 65, 504–509.

Wang, D., Guo, Z., Ma, X., Hu, Y., Huang, X., Fan, Y., et al., 2010. Effects of sulfated lentinan on cellular infectivity of avian infectious bronchitis virus. Carbohyd. Polym. 79, 461–465.

Wang, G., Zhang, J., Mizuno, T., Zhuang, C., Ito, H., Mayuzumi, H., et al., 1993. Antitumor active polysaccharides from the Chinese mushroom Songshan lingzhi, the fruiting body of *Ganoderma tsugae*. Biosci. Biotechnol. Biochem. 57, 894–900.

Wang, J., Xu, J., Zhang, J., Liu, Y., Sun, H., Zha, X., 2015. Physicochemical properties and antioxidant activities of polysaccharide from floral mushroom cultivated in Huangshan Mountain. Carbohyd. Polym. 131, 240–247.

Wang, Q., Wood, P.J., Cui, W., 2002. Microwave assisted dissolution of β-glucan in water—implications for the characterisation of this polymer. Carbohyd. Polym. 47, 35–38.

Wang, J., Yuan, Y., Yue, T., 2014a. Immunostimulatory activities of β-d-glucan from *Ganoderma lucidum*. Carbohyd. Polym. 102, 47–54.

Wang, Q., Chen, S., Han, L., Lian, M., Wen, Z., Jiayinaguli, T., et al., 2014b. Antioxidant activity of carboxymethyl (1→3)-β-d-glucan (from the sclerotium of *Poria cocos*) sulfate (in vitro). Int. J. Biol. Macromol. 69, 229–235.

Wang, S.Y., Hsu, M.L., Hsu, H.C., Lee, S.S., Shiao, M.S., Ho, C.K., 1997. The anti-tumor effect of *Ganoderma lucidum* is mediated by cytokines released from activated macrophages and T lymphocytes. Int. J. Cancer 70, 699–705.

Wang, X., Xu, X., Zhang, L., 2008. Thermally induced conformation transition of triple-helical lentinan in NaCl aqueous solution. J. Phys. Chem. B 112, 10343–10351.

Wang, Y., Zhang, L., Li, Y., Hou, X., Zeng, F., 2004. Correlation of structure to antitumor activities of five derivatives of a β-glucan from *Poria cocos* sclerotium. Carbohyd. Res. 339, 2567–2574.

Wang, Y., Yu, Y., Mao, J., 2009. Carboxymethylated β-glucan derived from *Poria cocos* with biological activities. J. Agr. Food Chem. 57, 10913–10915.

Wasser, S., 2002. Medicinal mushrooms as a source of antitumor and immunomodulating polysaccharides. Appl. Microbiol. Biot 60, 258–274.

Whyte, J., Englar, J.R., 1971. Polysaccharides of the Red Alga *Rhodymenia pertusa* Part II. cell--wall glucan; proton magnetic resonance studies on permethylated polysaccharides. Can. J. Chem. 49, 1302–1305.

Xiao, Z., Trincado, C.A., Murtaugh, M.P., 2004. β-Glucan enhancement of T cell IFNγ response in swine. Vet. Immunol. Immunop. 102, 315–320.

Yalin, W., Ishurd, O., Cuirong, S., Yuanjiang, P., 2005. Structure analysis and antitumor activity of (1→3)-beta-d-glucans (cordyglucans) from the mycelia of *Cordyceps sinensis*. Planta Med. 71, 381–384.

Yanaki, T., Ito, W., Tabata, K., Kojima, T., Norisuye, T., Takano, N., et al., 1983. Correlation between the antitumor activity of a polysaccharide schizophyllan and its triple-helical conformation in dilute aqueous solution. Biophys. Chem. 17, 337–342.

Yang, L., Zhang, L., 2009. Chemical structural and chain conformational characterization of some bioactive polysaccharides isolated from natural sources. Carbohyd. Polym. 76, 349–361.

You, Q., Yin, X., Zhang, S., Jiang, Z., 2014. Extraction, purification, and antioxidant activities of polysaccharides from *Tricholoma mongolicum* Imai. Carbohyd. Polym. 99, 1–10.

Yun, C., Estrada, A., van Kessel, A., Gajadhar, A.A., Redmond, M.J., Laarveld, B., 1997. β-(1→3, 1→4) Oat glucan enhances resistance to *Eimeria vermiformis* infection in immunosuppressed mice. Int. J. Parasitol. 27, 329–337.

Yun, C.H., Estrada, A., Kessel, A., Park, B.C., Laarveld, B., 2003. β-Glucan, extracted from oat, enhances disease resistance against bacterial and parasitic infections. FEMS Immunol. Med. Microbiol. 35, 67–75.

Yuri, C.H., Estrada, A., Van Kessel, A., Gajadhar, A., Redmond, M., Laarveld, B., 1998. Immunomodulatory effects of oat β-glucan administered intragastrically or parenterally on mice infected with *Eimeria vermiformis*. Microbiol. Immunol. 42, 457–465.

Zeković, D.B., Kwiatkowski, S., Vrvić, M.M., Jakovljević, D., Moran, C.A., 2005. Natural and modified (1→3)-β-D-glucans in health promotion and disease alleviation. Crit. Rev. Biotechnol. 25, 205–230.

Zhang, J., Tang, Q., Zimmerman-Kordmann, M., Reutter, W., Fan, H., 2002a. Activation of B lymphocytes by GLIS, a bioactive proteoglycan from *Ganoderma lucidum*. Life Sci. 71, 623–638.

Zhang, L., Li, X., Zhou, Q., Zhang, X., Chen, R., 2002b. Transition from triple helix to coil of lentinan in solution measured by SEC, viscometry, and 13C NMR. Polym. J. 34, 443–449.

Zhang, M., Cheung, P.C., Zhang, L., 2001. Evaluation of mushroom dietary fiber (nonstarch polysaccharides) from sclerotia of *Pleurotus tuber-regium* (Fries) Singer as a potential antitumor agent. J. Agr. Food Chem. 49, 5059–5062.

Zhang, M., Cheung, P.C.K., Zhang, L., Chiu, C., Ooi, V.E.C., 2004a. Carboxymethylated β-glucans from mushroom sclerotium of *Pleurotus tuber-regium* as novel water-soluble antitumor agent. Carbohyd. Polym. 57, 319–325.

Zhang, M., Cheung, P.C.K., Ooi, V.E.C., Zhang, L., 2004b. Evaluation of sulfated fungal β-glucans from the sclerotium of *Pleurotus tuber-regium* as a potential water-soluble antiviral agent. Carbohyd. Res. 339, 2297–2301.

Zhang, X., Zhang, L., Xu, X., 2004c. Morphologies and conformation transition of lentinan in aqueous NaOH solution. Biopolymers 75, 187–195.

Zhang, M., Cheung, P.C.K., Chiu, L.C.M., Wong, E.Y.L., Ooi, V.E.C., 2006. Cell-cycle arrest and apoptosis induction in human breast carcinoma MCF-7 cells by carboxymethylated β-glucan from the mushroom sclerotia of *Pleurotus tuber-regium*. Carbohyd. Polym. 66, 455–462.

Zhang, M., Cui, S.W., Cheung, P.C.K., Wang, Q., 2007. Antitumor polysaccharides from mushrooms: a review on their isolation process, structural characteristics and antitumor activity. Trends Food Sci. Tech 18, 4–19.

Zhang, Y., Li, S., Wang, X., Zhang, L., Cheung, P.C.K., 2011. Advances in lentinan: isolation, structure, chain conformation and bioactivities. Food Hydrocolloid. 25, 196–206.

Zhu, X., Chen, A., Lin, Z., 2007. *Ganoderma lucidum* polysaccharides enhance the function of immunological effector cells in immunosuppressed mice. J. Ethnopharmacol. 111, 219–226.

Chapter 4

Cordyceps Polysaccharides

Cordyceps, a well-known and valued traditional Chinese medicine, is also called "ChongCao" in Chinese. There are more than 350 species of *Cordyceps* all over world, out of which about 120 species have been reported in China (Zhong et al., 2009). The genus *Cordyceps* (Clavicipitaceae) is a group of entomopathogenic fungi that are widely used as a tonic food or invigorant with broad-spectrumed medicinal properties in China. Specifically, *Cordyceps sinensis*, also called "DongChongXiaCao" in Chinese, is a valued traditional Chinese medicine (TCM) that has been used in China for more than 700 years (Yan et al., 2014). *Cordyceps sinensis* was found in the prairie soil above 3500 m on the Qinghai–Tibetan plateau. In winter, it looks like an old silk worm in the soil, and in summer, hairs grow out of soil, and transform into grass. These need to be collected in the summer, if not, they will turn into the worm-like structure again (Fig. 4.1). The natural *Co. sinensis* needs a strict growing habitat, so its productivity is limited. On the other hand, excessive harvesting and environmental destruction have made the productivity of natural *Co. sinensis* decrease in recent decades. Since the natural *Co. sinensis* were so precious and could not meet the increasing demand, modern researchers in recent years focused on isolating the anamorphic fungi from *Co. sinensis*. Several other species such as *Paecilomyces sinensis*, *Cephalosporium sinensis*, *Tolypocladium sinensis*, and *Hirsutella sinensis* (Yin and Tang, 1995) have also been successfully cultivated. Later, fermentation technology was exploited for large-scale production of *Co. sinensis* fungal mycelia. *Cordyceps militaris*, another major entomopathogenic fungus, is commonly known as northern *Co. sinensis* or orange caterpillar fungus. Unlike *Co. sinensis*, *C. militaris* has a worldwide distribution, and its stromata can be artificially cultivated easily. Researchers have produced stromata of *C. militaris* on insects. Apart from stroma cultivation, production of *C. militaris* mycelia in submerged culture is also possible (Shrestha et al., 2012). Besides, *Cordyceps gunii* (Berk.), which is also called *Cordyceps hawkessi* Gray, belongs to the *Ascomycota*, *Pyrenomycetes*, *Sphaeriales*, *Clavicipitaceae*, and parasites on the larvae of *Hepialidae*. The anamorph of *Paecilomyces gunii* of *C. gunii* has been isolated and identified (Liang, 1985), and is widely distributed in Hunan and Guizhou province (Liu et al., 2008; Meng, 2001). *Cordyceps cicadae* Shing., also called "Chan Hua" in Chinese and "Senka" in Japanese, is a fungus parasitic on the larva of

Bioactive Polysaccharides. DOI: http://dx.doi.org/10.1016/B978-0-12-809418-1.00004-6

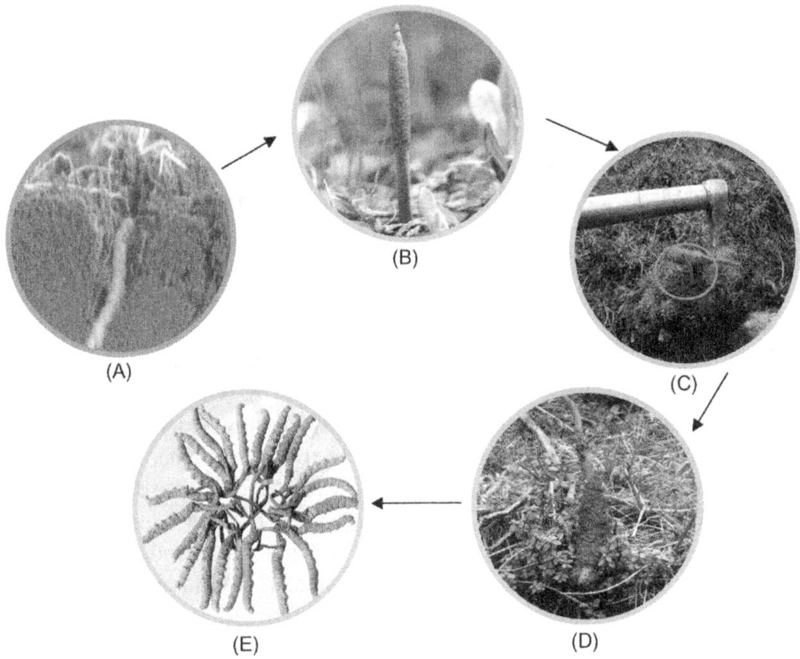

FIGURE 4.1 The growth and harvest of *Cordyceps sinensis*. In winter, it appears like an old silk worm in the soil (A), and in summer, the hairs grow out of soil and convert into grass (B). These need to be collected in the summer (C, D) and are stored and sold as a dried material (E).

Cicada flammata Dist. The dried material is composed of the fungal fruiting body and the insect body, and has been used as a medicine for childhood convulsions or palpitations in China.

In this chapter, we will emphasize the main bioactive components in *Cordyceps* polysaccharide, and provide a general overview of the latest developments in *Cordyceps* polysaccharide research concerning harvesting, extraction, and purification procedures, structural properties, and chain conformation, as well as a variety of bioactivities.

4.1 HARVESTING AND PREPARATION

Cordyceps encompasses several *Cordyceps* species which have been widely used for medicinal purposes for centuries, particularly in China, Japan, and other Asian countries. *Cordyceps* polysaccharide (CP) represents a class of structurally diverse biological macromolecules with wide-ranging physiochemical properties accounting for about 3%–8% of the total dry weight. They usually come from the fruiting bodies, the mycelium of solid fermentation submerged cultures, or the broth of *Cordyceps* fungi. Until now, many CPs and their components have been isolated, most of which have been isolated from cultured *Cordyceps* or *C. militaris*. There was a variation in the polysaccharide content among *Cordyceps* species according to previous literature. Wang et al. found that both *Cordyceps taii* (20.18%) and *Cordyceps jiangxiensis* (21.94%) had a higher total sugar content than *C. gunii* (17.1%), which had a content similar to the total sugar content of *Co. sinensis* (23.3%) (Wang et al., 2004). At present, different fermentation techniques have been developed successfully for the production of *Cordyceps*. Therefore, a large number of mycelium products are also studied in various research. Generally, the CP content in the fruiting bodies is low in any basic culture medium, but this could be increased through optimization of the medium composition and culturing conditions.

4.1.1 Polysaccharide Content in Cordyceps

4.1.1.1 Cordyceps sinensis

Wang et al. reported that the polysaccharide contents of three natural varieties of *Co. sinensis* that were harvested from Qinghai province in Golog, Yushu, and Xining were 8.32%, 8.90%, and 7.19%, respectively, but the polysaccharide content was significantly higher in cultured *Co. sinensis* (10.43%) (Wang et al., 2015a). On the other hand, Wu et al. figured out that the yield of crude polysaccharide extracted from natural *Co. sinensis* collected from Sichuan province was 3.75% (Wu et al., 2014b). However, the content of polysaccharide obtained from Gansu province was between 2.5% and 5.7% (Chen et al., 2009). The variation in polysaccharide content of natural *Co. sinensis* may be attributed to the differences in the growing environment such as geography, climate, and moisture. Besides, different parts of *Co. sinensis* also exhibited a variation in polysaccharide content. Hsu et al. compared the content of total sugar between the corpus (part of the insect) and fruiting body (ascoma) and found a slight difference, 24.2% and 24.9%, respectively, but this was much lower than the content in mycelium, at 39.4% (Hsu et al., 2002).

A novel fungal strain isolated from natural *Co. sinensis* is named UM01. Meng et al. revealed that the yield of polysaccharides from cultured UM01 mycelia was 7.4% (w/w), and the sugar content of this polysaccharide was 60.7% (w/w) (Meng et al., 2014). The production of intracellular selenium polysaccharide in a submerged culture of another strain, namely SU-02, was estimated at 11.52% under optimized extraction conditions (Wang et al., 2011a). Zhu et al. found that production of exopolysaccharide reached 29.182 mg/100 mL under fermentation conditions of original pH 7.0, inoculum quantity 11%, rotary speed 160 r/min, and temperature of 24°C (Zhu et al., 2013). Leung et al. revealed that the production of exopolysaccharide in mycelial liquid culture of *Co. sinensis* fungus Cs-HK1 had a rate proportional to the mycelial growth, reaching a maximum of about 2.5 g/L at day 7, which was slightly higher than the mass produced by *H. sinensis*-HS002 (2.16 g/L) reported by He et al. (He et al., 2013), and the total sugar content of exopolysaccharide showed a gradual increase from day 4 (46%) to day 6 (70%) (Leung et al., 2009). Determined by the phenol-sulfate method, the total sugar content of the crude polysaccharide from cultured *Co. sinensis* mycelium powder was 91.3%, with a yield of 4.2% (Akaki et al., 2009). However, the crude polysaccharide was obtained in a yield of 2.6% of fresh weight of the mycelium in Wu et al.'s report (Yalin et al., 2006). Wang et al. harvested a highest yield of crude polysaccharide (7.0%) from cultured *Co. sinensis* (Wang et al., 2010a). The polysaccharide content of mycelium and culture broth was also different, as reported by Kuo et al. (2007), since the mycelia and culture filtrate contained 14.3 mg/g and 197.1 mg/g of polysaccharide, respectively. Hsieh et al. (2005) optimized the production of polysaccharides using response surface methodology, and the composition of optimized medium calculated from the regression model of RSM was: sucrose 6.17%, corn steep powder 0.53%, $(NH_4)_2HPO_4$ 0.5%, and KH_2PO_4 0.15% at pH 4.44, with a predicted polysaccharide production at a maximum of 3.17 g/L and an actual production of 3.05 and 3.21 g/L in a shake flask and a 5-L jar fermenter, respectively. Particularly, both the cell growth and polysaccharide production were inhibited when the pH was controlled at a higher level such as pH 5.0, and a low pH (2.85) was required for maximum secretion of the polysaccharide.

Yang and Zhang (2016) optimized the fermentation parameters in production of intracellular selenium-enriched polysaccharides (ISPS) using response surface methodology (RSM) with: sodium selenite concentration 33.78 mg/L, incubation time 8.24 days, and incubation temperature 26.69°C, with a maximum yield of 197.35 mg/g.

4.1.1.2 *Cordyceps militaris*

Cordyceps militaris is the most common kind of *Cordyceps* spp., and can be cultivated well either in stroma cultivation or solid and liquid culture. Wen et al. reported that the content of polysaccharide in *C. militaris* was 86.49 mg/g (Wen et al., 2005). Through alcohol precipitation, 2.312 g of crude polysaccharide was obtained from 100 g of *C. militaris* mycelium, whose polysaccharide content was 28.1% determined by the anthrone–sulfuric acid method (Zhang et al., 2010). But the polysaccharide yield was much lower than extracting by alkaline solution (6%), as reported by Yu et al. (2009). Huang et al. found that the sugar content in the fruiting bodies and mycelia of *C. militaris* was 260.64 mg/g and 389.47 mg/g, respectively, but the former contained a higher reducing sugar content (Huang et al., 2006). Jiang et al. (2010) revealed that the polysaccharide yield of silkworm *C. militaris* was 25.75% for the sporophore and 12.85% for the medium. Zhang (2003) also found that the polysaccharide content of silkworm *Cordyceps* was highest in the fruiting bodies (86.94 mg/g), followed by stroma and sclerotium, while in the artificial medium, it was only 0.14% of dry matter.

Optimized liquid culture conditions and medium composition are reported to have a significant influence on the enhanced polysaccharide productivity, not only for exopolysaccharide but also for intracellular polysaccharide, which is also a necessary step for industrial application. Lv (2000) optimized polysaccharide production in single-factor and multi-factor design experiments. Results showed that the best production (10.11 mg/g) in a medium contained: sucrose 3%, peptone 10%, $MgSO_4 \cdot 7H_2O$ 0.5%, K_2HPO_4 1.0%, fermented at pH 7.5, 25°C, and rotated at 125 rpm for 5 days. Through an orthogonal experimental study, an optimal condition of incubation at 22°C, pH 6.0, and incubation time of 5 days with a medium containing: cornmeal 2%, sucrose 5%, yeast extract 0.03%, NH_4NO_3 0.04%, KNO_3 0.08%, $MgSO_4 \cdot 7H_2O$ 0.02%, K_2HPO_4 0.04%, and $FeSO_4 \cdot 7H_2O$ 0.02% was obtained, and this optimal condition yielded 170.3 mg polysaccharide per 100 mL of broth (Chen et al., 2005a). Using a desirability function approach and response surface methodology, Yang et al. (2014) obtained an optimal medium composition with a maximum intracellular polysaccharide yield of 0.6 g/L: yeast extract 10.33 g/L, sucrose 27.24 g/L, K_2HPO_4 5.60 g/L, and optimal culture conditions were initial pH 6, temperature 25°C, rotation speed 150 r/min, inoculum size 4% (v/v), and medium capacity 40 mL/250 mL. Cui and Jia (2010) applied a statistical experimental design strategy (SES) to optimize the medium for extracting exopolysaccharides (EPS) from *C. militaris* by submerged culture in a shaker flask, and found glucose and peptone had a significant influence on the EPS production. They applied a quadratic model to fit the EPS production and the optimum conditions were 48.67 g/L glucose, 12.56 g/L peptone, 1 g/L K_2HPO_4, 10 g/L yeast extract, and 0.5 g/L $MgSO_4 \cdot 7H_2O$. Under these optimized culture conditions, the EPS production was increased from 0.78 to 1.96 g/L, and earned a 2.5-fold increase compared with that of the original culture conditions. Kwon et al. (2009) investigated the effects of medium composition, environmental factors, and C/N ratio on exopolysaccharide secretion. Among these parameters, glucose 80 g, yeast extract 10 g, $MgSO_4 \cdot 7H_2O$ 0.5 g, and K_2HPO_4 0.5 g in 1 L distilled water were found to be the most suitable carbon, nitrogen, and mineral sources, respectively. Specifically, under uncontrolled pH, agitation at 200 rpm, aeration at 1.5 vvm, and cultivation at 24°C, polysaccharide content in shake flask cultures and 5 L jar bioreactors were 2.53 g/L and 6.74 g/L, respectively. *Cordyceps militaris* SU5-08 was derived from an initial strain (*C. militaris* SU5) by ultraviolet mutagenesis of protoplasts and under optimized submerged culture, the extraction rate of EPS was 1919.16 ± 165.27 mg/L, which was $120.38 \pm 11.36\%$ higher than that of *C. militaris* SU5 (Lin et al., 2012). Li et al. (2007) found that the polysaccharide yield of *C. militaris* was 4.84 g/L under the optimal medium conditions that were composed of 6% sucrose, 0.2% KNO_3, 2% yeast extract,

0.02% $MgSO_4 \cdot 7H_2O$, 0.04% K_2HPO_4, and 0.002% $FeSO_4 \cdot 7H_2O$, and a fermentation condition that was 6% of incubation volume, initial pH 7.0, and a fermentation temperature of 24°C.

4.1.1.3 Cordyceps gunii

Cordyceps gunii, another member of this genus, is a large *Cordyceps* hosted on hepialidae larvae and first reported in 1983 by Liang (1983), it was also called *Cordyceps hawkesii* (Gray). Mei found that the mycelial biomass of *P. gunii* 5–19 (teleomorph is *C. gunii*) contained 9.03 mg/g crude polysaccharide (Mei, 2006). But in Zhu's report (Zhu et al., 2012a), the yield of crude polysaccharide fraction was 11.32% of the dried mycelia, possibly due to the differences in fermented strain and culture conditions. Moreover, cultivation of *C. gunii* for different growth periods indicated that the polysaccharide accumulation was slow in the mycelia, but increased in the logarithmic growth phase and reached a maximum of 28.2 mg/g. Compared to the polysaccharide content in stroma, parasited worm, whole teleomorph, and mycelia, it showed the highest in mycelia, followed by stroma (Li et al., 1995; Meng et al., 2001). Through colorimetric assay using sulfuric acid anthrone, Bai et al. (2001) found that the content of polysaccharides in *Co. sinensis* and *C. hawkesii* (Gray) Cooke was 3.5% and 0.7%, respectively.

4.1.1.4 Cordyceps sobolifera

Cordyceps sobolifera is also known as chanhua, chanyongcao, chanrong, etc., in China. In the present study, Wen et al. (2006) revealed that the content of polysaccharide in *C. sobolifera* was 94.88 mg/g, significantly higher than that in *C. militaris* (32.30 mg/g), but slightly lower than that in *Co. sinensis* (125.68 mg/g).

4.1.1.5 Cordyceps pruinosa

Cordyceps pruinosa is an entomogenous fungus notable for its various bioactivities. The effect of synthetic medium and culture conditions on polysaccharide production in a shake flask culture was investigated. To get the maximum polysaccharide production from *C. pruinosa*, Xiao et al. (2004c) employed media and cultural conditions as follows: potato starch 2% (w/v), sucrose 2.5%, soybean 0.5%, beef extract 0.5%, yeast extract 0.1%, KCl 0.02%, K_2HPO_4 0.1%, $MgSO_4 \cdot 7H_2O$ 0.05%, pH 7.0, inoculum size 5%, medium capacity 50 mL/250 mL flask, dispersant 15 beads, culture time 7 days. On the other hand, the sucrose, beef extract, and yeast extract were replaced with molasses of sucrose, groundnut, and Vitamin B complex, respectively, in the fermentation medium. The resulting yield of polysaccharide production was 9.51 g/L after 54 h of fermentation under these optimized culture conditions, approximately twice as much (Xiao et al., 2004c).

4.1.1.6 *Cordyceps ophioglossoides*

Cordyceps ophioglossoides is a well-known fungus of the genus *Cordyceps* in the family of *Clavicipitaceae*. Xu et al. (2012) first investigated how to optimize the fermentation conditions in order to obtain a high yield of intracellular polysaccharide from *C. ophioglossoides*. Results showed that the optimal medium consisted of glucose 54.50 g/L, yeast powder 25.50 g/L, NaH_2PO_4 0.4 g/L, and K_2HPO_4 0.4 g/L, and the suggested culture conditions were 24°C, and initial pH 4.5 with a rotary speed of 120 r/min for 168 h. The yield of intracellular polysaccharide was 737.93 mg/L, which is 50% higher than that under the initial conditions.

4.1.1.7 *Cordyceps taii*

Cordyceps taii, a folk medicine mushroom native to south China, is recognized as an unparalleled resource of healthy foods and drug discovery. Xiao et al. optimized the yield of EPS from *C. taii* by systematic analysis so as to refine the fermentation parameters. It was found that the combination of the following variations: xylose 31.27 g/L, soybean steep liquor 4.85 g/L, and $(NH_4)_2SO_4$ 0.15 g/L resulted in optimum EPS production with a yield of 43.87 ± 0.28 g/L, accurately predicted by the model, and the figure was over eight times higher than that obtained from the preliminary experiments (Xiao et al., 2010).

4.1.1.8 Other *Cordyceps*

Xiao et al. found that the polysaccharide yield from *Cordyceps cephalosporium* mycelia by the hot-water extraction method was 3.07%, slightly lower than the ultrasonic method (Xiao et al., 2012b). The polysaccharide yields of *Cordyceps guangdongensis* from fruiting bodies, rice medium, mycelium, and submerged fermented liquid was 6.92%, 5.40%, 6.15%, and 3.42%, respectively, as Yan et al. (2009c) reported. Xiao et al. (2004a) investigated the optimal submerged culture requirements for mycelial growth and EPS production by *C. jiangxiensis* JXPJ 0109. The optimal temperature, initial pH, and incubation period for EPS production were 28°C, 7.0, and 10 days, respectively. The effect of medium ingredients was as follows: yeast extract > maltose > tryptone > glycerol > KH_2PO_4 > $CaCl_2$ > $MgSO_4$, and the optimal medium was (in g/L of distilled water) maltose (food-grade) 20, glycerol 8, tryptone 5, yeast extract 10, KH_2PO_4 1, and $CaCl_2$ 0.5. Under these optimal conditions, the maximum EPS and biomass production was 3.5 and 14.5 g/L after 10 and 8 days of fermentation, respectively. Sharma et al. (2015) optimized the mycelial culture to extract the polysaccharide, both the exopolysaccharide and intracellular polysaccharide, effectively under different factors and found the effect on EPS and IPS production in order as: temperature > incubation time > pH > rotary speed > medium capacity.

Particularly, research was carried out to investigate the effect of microelements on the biomass and content of polysaccharide in the fruiting body of

Cordyceps. Yang et al. (2006) studied the influence of lanthanum nitrate (La $(NO_3)_3$) concentrations on *Cordyceps brasiliensis* mycelium growth and polysaccharide content. Results showed that the polysaccharide content was highest, with a figure of 24.21 mg/g, 1.59 times higher than the control group, when 0.3 g/L was added. However, the polysaccharide content was then gradually decreased with an increase of La $(NO_3)_3$ concentration. On the other hand, Yu and Qian (2006) studied the influence of selenium on the polysaccharide content in *C. militaris*, and they found that supplementation of Se led to a significant increase in polysaccharide content, with a figure of 51.03 mg/g, almost twice that of the control sample without Se (26.71 mg/g).

4.1.2 Isolation and Purification of Polysaccharide in Cordyceps

Polysaccharides are mainly present in the cell wall of *Cordyceps*, and occur within the cytoplasm as a structural component. There are mainly two types of polysaccharide reported in the *Cordyceps*, based on their location in the fungal cells, i.e., intracellular polysaccharides (IPSs) and extracellular polysaccharides (EPSs). Intracellular polysaccharides mainly occur in the polysaccharide extracted from the fruiting bodies (or worm) and mycelium of *Cordyceps*, while EPS was commonly obtained from the culture broth of the fermentation liquid. Generally, IPS was extracted by different solvents, such as hot water, acid/alkali aqueous solution, aqueous buffers, etc., enabling isolation and purification of the polysaccharide according to their different solubility in water and organic solvents, or based on their different ionic properties and molecular weight distributions. According to previous research, extraction with hot or boiling water is the most common and convenient way to isolate water-soluble polysaccharide from *Cordyceps*. However, the main disadvantages for the water extraction method are low extraction efficiency and high energy consumption. As a result, other technologies, such as microwave, ultrasonic, or enzyme-assisted extraction methods, are often adopted to improve the extraction efficiency and enhance the polysaccharide productivity. On the other hand, for extraction of extracellular polysaccharide, the fermentation broth of *Cordyceps* was centrifuged and concentrated (Huang et al., 2013; Lee et al., 2010a; Yang et al., 2007). The resultant polysaccharide, collected either from water-extract of fruit bodies and mycelium or fermentation broth, was precipitated by organic solvent, usually using different concentrations of ethanol, and then centrifuged to harvest the crude polysaccharides.

4.1.2.1 *Hot-Water Extraction (HWE)*

Dong et al. (2009) employed a three-level Box–Behnken design, along with canonical and ridge analysis, to optimize the hot-water extraction parameters for polysaccharide from cultured mycelium of *Co. sinensis*. Results showed that the optimal conditions were extracting at 88.9°C three times, each for 110 min, and the corresponding polysaccharide production reach a maximum of 16.10%. The hot-water assisted extraction of polysaccharides from *C. militaris* was studied, and the effects of extraction time, solid–liquid ratio, and temperature on polysaccharide yield were investigated. An optimized condition was achieved through the orthogonal test: temperature 55°C, extraction time 4 h, solid–liquid ratio 1:40 (g/mL), and the resulting yield of the validated experiment was 34.14% (Gao et al., 2010c). The Soxhelt extraction method was also optimized by Gao et al., and the optimal conditions included an extraction time of 6 h and precipitation with 50% ethanol. Under this condition, the corresponding yield of polysaccharide from *C. militaris* was 31.95% (Gao et al., 2010a). In addition, Gao et al. (2011) investigated the reflux extraction process of polysaccharide from solid-state fermentation of *C. militaris*. Results showed that the optimal conditions were extraction temperature 80°C, extraction time 2 h, ratio of raw material to water 1:20, repeated twice, reached a maximum yield of 6.93%. Chen et al. optimized three extraction processes (decocting extraction, hydrothermal refluxing, and alkali extraction) of polysaccharide from the fruiting body of cultured *C. militaris*. Results showed that the optimum conditions of the decocting extraction method was a solid–liquid ratio of 1:40, extraction time of 3.0 h, and extraction 3 times, and the effects order of the three factors on the yield was solid–liquid ratio > decocting time > decocting times. The optimum conditions for the hydrothermal refluxing method were sample–water ratio of 1:20 and reflux twice at 80°C, each for 1.0 h, and the effects order was: number of extraction times > extraction time > extraction temperature > solid–liquid ratio. The optimum conditions of alkali extraction were a solid–liquid ratio of 1:8, NaOH concentration of 0.7 mol/L, extraction three times, and an extraction time of 3.0 h, and the effects order was number of extraction times > NaOH concentration > solid–liquid ratio > extraction time (Chen et al., 2010b).

Yu et al. determined the optimal condition to extract exopolysaccharide from *Co. sinensis* fermentation broth (fungus CS2) as: precipitation time 12 h, ethanol:fermentation broth 3.8:1, and pH 7.65. Under these conditions, 11.35 g/L of polysaccharide could be obtained from the fermentation broth (Yu et al., 2002b). Yang et al. (2007) could harvest 2.7734 g/L of polysaccharide from the fermentation broth of *C. brasiliensis* under the optimal extraction conditions of precipitating with 5 volumes of 95% ethanol for 10 h at pH 7.0.

4.1.2.2 Ultrasonic-Assisted Extraction (UAE)

Central composite design (CCD) was applied to investigate three critical extraction variables: temperature, duration time, and pH in the previous study. Results showed that a maximum extraction rate of 11.52% was obtained under the following conditions: ultrasonic treatment for 15.11 min, extraction temperature at 59.26°C, and pH 7.17 (Wang et al., 2011a).

Gao et al. (2010b) employed an orthogonal test to get the optimal condition of ultrasonic power 250 W, solid–liquid ratio 1:35 (g/mL), extraction time 13 min, and precipitation with 60% ethanol, and the resulting yield was 25.89%. Zhang and Sun (2013) optimized ultrasonic extraction conditions using a single factor experiment and an orthogonal test on the basis of polysaccharide yield. Results showed that the optimal conditions were ultrasonic power 93%, ultrasonic processing time of 40 min, ultrasonic temperature 50°C, and solid–liquid ratio 1:40, and the corresponding yield was 15.48%. The optimal ultrasonic extraction conditions for *C. gunii* were ultrasonic treatment for 214.4 s at a material–water ratio of 46.9 with a power of 413.7 W, and the predicted highest extraction yield was 2.2263% (Li, 2011). Liang and Shang (2010) found that the maximum yield of crude polysaccharide was 3.30% under conditions of solid–liquid ratio 1:30 (m/m), ethanol 15 mL, ultrasonic power 70 W, and ultrasonic time 90 min. Fan et al. (2011) investigated the process parameters such as ultrasonic treatment number, ultrasonic power, ethanol multiple, and extraction temperature for intracellular polysaccharide from *C. militaris* SU-08 extraction. Through Plackett–Burman (PB) tests and response surface methodology (RSM), four parameters that affect the extraction rate were determined. The optimum conditions were predicted to be: ultrasonic treatment number 61.45, ultrasonic power 543.64 W, extraction temperature 82.61°C, and ethanol multiple 3.28, and the extraction rate was estimated at 9.11%, with an actual value of 9.19. Qin and Li (2011) indicated that the polysaccharide yield could reach 5.57% under the extraction conditions of ultrasonic power 300 W, solid–liquid ratio 1:55, and extracting for 30 min at 35°C. Yan et al. (2009a) found that the parameters influencing extraction rate are in the following order: solid–liquid ratio > ultrasonic power > extraction temperature > extraction duration, with an optimal condition of 1:45, 140 W, 80°C, and 80 min, respectively, giving an optimal extraction rate of 3.98%. However, in a report by Zhang and Zhu (2012), the parameter order was extraction duration > extraction temperature > ultrasonic power, and the optimal condition was extraction for 15 min at 80°C under an ultrasonic power of 200 W, with a polysaccharide content of 10.61 mg/g.

4.1.2.3 Microwave-Assisted Extraction (MAE)

Microwave-assisted extraction was used to extract polysaccharides from cultured *C. militaris*. A Box–Behnken design and a quadratic polynomial model were applied to determine the effects of three independent parameters (microwave power, extraction time, and ratio of solution to solid) on the yield of polysaccharide from *C. militaris*. Results showed that the optimal conditions were as follows: microwave power 744.8 W, extraction time

4.2 min, and ratio of solution to solid 31.1 mL/g (Song et al., 2009). Zhiwei et al. (2010) found the extraction rate of polysaccharide was 3.78% under the conditions of microwave 700 W, extraction duration 2.0 min, and solid–liquid ratio1:6, slightly higher than the water extraction process. Shi et al. (2006) obtain optimal microwave-assisted extraction conditions of microwave power 80%, solid–liquid ratio1:20, and extraction for 20 min, the corresponding yield of polysaccharide was 10.97%.

4.1.2.4 Other Extraction Technologies

The application of high-power or high-intensity ultrasound or microwaves has been widely employed to extract polysaccharide from *Cordyceps*. Enzyme hydrolysis and high pressure are also used frequently in order to obtain a high yield of polysaccharide.

An and Zhu (2012) studied the enzyme extraction process of polysaccharide from a cultured medium of *C. militaris* using the response surface method (RSM). The optimum extraction conditions were as follows: extraction temperature of 39.89°C, solid–liquid ratio of 1:75.78, enzyme content of 2.39%, hydrolysis time of 4 h, and pH 3.12. Under these conditions, the extraction rate of polysaccharides was 9.96%. Chen et al. (2014b) combined ultrasonic wave treatment and enzyme hydrolysis to extract polysaccharide from *C. militaris*, and showed the optimal conditions were: the amount of cellulase 1.9%, enzymolysis time 1.1 h, ultrasonic wave temperature 50°C, and ultrasonic wave time 41 min. Under these conditions, the yield of *C. militaris* polysaccharides was up to 25.45%. Gao et al. (2009) employed ultra-high pressure (UHP) technology to extract polysaccharide from *C. militaris*. The optimal UHP extraction condition was extraction for 5 min using 45% as the extraction solvent at 400 MPa, with the material–liquid ratio of 1:35, and the resulting yield was 10.13%. Sun et al. (2012), using an ultrasonic-assisted aqueous two phase system to extract cultured *C. militaris*, found the highest extraction rate reaching 30.1%.

Zhao compared three extraction methods (water extraction, ultrasonic, and microwave) on the yield of polysaccharide from *C. militaris*. Results showed that ultrasonic method yield the highest amounts of polysaccharide, and the extraction rate of polysaccharide was 7.965 mg/mL under the conditions of ultrasonic wave frequency 40 kHz and extraction twice when the ultrasonic time was 30 min at 80°C, with ultrasonic power at 500 W (Zhao, 2008). Yi-Hong (2009) compared two extraction procedures, microwave and the hot-water method, on polysaccharide yield. The optimum microwave extraction procedure was microwave radiant intensity of 550 W, solid-to-solution ratio 1:30, and extraction time of 30 s once, while the optimum water extraction condition was 60°C for 3 h, and solid-to-solution ratio of 1:30 twice, and the yields were 3.67% and 3.35%, respectively. Weng et al. (2008) also compared boiling water, microwave, and ultrasonic extraction methods, which resulted in a yield of 3.50%, 2.23%, and 2.03%, respectively. However, Yin et al. (2014) found that the microwave-assisted method had the highest extraction yield compared to hot-water, ultrasonic-assisted, and Soxhelt extraction methods. All three pretreatment methods could increase the extraction rate, in an order of ultrasonic wave > microwave > water extraction. The extraction rate of polysaccharides was 7.965 mg/mL under the conditions of ultrasonic wave frequency 40 kHz, and extraction twice, each for 30 min, with an ultrasonic power of 500 W at 80°C (Zhao, 2008).

4.1.2.5 Purification Methods

After extracting the crude polysaccharide from *Cordyceps*, it was partially purified by deproteination, decoloration or dialysis against a certain molecular weight membrane to remove protein, pigment, and small molecular substances. Subsequently, *Cordyceps* polysaccharide was further purified by fractional precipitation with ethanol, acidic precipitation with acetic acid, absorption resin, ultra-filtration, dialysis, ion-exchange chromatography, gel filtration, and affinity chromatography methods, etc., (Izydorczyk, 2005). Among these, gradient ethanol precipitation, ion exchange chromatography, and size exclusion chromatography are frequently adopted in order to achieve a homogenous polysaccharide. Miyazaki et al. (1977) first isolated and purified a water-soluble polysaccharide by ethanol fractionation and gel filtration chromatography (1.5 × 90 cm) using phosphate buffer at the flow rate of 6 mL/h.

To isolate the protein-containing polysaccharide (CT-4N) the crude drug was pulverized, and extracted with acetone, hot methanol, and hot 70% aqueous ethanol successively, followed by exhaustive extraction with hot water. The residue was then soaked in 5% $NaCO_3$ for 12 h at room temperature. The alkaline suspension was filtered, neutralized with acetic acid, dialyzed against distilled water, and precipitated with three volumes of ethanol. It was then deproteinized using the Sevag procedure and purified by DEAE Sephadex A-25 column chromatography. The neutral fraction was lyophilized to give the crude polysaccharides in a 0.1% yield. These were separated using a column of Sephacryl S-300 to yield the minor CT-4N as colorless flakes in a 0.08% yield. Its molecular weight was estimated by gel filtration to be ~23,000 Da (Kiho et al., 1986).

Yu et al. (2004b) applied a couple of methods to isolate different homogenous polysaccharides from cultured *C. militaris*. Strains CSP50 and CSP70 were obtained by precipitating successively with a final ethanol concentration of 50% and 70%, respectively. The CSP70 strain was further subjected to a gel filtration column Sephadex G-100 (1.5 × 100 cm) to concentrate to a fraction CSP70-1, and it was further separated on a Sephadex G-200 column (1.1 × 60 cm) to result in two fractions, of which the first fraction was called CSP-1 (Yu et al., 2004b). The CSP50 was subjected to a Sephadex G-100 (1.5 × 100 cm) column as well, giving four main fractions named as CSP-2,

CSP-3, CSP-4, and CSP-5 (Yu et al., 2004c). On the other hand, CSP50 and CSP70 were injected into a DEAE-52 cellulose column first to obtain the respective polysaccharide. Then a gel filtration column of Sephacryl S-100 HR was applied to obtain three polysaccharides, namely P70-1, P70-2, and P50-1 (Yu et al., 2007). In recent years, the polysaccharide was first purified using XAD-7 macroporous adsorption resin (2.6 × 95 cm) with distilled water as the mobile phase at 2 mL/min. The fractions that contained the major peak detected by the phenol–sulfuric acid method, were collected, concentrated, and named as CM-1, and it was then fractionated on a DEAE-52 cellulose column (2.6 × 40 cm) and Sephadex G-25 (1.6 × 70 cm) subsequently, terminated in a polysaccharide CMP-1 (Jing et al., 2014).

Wu et al. combined the use of buffers and column chromatography and obtained a series of purified polysaccharides from the mycelium of *Co. sinensis*. Dried crushed mycelia were extracted successively with ethanol at 95% and 85%, to defat and decolorize, and then extracted with aqueous 75% ethanol for 12 h. After centrifugation (6700 rpm, 30 min), the residue was dried in air and then treated several times with 0.05 M phosphate buffer (pH 7.0) for 10 h at 80°C. The aqueous extracts were combined, the solvents removed under reduced pressure, and the residue dialyzed. The remainder was diluted with 95% ethanol, and the resulting precipitate was collected, washed three times with acetone, and dried. The yield of this crude polysaccharide fraction was 2.6% of the original weight, and was eluted from a column (2.6 × 60 cm) of DEAE-cellulose 52 with 0.02 M potassium phosphate buffer (400 mL, pH 7.0), followed by 0.10 M NaOH (400 mL). The molecular weight of the purified polysaccharide was estimated by gel filtration with a Sepharose CL-4B column to be 1.84×10^5 Da (Wu et al., 2006a,b). The mycelium residue was extracted with 0.05 M acetate buffer (pH ~ 6.0) for 5 h at 85°C, and the product deproteinated five times by the neutroenzyme–Sevag method to yield the crude polysaccharides. This was eluted from a column (2.6 × 40 cm) of DEAE Sepharose Fast Flow with 0.02 M sodium acetate–acetic acid buffer (pH 4.8, 400 mL), followed by distilled water (300 mL) to yield two fractions (I and II). Compound I was further purified using a column (1.6 × 60 cm) of Sephadex G-100 eluted with 0.1–1 M sodium chloride. Its molecular weight was estimated to be $\sim 7.7 \times 10^3$ Da using high-performance gel permeation chromatography (HPGPC) with two columns in series (Ultrahydrogel 250 and Ultrahydrogel 2000, Waters) (Wu et al., 2007).

The CSP1-2 fraction of *Co. sinensis* polysaccharide (CSP) was purified using DEAE-cellulose 52 and Sepharose CL-6B chromatography. The homogeneous polysaccharide CSP1-2 was mainly composed of mannose, glucose, and galactose in a molar ratio of about 2:2:1, with an average molecular weight of approximately 2.70×10^4 Da (Xiang et al., 2016).

Wu et al. (2012) isolated two acidic polysaccharide fractions (CM-jd-CPS2 and CM-jd(Y)-CPS2) from the fruiting bodies of cultured *C. militaris*

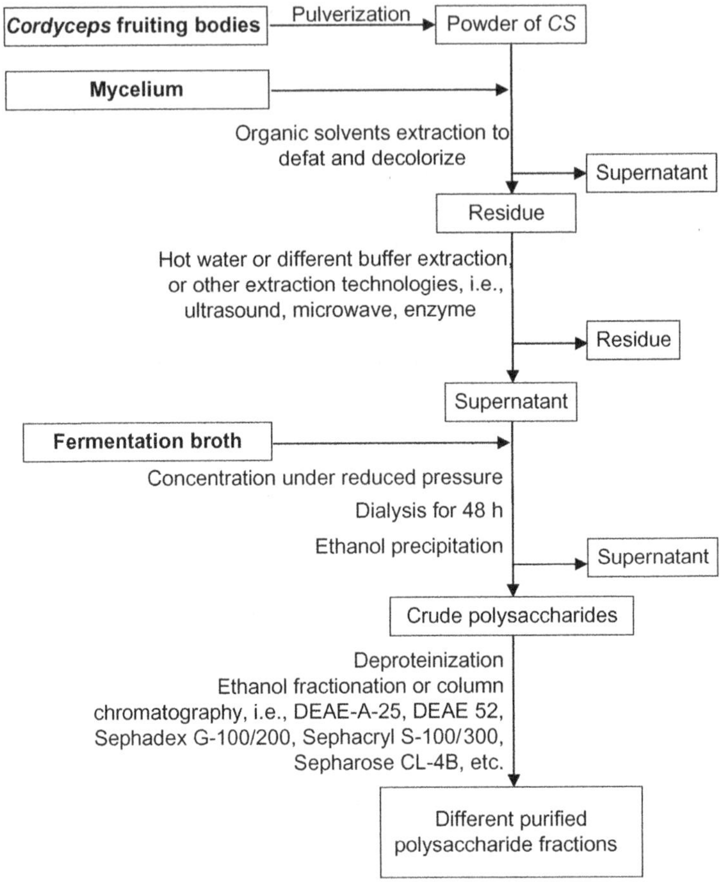

FIGURE 4.2 Flow chart for isolating and purifying polysaccharides from *Cordyceps*.

grown on solid rice medium and silkworm pupa, respectively, through water extraction, ethanol precipitation, and DEAE-cellulose-52 column followed by Sephadex G-100 column chromatography.

Other researchers used similar methods to obtain purified polysaccharides from fruiting bodies or mycelium of *Cordyceps* species, or the fermentation broth of *Cordyceps*. Fig. 4.2 summarizes the isolation and purification procedures used.

4.2 STRUCTURAL FEATURES

In order to characterize the primary structure of a polysaccharide, it is essential to figure out the following information: monosaccharide composition, molecular weight (Mw), configuration of glycosidic linkages, glycosidic linkage patterns, and sequence of monosaccharide and repeating units, as well as substitutions, especially some functional groups which will lead to a change of bioactivity compared with the native polysaccharide, and usually an enhancement was observed. The methods of characterizing polysaccharides have been comprehensively described (Athukorala et al., 2006; Cui, 2005; Cui et al., 1995; Dague et al., 2008; Dror et al., 2005; Harris et al., 1984; Mischnick et al., 1995; Needs and Selvendran, 1993; Nie et al., 2005; Poveda et al., 1998; Smiderle et al., 2006; Tanabe et al., 2000; Wang and Cui, 2005a,b; Yang and Zhang, 2009; Zhang et al., 2007). Although many polysaccharides, including IPS and EPS, either from natural and cultivated *Cordyceps* or fermented *Cordyceps* with different strains, have been obtained, most were mainly composed of glucose, mannose, and galactose in various mole ratios.

4.2.1 *Cordyceps sinensis*

The structures of polysaccharides from natural *Co. sinensis* were first studied by Miyazaki et al. (1977). A water-soluble polysaccharide (CS-I) was isolated and purified by ethanol fractionation and gel filtration. D-Galactose and D-mannose in a molar ratio of 1:1 were obtained by acid hydrolysis of CS-I. Periodate oxidation, Smith degradation, methylation analysis, partial acid hydrolysis, and 13C-NMR spectrometry showed that CS-I was a highly branched galactomannan. The mannan main chain contained mainly (1→2)-α-linked-D-mannopyranosyl residues, and the branches contained (1→3), (1→5), (1→6)-linked-D-galactofuranosyl, and (1→4)-linked-D-galactopyranosyl residues. The non-reducing ends were D-galactofuranose and D-mannopyranose, and the branching points were D-mannopyranosyl residues. Kiho et al (1986), thereafter, isolated a water-soluble, protein-containing galactomannan (CT-4N) from a 5% sodium carbonate extract of *Co. sinensis*. Its molecular weight by gel filtration was estimated to be ~23,000 Da. It was mainly composed of D-mannose and D-galactose in the molar ratio of 3:5, and contained a small proportion of protein. From the results of methylation analysis, Smith degradation, stepwise hydrolysis, and 13C-NMR spectroscopy, CT-4N was found to have a highly branched structure composed of (1→6)- and (1→2)-linked α-D-mannopyranosyl residues in the main chain, Some of the residues were present as branching points with (1→2,6) and (1→4,6) linkages. At these branches are mainly short chains, with a large proportion of (1→5)-linked β-D-galactofuranosyl residues and a small proportion of (1→6)-linked α-D-galactopyranosyl residues. The terminal groups are mainly β-D-galactofuranosyl, but also with a small proportion of α-D-mannopyranosyl groups. Additionally, Chen et al. (1997) reported a polysaccharide fraction from natural *Co. sinensis* (PSCS) with an apparent molecular weight of about 100 kDa by using gel filtration.

However, wild or natural *Co. sinensis* has become increasingly scarce in recent years, and it is for this reason that mycelial fermentation of *Cordyceps* fungal species is used as substitute.

There have been many studies reported in the literature about the structures of the polysaccharides from the mycelial fermentation of *Cordyceps* fungal species. Using antioxidation activity-guided fractionation, a polysaccharide (CSP-1) of molecular weight ~210 kDa was obtained by ion-exchange and size chromatography. The isolated product had strong antioxidant activity, and contained glucose, mannose, and galactose in a ratio of 1:0.6:0.75 (Li et al., 2003, 2006c). A neutral polysaccharide (CS-F30) was isolated containing galactose, glucose, and mannose (molar ratio, 62:28:10), with a molecular weight about 45,000 Da (Kiho et al., 1993, 2000). Based on activity-guided fractionation, a novel exopolysaccharide of molecular weight of ~82 kDa, namely cordysinocan, was obtained from cultured *Cordyceps* UST2000. The isolated polysaccharide contained glucose, mannose, and galactose in a ratio of 2.4:2:1 (Cheung et al., 2009). A water polysaccharide (CY) isolated from cultured *Cordyceps* with a molecular weight of 9.3×10^3 g/mol, was mainly composed of xylose, mannose, glucose, and galactose, and the backbone was connected with β-glucosidic bonds (Fu et al., 2014). A polysaccharide (CS-PS) was obtained from cultured *Co. sinensis* with an average molecular weight of 12 kDa determined by gel filtration. The polysaccharide was composed of mannose, rhamnose, arabinose, xylose, glucose, and galactose, and their corresponding mole percentages were 38.3 7%, 2.51%, 2.21%, 5.22%, 27.44%, and 24.25%, respectively (Zhang et al., 2011). A new galactoglucomannan (CS-F10) was purified from a hot water extract which was made up of galactose, glucose, and mannose in a molar ratio of 43:33:24 with a molecular weight about 15,000 Da. Chemical and spectroscopic investigations indicated that CS-F10 had a comb-type structure with α-D-glucopyranosyl residues at the terminal end of the side-chains. Nuclear magnetic resonance (NMR) spectroscopy indicated the presence of non-reducing terminal α-D-glucopyranosyl residues, (1→5 and/or 6)-linked β-D-galactofuranosyl residues, and (1→2) linked and branched α-D-mannopyranosyl residues (Kiho et al., 1999). HS002-II, a novel protein-bound polysaccharide with a molecular weight of 44 kDa, was isolated from submerged cultures of *H. sinensis* Liu, Guo, Yu, and Zeng by DEAE-Sepharose and Sephacryl S 200 column chromatography. Results showed that HS002-II consisted of 57.9% polysaccharide and 42.1% protein, with the existence of an N-type carbohydrate–protein linkage. The structure of HS002-II was mainly composed of a long backbone of (1→3)-linked α-D-ribofuranosyl residue, (1→4)-linked α-D-xylopyranosyl residue, and (1→4)-linked β-D-glucopyranosyl residue, substituted at C-6. Moreover, HS002-II had two branches of (1→6)-linked β-D-mannopyranosyl residue and (1→6)-linked β-D-galactopyranosyl residue, both of which were terminated with an α-L-arabinopyranosyl residue (He et al., 2013).

A D-glucan (SCP-I) with the molecular weight of about 1.84×10^5 Da contained an α-(1→4)-linked backbone and a branched α-(1→6)-linkage. This D-glucan gave a faint blue color with iodine (λ_{max} 564 nm), indicating the presence of α-(1→4)-linkages with short, exterior chains (Yalin et al., 2006). Another extracellular polysaccharide from the mycelia of *Co. sinensis* with an average molecular weight of 13,620 Da had a D-glucan backbone composed of (1→3)-β-D-glucosyl residues, and carried a single (1→4)-β-linked D-glucosyl residue. Spectra from NMR and IR showed that the sugar residues were β-glycosidically linked (Zhang et al., 2011).

A cell wall polysaccharide (cordyglucans) was obtained from fresh samples of the mycelia of *Co. sinensis* with an average molecular weight of 12,860 Da which consisted of a (1→3)-linked backbone with (1→6)-linked branches (Yalin et al., 2005). The structure was found for another purified polysaccharide (PS) from the mycelia of *Co. sinensis* using Sephadex G-100 (Pharmacia) column chromatography eluted with 0.2 M NaCl (500 mL). The polysaccharide was composed of D-Glc, D-Man, L-Ara, and D-Gal in a molar ratio of 8:90:1:1. The average molecular weight of PS was ~8.3 × 104 Da (Wu et al., 2006a,b).

A neutral mannoglucan with a molar mass of 7.7×10^3 Da was obtained from the 0.05 M acetate buffer extract of *Co. sinensis* mycelium purified by ion-exchange and gel-filtration chromatography which contained Man and Glc in the molar ratio of 1:9. This mannoglucan was reported to have a α-D-glucan backbone with (1→4)- and (1→3)-linkages, and side chains of α-D-(1→6)-Man*p* attached via O-6 of α-(1→3)-Glc*p* residues (Wu et al., 2007).

A polysaccharide (CHWp) was purified from the cultural mycelium of *Co. sinensis* by column chromatography on DEAE-TOYOPEARL 650 M after extraction with hot water and precipitation with ethanol. The CHWp consisted of D-Man, D-Gal, and D-Glc in a molar ratio of 1.0:2.7:1.8 and the molecular weight was $\sim 3.2 \times 10^4$ Da (Ji et al., 1993).

A polysaccharide (CS-81002) with a molecular weight of 43,000 Da was isolated from a fermentation medium of *Co. sinensis* (Berk) Sac having Man, Gal, and Glc in a molar ratio of 10.3:3.6:1. Methylation and partial acid hydrolysis indicated that CS-81002 was a heteropolysaccharide with branches. The backbone was composed of →6-)-Man-(1→. Six out of every ten Man residues of the backbone were substituted at the C3 position to form →3, 6-)-Man-(1→ side chains. Four out of every ten Man residues of backbone were substituted at the C2 position to form →2, 6-)-Man-(1→ side chains. Some other Man residues in the backbone were not substituted. Side chains contained →3-)-Gal-(1→, →1-)-Glc-(1→, 4-)-Man-(1→ and →2-)-Man-(1→. All three kinds of monosaccharides were present at the non-reducing end (Gong et al., 1990).

Four homogeneous polysaccharides were isolated from the mycelium of *Co. sinensis* by ethanol precipitation and separation using column chromatography with different eluants. The PC I consisted of mannose, galactose, and glucose in a molar ratio of 1:0.65:0.30, with a molecular weight of ~3.5×10^5 Da; PCA I consisted of mannose and galactose in a molar ratio of 1:1, with a molecular weight of ~5.56×10^5 Da; PCB I consisted of mannose and galactose in a molar ratio of 1:0.73, with a molecular weight of ~6×104 Da; PCB II consisted of mannose, galactose, and glucose in a molar ratio of 1:0.51:0.50, with a molecular weight of ~6×10^4 Da. IR, and NMR spectra suggested that the backbone of PCA I was composed of 1→4-linked Man*p*, and Gal*f* with Man*p* on the side chains linked to the backbone by 1→2, 1→3, and 1→6 linkages. The backbone of PCB I was also composed of 1→4-linked Man*p*, and Gal*f* with Man*p* on the side chains (Yuan et al., 2005).

Two other polysaccharides (CPS1 and CPS2) from the mycelium of *Co. sinensis* having molecular weights of 2.56×10^4 Da and 9.91×10^4 Da, respectively, were made up of Glu:Man:Gal:Ara = 46:36:18:1 and Glu:Man:Gal:GlaUA:Xyl:Ara:Rha = 30:25:14:4:3:3:1, respectively. Both had α-linkages with the backbone of CPS1 composed of 1,6-Man with glucose at the C2, C3, and C4 positions. The backbone of CPS2 was 1,6-Glu, 1,6-Man, and 1,6-Gal, with branches with different monosaccharide residues at the C3 positions (Fan et al., 2008).

An insoluble glucan-like polysaccharide (CS-Pp) was obtained from the mycelia of *Co. sinensis* with a monosaccharide composition ratio of Glc:Man:Gal = 21:2:1. The structure of CS-Pp is a 1,3-β-D-glucan with some 1,6-branched chain. Particle size analysis indicated that 80% of the particles were less than 5 μm, and the mean diameter was 1.5 μm (Akaki et al., 2009).

A water-soluble acid polysaccharide fraction (APSF) was purified by a DEAE-32 cellulose column eluting with a gradient NaCl solution (0–2 M) with phenol-sulfuric acid test monitoring. The APSF was proved to consist of mannose, glucose, and galactose in a ratio of 3.5:1:1.5 (Chen et al., 2010a,b).

An exopolysaccharide (EPS), obtained from cultivated *Co. sinensis*, had a β-D-glucan backbone and was purified by successively eluting a DEAE-32 column with water and a NaCl gradient of 0–2 M, and then was further filtered over a Sephadex G-200 column eluting with 0.05 M NaCl according to phenol-sulfuric acid monitoring. The EPS was reported to consist of mannose, glucose, and galactose in a ratio of 23:1:2.6, and its molecular weight was about 1.04×10^5 (Sheng et al., 2011). Another exopolysaccharide produced in a mycelial liquid culture of *Co. sinensis* fungus Cs-HK1, a *Tolypocladium* spp. fungus isolated from wild *Co. sinensis*, contained 65%–70% sugar and about 25% protein, suggesting that the polysaccharide was composed of polysaccharide–protein complexes. The GPC and gel electrophoresis results indicated that it was composed of biopolymers in a wide MW range from about 5 kDa to more than 200 kDa, and GC results suggested that the EPS molecules were homogenous β-glucans (Leung et al., 2009). A novel polysaccharide (EPS-1A) with an average molecular weight about 40 kDa was purified from the crude exopolysaccharide of a fermentation broth of cultured *Co. sinensis*. The EPS-1A was shown to be a slightly branched polysaccharide composed of glucose, mannose, and galactose in a molar ratio of 15.2:3.6:1.0. Its backbone was made up of (1→6)-α-D-glucose (~77%) and (1→6)-α-D-mannose residues (~23%), with branches at the O-3 position of (1→6)-α-D-mannose residues and terminated with D-galactose residues (Yan et al., 2010). In their recent study, EPS was purified by gradient ethanol precipitation, and the fraction obtained by precipitation with a 2/5 volume ratio of ethanol was characterized by FT-IR, GC, GC–MS, 1D and 2D NMR, together with periodate oxidation and Smith degradation, and partial acid hydrolysis, as well as methylation analysis. These data confirmed the structure of EPS with a linear backbone of (1→3)-β-D-glucopyranosyl residues and a single (1→6)-β-D-glucopyranosyl side-branching unit for every three β-D-glucopyranosyl residues. It showed a comb-like β-D-glucan with short and intensive branches, resulting in a high viscosity (Hu et al., 2016).

However, the exopolysaccharides produced by the Cs-HK1 mycelial fermentation had a high molecular weight with poor solubility and high viscosity in water. As a result, Chen et al. partially degraded the exopolysaccharide fraction by high-intensity ultra-sound (US) into a lower MW fraction EPS1U in order to improve the solutions properties and functional activities. It was worthy to note that EPSIU had good water solubility, with an average Mw of 730 kDa by light scattering analysis. It was composed of glucose (>95%) with minor or negligible mannose and galactose. Based on methylation analysis and NMR spectrometry, the structure of EPS1U was deduced as a (1→3)-β-D-glucan with glucose side chains attached to the O-6 position at the branching points (Chen et al., 2014a).

An acidic polysaccharide (AEPS-1) obtained from *Co. sinensis* Cs-HK1 in mycelial culture was mainly composed of glucopyranose (Glc*p*) and pyrano-glucuronic acid (GlcU*p*) in a molar ratio of 8:1, it also had a trace amount of mannose, with average molecular weight of about 36 kDa. The AEPS-1 had a linear main chain of (1→3)-linked α-D-Glc*p* and two branches, α-D-Glc*p* and α-D-GlcU*p*, were attached to the backbone through (1→6) glycosidic bonds at every seventh unit (Wang et al., 2011c).

A heteropolysaccharide (PS-A) isolated from water extracts of *Co. sinensis* by activity-guided fractionation had a molecular weight of 4.6×10^5 Da, and D-glucose, D-galactose, and D-mannose in a molar ratio of 2:1:1. The repeating backbone unit was →3-α-D-Glcp-1→3-β-D-Glcp-1→3-β-D-Galp-1→ with the branched residues (α-D-Manp-1→) linked at the O-2 position of residue 3-α-D-Glcp-1 (Kim, 2010).

Two novel polysaccharides (WIPS and AIPS) were isolated and fractionated from a mycelial biomass of *Co. sinensis* (strain Cs-HK1) by hot water and alkaline extraction, respectively. Both of the polysaccharides were characterized as α-D-glucans with a backbone of (1→4)-linked α-D-Glc*p* (>60%), and a very similar molecular weight (WIPS 1180 kDa; AIPS 1150 kDa). The WIPS had a short branch of (1→6)-linked α-D-Glc*p* (~14%), whereas the AIPS was a linear glucan, and the structure of branching, on the other hand, was rather distinctive from the reported glucans from other medicinal fungi. In alkaline solutions, both AIPS and WIPS exhibited as a random coil with similar conformational parameters, but a significantly different polydispersity index, 0.19 and 0.37, respectively (Yan et al., 2011).

The structure of a bioactive hydrophilic polysaccharide fraction obtained from *Co. sinensis* (CBHP) was studied. Monosaccharide composition analysis showed that it consisted mainly of glucose (95.19%), with trace amounts of mannose (0.91%) and galactose (0.61%). An α-1,4 linked Glcp is the main linkage type (65.7%), followed by *t*-Glcp (20.7%), 1,2,3,6-Glcp (4.1%), 1,2,4,6-Glcp (3.0%), 1,3,6-Glcp (2.0%), 1,4,6-Glcp (1.6%), 1,2-Manp (1.9%), and 1,3-Galp (1.0%). Based on NMR analysis, a preliminary structure is proposed: The backbone is composed of Glcp joined by 1→4 linkages and 1→3 linkages, with the branching points located at O-2 or O-6 of Glcp with α-terminal-D-Glcp as the side chain. The trace amounts of 1,2-Manp and 1,3-Galp linkages are probably located randomly in the side chains (Nie et al., 2011).

From structural information relating to different polysaccharides from various natural and cultured *Co. sinensis* which have been extensively studied over the past 30 years, it is evident that the polysaccharides from *Co. sinensis* are made up of glucose, mannose, and galactose that are consistent with the reports (Guan et al., 2010, 2011). The main chains of the polysaccharides from natural *Co. sinensis* consist essentially of mannose, but the main polysaccharide chains from cultured *Co. sinensis* contain mannose or/and glucose.

4.2.2 *Cordyceps militaris*

Yu et al. (2004c) obtained water-soluble crude polysaccharides from the fruiting bodies of cultured *C. militaris* by hot-water (60–70°C) and alkaline (0.3 M NaOH) extraction, followed by ethanol precipitation. The polysaccharides were successively purified by chromatography on anion-exchange and gel-permeation columns. Structural features of different purified fractions were investigated by a combination of chemical and instrumental analysis. Both CSP50 and CSP70 were obtained through fractionated precipitation with an ethanol concentration of 50% and 70%. The CSP50 was further purified by Sephadex G-100 to obtain four fractions, named CPS-2, CPS-3, CPS-4, and CPS-5. The CPS-2 (yield 0.12%) was eluted as a single symmetrical peak corresponding to an average molecular weight of 1.3×10^4, and was mainly composed of Rha, Glc, and Gal in a molar ratio of 1: 4.46: 2.43. The CPS-3 (0.08%) was a homogeneous polysaccharide having a molecular weight of 5.0×10^3, which was composed of D-glucose with a major linkage of α-D-glucose and the side chains were found at 6-O positions once in every eight glucose residues (Yu et al., 2004c). The CSP70 was further purified by Sephadex G-100 and Sephadex G-200 to get a polysaccharide called CSP-1, and the structure of CSP-1 was elucidated by sugar analysis, Smith degradation, IR, and 13C-NNR spectroscopy. Results showed that CSP-1 was mainly composed of Rha, Xyl, Man, Glc, and Gal in a molar ratio of 1:6.43:25.6:16.0:13.8, with a molecular weight of 2.3×10^4 Da. The chemical shift showed that CPS-1 might contain mannose bonded by (1→2) linkage, xylose bonded by (1→4) linkage, and rhamnose bonded with galactose by (1→2) or (1→3) linkage (Yu et al., 2004b). By using a DEAE-cellulose-52 column gradient eluted with NaCl aqueous solution (0–1 M), three polysaccharides, namely P70-1, P70-2, and P50-1, were obtained, all of which contained D-Man, D-Gal, and D-Glc in the respective proportions of 3.12:1.45:1.00, 1.75:1.35:1.00, and 2.80:2.62:1.00. The P70-1 had a backbone of (1→6)-β-D-Manp residues, with occasional branches at O-3, and the branches were mainly composed of (1→4)-α-D-Glcp and (1→6)-β-D-Galp residues, and terminated with β-D-Galp and α-D-Glcp residues (Yu et al., 2007). Moreover, CBP-1, isolated from alkaline extraction and subsequently purified through DEAE-Cellulose 52 and a Sephacryl S-100HR column, was a heteropolysaccharide and slightly branched. The molecular weight of CBP-1 was 17 kDa and was composed of D-Man, D-Gal, and D-Glc in a ratio of 2.81:4.01:1.00 determined by HPAEC-PAD, and 3.15:4.34:1.00 analyzed by GC–MS. Its backbone was (1→4)-α-D-mannose residues, with occasional branches at O-3, and the branches were mainly composed of (1→4)-α-D-glucose and (1→6)-β-D-galactose residues, terminating with β-D-galactose residues (Yu et al., 2009).

A novel low-molecular weight polysaccharide (CMP-1) with potent bioactivities (antioxidant, immunostimulatory, and anti-tumor activities) was isolated from cultured *C. militaris*. Results showed that CMP-1 had an average molecular weight of 4.3 kDa, and the structure of the main chain was composed of (1→4)-linked α-D-glucopyranosyl, (1→6)-linked β-D-glucopyranosyl, and (1→4)-linked β-D-glucopyranosyl residues, and branched at O-6. The side chain was (1→3)-linked α-L-rhamnopyranosyl residues terminated with (1→)-linked α-D-glucopyranosyl residues (Jing et al., 2014). Smiderle et al. isolated a homogeneous polysaccharide through 5% KOH extraction, purified by freeze-thawing treatment and dialysis (100 kDa) from the fungus *C. militaris*. The spectroscopy and spectrometry analysis indicated that the polysaccharide was a glucogalactomannan with a molecular weight of 23,000 Da, and consisted of mannose (56.7%), galactose (34.5%), and glucose (8.8%). It had a backbone of (1→2)-linked-α-D-Man*p* that can be substituted by (1→6)-linked-α-D-Man*p* or (1→2)-linked-β-D-Gal*f* fragments, and terminated at α-D-Man*p*, α-D-Gal*p*, α-D-Gal*f*, or β-D-Gal*f* units. The NMR analysis also indicated side chains of α-D-Man*p*-(1→2)- α-D-Man*p*-(1→6)- β-D-Glc*p*-(1→, and the β-D-Glc*p* may be also substituted at O-5 by non-reducing end units as α-D-Gal*p*, α-D-Galf or β-D-Gal*f* (Smiderle et al., 2013).

A water-soluble polysaccharide (named MCMP) was isolated from the mycelium with a high yield mutation *C. militaris* by hot-water extraction, deproteinization by Sevag, alcohol precipitation, anion-exchange, and gel filtration chromatography CL-6B. The polysaccharide contained mannose, rhamnose, galactose, and glucose in a molar ratio of 59.36:1:8.31:39.50, of which the average molecular weight is 8100 (Zhang et al., 2010). Zheng et al. investigated three chromatography columns (DEAE Cellulose, DEAE Sephadex A-50, and Sephadex G-200) to evaluate the purification process of polysaccharide from *C. guangdongensis* (CGP) in terms of molecular weight distribution, retention time, polydispersity, and proportion of each fraction. The results showed that the purity of the high-molecular weight fraction was improved, and small molecular polysaccharide fractions were obtained after gradient NaCl elution using different column chromatography (Zheng et al., 2009).

When purified by DEAE-52 cellulose anion exchange column and a Sepharose G-100 column, a homogenous polysaccharide, CMN1, was obtained from the aqueous extract of *C. militaris*, with a molecular weight of 37,842 Da. The CMN1 was composed of L-rhamnose, L-arabinose, D-mannose, and D-galactose, with a backbone comprised of (1→2) and (1→3) linkages, and branched (1→6) and (1→4) linkages (Dong et al., 2015).

It is worth noting that the configurations of polysaccharide produced from mycelia and broth of cultured *Cordyceps* are also not the same, and the difference may be attributed to the composition of the nutrient medium used for cultivation. Lee et al. isolated water-soluble polysaccharides from cultured mycelia and liquid culture broth of *C. militaris*, respectively, which were fractionated by DEAE cellulose and Sepharose CL-6B column chromatography. The CPMN Fr III, obtained from cultured mycelia, was able to stimulate the activation of macrophages effectively, as a result of further elucidating its structure features. Results showed that CPMN Fr III was a β-1,4-branched-β-1,6-galactolgucomannan that had a molecular weight of 210 kDa (Lee et al., 2010b). Similarly, CPSN Fr II, obtained from culture broth, was able to up-regulate the functional events mediated by activated macrophages as well, and was a 1,6-branched-glucogalactomannan that had a molecular weight of 36 kDa (Lee et al., 2010a). Wu et al. (2011) isolated a polysaccharide (CM-hs-CSP2) from the fruiting bodies of cultured *C. militaris* which was harvested from solid rice medium. It was shown that CM-hs-CSP2 was mainly composed of mannose, glucose, and galactose in a ratio of 1.35:8.34:1.00, and was linked by α-glycosidic bonds. Moreover, CM-hs-CSP2 was a kind of sulfated acidic polymer containing acetamido groups. Furthermore, they also isolated two kinds of acidic polysaccharide, CM-jd-CSP2 and CM-jd (Y)-CSP2, from the fruiting bodies of cultured *C. militaris* grown on solid rice medium and silkworm pupa, respectively. The monosaccharides of CM-jd-CPS2 and CM-jd(Y)-CPS2 both comprised mannose, glucose, and galactose, but in different proportions. The CM-jd-CPS2 was mainly composed of glucose, whereas CM-jd(Y)-CPS2 had a large percentage of mannose. The IR spectrum suggested that both of them were in a pyranose, but CM-jd-CPS2 was also linked by an α-glycosidic linkage, while

CM-jd(Y)-CPS2 was in a β-configuration form. Like CM-hs-CSP2, CM-jd-CPS2 was also a sulfated acidic polysaccharide having acetamido groups. However, CM-jd(Y)-CPS2 was found to be a type of carboxylated polysaccharide. These results indicated that different culture media resulted in totally different polysaccharide structures (Wu et al., 2012). Ohta et al. (2007) isolated an acidic polysaccharide (APS) from extract of *C. militaris* grown on germinated soybeans. The sugar composition analysis revealed that APS was composed of D-galactose, L-arabinose, D-xylose, L-rhamnose, and D-galacturonic acid, and mainly contained Ara*f*-(1→, →5)-Ara*f*-(1→, →4)-Gal*p*-(1→ and *f*→)-GalA*p*-(1→ linkage patterns based on methylation and GC-MS analysis. Zeng et al. (2015) isolated four polysaccharide fractions, i.e., P20, P40, P60, and P80, from the spent rice substrate of *C. militaris* through gradient ethanol precipitation. The P20 contained starch from the rice substrate, and the other three fractions had a main chain of α-1,4-glucan. The interpretation of structure variations of polysaccharide from *C. militaris* mycelium may be because of the different fermented strains. Otherwise, different culture media, such as silkworm pupa, solid rice medium, and broth, may contribute to these structural differences.

4.2.3 *Cordyceps gunii*

A polysaccharide with a high-molecular weight of 3.72×10^6 Da, named as CPS, was isolated from the mycelia of *C. gunii* and purified by DEAE-Sephadex A-25 and Sephadex G-100 (30 cm × 3 cm). The CPS had a sugar content of 92.84%, and mainly consisted of glucose with small amounts of rhamnose, arabinose, xylose, mannose, and galactose in a molar ratio of 3.0:2.6:1.0:1.3:106.0:2.8. The specific rotation was recorded at $[\alpha]_D^{25} = +134.2$ degrees, and CPS had a D-glucan containing α-(1→4)-linked backbone (Zhu et al., 2012b). Another fraction purified by graded ethanol precipitation, DEAE-Sephadex A-25, and Sephadex G-75 column obtained a homogenous water-soluble polysaccharide CPS50-I. Results showed that the molecular weight of CPS50-I was around 9874 Da, and was composed of xylose, mannose, glucose, and galactose in a proportion of 0.13:0.89:0.54:1. Periodate oxidation and Smith degradation results showed that CPS50-I contained (1→3)-linked-xylose and (1→2)-linked mannose and galactose, but did not have 1→4 linkages (Zhu et al., 2011). However, Xiao et al. (Zhu et al., 2014) reported a polysaccharide from *C. gunii* that was mainly D-glucopyranose containing an α-(1→3)-linked backbone, possibly owing to the difference in the original raw materials.

4.2.4 Other *Cordyceps* species

Wu et al. (2016) optimized the extraction conditions of polysaccharide from the mycelium of *Paecilomyces hepiali* HN1 (PHMPs) using a single factor and response surface methodology based on a Box–Behnken design as follows: extraction temperature 92°C, extracting time 190 min, and ratio of water to material 43 mL/g. The mean extraction yield of PHMPs under the optimum conditions was 11.08 ± 0.22% (Wu et al., 2016).

Lu (2011) obtained three polysaccharides from the culture broth, mycelium, and fruiting body of *Cordyceps japonica* and then purified them by Sephacryl S-300HR column (1.0 cm × 60 cm) to get four polysaccharides, named as EPS, IPS1, IPS2-1, and IPS2-2. All the polysaccharides were composed of mannose, galactose, and glucose, with a molar ratio of 12.9:2.4:1.0, 11.7:5.8:1.0, 1.3:1.0:2.9, and 1.5:1.1:1.0, respectively. The molecular weight of EPS and IPS2-2 was 74 and 25 kDa, much lower than IPS1 (Mw 233 kDa) and IPS2-1 (Mw 265 kDa). The IPS2-1 had an α configuration, while IPS2-2 was β (Lu, 2011). Zheng and He revealed that polysaccharide from *C. guangdongensis* was composed of three fractions, with molecular weights of 2.57×10^6, 3.84×10^4, and 5.00×10^3, respectively (Zheng and He, 2009).

Xiao et al. (2012b) carried out an ultrasonic extraction (USE) technique to isolate crude polysaccharide from *C. cephalosporium* mycelia. The polysaccharide was preliminarily purified by successively sub-fractionating with gradient concentrations of ethanol (40% and 70%), and applied to a Sephadex G-100 gel filtration column to obtain two major fractions (USEP40-1, USEP70-1). The molecular weights of USEP40-1 and USEP70-1 were 61.4 and 25.1 kDa. The USEP40-1 and USEP70-1 were composed of D-mannose, D-glucose, D-galactose, L-rhamnose, and L-arabinose with molar ratios of 11.52:5.54:8.75:2.45:2.59 and 11.50:6.74:5.75:4.46:2.39, respectively (Xiao et al., 2012b).

Wu et al. (2014a) first investigated polysaccharides from seven species of natural and cultured *Cordyceps* compared using saccharide mapping, partial acidic/enzymatic (α-amylase, β-glucanase, and pectinase) digestion, followed by polysaccharide analysis using carbohydrate gel electrophoresis (PACE) and high performance thin layer chromatography (HPTLC) analysis, respectively, to obtain a comprehensive profile of the polysaccharide hydrolysates and their characters. Results indicated that 1,4-α-D-glucosidic, 1,4-β-D-glucosidic, and 1,4-α-D-galactosidic linkages existed in natural and cultured *Co. sinensis*, cultured *C. militaris*, natural *Cordyceps gracilis*, and *Cordyceps ciecadae*. There was a relatively high similarity between polysaccharides from cultured *C. militaris* and natural *Co. sinensis*, and this might be contributed to by the rational use of *C. militaris*. Furthermore, different species of natural and cultured *Cordyceps* could be differentiated based on the saccharide mapping (Wu et al., 2014a).

Shi et al. (2007) isolated two water-soluble polysaccharides, BCPS-1 and BCPS-2, from cultured *C. brasiliensis* Henn. by DEAE-52 cellulose and Sephadex G-100 chromatography. Results showed that BCPS-1 mainly consisted of glucose, sorbose, galactose, and mannose, while BCPS-2 was mainly composed of glucose and sorbose. Their molecular weights were 41,000 and 33,000, respectively (Shi et al., 2007).

Xiao et al. (2016) isolated a purified alkali-soluble polysaccharide from *C. taii*, and the polysaccharide was found to mainly consist of glucose, galactose, and mannose in a molar ratio of 1.14:1.00:1.66. A series of α-1→4 glucosidic linkages and the mannose existed in the backbone of the polysaccharide structure (Xiao et al., 2016).

The monosaccharide composition of exopolysaccharide extracted from *Cordyceps bassiana* consisted of mannose and galactose in a molar ratio of 1.2:1.0, while the intracellular polysaccharide was composed of mannose, galactose, and glucose in a ratio of 5.4:4.0:1.0 (Yang, 2009).

A neutral heteropolysaccharide (PCIPS2) was obtained from mycelium of *Paecilomyces cicadae*, which was found to mainly consist of D-mannose, L-rhamnose, 3-*O*-methyl-D-galactose, D-glucose, and D-galactose with a molar ratio of 47.9:3.1:6.4:0.9:0.8. The backbone of PCIPS2 was 1,4-linked α-L-Rha*p* residues, and 1,6-linked α-D-Man*p* residues, and branched at O-3 of the α-D-Man*p* residues. The side chain was composed of minor terminal β-D-glucose and 1,4-linked α-3-*O*-Me-D-Gal*p* residues terminated at an α-D-galactose (Wei et al., 2016).

The anti-hypoxic effects of CMN1, a polysaccharide from *C. militaris*, were evaluated using sodium nitrite toxicosis analysis, acute cerebral ischemic/hypoxic assay, and normobaric hypoxia assay. Compared with the control group, CMN1 markedly improved the survival rate and enhanced gasping persistence. The CMN1 at a concentration of 0.5 g/kg possessed a similar anti-hypoxic function as rhodiola oral liquid (Dong et al., 2015).

Based on the above report, it is easy to figure out that not only the original source of material, but also the extraction process, for example, extraction treatment with different solvents, combination with new technologies, such as ultrasound and microwave, as well as multi-purification means, can influence the structural features and biological properties of polysaccharides.

4.3 MOLECULAR CHARACTERISTICS

The bioactivities of polysaccharides closely depend on their molecular weight, substitution of functional groups, chemical structures, and chain conformations. Generally, polysaccharides in aqueous solutions display different forms of chain conformations, such as random coil, single helix, double helix, triple helix, various helical forms, and aggregates (Yan et al., 2014). There is little literature available on the solution properties and chain conformations of *Cordyceps* polysaccharides. For example, a polysaccharide extracted from natural *Co. sinensis*, named cordysinan, was investigated for the solution properties through sizing exclusion chromatography coupled with multi-angle light scattering and triple detector array. The molecular weight, radius of gyration, and the intrinsic viscosity of cordysinan were reported as 22.45 ± 0.26 kDa and 2237 kDa, 15.4 ± 2.4 nm and 1.41 mL/g, respectively. Furthermore, the exponent υ and α value were calculated as

0.28 and 0.42, respectively, according to polymer solution theory, so as to reveal that cordysinan existed as a globular shape in 0.9% NaCl aqueous solution (Wu et al., 2014b). Atomic force microscopy (AFM) is a powerful tool in analyzing surface morphology and surface processing. It has been broadly used in biological morphology, but was not commonly reported in polysaccharide applications. Cai et al. (1998) first observed the morphology of *Cordyceps* polysaccharides through AFM. The results revealed that the polysaccharide had a multi-branched structure and various linkages between the adjacent monosaccharides, forming small rings or helical structures. He et al. (2013) reported that the topographical AFM image of HS002-II apparently showed that the polysaccharide formed a strong 3D interwoven network in which the main single chain and side chain extended in the dilute solution. The height of the polysaccharide clusters ranged from 0.6 to 2.5 nm, and this diameter was close to that of a single polysaccharide chain, indicating that the molecular aggregation was avoided. On the other hand, the diameter and length of the backbone were in the range of 5–15 nm and 40–160 nm. Recently, the morphological properties and chain information of exopolysaccharide isolated from the mycelial culture of *Co. sinensis* Cs-HK1 was also analyzed by AFM, together with the Cong red test, optical rotation, and dynamic light scattering. The result indicated that the exopolysaccharide formed large interwoven networks in water, and was primarily connected with a triple-helical conformation, as well as an occasional single-helical conformation. However, the intracellular polysaccharide exhibited as random coils in alkaline solutions. On the other hand, ultrasound treatment could lead to a notable reduction of the maximum molecular weight, and gave a more uniform molecular weight distribution, but did not change the primary structure of exopolysaccharide molecules. In contrast, the intrinsic viscosity of exopolysaccharide was reduced by only 20% in 1.0 M sulfuric acid at 50°C for 9 h (Wang et al., 2010b, 2011c; Yan et al., 2011). Likewise, Chen et al. (2014a) found that the exopolysaccharide chain existed in a multiple-helix conformation in aqueous solution, and high intensity ultrasound treatment did not cause conformation change of the polysaccharide from multiple to single helixes. However, alkali aqueous solution could disrupt the hydrogen bonds, resulting in a change from a multiple to a single helix chain, and finally to random coils. It was interesting to find that the random coils and aggregated networks of EPS-1 in aqueous solution were changed to single helices after sulfation (Yan et al., 2012a). Scanning electron microscope (SEM) observations revealed that the ultrastructure of CMP-1, a polysaccharide from cultured fruiting bodies of *C. militaris*, was mainly composed of a surface with a sheet-like appearance and randomly distributed ovoid-shape particles, and the diameter was approximately 2 m. These molecular characteristics might be attributed to the branches and network structures of this polysaccharide (Jing et al., 2014). In their study, ESP-1A, an exopolysaccharide isolated and purified from the fermentation broth of cultured *Co. sinensis* fungus Cs-HK1 was used to analyze the effects of temperature, metal ion (Ca^{2+}), denaturing agent (urea), and chemical modifier (congo red) on its conformation through ultraviolet (UV) spectroscopy and circular dichroism (CD) spectroscopy methods. The results indicated that EPS-1A had a random coil chain conformation, as revealed by Congo red reaction. Furthermore, external conditions could not alter the random coil chain of EPS-1A in aqueous solution, but could influence the inter- and intra-molecular hydrogen bond interactions of polysaccharide molecules to some extent, so as to become more loose and disordered (Yan et al., 2012b). By gradient precipitation with ethanol, a series of exopolysaccharide fractions, P1/5, P2/5, P1, P2, and P5 were isolated. According to the image of transmission electron microscopy (TEM), P2/5 existed as a worm-like chain shape with an approximate diameter in the range of 1000–2000 nm, corresponding to the results from DLS, and P2 exhibited as a globular shape and aggregated to compact clusters. Furthermore, it also strongly indicated that the high-molecular weight fractions (P1/5 and P2/5) were in a stiff chain conformation, whereas the low-molecular weight fractions (P1, P2, and P5) were in a globular geometry (Huang et al., 2013). Two polysaccharide fractions, CPSN Fr II and III, isolated and purified from the liquid culture broth of *C. militaris* by ethanol precipitation and column chromatography (DEAE cellulose and Sepharose CL-6B column), were both found to exhibit a random coil conformation similar to that of dextran, according to the results of the Cong red test (Lee et al., 2010a,b). Polysaccharide from *C. gunii*, on the other hand, expressed α-helix in its spatial structure, and its micro structure showed an aggregated state (Zhu et al., 2014). The chain information on the values of weight-average molar mass (Mw), root mean square radius, hydrodynamic radius (Rh), and intrinsic viscosity ($[\eta]$) for PCIPS2, polysaccharide from *P. cicadae*, were demonstrated to be 3.09×10^4 g/mol, 7.8 nm, 3.6 nm, and 8.5 mL/g, respectively, as revealed by Wei et al. The structural exponent α was calculated to be 0.57, indicating that PCIPS2 existed in a flexible chain conformation with a coil-like structure in 0.1 M $NaNO_3$ solution at 25°C. Based on the known theory for worm-like chains, the model parameters for PCIPS2 were predicted as follows: molar mass per unit contour length (ML) = 379 nm^{-1}, persistence length (q) = 0.74 nm, and hydrodynamic diameter of cylinder (d) = 0.82 nm, further confirmed by the results of AFM (Wei et al., 2016).

4.4 BIOACTIVITIES

Cordyceps species have been used as a popular medical remedy in China for over 300 years, and have been valued worldwide for their pharmaceutical properties. Numerous animal and clinical studies suggested that the polysaccharides from *Cordyceps* species played a crucial role in supporting human health. These benefits are now described. Previous studies on polysaccharides from various *Cordyceps* species have demonstrated many interesting biological activities, including anti-tumor, immunomodulating, hypoglycemic, antioxidant, and hypocholesterolemic effects, etc. The medical merits of *Cordyceps* polysaccharides have attracted attention.

4.4.1 Protective Effects on the Kidney

An early medical text lists the traditional usage of *Cordyceps* for "Kidney Improvement" (Li et al., 2006b; Zhu et al., 1998a,b). The protective effect on kidney cells of CPS1 and CPS2 purified from cultured *Co. sinensis* had been investigated. An obvious protection activity on the kidney cell (vero cell) was observed at a dose of 12.5 ~ 200 μg/mL (Fan et al., 2008). *Cordyceps* polysaccharide on gentamicin-induced acute renal failure in rats showed significant protection ($P < .01$ or $P < .05$) against renal failure, improved renal function, corrected metabolic disorder, and promoted regeneration of the nephron (Zhou et al., 2008).

The therapeutic effects of *Cordyceps* polysaccharide (CP) on adenine-induced chronic renal failure (CRF) in rats and occluding renal artery-induced acute kidney injury in dogs have been studied. Both renal index and content of serum TNF-α decreased significantly, indicating that CP had a therapeutic and preventive effect on adenine-induced CRF. It can prevent the generation and development of CRF, and hence improve renal function. Compared with the control group, the degree of injury of renal function in CP group dogs was reduced significantly ($P < .01$), and the renal pathological changes were alleviated significantly. The CP thus has a remarkably beneficial effect on renal ischemia-induced acute kidney injury. It has, therefore, potential to improve renal function, increase renal blood flow, correct metabolic disorder, and promote regeneration of the renal tubule (Yin et al., 2007a,b). Likewise, Hou et al. (2009) pointed out that *C. militaris*, especially the polysaccharide fraction, could obviously reduce the content of blood urea nitrogen and creatinine, as well as alleviate pathological lesion in adenine-induced CRF rats.

The therapeutic effects of *Cordyceps* polysaccharide on freezing kidney-induced investigation of CRF in rats showed that the serum BUN, SCr, Na^+, P, and MDA levels were significantly reduced, and the serum ALB and T-SOD levels significantly increased both in the positive control group and in all CP groups ($P < .01$ or $P < .05$) compared with the negative model group. Renal pathological injuries of rats were significantly improved in positive control groups and in the high-dose CP group, and to a lesser extent in mid- and low-dose CP groups. *Cordyceps* polysaccharide, it seems, can effectively prevent CRF, improve renal function, correct metabolic disorder, and promote regeneration of the nephron (Lv et al., 2007).

Wu et al. (2008) investigated the effect of *Cordyceps* polysaccharide on the transition of tubular epithelial mesenchyma in CD-1 male mice with unilateral ureteral obstruction. In the mice treated with CP, the degree of interstitial fibrosis was attenuated ($P < .01$). The expressions of α-SMA and ILK protein were alleviated significantly, and loss of E-cadherin expression was inhibited ($P < .01$). The CP might, therefore, suppress the loss of E-cadherin and the expression of α-SMA, and significantly inhibit the over-expression of ILK in the obstructed kidney. The CP thus could play an inhibitive role in the progression of EMT through down-regulating of mRNA and protein expressions of ILK.

The ureteric obstruction (UUO) model of renal fibrosis and the tubular epithelial cell line HK-2 were used to investigate the effects of polysaccharides from *Co. sinensis* on renal injury. The in vivo data supported an anti-fibrotic effect of CS on renal fibrosis. Furthermore, the data provided in vitro evidence showing that this effect was mediated by the soluble polysaccharide aggregate fraction of CS, which could significantly antagonize the effects of TGF-β1 through regulating the expression of the receptor, and therefore blunting the effect of TGF-β1 on driving epithelial injury which initiated a profibrotic response (Zhang et al., 2012).

A lipopolysaccharide (LPS)-induced acute endotoxemia rat model was used to illustrate the effect of polysaccharide from *C. sobolifera* (CS-P) on modulating nephro-functional damage. Results showed that pretreatment with CSP-P for 4 weeks could significantly improve the deleterious renal functions, and substantially reduce the elevated levels of blood urea nitrogen, creatinine, and urine protein, as well as improve LPS-induced pathological abnormalities. In addition, CS-P pretreatment could reduce the levels of NO, TNF-α, and IL-6, which were associated with significant down-regulation of iNOS expression. This ameliorative effect on renal damage was accompanied with up-regulation of antioxidant enzymes, such as superoxide dismutase, glutathione peroxidase, and catalase, as well as lower accumulation of lipid peroxidation in the kidney (Chiu et al., 2014).

A water-soluble polysaccharide (CPS-2) isolated from a cultured *Co. sinensis* polysaccharide has been proved to have significant therapeutic effect on relieving chronic renal failure caused by the fulgerizing kidney (Wang et al., 2010a). The underlying molecular mechanism was further studied on human mesangial cells (HMCs). Results showed that CSP-2 could inhibit PDGF-BB-induced HMC proliferation in a dose-dependent manner through the PDGF/ERK and TGF-β1/Smad pathways. More importantly, CSP-2 may have a bi-directional regulatory effect on the PDGF/ERK cellular signaling pathway (Wang et al., 2014).

4.4.2 Immunomodulatory Activity

Polysaccharide (CS-81002) purified from the fermentation medium of *Co. sinensis* showed a stimulating effect on the phagocytic function of macrophages in normal mice at dosages of 5 mg/kg (Gong et al., 1990).

Wu et al. (2006b) found that a purified polysaccharide (PS) from the edible mycelia of *Co. sinensis* initiated cellular and humoral immune responses in ICR mice against ovalbumin (OVA). They could identify significant immunomodulatory potentials by determining OVA-specific IgG, IgG1, and IgG2b antibody levels in serum. The effects of different components (I, II, III, IV) of *Co. sinensis* polysaccharides on the immune function of mice were investigated. Moreover, different components of *Co. sinensis* polysaccharides could enhance a similar immune response in the spleen index, thymus index, and the phagocytic function of monocyte-macrophages (Yu et al., 2004a).

An insoluble glucan-like polysaccharide (CS-Pp), extracted from the mycelia of *Co. sinensis*, was found to strongly induce TNF-α production. Further analysis of the effect of CS-Pp on TNF-α production of RAW 264.7 cells indicated time- and concentration-dependent production. Such TNF-α-inducing ability of CS-Pp was also confirmed in the presence of Polymyxin B (Akaki et al., 2009). It was also shown that CP could induce and stimulate the expression of IL-2R, but only selectively inhibit the production of IL-2 and IFN-γ. Both the stimulation and the inhibiting abilities were dose-dependent. This indicated that CP has a bi-directional action on PBLs in vitro (Qiu et al., 1995).

Carbon clearance, hemolysin, and dinitrofluorobenzene-induced delayed type hypersensitivity tests were used to examine the immunomodulating effects of PCM polysaccharides on non-specific and specific immunologic function. The PCM was found to increase the index of K, α, and the viscous index of spleen and thymus in the carbon clearance test. The PCM could also enhance murine ear swelling and hemolysin levels, which suggested that PCM could improve cellular and humoral immunologic functions (Gong et al., 1999). Another study also showed that polysaccharides from cultured *Co. sinensis* (PCCS) could increase the K and α indices in the carbon clearance test and PCCS enhanced the phagocytosis function of mononuclear macrophage in hydrocortisone-induced immunosuppressed mice. It was also found that PCCS could enhance murine ear swelling in immunosuppressed animals. The PCCS could improve the cellular and humoral immunologic function in immunosuppressed mice and significantly improve non-specific and specific immunologic abilities of immunosuppressed mice (Gong et al., 1999).

The effect of *Cordyceps* polysaccharide on the immune function in adjuvant arthritis (AA) rats in vitro was studied using an in vitro immune cell culture so as to detect the change of ConA-induced splenocyte proliferation, as well as cytokine (IL-1 and IL-2) synthesis. The CP (10 and 100 mg/L) not only enhanced ConA-induced splenocyte proliferation, but also improved the levels of IL-2 in AA rats (Jin et al., 2000). Another investigation found that crude polysaccharides from wild-type *Co. sinensis* and mycelia of *Co. sinensis* could induce macrophages from the mouse abdominal cavity to produce the tumor necrosis factor (TNF-α). It was found that crude polysaccharide from wild *Co. sinensis* was better than the mycelia of *Co. sinensis* in the ability to induce macrophage from the mouse abdominal cavity in vitro to produce TNF-α (Lin et al., 2003). However, the exopolysaccharide isolated from this strain of *Co. sinensis* also induced cell proliferation and the secretion of IL-2, IL-6, and IL-8 in cultured T-lymphocytes. In addition, the phosphorylation of extracellular signal-regulated kinases (ERK) was also induced transiently by treatment with this exopolysaccharide. This exopolysaccharide also increased the phagocytotic activity and the enzymatic activity of acid phosphatase in cultured macrophages. Together, these observations confirm the important role of *Cordyceps* polysaccharide in triggering such immune responses (Cheung et al., 2009).

The immunomodulatory activity of polysaccharides prepared from submerged cultured *Co. sinensis* was investigated in human peripheral blood. Exopolysaccharides (0.025 ~ 0.1 mg/mL) induced the production of TNF-α, IL-6, and IL-10 dose-dependently. Even a low concentration of exopolysaccharides (0.025 mg/mL) could significantly augment the surface expression of CD11b in monocytes and polymorphonuclear neutrophils (PMN). Functional assays revealed that EPS at a concentration of 0.05 mg/mL could elevate phagocytosis in monocytes, and PMN as well. However, the intracellular polysaccharides could only moderately induce TNF-α secretion, CD11b expression, and phagocytosis at the same concentrations. These results suggested that the immunomodulatory polysaccharide in submerged cultured *Co. sinensis* mainly existed in the culture broth (Kuo et al., 2007). Also, the modulating effects of the acid polysaccharide fraction (APSF) extracted from the mycelia of cultivated *Co. sinensis* on murine macrophage cell line RAW264.7 were evaluated. The APSF could stimulate the phagocytosis of macrophages, and promote both the mRNA and protein expressions of inducible nitric oxide synthase (iNOS). After APSF treatment, NF-κB levels increased, indicating that APSF probably stimulated macrophage activities by activating the IκB-NF-κB pathway (Chen et al., 2010a,b).

The conditioned medium from the polysaccharide fraction of *Co. sinensis* (PSCS) (10 μg/mL)-stimulated blood mononuclear cells (PSCS-MNC-CM) was found to have an activity that could significantly inhibit the proliferation of U937 cells. The PSCS-MNC-CM treatment induced about 50% of the cells to differentiate into mature monocytes/macrophages expressing nonspecific esterase activity and surface antigens of CD11b, CD14, and CD68. The levels of IFN-γ, TNF-α, and IL-1 can be greatly increased with PSCS stimulation. The results further revealed that the tumoricidal and differentiating effects of PSCS-MNC-CM were mainly derived from the elevated cytokines, particularly IFN-γ and TNF-α. These two cytokines acted synergistically on inhibiting cell growth and inducing differentiation of the target U937 cells (Chen et al., 1997).

The effects of an exopolysaccharide fraction (EPSF) from cultivated *Co. sinensis* on B16 melanoma-bearing mice were investigated. Results revealed that EPSF could significantly enhance the neutral red uptake ability of peritoneal macrophages (60 mg/kg, $P < .01$; 120 mg/kg, $P < .001$) and spleen lymphocyte proliferation rate (60 mg/kg, $P < .05$; 120 mg/kg, $P < .001$) in B16-bearing mice. Moreover, the metastasis of B16 melanoma cells to livers (30, 60, and 120 mg/kg) and lungs (120 mg/kg) was significantly inhibited. The EPSF could also reduce the quantity of Bcl-2 in livers (30, 60, and 120 mg/kg) and lungs (120 mg/kg). The results indicated that EPSF was a potent immunomodulatory and anti-tumor agent (Zhang et al., 2005).

The effect of an acidic polysaccharide (AEPS-1) from *Co. sinensis* Cs-HK1 on RAW 264.7 macrophage cells was studied. Data indicated that AEPS-1, at suitable doses between 25 and 250 μg/mL, significantly stimulated the release of several major cytokines (TNF-α, IL-1β, IL-6, and IL-10), demonstrating an immunomodulatory property (Wang et al., 2011c). On the other hand, the isolated exopolysaccharide was also demonstrated to significantly enhance immune organs and stimulate the release of major cytokines TNF-α and INF-γ in cyclophosphamide-induced immunosuppression mice (Hu et al., 2016).

The immunomodulatory effects of EPS from cultivated *Co. sinensis* were examined on the proliferation ability of lymphocytes from ICR mice when assayed by the MTT assay. Protein and mRNA expression levels of several cytokines in spleen and thymus cells were detected by ELISA and RT-PCR. The EPS elevated proliferation ability of spleen lymphocytes was found only at 100 g/mL after 48 h treatment. The TNF-α, IFN-γ, and IL-2 mRNA levels in splenocytes and thymocytes increased after EPS treatment for 2, 4, 8, or 20 h. The EPS also significantly elevated splenic TNF-α and IFN-γ protein expressions at 100 μg/mL, and increased thymic TNF-α and IFN-γ protein levels at 50 and 100 μg/mL. Thus, EPS may stimulate cytokine expressions of immunocytes (Sheng et al., 2011).

The effect of polysaccharide (CS-PS), from the fruiting bodies of cultured *Co. sinensis*, on the immune function of BALB/c mice exposed to ^{60}Co gamma radiation has been investigated. Mice were treated with doses of 50, 100, or 200 mg/kg body weight of CS-PS, and then exposed to ^{60}Co gamma radiation for four days. Lymphocyte proliferation, macrophage phagocytosis, DTH, and the total-SOD enzyme activity in CS-PS groups were significantly increased compared to the irradiated control group. On the contrary, lipid peroxidation levels were significantly decreased in CS-PS. Meanwhile, levels of cytokine (IL-4, IL-5, and IL-17) were also affected in the CS-PS group. These results indicated that CS-PS enhances the immunity activity in mice treated by ionizing radiation, through reducing oxidative injury and modulating the secretion of cytokine IL-4, IL-5, and IL-17 (Zhang et al., 2011).

Dendritic cells (DC) are professional antigen-presenting cells that are crucial to initiate tumor-specific T-cell responses, and defective DC maturation in tumor microenvironments is an important immunological issue. The effect of polysaccharide (cordlan) from *C. militaris* on DC maturation was investigated. Phenotypic maturation of DCs by cordlan was demonstrated in terms of the increased expression of CD40, CD80, CD86, MHC-I, and MHC-II molecules, and functional maturation was reported as enhancement of IL-12, IL-1β, TNF-α, and IFN-αβ expression, improved allogenic T-cell stimulation, and decreased endocytosis. It was important to note that cordlan-induced maturation of $tlr4^{+/+}$ DCs from C3H/HeN mice, instead of $tlr4^{-/-}$ DCs from C3H/HeJ mice, indicating the potential membrane receptor of cordlan. Furthermore, cordlan could significantly increase phosphorylation of ERK, p38, and JNK, and nuclear translocation of NF-κB p50/p65, suggesting that cordlan induced DC maturation through TLR4 signaling pathways (Kim et al., 2010). The effects of EPS from *Co. sinensis* on murine DCs, derived from the bone marrow of C57BL/6 mice, were also studied. The results showed that EPS could promote the levels of surface molecules (MHC II, CD40, CD80, and CD86) and decrease their ingestion capacity. The mRNA expressions of cytokines (IL-12 p40 and TNF-α) and inducible nitric oxide synthase were up-regulated as well. It was also found that EPS could significantly down-regulate the p-STAT3 level of DCs. All the findings indicated that EPS-induced DCs maturation and activation was probably related to the inhibition of STAT3 phosphorylation (Song et al., 2011). In their recent study, they further found that the mechanism may also be related to the inhibition of the JAK2/STAT3 signal pathway and promotion of the NF-κB pathway (Song et al., 2013). The activity of IL-12 and the effects of stimulation increased in chronic myeloid leukemia (CML) DCs pre-stimulated by the polysaccharide fraction of *Co. sinensis* (PSCS). Furthermore, the CML-DCs cultured with PSCS resulted in a rapid generation of CD86 and HLA-DR, an enhancement of IL-12 expression, and a stimulatory capacity in allogeneic mixed lymphocyte reaction (MLR). It seems, then, that PSCS could increase T-cell immunoresponse and represent a valuable traditional agent for the rapid generation of active DCs. It has potential as a vaccine against CML (Huang et al., 2011). *In vitro* assay showed that polysaccharide from cultured mycelia (UM01 PS) promoted RAW 264.7 macrophage proliferation, production of NO and multiple cytokines and chemokines, as well as phagocytic activity. In addition, UM01 PS could not only induce differentiation of RAW 264.7 macrophages into dendritic-like cells, but also promote phenotypic and functional maturation of mouse JAWS II dendritic cells. Therefore, it is believed that UM01 PS has the potential to activate macrophages and improve dendritic cell functions (Meng et al., 2014). Another polysaccharide, named as APSF, from an anamorph of *Co. sinensis*, was demonstrated to have stimulating activity on macrophages as well. Treated with APSF, the expression of TNF-α, IL-12, and iNOS was increased, and the expression of IL-10 was reduced in Ana-1 cells. In addition, the expression of SR and MR were down-regulated, and the level of NF-κB was reduced in M2 macrophages and up-regulated (Chen et al., 2012). He et al. demonstrated that the polysaccharide from *H. sinensis* could stimulate the expression of the pro-inflammatory cytokines, i.e., NO, TNF-α, IL-1β, and

NF-κB, in RAW264.7, indicating that the immunomodulatory effect of polysaccharide might be through the IkB-NF-κB pathway (He et al., 2013).

The effect of polysaccharide from *C. gunii* (PS) on the immunostimulatory activity, anti-tumor function, and expression of immune related cytokines was investigated in normal, immuno-suppressive, and H22-bearing mice, respectively, through assessing the macrophage phagocytic index, lymphocyte proliferation, natural killer (NK) cell activity, and related cytokine (IL-4, TNF-α, and IFN-γ) expression, as well as tumor inhibition index. Results showed that PS could significantly increase the thymus and spleen index, the macrophage phagocytosis, the splenic cells proliferation, and the level of IFN-γ and TNF-α, as well as remarkable inhibition effects, indicating that polysaccharide from *C. gunii* had a potent enhancement of immunostimulatory anti-tumor activity (Zhu et al., 2012a). Chen and An studied the immunomodulatory effects of polysaccharide from cultured *Co. sinensis* systematically in terms of phagocytosis, T and B lymphocyte proliferation, and immunoglobulin (IgM) production. Results showed that the polysaccharide can significantly stimulate LPS-induced B-cell proliferation and anti-CD3-induced T-cell proliferation from mouse spleen cells, as well as LPS-induced IgM secretion, but has no effect on IgA production. Additionally, the effect of polysaccharide on IgG subtypes was different. It could obviously suppress IgG2a production, whereas it has a positive effect on IgG1 generation at a concentration of 25 μg/mL, and a negative effect at concentrations above 50 μg/mL (Chen and An, 2009). Liu and Fei studied the effect of *C. taii* polysaccharide (CDP1) and *C. pruinosa* polysaccharide (CPP1) on the cellular immunological function of splenic T-cell and macrophage phagocytosis in mice. It was shown that both of the polysaccharides could not only enhance the proliferation of activated splenic T-cell in normal mice, but also in immunosupressed mice. An improvement of CTL cell activity under a proper concentration was observed, and a recovery of CTL activity in immunosupressed mice was discovered as well (Liu and Fei, 2000).

The CMP40 and CMP50 polysaccharide fraction from *C. militaris* by water extraction and stepwise ethanol precipitation purification was investigated for their immune-enhancing activities both in vivo and in vitro. Using in vitro assay, it was shown that both of the polysaccharides displayed a significant stimulation effect on lymphocyte proliferation, with CMP40 more obvious or numerically stronger. In an in vivo assay, 14-day-old chickens vaccinated with Newcastle disease vaccine were injected with polysaccharide at low, medium, and high concentrations, for 7, 14, 21, 28, 35, and 42 days after the first vaccination, and then the lymphocyte proliferation, serum antibody titer, IFNγ, and IL-4 were measured. Results indicated that CMP40 and CMP50 could significantly promote lymphocyte proliferation, increase serum antibody titer, as well as enhance serum IFN-γ and IL-4 secretion at a suitable dose (Wang et al., 2013). Wang et al. (2012) evaluated the immune activation activity of polysaccharide from *C. militaris* (CMP) in cyclophosphamide-induced immunosuppressed mice model. The lymphocyte proliferation, phagocytic index, and biochemical parameters were measured. Administration of CMP turned out to have a significant improvement against cyclophosphamide-induced immunosuppression, e.g., increased spleen and thymus index, as well as enhancement of spleen lymphocyte activity and macrophage function. Lee and Hong (2011) studied the signaling mechanism of macrophages immunostimulating properties induced by the polysaccharide from the fruiting body of *C. militaris* (CM). It was found that CM could up-regulate the expression of NO, ROS, TNF-α, and phagocytic uptake in both mouse peritoneal macrophages and RAW264.7 macrophages. Activation of macrophages seemed to be related to the activation of NF-κB and MAPKs pathways through Dectin-1 and TLR2 macrophage receptors. In addition, it was demonstrated that CM suppressed the growth of melanoma in an in vivo experimental mouse model (Lee and Hong, 2011). Furthermore, it is reported that polysaccharides from either cultured mycelia or liquid culture broth of *C. militaris* turned out to activate macrophages in terms of promoting NO production and cytokine (IL-1β and TNF-α) expression (Lee et al., 2010a,b).

In a recent study, they showed that the polysaccharide, namely PLCM from the culture broth of *C. militaris*, could enhance immunostimulatory activities such as releasing toxic molecules (nitric oxide and reactive oxygen species), secretion of cytokine TNF-α, and improving phagocytic uptake in RAW264.7 macrophages. The related mechanisms, therefore, were further elucidated.

After treating with PLCM, all the specific inhibitors against MAPK and NF-κB (SN50, BAY11-7082, PD98059, SP600125, and SB203580) markedly suppressed the production of nitric oxide and uptake of phagocytes. In addition, antibodies specific to the extracellular domain of TLR-2, TLR-4, or Dectin-1 significantly attenuated PLCM-induced secretion of TNF-α, indicating that the *C. militaris* polysaccharide activated macrophages through the MAPKs and NF-κB signaling pathways with Toll-like receptor 2, Toll-like receptor 4, and Dectin-1 as receptors (Lee et al., 2015). To investigate the immune activation activity of *C. militaris* polysaccharides (CMP) in vivo, 24 male and 24 female Kunming mice were randomly divided into four groups, with administration of CMP at 50, 100, or 200 mg/kg per day body weight *via* gavage. The viscera index, leukocyte, and differential leukocyte count, immunoglobulin (IgG) levels, and several biochemical parameters were measured after 30 days. The results demonstrated that CMP could significantly improve the immune function in mice with regard to increasing the spleen and thymus index, the spleen lymphocyte activity, the sum of white blood cells, as well as IgG function in mice serum, and up-regulate the expression of TNF-α, IFN-γ, and IL-1β mRNA in high-dose groups. Meanwhile, CMP exhibited significant antioxidant activity and decreased the level of malondialdehyde, suggesting that protection against oxidative stress was a crucial aspect when interpreting the immune function (Liu et al., 2016).

4.4.3 Antioxidant Activity

Yu et al. (2006) showed that extracts of *Co. sinensis* (CSE) protected lipid, protein, and low-density lipoprotein (LDL) against oxidative damage. The polysaccharide in CSE displayed antioxidant activity, which suggested that the activity might be derived partly from polysaccharides of CSE. Wang et al. (2005) also found that fractions of *Co. sinensis* with a higher concentration of total polysaccharides showed a stronger ability to scavenge free radicals. Furthermore, Dong and Yao (2008) also found that hot-water extracts containing polysaccharides from natural and cultured mycelia of *Co. sinensis* exhibited direct and potent antioxidant activities. They used several in vitro assays to study this behavior: inhibition of linoleic acid peroxidation, scavenging of DPPH, hydroxyl and superoxide anion radicals, reducing power, and the chelating ability of ferrous ions. There was no correlation between the antioxidant activity and the content of protein, polysaccharides, or mannitol. The antioxidant activity might, therefore, be attributed to a combination of these factors or some other compounds. But Li et al. (2001) found that the antioxidative activities increased 10–30-fold in partially purified polysaccharide fractions from the cultured *Cordyceps* mycelia compared with the water extracts. The antioxidative activity might then be partly derived from *Cordyceps* polysaccharides.

Li et al. (2003) first reported and identified a polysaccharide of molecular weight ~210 kDa from *Cordyceps* that protected against free radical-induced neuronal cell toxicity. A pretreatment of the polysaccharide on cultured rat pheochromocytoma PC12 cells showed a strong protective effect against H_2O_2-induced damage, with a significant increase in the survival of PC12 cell in medium over 60% at a concentration of 100 μg/mL. In parallel, the H_2O_2-induced production of malondialdehyde (MDA) in cultured cells was obviously reduced after treatment with the polysaccharide. Furthermore, the changes in glutathione peroxidase (GSH-Px) and superoxide dismutase (SOD) activities in H_2O_2-treated cells were significantly attenuated in a dose-dependent manner. Similarly, the protective capacity of an acid polysaccharide (APS) from cultivated *Cordyceps* mycelia against H_2O_2-induced oxidative damage in rat pheochromocytoma PC12 cells was assessed by measuring cell viability, lactate dehydrogenase (LDH) release, antioxidative enzyme activity, MDA levels, and intracellular accumulation of reactive oxygen species (ROS) and Ca^{2+}. Results revealed that pretreatment of APS to PC12 cells, followed by H_2O_2-induced damage, significantly promoted the survival of cells and the activities of GSH-Px, catalase (CAT), and SOD, as well as reducing the levels of LDH and MDA. Besides, the intracellular accumulation of ROS and Ca^{2+} were also inhibited. These results indicated that APS could improve the cellular antioxidant defense ability, and therefore, protect PC12 cells against oxidative stress (Shen et al., 2011).

The antioxidant activity of the polysaccharide (PS) from cultured *Co. sinensis* fungus on H22-bearing mice showed that the H22 tumor growth was significantly inhibited by PS. Moreover, PS could significantly enhance SOD activity in liver, brain, and serum, as well as GSH-Px activity of liver and brain in tumor-bearing mice. The PS could also significantly reduce the level of MDA in liver and brain of tumor-bearing mice (Chen et al., 2006b). In addition, the EPS from the mycelial liquid culture of *Co. sinensis* fungus showed moderate antioxidant activities with a Trolox equivalent antioxidant capacity of 35–40 μmol Trolox/g, and a ferric reducing ability of plasma of 50–52 μmol Fe(II)/g (Leung et al., 2009). But, Yan et al. (2009b) found that the hydrolyzed fractions of exopolysaccharides from *Co. sinensis* mycelial culture had much higher (30%–80%) antioxidant and radical-scavenging activities.

The antioxidant activity of intracellular selenium-enriched polysaccharide (ISPS) was studied systemically both in vitro and in vivo. The in vitro scavenging effects of ISPS on hydroxyl, superoxide, and DPPH radicals were $74.62 \pm 4.05\%$, $71.45 \pm 3.63\%$, and $79.48 \pm 4.75\%$, respectively, at a concentration of 1.0 mg/mL, and the reducing power of ISPS was 0.45 ± 0.01. On the other hand, its antioxidant effect on streptozotocin-induced rats was studied. After 4 weeks, malonaldehyde contents in heart, liver, and kidney tissues of the group administered with ISPS were significantly lower than those of model groups, but in pancreas tissue there was no significant difference. Total antioxidant capacities in liver, pancreas, and kidney tissues of the ISPS group were significantly higher than those of the model group, except for that in heart tissue, which was not significantly different. Serum total antioxidant capacity was also increased compared with that of the model group (Yang and Zhang, 2016). The in vivo antioxidant effects of ISPS from *Co. sinensis* SU-02 mycelium in submerged culture were studied. The results showed that the activities of GSH-Px and SOD in mice blood were 2.36 ± 0.17 U and 477.92 ± 35.14 U, respectively, and MDA level was 0.65 ± 0.04 μM/L. The in vitro inhibition effect of ISPS on superoxide anion radical and hydroxyl radical were $38.53 \pm 3.19\%$ and $43.22 \pm 3.29\%$, respectively, and the reducing power was 0.61 ± 0.04 (Wang et al., 2011a).

The studies on antioxidant activities of the polysaccharide from the fruiting bodies of *C. militaris* CM-hs, namely CM-hs-CPS2, were evaluated by various in vitro methods. Results showed that DPPH scavenging activity, reducing power, and ferrous ion chelating activity of CM-hs-CPS2 were 89%, 1.188, and 85%, respectively, at a concentration of 8 mg/mL (Wu et al., 2011). Yu et al. (2007, 2009) revealed that the polysaccharide P70-1 and CBP-1 from the fruiting bodies of *C. militaris* had hydroxyl radical-scavenging activity in vitro with an IC50 value of 0.548 and 0.638 mg/mL, respectively. Fan et al. (2011) showed that the inhibition effects of IPS, a polysaccharide from *C. militaris* SU-08, on hydroxyl, superoxide anion, and DPPH radical were $51.89 \pm 3.27\%$, $82.26 \pm 5.03\%$, and $74.59 \pm 4.53\%$, respectively, at a concentration of 5 g/L, $32.98 \pm 2.71\%$, $69.71 \pm 4.24\%$, and $41.64 \pm 3.28\%$ higher than that of the positive control, respectively. The reducing power of IPS was 0.79 ± 0.03, which was 11.27 ± 0.82% higher than the positive control. On the other hand, the exopolysaccharide from the culture broth of *C. militaris* SU5-08 was also investigated in vitro for antioxidant activity (Chen et al., 2013). The scavenging effects on hydroxyl, superoxide anion, and DPPH radicals were 63.64 ± 3.52%, 75.27 ± 5.16%, and 6.46 ± 5.03% at a concentration of 5 g/L, respectively, and the reducing power was 0.21 ± 0.01 (Lin et al., 2012). The results suggest that both the IPS and EPS of *C. militaris* SU5-08 can be explored as a potential antioxidant. Chen et al. revealed that polysaccharide fractions from cultured *C. militaris* C19, W-CBP50 II, are capable of scavenge DPPH, while another two fractions, W-CBP50 and W-CBP50I, could scavenge DPPH, hydroxyl, and superoxide radical.

The in vivo reactive oxygen species scavenging activity of *C. militaris* polysaccharides (CMP) was studied by Wang et al. (2012) in a CY-induced immunosuppression mice model. It was obvious to find that CMP can significantly increase the level of SOD, CAT, GSH-Px, and T-AOC, as well as decrease the level of MDA in heart, liver, and kidney.

Cordyceps jiangxiensis, also called "CaoMuWang", is a medicinal entomopathogenic macrofungus native to eastern China. Polysaccharide from *C. jiangxiensis* was evaluated for antioxidant activity through five in vitro assessments, i.e., DPPH, hydroxyl and superoxide anion radical scavenging activities, as well as reducing power and chelating ability on ferrous ions. Results showed that the polysaccharide presented higher scavenging abilities against superoxide anion radicals than the positive control. Compared with the positive control, the polysaccharide only had moderate scavenging effects on DPPH and hydroxyl radicals, and moderate reducing power and ferrous ion chelating ability (Xiao et al., 2012a, 2013). Simliarly, Zhu et al. (2011) pointed out that CPS50-I, a polysaccharide from *C. gunii*, could effectively protect against oxidation resistance in a D-galactose (D-Gal)-induced aging mice model.

In the previous study, Xiao et al. (2012a) investigated the antioxidant properties of polysaccharide from *C. taii* systematically. *In vitro* assays evaluated the scavenging activities of the aqueous extract and polysaccharides of *C. taii* against various radicals: DPPH, hydroxyl radical, and superoxide anion radical. The EC50 value for scavenging superoxide radical ranged from 2.04 to 2.49 mg/mL, at least 2.6 times higher than that of the antioxidant thiourea. The polysaccharides also significantly increased the antioxidant enzyme activities (SOD, CAT, and GSH-Px), and markedly decreased the MDA production of lipid peroxidation in a D-Gal-induced aging mice model. It was interesting to find that the immune function of the polysaccharide-treated groups was significantly boosted compared to the model group, indicating the close relationship between immune function and antioxidant activity. Recently, the in vitro antioxidant activities of alkali-soluble polysaccharide from *C. taii* were also evaluated in terms of DPPH, hydroxyl, and superoxide anion radical scavenging activities, the reducing power, and the Fe^{2+} chelating ability. Compared with the crude polysaccharide fraction (ACP), the refined polysaccharide fraction (ARP) exhibited more excellent scavenging abilities towards hydroxyl and superoxide anion radicals, with EC50 values of 2.72 ± 0.80 and 10.46 ± 1.97 mg/mL. On the contrary, ACP presented a slightly stronger scavenging activity against DPPH radicals. On the other hand, compared with the positive control, ARP showed a more significant superoxide anion radical scavenging effect, but moderate reducing ability (Xiao et al., 2016).

Polysaccharides from *P. hepiali* HN1 (PHMPs), composed of mannose, ribose, rhamnose, glucuronic acid, galacturonic acid, glucose, galactose, xylose, and arabinose in a molar ratio of 46.07:0.59:2.25:1.29:1.42:18.82:26.17:1.13:2.26, were further investigated to measure their protective effect against oxidative stress induced by D-galactose aging mice. As shown in the results, there were higher activities of SOD, CAT, GSH-Px, and level, of T-AOC, as well as lower levels of MDA in serums and livers in the PHMPs-treated group compared with the model group. The aging-related reduction in T-AOC level indicated that the decrease in the nonenzymatic antioxidant defense may also probably be attributed to endogenous lipid peroxidation, which needs further detailed research with regard to the mechanism so as to confirm the structure and functionality relationship group (Wu et al., 2016).

Cordyceps sobolifera polysaccharide (CSP), with a molecular weight of 1.53×10^4 kDa and composed of mannose, glucose, and galactose in a molar ratio of 1.7:8.9:1, was reported to possess potent antioxidant activity in vitro. It had reducing power with a value of 0.87 at 1.5 mg/mL, and scavenging abilities against hydroxyl radical (scavenging rate was 47.6% at 5 mg/mL) and DPPH radicals (scavenging rate was 55.6% at 1 mg/mL), both of which exhibited a concentration-dependent manner. (Lu et al., 2016).

Two polysaccharides isolated from the stroma of cultured *Cordyceps kyushuensis* by hot-water extraction and purified by ion-exchange and gel-permeation chromatography were tested for various antioxidant properties in in vitro systems. The data demonstrated that both of the polysaccharides possessed potent antioxidant activities in a dose-dependent manner, especially the scavenging effect on hydroxyl radicals. Both of the polysaccharides showed strong protective capacity against DNA damage (Zhang et al., 2015).

4.4.4 Hypoglycemic Effect

The polysaccharide-enriched fraction of *Cordyceps* can produce hypoglycemic effects in streptozotocin-induced diabetic rats. It is worthy of note that the polysaccharide-enriched fraction from *C. militaris* had a more potent effect (approximately 10-fold more) than that from *Co. sinensis*, in a dose-dependent manner (Zhang et al., 2006). Crude polysaccharide from the cultured mycelium of *Co. sinensis* showed significant hypoglycemic activities in normal mice, alloxan diabetes model mice, and streptozotocin diabetes model mice when given by intraperitoneal injection at a dose of 100 and 50 mg/kg. There was a dose-effect relationship to a certain extent (Ji et al., 1993).

Crude polysaccharides from cultured mycelium of *Co. sinensis* by hot-water extraction and from alkaline extraction showed significant activity in normal mice and in streptozotocin (STZ)-induced diabetic mice when delivered by intraperitoneal injection. The crude polysaccharide obtained by 5% sodium hydroxide extract slightly lowered the plasma glucose level in normal mice after oral administration. However, the neutral polysaccharide (CS-F30) exhibited higher hypoglycemic activity than its crude polysaccharide, and it significantly lowered the glucose level after oral administration (50 mg/kg). However, the same treatment hardly affected the plasma insulin level in normal mice (Kiho et al., 1993). Further work found that CS-F30 showed potent hypoglycemic activity in genetic diabetic mice after intraperitoneal administration, and the plasma glucose level was quickly decreased in normal and model groups after intravenous administration. However, administration of CS-F30 significantly increased the activities of hepatic glucokinase, hexokinase, and glucose-6-phosphate dehydrogenase, although the content of glycogen in liver was reduced. Besides, CS-F30 lowered the plasma triglyceride and cholesterol level as well (Kiho et al., 1996). The CS-F10 purified from a hot-water extract of the cultural mycelium of *Co. sinensis* could significantly lower the plasma glucose level in normal, STZ-induced diabetic, and epinephrine-induced hyperglycemic mice after intraperitoneal administration (50 mg/kg). Similar to CS-F30, the administration of CS-F10 could significantly improve the activity of hepatic glucokinase, reduce the hepatic glucose output. The CS-F10 also significantly reduced the protein content of glucose transporter isoform-2 from rat liver after intraperitoneal administration (Kiho et al., 1999).

The hypoglycemic effects of CSP-1 on normal and alloxan-diabetic mice and STZ-diabetic rats were also examined. The basal glucose level did not vary dramatically from that of normal mice. However, adminstration of CSP-1 at 200 and 400 mg/kg body weight per day for 7 days, p.o., significantly reduced the blood glucose level by $12.0 \pm 3.2\%$ and $22.5 \pm 4.7\%$ in normal mice, respectively ($P < 0.05$). When the dose was higher than 200 mg/kg body weight, CSP-1 produced a remarkable reduction in blood glucose level in both STZ- and alloxan-induced diabetic mice. The levels of serum insulin in diabetic mice were also increased ($P < 0.05$). The CSP-1 with hypoglycemic properties improved the circulating insulin level in diabetic mice, suggesting that CSP-1 might stimulate the pancreatic release of insulin and/or reduce insulin metabolism (Li et al., 2006c). Hypoglycemic effects were induced by polysaccharides from cultured mycelium of *Co. sinensis* (PCS), and a mechanism was proposed. Fasting serum glucose level and glycosylated serum protein concentration in diabetic mice were reduced significantly, but comparable effects on fasting serum glucose were not found with normal mice. The PCS was found also to reduce the increase in serum glucose level in diabetic mouse. Also, glucose taken into the insulin resistant (IR) adipocytes was raised by PCS. The results indicate that the hypoglycemic effects of PCS in diabetic mouse were, at least in part, due to an increase in glucose taken into IR adipocytes (Huang et al., 2002).

4.4.5 Anti-Tumor Activity

Wang et al found that fractions of *Co. sinensis* with higher concentrations of total polysaccharides had a strong inhibitory effect on the growth of colorectal and hepatocellular cancer cells by apoptosis (Wang et al., 2005). *Cordyceps* polysaccharide increased the proportion of ANAE+ cells in PBL, strengthened the DTH reaction and phagocytic activity of MΦ, and inhibited tumor growth in tumor-bearing mice. *Cordyceps* polysaccharide has, therefore, the potential to be used for immunotherapy of clinical tumors (Belli and Zudao, 1992). Furthermore, cordyglucans from cultured *Co. sinensis* were found to exhibit potent anti-tumor activity, and this activity can be correlated with their proportion of $(1 \rightarrow 3)$-β-D-glucan linkages (Wu et al., 2005).

A water-soluble polysaccharide, MCMP, obtained from the mycelium of *C. militaris*, was found to have a potent inhibitory effect on the Hep-G2 cell, Hela cell, and mesangial cell proliferation rate after 48 h treatment as determined by MTT assay (Zhang et al., 2010). Similarly, Yang et al. (2014) also revealed that the polysaccharide from *C. militaris* CGMCC 2909 showed a significant inhibitory activity against Hela and HepG2 cells in vitro, in a dose-dependent manner; in 72 h with a concentration of 1000 mg/mL the respective inhibition rate was 92.38% and 98.79%, and the IC50 was 70.91 μg/mL and 97.63 μg/mL, respectively. The tumor inhibition ratio on K562 cell by polysaccharide from *C. gunii* was 56.65% (Zhu et al., 2012b). An in vitro assay showed that polysaccharide from *Co. sinensis* mycelium exhibited weak cytotoxicity against an SPC-I (IC50 65 μg/mL) cancer line, but no obvious cytotoxic activity against another two cancer lines, BCAP37 (IC50 > 100 μg/mL) and SW480 (IC50 > 100 μg/mL) (Wu et al., 2007).

Exopolysaccharide fractions of cultivated *Co. sinensis* (EPSF) significantly lowered c-Myc, c-Fos, and VEGF levels in the lungs and livers of EPSF-treated mice compared with untreated mice ($P < 0.05$). The EPSF might then inhibit tumor growth in the lungs and livers of mice, and could be a potential adjuvant in cancer therapy (Yang et al., 2005). The effect of EPSF on immunocyte activity of H22 tumor-bearing mice was also evaluated, and the results indicated that EPSF not only significantly inhibited the growth of H22 tumor, but also obviously elevated the activity of immunocytes. It might thus enhance the phagocytotic capacity of peritoneal macrophages, and the proliferation ability of spleen lymphocytes at all the tested doses, and increase the TNF-α expression of macrophages, as well as improve the cytoxicity of spleen lymphocytes. The EPSF could also significantly enhance TNF-α and IFN-γ mRNA expression of splenic lymphocytes (Zhang et al., 2008).

The effect of polysaccharides from *Co. sinensis* (PSCS) on triptolide (TPL)-induced apoptosis in HL-60 cells was investigated. Results showed that PSCS could significantly inhibit the cell viability, enhance apoptosis rate of the HL-60 cells, decrease the protein expression of caspase-3, 6, 7, and 9, and NF-κB in a dose-dependent manner (Shen et al., 2009). The anti-tumor effects of two novel polysaccharides, WIPS and AIPS, respectively from hot-water and alkaline extracts of *Co. sinensis* (strain Cs-HK1) were explored. The AIPS was shown to have a much more significant anti-tumor effect than WIPS in animal tests using melanoma tumor-bearing mice (Yan et al., 2011).

A combination of the polysaccharide-rich fraction of *Co. sinensis* (CS) and cisplatin on H157 were used to treat non-small cell lung cancer (NSCLC) cells and detect the cell viability, cytotoxicity, and apoptosis, as well as expression of vascular endothelial growth factor (VEGF) and basic fibroblast growth factor (bFGF) in H157 cells. Results showed that cell viability was significantly reduced, and the levels of VEGF and bFGF protein expression were obviously decreased as well in the cells treated with a combination of CS and cisplatin, compared with the cells treated with cisplatin alone. This finding indicated that the polysaccharide could effectively inhibit tumor growth in NSCLC (Ji et al., 2011).

4.4.6 Protective Effect on Liver Disease

Cordyceps polysaccharides have been shown to protect against immunological liver injury in mice. Lipid peroxidation of immunological liver injury in mice was inhibited by all six types of *Cordyceps* polysaccharides, by decreasing the level of MDA and increasing the activity of SOD in the liver homogenates ($P < .05 \sim .01$) (Fan, 1999). The CP has been shown to improve the function of liver and peripheral blood lymphocytes in patients with chronic hepatitis C, and may prevent liver fibrosis (Ma et al., 1998).

Reactive oxygen species will cause hepatocyte apoptosis in hepatic fibrosis related to oxidative damage of the liver. As a result, Liu et al. studied the antioxidant compounds that may shed light on improvement in hepatic fibrosis. They isolated a major antioxidative polysaccharide of *C. militaris* (CPS), and investigated the effect of hydrogen peroxide (H_2O_2)-induced cell apoptosis. The data showed that CPS could obviously inhibit H_2O_2-induced mitochondrial dysfunction, reduce cell viability, improve the apoptotic rate, increase ROS production, reduce mitochondrial membrane potential (MMP), decrease the intracellular adenosine triphosphate (ATP) level, and enhance the Bax/Bcl-2 ratio, as well as increase cytochrome C (Cyt C) release. Therefore, it was indicated that CPS could protect HL-7702 cells against H_2O_2-induced mitochondrial dysfunction via decreasing the production of

ROS and regulating the mitochondrial apoptotic signaling that related to Cyt C, Bax, and Bcl-2 apoptosis (Liu et al., 2013).

Cordyceps polysaccharides attenuate liver fibrosis, decrease hepatic Hyp content and collagen production, reduce transforming growth factor β1 and its receptor expression, and decrease cell Dm expression. *Cordyceps* polysaccharides therefore have good potential against liver fibrosis, by inhibiting hepatic stellate cell activation and transforming growth factor β1 expression (Fang et al., 2000). At a dose of 1.0 ~ 40 μg/mL, CP inhibited hepatic stellate cell (HSC) proliferation and collagen synthesis in a drug concentration-dependent manner. The CP might then inhibit HSC proliferation and collagen synthesis, and it may be one of the routes leading to the depression of hepatic fibrosis by CP (Lu et al., 2003). Another study showed that CP reduced liver fibrosis by inhibiting the degradation of the deposited collagens and anti-lipid peroxidation (Li et al., 2006a).

Cordyceps polysaccharides also had a significant suppression effect on rat hepatic stellate cells and the activity of NF-κB, as well as down-regulating the expression of cytokine tumor necrosis factor in a dose-dependent manner. Besides, it also reduced the proliferation of hepatic stellate cells by inhibiting the NF-κB activity and TNF-α expression (Yan et al., 2003). The CP obviously inhibited the cell proliferation and collagen synthesis in a dose- and time-dependent manner, and reduced the expression of procollagen type I, III mRNA in vitro, which further confirmed the potent anti-fibrotic properties of CP (Jing et al., 1999).

Cordyceps polysaccharide coated with liposome and pure *Cordyceps* polysaccharide were used to treat patients with chronic active hepatitis B. The results indicated that there was an improvement in the liver functions of both groups. The HBeAg disappearance rates were 53.2% and 24%, respectively; and HBV-DNA disappearance rates were 68.4% and 30%, respectively. It was concluded that *Cordyceps* polysaccharide was useful in the treatment of chronic active hepatitis B, and that *Cordyceps* polysaccharide coated with liposome was more effective than simple *Cordyceps* polysaccharide (Chen et al., 1992). Yan et al. found polysaccharides from *C. militaris* with ultrasonic-assisted extraction could significantly reverse ethanol-induced hepatocyte injury in primary cultured rat hepatocytes, contributing to increased cell activity and enhancing the expression of the enzymatic system related to antioxidant function in a dose-dependent manner (Yan et al., 2008).

4.4.7 Anti-Platelet Effect

Lu et al. studied the antithrombotic effect of polysaccharide from *C. militaris* on platelet activation in vivo and in vitro, and the possible related mechanism, on the other hand, was also studied. Data showed that CME-1 had a potent antiplatelet effect initially on activating adenylate cyclase/cyclic AMP, and then on inhibiting intracellular signals (such as Akt and MAPKs), ultimately inhibiting platelet activation (Lu et al., 2014). In their recent study, the mechanisms of CME-1 in platelet activation were revealed, which may initially inhibit the PLCγ2-PKC-p47 cascade, suppress PI3-kinase/Akt and MAPK phosphorylation via adenylate cyclase/cyclic AMP activation, and then inhibit intracellular [Ca^{+2}] mobilization (Chang et al., 2015).

4.4.8 Antimutagenic Effect

The antimutagenic effects of *Cordyceps* polysaccharide were studied by testing the micronucleus of *Vicia faba* root tip cells, and this showed that it could effectively inhibit the mutation induced by cyclophosphomide in a dose-dependent manner. When the concentrations of *Cordyceps* intracellular polysaccharide and exopolysaccharide were 100 μg/mL, the respective micronucleus rates were 7.333‰ and 7.556‰, the inhibitory rates were 53.52% and 52.11%, respectively (Hongwei, 2005). *Cordyceps* polysaccharide on micronucleus of *V. faba* root tip cells caused by mitomycin-c (MMC) inhibited the yield of micronucleus of *V. faba* root tip cells caused by MMC (Chen et al., 2006a).

4.4.9 Longevity and Anti-Aging

Exopolysaccharide isolated from *Co. sinensis* can prolong the life-span of male fruit by 31.5% (Yu et al., 2002a). It is also able to protect against photoaging of the dermis layer of skin in mice (Li et al., 2008). The CP alleviated cell crimple could increase the activity of cells, decrease the content of MDA, and increase the level of HYP and the activity of SOD. The CP had a protective effect on photoaging skin fibroblast (Li et al., 2009).

Damage to DNA in UVB-irradiated human fibroblast cells was significantly lowered ($P < .01$) after 30 min and 24 h of *Cordyceps* pretreatment. There was a 27% reduction in cyclobutane-pyrimidine dimers (CPDs) in irradiated cells with 24 h pretreatment with 200 μg/mL of the hot-water *Cordyceps* extract, and a 34% reduction with 24 h pretreatment with 200 μg/mL of the exopolysaccharide extract. Clear evidence of protection against UVB-induced CPDs was seen with *Cordyceps* mycelial extracts. *Cordyceps* may thus offer photoprotection and lower the risk of basal cell carcinoma, the main skin cancer caused by CPDs. Further study is needed to identify protective mechanisms (Wong et al., 2011).

4.4.10 Other Activity

Deng et al. (2015) revealed that polysaccharide from mycelia of *Co. sinensis* (PMCS) has a favorable effect on pacific white shrimp (*Litopenaeus vannamei*) in terms of growth performance (including weight gain and feed conversion ratio), immunity, and antioxidant function. Hemolymph immunity indicators, including phenoloxidase (PO), alkaline phosphatase (ALP), acid phosphatase (ACP) and lysozyme (LSZ), were significantly higher in the PMCS group ($P < 0.05$). Additionally, antioxidative parameters, such as SOD, GSH, ROS, and TAC, were significantly higher in the treated groups. However, no significant differences in levels of MDA and GSH-Px were observed between the control and treatment groups. Moreover, the level of vibrio ichthyoenteri was lower in the PMCS group ($P < 0.05$). Similar findings had been reported previously (Chang et al., 2001). Jiang et al. (1998) also found that *Cordyceps* polysaccharides, as a feed additive, could enhance the immunity of *Macrobrachium nipponense* and *Penaeus chinensis*.

A heteropolysaccharide obtained from water extracts of *Co. sinensis* possessed a strong inhibitory activity against cholesterol esterase, with an IC50 value of 12.7 μg/mL in in vitro enzyme assay (Kim, 2010). Ohta et al. (2007) revealed that the polysaccharide from *C. militaris* (APS) could decrease virus titers in the bronchoalveolar lavage fluid and the lung of mice infected with influenza A virus, and improve the survival rate. Additionally, APS increased the levels of TNF-α and IFN-γ in mice compared with those of untreated mice. On the other hand, APS could promote NO production and induce iNOS mRNA and protein expression in RAW 264.7 cells. An increase of mRNA expression of cytokines including IL-1β, IL-6, IL-10, and TNF-α was also demonstrated. These results indicated that APS might have a beneficial therapeutic effect on influenza A virus infections which may in part be through the modulation of the immune function of macrophages. The CPS-1, another polysaccharide isolated from *C. militaris*, was shown to possess a significant anti-inflammatory activity, and suppressed the humoral immunity in mice, but had no significant effect on the cellular immunity and non-specific immunity (Yu et al., 2004b).

A polysaccharide from *Co. sinensis* mycelia (CME-1) was shown to have an anti-inflammatory effect in RAW 264.7 cells. The CME-1 could obviously protect RAW 264.7 cells against H_2O_2-induced injury through inhibition of sphingomyelinases (SMase) activity and a decrease in C16- and C18-ceramide levels. Pretreated with CME-1, the intracellular oxidative stress and mitochondrial membrane depolarization was diminished, and it was suggested that CME-1 acted as an intracellular antioxidant in a dose-dependent manner on H_2O_2-treated RAW264.7 cells through a mechanism other than scavenging activity (Wang et al., 2011b).

The CSP1 fraction of *Co. sinensis* polysaccharides was investigated for their effect on spontaneously hypertensive rats (SHR). Pharmacological tests showed that CSP1 could stimulate the secretion of vasodilator NO, decrease the level of ET-1, epinephrine, noradrenaline, and angiotensin II, inhibit the production of transforming growth factor β1 (TGF-β1), and reduce the level of inflammatory mediator of C-reactive protein (CRP), and these indicated that CSP1 possessed high potential in alleviating hypertension (Xiang et al., 2016).

Wang et al. (2015b) isolated residue polysaccharide from *C. militaris* SU-12 (RPS) and investigated its effect on lipid metabolism and oxidative stress. Data showed that RPS, with an average molecular weight of 2.86 × 103, and composed of glucose, arabinose, and mannose with a ratio of 62:1.6:1, obviously reduced the levels of blood and liver lipid, as well as improving the glutamate pyruvate transaminase and antioxidant effect. The histopathological results of mice livers clearly demonstrated that RPS could significantly attenuate liver cell injury (Wang et al., 2015b).

4.5 DISCUSSION: STRUCTURE–BIOACTIVITY RELATIONSHIP

It is important to interpret the connection between the polysaccharides structure and their biological activities. However, the relationships between bioactivity and structural features, as well as chain conformation are difficult to elucidate. Specifically, detailed information with regard to solution properties and chain conformation in aqueous solutions require further studies using modern technologies, such as TEM (transmission electron microscope), SEM (scanning electron microscopy), AFM, AFM-based single-molecule force spectroscopy, fluorescence correlation spectroscopy, static and dynamic light scattering, viscosity analysis based on the theory of dilute polymer solutions, circular dichroism analysis, computer modeling, and NMR spectroscopy (Yang and Zhang, 2009; Pérez et al., 1996; Striegel et al., 1999; Pol-Fachin et al., 2009; Brant and Burton, 1981).

According to previous research, it is not hard to figure out that structural characteristics, together with structure-related aspects, will lead to differences in bioactivities, such as anti-tumor, antioxidant, and immunomodulatory activities. Wu et al. compared the chemical structure and bioactivity of polysaccharides obtained from 39 batches of cultured *C. militaris* fruiting bodies. Using saccharide mapping, and partial acidic/enzymatic (α-amylase and β-glucanase) digestion followed by polysaccharide analysis using carbohydrate gel electrophoresis analysis (PACE), it was found that 1,4-α-D-glucosidic and 1,4-β-D-glucosidic linkages existed in cultured *C. militaris* and natural *Co. sinensis*. In addition, a relatively high similarity was observed among polysaccharides from different batches of *C. militaris*, as well as between *C. militaris* and natural *Co. sinensis*, suggesting that *C. militaris* might, at least partially, be used as a substitute for natural *Co. sinensis*. On the other hand, NO production by RAW 264.7 cells stimulated by polysaccharide from *C. militaris* and *Co. sinensis* turned out to be significantly correlated with the two kinds of linkages ($P < .01$) (Wu et al., 2014c). Besides, molecular weight was also reported to be an important factor that influences the bioactivity of polysaccharide. Yan et al. used an acidic solution to hydrolyze the exopolysaccharide from *Co. sinensis* to obtain a degraded exopolysaccharide that only cleaved the glycosidic linkages, but without any changes in the primary structure. The degraded polysaccharide, as expected, had much higher antioxidant activity (30%–80%) and radical-scavenging effect (Yan et al., 2009b).

Factors that will influence the structure of polysaccharides, such as the extraction process, for example, extraction treatment with different solvents, or in combination with new technologies, such as ultrasound and microwaves, as well as multi-purification methods, the originality of raw material and substitution of functional groups, tend to result in a change in pharmacological function. Xiao et al. carried out an ultrasonic extraction (USE) technology to isolate crude polysaccharide from *C. cephalosporium* mycelia. The polysaccharide was preliminarily purified by successive subfractionation with gradient concentrations of ethanol (40% and 70%), and applied to a Sephadex G-100 gel filtration column to obtain two major fractions (USEP40-1, USEP70-1). The molecular weights of USEP40-1 and USEP70-1 were 61.4 kDa and 25.1 kDa. The USEP40-1 and USEP70-1 were composed of D-mannose, D-glucose, D-galactose, L-rhamnose, and L-arabinose with a molar ratio of 11.52:5.54:8.75:2.45:2.59, and 11.50:6.74:5.75:4.46:2.39, respectively (Xiao et al., 2012b). Wu et al. (2012) evaluated the antioxidant activities of CM-jd-CPS2 and CM-jd(Y)-CPS2, isolated from the fruiting bodies of cultured *C. militaris* grown on solid rice medium and silkworm pupa, respectively, by various in vitro assessments. Results showed that both of them exhibited strong DPPH scavenging activity and ferrous ion chelating capacity, but moderate reducing power. The antioxidant activities of CM-jd (Y)-CPS2 were slightly higher than that of CM-jd-CPS2. In addition, these two acidic fractions were evaluated for in vitro proliferation of mouse splenocyte activity as well. Data showed that they both possessed mitogenic effects on mouse splenocytes in a dose-dependent manner, and synergistically promoted murine Con A-induced T- and LPS-induced B-proliferation. The CM-jd(Y)-CPS2 exhibited a stronger immunostimulation effect than CM-jd-CPS2. A series of EPS fractions (P1/5, P2/5, P1, P2, and P5) were isolated from the fermentation medium of *Co. sinensis* by gradient precipitation with ethanol with a proportion of 1/5, 2/5, 1, 2, and 5 volumes based on the resulting liquid medium. The P1/5 and P2/5 had a negligible protein content and a large molecular or particle size (intrinsic viscosity ($[\eta]$) 2025 mL/g; hydrodynamic radius (Rh) 905 nm). The fractions attained at a higher concentration of ethanol had a higher protein content but a lower carbohydrate content with a smaller molecular size. Particularly, P5 mainly consisted of protein with an average molecular weight of 16 kDa, $[\eta]$ of 4.3 mL/g, and Rh of 23.5 nm. The antioxidant capacity of EPS fractions exhibited a positive correlation with the protein content, since P1/5 and P2/5 showed negligible ability, P1 and P2 low or moderate, but P5 displayed very strong activity. Gradient ethanol precipitation was proven to be a convenient and feasible approach for fractionation of polysaccharide, protein, and their complexes with different molecular sizes (Huang et al., 2013).

Changes in conformation of polysaccharide also could result in an enhancement of bioactivity. Xiao et al. depolymerized polysaccharide from *C. gunii* by ultrasonic treatment, and further purified by a Sephadex A225 column. The effect of the polysaccharide (FP) and its depolymerization products (FP21, FP22, and FP23) was further evaluated for their effects on mouse spleen lymphocyte proliferation, mouse peritoneal macrophage (PMΦ) phagocytosis, and cytotoxin T lymphocyte (CTL) activity. Results showed that the depolymerizing fractions, especially FP2, had a higher immunocompetence than FP. Compared with control group, three kinds of depolymerizing fractions all had evident differences at doses of 1 μg/mL and 10 μg/mL (Xiao et al., 2004b). Zhu et al. (2014) investigated the immunocompetence changes in polysaccharide from *C. gunii* after it was depolymerized by ultrasonication treatment. The molecular weight and intrinsic viscosity was reduced, and the alpha-helicity was enhanced after the ultrasonic process. The main structure, on the other hand, remained unchanged after ultrasonic treatment in the tested conditions. The ultrasonically-treated polysaccharide with a lower molecular weight and intrinsic viscosity, exhibited a stronger anti-tumor activity. It was likely that molecular weight, intrinsic viscosity, and spatial structure had a significant effect on the anti-tumor activity of mycelial polysaccharide from *C. gunii* and ultrasonic treatment would possibly be an effective method for enhancing the anti-tumor activity of polysaccharide. The EPS-1 was produced by *Co. sinensis* fungus Cs-HK1 and sulfated with chlorosulfonic acid (CSA)-pyridine (Pyr) at different volume ratios. Four sulfated derivatives, SEPS-1A, B, C, and D were yielded with various degrees of substitution (DS, ranged from 0.25 to 1.38) and molecular weight (from 17.1 to 4.1 kDa). It was found that the sulfation of EPS-1 frequently happened at the C-6 hydroxyl groups attributing to the higher reactivity. In aqueous solution, the native EPS-1 had the conformation of random coils or aggregated networks, but the sulfated derivatives were changed to single helices. Obviously, the antioxidant activities of the sulfated derivatives were improved with the increasing DS and decreasing molecular weight, based on assessing the scavenging effect against hydroxyl and ABTS radicals. Therefore, sulfation was an effective and favorable process to improve the physicochemical properties and bioactivities of fungal polysaccharides (Yan et al., 2012a).

In a recent report, Zhu et al. (2016) showed the relationship between molecular conformation, physicochemical characteristics, and anti-tumor effects of polysaccharides extracted using five different extraction methods: room-temperature water extraction (RWE), HWE, MAE, UAE, and cellulase-assisted extraction (CAE). Different extraction methods, however, did not affect the primary structure of polysaccharide, although a change of intrinsic viscosity and spatial conformation were observed. Results showed that polysaccharides that had a clumpy structure exhibited better anti-tumor activity than those with a flaky appearance and helix structure, indicating the key role of intrinsic viscosity in exerting bioactivity. Microwave assisted extraction provides an available method to form this kind of structure, resulting the highest extraction rate and best anti-tumor activity.

The EPS and IPS collected from the mycelial liquid culture of the *Hirsutella spp.* liquid fermentation were precipitated with 50% ethanol (EPS-1, IPS-1), 65% ethanol (EPS-2, IPS-2), and 80% ethanol (EPS-3, IPS-3), respectively. The polysaccharide fractions precipitated at lower ethanol concentrations showed a lower content of neutral sugar and a larger molecular weight. The EPS-1, EPS-2, IPS-1, and IPS-2 were all composed of glucose, galactose, and mannose, but in EPS-3 and IPS-3 no galactose was detected. When evaluated by the scavenging ability of hydroxyl radicals, it was interesting to figure out that those polysaccharides with a higher protein content, lower neutral sugar content, and molecular weights between 10 and 20 kDa displayed better activity. Further analysis of significant correlations revealed that the antioxidant effect of the polysaccharides was correlated to the monosaccharide composition (mannose, $r = 0.942$; glucose, $r = -0.905$) (Meng et al., 2015).

REFERENCES

Akaki, J., Matsui, Y., Kojima, H., Nakajima, S., Kamei, K., Tamesada, M., 2009. Structural analysis of monocyte activation constituents in cultured mycelia of *Cordyceps sinensis*. Fitoterapia 80, 182–187.

An, D., Zhu, B., 2012. Optimization of enzyme extraction conditions of cordycepin polysaccharide by using response surface methodology. J. Anhui Agric. Sci. 25, 120.

Athukorala, Y., Jung, W.-K., Vasanthan, T., Jeon, Y.-J., 2006. An anticoagulative polysaccharide from an enzymatic hydrolysate of *Ecklonia cava*. Carbohydr. Polym. 66, 184–191.

Bai, Y., Wang, Y., Qi, X., Liu, X., 2001. Determination of polysaccharide in *Cordyceps sinensis* and *Cordyceps hawkesii*. Chin. J. Hosp. Pharm. 21, 19–20.

Belli, Z.Y.W.M.X., Zudao, W., 1992. Antitumor effect of cordyceptic polysaccharides in mice. Basic Med. Sci. Clin. 4, 013.

Brant, D.A., Burton, B., 1981. Solution Properties of Polysaccharides. ACS Publications.

Cai, L., Li, P., Lu, Z., 1998. Observation of the structure morphology of *Cordyceps* polysaccharide by atomic force microscope. J. Chin. Electron. Microsc. Soc. 18, 103–105.

Chang, M., Chen, X., Wu, Z., Hu, X., 2001. The effect of *Cordyceps* polysaccharide on *Macrobrachium nipponense* immune function. J. Huazhong Agric. Univ. 20, 275–278.

Chang, Y., Hsu, W.-H., Lu, W.-J., Jayakumar, T., Liao, J.-C., Lin, M.-J., et al., 2015. Inhibitory mechanisms of cme-1, a novel polysaccharide from the mycelia of *Cordyceps sinensis*, in platelet activation. Curr. Pharm. Biotechnol. 16, 451–461.

Chen, B., Tan, Y., Qiu, D., Zhang, Y., Yang, L., 1992. Clinical observation on *Cordyceps sinensis* (Berk.) polysaccharide liposome in the treatment of chronic active hepatitis B. Zhongguo Zhong Xi Yi Jie He Za Zhi. 3, 1–3.

Chen, H., Zhou, B., Zhu, Y., Liu, Q., 2005a. Study of the composition and optimum conditions for the submerged production of *Cordyceps* polysaccharide. J. Xuzhou Inst. Technol. 20, 78–83.

Chen, H.W., Zhu, Y.L., Hong, G.J., 2005b. Study on the inhibitory function of *Cordyceps* polysaccharide on micronucleus induced by cyclophoshomide. Food Ferment. Ind. 31, 9.

Chen, H., Zhu, Y., Wang, H., Fan, M., 2006a. Effects of *Cordyceps* polysaccharide on micronucleus formation in *Vicia faba* root tip cells caused by mitomycin-c. Food Sci. 27, 231–234.

Chen, J., Zhang, W., Lu, T., Li, J., Zheng, Y., Kong, L., 2006b. Morphological and genetic characterization of a cultivated *Cordyceps sinensis* fungus and its polysaccharide component possessing antioxidant property in H22 tumor-bearing mice. Life Sci. 78, 2742–2748.

Chen, J.-J., Wei, K.-L., Zhang, S.-X., Xie, F., 2009. Comparative studies on content of polysaccharide and mannitol from gansu *Cordyceps*. Guangzhou Chem. Ind. 7, 051.

Chen, L., An, L., 2009. Studies on the anti-tumour activity and immunomodulatory function of polysaccharide of Cordyceps sinensis. Shangdong Normal University (Natural Science).

Chen, W., Zhang, W., Shen, W., Wang, K., 2010a. Effects of the acid polysaccharide fraction isolated from a cultivated *Cordyceps sinensis* on macrophages in vitro. Cell. Immunol. 262, 69–74.

Chen, X.L., Wu, G.H., Gu, Q.X., Huang, Z.L., 2010b. Optimization of extraction process of polysaccharides from fruiting body of *Cordyceps militaris* L. Link by orthogonal test. Med. Plant 1, 92–94.

Chen, W., Yuan, F., Wang, K., Song, D., Zhang, W., 2012. Modulatory effects of the acid polysaccharide fraction from one of anamorph of *Cordyceps sinensis* on Ana-1 cells. J. Ethnopharmacol. 142, 739–745.

Chen, X., Wu, G., Huang, Z., 2013. Structural analysis and antioxidant activities of polysaccharides from cultured *Cordyceps militaris*. Int. J. Biol. Macromol. 58, 18–22.

Chen, X., Siu, K.-C., Cheung, Y.-C., Wu, J.-Y., 2014a. Structure and properties of a (1→3)-β-D-glucan from ultrasound-degraded exopolysaccharides of a medicinal fungus. Carbohydr. Polym. 106, 270–275.

Chen, X., Wang, Z., Hua, C., Zhou, F., Wang, R., Wang, X., et al., 2014b. Optimization of the extraction of *Cordyceps militaris* polysaccharides by combined ultrasonic wave treatment and enzyme hydrolysis. Food Sci. Technol. 39, 142–147.

Chen, Y.-J., Shiao, M.-S., Lee, S.-S., Wang, S.-Y., 1997. Effect of *Cordyceps sinensis* on the proliferation and differentiation of human leukemic U937 cells. Life Sci. 60, 2349–2359.

Cheung, J.K., Li, J., Cheung, A.W., Zhu, Y., Zheng, K.Y., Bi, C.W., et al., 2009. Cordysinocan, a polysaccharide isolated from cultured *Cordyceps*, activates immune responses in cultured T-lymphocytes and macrophages: signaling cascade and induction of cytokines. J. Ethnopharmacol. 124, 61–68.

Chiu, C.-H., Chyau, C.-C., Chen, C.-C., Lin, C.-H., Cheng, C.-H., Mong, M.-C., 2014. Polysaccharide extract of *Cordyceps sobolifera* attenuates renal injury in endotoxemic rats. Food Chem. Toxicol. 69, 281–288.

Cui, J.-D., Jia, S.-R., 2010. Optimization of medium on exopolysaccharides production in submerged culture of *Cordyceps militaris*. Food Sci. Biotechnol. 19, 1567–1571.

Cui, S.W., 2005. Food Carbohydrates: Chemistry, Physical Properties, and Applications. CRC Press.

Cui, W., Eskin, N., Biliaderis, C., 1995. NMR characterization of a water-soluble 1, 4-linked β-D-glucan having ether groups from yellow mustard (*Sinapis alba* L.) mucilage. Carbohydr. Polym. 27, 117–122.

Dague, E., Delcorte, A., Latgé, J.-P., Dufrêne, Y.F., 2008. Combined use of atomic force microscopy, X-ray photoelectron spectroscopy, and secondary ion mass spectrometry for cell surface analysis. Langmuir 24, 2955–2959.

Deng, B., Wang, Z., Tao, W., Li, W., Wang, C., Wang, M., et al., 2015. Effects of polysaccharides from mycelia of *Cordyceps sinensis* on growth performance, immunity and antioxidant indicators of the white shrimp *Litopenaeus vannamei*. Aquacult. Nutr. 21, 173–179.

Dong, C.-H., Yao, Y.-J., 2008. *In vitro* evaluation of antioxidant activities of aqueous extracts from natural and cultured mycelia of *Cordyceps sinensis*. LWT-Food Sci. Technol. 41, 669–677.

Dong, C.-H., Xie, X.-Q., Wang, X.-L., Zhan, Y., Yao, Y.-J., 2009. Application of Box-Behnken design in optimisation for polysaccharides extraction from cultured mycelium of *Cordyceps sinensis*. Food Bioprod. Process 87, 139–144.

Dong, Y., Hu, S., Liu, C., Meng, Q., Song, J., Lu, J., et al., 2015. Purification of polysaccharides from *Cordyceps militaris* and their anti-hypoxic effect. Mol. Med. Rep. 11, 1312–1317.

Dror, Y., Pyckhout-Hintzen, W., Cohen, Y., 2005. Conformation of polymers dispersing single-walled carbon nanotubes in water: a small-angle neutron scattering study. Macromolecules 38, 7828–7836.

Fan, K., Meng, G., Zhou, B., Deng, P., Liu, X., Jia, L., et al., 2011. Intracellular polysaccharide and its antioxidant activity by *Cordyceps militaris* SU-08. J. Appl. Polym. Sci. 120, 1744–1751.

Fan, M., 1999. Evaluation for the protective effect of *Cordyceps* polysaccharides on immunological liver injury in mice. Acta Univ. Med. Anhui. 34, 173–175.

Fan, W.-Q., Yin, H.-P., Zhou, C.-L., 2008. Separation, purification and activity of pharmacodynamic action of polysaccharide from *Cordyceps*. Chin. J. Bioproc. Eng. 1, 016.

Fang, B., Liu, P., Liu, C., Hu, Y., Xu, L., Li, F., et al., 2000. Effects and mechanism of *Cordyceps* polysaccharide on liver fibrosis induced by immunologic injury in rats. Shanghai J. Tradit. Chin. Med. 34, 37–40.

Fu, H.-T., Deng, C., Wang, M., Chen, J.-H., 2014. Extraction, structural identification and bioactivities of polysaccharide from fermented *Cordyceps*. Nat. Prod. Res. Dev. 1, 026.

Gao, F., Zhang, S., Liu, J., Yu, Y., 2009. Ultra high pressure extraction of polysaccharides from *Cordyceps militaris* (L. Fr) Link. Food Sci 16, 41–43.

Gao, F., Yu, Y.-L., Liu, J.-B., Zhang, S.-Q., 2010a. Study on extraction of polysaccharides from *Cordyceps militaris* by Soxhelt methods. Sci. Technol. Food Ind. 7, 047.

Gao, F., Yu, Y.-L., Liu, J.-B., Zhang, S.-Q., 2010b. Study on ultrasonic assisted extraction of polysaccharides from *Cordyceps militaris*. Food Res. Dev 31, 107–110.

Gao, F., Yu, Y., Liu, J., Zhang, S., 2010c. Study on extraction of polysaccharides from *Cordyceps militaris* assisted by hot water. Food Sci. Technol. 213–216.

Gao, Y., Cheng, J., He, L., Wu, X., Li, Y., Fu, L., et al., 2011. Research on extraction and separation of polysaccharides from solid-state fermentation substance of Cordyceps militaris. Forest By-Product and Speciality in China, pp. 1–4.

Gong, M., Zhu, Q., Wang, T., Wang, X., Ma, J., Zhang, W., 1990. Molecular structure and immunoactivity of the polysaccharide from *Cordyceps sinensis* (Berk.) Sacc. Chin. J. Biochem. Mol. Biol. 6, 486–492.

Gong, X., Ji, H., Lu, S., Li, S., Li, P., 1999. Effects of polysaccharides from cultured *Cordyceps sinensis* on murine immunologic function. J. China Pharm. Univ. 31, 53–55.

Guan, J., Yang, F.-Q., Li, S.-P., 2010. Evaluation of carbohydrates in natural and cultured *Cordyceps* by pressurized liquid extraction and gas chromatography coupled with mass spectrometry. Molecules 15, 4227–4241.

Guan, J., Zhao, J., Feng, K., Hu, D.-J., Li, S.-P., 2011. Comparison and characterization of polysaccharides from natural and cultured *Cordyceps* using saccharide mapping. Anal. Bioanal. Chem 399, 3465–3474.

Harris, P.J., Henry, R.J., Blakeney, A.B., Stone, B.A., 1984. An improved procedure for the methylation analysis of oligosaccharides and polysaccharides. Carbohydr. Res. 127, 59–73.

He, L., Ji, P., Cheng, J., Wang, Y., Qian, H., Li, W., et al., 2013. Structural characterization and immunostimulatory activity of a novel protein-bound polysaccharide produced by *Hirsutella sinensis* Liu, Guo, Yu, Zeng. Food Chem. 141, 946–953.

Hongwei, C., 2005. Study on the inhibitory function of *Cordyceps polysaccharide* on micronucleus induced by cyclophoshomide. Food Ferment. Ind. 31, 9.

Hou, A., Liu, Y., Meng, Q., Feng, Y., Zhu, K., Teng, L., 2009. Effect of *Cordyceps militaris* on adenine-induced rat chronic renal failure. Lishizhen Med. Mater. Med. Res. 20, 1103–1105.

Hsieh, C., Tsai, M.-J., Hsu, T.-H., Chang, D.-M., Lo, C.-T., 2005. Medium optimization for polysaccharide production of *Cordyceps sinensis*. Appl. Biochem. Biotechnol. 120, 145–157.

Hsu, T.-H., Shiao, L.-H., Hsieh, C., Chang, D.-M., 2002. A comparison of the chemical composition and bioactive ingredients of the Chinese medicinal mushroom DongChongXiaCao, its counterfeit and mimic, and fermented mycelium of *Cordyceps sinensis*. Food Chem. 78, 463–469.

Hu, T., Jiang, C., Huang, Q., Sun, F., 2016. A comb-like branched β-D-glucan produced by a *Cordyceps sinensis* fungus and its protective effect against cyclophosphamide-induced immunosuppression in mice. Carbohydr. Polym. 142, 259–267.

Huang, Q.-L., Siu, K.-C., Wang, W.-Q., Cheung, Y.-C., Wu, J.-Y., 2013. Fractionation, characterization and antioxidant activity of exopolysaccharides from fermentation broth of a *Cordyceps sinensis* fungus. Process Biochem. 48, 380–386.

Huang, S.-J., Tsai, S.-Y., Lee, Y.-L., Mau, J.-L., 2006. Nonvolatile taste components of fruit bodies and mycelia of *Cordyceps militaris*. LWT-Food Sci. Technol. 39, 577–583.

Huang, Z.-J., Ji, H., Li, P., Xie, L., Zhao, X.-C., 2002. Hypoglycemic effect and mechanism of polysaccharides from cultured mycelium of *Cordyceps sinensis*. J. China Pharm. Univ. 33, 51–54.

Huang, Z.-L., Jin, J., Tong, X.-M., Yang, Q.-Z., Gu, B.-L., Wang, J.-C., et al., 2011. The immunomodulatory effects of *Cordyceps sinensis* on dendritic cells derived from chronic myelogenous leukemia (CML). J. Med. Plants Res. 5, 5925–5932.

Izydorczyk, M., 2005. Understanding the Chemistry of Food Carbohydrates. CRC Press, Boca Raton, FL.

Ji, H., Tu, H., Li, N., 1993. Studies on extraction and hypoglycemic activity of polysaccharides from culture mycelium of *Cordyceps sinensis*. J. China Pharm. Univ. 24, 39–42.

Ji, N.-F., Yao, L.-S., Li, Y., He, W., Yi, K.-S., Huang, M., 2011. Polysaccharide of *Cordyceps sinensis* enhances cisplatin cytotoxicity in non-small cell lung cancer H157 cell line. Integr. Cancer Ther. 10, 359–367.

Jiang, M.Z., Yan, H., Wen, Y., Li, X.M., Qi, X., 2010. Extraction and purification of the polysaccharides of silkworm *Cordyceps militaris* fruiting body and rice culture medium. J. Anhui Agric. Sci. 18, 065.

Jiang, X., Liu, S., Mou, H., Wang, H., Guan, H., 1998. Effects of fungus polysaccharose on immune activities of serum and lymphocyte of *Penaeus chinensis*. Zool. Res. 20, 41–45.

Jin, Y., Li, J., Fan, M., Zhang, Y., Li, C., Xu, S., 2000. Effect of *Cordyceps* polysaccharides on immune function in adjuvant arthritis rats in vitro. Acta Univ. Med. Anhui. 35, 20–21.

Jing, D., Qiu, D., Xiao, S., Wang, X., Fan, J., 1999. Effects of cordycep polysaccharide on rat Ito cell proliferation and collagen gene expression in vitro. Chin. Hepatol. 4, 011.

Jing, Y., Cui, X., Chen, Z., Huang, L., Song, L., Liu, T., et al., 2014. Elucidation and biological activities of a new polysaccharide from cultured *Cordyceps militaris*. Carbohydr. Polym. 102, 288–296.

Kiho, T., Tabata, H., Ukai, S., Hara, C., 1986. A minor, protein-containing galactomannan from a sodium carbonate extract of *Cordyceps sinensis*. Carbohydr. Res. 156, 189–197.

Kiho, T., Hui, J., Yamane, A., Ukai, S., 1993. Polysaccharides in fungi. XXXII. Hypoglycemic activity and chemical properties of a polysaccharide from the cultural mycelium of *Cordyceps sinensis*. Biol. Pharm. Bull. 16, 1291–1293.

Kiho, T., Yamane, A., Hui, J., Usui, S., Ukai, S., 1996. Polysaccharides in fungi. XXXVI. Hypoglycemic activity of a polysaccharide (CS-F30) from the cultural mycelium of *Cordyceps sinensis* and its effect on glucose metabolism in mouse liver. Biol. Pharm. Bull. 19, 294–296.

Kiho, T., Ookubo, K., Usui, S., Ukai, S., Hirano, K., 1999. Structural features and hypoglycemic activity of a polysaccharide (CS-F10) from the cultured mycelium of *Cordyceps sinensis*. Biol. Pharm. Bull. 22, 966–970.

Kiho, T., Yamane, A., Hui, J., Usui, S., Ukai, S., 2000. Hypoglycemic activity of a polysaccharide (CS-F30) from the cultural mycelium of *Cordyceps sinensis* and its effect on glucose metabolism in mouse liver. Phytother. Res. 4, 647–649.

Kim, H.S., Kim, J.Y., Kang, J.S., Kim, H.M., Kim, Y.O., Hong, I.P., et al., 2010. Cordlan polysaccharide isolated from mushroom *Cordyceps militaris* induces dendritic cell maturation through toll-like receptor 4 signalings. Food Chem. Toxicol. 48, 1926–1933.

Kim, S.D., 2010. Isolation, structure and cholesterol esterase inhibitory activity of a polysaccharide, PS-A, from *Cordyceps sinensis*. J. Korean Soc. Appl. Biol. Chem. 53, 784–789.

Kuo, M.-C., Chang, C.-Y., Cheng, T.-L., Wu, M.-J., 2007. Immunomodulatory effect of exo-polysaccharides from submerged cultured *Cordyceps sinensis*: enhancement of cytokine synthesis, CD11b expression, and phagocytosis. Appl. Microbiol. Biotechnol. 75, 769–775.

Kwon, J.S., Lee, J.S., Shin, W.C., Lee, K.E., Hong, E.K., 2009. Optimization of culture conditions and medium components for the production of mycelial biomass and exo-polysaccharides with *Cordyceps militaris* in liquid culture. Biotechnol. Bioproc. Eng. 14, 756–762.

Lee, J.S., Hong, E.K., 2011. Immunostimulating activity of the polysaccharides isolated from *Cordyceps militaris*. Int. Immunopharmacol. 11, 1226–1233.

Lee, J.S., Kwon, J.S., Won, D.P., Lee, K.E., Shin, W.C., Hong, E.K., 2010a. Study on macrophage activation and structural characteristics of purified polysaccharide from the liquid culture broth of *Cordyceps militaris*. Carbohydr. Polym. 82, 982–988.

Lee, J.S., Kwon, J.S., Yun, J.S., Pahk, J.W., Shin, W.C., Lee, S.Y., et al., 2010b. Structural characterization of immunostimulating polysaccharide from cultured mycelia of *Cordyceps militaris*. Carbohydr. Polym. 80, 1011–1017.

Lee, J.S., Kwon, D.S., Lee, K.R., Park, J.M., Ha, S.-J., Hong, E.K., 2015. Mechanism of macrophage activation induced by polysaccharide from *Cordyceps militaris* culture broth. Carbohydr. Polym. 120, 29–37.

Leung, P.H., Zhao, S., Ho, K.P., Wu, J.Y., 2009. Chemical properties and antioxidant activity of exopolysaccharides from mycelial culture of *Cordyceps sinensis* fungus Cs-HK1. Food Chem. 114, 1251–1256.

Li, D., 2011. Liquid fermentation of *Cordyceps gunnii* and extraction of polysaccharides and its anti-tumor activity. Jinlin University.

Li, H., Ye, T.-J., Li, B.-Q., Ying, X.-P., 2008. Protective action of *Cordyceps* polysaccharide on photoaging skin in mice. Lishizhen Med. Mater. Med. Res. 11, 048.

Li, H., Ye, T.-J., Li, B.-Q., Ying, X.-P., 2009. Antagonism of *Cordyceps* polysaccharides to photoaging fibroblasts induced by 8-MOP/UVA. Lishizhen Med. Mater. Med. Res. 5, 024.

Li, N., Song, J., Liu, J., Zhang, H., 1995. Comparison of chemical composition of *Cordyceps militaris* and *Cordyceps sinensis*. J. Jilin Agric. Univ. 17, 80–83.

Li, S., Li, P., Dong, T., Tsim, K., 2001. Anti-oxidation activity of different types of natural *Cordyceps sinensis* and cultured *Cordyceps* mycelia. Phytomedicine 8, 207–212.

Li, F., Liu, P., Xiong, W., Xu, G., 2006a. Effects of corydyceps polysaccharide on liver fibrosis induced by DMN in rats. China J. Chin. Mater. Med. 31, 1968–1971.

Li, S., Yang, F., Tsim, K.W., 2006b. Quality control of *Cordyceps sinensis*, a valued traditional Chinese medicine. J. Pharm. Biomed. Anal. 41, 1571–1584.

Li, S., Zhang, G., Zeng, Q., Huang, Z., Wang, Y., Dong, T., et al., 2006c. Hypoglycemic activity of polysaccharide, with antioxidation, isolated from cultured *Cordyceps* mycelia. Phytomedicine 13, 428–433.

Li, S.P., Zhao, K.J., Ji, Z.N., Song, Z.H., Dong, T.T., Lo, C.K., et al., 2003. A polysaccharide isolated from *Cordyceps sinensis*, a traditional Chinese medicine, protects PC12 cells against hydrogen peroxide-induced injury. Life Sci. 73, 2503–2513.

Li, Y.-L., Su, Y.-Y., Miao, M., Zhang, X.-Z., Shi, R.-J., Liu, J.-H., 2007. Studies on fermentation conditions for exopolysaccharides production by *Cordyceps militaris*. J. Taishan Med. Coll. 12, 008.

Liang, C., Shang, H., 2010. Extraction and antioxidant performance of polysaccharide from culture medium of *Cordyceps militaris*. Food Res. Dev. 31, 26–29.

Liang, Z., 1983. The first report on *Cordyceps* in China–*Cordyceps gunnii*. Acta Mycol. Sin 2, 258–259.

Liang, Z., 1985. Isolation and identification of the conidial stage of *Cordyceps gunnii*. Acta Mycol. Sin. 4, 162–166.

Lin, R., Liu, H., Wu, S., Pang, L., Jia, M., Fan, K., et al., 2012. Production and in vitro antioxidant activity of exopolysaccharide by a mutant, *Cordyceps militaris* SU5-08. Int. J. Biol. Macromol. 51, 153–157.

Lin, X., Zheng, Y., Chen, J., Arai, M., 2003. Effects of crude polysaccharides from three aweto (*Cordyceps sinensis*) strains on induction of mouse macrophage from abdominal cavity to produce TNF-alpha (tumor necrosis factor) in vitro. Wei Sheng Wu Xue Za Zhi 24, 22–23.

Liu, A.-J., Zhong, Y.-R., Zhu, Z.-Y., Zhu, C.-M., Ma, L., Ma, Z.-M., 2008. Extraction, isolation and analysis of the polysaccharides from *Cordyceps gunnii* (Berk.) Berk. Mod. Food Sci. Technol. 1, 012.

Liu, J., Fei, Y., 2000. Enhancement of *Cordyceps taii* polysaccharide and *Cordyceps pruinosa* polysaccharide on cellular immune function in vitro. J. Immunol 17, 189–191.

Liu, J.-Y., Feng, C.-P., Li, X., Chang, M.-C., Meng, J.-L., Xu, L.-J., 2016. Immunomodulatory and antioxidative activity of *Cordyceps militaris* polysaccharides in mice. Int. J. Biol. Macromol. 86, 594–598.

Liu, Y., Zuo, J., Tao, Y., Liu, W., 2013. Protective effect of *Cordyceps* polysaccharide on hydrogen peroxide-induced mitochondrial dysfunction in HL-7702 cells. Mol. Med. Rep 7, 747–754.

Lu, C., Liu, L., Yang, L., Wang, S., Tao, F., 2003. Effects of cordycep polysaccharide on rat hepatic stellate cell proliferation and collagen gene expression in vitro. Heilongjiang Med. Pharm. 27, 17–18.

Lu, J., Gu, G., Hao, L., Jin, Z., Wang, X., 2016. Characterization and in vitro antioxidant activity of a polysaccharide from *Cordyceps sobolifera*. J. Food Process. Preserv. 40, 447–452.

Lu, L., 2011. Study on biological characteristics and some active metabolites of *Cordyceps japonica*. MSc Thesis, Ocean University of China.

Lu, W.-J., Chang, N.-C., Jayakumar, T., Liao, J.-C., Lin, M.-J., Wang, S.-H., et al., 2014. Ex vivo and in vivo studies of CME-1, a novel polysaccharide purified from the mycelia of *Cordyceps sinensis* that inhibits human platelet activation by activating adenylate cyclase/cyclic AMP. Thromb. Res. 134, 1301–1310.

Lv, Q., 2000. The selection of the optimum conditions for the production of polysaccharides by *Cordyceps militaris* and the bioactivity of the polysaccharides. MSc Thesis, Ocean University of China.

Lv, X., Yin, H., Li, H., Chen, X., 2007. Therapeutic effects of *Cordyceps* polysaccharide on freezing-kidney-induced chronic renal failure in rats. Yao Xue Jin Zhan 31, 314–319.

Ma, X., Qiu, D., Xu, J., Zeng, M., 1998. Effects of *Cordyceps* polysaccharides in patients with chronic hepatitis C. Huaren Xiaohua Zazhi 6, 582–584.

Mei, D., 2006. Optimization of submerged culture conditions for mycelia production in *Cordyceps gunnii* and the effect of polysaccharide on immunosuppressive factors from tumor. MSc Thesis, Guizhou University.

Meng, L., Sun, S., Li, R., Shen, Z., Wang, P., Jiang, X., 2015. Antioxidant activity of polysaccharides produced by *Hirsutella* sp. and relation with their chemical characteristics. Carbohydr. Polym. 117, 452–457.

Meng, L.-Z., Feng, K., Wang, L.-Y., Cheong, K.-L., Nie, H., Zhao, J., et al., 2014. Activation of mouse macrophages and dendritic cells induced by polysaccharides from a novel *Cordyceps sinensis* fungus UM01. J. Funct. Foods 9, 242–253.

Meng, X., 2001. Study on biological characteristics and some active component of *Cordyceps gunnii*. MSc Thesis, Hunan Normal University.

Meng, X., Zhang, Z., Liu, X., Chen, Z., Meng, X., 2001. Detection and analysis of some active components of *Cordyceps gunnii*. J. Hunan Univ. (Nat. Sci. Ed.) 29, 13–17, 23.

Mischnick, P., Lange, M., Gohdes, M., Stein, A., Petzold, K., 1995. Trialkylsilyl derivatives of cyclomaltoheptaose, cellulose, and amylose: rearrangement during methylation analysis. Carbohydr. Res. 277, 179–187.

Miyazaki, T., Oikawa, N., Yamada, H., 1977. Studies on fungal polysaccharides. XX. Galactomannan of *Cordyceps sinensis*. Chem. Pharm. Bull. 25, 3324–3328.

Needs, P., Selvendran, R., 1993. Avoiding oxidative degradation during sodium hydroxide/methyl iodide-mediated carbohydrate methylation in dimethyl sulfoxide. Carbohydr. Res. 245, 1–10.

Nie, S., Xie, M., Wang, Y., 2005. Preparation of tea glycoprotein and its application as a calibration standard for the quantification and molecular weight determination of tea glycoprotein in different tea samples by high-performance gel-permeation chromatography. Anal. Bioanal. Chem. 383, 680–686.

Nie, S.-P., Cui, S.W., Phillips, A.O., Xie, M.-Y., Phillips, G.O., Al-Assaf, S., et al., 2011. Elucidation of the structure of a bioactive hydrophilic polysaccharide from *Cordyceps sinensis* by methylation analysis and NMR spectroscopy. Carbohydr. Polym. 84, 894–899.

Ohta, Y., Lee, J.-B., Hayashi, K., Fujita, A., Park, D.K., Hayashi, T., 2007. *In vivo* anti-influenza virus activity of an immunomodulatory acidic polysaccharide isolated from *Cordyceps militaris* grown on germinated soybeans. J. Agric. Food Chem. 55, 10194–10199.

Pérez, S., Kouwijzer, M., Mazeau, K., Engelsen, S.B., 1996. Modeling polysaccharides: present status and challenges. J. Mol. Graphics 14, 307–321.

Pol-Fachin, L., Fernandes, C.L., Verli, H., 2009. GROMOS96 43a1 performance on the characterization of glycoprotein conformational ensembles through molecular dynamics simulations. Carbohydr. Res. 344, 491–500.

Poveda, A., Martin-Pastor, M., Bernabe, M., Leal, J.A., Jimenez-Barbero, J., 1998. Solution conformation and dynamics of a fungal cell wall polysaccharide isolated from *Microsporum gypseum*. Glycoconj. J. 15, 309–321.

Qin, X., Li, F., 2011. Optimization of the extraction of *Cordyceps militaris* polysaccharides by combined ultrasonic wave treatment and enzyme hydrolysis. Jiangsu Agric. Sci 5, 378–380.

Qiu, D., Xiao, S., Zeng, M., 1995. Effects of *Cordyceps* polysaccharides on the production of IL-2 and IFN-γ and the Expression of IL-2R in human peripheral blood lymphocytes. Shanghai Mian Yi Xue Za Zhi 15, 321–323.

Sharma, S.K., Gautam, N., Atri, N.S., 2015. Optimized extraction, composition, antioxidant and antimicrobial activities of exo and intracellular polysaccharides from submerged culture of *Cordyceps cicadae*. BMC Complement. Alternat. Med. 15, 1.

Shen, W., Song, D., Wu, J., Zhang, W., 2011. Protective effect of a polysaccharide isolated from a cultivated *Cordyceps* mycelia on hydrogen peroxide-induced oxidative damage in PC12 cells. Phyto. Res. 25, 675–680.

Shen, Y., Shao, X., Ni, Y., Xu, H., Tong, X., 2009. *Cordyceps sinensis* polysaccharide enhances apoptosis of HL-60 cells induced by triptolide. J. Zhejiang Univ. Med. Sci. 38, 158–162.

Sheng, L., Chen, J., Li, J., Zhang, W., 2011. An exopolysaccharide from cultivated *Cordyceps sinensis* and its effects on cytokine expressions of immunocytes. Appl. Biochem. Biotechnol. 163, 669–678.

Shi, Y., Wu, Y., Liao, S., Liu, X., Zou, Y., 2006. Microwave-assisted extraction of polysaccharide from *Cordceps militaris*. Guangdong Agric. Sci 11, 41–42.

Shi, Y., Wu, Y., Liu, X., Liao, S., Chen, Y., Chen, W., et al., 2007. Separation, purification and structural analysis of polysaccharides in artifical Brazil *Cordyceps sinensis*. Sci. Seric 3, 493–495.

Shrestha, B., Zhang, W., Zhang, Y., Liu, X., 2012. The medicinal fungus *Cordyceps militaris*: research and development. Mycol. Prog. 11, 599–614.

Smiderle, F.R., Carbonero, E.R., Mellinger, C.G., Sassaki, G.L., Gorin, P.A., Iacomini, M., 2006. Structural characterization of a polysaccharide and a β-glucan isolated from the edible mushroom *Flammulina velutipes*. Phytochemistry 67, 2189–2196.

Smiderle, F.R., Sassaki, G.L., Van Griensven, L.J., Iacomini, M., 2013. Isolation and chemical characterization of a glucogalactomannan of the medicinal mushroom *Cordyceps militaris*. Carbohydr. Polym. 97, 74–80.

Song, D., Lin, J., Yuan, F., Zhang, W., 2011. Ex vivo stimulation of murine dendritic cells by an exopolysaccharide from one of the anamorph of *Cordyceps sinensis*. Cell Biochem. Funct. 29, 555–561.

Song, D., He, Z., Wang, C., Yuan, F., Dong, P., Zhang, W., 2013. Regulation of the exopolysaccharide from an anamorph of *Cordyceps sinensis* on dendritic cell sarcoma (DCS) cell line. Eur. J. Nutr. 52, 687–694.

Song, J.F., Li, D.J., Liu, C.Q., 2009. Response surface analysis of microwave-assisted extraction of polysaccharides from cultured *Cordyceps militaris*. J. Chem. Technol. Biotechnol. 84, 1669–1673.

Striegel, A.M., Plattner, R.D., Willett, J., 1999. Dilute solution behavior of dendrimers and polysaccharides: SEC, ESI-MS, and computer modeling. Anal. Chem. 71, 978–986.

Sun, S.Q., Wang, Y.J., Zhu, C.J., Xu, W., Liu, X.X., 2012. Extraction conditions of polysaccharide from cultured *Cordyceps militaris* by ultrasonic assisted aqueous two phase system. J. Anhui Agric. Sci. 33, 130.

Tanabe, T., Kawase, T., Watanabe, T., Uchida, Y., Mitsutomi, M., 2000. Purification and characterization of a 49-kDa chitinase from *Streptomyces griseus* HUT 6037. J. Biosci. Bioeng. 89, 27–32.

Wang, B.-J., Won, S.-J., Yu, Z.-R., Su, C.-L., 2005. Free radical scavenging and apoptotic effects of *Cordyceps sinensis* fractionated by supercritical carbon dioxide. Food Chem. Toxicol. 43, 543–552.

Wang, J., Kan, L., Nie, S., Chen, H., Cui, S.W., Phillips, A.O., et al., 2015a. A comparison of chemical composition, bioactive components and antioxidant activity of natural and cultured *Cordyceps sinensis*. LWT- Food Sci. Technol. 63, 2–7.

Wang, L., Xu, N., Zhang, J., Zhao, H., Lin, L., Jia, S., et al., 2015b. Antihyperlipidemic and hepatoprotective activities of residue polysaccharide from *Cordyceps militaris* SU-12. Carbohydr. Polym. 131, 355–362.

Wang, L., Wang, G., Zhang, J., Zhang, G., Jia, L., Liu, X., et al., 2011a. Extraction optimization and antioxidant activity of intracellular selenium polysaccharide by *Cordyceps sinensis* SU-02. Carbohydr. Polym. 86, 1745–1750.

Wang, S.-H., Yang, W.-B., Liu, Y.-C., Chiu, Y.-H., Chen, C.-T., Kao, P.-F., et al., 2011b. A potent sphingomyelinase inhibitor from *Cordyceps* mycelia contributes its cytoprotective effect against oxidative stress in macrophages. J. Lipid Res. 52, 471–479.

Wang, Z.-M., Peng, X., Lee, K.-L.D., Tang, J.C.-O., Cheung, P.C.-K., Wu, J.-Y., 2011c. Structural characterisation and immunomodulatory property of an acidic polysaccharide from mycelial culture of *Cordyceps sinensis* fungus Cs-HK1. Food Chem. 125, 637–643.

Wang, M., Meng, X.Y., Le Yang, R., Qin, T., Wang, X.Y., Zhang, K.Y., et al., 2012. *Cordyceps militaris* polysaccharides can enhance the immunity and antioxidation activity in immunosuppressed mice. Carbohydr. Polym. 89, 461–466.

Wang, M., Meng, X., Yang, R., Qin, T., Li, Y., Zhang, L., et al., 2013. *Cordyceps militaris* polysaccharides can improve the immune efficacy of Newcastle disease vaccine in chicken. Int. J. Biol. Macromol. 59, 178–183.

Wang, Q., Cui, S.W., 2005a. Understanding the conformation of polysaccharides. Food Carbohydrates: Chemistry, Physical Properties, and Applications. CRC Press, p. 219.

Wang, Q., Cui, S.W., 2005b. Understanding the physical properties of food polysaccharides. Food Carbohydrates: Chemistry, Physical Properties, and Applications. CRC Press, pp. 161–217.

Wang, Y., Yin, H., Lv, X., Wang, Y., Gao, H., Wang, M., 2010a. Protection of chronic renal failure by a polysaccharide from *Cordyceps sinensis*. Fitoterapia 81, 397–402.

Wang, Z.-M., Cheung, Y.-C., Leung, P.-H., Wu, J.-Y., 2010b. Ultrasonic treatment for improved solution properties of a high-molecular weight exopolysaccharide produced by a medicinal fungus. Bioresour. Technol. 101, 5517–5522.

Wang, Y., Liu, D., Zhao, H., Jiang, H., Luo, C., Wang, M., et al., 2014. *Cordyceps sinensis* polysaccharide CPS-2 protects human mesangial cells from PDGF-BB-induced proliferation through the PDGF/ERK and TGF-β1/Smad pathways. Mol. Cell. Endocrinol. 382, 979–988.

Wang, Z., Gu, Y., Zhou, L., Yuan, Q., Yu, Y., 2004. Chemical constituents of *Cordyceps sinensis* mycelial fermentation preparations in solid media. Nat. Prod. Res. Dev. 17, 331–336.

Wei, C.-Y., Li, W.-Q., Shao, S.-S., He, L., Cheng, J., Han, S., et al., 2016. Structure and chain conformation of a neutral intracellular heteropolysaccharide from mycelium of *Paecilomyces cicadae*. Carbohydr. Polym. 136, 728–737.

Wen, L., Tang, Y., Yin, Q., Xia, M., Yang, Y., 2005. Assays on nutrient and effective ingredients in different parts of *Cordyceps militaris*. Zhongguo Zhong Yao Za Zhi 30, 659–661.

Wen, L., Tang, Y., Zhang, P., 2006. Assays on effective ingredients in different parts of *Cordyceps sobolifera*. Jiangsu J. Tradit. Chin. Med. 27, 45–46.

Weng, L., Wen, L., Yang, F., Wang, J.-J., 2008. Empirical study of different extraction methods influence on the antioxidation of polysaccharide in *Cordyceps militaris*. Food Sci. Technol. 11, 076.

Wong, W., Wu, J., Benzie, I., 2011. Photoprotective potential of *Cordyceps* polysaccharides against ultraviolet B radiation-induced DNA damage to human skin cells. Br. J. Dermatol. 164, 980–986.

Wu, D.-T., Cheong, K.-L., Wang, L.-Y., Lv, G.-P., Ju, Y.-J., Feng, K., et al., 2014a. Characterization and discrimination of polysaccharides from different species of *Cordyceps* using saccharide mapping based on PACE and HPTLC. Carbohydr. Polym. 103, 100–109.

Wu, D.-T., Meng, L.-Z., Wang, L.-Y., Lv, G.-P., Cheong, K.-L., Hu, D.-J., et al., 2014b. Chain conformation and immunomodulatory activity of a hyperbranched polysaccharide from *Cordyceps sinensis*. Carbohydr. Polym. 110, 405–414.

Wu, D.-T., Xie, J., Wang, L.-Y., Ju, Y.-J., Lv, G.-P., Leong, F., et al., 2014c. Characterization of bioactive polysaccharides from *Cordyceps militaris* produced in China using saccharide mapping. J. Funct. Foods 9, 315–323.

Wu, F., Yan, H., Ma, X., Jia, J., Zhang, G., Guo, X., et al., 2011. Structural characterization and antioxidant activity of purified polysaccharide from cultured *Cordyceps militaris*. Afr. J. Microbiol. Res. 5, 2743–2751.

Wu, F., Yan, H., Ma, X., Jia, J., Zhang, G., Guo, X., et al., 2012. Comparison of the structural characterization and biological activity of acidic polysaccharides from *Cordyceps militaris* cultured with different media. World J. Microbiol. Biotechnol. 28, 2029–2038.

Wu, J., Liu, B., Xia, H., Liu, D., 2008. Effect of *Cordyceps* polysaccharide (Cp) on the transition of tubular epithelial mesenchymal in mice with unilateral ureteral obstructive. Chin. J. Intergr. Tradit. West. Nephrol. 9, 400–402.

Wu, Y., Hu, N., Pan, Y., Zhou, L., Zhou, X., 2007. Isolation and characterization of a mannoglucan from edible *Cordyceps sinensis* mycelium. Carbohydr. Res. 342, 870–875.

Wu, Y.L., Omar, I., Sun, C.R., Pan, Y.J., 2005. Structure analysis and antitumor activity of (1→3)-β-D-glucans (cordyglucans) from the mycelia of *Cordyceps sinensis*. Planta Med. 71, 381–384.

Wu, Y., Sun, H., Qin, F., Pan, Y., Sun, C., 2006a. Effect of various extracts and a polysaccharide from the edible mycelia of *Cordyceps sinensis* on cellular and humoral immune response against ovalbumin in mice. Phytother. Res. 20, 646–652.

Wu, Y.L., Sun, C.R., Pan, Y.J., 2006b. Studies on isolation and structural features of a polysaccharide from the mycelium of an Chinese edible fungus (*Cordyceps sinensis*). Carbohydr. Polym. 63, 251–256.

Wu, Z., Zhang, M., Xie, M., Dai, Z., Wang, X., Hu, B., et al., 2016. Extraction, characterization and antioxidant activity of mycelial polysaccharides from *Paecilomyces hepiali* HN1. Carbohydr. Polym. 137, 541–548.

Xiang, F., Lin, L., Hu, M., Qi, X., 2016. Therapeutic efficacy of a polysaccharide isolated from *Cordyceps sinensis* on hypertensive rats. Int. J. Biol. Macromol. 82, 308–314.

Xiao, D.-M., Yu, S., Xiao, J.-H., 2016. Antioxidant activities of alkali-soluble polysaccharides from medicinal mushroom *Cordyceps taii* and its chemical characteristics. Biomed. Res.

Xiao, J., Chen, D., Liu, J., Liu, Z., Wan, W., Fang, N., et al., 2004a. Optimization of submerged culture requirements for the production of mycelial growth and exopolysaccharide by *Cordyceps jiangxiensis* JXPJ 0109. J. Appl. Microbiol. 96, 1105–1116.

Xiao, J., Liang, Z., Hu, X., Xiao, Y., Fang, N., Wan, W., 2004b. Immunocompetence of *Cordyceps gunnii* polysaccharide and its depolymeerizing substance. J. Immunol. 21, 51–53, 58.

Xiao, J.H., Xiao, Y., Liu, J.W., Wan, W.H., Fang, N., Tan, B.B., et al., 2004c. Optimization of submerged culture conditions for mycelial polysaccharide production in *Cordyceps pruinosa*. Process Biochem. 39, 2241–2247.

Xiao, J.-H., Xiao, D.-M., Xiong, Q., Liang, Z.-Q., Zhong, J.-J., 2010. Nutritional requirements for the hyperproduction of bioactive exopolysaccharides by submerged fermentation of the edible medicinal fungus *Cordyceps taii*. Biochem. Eng. J. 49, 241–249.

Xiao, J.-H., Xiao, D.-M., Chen, D.-X., Xiao, Y., Liang, Z.-Q., Zhong, J.-J., 2012a. Polysaccharides from the medicinal mushroom *Cordyceps taii* show antioxidant and immunoenhancing activities in a D-galactose-induced aging mouse model. Evid. Based Complement. Alternat. Med. 2012, 273435.

Xiao, J., Sun, J., Yao, L., Zhao, Q., Wang, L., Wang, X., et al., 2012b. Physicochemical characteristics of ultrasonic extracted polysaccharides from *Cordyceps cephalosporium* mycelia. Int. J. Biol. Macromol. 51, 64–69.

Xiao, J.-H., Xiao, D.-M., Sun, Z.-H., Xiao, Y., Zhong, J.-J., 2013. Antioxidative potential of polysaccharide fractions produced from traditional Chinese medicinal macrofungus *Cordyceps jiangxiensis in vitro*. Afr. J. Biotechnol. 10, 6607–6615.

Xu, Q.Q., Liu, Z.H., Sun, Y.S., Ding, Z.J., Lv, L.X., Li, Y.Q., 2012. Optimization for production of intracellular polysaccharide from *Cordyceps ophioglossoides* L2 in submerged culture and its antioxidant activities in vitro. Chin. J. Chem. Eng. 20, 294–301.

Yalin, W., Ishurd, O., Cuirong, S., Yuanjiang, P., 2005. Structure analysis and antitumor activity of (1→3)-beta-D-glucans (cordyglucans) from the mycelia of *Cordyceps sinensis*. Planta Med. 71, 381–384.

Yalin, W., Cuirong, S., Yuanjiang, P., 2006. Studies on isolation and structural features of a polysaccharide from the mycelium of a Chinese edible fungus (*Cordyceps sinensis*). Carbohydr. Polym. 63, 251–256.

Yan, H., Zhu, D., Xu, D., Wu, J., Bian, X., 2008. A study on *Cordyceps militaris* polysaccharide purification, composition and activity analysis. Afr. J. Biotechnol. 7, 4004–4009.

Yan, H., Zhu, D., Wu, J., 2009a. Study on ultrasonic wave assisted extraction of polysacchride from *Cordyceps*. Jiangsu Agric. Sci 1, 255–257.

Yan, J.K., Li, L., Wang, Z.M., Leung, P.H., Wang, W.Q., Wu, J.Y., 2009b. Acidic degradation and enhanced antioxidant activities of exopolysaccharides from *Cordyceps sinensis* mycelial culture. Food Chem. 117, 641–646.

Yan, W.J., Li, T.H., Tang, F.Y., Zheng, B.S., Jiang, Z.D., 2009c. Extraction and contents determination of polysaccharide in *Cordyceps guangdongensis*. Hua Nan Nong Ye Da Xue Xue Bao 30, 53–56.

Yan, J.-K., Li, L., Wang, Z.-M., Wu, J.-Y., 2010. Structural elucidation of an exopolysaccharide from mycelial fermentation of a *Tolypocladium* sp. fungus isolated from wild *Cordyceps sinensis*. Carbohydr. Polym. 79, 125–130.

Yan, J.-K., Wang, W.-Q., Li, L., Wu, J.-Y., 2011. Physiochemical properties and antitumor activities of two α-glucans isolated from hot water and alkaline extracts of *Cordyceps* (Cs-HK1) fungal mycelia. Carbohydr. Polym 85, 753–758.

Yan, J.-K., Wang, W.-Q., Ma, H.-L., Wu, J.-Y., 2012a. Sulfation and enhanced antioxidant capacity of an exopolysaccharide produced by the medicinal fungus *Cordyceps sinensis*. Molecules 18, 167–177.

Yan, J., Li, L., Wu, J., 2012b. Influence of external conditions on conformation of exopolysaccharide from *Cordyceps sinensis*. Food Sci. 33, 62–65.

Yan, J.-K., Wang, W.-Q., Wu, J.-Y., 2014. Recent advances in *Cordyceps sinensis* polysaccharides: mycelial fermentation, isolation, structure, and bioactivities: a review. J. Funct. Foods 6, 33–47.

Yan, J.-L., Li, H., Fan, Y., 2003. Effect of *Cordycep* polysaccharide on expression of tumor necrosis factor-alpha and activity of nuclear factor-kappaB of hepatic stellate cells in rat. Fudan Univ. J. Med. Sci. 30, 27–29.

Yang, J., Zhang, W., Shi, P., Chen, J., Han, X., Wang, Y., 2005. Effects of exopolysaccharide fraction (EPSF) from a cultivated *Cordyceps sinensis* fungus on c-Myc, c-Fos, and VEGF expression in B16 melanoma-bearing mice. Pathol. Res. Pract. 201, 745–750.

Yang, L., Zhang, L.-M., 2009. Chemical structural and chain conformational characterization of some bioactive polysaccharides isolated from natural sources. Carbohydr. Polym. 76, 349–361.

Yang, R., Zhao, X., Xiao, G., Wu, X., Xu, Y., Liu, X., 2006. Effect of lanthanum nitrate on the growth of *Cordyceps brasiliensis* Henn. Acta Ed. Fungi 13, 33–36.

Yang, R., Chen, W., Wu, X., Liu, X., 2007. Extraction technology optimization for extra-cellular polysaccharide (EPS) of *Cordyceps brasiliensis*. Food Sci. 28, 91–96.

Yang, S., Zhang, H., 2016. Production of intracellular selenium-enriched polysaccharides from thin stillage by *Cordyceps sinensis* and its bioactivities. Food Nutr. Res. 60, 30153.

Yang, S., Jin, L., Ren, X., Lu, J., Meng, Q., 2014. Optimization of fermentation process of *Cordyceps militaris* and antitumor activities of polysaccharides in vitro. J. Food Drug Anal. 22, 468–476.

Yang, Z., 2009. Study on biological characteristics and some active component of *Cordyceps bassiana*. Master Thesis, Ocean University of China.

Yi-Hong, Z., 2009. Comparative study on polysaccharide extraction from northern *Cordyceps* by Microwave Method and Traditional Technology. Inn. Mong. Agric. Sci. Technol. 3, 043.

Yin, D., Tang, X., 1995. Progresses of cultivation research of *Cordyceps sinensis*. Zhongguo Zhong Yao Za Zhi 20, 707–709.

Yin, D., Wang, R., Duan, H., 2014. Optimization of extraction process of polysaccharide from *Cordyceps militaris*. Jiangsu Agric. Sci. 42, 277–279.

Yin, H.-P., Lv, X.-B., Chen, X., 2007a. The therapeutic effects of *Cordyceps* polysaccharide on occluding renal artery-induced acute kidney injury in dog. Chin. J. Bioproc. Eng. 4, 013.

Yin, H., Lv, X., Chen, T., 2007b. Therapeutic effects of *Cordyceps* polysaccharide on adenine-induced chronic renal failure in rats. Tradit. Chin. Drug Res. Clin. Pharmacol. 6, 014.

Yu, C., Huang, B., Fan, M., 2002a. Studies of extracellular polysaccharide produced from the filtrate of fermention of *Cordyceps*. J. Anhui Agric. Univ. (China) 3, 260–264.

Yu, X., Luo, C., Liao, J., 2002b. Optimation of the extraction process of *Cordyceps* polysaccharide. Zhong Cao Yao 33, 1086–1087.

Yu, H.M., Wang, B.-S., Huang, S.C., Duh, P.-D., 2006. Comparison of protective effects between cultured *Cordyceps militaris* and natural *Cordyceps sinensis* against oxidative damage. J. Agric. Food Chem. 54, 3132–3138.

Yu, L., Zhang, B., Ruan, Y., Yao, G., Xie, L., Lu, K., 2004a. Effects of different components of *Cordyceps sinensis* polysaccharide on immune function. J. Zhejiang Coll. Tradit. Chin. Med. 28, 49–50.

Yu, R., Song, L., Zhao, Y., Bin, W., Wang, L., Zhang, H., et al., 2004b. Isolation and biological properties of polysaccharide CPS-1 from cultured *Cordyceps militaris*. Fitoterapia 75, 465–472.

Yu, R., Wang, L., Zhang, H., Zhou, C., Zhao, Y., 2004c. Isolation, purification and identification of polysaccharides from cultured *Cordyceps militaris*. Fitoterapia 75, 662–666.

Yu, R., Yang, W., Song, L., Yan, C., Zhang, Z., Zhao, Y., 2007. Structural characterization and antioxidant activity of a polysaccharide from the fruiting bodies of cultured *Cordyceps militaris*. Carbohydr. Polym. 70, 430–436.

Yu, R., Yin, Y., Yang, W., Ma, W., Yang, L., Chen, X., et al., 2009. Structural elucidation and biological activity of a novel polysaccharide by alkaline extraction from cultured *Cordyceps militaris*. Carbohydr. Polym. 75, 166–171.

Yu, T., Qian, H., 2006. Effect of selenium on the chemical components of *Cordyceps militaris*. Food Sci. Technol. 1, 133–135.

Yuan, J., Cheng, X., Hou, Y., 2005. Studies on the components and pharmacological action of polysaccharide from *Cordyceps sinensis*. Food Drug 7, 45–48.

Zeng, Y., Zhang, Y., Zhang, L., Cui, S., Sun, Y., 2015. Structural characterization and antioxidant and immunomodulation activities of polysaccharides from the spent rice substrate of *Cordyceps militaris*. Food Sci. Biotechnol. 24, 1591–1596.

Zhang, A., Lu, J., Zhang, N., Zheng, D., Zhang, G., Teng, L., 2010. Extraction, purification and anti-tumor activity of polysaccharide from mycelium of mutant *Cordyceps militaris*. Chem. Res. Chin. Univ. 26, 798–802.

Zhang, G., Huang, Y., Bian, Y., Wong, J.H., Ng, T., Wang, H., 2006. Hypoglycemic activity of the fungi *Cordyceps militaris*, *Cordyceps sinensis*, *Tricholoma mongolicum*, and *Omphalia lapidescens* in streptozotocin-induced diabetic rats. J. Microbiol. Biotechnol. 72, 1152–1156.

Zhang, G., Yin, Q., Han, T., Zhao, Y., Su, J., Li, M., et al., 2015. Purification and antioxidant effect of novel fungal polysaccharides from the stroma of *Cordyceps kyushuensis*. Ind. Crops Prod. 69, 485–491.

Zhang, J., Sun, Y., 2013. Ultrasonic extraction and anti-oxidation of polysaccharide from mycelium of *Cordyceps militaris*. Food Sci. Technol. 38, 203–207.

Zhang, J., Yu, Y., Zhang, Z., Ding, Y., Dai, X., Li, Y., 2011. Effect of polysaccharide from cultured *Cordyceps sinensis* on immune function and anti-oxidation activity of mice exposed to 60 Co. Int. Immunopharmacol. 11, 2251–2257.

Zhang, M., Zhu, W., 2012. Ultasonic-assisted extraction of the polysaccharide from *Cordyceps militaris*. Farm Mach 30, 050.

Zhang, M., Cui, S., Cheung, P., Wang, Q., 2007. Antitumor polysaccharides from mushrooms: a review on their isolation process, structural characteristics and antitumor activity. Trends Food Sci. Technol. 18, 4–19.

Zhang, W., Yang, J., Chen, J., Hou, Y., Han, X., 2005. Immunomodulatory and antitumour effects of an exopolysaccharide fraction from cultivated *Cordyceps sinensis* (Chinese caterpillar fungus) on tumour-bearing mice. Biotechnol. Appl. Biochem. 42, 9–15.

Zhang, W., Li, J., Qiu, S., Chen, J., Zheng, Y., 2008. Effects of the exopolysaccharide fraction (EPSF) from a cultivated *Cordyceps sinensis* on immunocytes of H22 tumor bearing mice. Fitoterapia 79, 168–173.

Zhang, X.L., Bi-Cheng, L., Al-Assaf, S., Phillips, G.O., Phillips, A.O., 2012. *Cordyceps sinensis* decreases TGF-β1 dependent epithelial to mesenchymal transdifferentiation and attenuates renal fibrosis. Food Hydrocoll. 28, 200–212.

Zhang, X.-Z., 2003. The artificial culture and its nutrition composition of *Cordyceps militaris* (L. Fr) Link. Shi Yong Jun 22, 19–21.

Zhao, F., 2008. Effect of pretreatment on extraction of cordycepic polysaccharides by water extraction and alcohol precipitation. Sci. Technol. Food Ind. 1, 140–142.

Zheng, B., He, L., 2009. Study on deproteinization of *Cordyceps guangdongensis* polysaccharide and its molecular weight distribution by GPC. Mod. Food Sci. Technol. 25, 707–710.

Zheng, B.-S., Tang, F.-Y., He, L.-Y., Yan, W.-J., Li, T.-H., 2009. Study on the effect of column chromatography on the purification of *Cordyceps guangdongensis* polysaccharide by GPC. Food Sci. Technol. 5, 090.

Zhiwei, Y., Li, Z., Lu, Y., 2010. Polysaccharide extracting from *Cordyceps militaris* (L.) Link by microwave method. J. Wuhan Bioeng. Inst. 4, 010.

Zhong, S., Pan, H., Fan, L., Lv, G., Wu, Y., Parmeswaran, B., et al., 2009. Advances in research of polysaccharides in *Cordyceps* species. Food Technol. Biotechnol. 47, 304–312.

Zhou, F., Chen, T., Yin, H., 2008. Protective role of *Cordyceps* polysaccharide on gentamicin-induced acute renal failure in rats. Pharm. Clin. Res. 16, 432–435.

Zhu, C., Liu, G., Kuang, S.-M., Zhou, G.-Y., 2013. Optimization of fermentation conditions of exo-polysaccharides by Ophio *Cordyceps sinensis* in liquid fermentation. Food Sci. Technol. 38, 18–20.

Zhu, J.-S., Halpern, G.M., Jones, K., 1998a. The scientific rediscovery of a precious ancient Chinese herbal regimen: *Cordyceps sinensis* part II. J. Alternat. Complement. Med. 4, 429–457.

Zhu, J.-S., Halpern, G.M., Jones, K., 1998b. The scientific rediscovery of an ancient Chinese herbal medicine: *Cordyceps sinensis* part I. J. Alternat. Complement. Med. 4, 289–303.

Zhu, Z.Y., Si, C.L., Zhong, Y.R., Zhu, C.M., Zhou, J.P., Liu, A.J., et al., 2011. The purification and antioxidant activities in D-galactose-induced aging mice of a water-soluble polysaccharide from *Cordyceps gunnii* (Berk.) Berk. mycelium. J. Food Biochem. 35, 303–322.

Zhu, Z.-Y., Chen, J., Si, C.-L., Liu, N., Lian, H.-Y., Ding, L.-N., et al., 2012a. Immunomodulatory effect of polysaccharides from submerged cultured *Cordyceps gunnii*. Pharm. Biol. 50, 1103–1110.

Zhu, Z.-Y., Liu, N., Si, C.-L., Liu, Y., Ding, L.-N., Jing, C., et al., 2012b. Structure and antitumor activity of a high-molecular-weight polysaccharide from cultured mycelium of *Cordyceps gunnii*. Carbohydr. Polym. 88, 1072–1076.

Zhu, Z.-Y., Pang, W., Li, Y.-Y., Ge, X.-R., Chen, L.-J., Liu, X.-C., et al., 2014. Effect of ultrasonic treatment on structure and antitumor activity of mycelial polysaccharides from *Cordyceps gunnii*. Carbohydr. Polym. 114, 12–20.

Zhu, Z.-Y., Dong, F., Liu, X., Lv, Q., Liu, F., Chen, L., et al., 2016. Effects of extraction methods on the yield, chemical structure and anti-tumor activity of polysaccharides from *Cordyceps gunnii* mycelia. Carbohydr. Polym. 140, 461–471.

Chapter 5

Complex Glucomannan From *Ganoderma atrum*

The genus *Ganoderma* is one of the most valued Chinese traditional medicines. It is well-known as "Lingzhi" in Chinese, "Reishi" in Japanese, and "Youngzhi" in Korean. The fruiting bodies, cultured mycelia, and spores were reported to be effective in the treatment of chronic hepatopathy, hypertension, hyperglycemia, and neophasia (Bao et al., 2002; Franz, 1989; Furusawa et al., 1992; Liu et al., 2007; Shiao et al., 1994). This fungus has attracted considerable attention because its polysaccharides have been demonstrated by recent research to possess diverse and potentially significant pharmacological activities such as anti-tumor, anti-aging (Miyazak and Nishijima, 1981; Wang et al., 1997), hypoglycemic (Hikino et al., 1985; Tomoda et al., 1986), and anti-microbial/viral activities (Eo et al., 2000; Yoon et al., 1994), including anti-human immunodeficiency virus (HIV) (El-Mekkawy et al., 1998; Kim et al., 1997) activities.

In China, the genus *Ganoderma* was first classified into six kinds, according to the difference in color and luster in "Shen Nong's Herbal Classic", but most of the previous studies were just concentrated on *Ganoderma lucidum* (red) and *Ganoderma tsugae* (red-brown), which were commonly known to possess medicinal and/or nutritional values. Up to now, most of the previous studies were concentrated on just these two members. It has been reported that three anti-tumor heteropolysaccharide–protein complexes with a molecular mass between 1.0 and 1.6×10^4 have been isolated from the mycelium of *G. tsugae* (Zhang et al., 1994). Chen et al. (2000) reported that a β-D-glucanprotein complex with a molecular mass of 64.6×10^4 was extracted from *G. lucidum* mycelium. Bao et al. (2001) isolated a polysaccharide with a molecular weight of 1.26×10^5 from the sporoderm-broken spores of *G. lucidum*, and found that the polysaccharide showed a strong antibody and lymphocyte suppressive activity. In 2005, Peng et al. (2005) obtained six branched (1→3)-β-D-glucan and (1→4)-α-D-glucan water-soluble polysaccharides with high anti-tumor activity from the crude extracellular polysaccharide of *G. tsugae* mycelium. However, more and more studies have shown that another member, *Ganoderma atrum* (black) could also be used to promote health and longevity (Gao et al., 2004, 2005). However, to our knowledge, there were few reports about the structure and

Bioactive Polysaccharides. DOI: http://dx.doi.org/10.1016/B978-0-12-809418-1.00005-8

bioactivities of polysaccharide from *G. atrum*. In our laboratory, we focused on the structure and bioactivities of polysaccharide from *G. atrum* over the past 10 years. In this chapter, we take this chance to summarize our works.

5.1 PREPARATION AND STRUCTURAL CHARACTERIZATION

5.1.1 Isolation of PSG-1

The crude polysaccharide of PSG was extracted as a water-insoluble dust-colored powder by hot-water extraction from fruiting bodies of *G. atrum* which were cultivated in Ganzhou, Jiangxi Province, China. The total yield of water-soluble polysaccharide was 3.9% by this isolation procedure, higher than the yield of 1.8% from *G. tsugae* mycelium extracted by the same procedure (Zhang et al., 1994), and almost equal to the yield of 3.8% extracted from *G. lucidum* fruiting bodies using a similar process (Bao et al., 2002). This yield of 3.9% was also much higher than the yield of 0.15% from the fruiting bodies of *G. lucidum* extracted by sodium phosphate buffer (Zhang and Chen, 1997), and the yield of 0.82% from the fruiting bodies of *G. lucidum* extracted by chloroform followed by hot water (Han et al., 2006). Therefore, the conditions of the isolation procedure used here were helpful in obtaining water-soluble polysaccharides from *G. atrum* fruiting bodies in a higher yield.

The crude polysaccharide was separated and sequentially purified through Sephadex G-200 (2.6 × 60 cm), giving two large overlapping elution peaks: PSG-1 and PSG-2 (eluted with water) (Fig. 5.1), as detected by

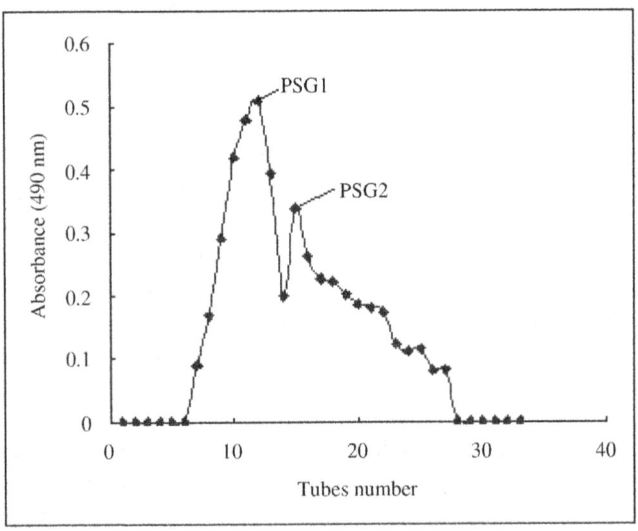

FIGURE 5.1 THE Sephadex G-200 column elution profile of the polysaccharides present in the aqueous extract of *Ganoderma atrum*. Eluent: H_2O; flow rate: 2 mL/min; 8 mL/tube.

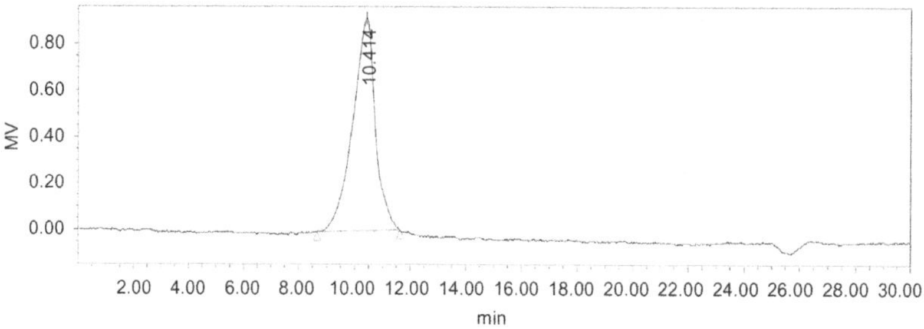

FIGURE 5.2 Profile of PSG-1 in HPGPC on ultrahydrogel-500 column, with 0.1 mol/L NaCl at 0.6 mL/min.

TABLE 5.1 Yields, Protein Contents, Sugar Contents, Uronic Acid Contents, and M_w of PSG-1

Sample	Appearance	Yield[a] (%)	Protein (%)	Uronic Acid (%)	M_w (kDa)	Sugar Component (%)		
						D-Glu	D-Man	D-Gala
PSG-1	Dust-color, fluffy	2.1	10.11	15.57	1013	68.3	13.9	17.8

[a]*Calculated as weight ratio of PSG-1/PSG.*

phenol–sulfuric acid assay. Its homogeneity was determined by high performance gel permeation chromatography (HPGPC) on a Waters HPLC system which showed as a single and symmetrically sharp peak on an Ultrahydrogel-500 column (Fig. 5.2).

5.1.2 Chemical Analysis of PSG-1

The results of HPGPC (Fig. 5.2) indicated that PSG-1 was a homogeneous protein-bound polysaccharide, with a weight-average molecular weight based a on column calibration of 1013 kDa. It had a positive response to the Lowry test, and absorption at 280 nm or 260 nm in the UV spectrum, indicating the existence of protein and nucleic acid. The yields, average molecular weight, protein content, uronic acid contents, and sugar compositions of PSG-1 were determined and are given in Table 5.1. The total sugar content of the polysaccharide was determined to be 89.1% with a purity of >99.8% in PSG-1, using the phenol–sulfuric acid method. Protein was estimated by the Lowry method. Though the protein content of PSG-1 was a little high, it could be considered to be protein-bound polysaccharide, because the Sevag method was repeated many times to remove free proteins. Generally, the pure protein exists as a globular shape in aqueous solution, but protein-bound polysaccharide exhibited a relatively expanded flexible chain, such as proteoglycan monomers (Ghosh and Reed, 1995). Therefore, we can presume that PSG-1 exhibits a more expanded flexible coil rather than a compact coil chain.

On hydrolysis by 2 M TFA, the presence of D-glucose, D-mannose, and D-galactose was detected by GC analysis. Usually, GC analysis could give an accurate content of the sugars in the polysaccharides. The experimental results from GC are summarized in Table 5.1. These sugars (mannose, galactose, and glucose) were found to be present in a molar ratio of 1:1.28:4.91. The absolute configuration of the monosaccharide was determined by GC analysis of acetylated (+)-2-octyl glycosides, and showed that all have D configurations (Gerwig et al., 1979). The sugar composition of this fraction (PSG-1) is similar to the glycosyl residue compositions of PL-1 isolated from *G. lucidum* by Bao et al. (2002), except that the latter was composed of Rha, Gal, and Glc in the molar ratios of 1:4:13.

The amino acid composition of PSG-1 was analyzed (Table 5.2). The PSG-1 contains 17 general amino acids and is rich in glutamic acid, asparaginic acid, alanine, glycine, threonine, and serine, while lacking tryptophan, asparagine, and glutamine. The high concentration of threonine and serine indicates the possibility of the existence of *O*-glycosidic linkages.

5.1.3 Structural Features of PSG-1

5.1.3.1 Infrared (IR) Spectroscopy

The spectra were recorded in the absorbance mode from 4000 to 400 cm^{-1} (mid infrared region) at a resolution of 4 cm^{-1} with 128 co-added scans. There are two types of end carbon–glucoside bonds: α- and β-styles, which can be measured by IR. In IR spectra, the C–H bond in α-style has an absorption peak near 844 cm^{-1}, while that of the C–H bond is in β-style near 891 cm^{-1} (Barker et al., 1954). Fig. 5.3 shows the IR spectra of the polysaccharide fractions (PSG-1) of *G. atrum*. The FTIR spectrum of PSG-1 showed a strong band between 950 and 1160 cm^{-1}, attributed to the stretching vibrations of the pyranose ring. In the anomeric region (950–700 cm^{-1}), the fraction exhibited the obvious characteristic absorption at 920 and 808.6 cm^{-1}, corresponding to the existence of mannose (Mathlouthi and Koenig, 1987). A characteristic absorption at 899 cm^{-1} was also observed, indicating the β-configuration of the sugar units. There was no absorption at 850 cm^{-1} for the α-configuration.

The strong absorption at 1650.4 and 1548.2 cm^{-1}, corresponding to the stretching vibration of the carbonyl bond of the amide group and the bending vibration of the N–H bond, respectively, shows the existence of protein.

TABLE 5.2 The Amino Acid Composition of PSG-1

Peak	Name of Amino Acid	Retention Time (min)	Area	Concentration (%)
1	Asp	4.77	14,1779	0.691
2	Thr	5.55	160,457	0.674
3	Ser	6.23	186,334	0.662
4	Glu	7.03	158,512	0.858
5	Gly	10.51	187,606	0.513
6	Ala	11.72	168,857	0.638
7	Cys	12.57	6653	0.057
8	Val	13.16	77,713	0.358
9	Met	14.39	201,113	1.149
10	Ile	16.52	33,932	0.186
11	Leu	17.48	81,428	0.390
12	Tyr	18.07	16,090	0.101
13	Phe	18.84	37,273	0.195
14	Lys	21.11	68,379	0.281
16	His	23.48	3421	0.017
17	Arg	27.72	30,252	0.165
18	Pro	7.76	37,633	0.175
Total concentration of the 17 general amino acids				7.103
Proportion of the essential amino acids				45.52

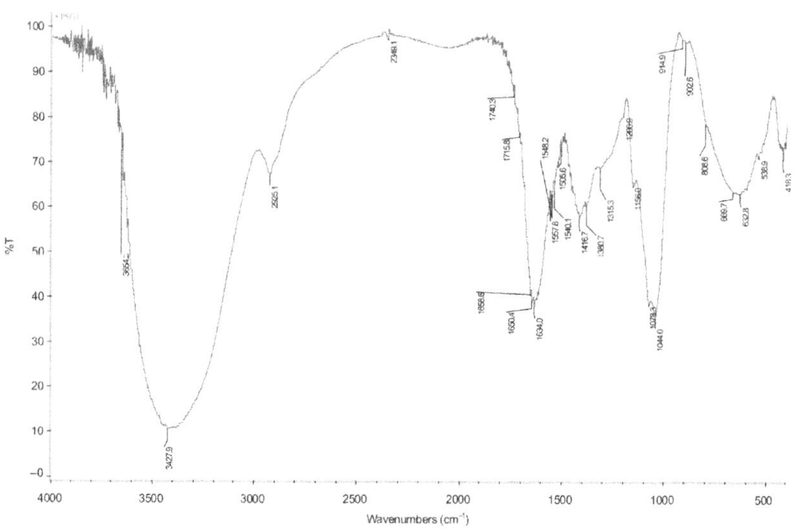

FIGURE 5.3 The IR spectra of polysaccharide fractions (PSG-1) of *Ganoderma atrum*.

Besides, absorption band over 1700 cm^{-1} indicates the trace of uronic acid in the samples. The absorption peak at c.1740.3 cm^{-1} is related to the free carboxyl group without coupling with benzylamine.

The bands in the region of 3427.9 cm^{-1} are due to the hydroxyl stretching vibration of the polysaccharides. The bands in the region of 2925.1 cm^{-1} are due to the C–H stretching vibration. Absorptions at 914.9 cm^{-1} are typical for D-Glc in the pyranose form (Barker et al., 1954).

5.1.3.2 *Linkage Analysis*

Linkages in the structure of glycoproteins can be divided into two types on the basis of their stability to alkali: *O*-glycosidic linkages and *N*-glycosidic linkages. The alkali-sensitive *O*-glycosidic linkages (involving Xyl, GlcNAc, Gal or Man, and Ser or Thr) are easily split in relatively mild conditions by a β-elimination mechanism resulting in the release of the carbohydrate moiety. This method has been widely used to analyze the type of linkages in glycoproteins (Greis and Hart, 1998). By comparing the UV scanning spectra of the samples with and without alkali treatment, the sample with alkali treatment had a distinctly higher absorbance at 241 nm than that without alkali treatment, showing that a β-elimination reaction had taken place, which demonstrated that the protein and carbohydrate were linked by *O*-linkage in PSG-1.

5.1.3.3 *Methylation Analysis*

From the results of the chemical composition of PSG-1, we knew the content of uronic acid in PSG-1 was high. The considerable proportion of uronic acid in PSG-1 would make it difficult to carry out the methylation experiment (Cui, 2005), so the carboxyl group of PSG-1 should be reduced before methylation analysis. The individual peaks of the partially methylated alditol acetate (PMAA) from the GC analysis and mass spectroscopy of each peak were identified by their retention time, and by comparison with mass spectrum patterns from literature (Carpita and Shea, 1989). The linkage patterns and corresponding molar ratio of PSG-1 are shown in Table 5.3.

The proportion of the non-reducing T-D-Glc*p* residue was 18.68%. The majority of the branched sugar residue was 1,3,6-D-Glc*p*, which accounted for 12.97% of the total sugar residues, followed by 1,4,6-D-Glc*p* (3.24%). These data indicated that the ratio between terminal units and branching point was 1.15, which was consistent with the fact that the number of polysaccharide branching points approximately equals the number of terminal units. The unsubstituted residues were shown to be 1,3-D-Glc*p*, 1,6-D-Glc*p*, 1,4-D-Gal*p*, 1,2-D-Man*p*, and 1,4-D-Man*p* in the molar ratio of 21.08%, 12.33%, 12.70%, 8.06%, and 4.55%, respectively. The galacturonic acid was found to be as 1,4-D-Gal*p*A with an amount of 6.15%. The data from the methylation analysis suggested that PSG-1 was mainly composed

TABLE 5.3 Linkage Patterns and Corresponding Percentage of PSG-1 by Methylation Analysis

RT[a] (min)	PMAA	Linkage Pattern	Peak Area Percentage
14.70	2,3,4,6-Me_4-Glc*p*	T-Glc*p*	18.68
20.06	2,4,6-Me_3-Glc*p*	1,3-Linked-Glc*p*	21.08
22.24	2,3,4-Me_3-Glc*p*	1,6-Linked-Glc*p*	12.33
28.15	2,4-Me_2-Glc*p*	1,3,6-Linked-Glc*p*	12.97
28.82	2,3-Me_2-Glc*p*	1,4,6-Linked-Glc*p*	3.24
20.31	3,4,6-Me_3-Man*p*	1,2-Linked-Man*p*	8.06
22.51	2,3,6-Me_3-Man*p*	1,4-Linked-Man*p*	4.55
23.09	2,3,6-Me_3-Gal*p*	1,4-Linked-Gal*p*	12.70
23.09	2,3,6-Me_3-Gal*p*A	1,4-Linked-Gal*p*A	6.15

[a]*RT: Retention time.*

of →1)-D-Glc*p*-(3→ and →1)-D-Glc*p*-(6→ as the backbone, with some →1)-D-Gal*p*-(4→, →1)-D-Man*p*-(2→, and →1)-D-Man*p*-(4→ which could exist in the backbone or side chains. The major branching points were at *O*-4 and/or *O*-6 positions of the Glc*p* chain, with β-D-Glc*p* as the terminal residue.

To our knowledge, this was the first report on the structural characteristics of polysaccharide from *G. atrum*. Compared with the previous study of polysaccharide from *G. lucidum*, our results were similar to Liu et al. (2010) which assigned →3)-Glc*p*-(1→, →4)-Glc*p*-(1→, and →6)-Glc*p*-(1→ in the backbone, and *O*-4 and/or *O*-6 as the branching points. However, some discrepancies have also been published in the literature. For example, Bao et al. (2002) claimed a polysaccharide from *G. lucidum* (PL-1) with immunocompetence had a backbone consisting of 1,4-α-D-Glc*p* and 1,6-β-D-Gal*p* with branches at *O*-6 of glucose residues and *O*-2 of galactose residues. The discrepancies in these studies cannot be easily resolved, because of the differences in species and extraction methods used for sample preparation, and the differences highlighted the uniqueness of PSG-1. The subsequent NMR spectroscopy confirmed the conclusions drawn from methylation analysis, and provided us with more details of the structure of PSG-1.

5.1.3.4 Nuclear Magnetic Resonance (NMR) Analysis

The structural features of PSG-1 were further identified by 1D and 2D NMR spectrum. The ^{1}H NMR spectrum (Fig. 5.4A) showed more than 15 peaks in the anomeric region (4.3–5.8 ppm), but only 9 peaks were found to be significant and used for the analysis. They were designated A, B, C, ..., I

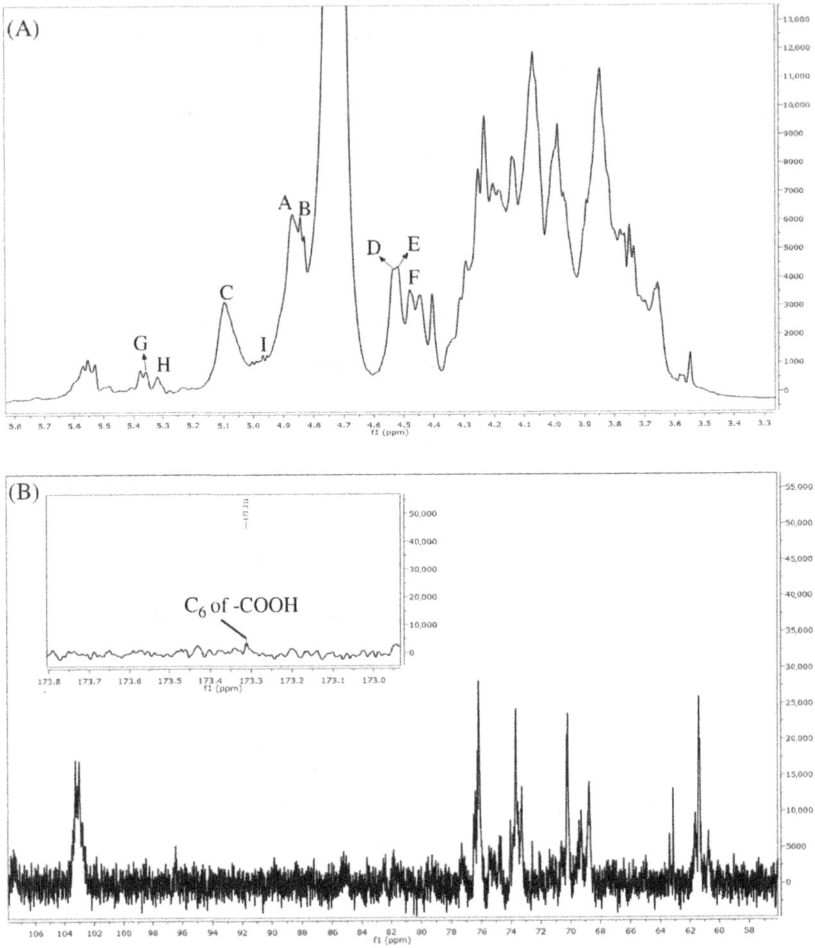

FIGURE 5.4 One-dimensional NMR spectrum of PSG-1 recorded at 60°C. (A) ^{1}H NMR spectrum (600.1 MHz); (B) ^{13}C NMR spectrum (151.0 MHz).

according to the decreasing order of peak intensity. In the ^{13}C NMR spectrum (Fig. 5.4B), a small peak at δ 173.31 ppm was a typical signal of uronic acid, which was indicated as 1,4-D-Gal*p*A according to methylation and GC-MS results.

The HSQC spectrum (Fig. 5.5C) showed nine cross peaks in the anomeric region, which indicated that nine spin systems were present in the PSG-1. They were marked as A1, B1, C1, ..., I1, in accordance with the ^{1}H spectrum labels. All the ^{1}H and ^{13}C signals of residues from A to I were assigned completely using DQF-COSY, TOCSY, and HSQC, and the linkage sites and sequence among residues were identified by HMBC.

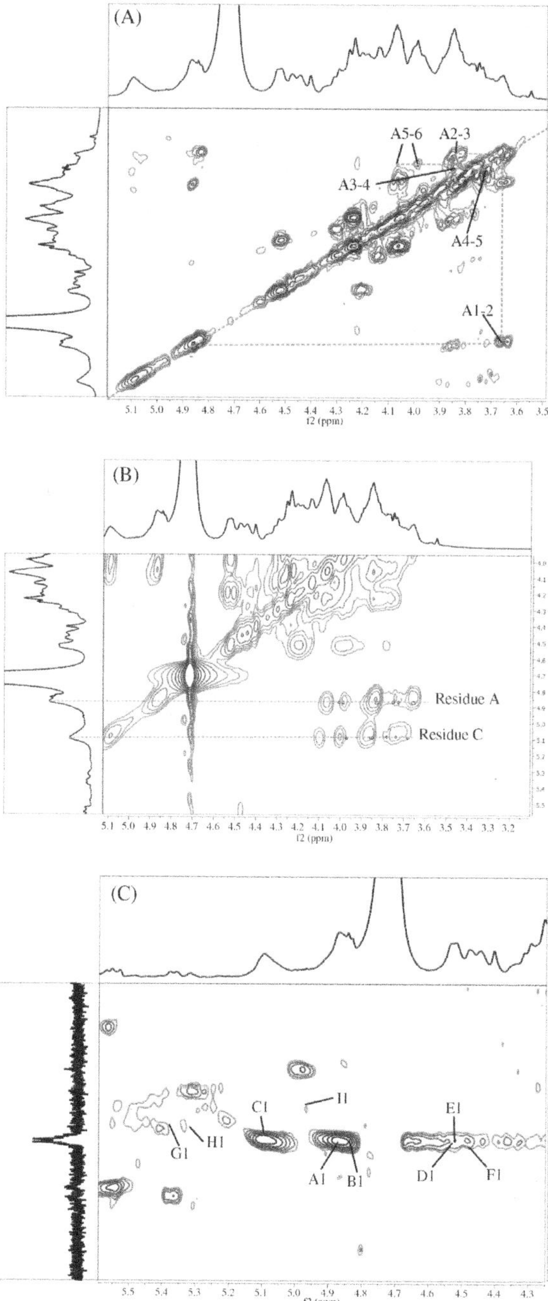

FIGURE 5.5 ^{1}H/^{1}H and ^{1}H/^{13}C correlation spectrum of PSG-1. (A) ^{1}H/^{1}H DQF-COSY correlation spectrum (the assignments are the correlations of protons of β-1,3-D-Glc*p* (residue **A**)); (B) ^{1}H/^{1}H TOCSY correlation spectrum (the assignments are the correlations of protons residue **A** and residue **C**); (C) part of ^{1}H/^{13}C HSQC correlation spectrum (anomeric region).

β-1,3-Glc*p* (residue A): According to the methylation analysis, 1,3-Glc*p* was expected to be the most abundant residue. In the ^{1}H spectrum of PSG-1, the peak at 4.87 ppm was the most intense in the anomeric region (Fig. 5.4A). By comparing previous studies (Laws et al., 2008; Perepelov et al., 2007; Ye et al., 2010) in combination with the peak intensity, this peak was tentatively assigned as the anomeric proton of 1,3-Glc*p*. The H-1 chemical shift of residue **A** at δ 4.87 ppm indicated that it was β-linked. A cross peak at δ 4.87/3.66 ppm was easily found in the DQF-COSY spectrum (Fig. 5.5A), implying that the chemical shift of H-2 was δ 3.66 ppm. The other proton signals for H-3, H-4, and H-5 of this residue were assigned according to the DQF-COSY spectrum using the same method, and H-6 was confirmed by means of TOCSY (Fig. 5.5b) combined with DQF-COSY. On the basis of the proton assignments, the chemical shifts in the carbon atoms of residue **A** were readily obtained from the HSQC spectrum (Fig. 5.5C). Both ^{1}H and ^{13}C chemical shifts were consistent with previous literature (Laws et al., 2008; Maity et al., 2011; Perepelov et al., 2007; Ye et al., 2010), and the downfield shifts of the C-1 (δ 103.25 ppm) and C-3 (δ 84.92 ppm) carbon signals indicated that residue **A** was a β-1,3-D-glucopyranose.

β-T-Glc*p* (residue B): As the second highest peak in ^{1}H spectrum, residue **B** at 4.84 ppm in the anomeric region was tentatively assigned as T-Glc*p*, according to the methylation analysis result. The anomeric signal at δ 4.84 ppm indicated that residue **B** was a β-linked residue. The chemical shifts of H-2, H-3, H-4, H-5, and H-6 were obtained from DQF-COSY (Fig. 5.5A), and confirmed by the TOCSY spectrum (Fig. 5.5B). The carbon signals from C-1 to C-6 of residue **B** were identified from the HSQC spectrum (Fig. 5.5C). Both carbon and proton chemical shifts were compared with previous studies (Agrawal, 1992; Bhunia et al., 2010; Ge et al., 2009; Ghosh et al., 2008) to confirm that residue **B** was a β-T-D-glucopyranose.

α-1,4-Gal*p* (residue C): The anomeric signal at δ 5.10 ppm indicated that residue **C** was an α-linked residue, which was further corroborated by the C-1 chemical shift at δ 102.32 ppm. Residue **C** was tentatively assigned as 1,4-Gal*p* by comparison with literature values of the anomeric chemical shifts (Das et al., 2009). With the help of the DQF-COSY spectrum, the chemical shifts from H-2 to H-6 of residue **C** were assigned, and the results were confirmed by the TOCSY spectrum combined with literature (Chandra et al., 2009; Luo et al., 2009; Ojha et al., 2008). The carbon signals from C-1 to C-6 of residue **C** were identified from the HSQC spectrum (Fig. 5.5C). The downfield shifts of the C-1 (δ 102.32 ppm) and C-4 (δ 76.55 ppm) carbon signals indicated that residue **C** was an α-1,4-linked-D-galactopyranose.

α-1,4-Gal*p*A (residue I): According to the methylation analysis results (Table 5.3), the content of 1,4-Gal*p*A was approximately 6.15% in PSG-1. The anomeric proton chemical shift of residue **I** at δ 4.97 ppm and the C-1 identified from HSQC at δ 99.77 ppm indicated it was α-linked. This residue showed only five proton signals, and a high chemical shift of H-5 (δ 4.60 ppm) was observed from the DQF-COSY spectrum, and the resonance at 173.31 ppm in ^{13}C spectrum (Fig. 5.4b) was attributed to the −COOH group, so residue **I** was deduced as a Gal*p*A. All the ^{1}H and ^{13}C chemical shifts of residue **I** assigned by DQF-COSY and HSQC in this study were consistent with previous studies (Maiti et al., 2008), and the downfield shift of C-4 (δ 79.50 ppm) carbon signals with respect to standard values of methyl glycosides indicated that residue **I** was an α-1,4-linked-D-Gal*p*A.

β-1,6-Glc*p* (residue E): For residue **E**, the anomeric proton signal appeared at δ 4.52 ppm and the anomeric carbon signal at δ 103.12 ppm indicated that it was a β-linked residue. The assignment of proton signals from H-2 to H-6 was conducted by DQF-COSY and TOCSY spectra combined with previous literature (Das et al., 2010b; Mandal et al., 2010). The corresponding carbon chemical shifts were obtained from the cross peaks of H-1-C-1, H-2-C-2, …, H-6-C-6 in the HSQC spectrum. The downfield shift of C-6 (69.52 ppm) and a small upfield shift for C-5 (73.19 ppm) with respect to standard values indicated that residue **E** was linked at C-6. Based on the above results and previous reports (Das et al., 2010a,b; Mandal et al., 2010), residue **E** was determined as a β-1,6-D-glucopyranose.

β-1,4,6-Glc*p* (residue F): Residue **F** had an anomeric proton signal at δ 4.48 ppm and an anomeric carbon signal at δ 103.42 ppm which indicated it was a β-linked moiety. It was tentatively assigned as 1,4,6-Glc*p* by comparison with literature values of the anomeric chemical shifts (Roy et al., 2008). The proton and carbon signals of residue **F** were assigned in the same way as described above. The downfield shifts of C-4 (δ 79.21 ppm) and C-6 (δ 69.57 ppm) with respect to the values of standard methyl glycosides indicated that residue **F** was linked at C-4 and C-6. Based on the above results, residue **F** was deduced to be a β-1,4,6-glucopyranose.

The proton and carbon signals of residue **D**, **G**, and **H** were assigned in the same way as described above, with the help of DQF-COSY and HSQC spectrum along with previous literature (Ahrazem et al., 2002; Bhunia et al., 2010). Thus, combining the information from the ^{1}H, ^{13}C, HSQC, DQF-COSY, and TOCSY spectra with methylation analysis results, a complete assignment of all the linkage patterns was identified (Table 5.4).

Linkage sites and sequence among residues: The sequences of sugar residues were determined from a long-range HMBC experiment, and the following connectivities were observed. The cross peaks of both anomeric protons and carbons of each sugar residue of PSG-1 were examined from the HMBC experiment, and both intra- and inter-residual connectivities were observed (figures from HMBC results have not been shown). Cross peaks were found between H-1 of residue **A** (δ 4.87 ppm) and C-3 of residue **D** (**A** H-1/**D** C-3), C-1 of residue **A** (δ 103.25 ppm) and H-3 of residue **D** (**A** C-1/**D** H-3), H-6 of residue **E** (**A** C-1/**E** H-6), along with intra-residual coupling between H-1 of residue **A** with its own C-3 (**A** H-1/**A** C-3). Similar

TABLE 5.4 ^{1}H and ^{13}C NMR Chemical Shifts of PSG-1 Recorded in D_2O at 60°C

Residue		Proton or Carbon						
		1	2	3	4	5	6a	6b
β-T-Glc*p* (**B**)	H	4.84	3.62	3.81	3.68	3.76	3.87	4.01
	C	102.99	73.48	76.23	70.31	75.82	63.18	
β-1,3-Glc*p* (**A**)	H	4.87	3.66	3.84	3.75	3.73	3.99	4.07
	C	103.25	73.06	**84.92**	70.18	75.65	61.48	
α-1,4-Gal*p* (**C**)	H	5.10	3.74	3.84	3.97	3.85	3.68	3.77
	C	102.32	70.22	69.35	**76.55**	73.19	61.21	
β-1,3,6-Glc*p* (**D**)	H	4.53	3.49	3.63	3.57	3.61	3.84	4.14
	C	103.17	73.19	**83.34**	70.43	75.81	**69.28**	
β-1,6-Glc*p* (**E**)	H	4.52	3.54	3.67	3.58	3.65	3.86	4.20
	C	103.12	73.65	76.11	71.03	73.19	**69.52**	
β-1,4,6-Glc*p* (**F**)	H	4.48	3.49	3.67	3.80	3.87	3.78	4.09
	C	103.43	73.76	76.23	**79.21**	72.59	**69.57**	
α-1,2-Man*p* (**G**)	H	5.38	4.13	3.87	3.76	3.85	3.74	4.06
	C	101.52	**79.42**	71.42	69.55	70.86	62.79	
α-1,4-Man*p* (**H**)	H	5.32	4.18	3.67	3.82	3.75	3.84	4.02
	C	101.78	69.25	73.21	**80.00**	75.98	63.44	
α-1,4-Gal*p*A (**I**)	H	4.97	3.56	4.12	4.29	4.60		
	C	99.77	71.05	73.39	**79.50**	71.84	**173.31**	

Numbers in bold indicate linkage positions.

cross peaks were also found between H-1 of residue **D** (δ 4.53 ppm) and C-6 of residue **E** (**D** H-1/**E** C-6), C-1 of residue **D** (δ 103.17 ppm) and H-6 of residue **E** (**D** C-1/**E** H-6), along with intra-residual cross couplings between H-1 of residue **D** and C-6 of its own (**D** H-1/**D** C-6). With the same method, the cross peaks of other residues were found and are summarized in Table 5.5.

According to the results stated above, PSG-1 had been proved to be a hyteropolysaccharide comprised a backbone of 1,3-linked and 1,6-linked β-Glc*p* residues substituted at the *O*-3 and *O*-6 position as the branch points. The residues of α-1,4-Gal*p*, α-1,2-Man*p*, and α-1,4-Man*p* were also found in the backbone. Side chains were terminated by β-Glc*p*, with the composition of α-1,4-Gal*p*, α-1,4-Gal*p*A, β-1,3-Glc*p*, and β-1,6-Glc*p*. Based on all

TABLE 5.5 Significant $^3J_{H,C}$ Connectivities Observed in HMBC Spectrum for the Anomeric Protons/Carbons of the Sugar Residues of PSG-1

Sugar Residue	Sugar Linkage	H-1/C-1		Observed Connectivities	
		δ_H/δ_C	δ_H/δ_C	Residue	Atom
A	β-1,3-Glc*p*	4.87	84.92	A	C-3
			83.34	D	C-3
		103.25	3.84	A	H-3
			3.63	D	H-3
			3.86	E	H-6
B	β-T-Glc*p*	4.84	84.92	A	C-3
			76.55	C	C-4
		102.99	3.84	A	H-3
			3.97	C	H-4
C	α-1,4-Gal*p*	5.10	84.92	A	C-3
			69.28	D	C-6
			69.57	F	C-6
		102.32	3.84	A	H-3
			3.84	D	H-6
D	β-1,3,6-Glc*p*	4.53	69.28	D	C-6
			69.52	E	C-6
		103.17	3.84	D	H-6
			3.86	E	H-6
E	β-1,6-Glc*p*	4.52	79.21	F	C-4
			80.00	H	C-4
			79.50	I	C-4
		103.12	3.80	F	H-4
			3.82	H	H-4
			4.29	I	H-4
F	β-1,4,6-Glc*p*	4.48	79.42	G	C-2
G	α-1,2-Man*p*	5.38	83.34	D	C-3
		101.52	3.63	D	H-3
H	α-1,4-Man*p*	5.32	76.55	C	C-4
I	α-1,4-Gal*p*A	4.97	76.55	C	C-4

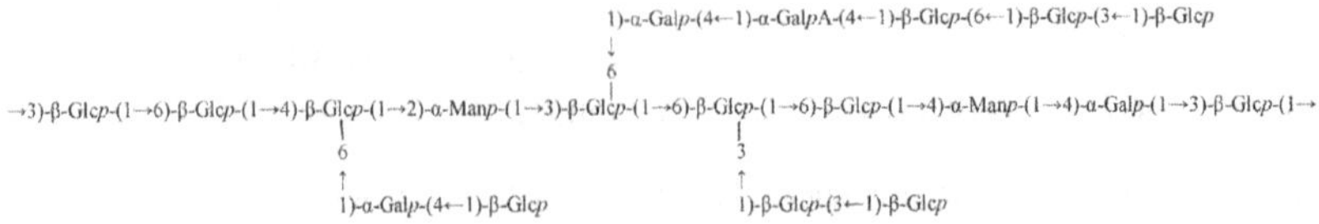

FIGURE 5.6 Possible molecular structure of the repeating unit of PSG-1.

these chemical and spectroscopic data, a possible molecular structure of the repeating unit of PSG-1 was established as is shown in Fig. 5.6.

5.2 BIOACTIVITIES

5.2.1 Immunomodulatory

5.2.1.1 Macrophage Immunomodulatory Activity

In recent decades, a wide range of polysaccharides from natural products were isolated and structurally characterized. Due to their broad spectrum of therapeutic properties and relatively low toxicity, these polysaccharides have attracted increasing attention (Chang et al., 2008; Chen et al., 2010a,b; Chen et al., 2011; Zhang et al., 2011; Zhao et al., 2010). Among them, enormous interest has been focused on the beneficial effects of fungal polysaccharides on immune function. Indeed, one of the primary mechanisms of immune system potentiation by fungal polysaccharides is thought to be the influence on innate immunity through activating macrophages.

Macrophages exist in almost all tissues of the body, and play a central role in innate and adaptive immune responses (Leiro et al., 2007). The hallmark function of macrophages is their potent phagocytic ability. Macrophages can engulf microorganisms, proteins, and other smaller cells. The macrophage is a highly secretory cell, when the body is stimulated by pathologic stimuli or injury, the macrophage protects the host by releasing numerous cytokines, such as tumor necrosis factor (TNF)-α and interleukin (IL)-1β, and cytotoxic molecules, such as nitric oxide (NO) and reactive oxygen species (ROS) (Cheng et al., 2008; Lee and Jeon, 2005).

Activation of macrophages is a key event in innate and adaptive immunity for the initiation and propagation of defensive reactions against pathogens (Cheng et al., 2008; Lee and Jeon, 2005). Activation of macrophages is possibly mediated through a series of signal transduction pathways (Schepetkin and Quinn, 2006). However, the signal transduction pathways responsible for PSG-1-induced macrophage activation are not fully understood yet.

The transmission of extracellular signals to their intracellular targets is mediated by a network of interacting proteins that govern a large number of cellular processes (Karin and Hunter, 1995). Phosphoinositide 3-kinases

(PI3K)/Akt, and mitogen-activated protein kinases (MAPKs), as well as transcription factors such as nuclear factor (NF)-κB, are major components of the signaling machinery (Cantley, 2002; Seger and Krebs, 1995). To examine PSG-1-mediated immunomodulating activities, a study was conducted to characterize the signaling pathways involved in the activation of macrophages by PSG-1. Specifically, the effects of PSG-1 on PI3K/Akt, MAPKs, and NF-κB pathways were assessed, and the molecular mechanism of PSG-1-induced signal transduction in the regulation of TNF-α secretion was elucidated.

5.2.1.1.1 PSG-1 is Free of Endotoxin (LPS) Contamination

Endotoxin (LPS), a known immunomodulator, is often a contaminant in biological preparations. Many previous reports have demonstrated the immunostimulating activities of several polysaccharides (Wong et al., 2011); the effect of these polysaccharides on the immune system is quite similar to that of LPS. Thus, one of the principal concerns is that the immune stimulating properties of polysaccharides might be due to contamination from LPS. The results showed that the quantity of endotoxin in PSG-1 was less than 0.015 EU/mg (negative), as measured by LAL assay. To further ensure that the effects of PSG-1 were not due to endotoxin contamination, the effect of polymyxin B (PMB) on PSG-1-induced TNF-α production was examined. Polymyxin B has been known to inhibit macrophage activation by LPS through the binding of lipid A (Morrison and Jacobs, 1976). As shown in Fig. 5.7, the PMB treatment did not affect PSG-1-induced TNF-α generation, while it abolished TNF-α production in LPS-induced macrophages. These results demonstrated that PSG-1 is free of LPS contamination, and the activation of macrophages was attributed to the stimulation of PSG-1.

5.2.1.1.2 Effect of PSG-1 on FITC-Dextran Internalization

Phagocytosis is the first step in the macrophage response to invading microorganisms (Kim et al., 2007). After phagocytic uptake, macrophages turn their role into antigen-presenting cells with expression of higher levels of costimulatory molecules, and then mediate an interaction between T cells and macrophages (Fidler and Kleinerman, 1993). To confirm the effect of PSG-1 on the phagocytosis of RAW264.7 cells, the FITC-dextran internalization of cells was analyzed by flow cytometry. As is shown in Fig. 5.8, the phagocytosis increased when incubated with PSG-1 for 24 h, the majority of the cells (>90%) of each group phagocytosed within 1 h of incubation. The PSG-1 treatment increased the phagocytic activity of the RAW264.7 cells with regard to the dose of PSG-1. Since phagocytes act as regulatory and effector cells in the immune system, the enhancement of the phagocytic function by PSG-1 means elevation of the innate immune response.

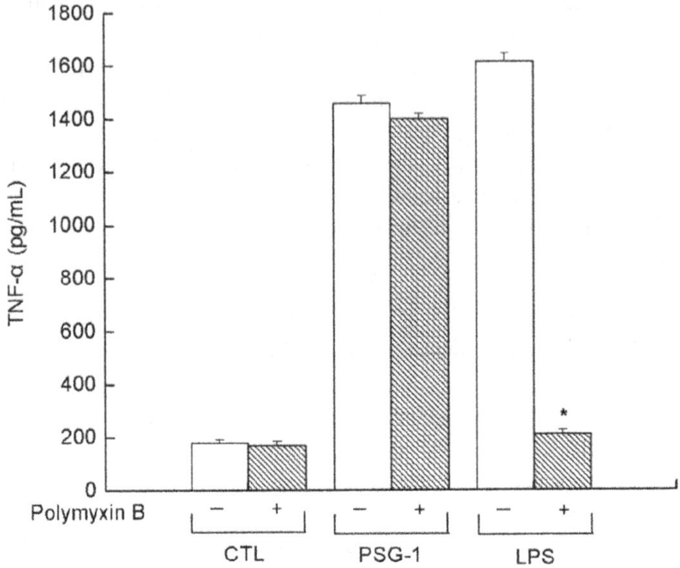

FIGURE 5.7 Effect of polymyxin B on PSG-1 or LPS-induced TNF-α production. PSG-1 or LPS was incubated with polymyxin B (1000 U/mL) for 1 h before challenging RAW264.7 cells. Cells were treated with PSG-1 (160 μg/mL) or LPS (1 μg/mL) for 24 h. TNF-α generation was determined from the culture supernatant. Results are representative of three separate experiments. $^*P < .05$ compared with the corresponding group without PMB treatment.

5.2.1.1.3 Effect of PSG-1 on Nitric Oxide Production and mRNA Expression of iNOS

Nitric oxide (NO) has been identified as a major effector molecule produced by macrophages, and is involved in the regulation of apoptosis and in host defenses against microorganisms and tumor cells. The synthesis of NO by activated macrophages is an important cytotoxic/cytostatic mechanism of non-specific immunity (Brüne, 2003). The effect of PSG-1 on the NO production of RAW264.7 cells was determined by Griess assay. The nitrite concentration of the culture supernatant was significantly increased by PSG-1 (20 μg/mL, $P < .05$; 40, 80, and 160 μg/mL, $P < .01$) (Fig. 5.9A). This suggested that PSG-1 may stimulate the NO production of RAW264.7 cells. To confirm the increase in NO production, RT-PCR was used to analyze the gene expression of iNOS in RAW264.7 cells, which is the most crucial enzyme in the synthesis of NO after the activation of macrophages. As is shown in Fig. 5.9B, the transcription of iNOS mRNA was dose-dependently induced by PSG-1 in macrophages. The result indicated that the increased production of NO in macrophages may be mediated by the expression of the iNOS gene.

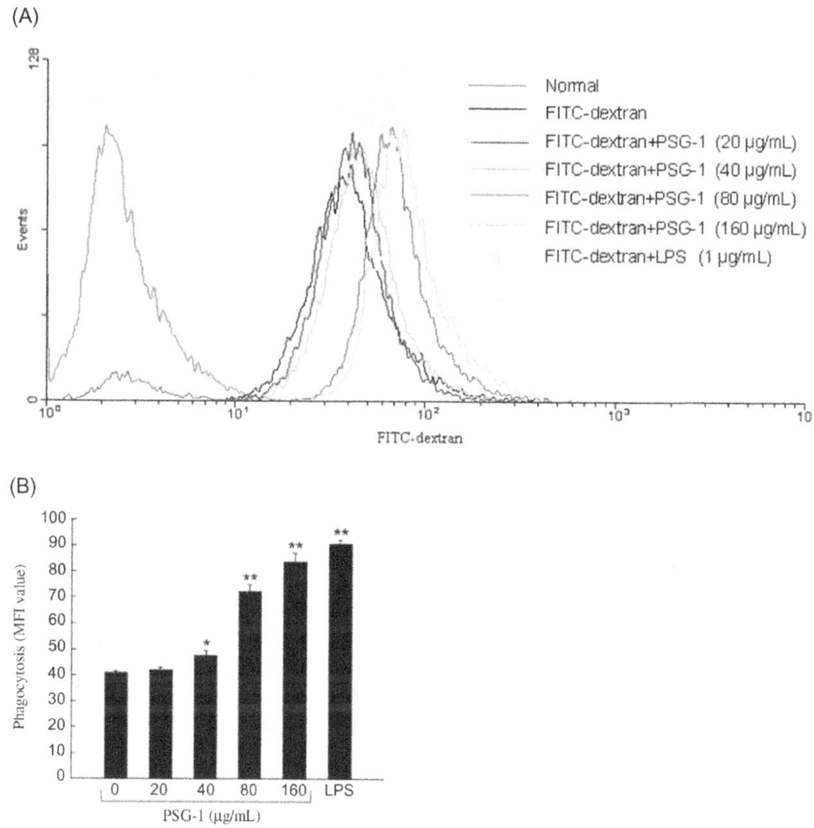

FIGURE 5.8 Effect of PSG-1 on the phagocytosis of RAW264.7 cells by flow cytometry. (A) The extent of the phagocytic uptake determined. (B) Statistic data of MFI value. $*P<.05$, $**P<.01$, $n=3$.

5.2.1.1.4 Effect of PSG-1 on TNF-α and IL-1β Release

Cytokines are intercellular signaling proteins released by both immune and non-immune cells. They play important roles in controlling homeostasis of the whole organism by the induction of cell differentiation, proliferation, and apoptosis, as well as defense functions such as immune responses and inflammatory reactions (Haddad, 2002). When stimulated by pathologic stimuli or injury, activated macrophages play an important role by producing cytokines, such as TNF-α and IL-1β. The induction of cytokine synthesis is one of the methods used to evaluate the augmentation activity of innate immunity. The secretory effects of PSG-1 on the production of TNF-α and IL-1β by macrophages were determined by ELISA. The RAW264.7 cells were treated with various concentrations of PSG-1 for 24 h, and culture supernatants were assayed for cytokines. Fig. 5.9C shows that treatment of

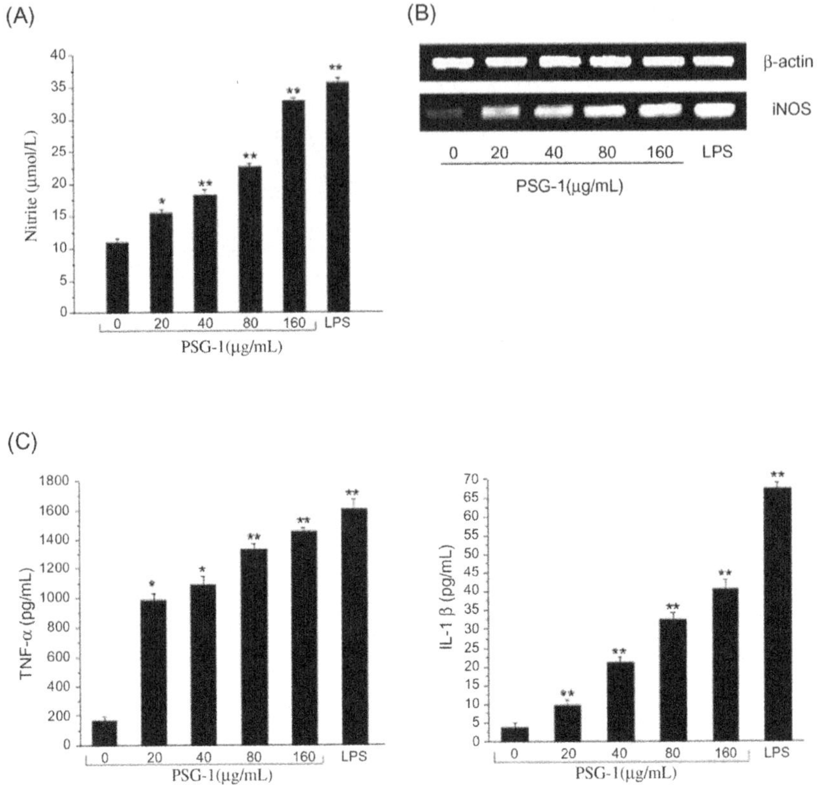

FIGURE 5.9 Effect of PSG-1 on NO and cytokine production in RAW264.7 cells. (A) Cells were treated with PSG-1(20, 40, 80, and 160 μg/mL) or LPS (1 μg/mL) for 24 h. The supernatant nitrite levels were determined using Griess reagent. (B) Cells were incubated with PSG-1 or LPS for 24 h. Total RNA was isolated and analyzed for the magnitude of mRNA expression of iNOS using RT-PCR. One of three representative experiments is shown. (C) Cells were pretreated with PSG-1 (20, 40, 80, and 160 μg/mL) or LPS (1 μg/mL) for 24 h. The supernatant cytokine levels were determined using ELISA kits. *$P<.05$, **$P<.01$ compared with untreated group (0 μg/mL), $n=8$.

PSG-1 for 24 h dose-dependently induced TNF-α and IL-1β release in RAW264.7 cells. When the concentration was 160 μg/mL, the TNF-α and IL-1β production reached up to 1454.3 ± 25.6 pg/mL and 40.7 ± 4.3 pg/mL, respectively. On the other hand, the levels of TNF-α and IL-1β from the positive control, LPS (1 μg/mL)-stimulated RAW264.7 cells, were much higher (1605.0 ± 66.2 pg/mL and 67.3 ± 3.6 pg/mL, respectively).

5.2.1.1.5 Effect of PSG-1 on Reactive Oxygen Species Generation

The reactive oxidants play key roles in host defense and other physiological processes, and are also involved in the regulation of apoptosis and immune homeostasis (Knight, 2000). The generation of ROS was assessed using the cell-permeable probe cDCFH-DA. This dye is cleaved by cellular esterases to nonfluorescent 2′,7′-dichlorofluorescin (DCFH), and oxidized by ROS to a fluorescent product dichlorofluorescein (DCF). The intracellular ROS level (dye oxidation) was measured by flow cytometry. To evaluate whether PSG-1 influences the ROS generation of RAW264.7 cells, the cells were cultured with PSG-1 or LPS for 24 h and then analyzed for DCF by flow cytometry. As shown in Fig. 5.10A, compared with the untreated control, RAW264.7 cells treated with PSG-1 exhibited an increase in ROS production. The concentration-dependent relationship is shown in Fig. 5.10B, with the four doses (from 20 to 160 μg/mL) tested.

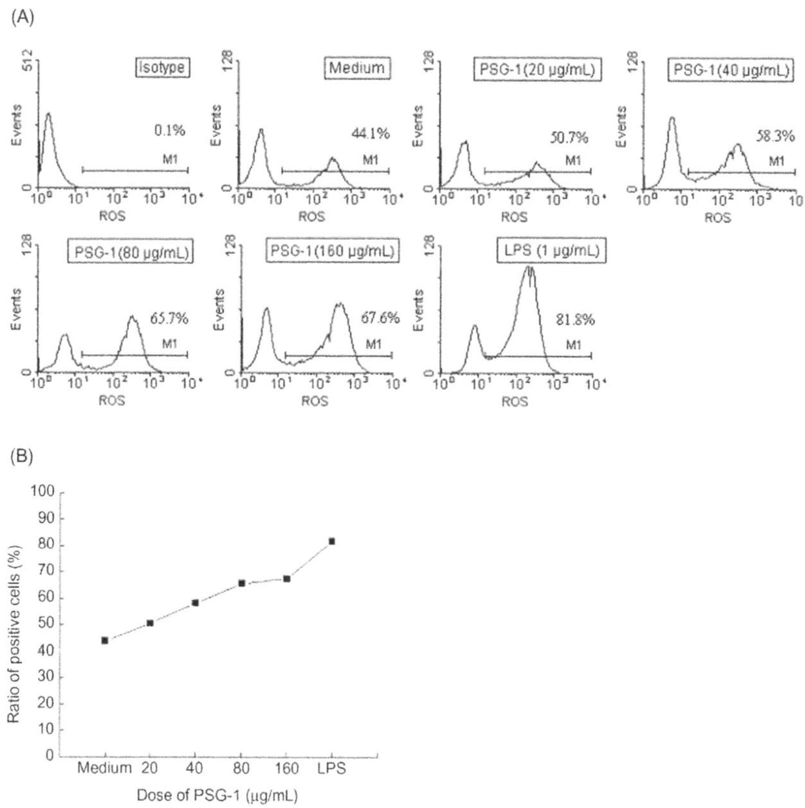

FIGURE 5.10 Effect of PSG-1 on ROS generation in RAW264.7 cells. (A) The expression of fluorescence of 2′,7′-dichlorofluorescein (DCF) in RAW264.7 cells was analyzed by flow cytometer after treatment of PSG-1 (20, 40, 80, or 160 μg/mL) or 1 μg/mL LPS for 24 h. (B) The upregulation effect of PSG-1 on ROS generation was dose dependent. The results shown were from one representative experiment of three independent experiments performed.

5.2.1.2 The Signal Pathways Involved in the Macrophage Immunomodulatory by PSG-1

5.2.1.2.1 PSG-1 Induces Activation of MAPKs Pathway

MAPKs are a family of serine/threonine protein kinases that include three major groups in mammalian cells, i.e., ERK1/2, JNK1/2, and p38. All the three families of MAPKs have been reported to be activated in macrophages initiated by a variety of extracellular stimuli, such as LPS, cytokines, polysaccharides, and so on (Rao, 2001). Many studies have demonstrated that polysaccharides derived from mushrooms, including *Lentinus edodes* and *G. lucidum* could stimulate MAPKs phosphorylation. Lee et al. (2008) showed that fraction II (F-II), an exopolysaccharide prepared from a liquid culture of *L. edodes*, induced rapid MAPKs phosphorylation in the RAW264.7 murine macrophage cell line. Therefore, it was crucial to demonstrate the activation of all the three MAPKs stimulated by PSG-1 in RAW264.7 cells.

Western blot analysis was used to investigate whether the MAPKs signaling pathway was involved in macrophage activation by PSG-1. The phosphorylation of three major groups of MAPKs, including ERK1/2, JNK1/2, and p38, in response to PSG-1 in controlled time-dependent experiments was examined. The phosphorylation state-independent MAPKs were shown to verify semiquantitative loading for gel analysis per loading lane. As shown in Fig. 5.11, phosphorylated ERK1/2 in RAW264.7 cells was detected in 5 min, and the level became sevenfold higher when it reached the maximum in 30 min, after 120 min, ERK1/2 phosphorylation returned to threefold. Next was to investigate the possibility of PSG-1 activation of the JNK1/2 pathway. Cells incubated with PSG-1 showed JNK1/2 phosphorylation, as determined by Western blot analysis, with anti-phospho-JNK1/2 antibodies. The JNK1/2 phosphorylation gradually increased in 15 min, and reached the maximal level (fivefold increase) at 30 min, after 120 min, the induced JNK1/2 phosphorylation gradually returned to the basal level. Then, it was examined whether PSG-1 could induce p38 phosphorylation, another important MAPKs member. A time course study of PSG-1-induced p38 phosphorylation showed that after 30 min, p38 phosphorylation increased fourfold compared with that of control cells, after 120 min, it returned to twofold. In summary, the time course study showed that the phosphorylation levels of all three MAPKs reached a maximum after 30 min.

5.2.1.2.2 Activation of PI3K/Akt Pathway by PSG-1

The PI3K family belongs to an evolutionarily conserved family of signal transducing enzymes, which control several cellular responses including cell growth, survival, cytoskeletal remodeling, and the trafficking of intracellular organelles in many different types of cell (Carpenter and Cantley, 1996). The PI3K and downstream serine/threonine kinase Akt have recently received much attention due to their particularly important functions in the immune system (Fruman and Cantley, 2002; Koyasu, 2003; Martin, 1998;

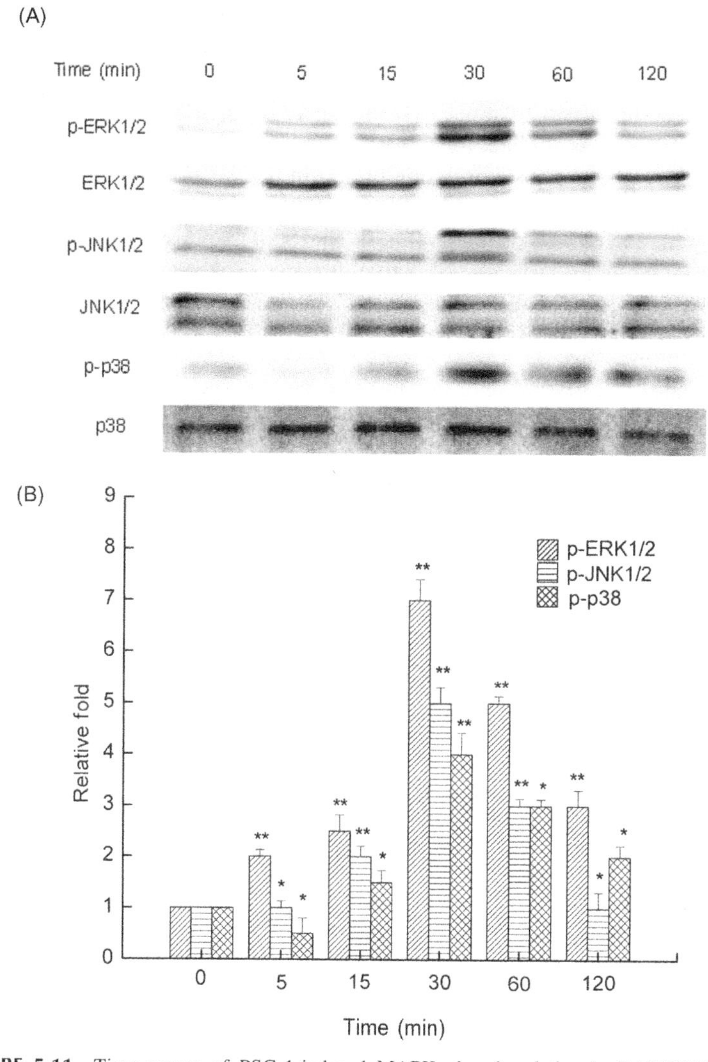

FIGURE 5.11 Time course of PSG-1-induced MAPK phosphorylation in RAW264.7 cells. (A) RAW264.7 cells were exposed to PSG-1 (160 μg/mL) at the indicated time periods. The protein levels of total ERK1/2, JNK1/2, and p38 and its phosphorylated form in the cytoplasmic extracts were analyzed using Western blot. The images are a representative of three independent experiments. (B) Histogram represents quantification by PhosphorImager of PSG-1-stimulated phosphor-ERK1/2, phosphor-JNK1/2, and phosphor-p38 in RAW264.7 cell sample using Quantity one software. All data of relative activity are expressed in comparison with untreated cells ($t = 0$; phosphorylation of control cells defined as 1). $^*P < .05$, $^{**}P < .01$ compared with control group. Similar experiments were repeated three times.

Okkenhaug and Vanhaesebroeck, 2003). However, there is rare evidence on the role of the PI3K/Akt pathway in intracellular mechanisms of macrophage activation. To study the putative role of the PI3K/Akt pathway in PSG-1-induced macrophage activation, first the capability of PSG-1 to activate Akt in macrophages was studied.

Furthermore, the phosphorylation of Akt in RAW264.7 cells was detected to investigate the effect of PSG-1 on the PI3K/Akt pathway. Cells were treated with PSG-1 for different times. The Ser473 phosphorylation of Akt was measured as an indicator of active enzyme (Toker and Newton, 2000). As shown in Fig. 5.12, there was a noticeable increase in Ser473

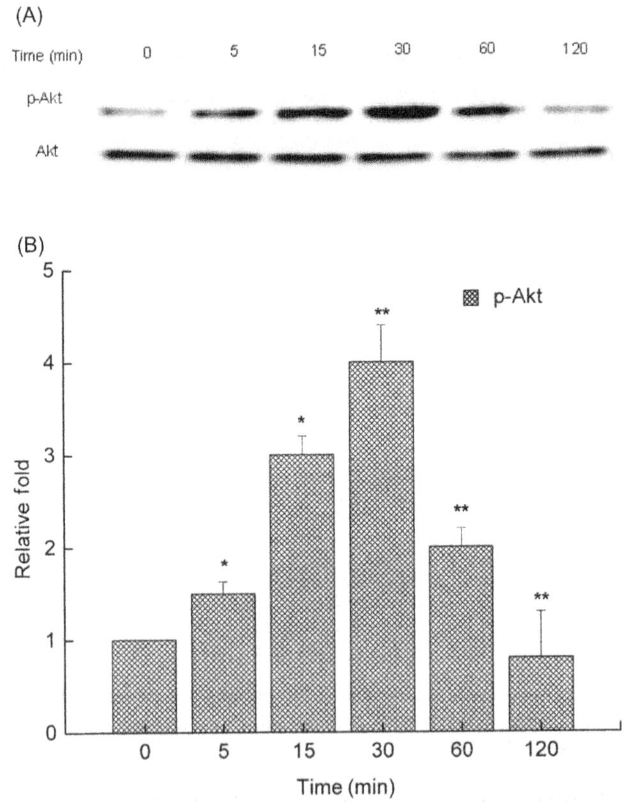

FIGURE 5.12 Time course of PSG-1-induced Akt phosphorylation in RAW264.7 cells. (A) RAW264.7 cells were treated with PSG-1 (160 μg/mL) at the indicated time periods. After treatment, total cell extract was prepared, and the protein levels of phosphorylated Akt (p-Akt) and total Akt (Akt) were determined by Western blot analysis as described in Materials and Methods. Data are representative of three experiments. (B) Histogram represents quantification by PhosphorImager of PSG-1-stimulated phospho-Akt in RAW264.7 cell sample using Quantity one software. All data of relative activity are expressed in comparison with untreated cells ($t = 0$; phosphorylation of control cells defined as 1). $^{*}P < .05$, $^{**}P < .01$ compared with control group. Similar experiments were repeated three times.

phosphorylation of Akt, which reached its peak at 30 min. The result showed that the Akt phosphorylation was activated and reached its peak at 30 min, a similar pattern to MAPKs phosphorylation.

5.2.1.2.3 Inhibition of PI3K/Akt Prevented PSG-1-Induced Activation of MAPKs

Various studies have shown that the activation of MAPKs was mediated via the PI3K/Akt signaling pathway, however, there are some controversial reports. For example, Hsu et al. (2004a,b) reported that inhibition of PI3K by LY294002 did not alter EORP, a polysaccharide extract from *G. lucidum* (also called reishi), but induced JNK1/2 and p38 phosphorylation in the J774.1 murine macrophage cell line. Examination of the phosphorylation of ERK1/2, JNK1/2 and p38 in PSG-1-stimulated macrophages treated with increasing concentrations of LY294002 can address this issue. Specifically, RAW264.7 cells were pre-incubated with and without LY294002 for 30 min and then treated with PSG-1 (160 μg/mL) for an additional 30 min. Cell lysates were analyzed by Western blot with anti-phosphorylated ERK1/2, anti-phosphorylated JNK1/2, or anti-phosphorylated p38 antibody. The results showed a dose-dependent decrease of ERK1/2, JNK1/2, and p38 phosphorylation (Fig. 5.13 samples 3, 4, 5, and 6 vs 2), indicating that the PI3K/Akt pathway was associated with PSG-1 activation of the MAPKs pathway. The data showed that inhibition of PI3K by LY294002 diminished PSG-1-induced ERK1/2, JNK1/2, and p38 phosphorylation in a dose-dependent manner. These results suggested that the PI3K/Akt signaling pathway acted upstream of the MAPKs pathway in PSG-1-stimulated macrophages.

5.2.1.2.4 Effect of PSG-1 on Activation of NF-κB

The activation of macrophages is associated with transcriptional up-regulation of numerous genes related to immune defense functions (Chen et al., 2009). It is well documented that NF-κB is a ubiquitous transcription factor that plays a crucial role in macrophage activation via the transcription of downstream inflammatory mediators of multiple types of stimulatory events (Li et al., 2011a,b). Therefore, the ability of PSG-1 to activate NF-κB was tested to identify the nuclear factors that contribute to the activation of macrophages. Since the p65 subunit of NF-κB has been demonstrated to exert critical activity in the transcription of many inflammatory genes (Li and Verma, 2002), a Western blot assay was performed to detect expression of the NF-κB p65 protein. As shown in Fig. 5.14, NF-κB p65 was significantly elevated ($*P < .05$ or $**P < .01$). The results indicated that stimulation of macrophages with PSG-1 resulted in increased protein synthesis of nuclear NF-κB, suggesting that NF-κB may play a central role in macrophage-mediated cellular activation by PSG-1. Detailed intracellular signal pathways will be further explored in the future.

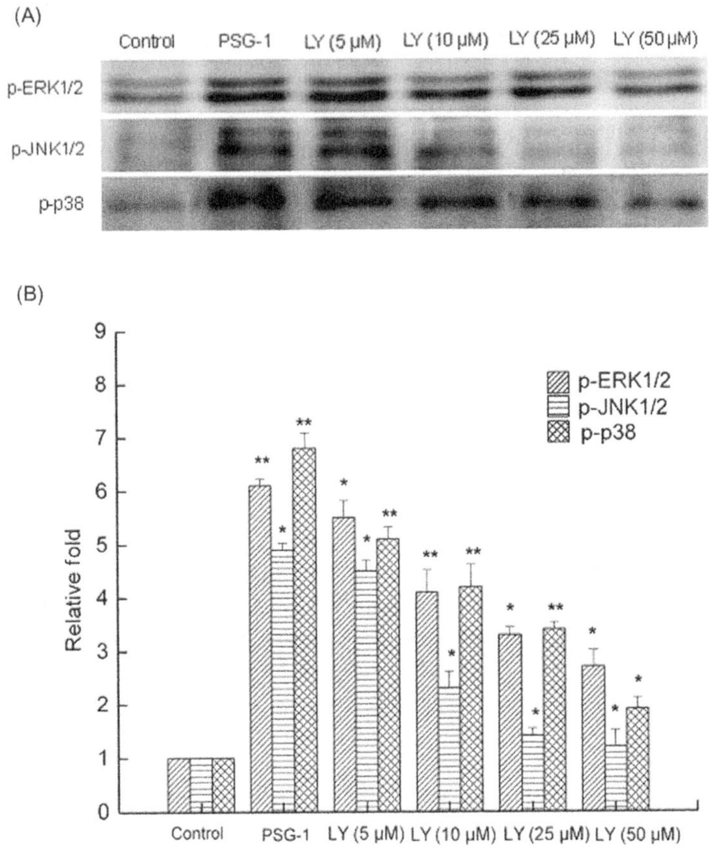

FIGURE 5.13 Effect of LY294002 on PSG-1-induced phosphorylation of ERK1/2, JNK1/2, and p38 in RAW264.7 cells. (A) RAW264.7 cells were pretreated with various concentrations of LY294002 (5, 10, 25, and 50 μM) for 30 min before stimulation with PSG-1 (160 μg/mL) for an additional 30 min. The cell lysates were collected and analyzed by Western blot with antiphosphorylated ERK1/2, antiphosphorylated JNK1/2, or antiphosohorylated p38 monoclonal antibody, respectively. Similar results were obtained in three separate experiments. (B) Histogram represents quantification by PhosphorImager of phospho-ERK1/2, phospho-JNK1/2, and phospho-p38 in RAW264.7 cell sample using Quantity one software. All data of relative activity are expressed in comparison with untreated cells ($t = 0$; phosphorylation of control cells defined as 1). $^{*}P < .05$, $^{**}P < .01$ compared with PSG-1 alone. Similar experiments were repeated three times.

5.2.1.2.5 PI3K/Akt and MAPKs Pathways Were Involved in the Activation of NF-κB

Several major pathways are currently considered to be relevant to the activation of the transcription factor; PI3K/Akt and MAPKs are the examples of the pathways (Fishman et al., 2006). Whether NF-κB activation by PSG-1 was triggered by these upstream signaling cascades was carefully explored

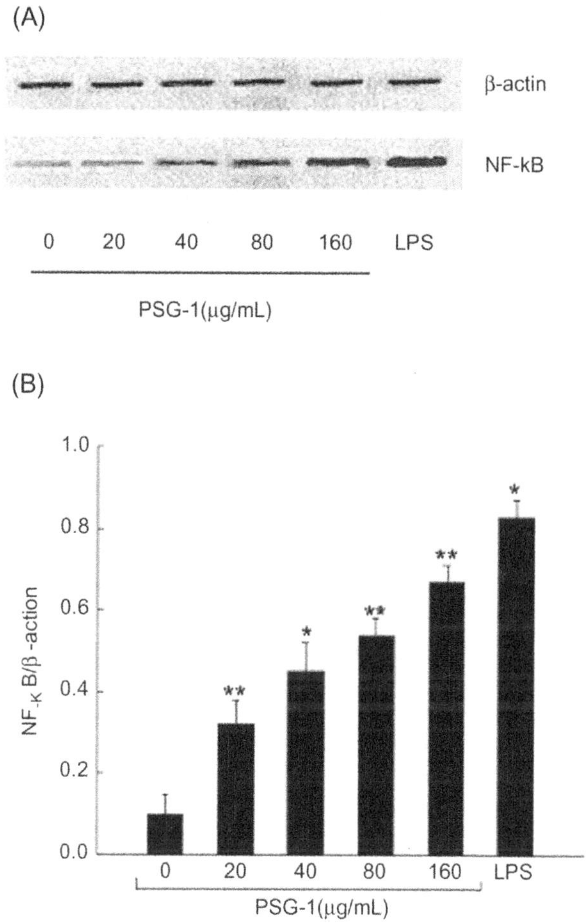

FIGURE 5.14 Effect of PSG-1 on the activation of NF-κB detected by Western blotting. (A) The nuclear protein was isolated after cells were cultured with PSG-1 or LPS for 24 h, then analyzed by Western blotting with anti-β-actin and anti-NF-κB p65 antibodies. (B) Blots were scanned and expression of β-actin and NF-κB p65 was quantified by densitometric analysis. The ratios for these proteins are shown. Eight independent experiments were done and all gave similar results. $^{*}P<.05$, $^{**}P<.01$ compared with untreated group (0 μg/mL).

using a pharmacological assay. To elucidate the relationship between the PI3K/Akt/MAPKs pathway and NF-κB, a further study was conducted to investigate whether PI3K and MAPKs pathways were involved in regulating the DNA binding activity of NF-κB. A gel mobility shift assay was used to examine the ability of PI3K and MAPKs inhibitors to block PSG-1-induced DNA binding of NF-κB. To this end, RAW264.7 cells were stimulated with PSG-1 after pretreatment with LY294002, PD98059, SP600125, and SB203580. Nuclear extracts were prepared, and NF-κB EMSA was

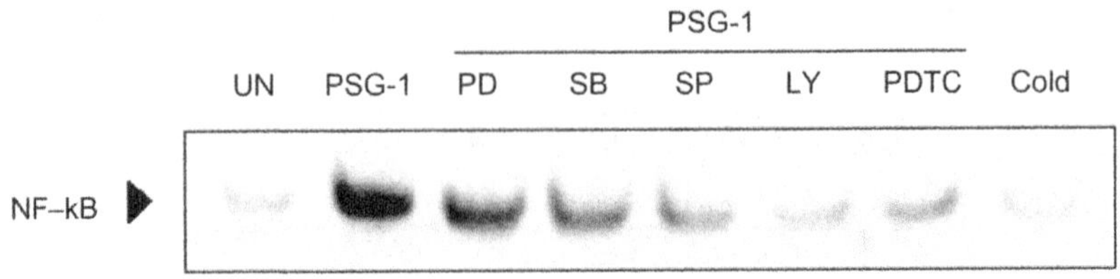

FIGURE 5.15 Effect of PI3K and MAPK inhibitors on PSG-1-induced DNA binding of NF-κB in RAW 264.7 cells. RAW 264.7 cells were treated with PD98059 (30 μM), SB203580 (30 μM), SP600125 (30 μM), LY294002 (25 μM), or PDTC (30 μM) for 30 min before being incubated with PSG-1 (160 μg/mL). Nuclear extracts were prepared and analyzed for the binding activity of NF-κB by EMSA. The results presented are representative of three independent experiments.

performed. As shown in Fig. 5.15, a low level of NF-κB DNA binding activity was observed in nuclear extracts from untreated cells. The PSG-1 challenge robustly enhanced NF-κB DNA binding. Pretreatment with LY294002, PD98059, SP600125, or SB203580 was able to significantly reduce PSG-1-induced DNA binding of NF-κB. These results showed that PSG-1-induced DNA binding activity of NF-κB was abrogated by the treatment of specific inhibitors of PI3K/Akt and MAPKs pathways, suggesting that the induction of NF-κB DNA binding activity by PSG-1 was dependent on the PI3K/Akt/MAPKs pathway in RAW 264.7 cells.

Activation of macrophages is a key event in innate and adaptive immunity for the initiation and propagation of defensive reactions against pathogens (Lee et al., 2007). Thus, the identification of agents that can modulate macrophages is of significant interest. The present study demonstrates that PSG-1 from *G. atrum* may promote the phagocytosis and stimulate production of NO, ROS, and cytokines (TNF-α and IL-1β). In addition, the results presented in this report suggested that the PI3K/Akt/MAPKs/NF-κB signaling pathway was involved in macrophage activation by PSG-1.

5.2.1.3 The Membrane Receptor and Signaling Pathway Involved in the Activation of Macrophages by PSG-1

It was found that PSG-1 possesses immunomodulatory effects on macrophages (Yu et al., 2013), and the PI3K/Akt/MAPKs/NF-κB signaling pathway plays a key role in macrophage activation by PSG-1 (Yu et al., 2012). However, the membrane receptor and detailed signaling pathway involved in the activation of macrophages in response to PSG-1 have not yet been completely clarified.

Emerging studies have established the essential role of specific cellular receptors on macrophages in response to a wide variety of microbial pathogens with subsequent induction of intracellular signaling cascade (Ehlers, 2000; Medzhitov and Janeway, 2000; Netea et al., 2006). Toll-like receptors are of a family that has been found to be crucial for innate immunity (Medzhitov, 2001; Trinchieri and Sher, 2007). To date, 10 and 12 functional TLRs have been identified in humans and mice, respectively, with TLR1–TLR9 being conserved in both species. Mouse TLR10 is not functional because of a retrovirus insertion, and TLR11, TLR12, and TLR13 have been lost from the human genome (Akira et al., 2001). Binding of ligands to TLR leads to the formation of a complex between the cytoplasmic region of TLR, the adaptor protein Myd88, and the IRAK. This is followed by the activation of TRAF6, and TRAF6 in turn activates the IKK complex. The activation of the IKK complex leads to the degradation of IκB and the activation of NF-κB, which regulates a wide spectrum of target genes (Karin and Ben-Neriah, 2000). Toll-like receptors are expressed by various cells of the immune system. Among them, TLR4 has been identified as an important membrane receptor of macrophages (Moresco et al., 2011), it mediates macrophage activation by transmitting a variety of extracellular signals. Endotoxin is one of the best studied components that can interact with TLR4 (Anderson, 2000); however, the role of TLR4 in response to PSG-1 is unclear.

ROS have been traditionally regarded as toxic by-products of metabolism, which could cause damage to lipids, proteins, and DNA (Freeman and Crapo, 1982). However, more and more evidence shows that ROS are not only injurious by-products of cellular metabolism, but are also involved in cell signaling and regulation (Finkel, 1998; Rhee, 1999). For example, numerous studies have implied that ROS are involved in MAPKs activation after cell stimulation with various agents (Iwaoka et al., 2006). Lots of cellular stimuli that induce ROS production are capable of simultaneously activating MAPKs pathways in various cell types (McCubrey et al., 2006; Rhee, 1999). The prevention of ROS accumulation by antioxidants inhibits MAPKs activation, indicating the involvement of ROS in the activation of MAPKs pathways. Besides MAPKs, other signaling molecules, such as protein serine/threonine kinase, and transcriptional factors can also be activated by ROS (Thannickal and Fanburg, 2000). The PI3K, as well as its downstream serine/threonine kinase Akt, which have been implicated in a number of cellular responses, including cell migration, phagocytosis, and apoptosis (Carpenter and Cantley, 1996), were reported to be activated by ROS (Ushio-Fukai et al., 1999). Moreover, NF-κB is well-known as a critical transcription factor in the induction of a wide variety of genes that are involved in the regulation of immune and inflammatory responses (Hayden and Ghosh, 2004). A number of studies indicate that ROS may serve as common intracellular agents that contribute to the process of NF-κB activation in response to a diverse range of stimuli (Baeuerle and Henkel, 1994).

Many reports indicate that one of the important sources of ROS in macrophages is NADPH oxidase activation during phagocytosis (Bokoch, 1995). The NADPH oxidase complex of the macrophages consists of four core protein subunits, which are cytochrome b558, along with the cytosolic components p67 phox, p47 phox, and p40 phox. Activation of the oxidase, however, requires the additional participation of Rac-1, a member of the Ras superfamily of small GTP-binding proteins. During activation, Rac-1 binds GTP and migrates to the plasma membrane along with the cytosolic components, to form the active oxidase complex (Bosco et al., 2009). However, little is known about the potential role of ROS in controlling PSG-1-mediated intracellular signaling pathways and the mechanism by which PSG-1 induces ROS generation in macrophages.

In this study, the possibility of TLR4 as a candidate receptor for PSG-1-mediated signaling was examined. Moreover, the role of ROS in PSG-1-mediated signal transduction involved in the regulation of TNF-α production, as well as the mechanism for PSG-1 induction ROS in macrophages, was investigated.

5.2.1.3.1 Fluorescence Labeling of PSG-1 and Dextran

A number of studies have well documented that polysaccharides cannot penetrate cells due to their large molecular mass, so the first step in the modulation of cellular activity would be binding to immune cell receptors. Therefore, it was hypothesized that macrophage activation by PSG-1 starts with specific binding to cells. In order to identify cells expressing specific receptor(s) for PSG-1, PSG-1 was conjugated with fluoresceinamine using the CNBr-activation method; f-dextran was constructed as a control using the same method. As shown in Fig. 5.16, a small amount of fluoresceinamine co-eluted with polysaccharides or dextran, while free fluoresceinamine eluted much more slowly. Fluorescence-labeled polysaccharide or dextran fractions were mixed and stored at $-20°C$ for future use.

5.2.1.3.2 TLR4 is the Major Receptor Involved in Specific Binding of PSG-1 to Macrophages

The innate immune system identifies infectious agents or compounds by using pattern-recognition receptors (PRRs). The best characterized group of PRRs is the evolutionarily conserved family of TLRs (Barton and Medzhitov, 2003). It is now evident that mammalian TLRs play a prominent role in the direct activation of host defense mechanisms. The activation of TLRs stimulates an innate immune response, which involves the production of direct antimicrobial effector molecules, including ROS, and increases an adaptive immune response by inducing the production of TNF-α and IL-1β that augment both cell-mediated and humoral immune responses. A founding member of the TLR family, TLR4, is mainly expressed in macrophages, monocytes, neutrophils, and dendritic cells (Beutler, 2000). Recently, many studies have identified that TLR4 is an important membrane receptor of polysaccharides. Polysaccharides from *G. lucidum* and *Paecilomyces cicadae* interact with TLR4 on macrophages (Kim et al., 2012; Shao et al., 2004). In

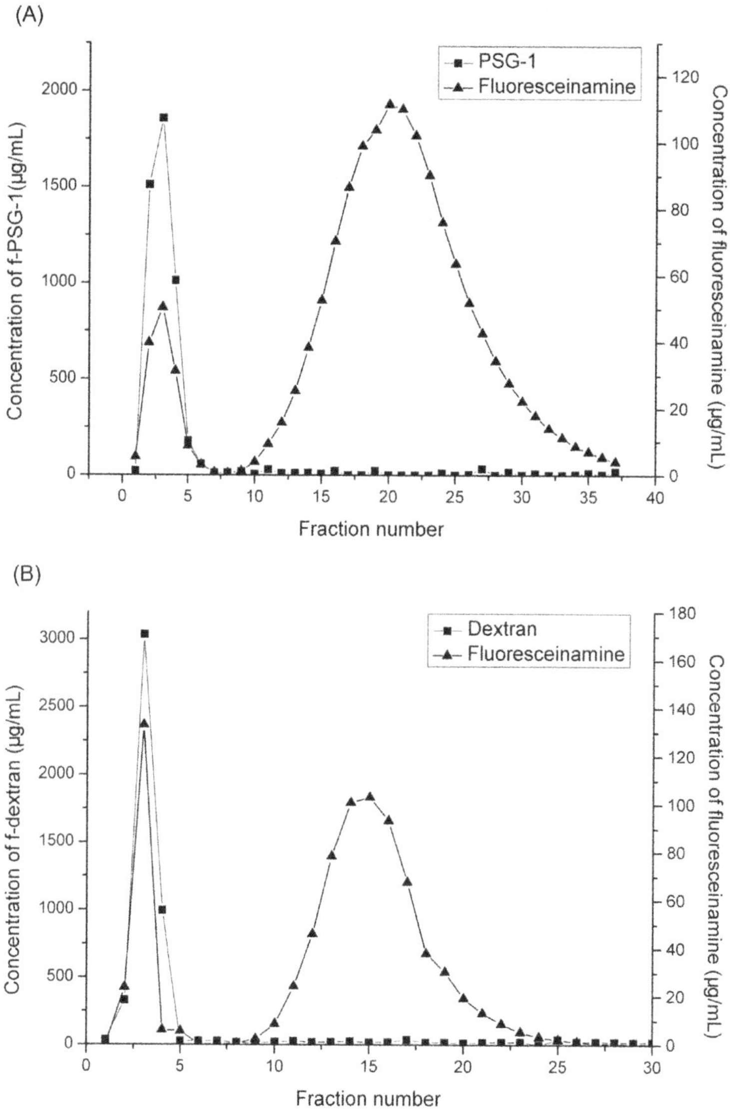

FIGURE 5.16 Polysaccharide labeling with fluoresceinamine. CNBr-activated PSG-1 (A) and dextran (B) were reacted with fluoresceinamine for 10 h at room temperature in the dark. The mixtures were fractionated on a Sephadex-G 25 column, and concentrations (μg/mL) of carbohydrate (■) and fluoresceinamine (▲) in each fraction were determined.

addition, polysaccharides from *Polyporus umbellatus* (Per) Fr, *Cordyceps militaris*, have been reported to activate DC cells through TLR4 (Kim et al., 2010; Li et al., 2010a,b). This evidence prompted us to study TLR4 as the candidate receptor for PSG-1.

To demonstrate the specific binding of PSG-1 to macrophages, RAW264.7 cells were stained with f-PSG-1 or f-dextran for 1 h at 4°C, followed by flow cytometric analysis and confocal laser-scanning microscopy. As shown in Fig. 5.17A–C, both methods could detect positive staining. However, the staining of f-PSG-1 could be noticeably inhibited by the addition of eightfold molar excess of unlabeled PSG-1 (Fig. 5.17D), which demonstrated that f-PSG-1 specifically bound to macrophages. To explore which cell receptor was involved in the specific binding of PSG-1 to macrophages, RAW264.7 cells were incubated with anti-TLR2, anti-TLR4, or anti-CR3 mAb for 1 h, and then with f-PSG-1 (20 μg/mL) for another hour at 4°C. As shown in Fig. 5.17E and F, anti-TLR4 mAb significantly inhibited the ability of f-PSG-1 to bind to macrophages; however, anti-TLR2 and anti-CR3 mAb did not exhibit any significant inhibitory effect, suggesting that TLR4 is the major receptor involved in PSG-1 binding to macrophages.

5.2.1.3.3 TLR4-Dependent Activation of Macrophages by PSG-1

Although TLRs are essential for protective immunity against infection, inappropriate TLR responses contribute to acute and chronic inflammation, as well as to systemic autoimmune diseases. For example, TLR4 responds to bacterial LPS, which is considered to play a key role in septic shock syndrome (Kawai and Akira, 2006). More and more researchers speculate that natural polysaccharides can directly activate the innate immune response by mediating TLR signaling, since natural polysaccharides have been safely used as ingredients in traditional medicines for a long time in oriental countries, and have not been associated with any detrimental tissue injuries caused by LPS. Recently, several reports have demonstrated the TLR4-dependent activation of macrophages by mushroom polysaccharides. For instance, Li and Xu (2011) reported that the polysaccharide fraction from *P. umbellatus* (pers.) Fries stimulated TNF-α and IL-1β production via TLR4 activation of signaling pathway in macrophages. Especially, Zhang et al. (2013) reported that PSG-1 may elicit its anti-tumor effect through TLR4. However, the authors only preliminarily investigated the role of TLR4 by measuring the protein expression of TLR4, and then suggested that it may be a possible receptor for PSG-1. This prompted us to conduct more systematic and convincing research to illustrate the role of TLR4 on the activation of macrophage by PSG-1. Specifically, first Fluorescein-labeled PSG-1(f-PSG-1) was synthesized, and the specific binding of f-PSG-1 to macrophages by flow cytometric analysis and confocal laser-scanning microscopy was demonstrated. Then, anti-TLR4 mAb, anti-TLR2 mAb, and anti-CR3 mAb were used to inhibit f-PSG-1 binding, and it was found that TLR4 is the major receptor involved in PSG-1 binding to macrophages. Furthermore, the results demonstrated that PSG-1 activated peritoneal macrophages from

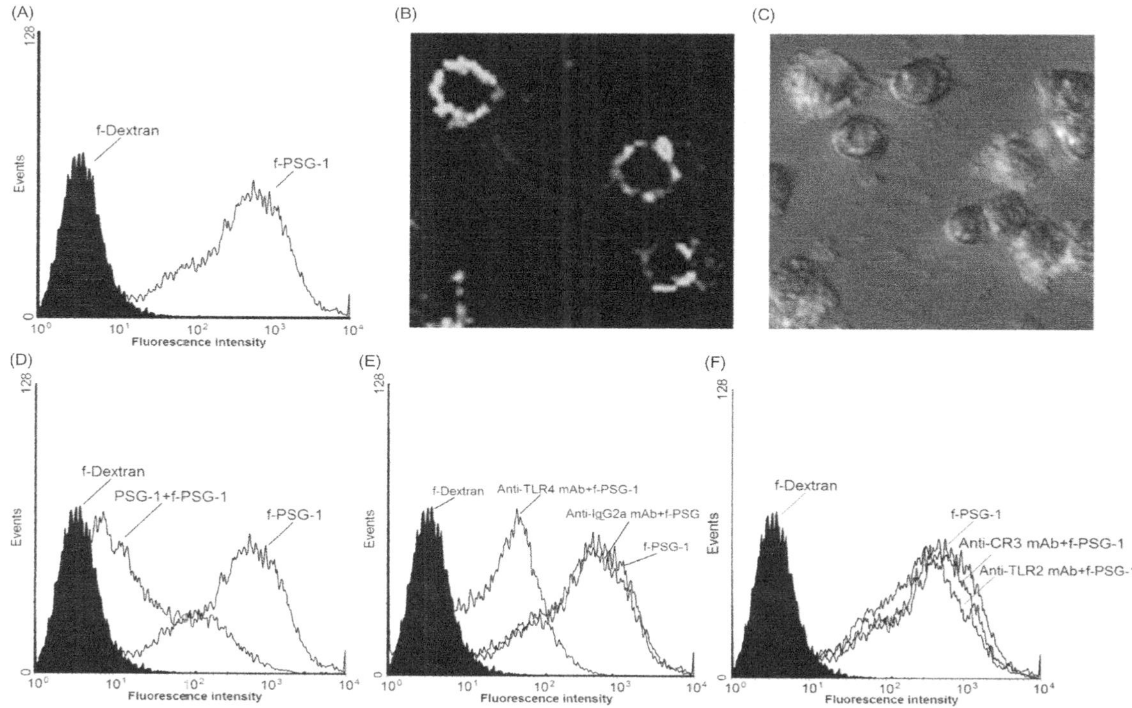

FIGURE 5.17 TLR4-dependent staining of macrophages with f-PSG-1. (A) RAW264.7 cells were stained with f-dextran or f-PSG-1 for 1 h for flow cytometric analysis. (B, C) RAW264.7 cells stained with f-PSG-1 were observed using a confocal laser-scanning microscope. (D) Competitive inhibition of f-PSG-1 binding to cells by unlabeled PSG-1. RAW264.7 cells were stained with f-PSG-1 (20 μg/mL) in the presence or absence of unlabeled PSG-1 at 160 μg/mL for 1 h at 4°C for flow cytometric analysis. (E, F) Inhibition of f-PSG-1 binding to macrophages by anti-TLR4 mAb but not by anti-TLR2 or anti-CR3 mAb. RAW264.7 cells were treated with anti-TLR4, anti-TLR2, or anti-CR3 mAb (20 μg/mL) for 1 h at 4°C followed by f-PSG-1 (20 μg/mL) staining for another hour, then measured by flow cytometry.

wild-type, but not TLR4-deficient, mice to produce TNF-α. Moreover, anti-TLR4 mAb could significantly block PSG-1-mediated TNF-α secretion.

To verify whether TLR4 was required for PSG-1 activation of macrophages, the effect of PSG-1 on TNF-α production by peritoneal macrophages from wild-type C3H/HeN mice and functional TLR4-deficient C3H/HeJ mice was compared. It was found that C3H/HeJ peritoneal macrophages did not respond to PSG-1 stimulation in terms of TNF-α production (Fig. 5.18A). However, PSG-1 could activate macrophages from C3H/HeN mice. It is likely that TLR4 plays a pivotal role in PSG-1-mediated macrophage activation. To further confirm, the effect of anti-TLR4 antibody on PSG-1-induced TNF-α production by RAW 264.7 cells was assessed. The results showed that pretreatment of anti-TLR4 antibody significantly blocked PSG-1-induced TNF-α production (Fig. 5.18B). These results can lead to a conclusion that TLR4 is the major receptor involved in PSG-1-mediated activation of macrophages.

5.2.1.3.4 The Role of ROS in Regulating PSG-1-Induced PI3K/Akt, MAPKs, NF-κB Pathways

Increasing evidence demonstrated that ROS can not only play key roles in host defense, but also can act as a cellular secondary messenger in many biological functions, such as protein phosphorylation (Ito et al., 2006), transcription factor activation (Kaul et al., 1998), and cytokine production (Hong et al., 1997). The present study showed that ROS was involved in the regulation of cytokine TNF-α production stimulated by PSG-1. Furthermore, TLR4 was demonstrated as the critical receptor for ROS production stimulated by PSG-1, evidenced by the inhibition effect of anti-TLR4 mAb on ROS production. These facts inspired us to explore the role of ROS in TLR4-mediated downstream signaling pathway in response to PSG-1.

Extensive literature has documented that ROS can mediate the activation of MAPK pathways by a number of cellular stimuli in several cell types, although a recent paper (Yang et al., 2011) challenges this concept, the capsular polysaccharide of pyrogenic liver abscess *Klebsiella pneumonia* (PLAK. pneumoniae CPS)-mediated activation of ERK1/2, JNK1/2, and p38 is independent of ROS. However, little information is available regarding whether ROS is involved in MAPK activation stimulated by PSG-1 in macrophages. To examine the potential role of ROS in PSG-1-mediated activation of MAPKs pathways, RAW264.7 cells were pre-incubated with and without various concentrations of NAC for 30 min, and then treated with PSG-1 (160 μg/mL) for an additional 30 min. Cell lysates were analyzed by Western blot with anti-p-ERK1/2, anti-p-JNK1/2, or anti-p-p38 antibody. The phosphorylation state-independent ERK1/2, JNK1/2, and p38 were shown to verify semiquantitative loading for gel analysis per loading lane.

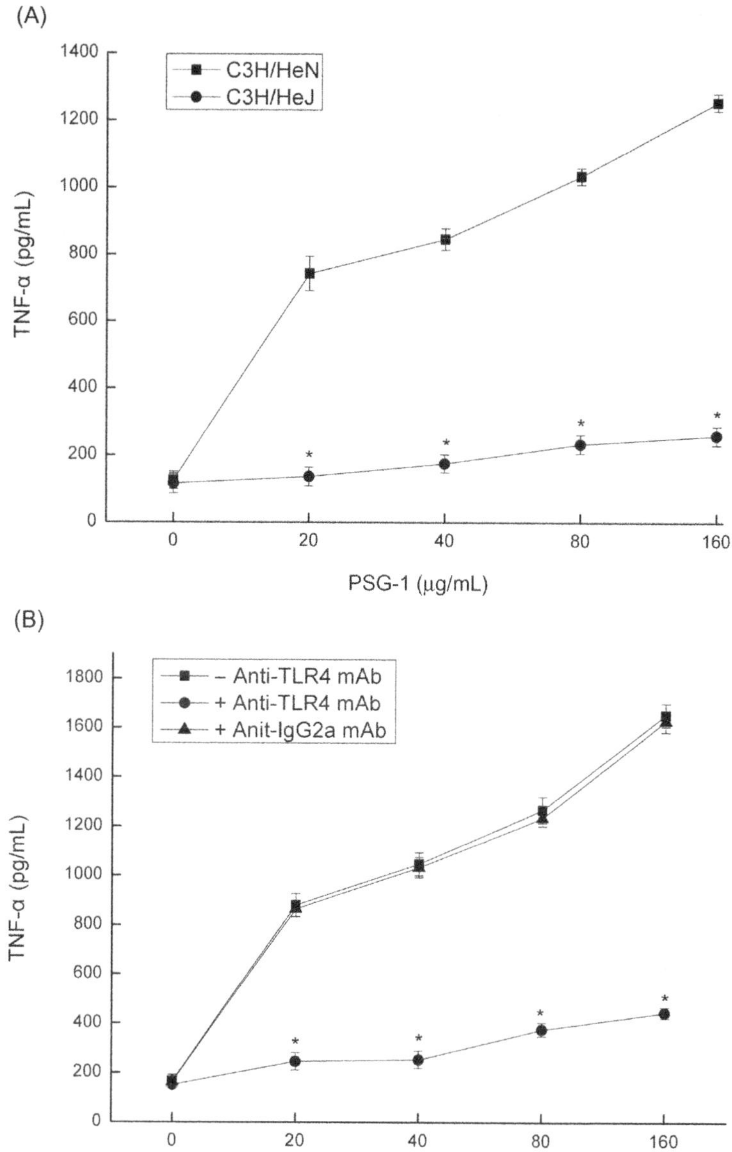

FIGURE 5.18 Involvement of TLR4 in the activation of TNF-α secretion by PSG-1. (A) Peritoneal macrophages from C3H/HeN and C3H/HeJ mice were stimulated with indicated concentrations of PSG-1 for 6 h. Supernatant TNF-α levels were determined using ELISA. (B) RAW 264.7 cells were preincubated with anti-TLR4 mAb (20 μg/mL) or isotype anti-IgG2a mAb for 1 h, and then treated with indicated concentrations of PSG-1 for 6 h. After incubation, the released TNF-α in conditioned media was measured by ELISA. These data are the means ± SEM of triplicate determinations. *$P < .05$ compared with PSG-1 alone.

As shown in Fig. 5.19C, upon PSG-1 stimulation, the level of phosphorylated ERK1/2 in RAW264.7 cells detected in 30 min became about sevenfold higher. In contrast, ERK1/2 phosphorylation gradually reduced by pretreatment with increasing concentration of NAC (Lane 3, 4, and 5 vs Lane 2). When cells were pre-incubated with NAC (10 mM), the level of phosphorylated ERK1/2 decreased to 2.3-fold, as compared with the control group. Cells incubated with PSG-1 alone showed a 3.5-fold JNK1/2 phosphorylation increase. However, a dose-dependent decrease of JNK1/2 phosphorylation by pretreatment of NAC (Lane 3, 4, and 5 vs Lane 2) was shown, and the induced JNK1/2 phosphorylation returned to basal level when it was pretreated with 10 mM NAC. Then, it was examined whether ROS could mediate p38 phosphorylation, another important MAPKs member. The PSG-1-induced p38 phosphorylation increased about 5.8-fold compared with that of control cells. By contrast, p38 phosphorylation was dose-dependently inhibited by pretreatment with NAC, it became to almost onefold by block of NAC (10 mM). The data showed that inhibition of ROS by NAC diminished PSG-1-induced ERK1/2, JNK1/2, and p38 phosphorylation in a dose-dependent manner, which suggested that PSG-1-mediated activation of ERK1/2, JNK1/2, and p38 is dependent on ROS.

In addition, a previous study demonstrated that the PI3K/Akt signaling pathway acted upstream of the MAPKs pathway in PSG-1-stimulated macrophages. Given the above results demonstrating that the activation of MAPKs is mediated by ROS, we next sought to determine whether ROS also induces the PI3K/Akt signal transduction pathway. To address this issue, the effect of NAC on Akt phosphorylation was tested, and it was found that PSG-1-induced Akt phosphorylation was inhibited by pretreatment of NAC in a dose-dependent manner. As shown in Fig. 5.19A, in the absence of NAC, PSG-1 induced the phosphorylation of Akt by 5.3-fold, compared with the control (Lane 1). In contrast, cells pre-incubated with NAC, at a concentration as high as 10 mM, followed by treatment with PSG-1, showed an almost complete inhibition. These results demonstrated that ROS was involved in the PI3K/Akt pathway in PSG-1 stimulated macrophages.

Moreover, NF-κB has been reported to be regulated by ROS (Asehnoune et al., 2004; Meyer et al., 1993; Schreck et al., 1992). Based on a previous study, which showed that NF-κB activation by PSG-1 was triggered by upstream signaling cascades, PI3K/Akt and MAPKs, it was hypothesized that ROS might be involved in PSG-1-mediated NF-κB activation. Transient transfection and luciferase reporter assay was used to examine the ability of NAC to block the PSG-1-induced transactivation activity of NF-κB. As shown in Fig. 5.19E, the PSG-1 challenge robustly enhanced luciferase activity, while pretreatment with NAC could significantly reduce luciferase activity, which suggested that the induction transactivation activity of NF-κB by PSG-1 was dependent on ROS in RAW 264.7 cells. These results support the hypothesis that NF-κB activation by PSG-1 was triggered by ROS. These

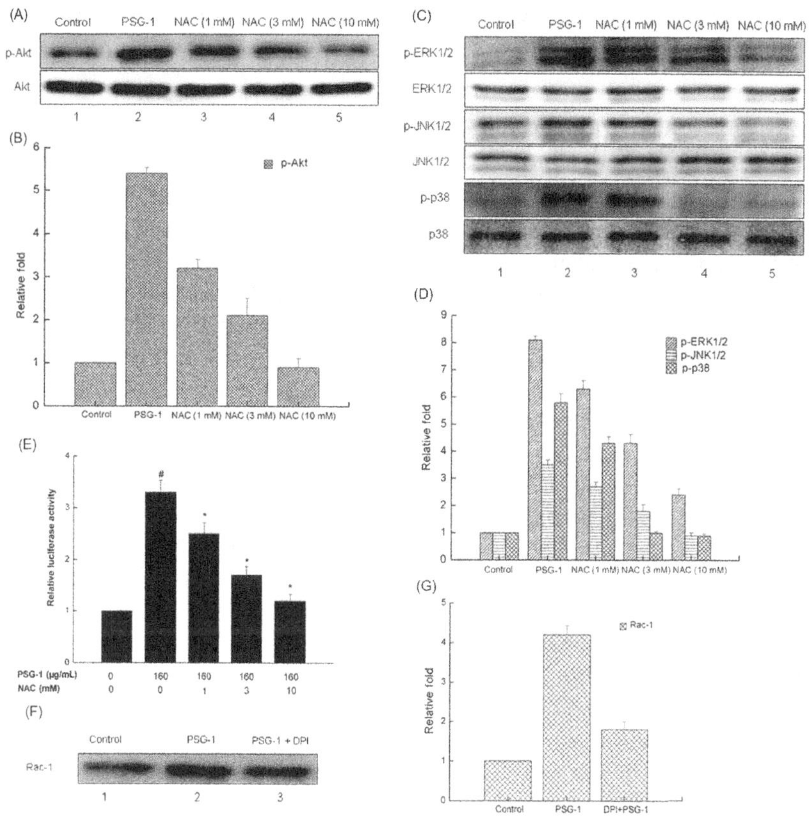

FIGURE 5.19 NADPH oxidase-derived ROS acts upstream of the PI3K/Akt/MAPKs/NF-κB signaling pathway. (A, C) Effect of NAC on PSG-1-induced phosphorylation of Akt, ERK1/2, JNK1/2, and p38 in RAW264.7 cells. RAW264.7 cells were pretreated with various concentrations of NAC (1, 3, and 10 mM) for 30 min before stimulation with PSG-1 (160 μg/mL) for an additional 30 min. The cell lysates were collected and analyzed by Western blot with anti-phosohorylated Akt, anti-phosphorylated ERK1/2, anti-phosphorylated JNK1/2, or anti-phosohorylated p38 antibody respectively. Similar results were obtained in three separate experiments. (E) Effects of NAC on the NF-κB activation in PSG-1-stimulated RAW 264.7 cells. RAW264.7 cells were transiently transfected with luciferase reporter plasmid NF-κB-luc for 48 h, then cells were pretreated with various concentrations of NAC for 30 min before incubation with PSG-1 (160 μg/mL) for 6 h. Luciferase activity was measured by luciferase assay. Results are expressed as fold induction of relative light units (RLU) of treated cells over that of untreated cells. Values are means ± SEM with three separate experiments. $^{\#}P < .05$ compared with untreated cells, $^{*}P < .05$ compared with PSG-1 alone. (F) The role of NADPH oxidase in regulation of PSG-1-mediated Rac1 activity. Cells were pretreated with NADPH oxidase inhibitor, DPI (5 μM) for 30 min before stimulation of PSG-1 for additional 30 min. The Rac1-GTP complex was coprecipitated from whole cell lysate by PAK-1 PBD as well as active Rac1. Then the immunoprecipitant was quantitatively measured by Western blot using the anti-Rac antibody. (B), (D), and (G) Bar graph represents quantification by Quantity one software. All data of relative activity are expressed in comparison with untreated cells (control cells defined as 1). Similar experiments were repeated three times.

results indicated that ROS acted upstream of PSG-1 activation of PI3K/Akt, MAPKs, and NF-κB pathways.

5.2.1.3.5 PSG-1-Induced Rac-1 Activity and the Effect of DPI in Rac-1 Activation

Mounting evidence supports one of the important sources of ROS in macrophages to be NADPH oxidase activation during phagocytosis (Bokoch, 1995), so the most likely origin of ROS following PSG-1 stimulation in macrophages would be expected to come from NADPH oxidase. Given that Rac1 is known as an important component of functional NADPH oxidase (DeLeo et al., 1998), it was tested whether Rac1 activity was elevated following PSG-1 treatment. The results showed that Rac-1 activity was elevated 4.2-fold (Fig. 5.19F and G, Lane 2 vs Lane 1). In contrast, DPI, an NADPH oxidase effective inhibitor, significantly reduced PSG-1-triggered Rac1 activation by 60%, as compared with the PSG-1 alone (Fig. 5.19F and G, Lane 3 vs Lane 2). The significant increase of GTP-bound Rac1 after treatment of RAW264.7 cells with PSG-1, and the DPI reduction of PSG-1-triggered Rac1 activation, clearly demonstrate that the origin of PSG-1-induced ROS as a secondary messenger is mainly derived from NADPH oxidase.

In summary, the present study demonstrated that TLR4 is the major receptor involved in the interaction of PSG-1 and macrophages. Moreover, NADPH oxidase-derived ROS acts upstream of the PI3K/Akt/MAPKs/NF-κB signaling pathway in the regulation of PSG-1 stimulated TNF-α production. Therefore, it was established that a major pathway by which PSG-1 exerted its effects on macrophage activation was via a TLR4/ROS/PI3K/Akt/MAPKs/NF-κB signal pathway (Fig. 5.20).

5.2.2 Anti-Tumor

Cancer remains one of the most deadly diseases in the world, and the number of new cases increases each year (Boyle and Levin, 2008). For many decades, anti-cancer drugs have been widely used in chemotherapy for cancer patient treatment. Yet, despite the fact that the effectiveness of chemotherapy for many types of common cancer has improved in previous decades, the high-toxic effects of chemotherapeutic agents leading to a severe drop in quality of life are still huge challenges in clinical trials (Rein and Kurbacher, 2001).

Plants have made important contributions to anti cancer drug development. More than half of anti-tumor agents approved by the United States Food and Drug Administration (FDA) since 1960 were derived from natural resources in the world, such as terrestrial plants and marine organisms (Kim and Park, 2002). They provide a unique and yet significant undeveloped source for formidable novel pharmaceutical products. In recent years, a large number of polysaccharides with potent anti tumor activity and low toxicity

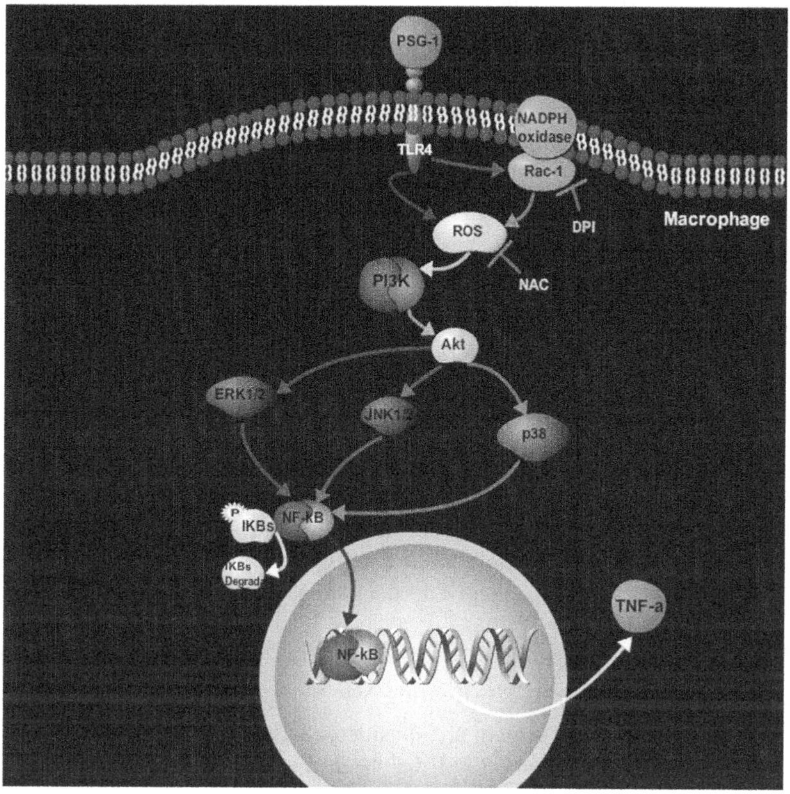

FIGURE 5.20 The proposed model for PSG-1-mediated ROS and downstream signal transduction pathways in the regulation of TNF-α production in macrophages.

have been reported, such as *L. edodes*, *Panax ginseng*, *Agaricus blazei*, and *Corious versicolor* (Chihara et al., 1969; Kobayashi et al., 2005; Nakazato et al., 1994; Shin et al., 2002).

Apoptosis, well-known as the regulated destruction of cells, is a controlled biological strategy to remove unwanted, injured, or virus-infected cells from a given tissue, and may account for much of the spontaneous cell loss in many tumors. Apoptosis is triggered by many upstream signaling pathways, and requires the coordinated activation and execution of multiple subprogrammes (Hengartner, 2000; Hsu et al., 2004a,b). One critical hallmark of cancer cells is their resistance to apoptosis induction (Hanahan and Weinberg, 2000). Therefore, most of the anti-neoplastic drugs kill cancer cells by inducing apoptotic death to block or suppress the growth of cancer cells.

Although PSG-1 has been acknowledged as one of the most active antineoplastic agents in the past decades, the exact molecular mechanisms that integrate its success have remained largely unclear. Therefore, research and development of PSG-1 as a novel agent with anti-tumor activity is of great importance. To further identify the bioactivity and molecular mechanism of PSG-1, the abilities of PSG-1 to inhibit tumor growth in CT26-bearing mice, and to activate immune responses were analyzed. In addition, the molecular mechanism of PSG-1-inhibited tumor growth was elucidated.

5.2.2.1 Anti-Tumor Activity of PSG-1 on CT26-Bearing Mice

5.2.2.1.1 Effect of PSG-1 on the Proliferation of CT26 Vells *In Vitro*

To investigate the effects of PSG-1 on the proliferation of CT26 cells, the cells were cultivated with different concentrations of PSG-1. The MTT assays showed that PSG-1 exhibited no growth inhibitory activity on CT26 cells at concentrations of 20–320 μg/mL (Fig. 5.21A).

5.2.2.1.2 PSG-1 Inhibited the Growth of Transplantable Tumors

After 14 day-treatment, PSG-1 (50, 100, and 200 mg/kg) and 5-Fu (20 mg/kg) had significant inhibitory effects on the growth of inoculated CT26 in mice. Fig. 5.21B shows the inhibition ratio of tumor growth for a different treatment in tumor-bearing mice. The data suggest that PSG-1 could inhibit tumor growth *in vivo*.

5.2.2.1.3 Apoptosis in CT26-Bearing Mice

Tumor development and growth are considered to be a result of the high proliferative capacity of tumor cells. It has been reported that apoptosis and its related signaling pathways contributed to the high rate of cell loss in malignant tumors and, moreover, could provide a defense against tumor progression (Lowe and Lin, 2000). Apoptosis is the process of cell death occurring in various multicellular organisms, and is well-known to induce characteristic changes of cell morphology, such as membrane blebbing, cell shrinkage, chromatin condensation, and chromosomal DNA fragmentation (Doonan and Cotter, 2008). It is now well-established that anti-cancer agents induce apoptosis, and that disruption of apoptotic programs can promote tumor initiation, progression, and treatment resistance (Schmitt and Lowe, 1999). Flow cytometry assays have shown marked changes in cell profiles after treatment with 50, 100, and 200 mg/kg PSG-1. The apoptotic rates ranged from 1.68% to 41.04%, and the 200 mg/kg PSG-1 group was the highest among them, indicating that administration of PSG-1 induced apoptosis (Fig. 5.22A–E).

5.2.2.1.4 Effect of PSG-1 on the Mitochondrial Transmembrane Potential ($\Delta\Psi_m$) in Tumors of CT26-Bearing Mice

Mitochondrial outer membrane permeabilization is a decisive event in apoptosis which is mediated by the Bcl-2 family of proteins (Cui et al., 2014; Pinton et al., 2011). At present, existing data demonstrate that elimination of

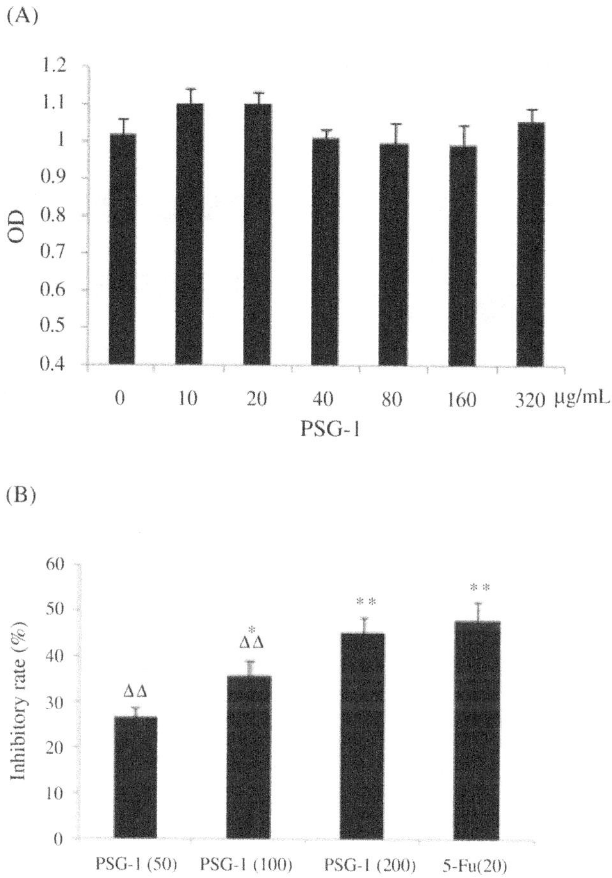

FIGURE 5.21 Anti-tumor effects of PSG-1 *in vitro* and *in vivo*. (A) Growth inhibitory effect of PSG-1 on CT26 cells. The effects of PSG-1 on the growth of CT26 cells were determined by the MTT method. The data presented are averages derived from at least triplicate experiments. (B) The tumor growth inhibitory rates of 5-Fu (20 mg/kg) and PSG-1 (50, 100, and 200 mg/kg) treated groups. Each data point is the average from 8 mice ($n = 8$). $^{*}P < .05$ and $^{**}P < .01$ versus control group; ΔΔ $P < .01$ versus 5-Fu group.

potentially malignant cells often relays on the mitochondrial apoptotic pathway (Ghosh et al., 2006). The loss of mitochondrial membrane potential, $\Delta\psi_m$, is one of the crucial parameters of mitochondrial dysfunction, and could lead to an irreversible event during the process of cell apoptosis (Zhou et al., 2012). The disruption of $\Delta\psi_m$ could result in the release of some apoptogenic proteins, activating a cascade of events that lead to the activation of caspases in the mitochondrial-dependent apoptotic pathway (Nakazato et al., 2005). To assess whether PSG-1 affects the function of mitochondria, the mitochondrial transmembrane potential was investigated

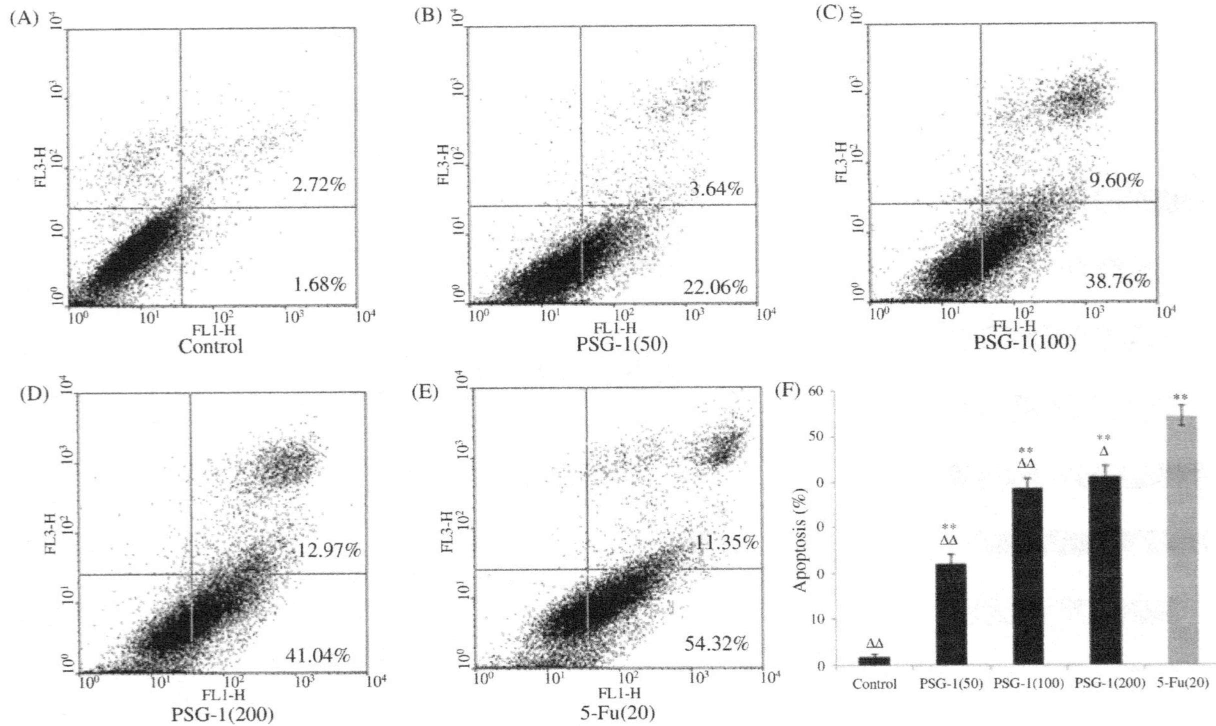

FIGURE 5.22 (A–E) AnnexinV-FITC/PI double-staining assay of CT26 cells from tumor-bearing mice. (F) Bar graph represents the percentage of CT26 cell apoptosis. Each value represents the mean of three experiments. $^{**}P < .01$ versus control group, $\Delta P < .05$, $\Delta\Delta P < .01$ versus 5-Fu group.

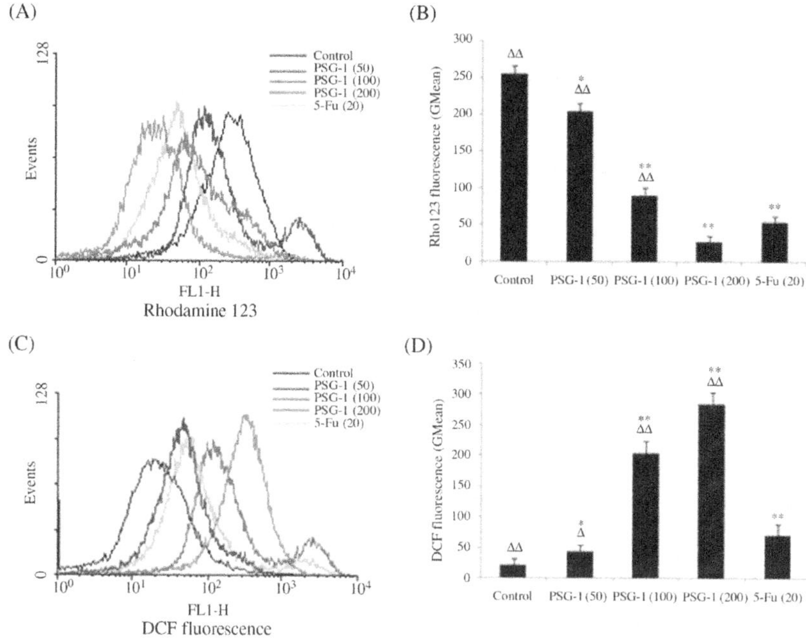

FIGURE 5.23 Effects of PSG-1 on the loss of $\Delta\Psi_m$ and generation of ROS. (A) The loss of $\Delta\Psi_m$ was evaluated by Rh123 staining. (B) Column chart of cell fluorescence for Rh123. (C) Intracellular ROS levels were measured by flow cytometry using an oxidation-sensitive fluorescent probe, DCFH-DA, which is oxidized to DCF in the presence of ROS. (D) Column chart of cell fluorescence for DCF. Values are mean ± SEM of eight mice. **P* < .05 and ***P* < .01 versus control group; Δ*P* < .05 and ΔΔ*P* < .01 versus 5-Fu group.

using the mitochondria probe, Rhodamine 123. With the increase of PSG-1 concentration, a remarkable attenuation of $\Delta\Psi_m$ was observed (Fig. 5.23A and B). PSG-1 administration led to collapse of the $\Delta\Psi_m$.

5.2.2.1.5 PSG-1 Increased Release of Cytochrome c and Apoptosis-Associated Protein Expression in the Tumor of CT26-Bearing Mice

The disruption of $\Delta\psi_m$ in mitochondria could cause not only an uncoupling of the respiratory chain, but also the efflux of some proteins as well as small molecules (e.g., cytochrome c and calcium). When cytochrome c is released into the cytosol, cytochrome c would bind to Apaf-1 to produce apoptosome, accelerating the activation of caspase-9 and then caspase-3 for caspase-dependent apoptosis (Kim, 2005). The dissipation of $\Delta\psi_m$ prompted us to investigate the effect of PSG-1 on the release of cytochrome c. Western-blotting results revealed that the level of cytosolic cytochrome c protein was unregulated dose-dependently (Fig. 5.24B). In contrast, PSG-1 caused a slight decrease in the protein level of cytochrome c in the mitochondrial

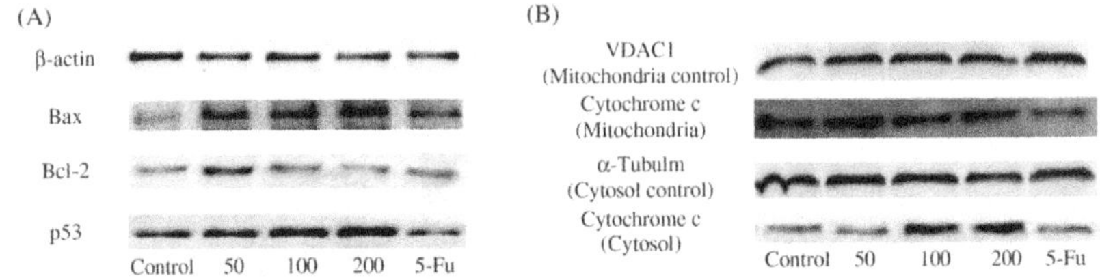

FIGURE 5.24 Western blot analysis of cytochrome c release and the expression of the Bcl-2 family in tumor. (A) Total protein extracts were prepared and then analyzed by Western blotting with antibodies to p53, Bax, Bcl-2, and β-actin. (B) The cytosolic and mitochondrial proteins were analyzed by Western blotting with anti-cytochrome c, anti-VDAC1, and anti-α-tubulin antibodies. Western blots were representative of three independent experiments.

fraction in 100 and 200 mg/kg body weight treated mice. These results indicated that PSG-1 led to the release of cytochrome c from the mitochondria to the cytosol.

The mitochondria-mediated apoptotic pathway was under the control of multiple layers of regulation, one of the most important players being members of the Bcl-2 family (Gogvadze et al., 2006). Two most important members of Bcl-2 family, the anti-apoptotic protein Bcl-2 and the pro-apoptotic protein Bax, are key regulators of this process. Their main function is to control mithochondrial permeability. Bax can promote cell death. Conversely, Bcl-2 proteins cause resistance to chemotherapeutic agents and radiation therapy, while reducing Bcl-2 expression could promote apoptotic responses to anti-tumor agents (Portt et al., 2011). Triterpene extract was found to be able to modulate the expression of Bcl-2 and Bax in many types of tumor cell lines (Choudhuri et al., 2002). In addition, tumor suppressor p53 has been identified as an effector of apoptosis signals. Reactivating p53 provides a general strategy in conquering cancer specifically (Chen et al., 2010a,b). It is a regulator of cell cycle arrest progression, and the mediator of apoptosis in various cases. In order to clarify the mechanism of PSG-1-induced apoptosis, this study examined the expression of p53, Bax, and Bcl-2 proteins. Western blotting testing revealed that protein levels of both Bax and p53 increased dramatically in the 100 and 200 mg/kg PSG-1-treated tumor-bearing groups, compared with control group, whereas the protein expression level of Bcl-2 was decreased in a dose-dependent manner (Fig. 5.24A).

5.2.2.1.6 Effect of PSG-1 on Caspase Activity in the Tumor of CT26-Bearing Mice

Caspases are a ubiquitous family of cysteine proteases that modulate the apoptotic response. Particularly, the activation of caspase-3 is one of the most commonly involved in the apoptosis of various cell types (Mancini et al., 1998). Caspase-3, which might be activated by caspase-9, typically functions downstream of other caspases, and directly activates enzymes that are

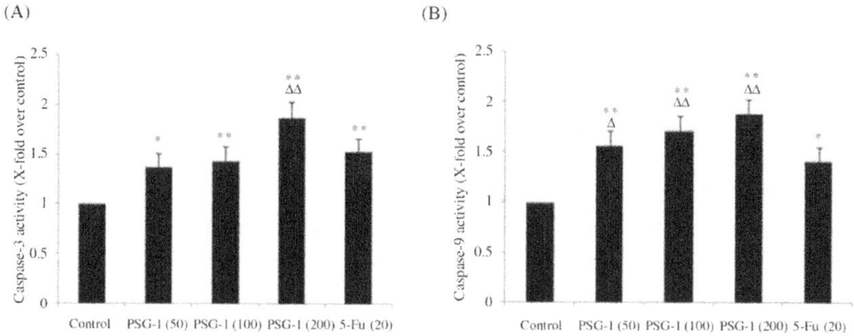

FIGURE 5.25 Effects of PSG-1 on the activation of caspase in tumors of CT26-bearing mice. Quantitative analysis of caspase-3 (A) and caspase-9 (B) activities in the tumor cells of CT26-bearing mice. Values are mean ± SEM of eight mice. **P* < .05, ***P* < .01 versus control group; ΔP < 0.05 and ΔΔ*P* < 0.01 versus 5-Fu group.

responsible for DNA fragmentation in the mitochondrial apoptosis pathway. Lots of microtubule inhibitors such as vincristine, paclitaxel, colchicines, and vinblastine induce growth arrest, subsequent inactivation of Bcl-2, and finally apoptotic cell death through the caspase-9-dependent pathway (Wang et al., 1999).

According to the results of the apoptosis assay, it is hypothesized that PSG-1 might promote the activation of the caspase-dependent cell apoptosis pathway. Both caspase-3 and caspase-9 activities were elevated by treatment of PSG-1 in CT26-bearing mice (Fig. 5.25A and B). Taken together, all these lines of evidence suggest that PSG-1 could activate caspase-3 and caspase-9, and induce caspase-dependent apoptosis in tumors of CT26-bearing mice.

5.2.2.1.7 Effect of PSG-1 on Reactive Oxygen Species Production in Tumors of CT26-Bearing Mice

In cancer cells, ROS have been broadly reported to play a crucial part in apoptosis induction in both pathological and physiological processes. It has been detected as an early signal that initiates and mediates cellular apoptosis (Cort et al., 2012). Interestingly enough, mitochondria are both the major source of, and target of, ROS. It could lead to the breakdown of membrane phospholipids, inactivation of enzymes, and destruction of mitochondrial function, which contribute to the intermembrane proteins, such as cytochrome c, to be released out of the mitochondria and eventually initiate caspase-3 activation. The level of ROS was detected to test whether oxidative stress went into effect in PSG-1-induced apoptosis in tumors of CT26-bearing mice. As shown in Fig. 5.23C and D, PSG-1 led to a remarkable increase in intracellular ROS level compared with the control group. Therefore, the elevation of ROS production may be a major cause of PSG-1-induced apoptosis.

5.2.2.1.8 Distinct Changes of Intracellular cAMP and cGMP Concentrations in Tumors

The rate of cAMP production and degradation is sensitive to a wide range of extracellular and intracellular signals, such that cAMP can directly regulate a variety of cell functions, from metabolism to ion channel activation, cell growth and differentiation, to gene expression and apoptosis (Chin et al., 2002). Previous studies have indicated that the cAMP signaling pathway and cGMP mediator are known to be of paramount importance in PSG-1-induced apoptosis of S180 cells. To investigate the correlation between intracellular cAMP and cGMP levels and the percentage of apoptosis, the intracellular cAMP and cGMP contents in CT26 cells from tumor-bearing mice treated with different doses of PSG-1 were measured. A significant improvement in the intracellular cAMP concentration was observed. In contrast, the intracellular cGMP level decreased in the CT26 cells.

5.2.2.1.9 Effect of PSG-1 Treatment on AC and DAG Content

Since the intracellular cAMP level is mainly regulated by the AC isoform, the activity of the AC system might be related to the alteration of cAMP levels triggered by PSG-1. DAG is a physiological activator of PKC. The production of DAG in the membrane facilitates translocation of PKC from the cytosol to the plasma membrane. The present study thus determined the levels of intracellular AC and DAG. As shown in Fig. 5.26C and D, a great change in the activity of AC was observed. The AC activity was enhanced along with an increase in the dose. However, the level of DAG was dramatically down-regulated in the PSG-1 groups.

5.2.2.1.10 Effect of PSG-1 on PKA and PKC in Tumors During PSG-1-Induced Apoptosis

The cAMP dependent kinase, PKA, is a central component of the signal transduction pathway. The activation of PKA is one of the primary elements in the cAMP signal transduction downstream cascade. The cAMP/PKA pathway in cancer starts with AC activation, which leads to the formation of cAMP. The PKA is activated by the newly minted cAMP, and causes reversible phosphorylation of protein substrates that regulate a number of cellular processes (Leadsham and Gourlay, 2010). Excessive PKA activity in yeast could trigger mitochondria-dependent apoptosis *via* a distinct signaling mechanism (Leadsham et al., 2009). Chen et al. (1998) reported that up-regulation of the cAMP/PKA pathway decreased proliferation, induced differentiation, and lead to apoptosis in malignant gliomas. This is consistent with a report of Hanson; the growth of neuroblastoma cells is inhibited by an increase in cAMP (Hanson et al., 2005). As a crucial step towards understanding the mechanism of PSG-1-induced apoptosis, intracellular PKA activity in the CT26 cells from tumor-bearing mice was measured. It was found that in

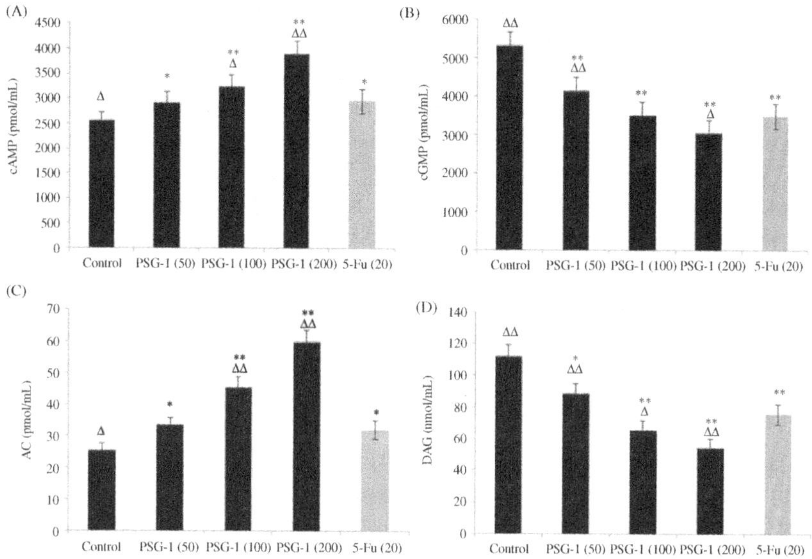

FIGURE 5.26 Effect of PSG-1 on intracellular cAMP, cGMP, AC, and DAG production in the tumor. The production of cAMP(A),cGMP(B),AC(C),DAG(D) was measured according to ELISA Kit instructions. Bar graphs show mean ± SEM of three independent experiments performed. *$P < .05$, **$P < .01$ versus control group; $\Delta P < .05$, $\Delta\Delta P < .01$ versus 5-Fu group.

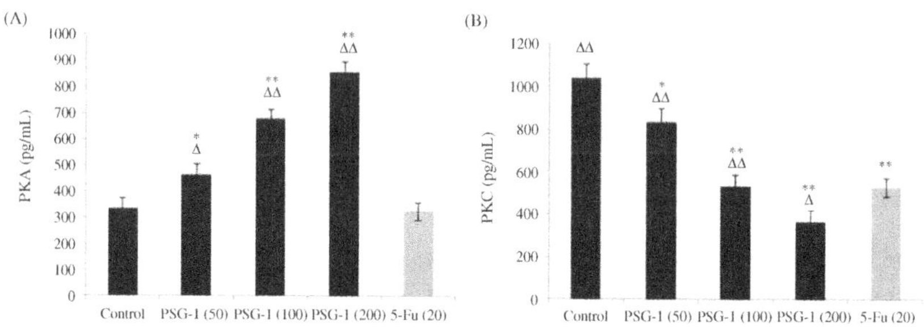

FIGURE 5.27 Effect of PSG-1 on the levels of PKA and PKC in tumor cells from tumor-bearing mice. (A) PKA and (B) PKC were detected by kinase activity kits by *$P < .05$,**$P < .01$ versus control group; $\Delta P < .05$, $\Delta\Delta P < .01$ versus 5-Fu group.

parallel with a significant increase in cAMP level, the activity of PKA was increased by PSG-1 treatment (Fig. 5.27A), implying the importance of the cAMP–PKA pathway in PSG-1-induced apoptosis of CT26 cells.

A family of serine/threonine kinases, PKC, is involved in the transduction of signals for cell proliferation and differentiation (Caponigro et al., 1997). The important role of PKC in processes relevant to neoplastic transformation, carcinogenesis, and tumor cell invasion renders it a potentially suitable target for anti-cancer therapy (Mackay and Twelves, 2003). Early studies suggested a role for PKC isozymes in tumor promotion. A direct correlation between the ability of individual phorbol esters to promote tumors and the activation of PKC has been demonstrated (Nishizuka, 1984). Increased levels of PKC have been associated with malignant transformation in a number of cell lines including breast, lung, and gastric carcinomas (O'Brian et al., 1989; Schwartz et al., 1993; Takenaga and Takahashi, 1986). In colon cancer, PKC β II appears to be overexpressed early in the colon carcinogenic process (Gökmen-Polar et al., 2001). A decrease in the content of PKC α by antisense or phorbol ester led to apoptosis in many different cell types, such as glioma cells, endothelial cells, and promyelocytic U937 cells (Dooley et al., 1998; Haimovitz-Friedman et al., 1994; Whelan and Parker, 1998). Dean et al. (1996) have reported that the antisense PKC α oligonucleotides inhibited tumor growth in nude mice bearing implanted human glioblastomas and human bladder, lung, and colon carcinomas. In addition, PKC δ has been implicated in apoptotic response and cytochrome *c* release (Fujii et al., 2000; Majumder et al., 2000). Hence, inhibition of PKC may have an impact on the invasive and metastatic potential of malignant cells, resulting in tumor cell apoptosis. In the present study, it was found that intracellular PKC activity of the PSG-1-treated group decreased significantly (Fig. 5.27B).

To further confirm this potential role of PKA and PKC, a Western blot experiment was conducted to examine the expression of PKA and PKC proteins. As shown in Fig. 5.28, PSG-1 caused a marked increase in the expression of PKA protein. In contrast, PKC level decreased in the PSG-1 treated groups compared to the control group.

Taken together, the present study has, for the first time, provided unequivocal evidence that polysaccharide from *G. atrum* (PSG-1) possesses potent anti-tumor activity. In this study, the most important observations were that PSG-1 markedly suppressed the tumor growth, and induced tumor apoptosis through mitochondrial pathways, and the cAMP/PKA signaling pathway, as well as the PKC pathway in CT26-bearing mice. In conclusion, PSG-1 may have potential as a novel therapeutic agent to replace or augment more cytotoxic agents currently used to treat carcinoma patients.

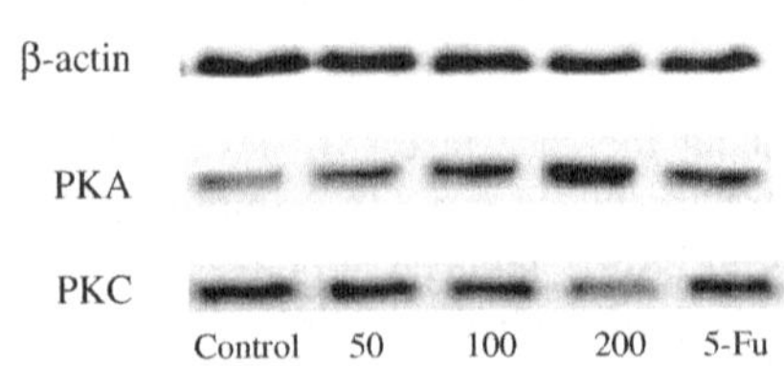

FIGURE 5.28 Expressions of PKA and PKC in tumor-bearing mice by Western blot. Total cellular proteins were separated by SDS-PAGE, and analyzed by Western blot with specific antibodies.

5.2.2.2 *Immunomodulatory Effect of* G. atrum *Polysaccharide on CT26 Tumor-Bearing Mice*

Cancer is one of the most serious diseases in developed countries, and the second largest deadly disease in developing countries (Jemal et al., 2011). There is a close relationship between the occurrence, growth, and decline of tumor and immune states. Immunomodulation may have a great contribution to make to the prevention and cure of neoplastic diseases (Mitchell, 2003). Anti-cancer treatments based on exploiting the host's own anti-tumor defense mechanisms can be beneficial. The enhancement or potentiation of the host immune response has been recognized as a possible way of inhibiting tumor growth without harming the host (Yuan et al., 2006). Therefore, it is very important to develop novel anti-tumor drugs with improved immunity potential.

Previous studies have shown that natural product-derived polysaccharides have an anti-tumor effect in tumor-bearing mice. However, the underlying mechanisms of this inhibition activity remain unclear (Yuan et al., 2006). The enhancement or potentiation of host defense mechanisms has been recognized as a possible means of inhibiting tumor growth without harming the host.

The present study was aimed at investigating its anti-tumor and immunomodulatory activities in CT26-bearing mice and S180 tumor-bearing mice, and providing insights into the underlying mechanisms.

5.2.2.2.1 Immune Function of PSG-1 in CT26-Bearing Mice

The host immune function is another key factor for cancer treatment and recovery. Clinical verification is being investigated into the maximal potential of anti-cancer treatments based on exploiting the host's own anti-tumor defense mechanisms (Ehrke, 2003). Immunostimulation is a crucial strategy to improve the body's defense mechanism. The anti-tumor immune response can be divided into specific and nonspecific immunity. The thymus and spleen are two important immune organs. The thymus and spleen indexes are an important index for non-specific immunity. As shown in Table 5.6, compared with the control group, the thymus and spleen indexes of tumor-bearing mice treated with 5-Fu at a dose of 20 mg/kg BW decreased significantly. Treatment with PSG-1 at different doses produced a marked and dose-dependent increase in both spleen and thymus indexes ($P < .05$), especially in the groups with high and medium doses. This finding suggested that PSG-1 enhanced the immunization of tumor-bearing mice.

5.2.2.2.2 Effect of PSG-1 on Spleen Lymphocyte Proliferation in CT26-Bearing Mice

To elucidate the mechanism of the immunoregulatory activity of PSG-1, the effect on spleen lymphocyte proliferation was examined. Compared with the

TABLE 5.6 Effects of PSG-1 on Spleen Index and Thymus Index in CT26-Bearing Mice

Group	Dose (mg/kg)	Tumor Weight (g)	Thymus Index (mg/g)	Spleen Index (mg/g)
CT26 control	–	1.32 ± 0.34^{a}	1.67 ± 0.12^{b}	5.54 ± 1.01^{a}
5-Fu	20	0.34 ± 0.10^{c}	1.01 ± 0.10^{d}	3.28 ± 0.85^{c}
PSG-1	50	$0.86 \pm 0.37^{a,d}$	1.84 ± 0.34^{a}	$6.57 \pm 1.14^{a,d}$
	100	$0.58 \pm 0.33^{b,c}$	$2.32 \pm 0.25^{a,c}$	$7.24 \pm 1.06^{a,c}$
	200	0.46 ± 0.31^{c}	$3.16 \pm 0.36^{a,c}$	$8.53 \pm 1.23^{a,c}$

Values are mean ± SEM of eight mice.
[a]*Means significantly different from 5-Fu group at P < .01.*
[b]*Means significantly different from 5-Fu group at P < .05.*
[c]*Means significantly different from CT26 control group at P < .01.*
[d]*Means significantly different from CT26 control group at P < .05.*

TABLE 5.7 Effect of PSG-1 on the Immune Function of the CT26-Bearing Mice

Group	Dose (mg/kg)	Phagocytosis Ratio (%)	Proliferation Ratio (%)	
			Con A (1 mg/L)	LPS (5 mg/L)
Normal	–	100	100	100
CT26 control	–	74.03 ± 4.25^{a}	78.04 ± 4.34^{b}	80.26 ± 6.92^{b}
5-Fu	20	45.33 ± 5.21^{c}	61.67 ± 3.53^{d}	67.92 ± 4.64^{d}
PSG-1	50	$133.25 \pm 8.78^{a,c}$	$93.68 \pm 4.72^{a,d}$	$96.21 \pm 5.73^{a,d}$
	100	$151.44 \pm 10.65^{a,c}$	$114.22 \pm 8.81^{a,c}$	$136.56 \pm 6.78^{a,c}$
	200	$182.42 \pm 17.93^{a,c}$	$132.31 \pm 6.38^{a,c}$	$149.33 \pm 5.87^{a,c}$

Values are mean ± SEM of eight mice.
[a]*Means significantly different from 5-Fu group at P < .01.*
[b]*Means significantly different from 5-Fu group at P < .05.*
[c]*Means significantly different from CT26 control group at P < .01.*
[d]*Means significantly different from CT26 control group at P < .05.*

control group, proliferation of spleen lymphocyte induced by Con A and LPS was increased significantly in PSG-1-treated CT26-bearing mice. In contrast, 5-Fu administration decreased the proliferation of spleen lymphocytes markedly in CT26-bearing mice (Table 5.7).

5.2.2.2.3 Effect of PSG-1 on Peritoneal Macrophages Phagocytosis in CT26-Bearing Mice

To verify the effect of PSG-1 on the activation of immune cells, the phagocytosis of macrophages from each group was estimated. As shown in Table 5.7, PSG-1 (50, 100, and 200 mg/kg) augmented the percentage of phagocytosis of NR by peritoneal macrophages in CT26-bearing mice, compared with the control group. It is speculated that PSG-1 restored impaired peritoneal macrophage phagocytosis.

5.2.2.2.4 Effects of PSG-1 on NO, TNF-α, and IL-1β Production by Macrophages in CT26-Bearing Mice

Levels of NO, TNF-α, and IL-1β in animals treated with 50–200 mg/kg BW PSG-1 were significantly increased compared with the positive control group (Table 5.8), suggesting that PSG-1 promoted NO, TNF-α, and IL-1β production, and improved immunity. The anti-tumor effect of PSG-1 may be related to their potentiation of the production of TNF-α and IL-1β in tumor-bearing mice.

5.2.2.2.5 Anti-TLR4 Antibody Partly Abrogates TNF-α Production in Peritoneal Macrophages From PSG-1-Treated Tumor-Bearing Mice

Activation of TLR4 by PSG-1 is crucial in the production of inflammatory cytokines. To investigate whether TLR4 takes part in the activating effects of PSG-1 on macrophages, the effect of anti-TLR4 antibody on TNF-α secretion was studied. Macrophages were pre-incubated with anti-TLR4

TABLE 5.8 Effects of PSG-1 on NO, TNF-α, and IL-1β Production by Macrophages in CT26-Bearing Mice

Group	Concentration (mg/kg)	NO (μmol/L)	TNF-α (pg/mL)	IL-1β (pg/mL)
Model control	0	3.13 ± 0.52	30.27 ± 1.71	51.20 ± 3.51
5-Fu	20	4.24 ± 0.18[a]	32.80 ± 0.97	56.81 ± 3.01[b]
PSG-1	50	5.12 ± 0.46[a,c]	40.72 ± 1.56[a,d]	74.21 ± 4.27[a,d]
	100	7.86 ± 0.34[a,d]	42.31 ± 2.48[a,d]	83.63 ± 6.37[a,d]
	200	9.97 ± 0.50[a,d]	58.26 ± 2.52[a,d]	87.55 ± 5.42[a,d]

Values are mean ± SEM, $n = 8$.
[a]*Means significantly different from model control group at* P < *.01.*
[b]*Means significantly different from model control group at* P < *.05.*
[c]*Means significantly different from 5-Fu control group at* P < *.05.*
[d]*Means significantly different from 5-Fu control group at* P < *.01.*

TABLE 5.9 Anti-TLR4 Antibody Partly Abrogates TNF-α Production by Macrophages in PSG-1 Treated Tumor-Bearing Mice

Group	Concentration (μg/mL + μg/mL)	TNF-α (pg/mL)
Untreated		93.93 ± 4.43
PSG-1	200	167.00 ± 8.45[a]
PSG-1 + AT4A	200 + 2.5	148.50 ± 6.28[b]
PSG-1 + AT4A	200 + 5.0	134.64 ± 6.01[c]
PSG-1 + AT4A	200 + 10.0	128.93 ± 5.37[c]
PSG-1 + AT4A	200 + 20.0	116.78 ± 5.63[c]
AT4A	20	85.36 ± 3.52[d]
PSG-1 + NSA	200 + 20	153.57 ± 7.57[e]

Values are mean ± SEM, $n = 8$.
AT4A, anti-TLR4; NSA, non-specific antibody IgG.
[a]*Means significantly different from untreated group at* $P < .01$.
[b]*Means significantly different from group PSG-1 at* $P < .05$.
[c]*Means significantly different from group PSG-1 at* $P < .01$.
[d]$P = .355$ *compared with untreated group.*
[e]$P = .468$ *compared with PSG-1 group.*

blocking antibody for 30 min before and during PSG-1 (200 μg/mL) treatments, and the levels of TNF-α were measured in the cell supernatant. The results (Table 5.9) showed that the blocking TLR4 receptor abolished the production of TNF-α induced by PSG-1, suggesting that TLR4 is required for PSG-1-induced TNF-α production.

The results suggested that anti-TLR4 antibody showed marked, although not complete, suppression of this promotion of PSG-1 in a dose-dependent manner. But there is no significant difference between the non-specific antibody IgG group and PSG-1 group ($P = .355$). In addition, only added anti-TLR4 antibody had no significant effect on TNF-α secretion (ALR4 group compared with untreated group, $P = .468$). It suggested that anti-TLR4 antibody has a specific effect on PSG-1, but it could not completely suppress the promotion. For example, when 20 μg/mL of anti-TLR4 antibody was added, there still remained approximately 24.3% promotion.

5.2.2.2.6 Specific Blockers of p38MAPK and NF-κB Abrogate TNF-α Production in Peritoneal Macrophages in PSG-1-Treated Tumor-Bearing Mice

The results (Fig. 5.29) showed that 20 μM of SB203580 could completely abolish the stimulatory effect of PSG-1 on the production of TNF-α from tumor-bearing mouse peritoneal macrophages. In addition, PDTC could

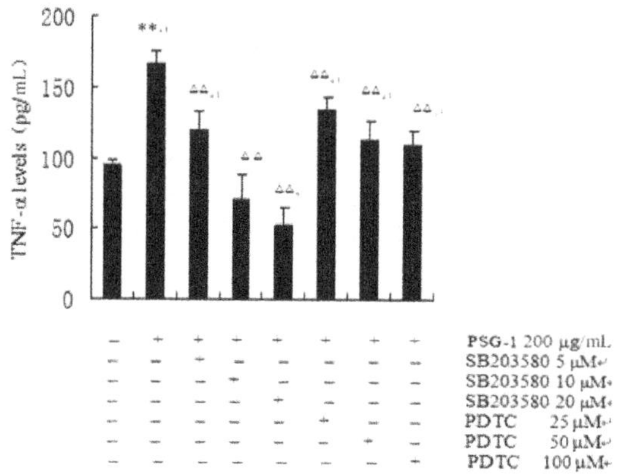

FIGURE 5.29 Specific blockers of p38MAPK and NF-κB abrogate TNF-α production. Peritoneal macrophages were pretreated with 5, 10, and 20 μM of SB203580 (a p38 MAPK inhibitor) or 25, 50, and 100 μM of PDTC for 30 min, and then treated with PSG-1 for 48 h. The levels of TNF-α in the culture supernatants were analyzed by ELISA. Values are means ± SEM, $n = 8$. ** Means significantly different from untreated group at $P < .01$. ΔΔ Means significantly different from group PSG-1 at $P < .01$.

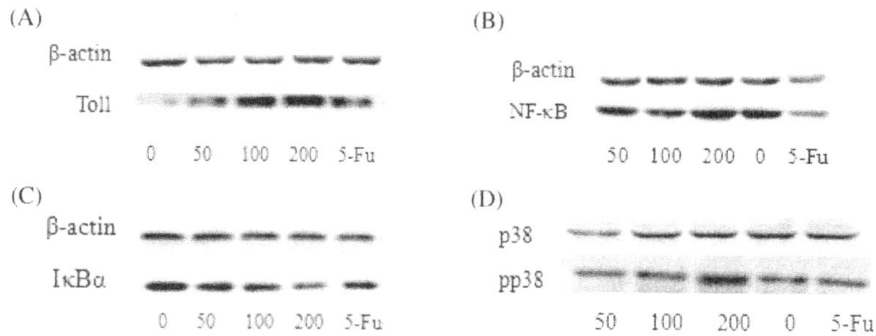

FIGURE 5.30 Western blot analyses of TLR4-mediated and p38 MAPK signaling pathways. Peritoneal macrophages from each tumor-bearing mouse were collected. (A) Total cell extracts immunoblots were prepared and Western blotted for TLR4 using specific TLR4 antibodies. (B) Nuclear extracts were prepared for the Western blotting of p65 of NF-κB using a specific anti-p65 monoclonal antibody. (C) Cytosolic extracts of the cells were then prepared and Western blotted for IκBα using specific IκBα antibodies. (D) Whole cell lysates from peritoneal macrophages were analyzed using antibodies raised against the activated MAPK. This experiment was repeated three times, producing similar results.

significantly suppress the stimulatory effect of PSG-1 on the production of TNF-α.

5.2.2.2.7 PSG-1 Stimulates TLR4-Mediated NF-κB Pathways

As shown in Fig. 5.30A, PSG-1 caused a relative increase in TLR4 expression. In contrast, compared with the model control group, 5-Fu, a positive control, has an adverse effect on TLR4 expression. Next, the levels of nuclear proteins in macrophages from tumor-bearing mice were analyzed, the results showed that PSG-1 increased p65 levels in nuclear fractions as compared with the model control group (Fig. 5.30B). The cytoplasmic protein levels of IκBα were also investigated, because IκBα degradation induces NF-κB activation (Baeuerle and Baltimore, 1988). As expected, the degradation of IκBα was significantly increased in macrophages from PSG-1-treated tumor-bearing mice (Fig. 5.30C).

5.2.2.2.8 Involvement of the p38 MAPK Pathway in PSG-1-Induced Anti-Tumor Responses

To investigate whether the activation of macrophages by PSG-1 is mediated through the MAPK pathway, the effect of PSG-1 on the phosphorylation of p38 MAPK in macrophages was examined by Western blotting with phospho-specific antibodies. As shown in Fig. 5.30D, PSG-1 induced the activation of p38 MAPK in a dose-dependent manner. Interestingly, the level of non-phosphorylated p38 MAPK was unaffected by PSG-1. These results suggest that PSG-1 treatment may activate the phosphorylation of p38 in macrophages.

In conclusion, these results suggest that PSG-1 could enhance immunity, and the anti-tumor effects of PSG-1 may be related to their potentiation of the production of TNF-α in tumor-bearing mice. The signaling mechanism by which PSG-1 promotes TNF-α production may be as follows: PSG-1 acts on the TLR4 receptors on macrophages, signals through the p38 MAPK pathway, and then activates NF-κB. The activated NF-κB in turn initiates the release of TNF-α. In addition, PSG-1 has low toxicity and side-effects compared to normal cells, and may have potential as a novel therapeutic agent. However, further clinical studies are needed to fully support our strategy for the treatment of patients with tumor.

5.2.3 Anti-Diabetes

Diabetes mellitus is the most common metabolic disorder caused by absolute or relative insulin deficiency, it is characterized by hyperglycemia, as well as impaired metabolism of carbohydrate, fat, and protein (Pooja and Samanta, 2011). Type 1 diabetes and type 2 diabetes are the two main types of diabetes, while type 2 diabetes accounts for approximately 90% of all diagnosed cases of diabetes (Pham et al., 2007; Zhang et al., 2012c,d). Over the past half-century, changes in diet and other aspects of lifestyle have resulted in a dramatic rise in the prevalence and incidence of type 2 diabetes in virtually every country in the world. Meanwhile, reduction in physical activity, increase in energy intake, and the aging of the population play pivotal roles in bringing about this rapid change (Feinglos and Bethel, 2008).

To our knowledge, there has been little investigation into the hypoglycemic and hypolipidemic effects of polysaccharides from *G. atrum* in type 2 diabetes. Therefore, the aim of this study was to evaluate the hypoglycemic and hypolipidemic effects of PSG-1 in an established animal model of type 2 diabetes. Our results demonstrated that PSG-1 may have potential in the treatment of hyperglycemia, hyperlipidemia, hyperinsulinemia, and insulin resistance in type 2 diabetics.

5.2.3.1 PSG-1 Attenuates Hyperglycemia and Hyperlipidemia

5.2.3.1.1 Effect of PSG-1 on FBG Levels of Type 2 Diabetic Rats

During the experimental period, the effect of PSG-1 on glucose control was assessed by measurement of FBG levels weekly. Fig. 5.31A showed that the FBG levels increased significantly in HFD/STZ-induced diabetic groups in comparison to the non-diabetic control group ($P<.01$), but were significantly inhibited by both doses of PSG-1 and 1,1-dimethylbiguanide hydrochloride ($P<.05$) compared to the untreated diabetic group. In addition, the anti-diabetic effect of PSG-1 was dose-dependent, the higher dose tended to show a consistently stronger hypoglycemic effect than the lower dose, but there were no significant differences between these two doses.

5.2.3.1.2 Effect of PSG-1 on OGTT Blood Glucose Levels in Type 2 Diabetic Rats

Blood concentration–time curves of glucose in rats before and after oral administration of glucose are shown in Fig. 5.31B. Diabetic rats exhibited severely impaired glucose tolerance compared to non-diabetic controls. Blood glucose levels in the diabetic group were higher than the non-diabetic rats at 30, 60, 120, and 180 min. The FBG levels declined gradually as the dose of PSG-1 increased. Blood glucose levels of 400 mg/kg BW group 30 min after glucose administration were significantly ($P<.01$) lower than that of the untreated diabetic group, but were higher in the positive control group.

Type 2 diabetes is a group of complex metabolic diseases, characterized by hyperglycemia due to peripheral insulin resistance including skeletal muscle, liver, and adipose tissue. An animal model of type 2 diabetes has been developed by feeding a high fat diet to rats for a relatively long period, followed by an injection of low dose of STZ (Reed et al., 2000; Srinivasan et al., 2005). The characteristic dyslipidemia of HFD/STZ-induced type 2 diabetic rats consists of elevated levels of TC, TG, FFA, and LDL-C, along with a decrease in HDL-C (Lu et al., 2010). Using a rat model, it can be shown that the characteristics of type 2 diabetic rats, including hyperglycemia, hyperlipidemia, hyperinsulinemia, and insulin resistance compared with non-diabetic controls, were similar to those in patients with type 2 diabetes (Petersen et al., 2005). The results also showed that PSG-1 evoked a marked

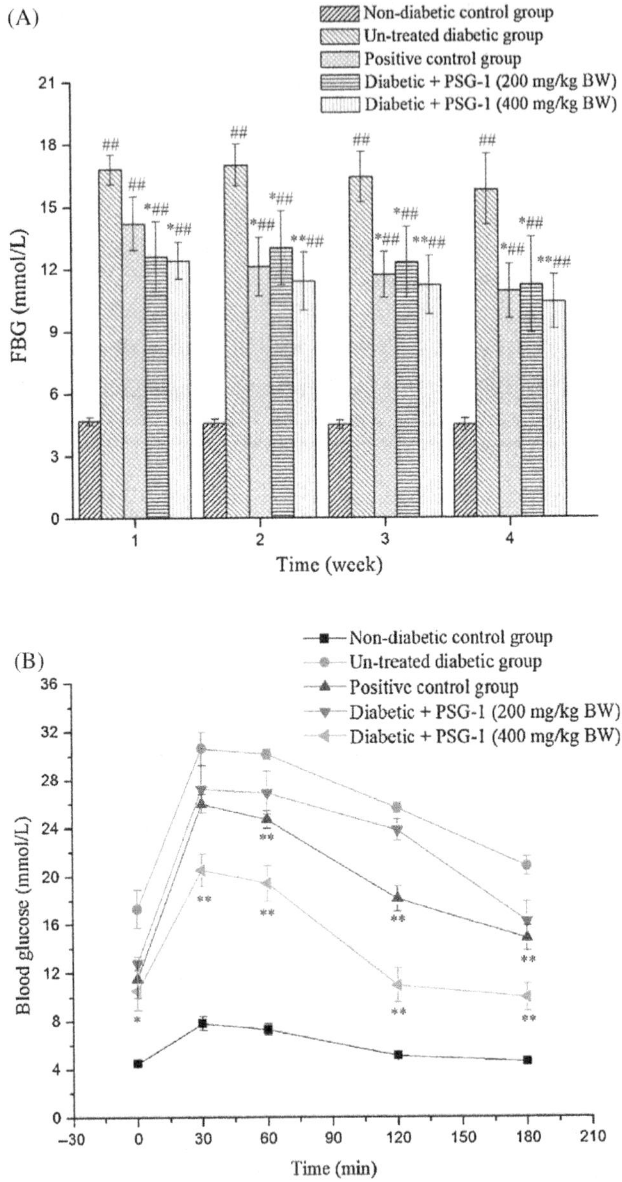

FIGURE 5.31 (A) Effect of PSG-1 on FBG levels of type 2 diabetic rats. Values are expressed as means ± SEM ($n = 10$). The FBG levels were measured once a week. The FBG levels increased significantly in HFD/STZ-induced diabetic groups in comparison to the non-diabetic control group ($P < .01$), but was significantly inhibited by both doses of PSG-1 ($P < .05$) in a dose-dependent manner. $^{\#}P < .05$, $^{\#\#}P < .01$ compared with the non-diabetic control group. $^{*}P < .05$, $^{**}P < .01$ compared with the untreated diabetic group. (B) Effect of PSG-1 on OGTT blood glucose levels in type 2 diabetic rats. Values are expressed as means ± SEM ($n = 10$). The diabetic group showed an increase in blood glucose level at 30 min, and remained high over the next 60 min after oral glucose administration compared to the non-diabetic control group. PSG-1 supplementation evoked a decrease in the blood glucose level at 60 and 120 min in the diabetic group. $^{**}P < .01$ compared with the untreated diabetic group.

improvement in hyperglycemia and hyperlipidemia, as well as glucose tolerance and insulin resistance, after four weeks of administration.

5.2.3.1.3 Effect of PSG-1 on Insulin Levels and Insulin Sensitivity in Type 2 Diabetic Rats

Until recently, insulin was the sole pancreatic β-cell hormone known to lower blood glucose levels (Brezar et al., 2011). Healthy individuals maintain a stable blood glucose levels by basal insulin secretion (Raghavan and Garber, 2008), but hyperglycemia in type 2 diabetes is a main symptom caused by a lack of insulin or insulin resistance, or both (Ke et al., 2009). Fasting insulin levels of diabetic groups were significantly higher than a non-diabetic control group. Decreased insulin levels were observed after oral administration of PSG-1 for 4 weeks. A maximum ($P < .05$) amelioration was shown in diabetic rats at a dose of 400 mg/kg BW. HOMA-IR and QUICKI were used to estimate insulin resistance/sensitivity (Wallace et al., 2004). The experimentally-induced diabetes significantly increased the value of HOMA-IR by nearly six times the control level in diabetic rats. However, the HOMA-IR values of diabetic rats treated with PSG-1 were significantly decreased, compared to the untreated diabetic group. The QUICKI values were significantly lower in diabetic rats than in non-diabetic control rats. Interestingly, PSG-1 increased QUICKI values in diabetic rats (Table 5.10). These results suggest the beneficial effect of PSG-1 supplementation on insulin resistance of the diabetic rats.

TABLE 5.10 Effect of PSG-1 on Insulin Levels and Insulin Sensitivity in Type 2 Diabetic Rats[a,b,c]

Groups	Fasting Insulin (μIU/mL)	HOMA-IR	QUICKI
Non-diabetic control	9.76 ± 1.34	1.94 ± 0.28	0.349 ± 0.019
Untreated diabetic	18.74 ± 1.05$^{\#\#}$	13.17 ± 1.60$^{\#\#}$	0.269 ± 0.009$^{\#\#}$
Positive control	15.68 ± 0.66$^{*,\#\#}$	7.53 ± 1.32$^{**,\#\#}$	0.288 ± 0.006$^{*,\#}$
Diabetic + PSG-1 (200 mg/kg BW)	15.81 ± 1.31$^{\#\#}$	8.81 ± 1.97$^{*,\#\#}$	0.285 ± 0.017$^{\#}$
Diabetic + PSG-1 (400 mg/kg BW)	15.60 ± 1.43$^{*,\#\#}$	6.24 ± 1.32**	0.292 ± 0.018**

[a]*Values are expressed as means ± SEM (n = 10).*
[b]**P < .05, **P < .01 compared with un-treated diabetic group.*
[c]*#P < .05, ##P < .01 compared with non-diabetic control group.*

5.2.3.1.4 Effect of PSG-1 on Serum Lipids in Type 2 Diabetic Rats

Since dyslipidemia is a potent cardiovascular risk factor for the evolution of cardiovascular disease in diabetic patients (Mooradian, 2009), the ideal treatment for patients with diabetes, in addition to glycemic control, should produce a more favorable effect on lipid profile. In the present study, a marked increase in TC, TG, and decreased HDL-C were observed in diabetic rats (Table 5.11). These results are in agreement with other studies in which a high fat diet combined with an injection of a low dose of STZ resulted in an elevation of plasma or serum cholesterol. The results also have shown that successive administration of PSG-1 produced an obvious reduction in serum TC, TG, and an increase in HDL-C. Results showed that PSG-1 attenuated the disorder of lipid metabolism.

Up-regulation of LDL-C may cause excess cholesterol deposition in the arteries and aorta, and act as a direct risk factor for coronary heart disease (Rosenson et al., 2002). The ratios of TC/HDL-C and LDL-C/HDL-C were calculated as indices of atherogenic risk, both of them are strong and valid indicators for monitoring and evaluating metabolic situations of cholesterol, i.e., deposited into tissue or metabolized and excreted (Chang et al., 2011). In these studies, administration of PSG-1 significantly reduced LDL-C in HFD/STZ-induced type 2 diabetic rats. The results have shown that PSG-1 at a high dose (400 mg/kg BW) produced a significant ameliorative action on the ratios of TC/HDL-C and LDL-C/HDL-C (1.76 ± 0.19 and 0.48 ± 0.11, respectively) compared to those of diabetic rats (3.85 ± 0.67 and 2.35 ± 0.17, respectively), suggesting that PSG-1 may possess cardioprotective potential.

Free fatty acids (FFA) play important roles in many tissues, including heart, skeletal muscle, pancreas, and liver, and FFA levels are commonly elevated in obese individuals and patients with type 2 diabetes (Boden and Shulman, 2002). Furthermore, studies have shown that elevated circulating levels of FFA may directly contribute to the development of insulin resistance, both in the liver and peripheral cells (Shulman, 2000). In a recent study, FFA levels were elevated two to threefold in diabetic rats compared to non-diabetic controls. Interestingly, treatment with PSG-1 at 200 or 400 mg/kg BW was found to significantly reduce the FFA levels. Thereby, it can be hypothesized that the ameliorating effect of PSG-1 in insulin resistance in type 2 diabetes may be associated with reducing serum FFA levels and attenuating the disorder of lipid metabolism.

5.2.3.1.5 Histopathological Observation

Fig. 5.32A depicts the effect of PSG-1 on the histology of the pancreas in HFD/STZ-induced type 2 diabetic rats. Diabetic rats showed shrunken islets and the presence of pancreatic acini. Treatment of diabetic rats with PSG-1 at a dose of 200 mg/kg BW or 400 mg/kg BW resulted in expansion of the

TABLE 5.11 Effect of PSG-1 on Serum Lipids in Type 2 Diabetic Rats[a,b,c]

Groups	TC (mmol/L)	TG (mmol/L)	HDL-C (mmol/L)	LDL-C (mmol/L)	FFA (μmol/L)
Non-diabetic control	2.46 ± 0.23	0.61 ± 0.04	1.68 ± 0.13	0.48 ± 0.02	447.84 ± 9.80
Untreated diabetic	$3.78 \pm 0.91^{\#\#}$	$1.10 \pm 0.05^{\#\#}$	$0.99 \pm 0.06^{\#\#}$	$2.32 \pm 0.98^{\#\#}$	$1158.83 \pm 33.76^{\#\#}$
Positive control	2.72 ± 0.29	$1.09 \pm 0.06^{\#}$	$1.05 \pm 0.16^{\#}$	$1.19 \pm 0.45^{\#}$	$876.41 \pm 55.73^{\#\#}$
Diabetic + PSG-1 (200 mg/kg BW)	2.85 ± 0.25	$0.97 \pm 0.16^{*}$	$1.22 \pm 0.36^{*,\#}$	$1.20 \pm 0.44^{\#}$	$772.09 \pm 43.75^{*,\#}$
Diabetic + PSG-1 (400 mg/kg BW)	$2.50 \pm 0.13^{*}$	$0.88 \pm 0.10^{**}$	$1.47 \pm 0.13^{**}$	$0.68 \pm 0.22^{*}$	$689.38 \pm 23.55^{**}$

[a] *Values are expressed as means ± SEM (n = 10).*
[b] *$^{*}P < .05$, $^{**}P < .01$ compared with untreated diabetic group.*
[c] *$^{\#}P < .05$, $^{\#\#}P < .01$ compared with non-diabetic control group.*

(A)

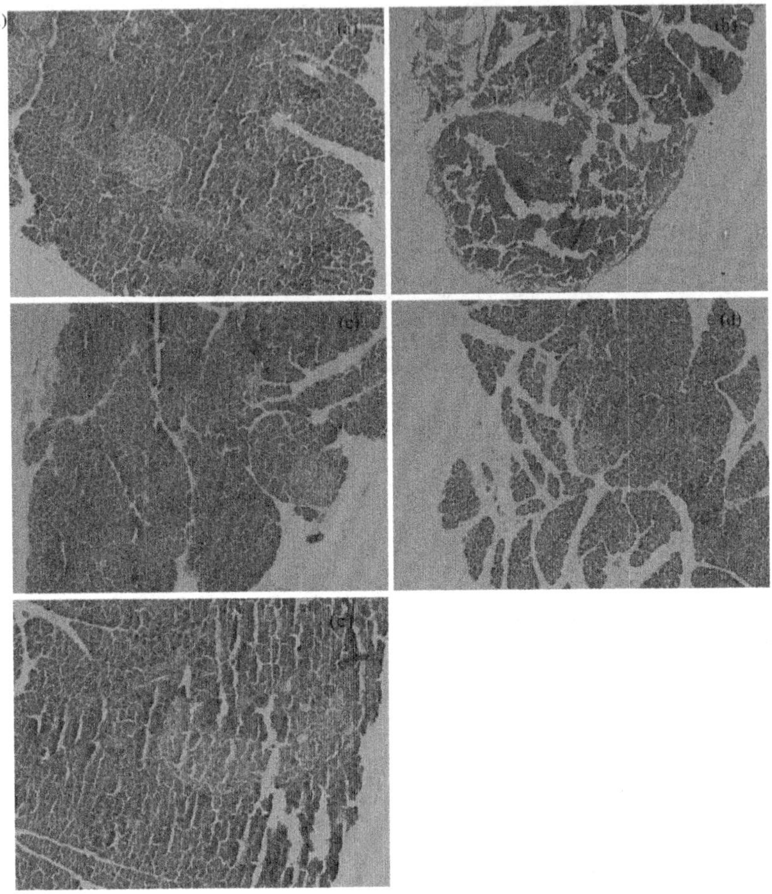

FIGURE 5.32 (A) The protective effect of PSG-1 treatment on the histological structure of pancreas in diabetic rats. (a) Non-diabetic control group; (b) Untreated diabetic group; (c) Positive control group; (d) Diabetic + PSG-1 (200 mg/kg BW); (e) Diabetic + PSG-1 (400 mg/kg BW), $n = 10$ in all groups. Original magnification 200X. (B) Effect of PSG-1 on the expression of Bax in pancreatic tissues of type 2 diabetic rats. (a) Non-diabetic control group; (b) Untreated diabetic group; (c) Positive control group; (d) Diabetic + PSG-1 (200 mg/kg BW); (e) Diabetic + PSG-1 (400 mg/kg BW), $n = 10$ in all groups. Original magnification 200 X; (f) IOD of immunostaining on Bax in pancreatic tissues of type 2 diabetic rats. $^{\#}P < .05$, $^{\#\#}P < .01$ compared with non-diabetic control group. $^{*}P < .05$, $^{**}P < .01$ compared with untreated diabetic group. (C) Effect of PSG-1 on the expression of Bcl-2 in pancreatic tissues of type 2 diabetic rats. (a) Non-diabetic control group; (b) Untreated diabetic group; (c) Positive control group; (d) Diabetic + PSG-1 (200 mg/kg BW); (e) Diabetic + PSG-1 (400 mg/kg BW), $n = 10$ in all groups. Original magnification 200X; (f) IOD of immunostaining on Bcl-2 in pancreatic tissues of type 2 diabetic rats. $^{\#}P < .05$, $^{\#\#}P < .01$ compared with non-diabetic control group. $^{*}P < .05$, $^{**}P < .01$ compared with untreated diabetic group.

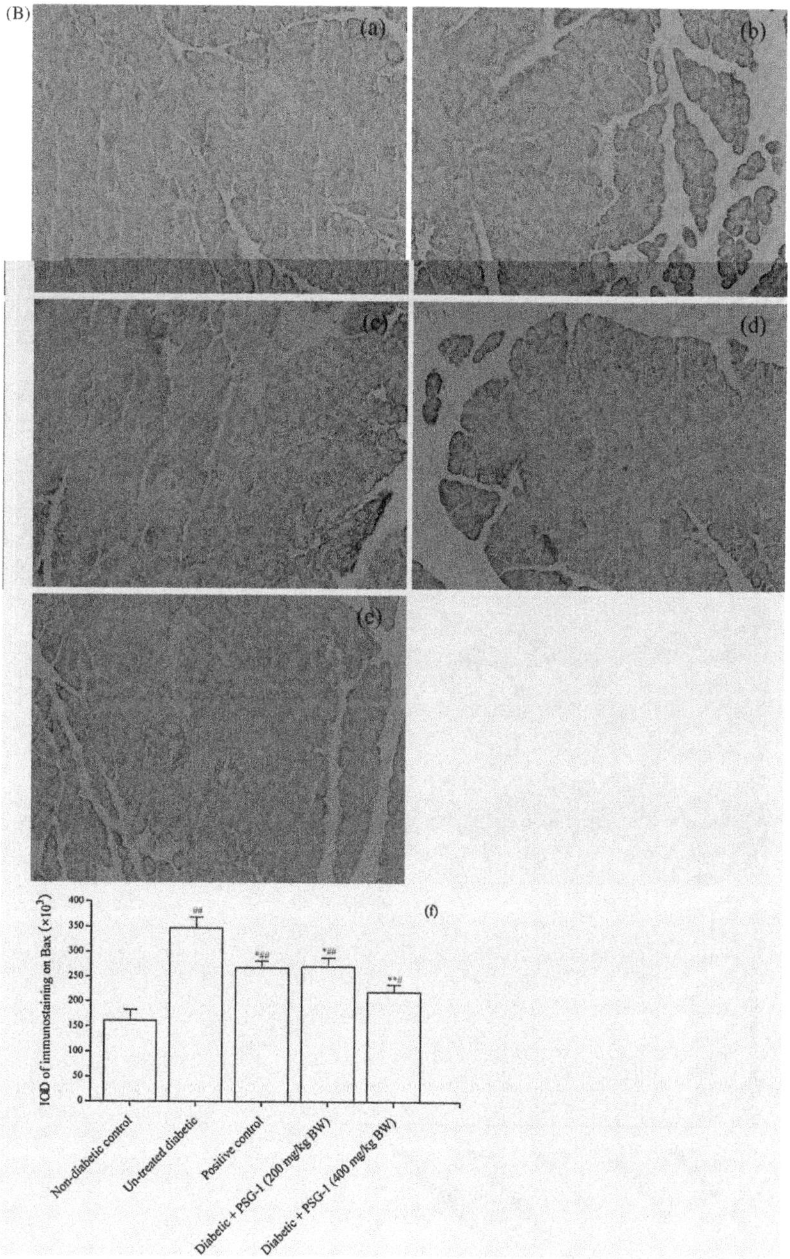

FIGURE 5.32 (Continued)

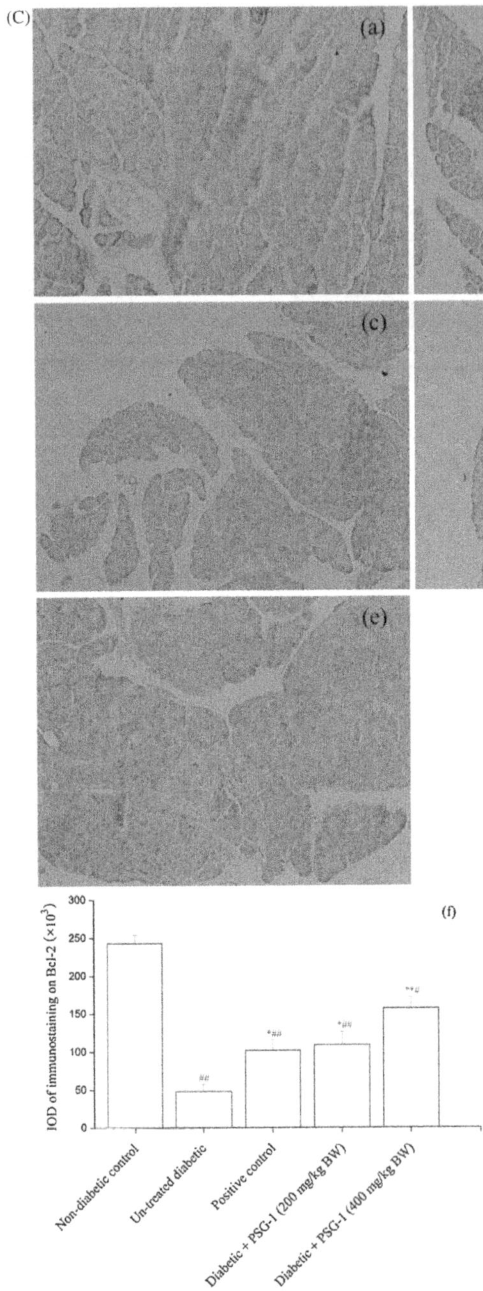

FIGURE 5.32 (Continued)

pancreatic islets, suggesting the protective role of PSG-1 in the pancreas of diabetic rats.

5.2.3.1.6 Effect of PSG-1 on the Expression of Bax and Bcl-2 in the Pancreatic Tissues of Type 2 Diabetic Rats

Immunohistochemical staining of Bax in pancreatic tissues of type 2 diabetic rats and image analysis are shown in Fig. 5.32B (a–f). Bax protein expression levels are presented as the value of IOD. The expression of Bax in untreated diabetic rats is significantly higher than that in non-diabetic control rats. There is a statistically significant difference: the value of IOD on bax was significantly higher in the diabetic groups. However, the values of IOD in the other three groups were significantly decreased ($P < .05$) compared with the untreated diabetic group after four weeks of treatment. The results demonstrated that PSG-1 inhibited Bax protein levels and increased the levels of antiapoptotic Bcl-2 protein, suggesting that the anti-apoptotic effect of PSG-1 worked through inhibiting expression of Bax and improving the expression of Bcl-2 protein in pancreatic tissues of type 2 diabetic rats.

In conclusion, the present study demonstrates that PSG-1 may have potential for the treatment of hyperglycemia and hyperlipidemia. The results also indicate that PSG-1 may have an anti-apoptotic effect through inhibiting the expression of Bax and improving the expression of Bcl-2 protein in pancreatic tissues of type 2 diabetic rats. In addition, PSG-1 may exert a beneficial effect on the prevention of cardiovascular diseases, via a favorable effect on the lipid profile. However, the effect of PSG-1 in the prevention of atherogenic cardiovascular diseases needs to be determined in the future.

5.2.3.2 Effect of Polysaccharide From G. atrum *on the Serum Metabolites of Type 2 Diabetic Rats*

Metabolomics is a newly emergent "omics" which provides major insights into the similarities and differences related to the prognosis or diagnosis of diseases by measuring the dynamic range of small molecule metabolites from biological samples (Kim et al., 2011). Small molecule metabolites including lipids, amino acids, peptides, nucleic acids, organic acids, vitamins, etc., give metabolomics potential advantages over endogenous metabolites in biological samples (Zhang et al., 2012a,b). With large-scale metabolomics data, multivariate analysis techniques, including principal components analysis (PCA) and partial-least-square discriminant analysis (PLS-DA), were employed to determine the distribution of samples and identify metabolites that differentiate between different samples (Want and Masson, 2011; Worley et al., 2013).

Diabetes, as the most prevalent metabolic disease, is caused by absolute or relative insulin deficiency, and is characterized by hyperglycemia as well as impaired metabolism of carbohydrate, fat, and protein (Sharma, 2011).

Thus, it is a formidable task to fully understand the complicated mechanisms of diabetes on the basis of metabolomics. Griffin and Nicholls (2006) reported that metabolomics was used to study diseases related to the metabolic syndrome, including type 2 diabetes, obesity, cardiovascular disease, and fatty liver disease.

This study aimed to investigate the metabolites of polysaccharides from *G. atrum* (PSG-1) in serum samples from high fat diet- and streptozotocin-induced type 2 diabetic rats by employing ultra-performance liquid chromatography quadrupole time-of-flight mass spectrometry (UPLC-Q-TOF/MS). Then, *t* test unpaired statistical analysis and PLS-DA were carried out to determine the number of significantly differential metabolites, and the distribution of samples from different groups.

5.2.3.2.1 Analysis of Metabolite Profiles

Fig. 5.33 shows the distribution of the retention times for the molecular features detected using UPLC-Q-TOF/MS in the positive and negative ion mode. The two chromatograms reporting wider information in terms of number of peaks were obtained with a method based on a C_{18} column in both

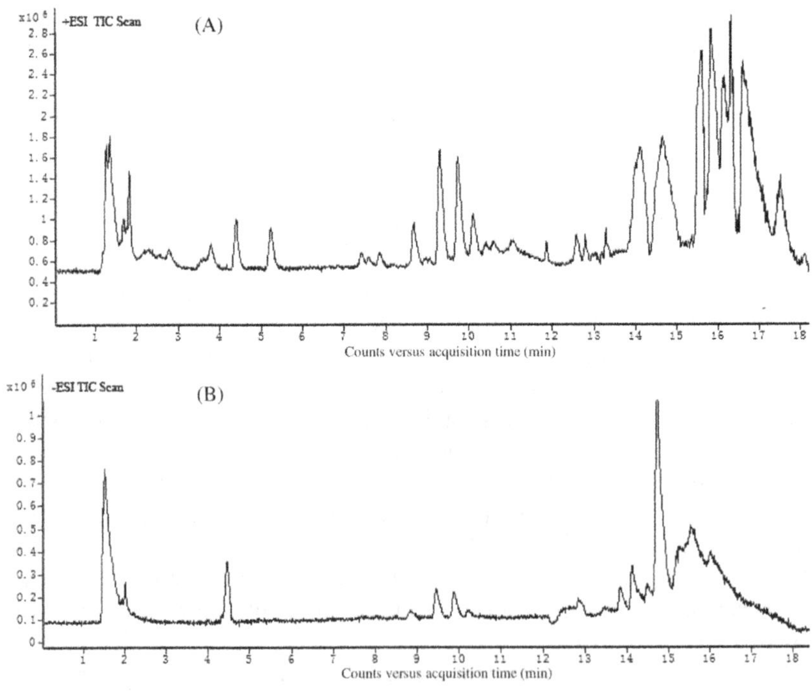

FIGURE 5.33 The positive (A) and negative (B) ion total ion chromatogram of the serum sample from a type 2 diabetic rat.

positive and negative ionization modes. A reference mass solution containing reference ions 121.0508, and 922.0097 in positive ionization mode, and 112.39855, 1034.9881 in negative ionization mode was used to ensure mass accuracy and reproducibility.

5.2.3.2.2 Effect of PSG-1 on Serum Metabolic Profiling

Partial least squares-discriminant analysis (PLS-DA) is one of the most popular pattern recognition methods to gain information for classification and to identify metabolites (Liu et al., 2012). As shown in Fig. 5.34, good separation of different groups was achieved, suggesting that biochemical changes happened in the serum of untreated diabetic and PSG-1 treated groups. After

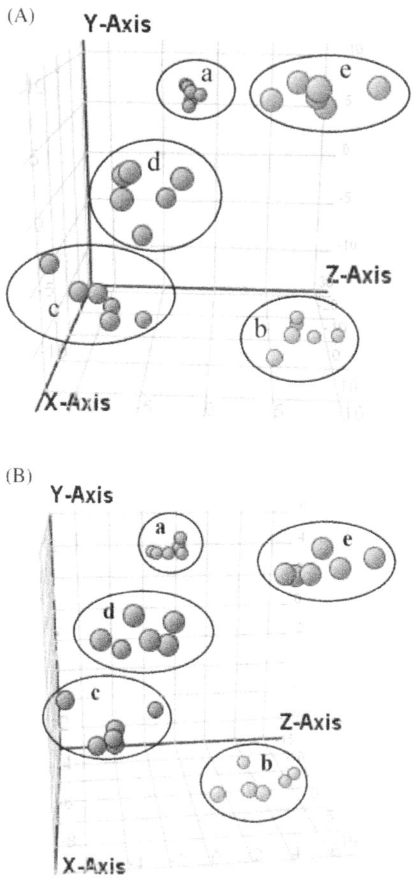

FIGURE 5.34 Score plots (A) in positive ion mode; and (B) negative ion mode from 3D PLS-DA model of different groups. (a) Non-diabetic control group; (b) Untreated diabetic group; (c) Positive control group; (d) Diabetic + PSG-1 (200 mg/kg BW); (e) Diabetic + PSG-1 (400 mg/kg BW), $n = 6$ in all groups.

treatment for 1 week, the PSG-1 treated groups are closer to the non-diabetic control group than the untreated diabetic group in the direction of the principal component, which implied that PSG-1 intervened in the metabolic process of diabetic rats to some degree. The results were consistent with the previous report that PSG-1 had a potential for the treatment of hyperglycemia, hyperlipidemia, hyperinsulinemia, and insulin resistance in type 2 diabetic rats (Zhu et al., 2013).

5.2.3.2.3 Statistical Analyses and Potential Biomarkers

The statistical analysis of data and database search systems are both important methods for the identification of potential biomarkers (Tsutsui et al., 2011). The differential metabolites were validated using *t* test unpaired statistical analysis. As shown in Table 5.12, there are 255 significantly differential metabolites in positive ion mode, and 94 metabolites in negative ion mode from the untreated diabetic group compared to the non-diabetic control group. Compared with the untreated diabetic group, there are 50, 161, and 196 metabolites in positive ion mode from the positive control group, PSG-1 (200 mg/kg BW), and PSG-1 (400 mg/kg BW), respectively. However, less metabolites were found in the negative ion model, only 30, 49, and 74 metabolites from the positive control group, PSG-1 (200 mg/kg BW) and PSG-1 (400 mg/kg BW), respectively.

5.2.3.2.4 Identification of Potential Biomarkers and Biological Interpretation

Biological pathways and processes that are significantly represented after PSG-1 treatment were analyzed using Mass Profiler Professional. Then, UPLC-Q-TOF MS/MS was used to confirm the identities of metabolites. As shown in Table 5.13, eight potential biomarkers, including methionine,

TABLE 5.12 The Results of Unpaired *t* Test in Positive Ion Mode ($n = 6$)

Ion Model / Groups' Comparison	Number of Significantly Differential Metabolites	
	Positive Ion Mode	Negative Ion Mode
Untreated diabetic group vs non-diabetic control group	255	94
Positive control group vs untreated diabetic group	50	30
PSG-1 (200 mg/kg BW) vs untreated diabetic group	161	49
PSG-1 (400 mg/kg BW) vs untreated diabetic group	196	74

TABLE 5.13 Potential Biomarkers Identified in Positive/Negative Ion Mode and the Related Pathways

Ion Model	Retention Time_M^+	MS $[M+H]^+$	Actual_M	Proposed Compound	MS/MS	Proposed Structure	Related Pathway
Positive	2.04_150.0582	150.0582	149.0509	Methionine	133.0316, 121.0314, 105.0366, 104.0530, 102.0553	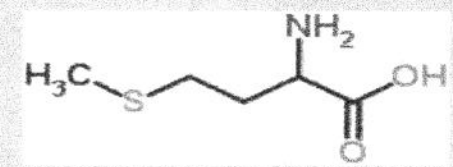	Methylation pathways, vitamin B_{12} metabolism, TCA and urea cycles, metabolism of water-soluble vitamins and cofactors
	2.08_323.0650	323.0650	322.0577	dTMP	233.0410, 210.0411, 132.1007, 126.0535, 115.0374	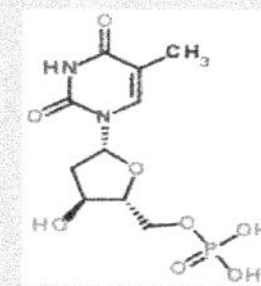	*Trans*-sulfuration and one carbon metabolism
	5.41_205.0971	205.0971	204.0898	D-Tryptophan	188.0706, 170.0603, 160.0748, 159.0915, 144.0805, 132.0803, 118.0648, 115.0545	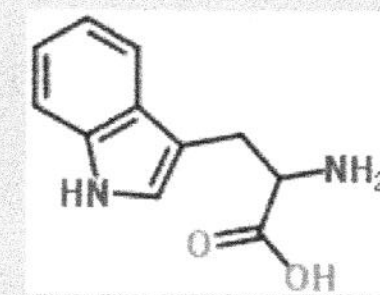	Tryptophan metabolism, selenium micronutrient network, TCA and urea cycles
	11.16_176.0705	176.0705	175.0632	5-Hydroxy-indoleacetaldehyde	159.0440, 133.0486, 130.0661, 117.0696, 110.0602	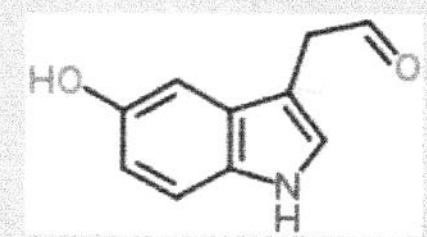	Tryptophan metabolism

(*Continued*)

TABLE 5.13 (Continued)

Ion Model	Retention Time_M$^+$	MS $[M+H]^+$	Actual_M	Proposed Compound	MS/MS	Proposed Structure	Related Pathway
	13.52_466.3158	466.3162	465.3085	Glycocholic acid	448.3054, 431.2949, 355.2634, 339.2528, 240.1572, 158.0801, 123.1155	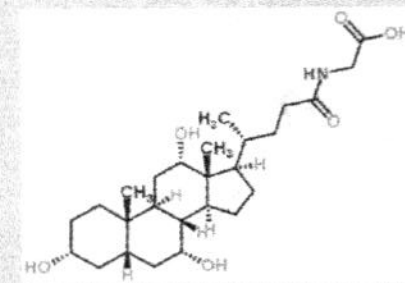	Fatty acid biosynthesis, glycolysis and gluconeogenesis, metabolism of lipids and lipoproteins
	14.58_583.2545	583.2545	582.2472	Biliverdin IX	523.2447, 469.2263, 297.1231, 243.1223, 133.0518, 114.0907, 100.0750	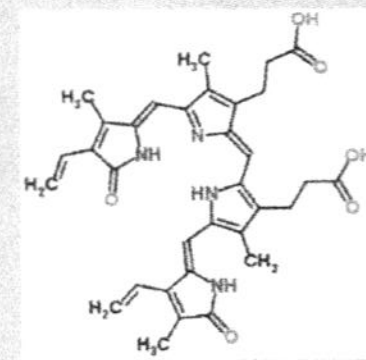	Heme biosynthesis, metabolism of porphyrins
Negative	1.88_133.0139	133.0139	134.0212	Malate	115.0040, 71.0135	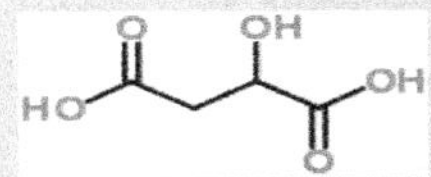	Metabolism of carbohydrates, fatty acid, triacylglycerol, and ketone body metabolism, TCA and urea cycles
	11.74_514.2839	514.2835	515.2912	Taurocholic acid	433.3179, 350.2828, 303.1931, 295.2537, 195.0995, 175.0644, 164.0763, 163.0977, 131.0037	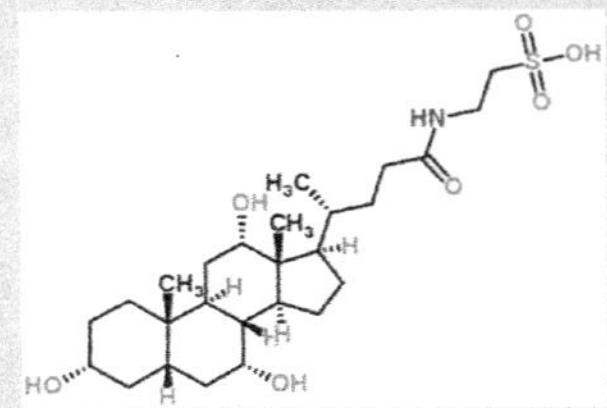	Glycolysis and gluconeogenesis, triacylglyceride synthesis

dTMP, D-tryptophan, 5-hydroxy-indoleacetaldehyde, glycocholic acid, biliverdin IX, malate, and taurocholic acid have been identified.

Methionine, as a sulfur-containing amino acid, is necessary for the growth and development of mammals (Finkelstein, 1990). Methionine is converted to homocysteine by *trans*-methylation and reformed by methylation of homocysteine (Klepacki et al., 2013). Moreover, homocysteine induces endothelial injury and arteriosclerosis *in vivo*. Tessari et al. (2005) found that type 2 diabetic rats showed a lower concentration of methionine and greater homocysteine content. Methionine was reported to restore the levels of antioxidants to some degree, and to reduce the peroxidation of plasma lipid in diabetic rats (Anuradha and Selvam, 1993; Özkaya et al., 2011). The present study found that PSG-1 influenced the concentration of methionine, which may be related to methylation pathways, Vitamin B_{12} metabolism, TCA and urea cycles, metabolism of water-soluble vitamins, and cofactors in type 2 diabetic rats. A previous study reported that metformin, an anti-diabetic drug, exerted an effect on serum concentrations of homocysteine, folate, and vitamin B_{12} in type 2 diabetes (Palomba et al., 2010; Wulffele et al., 2003). These results may in part explain why PSG-1 protected against diabetes-induced endothelial dysfunction in type 2 diabetic rats (Zhu et al., 2014).

The *trans*-sulfuration pathway is functional in the liver, kidney, and pancreas (Guéant et al., 2013), while one-carbon metabolism is essential for thymidylate (dTMP) synthesis and remethylation of homocysteine to methionine (Beaudin et al., 2011). A recent study found that PSG-1 treatment influenced dTMP content in the serum of type 2 diabetic rats, suggesting that PSG-1 may exert beneficial effects on diabetic nephropathy and diabetic cardiovascular disease through regulating *trans*-sulfuration and one-carbon metabolism.

Tryptophan is one of eight essential amino acids, and plays an important role in the process of protein synthesis (Nikolić et al., 2014). Trulson et al. (1986) found that total levels of plasma tryptophan and brain tryptophan were reduced in streptozotocin-induced diabetic rats. Studies have shown that PSG-1 influenced the concentration of tryptophan, which may be through regulation of several related pathways, including tryptophan metabolism, selenium micronutrient network, and TCA and urea cycles. Here, 5-hydroxyindoleacetaldehyde is the oxidized metabolite of 5-hydroxytryptamine by monoamine oxidase. MacKenzie and Trulson (1978) studied the effects of diabetes on the accumulation of tryptophan and the responsivity of the 5-HT metabolic pathway to an acute tryptophan load. This finding was confirmed in the present study, that the change of 5-hydroxyindoleacetaldehyde concentration is related to tryptophan metabolism.

In humans, taurocholic acid and glycocholic acid (derivatives of cholic acid) represent approximately 80% of all bile salts (Roberts et al., 2011). Hansen (2001) reported that excretion of cholesterol from the body may be much more likely to occur through taurocholic acid than glycocholic acid and unconjugated bile acids. The bile pigment biliverdin IX is one of the endogenously derived substances generated during enzymatic heme degradation (Choudhary, 2012). Then, they may affect fatty acid biosynthesis, glycolysis and gluconeogenesis, and metabolism of lipids and lipoproteins. A study found that PSG-1 influenced the concentrations of taurocholic acid and glycocholic acid, which was consistent with the previous study showing that PSG-1 attenuated the disorder of lipid metabolism.

A series of mitochondrial and/or cytosolic reactions, including the generation of malate, occurred while dysregulated pyruvate cycling mediates lipid-induced β-cell dysfunction in type 2 diabetes (Muoio and Newgard, 2008). Studies have shown that PSG-1 ameliorated malate content in type 2 diabetic rats through influencing the metabolism of carbohydrates, fatty acid, triacylglycerol, and ketone body metabolism.

An analysis method based on UPLC-Q-TOF/MS has been developed to study the metabolites of PSG-1 on serum from STZ-induced type 2 diabetic rats. With T test unpaired statistical analysis, a clear separation among different groups was achieved. Eight potential biomarkers have been found and identified in serum samples from type 2 diabetic rats. Biological pathways and processes, including metabolism of carbohydrates, fatty acid biosynthesis, glycolysis and gluconeogenesis, and metabolism of lipids and lipoproteins, are significantly changed by PSG-1 treatment using the pathway analysis features in Mass Profiler Professional.

5.3 STRUCTURE–BIOACTIVITY RELATIONSHIP

Antioxidant supplements or foods containing high concentrations of antioxidants may help reduce oxidative damage. Identification of new antioxidants remains a hot topic for researchers. Generally speaking, the antioxidant activity of polysaccharide depends on several structural parameters, such as degree of substitution (DS), degree of branching, molecular weight (MW), monosaccharide composition, and functional groups (Bohn and BeMiller, 1995). Therefore, increasing attention has been paid to molecular modification and the structure–activity relationship of polysaccharides.

Ganoderma atrum was found to contain a high level of crude polysaccharide (PSG, which mainly consisted of PSG-1) with potential antioxidant and immunomodulating activities both *in vitro* and *in vivo* (Li et al., 2010a,b; Li et al., 2011a,b). However, the structure–antioxidant activity relationship for this polysaccharide has not been elucidated.

For further research and utilization of polysaccharides from *G. atrum*, we report here the preparation, structural characterization, and some bioactivities *in vitro* of different derivatives of PSG. Firstly, different derivatives from PSG were prepared by means of acetylation and carboxylmethylation. Then, the structures of these derivatives were characterized by chemical analysis, FT-IR spectroscopy, and NMR spectroscopy. Finally, their antioxidant and immunomodulating activities *in vitro* were evaluated and compared, and the relationship between the chemical structure and antioxidant, immunomodulating activities of PSG was discussed.

5.3.1 Acetylation and Carboxymethylation of the Polysaccharide From PSG-1

5.3.1.1 Chemical Analysis of Ac-PSG and CM-PSG

The polysaccharide we used here is the crude polysaccharide, so it is not a pure and single polysaccharide. Thus, correspondingly, the profiles of CM-PSG and Ac-PSG in HPGPC (Fig. 5.35) were not very sharp and symmetrical. However, its distribution or the range of elution time was comparatively centralized. This indicated that CM-PSG or Ac-PSG was constituted of some polysaccharides with similar or close molecular weights, which made it possible to determine the average molecular weight of CM-PSG and Ac-PSG. The yield, DS, carbohydrate content, and molecular weight of PSG and its derivatives are summarized in Table 5.14. Compared with the native sample, not only did the total sugar content of CM-PSG and Ac-PSG decrease significantly ($P < .05$), but also the molecular weight decreased, which was in accordance with the results of Das et al. (2010a,b). The DS of the PSG derivatives were listed in decreasing order as follows: Ac-PSG-3 > Ac-PSG-2 > Ac-PSG-1, CM-PSG-2 > CM-PSG-1. However, their carbohydrate content and molecular weight listed in a converse order, which indicated that a slight degradation might occur in the process of Ac-PSG and CM-PSG preparation.

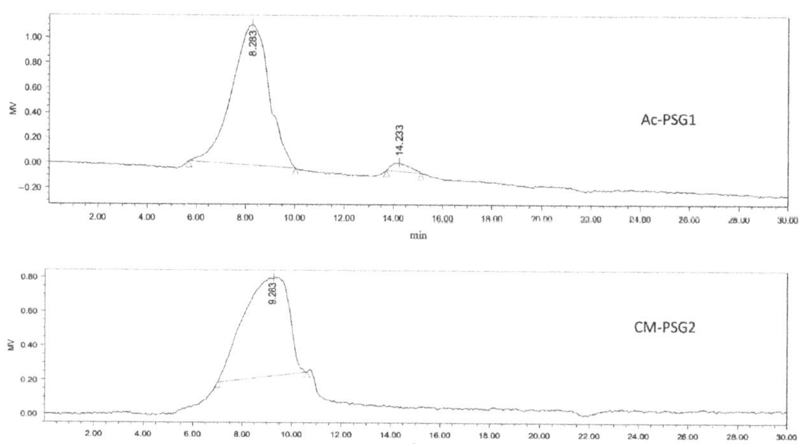

FIGURE 5.35 Profiles of CM-PSG and Ac-PSG in HPGPC.

TABLE 5.14 Molecular Structure Parameters of PSG and Its Derivatives

Samples	Yield (%)	Total Sugar (%)	DS (Carboxylmethyl acetyl)	Molecular Weight ($\times 10^{-6}$)
PSG	–	90.7	–	8.89
CM-PSG-1	69.3	80.6	0.37	4.56
CM-PSG-2	54.1	72.3	0.53	2.68
Ac-PSG-1	80.2	82.1	0.71	6.34
Ac-PSG-2	74.2	74.5	0.89	5.18
Ac-PSG-3	43.6	64.4	1.04	1.87

5.3.1.2 *Infrared Spectra of PSG and Its Derivatives*

The FT-IR spectra of PSG and its derivatives are shown in Fig. 5.36. The general profile of the spectra for PSG, AC-PSG, and CM-PSG were similar, and exhibit the typical signals of a polysaccharide in the range from 4000 to 500 cm^{-1}. However, if we go further into the details, many differences in specific spectra range could be found. As shown in Fig. 5.36A, for acetylation, a new absorption appeared at 1735 cm^{-1} in Ac-PSG, which was assigned to the C = O stretching vibration, and the increased absorption intensity at 1247 cm^{-1} showed that the acetylated derivative was successfully synthesized (Zhang et al., 2010). Acetylated PSG was also confirmed by strong absorption bands in the range 1365–1380 cm^{-1} (δS_{CH3}), which became stronger as the degree of modification increased.

As shown in Fig. 5.36B, carboxyl of CM-PSG was confirmed by two new strong absorption bands in the 1601 region [γsym(COO$^-$)] and 1420 cm^{-1} [γas(COO$^-$)]. The bands, which were attributed to C–O–C symmetrical stretching vibrations, appeared at about 1070 cm^{-1}, and asymmetrical stretching vibrations appeared at about 1247 cm^{-1}. The spectra are dominated by a broad band at about 3384 cm^{-1}, which was assigned to the stretching vibration modes of OH groups. This band tends to shift to a higher wave number, up to 3422 cm^{-1}, after carboxymethylation. The bands in the region of 2924 cm^{-1} were assigned to the CH stretching of the CH_2 groups.

5.3.1.3 *^{13}C Nuclear Magnetic Resonance Analyses*

The ^{13}C NMR spectrum of PSG and its derivatives provided more detailed structural information, as shown in Fig. 5.37. In the ^{13}C NMR spectrum of PSG, according to a comparison with β-D-glucan (Storseth et al., 2004), the peaks at around 102.88 (C-1), 75.11 (C-2), 76.23 (C-3), 70.31(C-4), 75.82

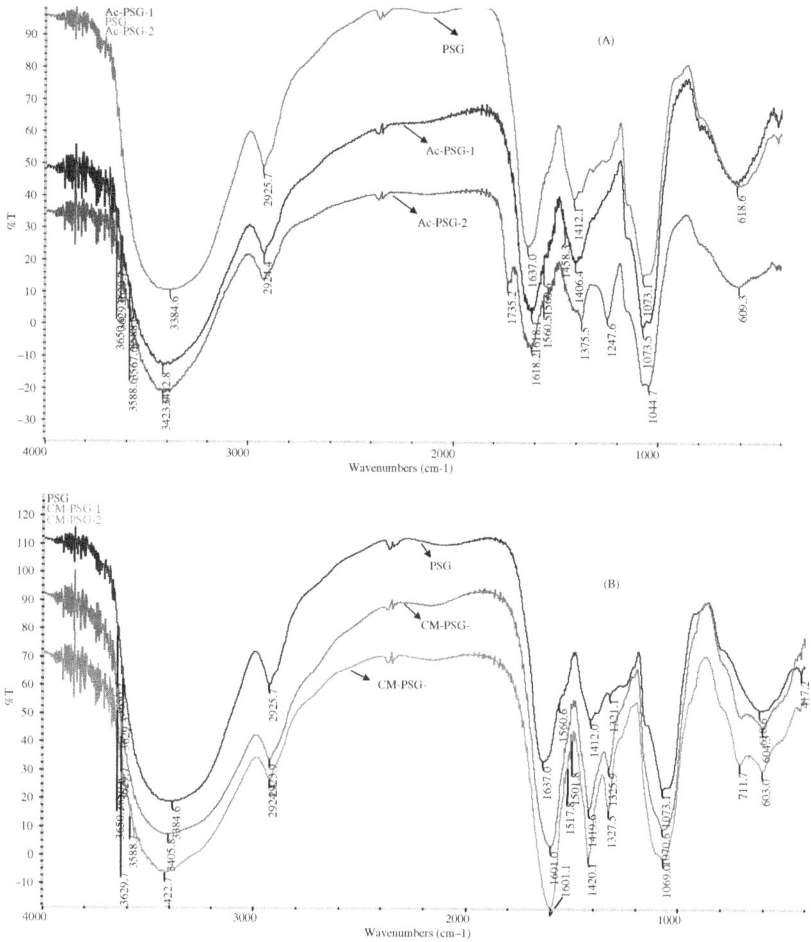

FIGURE 5.36 FTIR spectra of PSG and its derivatives: (A) CM-PSG; (B) Ac-PSG.

(C-5), and 60.54 (C-6) ppm were assigned to the signals of the backbone chain. In the ^{13}C-NMR spectrum (Fig. 5.37A), the signal at 84.88 ppm was attributed to substitution at O-3, while that at 81.30 ppm arose from 3,6-di-*O*-substitution units. The shift of C-6 from 60.54–69.96 ppm indicated that a (1–6)-linked glycosidic bond may exist, which was in agreement with Chakraborty et al.(2004). Thus, PSG probably had a backbone chain mainly composed of β-(1,3), β-(1,6), and β-(1,3,6)-D-glucan. In addition, the peaks at 98.33, 71.57, 73.56, 66.98, 76.02, and 61.02 ppm for PSG were attributed to α-D-galactose bound to glucose (Zhang et al., 2012c,d).

After chemical modification, the signals of the backbone chain of PSG mentioned above still existed, which indicated that the main structure of

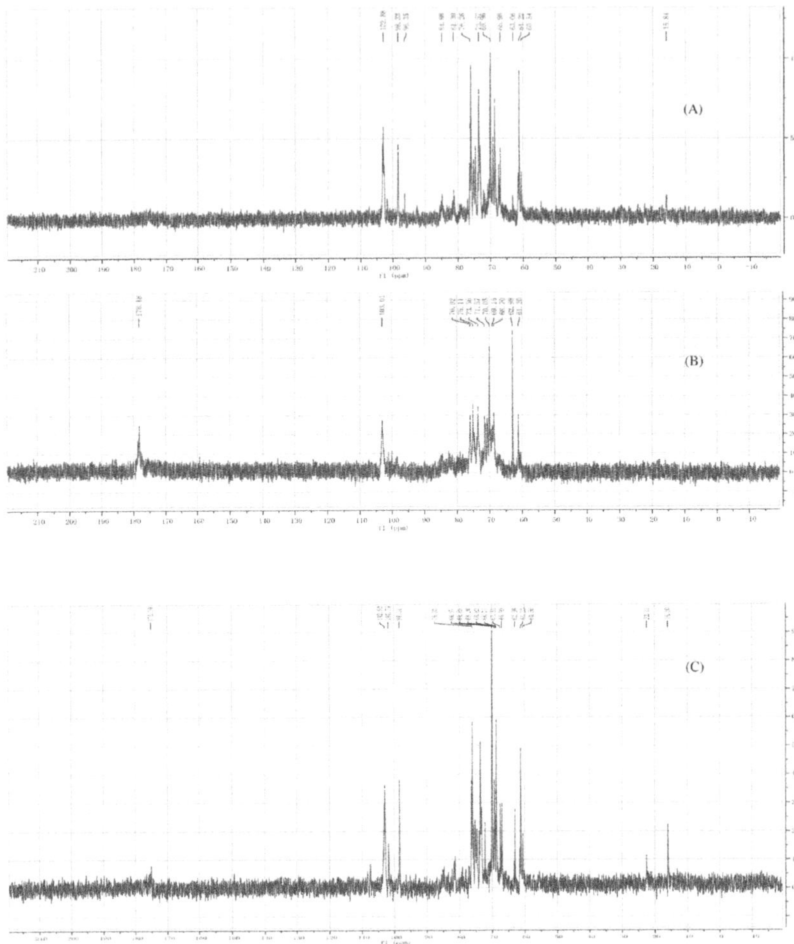

FIGURE 5.37 ^{13}C NMR spectra of (A) GLP; (B) CM-GLP; (C) Ac-PSG.

PSG in its chemically modified derivatives was reserved. For carboxymethylation (Fig. 5.37B), an intense group of peaks belong to the carbonyl groups at 178.9, 178.18, and 177.9 ppm could be seen as evidence of a carboxymethylation reaction, which was in agreement with IR results. And the signals at 71.57 and 73.56 ppm were assigned to the methylene carbon atoms of the carboxymethyl substituents (Ren et al., 2008). The increased signal at 70.8 ppm, and the decreased signal at 61.2 ppm, revealed that the carboxymethyl substitution occurred at the C-6 position (Silva et al., 2004). Moreover, the shifts of C-2 from 75.11 to 73.56 ppm, and C-4 from 70.31 to 75.1 ppm, indicated that the residues in CM-PSG are also substituted at the O-2 and O-4 positions by the substitution of $-CH_2COOH$ (Tao et al., 2009).

Based on the above information from NMR spectra, we proposed that the substitutions may nonselectively occur at C-6, C-4, or C-2. However, a further structural determination needs to be carried out in a future study.

Similarly, as shown in Fig. 5.37C, the signal at 175.10 ppm in the ^{13}C-NMR spectrum of Ac-PSG confirmed the presence of an acetyl group. By comparison with the ^{13}C-NMR spectrum of PSG (Fig. 5.37A), the signal at 62.98 ppm was assigned to an O-6 substituted carbon. The signal at 22.4 ppm gave further evidence for the presence of acetyl group.

5.3.2 Antioxidant Activity

5.3.2.1 DPPH Radical-Scavenging Assay

As shown in Fig. 5.38A, all the samples were found to possess DPPH radical scavenging activity in a concentration-dependent manner. After chemical derivative processing, the DPPH free radical scavenging ability of PSG had been changed.

The acetylated PSG, Ac-PSG-1 with low acetyl substitution showed a slightly decreased scavenging effect compared to the native polysaccharide; while with the acetyl substitution degree increased, the scavenging ability was greatly enhanced. However, for all the samples tested, the reducing powers were lower than that of V_C and BHT. At a concentration of 2 mg/mL, the scavenging rates of PSG, Ac-PSG-1, Ac-PSG-2, V_C, and BHT were 47.64%, 32.58%, 71.36%, 96.33%, and 91.05% respectively. These results indicated that the insertion of the acetyl group has a noticeable effect on the scavenging abilities of DPPH free radicals, especially at high substitution degrees. We proposed that the introduction of acetyl groups may affect the polarity, conformation, or charge density of native polysaccharide, which may lead to the various changes of scavenging activity. However, further investigation will be required to determine the possible antioxidant mechanism.

Similarly, the scavenging activity of carbomethylated derivatives was correlated well with the increase of substitution degree, while in contrast, the carbomethylated derivatives exhibited weaker activity to scavenge the DPPH radicals than the native one. The IC_{50} of the CM-PSG-1 and CM-PSG-2 derivatives were determined to be 4.24 and 6.07 mg/mL, while that of the native polysaccharides was 2.04 mg/mL. These results are consistent with the results reported by other researchers (Shin et al., 2007). Therefore, the hydroxyl groups in the polysaccharides may play an important role in maintaining its hydrogen donating ability. These results revealed that not only the charge density, but also the chemical properties of the function groups on derivatized PSG affected its antioxidant activities.

5.3.2.2 Ferric Reducing Ability of Plasma (FRAP) Assay

The reduction of Fe^{3+} is often used as an indicator of electron-donating activity. In the FRAP assay, the presence of antioxidants in the samples would

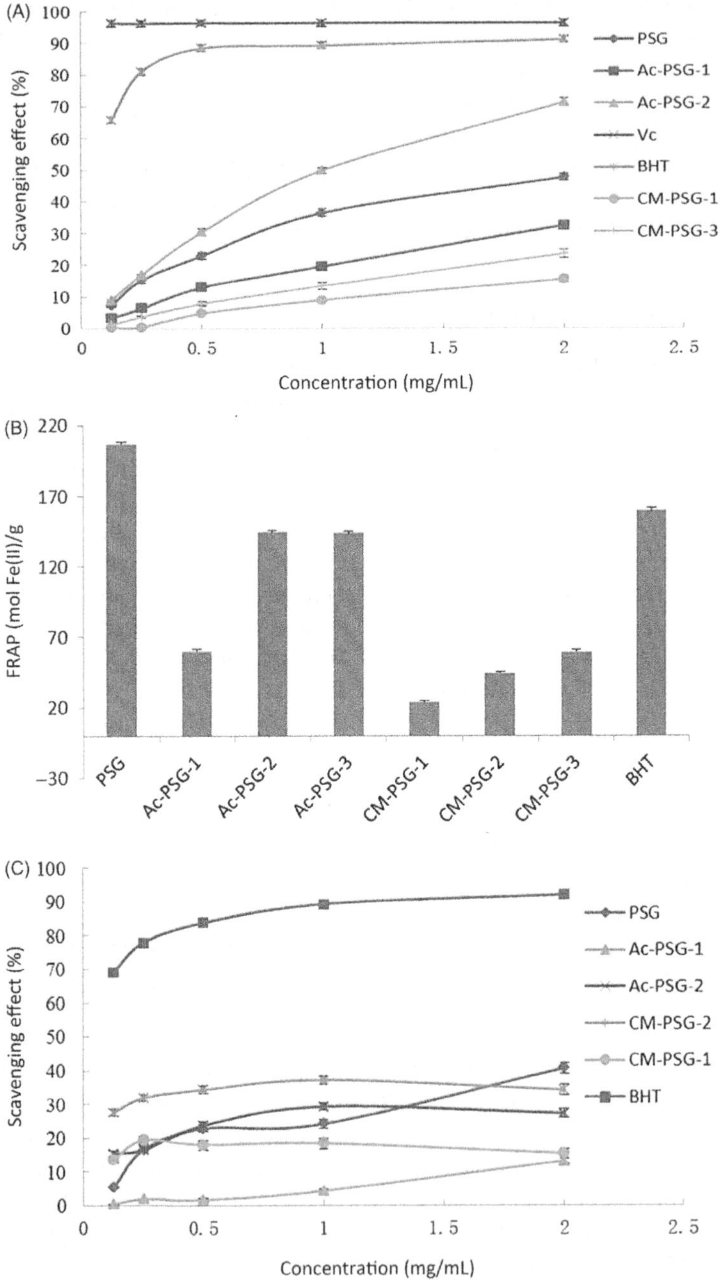

FIGURE 5.38 Antioxidant activity analysis of PSG and its derivatives with various methods: (A) Scavenging effects of PSG and its derivatives on DPPH free radicals; (B) Ferric reducing antioxidant property of the PSG and its derivatives; (C) Inhibitory abilities of PSG and its derivatives on β-carotene and linoleic acid system.

result in the reduction of TPRZ-Fe (III) complex to TPTZ-Fe (II) by donating an electron. This reducing capacity of compounds could be used as an index to evaluate potential antioxidant properties (Benzie and Strain, 1996).

The FRAP values of all the samples are summarized in Fig. 5.38B. All samples showed some degree of reducing power. The reducing power for Ac-PSG increased with the increasing substitution degree, and reached a plateau of 143.6 μm Fe (II)/g; however, it was still much lower than that of the native polysaccharide (206.6 μm Fe (II)/g), but comparable with that of BHT. For CM-PSG, although the reducing power increased in a DS-dependent manner, it exhibited the lowest reducing power among all those samples tested. Even the highest value of 59.3 μm Fe (II)/g for CM-PSG-3 was still inferior to that of AC-PSG-1 with the lowest acetyl DS. This trend was in accordance with the results in Section 5.3.3.1. Therefore, the chemical properties of the function groups on derivatized PSG may also play an important role in its antioxidant activities.

It has been previously reported that reducing properties are generally associated with the presence of reductones, which can break the free radical chain by donating a hydrogen atom (Tseng et al., 2008). Reductones are also reported to react with certain precursors of peroxide, thus preventing peroxide formation. In this assay, PSG with high donating-hydrogen abilities showed excellent reducing power, while the introduction of the acetyl and carbomethyl group might lead to the diminution of hydroxyl groups and the steric conformation of PSG, which decreased the electron cloud density of active hydroxyl groups, and prevented some active carboxymethyl and acetyl groups binding to the metal ion. Thus, it resulted in a decrease in the reducing power. Tao et al. (2009) reported that the reducing capacity of two carboxymethylated polysaccharide–protein derivatives from *Pleurotus tuber-regium* was enhanced, which could not be found in the present study. This might be due to the different physicochemical properties of different kinds of polysaccharides.

5.3.2.3 Beta-Carotene–Linoleic Acid Assay

The bleaching mechanism of β-carotene is a free radical mediated phenomenon resulting from the hydroperoxides formed from linoleic acid (Jayaprakasha et al., 2001). Beta-carotene undergoes rapid discoloration in the absence of antioxidant. Linoleic acid becomes a free radical with a hydrogen atom abstracted from one of its diallylic methylene groups. The radical formed then attacks the highly unsaturated β-carotene molecules. As a result, β-carotene molecules lose their double bonds by oxidation, and the compound loses its chromophore and characteristic orange color, which can be monitored spectrophotometrically. The presence of different extracts can hinder the extent of β-carotene bleaching by neutralizing the linoleate free radical and other free radicals formed in the system.

The antioxidant activity of PSG and its derivatives determined in terms of percent inhibition in the β-carotene–linoleic acid system is presented in Fig. 5.38C. Among all the detected derivatives, only Ac-PSG-2 and CM-PSG-2, with a higher substitution degree of acetyl and carboxymethyl groups, showed stronger inhibition effects than that of PSG, with 29.3% and 37.12% at 1 mg/mL. These results indicated that acetylated and carboxymethylated derivatives of PSG, with proper substitution, could be used as natural antioxidants to inhibit lipid peroxidation during food processing and preserving.

5.3.3 Immunomodulatory Activity

5.3.3.1 Pinocytic Activity

As shown in Fig. 5.39A, pinocytic activity was enhanced in macrophages after treatment with PSG and all its acetylated derivatives at doses of 6.25–400 μg/mL. Moreover, compared with PSG, all the three acetylated

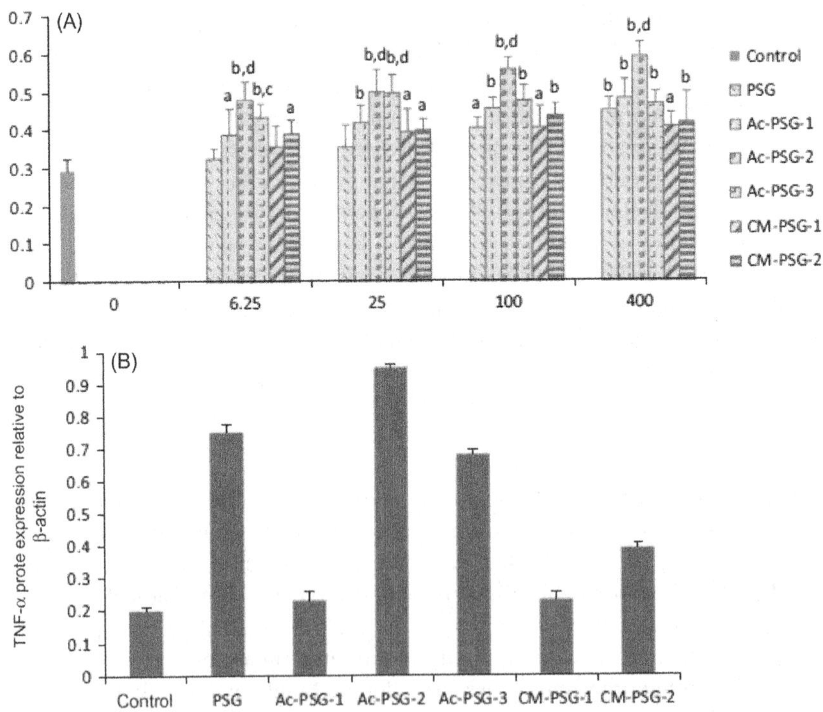

FIGURE 5.39 (A) Effects of PSG and its derivatives on the pinocytic activity of macrophages. Significance was determined using ANOVA. $^{a}P<.05$ and $^{b}P<.01$, compared with the normal control group (PSG 0 mg/kg). $^{c}P<.05$ and $^{d}P<.01$, compared with the PSG treated model group. (B) Effects of PSG and its derivatives on macrophage TNF-a production.

derivatives showed much better effects on the pinocytic activity of mouse peritoneal macrophages. Notably, among them, Ac-PSG-2 exhibited the strongest enhancement effect in a dose-dependent manner. It indicated that the substitution degree of the acetyl group in PSG may be one of the important factors affecting its immunomodulatory activity, and the prime activity was observed for polysaccharides with moderate substitution. Since the bioactivity of polysaccharide is also closely related to their solubility and conformation in solution, we proposed that the introduction of acetyl groups into PSG probably may change its configuration and orientation, and induce different exposure of hydroxyl groups. The moderate substitution may cause better effects to its solubility and conformation in water, thus improving its interaction with specific receptors and stimulating the activity of macrophages.

The pinocytic activities of carboxymethylated samples were also shown in Fig. 5.39A. All the samples exerted a concentration-dependent enhancement effect on the pinocytic activity of macrophages. In addition, the enhancement rates in groups treated with moderate and low doses ($\leq 100\ \mu g/mL$) of CM-PSG samples were higher than those of the PSG group, but no significant difference was observed ($P \geq .05$); while at a higher concentration ($400\ \mu g/mL$), they showed slightly decreased effects compared to the native polysaccharide. These results showed that, similar to Ac-PSG, introduction of a proper number of carboxymethyl groups also contributed to its immunomodulatory activity.

5.3.3.2 *TNF-α Secretion*

As shown in Fig. 5.39B, compared with the native polysaccharide, the higher stimulating effects on secretion of TNF-α protein were only observed for Ac-PSG-2 with a middle substitution degree of acetyl groups, where the level of TNF-α was increased by 25%. However, for other Ac-PSG with low or high substitution degrees of acetyl groups, TNF-α secretion was decreased. These results suggested that a proper substitution degree of acetyl groups should be approximately 0.89, which was consistent with our results from the pinocytic activity assay.

For carboxymethyl-modified polysaccharides, with an increase in the substitution degree of carboxymethyl groups, the TNF-α protein expression of macrophages increased gradually, but was still much lower than that of the native polysaccharide. This might be due to the low substitution degree of carboxymethylation in this study.

5.3.4 Relationship Between Structure and Bioactivity of PSG

Chemical modification of polysaccharide is an important way of obtaining new antioxidants and immunomodulatory agents. In general, it has been suggested that the bioactivities of polysaccharide derivatives might be attributed to their hydroxyl groups and the introduction of substituting groups such as –SH, –COOH, –C–O–, and $-NH_2$, which can change the structure of native polysaccharide (including molecular weight, monosaccharide composition, and configuration) and decrease the intermolecular/intramolecular hydrogen bond. Here, we demonstrated that the introduction of chemically modified groups, acetyl and carboxymethyl groups, into the PSG molecule could significantly change its antioxidant and immunomodulatory activities. Furthermore, Ac-PSG and CM-PSG exhibited some differences in antioxidant and immunomodulatory activities, indicating that the substituting groups may play an essential role in the bioactivities of the derivatives. The introduction of acetyl groups significantly increased the original polysaccharide's water solubility, so that the DPPH radical-scavenging activity and immunomodulatory activity were increased. However, reduced DPPH radical-scavenging activity and reducing power were found for the polysaccharides substituted with a carboxymethyl group.

Except for the substituted groups, some other structural characteristics such as DS, molecular weight, and chain conformation of polysaccharide derivatives are also important factors affecting their bioactivities. In this study, the samples of Ac-PSG-1 and CM-PSG-1 with lower DS showed lower activities than the raw material in the β-carotene–linoleic acid assay. However, higher activities were shown for samples Ac-PSG-2 and CM-PSG-2 with higher and approximate DS. This suggested that although the substitution group was essential, the certain DS was also necessary for the activity. Besides, our present studies also indicated that the molecular weight was another important parameter influencing bioactivities. The molecular weight of the derivatives was decreased due to the slight degradation during the reaction, while higher antioxidant activities were found for the derivatives with lower molecular weight. This might relate to the higher solubility of the derivatives as well. Moreover, research on the effects of conformation of polysaccharide on its bioactivities is also very important. A previous study showed that the relative extended chain conformation of carboxymethylated derivatives was beneficial for enhancing anti-tumor activity (Falch et al., 2000). Thus, in order to understand the relationship between structure and bioactivity of polysaccharide derivatives better, further work on conformation and other structural features should be considered.

These results indicated that the introduction of different functional groups and their DS could affect the molecular weight, polarity, solubility, charge density, or conformation of native polysaccharide, which might lead to a change in bioactivities. This study has laid the foundation for analysis of the structure–activity relationship of polysaccharides. Moreover, the results in turn can also guide the direction of molecular modification, and provide theoretical support for the development of novel food additives.

REFERENCES

Agrawal, P.K., 1992. NMR spectroscopy in the structural elucidation of oligosaccharides and glycosides. Phytochemistry 31, 3307–3330.

Ahrazem, O., Prieto, A., Leal, J.A., Jiménez-Barbero, J., Bernabé, M., 2002. Fungal cell-wall galactomannans isolated from *Geotrichum* spp. and their teleomorphs, *Dipodascus* and *Galactomyces*. Carbohydr. Res. 337, 2347–2351.

Akira, S., Tokeda, K., Kaisho, T., 2001. Toll-like receptors: critical proteins linking innate and acquired immunity. Nat. Immunol. 2, 675–680.

Anderson, K.V., 2000. Toll signaling pathways in the innate immune response. Curr. Opin. Immunol. 12, 13–19.

Anuradha, C., Selvam, R., 1993. Effect of oral methionine on tissue lipid peroxidation and antioxidants in alloxan-induced diabetic rats. J. Nutr. Biochem. 4, 212–217.

Asehnoune, K., Strassheim, D., Mitra, S., Kim, J.Y., Abraham, E., 2004. Involvement of reactive oxygen species in Toll-like receptor 4-dependent activation of NF-κB. J. Immunol. 172, 2522–2529.

Baeuerle, P.A., Baltimore, D., 1988. IKB: a specific inhibitor of the NF-κB transcription factor. Science 242, 540–546.

Baeuerle, P.A., Henkel, T., 1994. Function and activation of NF-kappaB in the immune system. Annu. Rev. Immunol. 12, 141–179.

Bao, X., Liu, C., Fang, J., Li, X., 2001. Structural and immunological studies of a major polysaccharide from spores of *Ganoderma lucidum* (Fr.) Karst. Carbohydr. Res. 332, 67–74.

Bao, X.F., Wang, X.S., Dong, Q., Fang, J.N., Li, X.Y., 2002. Structural features of immunologically active polysaccharides from *Ganoderma lucidum*. Phytochemistry 59, 175–181.

Barker, S., Bourne, E., Stacey, M., Whiffen, D.H., 1954. Infra-red spectra of carbohydrates. Part I. Some derivatives of D-glucopyranose. J. Chem. Soc. (Resumed)171–176. Available from: http://dx.doi.org/10.1039/JR9540000171.

Barton, G.M., Medzhitov, R., 2003. Toll-like receptor signaling pathways. Science 300, 1524.

Beaudin, A.E., Abarinov, E.V., Noden, D.M., Perry, C.A., Chu, S., Stabler, S.P., et al., 2011. Shmt1 and de novo thymidylate biosynthesis underlie folate-responsive neural tube defects in mice. Am. J. Clin. Nutr. 93, 789–798.

Benzie, I.F., Strain, J., 1996. The ferric reducing ability of plasma (FRAP) as a measure of "antioxidant power": the FRAP assay. Anal. Biochem. 239, 70–76.

Beutler, B., 2000. Tlr4: central component of the sole mammalian LPS sensor. Curr. Opin. Immunol. 12, 20–26.

Bhunia, S.K., Dey, B., Maity, K.K., Patra, S., Mandal, S., Maiti, S., et al., 2010. Structural characterization of an immunoenhancing heteroglycan isolated from an aqueous extract of an edible mushroom, *Lentinus squarrosulus* (Mont.) Singer. Carbohydr. Res. 345, 2542–2549.

Boden, G., Shulman, G., 2002. Free fatty acids in obesity and type 2 diabetes: defining their role in the development of insulin resistance and β-cell dysfunction. Eur. J. Clin. Invest. 32, 14–23.

Bohn, J.A., BeMiller, J.N., 1995. (1→3)-β-D-Glucans as biological response modifiers: a review of structure-functional activity relationships. Carbohydr. Polym. 28, 3–14.

Bokoch, G.M., 1995. Regulation of the phagocyte respiratory burst by small GTP-binding proteins. Trends Cell Biol. 5, 109–113.

Bosco, E.E., Mulloy, J.C., Zheng, Y., 2009. Rac1 GTPase: a "Rac" of all trades. Cell. Mol. Life Sci. 66, 370–374.

Boyle, P., Levin, B., 2008. World Cancer Report 2008. IARC Press, International Agency for Research on Cancer.

Brezar, V., Carel, J.C., Boitard, C., Mallone, R., 2011. Beyond the hormone: insulin as an autoimmune target in type 1 diabetes. Endocr. Rev. 32, 623–669.

Brüne, B., 2003. Nitric oxide: NO apoptosis or turning it ON? Cell Death Differ. 10, 864–869.

Cantley, L.C., 2002. The phosphoinositide 3-kinase pathway. Science 296, 1655.

Caponigro, F., French, R.C., Kaye, S.B., 1997. Protein kinase C: a worthwhile target for anticancer drugs? Anticancer Drugs 8, 26–33.

Carpenter, C.L., Cantley, L.C., 1996. Phosphoinositide kinases. Curr. Opin. Cell Biol. 8, 153–158.

Carpita, N.C., Shea, E.M., 1989. Linkage structure of carbohydrates by gas chromatography-mass spectrometry (GC-MS) of partially methylated alditol acetates. Analysis of Carbohydrates by GLC and MS, pp. 157–216.

Chakraborty, I., Mondal, S., Pramanik, M., Rout, D., Islam, S.S., 2004. Structural investigation of a water-soluble glucan from an edible mushroom, *Astraeus hygrometricus*. Carbohydr. Res. 339, 2249–2254.

Chandra, K., Ghosh, K., Ojha, A.K., Islam, S.S., 2009. Chemical analysis of a polysaccharide of unripe (green) tomato (*Lycopersicon esculentum*). Carbohydr. Res. 344, 2188–2194.

Chang, C.J., Tzeng, T.F., Liou, S.S., Chang, Y.S., Liu, I.M., 2011. Regulation of lipid disorders by ethanol extracts from *Zingiber zerumbet* in high-fat diet-induced rats. Food Chem. 132, 460–467.

Chang, Z.Q., Oh, B.C., Lee, S.P., Rhee, M.H., Park, S.C., 2008. Comparative immunomodulating activities of polysaccharides isolated from *Phellinus* spp. on cell-mediated immunity. Phytother. Res. 22, 1396–1399.

Chen, F., Wang, W., El-Deiry, W.S., 2010a. Current strategies to target p53 in cancer. Biochem. Pharmacol. 80, 724–730.

Chen, X., Lin, Z., Ye, Y., Zhang, R., Yin, J., Jiang, Y., et al., 2010b. Suppression of diabetes in non-obese diabetic (NOD) mice by oral administration of water-soluble and alkali-soluble polysaccharide conjugates prepared from green tea. Carbohydr. Polym. 82, 28–33.

Chen, J., Zhang, L., Yu, D., Zhu, R., 2000. The chemical structure and solution properties of polysaccharides from *Ganoderma lucidum* mycelium. Chem. J. Chin. Univ. 21, 961–964.

Chen, T.C., Hinton, D.R., Zidovetzki, R., Hofman, F.M., 1998. Up-regulation of the cAMP/PKA pathway inhibits proliferation, induces differentiation, and leads to apoptosis in malignant gliomas. Lab. Invest. 78, 165–174.

Chen, X., Wang, Y., Wu, Y., Han, B., Zhu, Y., Tang, X., et al., 2011. Green tea polysaccharide-conjugates protect human umbilical vein endothelial cells against impairments triggered by high glucose. Int. J. Biol. Macromol. 49, 50–54.

Chen, Z., Soo, M.Y., Srinivasan, N., Tan, B.K., Chan, S.H., 2009. Activation of macrophages by polysaccharide-protein complex from *Lycium barbarum* L. Phytother. Res. 23, 1116–1122.

Cheng, A., Wan, F., Wang, J., Jin, Z., Xu, X., 2008. Macrophage immunomodulatory activity of polysaccharides isolated from *Glycyrrhiza uralensis* fish. Int. Immunopharmacol. 8, 43–50.

Chihara, G., Maeda, Y., Hamuro, J., Sasaki, T., Fukuoka, F., 1969. Inhibition of mouse sarcoma 180 by polysaccharides from *Lentinus edodes* (Berk.) sing. Nature 222, 687–688.

Chin, K.V., Yang, W.L., Ravatn, R., Kita, T., Reitman, E., Vettori, D., et al., 2002. Reinventing the wheel of cyclic AMP. Ann. N.Y. Acad. Sci. 968, 49–64.

Choudhary, A.K., 2012. Role of Heme Arginate in Modulation of Inflammation and Type 2 Diabetes. University of Edinburgh.

Choudhuri, T., Pal, S., Agwarwal, M.L., Das, T., Sa, G., 2002. Curcumin induces apoptosis in human breast cancer cells through p53-dependent Bax induction. FEBS Lett. 512, 334–340.

Cort, A., Timur, M., Dursun, E., Kucuksayan, E., Aslan, M., Ozben, T., 2012. Effects of N-acetylcystein on bleomycin-induced apoptosis in malignant testicular germ cell tumors. J. Physiol. Biochem. 68, 555–562.

Cui, S.W., 2005. Food Carbohydrates: Chemistry, Physical Properties, and Applications. CRC Press.

Cui, Z.G., Piao, J.L., Rehman, M.U., Ogawa, R., Li, P., Zhao, Q.-L., et al., 2014. Molecular mechanisms of hyperthermia-induced apoptosis enhanced by withaferin A. Eur. J. Pharmacol. 723, 99–107.

Das, A.B., Singh, G., Singh, S., Riar, C.S., 2010a. Effect of acetylation and dual modification on physico-chemical, rheological and morphological characteristics of sweet potato (*Ipomoea batatas*) starch. Carbohydr. Polym. 80, 725–732.

Das, D., Mondal, S., Roy, S.K., Maiti, D., Bhunia, B., Maiti, T.K., et al., 2010b. A (1→6)-β-glucan from a somatic hybrid of *Pleurotus florida* and *Volvariella volvacea*: isolation, characterization, and study of immunoenhancing properties. Carbohydr. Res. 345, 974–978.

Das, D., Mondal, S., Roy, S.K., Maiti, D., Bhunia, B., Maiti, T.K., et al., 2009. Isolation and characterization of a heteropolysaccharide from the corm of *Amorphophallus campanulatus*. Carbohydr. Res. 344, 2581–2585.

Dean, N., McKay, R., Miraglia, L., Howard, R., Cooper, S., Giddings, J., et al., 1996. Inhibition of growth of human tumor cell lines in nude mice by an antisense oligonucleotide inhibitor of protein kinase C-α expression. Cancer Res. 56, 3499–3507.

DeLeo, F.R., Renee, J., McCormick, S., Nakamura, M., Apicella, M., Weiss, J.P., et al., 1998. Neutrophils exposed to bacterial lipopolysaccharide upregulate NADPH oxidase assembly. J. Clin. Invest. 101, 455.

Dooley, N.P., Baltuch, G.H., Groome, N., Villemure, J.G., Yong, V.W., 1998. Apoptosis is induced in glioma cells by antisense oligonucleotides to protein kinase Cα and is enhanced by cycloheximide. NeuroReport 9, 1727–1733.

Doonan, F., Cotter, T.G., 2008. Morphological assessment of apoptosis. Methods 44, 200–204.

Ehlers, M.R., 2000. CR3: a general purpose adhesion-recognition receptor essential for innate immunity. Microbes Infect. 2, 289–294.

Ehrke, M.J., 2003. Immunomodulation in cancer therapeutics. Int. Immunopharmacol. 3, 1105–1119.

El-Mekkawy, S., Meselhy, M.R., Nakamura, N., Tezuka, Y., Hattori, M., Kakiuchi, N., et al., 1998. Anti-HIV-1 and anti-HIV-1-protease substances from *Ganoderma lucidum*. Phytochemistry 49, 1651–1657.

Eo, S.K., Kim, Y.S., Lee, C.K., Han, S.S., 2000. Possible mode of antiviral activity of acidic protein bound polysaccharide isolated from *Ganoderma lucidum* on herpes simplex viruses. J. Ethnopharmacol. 72, 475–481.

Falch, B.H., Espevik, T., Ryan, L., Stokke, B.T., 2000. The cytokine stimulating activity of (1→3)-β-D-glucans is dependent on the triple helix conformation. Carbohydr. Res. 329, 587–596.

Feinglos, M.N., Bethel, M.A., 2008. Type 2 Diabetes Mellitus: An Evidence-Based Approach to Practical Management. Humana Press Inc.

Fidler, I., Kleinerman, E., 1993. Therapy of cancer metastasis by systemic activation of macrophages: from the bench to the clinic: macrophages and cancer. Res. Immunol. 144, 284–287.

Finkel, T., 1998. Oxygen radicals and signaling. Curr. Opin. Cell Biol. 10, 248–253.

Finkelstein, J.D., 1990. Methionine metabolism in mammals. J. Nutr. Biochem. 1, 228–237.

Fishman, P., Bar-Yehuda, S., Madi, L., Rath-Wolfson, L., Ochaion, A., Cohen, S., et al., 2006. The PI3K-NF-κB signal transduction pathway is involved in mediating the anti-inflammatory effect of IB-MECA in adjuvant-induced arthritis. Arthritis Res. Ther. 8, 33.

Franz, G., 1989. Polysaccharides in pharmacy: current applications and future concepts. Planta Med. 55, 493–497.

Freeman, B.A., Crapo, J.D., 1982. Biology of disease: free radicals and tissue injury. Lab. Invest. 47, 412.

Fruman, D.A., Cantley, L.C., 2002. Phosphoinositide 3-kinase in immunological systems. Semin. Immunol. 14, 7–18.

Fujii, T., García-Bermejo, Ma.L., Bernabó, J.L., Caamaño, J., Ohba, M., Kuroki, T., et al., 2000. Involvement of protein kinase C δ (PKCδ) in phorbol ester-induced apoptosis in LNCaP prostate cancer cells. Lack of proteolytic cleavage of PKCδ. J. Biol. Chem. 275, 7574–7582.

Furusawa, E., Chou, S., Furusawa, S., Hirazumi, A., Dang, Y., 1992. Antitumour activity of *Ganoderma lucidum*, an edible mushroom, on intraperitoneally implanted Lewis lung carcinoma in synergenic mice. Phytother. Res. 6, 300–304.

Gao, Y., Chan, E., Zhou, S., 2004. Immunomodulating activities of *Ganoderma*, a mushroom with medicinal properties. Food Rev. Int. 20, 123–161.

Gao, Y., Tang, W., Gao, H., Chan, E., Lan, J., Li, X., et al., 2005. Antimicrobial activity of the medicinal mushroom *Ganoderma*. Food Rev. Int. 21, 211–229.

Ge, Q., Zhang, A.-q, Sun, Pl, 2009. Structural investigation of a novel water-soluble heteropolysaccharide from the fruiting bodies of *Phellinus baumii* Pilát. Food Chem. 114, 391–395.

Gerwig, G.J., Kamerling, J.P., Vliegenthart, J.F., 1979. Determination of the absolute configuration of monosaccharides in complex carbohydrates by capillary GLC. Carbohydr. Res. 77, 1–7.

Ghosh, J.C., Dohi, T., Raskett, C.M., Kowalik, T.F., Altieri, D.C., 2006. Activated checkpoint kinase 2 provides a survival signal for tumor cells. Cancer Res. 66, 11576–11579.

Ghosh, K., Chandra, K., Roy, S.K., Mondal, S., Maiti, D., Das, D., et al., 2008. Structural investigation of a polysaccharide (Fr. I) isolated from the aqueous extract of an edible mushroom, *Volvariella diplasia*. Carbohydr. Res. 343, 1071–1078.

Ghosh, S., Reed, W.F., 1995. New characteristic signatures from time-dependent static light scattering during polymer depolymerization, with application to proteoglycan subunit degradation. Biopolymers 35, 435–450.

Gogvadze, V., Orrenius, S., Zhivotovsky, B., 2006. Multiple pathways of cytochrome c release from mitochondria in apoptosis. BBA-Bioenergetics 1757, 639–647.

Gökmen-Polar, Y., Murray, N.R., Velasco, M.A., Gatalica, Z., Fields, A.P., 2001. Elevated protein kinase C βII is an early promotive event in colon carcinogenesis. Cancer Res. 61, 1375–1381.

Greis, K.D., Hart, G.W., 1998. Analytical methods for the study of O-GlcNAc glycoproteins and glycopeptides. Methods Mol. Biol. 76, 19–33.

Griffin, J.L., Nicholls, A.W., 2006. Metabolomics as a functional genomic tool for understanding lipid dysfunction in diabetes, obesity and related disorders. Pharmacogenomics 7, 1095–1107.

Guéant, J.L., Namour, F., Guéant-Rodriguez, R.M., Daval, J.L., 2013. Folate and fetal programming: a play in epigenomics? Trends Endocrinol. Metab. 24, 279–289.

Haddad, J.J., 2002. Cytokines and related receptor-mediated signaling pathways. Biochem. Biophys. Res. Commun. 297, 700–713.

Haimovitz-Friedman, A., Balaban, N., McLoughlin, M., Ehleiter, D., Michaeli, J., Vlodavsky, I., et al., 1994. Protein kinase C mediates basic fibroblast growth factor protection of endothelial cells against radiation-induced apoptosis. Cancer Res. 54, 2591–2597.

Han, H.F., Nakamura, N., Hattori, M., 2006. Protective effects of an acidic polysaccharide isolated from fruiting bodies of *Ganoderma lucidum* against murine hepatic injury induced by *Propionibacterium acnes* and lipopolysaccharide. J. Nat. Med. 60, 295–302.

Hanahan, D., Weinberg, R.A., 2000. The hallmarks of cancer. Cell 100, 57–70.

Hansen, S.H., 2001. The role of taurine in diabetes and the development of diabetic complications. Diabetes Metab. Res. 17, 330–346.

Hanson, A.J., Nahreini, P., Andreatta, C., Yan, X.D., Prasad, K.N., 2005. Role of the adenosine 3′, 5′-cyclic monophosphate (cAMP) in enhancing the efficacy of siRNA-mediated gene silencing in neuroblastoma cells. Oncogene 24, 4149–4154.

Hayden, M.S., Ghosh, S., 2004. Signaling to NF-κB. Genes Dev. 18, 2195–2224.

Hengartner, M.O., 2000. The biochemistry of apoptosis. Nature 407, 770–776.

Hikino, H., Konno, C., Mirin, Y., Hayashi, T., 1985. Isolation and hypoglycemic activity of ganoderans A and B, glycans of *Ganoderma lucidum* fruit bodies. Planta Med. 51, 339–340.

Hong, Y.H., Peng, H.B., La Fata, V., Liao, J.K., 1997. Hydrogen peroxide-mediated transcriptional induction of macrophage colony-stimulating factor by TGF-beta1. J. Immunol. 159, 2418–2423.

Hsu, H.Y., Hua, K.F., Lin, C.C., Lin, C.H., Hsu, J., Wong, C.H., 2004a. Extract of Reishi polysaccharides induces cytokine expression via TLR4-modulated protein kinase signaling pathways. J. Immunol. 173, 5989.

Hsu, Y.L., Kuo, P.L., Liu, C.F., Lin, C.C., 2004b. Acacetin-induced cell cycle arrest and apoptosis in human non-small cell lung cancer A549 cells. Cancer Lett. 212, 53–60.

Ito, K., Hirao, A., Arai, F., Takubo, K., Matsuoka, S., Miyamoto, K., et al., 2006. Reactive oxygen species act through p38 MAPK to limit the lifespan of hematopoietic stem cells. Nat. Med. 12, 446–451.

Iwaoka, S., Nakamura, T., Takano, S., Tsuchiya, S., Aramaki, Y., 2006. Cationic liposomes induce apoptosis through p38 MAP kinase–caspase-8-Bid pathway in macrophage-like RAW264.7 cells. J. Leukoc. Biol. 79, 184–191.

Jayaprakasha, G., Singh, R., Sakariah, K., 2001. Antioxidant activity of grape seed (*Vitis vinifera*) extracts on peroxidation models in vitro. Food Chem. 73, 285–290.

Jemal, A., Bray, F., Center, M.M., Ferlay, J., Ward, E., Forman, D., 2011. Global cancer statistics. CA-Cancer J. Clin. 61, 69–90.

Karin, M., Ben-Neriah, Y., 2000. Phosphorylation meets ubiquitination: the control of NF-κB activity. Annu. Rev. Immunol. 18, 621–663.

Karin, M., Hunter, T., 1995. Transcriptional control by protein phosphorylation: signal transmission from the cell surface to the nucleus. Curr. Biol. 5, 747–757.

Kaul, N., Gopalakrishna, R., Gundimeda, U., Choi, J., Forman, H.J., 1998. Role of protein kinase C in basal and hydrogen peroxide-stimulated NF-κB activation in the murine macrophage J774A.1 cell line. Arch. Biochem. Biophys. 350, 79–86.

Kawai, T., Akira, S., 2006. TLR signaling. Cell Death Differ. 13, 816–825.

Ke, Y.D., Delerue, F., Gladbach, A., Götz, J., Ittner, L.M., 2009. Experimental diabetes mellitus exacerbates tau pathology in a transgenic mouse model of Alzheimer's disease. PLoS One. 4, e7917.

Kim, A.J., Kim, Y.O., Shim, J.S., Hwang, J.K., 2007. Immunostimulating activity of crude polysaccharide extract isolated from *Curcuma xanthorrhiza* Roxb. Biosci. Biotechnol. Biochem. 71, 1428–1438.

Kim, H.S., Kim, J.Y., Kang, J.S., Kim, H.M., Kim, Y.O., Hong, I.P., et al., 2010. Cordlan polysaccharide isolated from mushroom *Cordyceps militaris* induces dendritic cell maturation through toll-like receptor 4 signalings. Food Chem. Toxicol. 48, 1926–1933.

Kim, H.S., Kim, Y.J., Lee, H.K., Ryu, H.S., Kim, J.S., Yoon, M.J., et al., 2012. Activation of macrophages by polysaccharide isolated from *Paecilomyces cicadae* through toll-like receptor 4. Food Chem. Toxicol. 50, 3190–3197.

Kim, H.W., Shim, M.J., Choi, E.C., Kim, B.K., 1997. Inhibition of cytopathic effect of human immunodeficiency virus-1 by water-soluble extract of *Ganoderma lucidum*. Arch. Pharm. Res. 20, 425–431.

Kim, J., Park, E., 2002. Cytotoxic anticancer candidates from natural resources. Curr. Med. Chem. Anticancer Agents 2, 485–537.

Kim, N., Kim, K., Choi, B.Y., Lee, D., Shin, Y.S., Bang, K.H., et al., 2011. Metabolomic approach for age discrimination of *Panax ginseng* using UPLC-Q-Tof MS. J. Agric. Food Chem. 59, 10435–10441.

Kim, R., 2005. Recent advances in understanding the cell death pathways activated by anticancer therapy. Cancer 103, 1551–1560.

Klepacki, J., Brunner, N., Schmitz, V., Klawitter, J., Christians, U., Klawitter, J., 2013. Development and validation of an LC-MS/MS assay for the quantification of the *trans*-methylation pathway intermediates S-adenosylmethionine and S-adenosylhomocysteine in human plasma. Clin. Chim. Acta 421, 91–97.

Knight, J.A., 2000. Review: free radicals, antioxidants, and the immune system. Ann. Clin. Lab. Sci. 30, 145–158.

Kobayashi, H., Yoshida, R., Kanada, Y., Fukuda, Y., Yagyu, T., Inagaki, K., et al., 2005. Suppressing effects of daily oral supplementation of beta-glucan extracted from *Agaricus blazei* Murill on spontaneous and peritoneal disseminated metastasis in mouse model. J. Cancer Res. Clin. 131, 527–538.

Koyasu, S., 2003. The role of PI3K in immune cells. Nat. Immunol. 4, 313–319.

Laws, A.P., Chadha, M.J., Chacon-Romero, M., Marshall, V.M., Maqsood, M., 2008. Determination of the structure and molecular weights of the exopolysaccharide produced by *Lactobacillus acidophilus* 5e2 when grown on different carbon feeds. Carbohydr. Res. 343, 301–307.

Leadsham, J.E., Gourlay, C.W., 2010. cAMP/PKA signaling balances respiratory activity with mitochondria dependent apoptosis via transcriptional regulation. BMC Cell Biol. 11, 1.

Leadsham, J.E., Miller, K., Ayscough, K.R., Colombo, S., Martegani, E., Sudbery, P., et al., 2009. Whi2p links nutritional sensing to actin-dependent Ras-cAMP-PKA regulation and apoptosis in yeast. J. Cell Sci. 122, 706–715.

Lee, J.Y., Kim, J.Y., Lee, Y.G., Rhee, M.H., Hong, E.K., Cho, J.Y., 2008. Molecular mechanism of macrophage activation by exopolysaccharides from liquid culture of *Lentinus edodes*. J. Microbiol. Biotechnol. 18, 355–364.

Lee, K.Y., Jeon, Y.J., 2005. Macrophage activation by polysaccharide isolated from Astragalus membranaceus. Int. Immunopharmacol. 5, 1225–1233.

Lee, Y.G., Byeon, S.E., Kim, J.Y., Lee, J.Y., Rhee, M.H., Hong, S., et al., 2007. Immunomodulatory effect of *Hibiscus cannabinus* extract on macrophage functions. J. Ethnopharmacol. 113, 62–71.

Leiro, J.M., Castro, R., Arranz, J.A., Lamas, J., 2007. Immunomodulating activities of acidic sulphated polysaccharides obtained from the seaweed *Ulva rigida* C. Agardh. Int. Immunopharmacol. 7, 879–888.

Li, H., Kim, J.Y., Hyeon, J., Lee, H.J., Ryu, J.H., 2011a. In vitro antiinflammatory activity of a new sesquiterpene lactone isolated from *Siegesbeckia glabrescens*. Phytother. Res. 25, 1323–1327.

Li, W.J., Chen, Y., Nie, S.P., Xie, M.Y., He, M., Zhang, S.S., et al., 2011b. *Ganoderma atrum* polysaccharide induces anti-tumor activity via the mitochondrial apoptotic pathway related to activation of host immune response. J. Cell. Biochem. 112, 860–871.

Li, Q., Verma, I.M., 2002. NF-κB regulation in the immune system. Nat. Rev. Immunol. 2, 725–734.

Li, W.J., Nie, S.P., Chen, Y., Xie, M.Y., He, M., Yu, Q., et al., 2010a. *Ganoderma atrum* polysaccharide protects cardiomyocytes against anoxia/reoxygenation-induced oxidative stress by mitochondrial pathway. J. Cell. Biochem. 110, 191–200.

Li, X., Xu, W., Chen, J., 2010b. Polysaccharide purified from *Polyporus umbellatus* (Per) Fr induces the activation and maturation of murine bone-derived dendritic cells via toll-like receptor 4. Cell. Immunol. 265, 50–56.

Li, X.Q., Xu, W., 2011. TLR4-mediated activation of macrophages by the polysaccharide fraction from *Polyporus umbellatus* (pers.) Fries. J. Ethnopharmacol. 135, 1–6.

Liu, J., Shimizu, K., Konishi, F., Noda, K., Kumamoto, S., Kurashiki, K., et al., 2007. Anti-androgenic activities of the triterpenoids fraction of *Ganoderma lucidum*. Food Chem. 100, 1691–1696.

Liu, W., Wang, H., Pang, X., Yao, W., Gao, X., 2010. Characterization and antioxidant activity of two low-molecular-weight polysaccharides purified from the fruiting bodies of *Ganoderma lucidum*. Int. J. Biol. Macromol. 46, 451–457.

Liu, X., Zhang, S., Lu, X., Zheng, S., Li, F., Xiong, Z., 2012. Metabonomic study on the anti-osteoporosis effect of Rhizoma Drynariae and its action mechanism using ultra-performance liquid chromatography–tandem mass spectrometry. J. Ethnopharmacol. 139, 311–317.

Lowe, S.W., Lin, A.W., 2000. Apoptosis in cancer. Carcinogenesis 21, 485–495.

Lu, Y.X., Zhang, Q., Li, J., Sun, Y.X., Wang, L.Y., Cheng, W.M., et al., 2010. Antidiabetic effects of total flavonoids from Litsea Coreana leve on fat-fed, streptozotocin-induced type 2 diabetic rats. Am. J. Chinese Med. 38, 713–725.

Luo, A., He, X., Zhou, S., Fan, Y., He, T., Chun, Z., 2009. In vitro antioxidant activities of a water-soluble polysaccharide derived from *Dendrobium nobile* Lindl. extracts. Int. J. Biol. Macromol. 45, 359–363.

Mackay, H., Twelves, C., 2003. Protein kinase C: a target for anticancer drugs? Endocr. Relat. Cancer 10, 389–396.

MacKenzie, R., Trulson, M., 1978. Regional accumulation of tryptophan and serotonin metabolism following tryptophan loading in diabetic rats. J. Neurochem. 31, 157–160.

Maiti, D., Chandra, K., Mondal, S., Ojha, A.K., Das, D., Roy, S.K., et al., 2008. Isolation and characterization of a heteroglycan from the fruits of *Astraeus hygrometricus*. Carbohydr. Res. 343, 817–824.

Maity, K.K., Patra, S., Dey, B., Bhunia, S.K., Mandal, S., Das, D., et al., 2011. A heteropolysaccharide from aqueous extract of an edible mushroom, *Pleurotus ostreatus* cultivar: structural and biological studies. Carbohydr. Res. 346, 366–372.

Majumder, P.K., Pandey, P., Sun, X., Cheng, K., Datta, R., Saxena, S., et al., 2000. Mitochondrial translocation of protein kinase C δ in phorbol ester-induced cytochrome c release and apoptosis. J. Biol. Chem. 275, 21793–21796.

Mancini, M., Nicholson, D.W., Roy, S., Thornberry, N.A., Peterson, E.P., Casciola-Rosen, L.A., et al., 1998. The caspase-3 precursor has a cytosolic and mitochondrial distribution: implications for apoptotic signaling. J. Cell Biol. 140, 1485–1495.

Mandal, S., Maity, K.K., Bhunia, S.K., Dey, B., Patra, S., Sikdar, S.R., et al., 2010. Chemical analysis of new water-soluble (1→6)-, (1→4)-α, β-glucan and water-insoluble (1→3)-, (1→4)-β-glucan (Calocyban) from alkaline extract of an edible mushroom, *Calocybe indica* (Dudh Chattu). Carbohydr. Res. 345, 2657–2663.

Martin, T., 1998. Phosphoinositide lipids as signaling molecules: common themes for signal transduction, cytoskeletal regulation, and membrane trafficking. Annu. Rev. Cell Dev. Biol. 14, 231–264.

Mathlouthi, M., Koenig, J.L., 1987. Vibrational spectra of carbohydrates. Adv. Carbohydr. Chem. Biochem. 44, 7–89.

McCubrey, J.A., LaHair, M.M., Franklin, R.A., 2006. Reactive oxygen species-induced activation of the MAP kinase signaling pathways. Antioxid. Redox Signal. 8, 1775–1789.

Medzhitov, R., 2001. Toll-like receptors and innate immunity. Nat. Rev. Immunol. 1, 135–145.

Medzhitov, R., Janeway Jr, C., 2000. The Toll receptor family and microbial recognition. Trends Microbiol. 8, 452–456.

Meyer, M., Schreck, R., Baeuerle, P.A., 1993. H_2O_2 and antioxidants have opposite effects on activation of NF-kappa B and AP-1 in intact cells: AP-1 as secondary antioxidant-responsive factor. EMBO J. 12, 2005.

Mitchell, M.S., 2003. Immunotherapy as part of combinations for the treatment of cancer. Int. Immunopharmacol. 3, 1051–1059.

Miyazak, T., Nishijima, M., 1981. Structual examination of a water soluble antitumor polysaccharide of *Ganoderma lucidum*. Chem. Pharm. Bull. 29, 3611–3616.

Mooradian, A.D., 2009. Dyslipidemia in type 2 diabetes mellitus. Nat. Clin. Pract. Endocrinol. Metab. 5, 150–159.

Moresco, E.M.Y., LaVine, D., Beutler, B., 2011. Toll-like receptors. Curr. Biol. 21, R488–R493.

Morrison, D.C., Jacobs, D.M., 1976. Binding of polymyxin B to the lipid A portion of bacterial lipopolysaccharides. Immunochemistry 13, 813–818.

Muoio, D.M., Newgard, C.B., 2008. Molecular and metabolic mechanisms of insulin resistance and β-cell failure in type 2 diabetes. Nat. Rev. Mol. Cell Biol. 9, 193–205.

Nakazato, H., Koike, A., Saji, S., Ogawa, N., Sakamoto, J., Nakazato, H., et al., 1994. Efficacy of immunochemotherapy as adjuvant treatment after curative resection of gastric cancer. Lancet 343, 1122–1126.

Nakazato, T., Ito, K., Ikeda, Y., Kizaki, M., 2005. Green tea component, catechin, induces apoptosis of human malignant B cells via production of reactive oxygen species. Clin. Cancer Res. 11, 6040–6049.

Netea, M.G., Ferwerda, G., van der Graaf, C.A., Van der Meer, J.W., Kullberg, B.J., 2006. Recognition of fungal pathogens by toll-like receptors. Curr. Pharm. Des. 12, 4195–4201.

Nikolić, D., Li, J., Breemen, R.B., 2014. Metabolism of Nω-methylserotonin, a serotonergic constituent of black cohosh (*Cimicifuga racemosa*, L.(Nutt.)), by human liver microsomes. Biomed. Chromatogr. 28, 1647–1651.

Nishizuka, Y., 1984. The role of protein kinase C in cell surface signal transduction and tumour promotion. Nature 693–698.

O'Brian, C.A., Vogel, V.G., Singletary, S.E., Ward, N.E., 1989. Elevated protein kinase C expression in human breast tumor biopsies relative to normal breast tissue. Cancer Res. 49, 3215–3217.

Ojha, A.K., Maiti, D., Chandra, K., Mondal, S., Roy, D.D.S.K., Ghosh, K., et al., 2008. Structural assignment of a heteropolysaccharide isolated from the gum of Cochlospermum religiosum (Katira gum). Carbohydr. Res. 343, 1222–1231.

Okkenhaug, K., Vanhaesebroeck, B., 2003. PI3K in lymphocyte development, differentiation and activation. Nat. Rev. Immunol. 3, 317–330.

Özkaya, D., Nazıroğlu, M., Armağan, A., Demirel, A., Köroglu, B.K., Çolakoğlu, N., et al., 2011. Dietary vitamin C and E modulates oxidative stress induced-kidney and lens injury in diabetic aged male rats through modulating glucose homeostasis and antioxidant systems. Cell Biochem. Funct. 29, 287–293.

Palomba, S., Falbo, A., Giallauria, F., Russo, T., Tolino, A., Zullo, F., et al., 2010. Effects of metformin with or without supplementation with folate on homocysteine levels and vascular endothelium of women with polycystic ovary syndrome. Diabetes Care 33, 246–251.

Peng, Y., Zhang, L., Zeng, F., Kennedy, J.F., 2005. Structure and antitumor activities of the water-soluble polysaccharides from *Ganoderma tsugae* mycelium. Carbohydr. Polym. 59, 385–392.

Perepelov, A.V., Han, W., Sof'ya, N.S., Shevelev, S.D., Shashkov, A.S., Feng, L., et al., 2007. Structure of the O-polysaccharide of *Escherichia coli* O150 containing 2-acetamido-4-O-[(S)-1-carboxyethyl]-2-deoxy-D-glucose. Carbohydr. Res. 342, 648–652.

Petersen, K.F., Dufour, S., Befroy, D., Lehrke, M., Hendler, R.E., Shulman, G.I., 2005. Reversal of nonalcoholic hepatic steatosis, hepatic insulin resistance, and hyperglycemia by moderate weight reduction in patients with type 2 diabetes. Diabetes 54, 603–608.

Pham, P.C.T., Pham, P.M.T., Pham, S.V., Miller, J.M., Pham, P.T.T., 2007. Hypomagnesemia in patients with type 2 diabetes. Clin. J. Am. Soc. Nephrol. 2, 366–373.

Pinton, P., Giorgi, C., Pandolfi, P., 2011. The role of PML in the control of apoptotic cell fate: a new key player at ER–mitochondria sites. Cell Death Differ. 18, 1450–1456.

Pooja, V.S., Samanta, K., 2011. Hypoglycemic activity of methanolic extract of *Tectona grandis* linn. Root in alloxan induced diabetic rats. J. Appl. Pharm. Sci. 1, 106–109.

Portt, L., Norman, G., Clapp, C., Greenwood, M., Greenwood, M.T., 2011. Anti-apoptosis and cell survival: a review. BBA-Mol. Cell Res. 1813, 238–259.

Raghavan, V., Garber, A.J., 2008. Postprandial Hyperglycemia. Humana Press, Durham, NC.

Rao, K., 2001. MAP kinase activation in macrophages. J. Leukoc. Biol. 69, 3–10.

Reed, M.J., Meszaros, K., Entes, L.J., Claypool, M.D., Pinkett, J.G., Gadbois, T.M., et al., 2000. A new rat model of type 2 diabetes: the fat-fed, streptozotocin-treated rat. Metabolism 49, 1390–1394.

Rein, D.T., Kurbacher, C.M., 2001. The role of chemotherapy in invasive cancer of the cervix uteri: current standards and future prospects. Anticancer Drugs 12, 787–795.

Ren, J.L., Sun, R.C., Peng, F., 2008. Carboxymethylation of hemicelluloses isolated from sugarcane bagasse. Polym. Degrad. Stab. 93, 786–793.

Rhee, S.G., 1999. Redox signaling: hydrogen peroxide as intracellular messenger. Exp. Mol. Med. 31, 53.

Roberts, R.E., Alaghband-Zadeh, J., Le Roux, C.W., 2011. The role of bile acids in gut-hormone-induced weight loss after bariatric surgery: implications for appetite control and diabetes. Handbook of Behavior, Food and Nutrition. Springer, pp. 1317–1330.

Rosenson, R.S., Otvos, J.D., Freedman, D.S., 2002. Relations of lipoprotein subclass levels and low-density lipoprotein size to progression of coronary artery disease in the Pravastatin Limitation of Atherosclerosis in the Coronary Arteries (PLAC-I) trial. Am. J. Cardiol. 90, 89–94.

Roy, S.K., Maiti, D., Mondal, S., Das, D., Islam, S.S., 2008. Structural analysis of a polysaccharide isolated from the aqueous extract of an edible mushroom, *Pleurotus sajor-caju*, cultivar Black Japan. Carbohydr. Res. 343, 1108–1113.

Schepetkin, I.A., Quinn, M.T., 2006. Botanical polysaccharides: macrophage immunomodulation and therapeutic potential. Int. Immunopharmacol. 6, 317–333.

Schmitt, C.A., Lowe, S.W., 1999. Apoptosis and therapy. J. Pathol. 187, 127–137.

Schreck, R., Albermann, K.A.J., Baeuerle, P.A., 1992. Nuclear factor kB: an oxidative stress-responsive transcription factor of eukaryotic cells (a review). Free Radic. Res. 17, 221–237.

Schwartz, G.K., Jiang, J., Kelsen, D., Albino, A.P., 1993. Protein kinase C: a novel target for inhibiting gastric cancer cell invasion. J. Natl. Cancer Inst. 85, 402–407.

Seger, R., Krebs, E., 1995. The MAPK signaling cascade. FASEB J. 9, 726.

Shao, B.M., Dai, H., Xu, W., Lin, Z.B., Gao, X.M., 2004. Immune receptors for polysaccharides from *Ganoderma lucidum*. Biochem. Biophys. Res. Commun. 323, 133–141.

Sharma, V., 2011. Hypoglycemic activity of methanolic extract of *Tectona grandis* linn. Root in alloxan induced diabetic rats. J. Appl. Pharm. Sci. 1, 16.

Shiao, M.S., Lee, K.R., Lin, L.J., Wang, C.T., 1994. Natural products and biological activities of the Chinese medicinal fungus *Ganoderma lucidum*, ACS symposium series (USA).

Shin, J.Y., Song, J.Y., Yun, Y.S., Yang, H.O., Rhee, D.K., Pyo, S., 2002. Immunostimulating effects of acidic polysaccharides extract of *Panax ginseng* on macrophage function. Immunopharmacol Immunotoxicol. 24, 469–482.

Shin, J.Y., Lee, S., Bae, I.Y., Yoo, S.H., Lee, H.G., 2007. Structural and biological study of carboxymethylated *Phellinus linteus* polysaccharides. J. Agric. Food Chem. 55, 3368–3372.

Shulman, G.I., 2000. Cellular mechanisms of insulin resistance. J. Clin. Invest. 106, 171–176.

Silva, D.A., de Paula, R.C., Feitosa, J.P., de Brito, A.C., Maciel, J.S., Paula, H.C., 2004. Carboxymethylation of cashew tree exudate polysaccharide. Carbohydr. Polym. 58, 163–171.

Srinivasan, K., Viswanad, B., Asrat, L., Kaul, C., Ramarao, P., 2005. Combination of high-fat diet-fed and low-dose streptozotocin-treated rat: a model for type 2 diabetes and pharmacological screening. Pharmacol. Res. 52, 313–320.

Storseth, T.R., Hansen, K., Skjermo, J., Krane, J., 2004. Characterization of a β-D-(1→3)-glucan from the marine diatom Chaetoceros mulleri by high-resolution magic-angle spinning NMR spectroscopy on whole algal cells. Carbohydr. Res. 339, 421–424.

Takenaga, K., Takahashi, K., 1986. Effects of 12-O-tetradecanoylphorbol-13-acetate on adhesiveness and lung-colonizing ability of Lewis lung carcinoma cells. Cancer Res. 46, 375–380.

Tao, Y., Zhang, Y., Zhang, L., 2009. Chemical modification and antitumor activities of two polysaccharide–protein complexes from *Pleurotus tuber-regium*. Int. J. Biol. Macromol. 45, 109–115.

Tessari, P., Coracina, A., Kiwanuka, E., Vedovato, M., Vettore, M., Valerio, A., et al., 2005. Effects of insulin on methionine and homocysteine kinetics in type 2 diabetes with nephropathy. Diabetes 54, 2968–2976.

Thannickal, V.J., Fanburg, B.L., 2000. Reactive oxygen species in cell signaling. Am. J. Physiol. Lung Cell. Mol. Physiol. 279, L1005–L1028.

Toker, A., Newton, A.C., 2000. Akt/protein kinase B is regulated by autophosphorylation at the hypothetical PDK-2 site. J. Biol. Chem. 275, 8271.

Tomoda, M., Gonda, R., Kasahara, Y., Hikino, H., 1986. Glycan structures of ganoderans B and C, hypoglycemic glycans of *Ganoderma lucidum* fruit bodies. Phytochemistry 25, 2817–2820.

Trinchieri, G., Sher, A., 2007. Cooperation of Toll-like receptor signals in innate immune defence. Nat. Rev. Immunol. 7, 179–190.

Trulson, M.E., Jacoby, J.H., MacKenzie, R.G., 1986. Streptozotocin-induced diabetes reduces brain serotonin synthesis in rats. J. Neurochem. 46, 1068–1072.

Tseng, Y.H., Yang, J.H., Mau, J.L., 2008. Antioxidant properties of polysaccharides from *Ganoderma tsugae*. Food Chem. 107, 732–738.

Tsutsui, H., Maeda, T., Min, J.Z., Inagaki, S., Higashi, T., Kagawa, Y., et al., 2011. Biomarker discovery in biological specimens (plasma, hair, liver and kidney) of diabetic mice based upon metabolite profiling using ultra-performance liquid chromatography with electrospray ionization time-of-flight mass spectrometry. Clin. Chim. Acta 412, 861–872.

Ushio-Fukai, M., Alexander, R.W., Akers, M., Yin, Q., Fujio, Y., Walsh, K., et al., 1999. Reactive oxygen species mediate the activation of Akt/protein kinase B by angiotensin II in vascular smooth muscle cells. J. Biol. Chem. 274, 22699–22704.

Wallace, T.M., Levy, J.C., Matthews, D.R., 2004. Use and abuse of HOMA modeling. Diabetes Care 27, 1487–1495.

Wang, L.G., Liu, X.M., Kreis, W., Budman, D.R., 1999. The effect of antimicrotubule agents on signal transduction pathways of apoptosis: a review. Cancer Chemother. Pharmacol. 44, 355–361.

Wang, S.Y., Hsu, M.L., Hsu, H.C., Lee, S.S., Shiao, M.S., Ho, C.K., 1997. The anti-tumor effect of *Ganoderma lucidum* is mediated by cytokines released from activated macrophages and T lymphocytes. Int. J. Cancer 70, 699–705.

Want, E., Masson, P., 2011. Processing and analysis of GC/LC-MS-based metabolomics data. Methods Mol. Biol. 708, 277–298.

Whelan, R.D., Parker, P.J., 1998. Loss of protein kinase C function induces an apoptotic response. Oncogene 16, 1939–1944.

Wong, K.H., Lai, C.K.M., Cheung, P.C.K., 2011. Immunomodulatory activities of mushroom sclerotial polysaccharides. Food Hydrocolloids 25, 150–158.

Worley, B., Halouska, S., Powers, R., 2013. Utilities for quantifying separation in PCA/PLS-DA scores plots. Anal. Biochem. 433, 102–104.

Wulffele, M., Kooy, A., Lehert, P., Bets, D., Ogterop, J., Burg, B., et al., 2003. Effects of short-term treatment with metformin on serum concentrations of homocysteine, folate and vitamin B12 in type 2 diabetes mellitus: a randomized, placebo-controlled trial. J. Intern. Med. 254, 455–463.

Yang, F.L., Yang, Y.L., Liao, P.C., Chou, J.C., Tsai, K.C., Yang, A.S., et al., 2011. Structure and immunological characterization of the capsular polysaccharide of a pyrogenic liver abscess caused by *Klebsiella pneumoniae* activation of macrophages through toll-like receptor 4. J. Biol. Chem. 286, 21041–21051.

Ye, L., Li, J., Zhang, J., Pan, Y., 2010. NMR characterization for polysaccharide moiety of a glycopeptide. Fitoterapia 81, 93–96.

Yoon, S.Y., Eo, S.K., Kim, Y.S., Lee, C.K., Han, S.S., 1994. Antimicrobial activity of *Ganoderma lucidum* extract alone and in combination with some antibiotics. Arch. Pharm. Res. 17, 438–442.

Yu, Q., Nie, S.P., Wang, J.Q., Yin, P.F., Li, W.J., Xie, M.Y., 2012. Polysaccharide from *Ganoderma atrum* induces tumor necrosis factor-α secretion via Phosphoinositide 3-kinase/Akt, mitogen-activated protein kinases and nuclear factor-κB signaling pathways in RAW264.7 cells. Int. Immunopharmacol. 14, 362–368.

Yu, Q., Nie, S.P., Li, W.J., Zheng, W.Y., Yin, P.F., Gong, D.M., et al., 2013. Macrophage immunomodulatory activity of a purified polysaccharide isolated from *Ganoderma atrum*. Phytother. Res. 27, 186–191.

Yuan, H., Song, J., Li, X., Li, N., Dai, J., 2006. Immunomodulation and antitumor activity of κ-carrageenan oligosaccharides. Cancer Lett. 243, 228–234.

Zhang, A., Sun, H., Wang, P., Han, Y., Wang, X., 2012a. Modern analytical techniques in metabolomics analysis. Analyst 137, 293–300.

Zhang, A., Sun, H., Wang, P., Han, Y., Wang, X., 2012b. Recent and potential developments of biofluid analyses in metabolomics. J. Proteomics 75, 1079–1088.

Zhang, H., Li, W.J., Nie, S.P., Chen, Y., Wang, Y.X., Xie, M.Y., 2012c. Structural characterisation of a novel bioactive polysaccharide from *Ganoderma atrum*. Carbohydr. Polym. 88, 1047–1054.

Zhang, W., Zheng, L.J., Zhang, Z.M., Hai, C.X., 2012d. Protective effect of a water-soluble polysaccharide from *Salvia miltiorrhiza* Bunge on insulin resistance in rats. Carbohydr. Polym. 89, 890–898.

Zhang, H., Wang, Z.Y., Zhang, Z., Wang, X., 2011. Purified *Auricularia auricular-judae* polysaccharide (AAP I-a) prevents oxidative stress in an ageing mouse model. Carbohydr. Polym. 84, 638–648.

Zhang, J., Wang, G., Li, H., Zhuang, C., Mizuno, T., Ito, H., et al., 1994. Antitumor active protein-containing glycans from the Chinese mushroom songshan lingzhi, *Ganoderma tsugae* mycelium. Biosci. Biotechnol. Biochem. 58, 1202–1205.

Zhang, L., Chen, J., 1997. Isolation and determination of molecular weights of water-soluble polysaccharides from fruiting body of *Ganoderma lucidum*. Acta Polym. Sin. 1, 71–72.

Zhang, S.S., Nie, S.P., Huang, D.F., Huang, J.Q., Wang, Y.X., Xie, M.Y., 2013. Polysaccharide from *Ganoderma atrum* evokes antitumor activity via toll-like receptor 4-mediated NF-κB and mitogen-activated protein kinase signaling pathways. J. Agric. Food Chem. 61, 3676–3682.

Zhang, Z., Zhang, Q., Wang, J., Song, H., Zhang, H., Niu, X., 2010. Chemical modification and influence of function groups on the in vitro-antioxidant activities of porphyran from *Porphyra haitanensis*. Carbohydr. Polym. 79, 290–295.

Zhao, L., Dong, Y., Chen, G., Hu, Q., 2010. Extraction, purification, characterization and antitumor activity of polysaccharides from *Ganoderma lucidum*. Carbohydr. Polym. 80, 783–789.

Zhou, Q., Li, Y., Jin, J., Lang, L., Zhu, Z., Fang, W., et al., 2012. Lx2-32c, a novel taxane derivative, exerts anti-resistance activity by initiating intrinsic apoptosis pathway in vitro and inhibits the growth of resistant tumor in vivo. Biol. Pharm. Bull. 35, 2170–2179.

Zhu, K.X., Nie, S.P., Li, C., Lin, S.L., Xing, M.M., Li, W.J., et al., 2013. A newly identified polysaccharide from *Ganoderma atrum* attenuates hyperglycemia and hyperlipidemia. Int. J. Biol. Macromol. 57, 142–150.

Zhu, K.X., Nie, S.P., Li, C., Gong, D.M., Xie, M.Y., 2014. *Ganoderma atrum* polysaccharide improves aortic relaxation in diabetic rats via PI3K/Akt pathway. Carbohydr. Polym. 103, 520–527.

Chapter 6

Glucomannans From *Dendrobium officinale* and Aloe

6.1 INTRODUCTION

6.1.1 Dendrobium Herbs and *Dendrobium officinale*

Dendrobium, a genus identified by Olaf Swartz in 1799 AD, is the second largest genus in the family Orchidaceae (Puchooa, 2004). It contains approximately 1100 species which are mainly distributed in the subtropical and tropical regions of Asia and Oceania (Xing et al., 2013). Seventy-four Dendrobium species have been found in the tropical and subtropical provinces of China, 15 of which grow in Taiwan (Xing et al., 2013). Since ancient times, many Dendrobium plants have been used as ingredients for nutraceutical beverages and food products (Bao et al., 2009). For example, Dendrobium plants have been used by traditional Chinese medical practitioners to alleviate diabetes, obesity, rheumatoid arthritis, and many other diseases, which have been recorded in at least 18 ancient Chinese medical works including the *Compendium of Materia Medica*. Among all the Dendrobium herbs in China, *D. officinale* stems and leaves have been believed by traditional medical practitioners to have the best medical properties. Consequently, this herb is in high demand by the Chinese.

According to differences in morphology, *Dendrobium* can be divided into dozens of varieties, including *Dendrobium officinale*, *Dendrobium nobile*, *Dendrobium huoshanense*, *Dendrobium chrysanthum*, *Dendrobium loddigesii*, *Dendrobium fimbriatum*, and so on. Among them, *D. officinale* is considered to be the most precious (Lin et al., 2011). The wild herbal plant with the taxonomic name "*D. officinale* Kimura et Migo," called "Tie-Pi- Shi-Hu" in Chinese, grows mainly on the semishaded, humid rocks at an altitude of around 1600 m in the mountain areas of Southern Anhui, Eastern Zhejiang, Western Fujian, North-western Guangxi, Sichuan, and Southeastern Yunnan in China (Xing et al., 2013). Fig. 6.1 is a picture of *D. officinale* herb.

The chemical composition of *D. officinale* is complex, including polysaccharide, phenanthrene and bibenzyls, amino acids, trace elements, lignans, phenolic acids, phenylpropanoid, other compounds such as ocean dihuang

Bioactive Polysaccharides. DOI: http://dx.doi.org/10.1016/B978-0-12-809418-1.00006-X

FIGURE 6.1 *Dendrobium officinale* herbs.

lactones, naringenin, β-sitosterol, palmitic acid, and so on. The main bioactive constituents are polysaccharide alkaloids, amino acids, and trace elements.

Polysaccharides are the most important bioactive substance in *D. officinale*. It is set forth in the *Chinese Pharmacopoeia* published in 2010 that the mass fraction of polysaccharide contained in *D. officinale* should not be less that 25% (Lv et al., 2013). However, the polysaccharide content in *D. officinale* is affected by various factors, such as planting patterns, illumination intensity and illumination time, harvesting time, and the part of the plant. Moisture, light intensity, and their interactions can largely affect the water-soluble polysaccharide content in *D. officinale*. Under the same light intensity, the water-soluble polysaccharide content increases with a decrease in moisture level. However, when the moisture is similar, the water-soluble polysaccharide content significantly increases with increasing illumination intensity (Kong et al., 2015). In addition, the polysaccharide content has a close relationship with flowering. Flowering significantly reduces the polysaccharide content in herbs, especially the amounts of mannose, glucose, and galacturonic acid (Yu et al., 2014). However, the flowering process might possibly promote the accumulation of xylose and arabinose. The polysaccharide content in biennial plants is obviously higher than that in 1-year-old plants (Ling and Fang, 2012). Therefore, biennial herbs harvested before flowering should have the highest amount of polysaccharide (Yu et al., 2014). Li and Zhu (2010) compared the polysaccharide content in Dendrobium harvested in different seasons, and found that those plants harvested in winter contained the highest sugar content. Furthermore, the sugar content in the stem was higher than that in roots (Hua, 2005).

Dendrobium alkaloid is the compound that was first isolated and identified in Dendrobium herbs (Nie and Cai, 2012). The commonly used method in the determination of total alkaloids is the acid dye colorimetry method (Xu et al., 2014; Yang and Wang, 2015). In general, there is an obvious range in the content of alkaloid in *D. officinale* (Nie and Cai, 2012), and this range is approximately from 0.019% to 0.043% (Lv et al., 2013), far lower than that in *D. nobile* (Nie and Cai, 2012). This range can be used to distinguish fake *D. officinale*. If the total alkaloid content in a *D. officinale* sample largely deviates from this range, it can be virtually confirmed that this *D. officinale* sample is a fake one (Nie and Cai, 2012). The alkaloid content is higher in wild-type than in cultured herb (Nie and Cai, 2012). For artificially cultured Dendrobium, the alkaloid content is higher in the annual plant than the biennial plant (Nie and Cai, 2012). In addition, the plants in Yunan contained more alkaloid than those in Zhejiang. So far, five types of alkaloids with different structural skeletons have been reported from Dendrobium, namely sesquiterpenoids, indolizidine, pyrrolidines, phthalides, and imidazoles (Xu et al., 2013). The total alkaloid content of *D. officinale* with different germplasms and physiological ages exhibited a significant discrepancy, and it increased with the age of growth (Yang and Wang, 2015).

D. officinale contains aspartic acid, glutamic acid, glycine, valine, leucine, threonine, serine, alanine, cystine, methionine, isoleucine, lysine, histidine, arginine, proline, and tyrosine, etc. The former five kinds of amino acid comprised 53% of the total amino acid. *D. officinale* contains all the essential amino acids, except tryptophan (Li et al., 2011a,b). Prolonging the growing years is beneficial to the enrichment of amino acids in *D. officinale*, especially to the content of glycine, threonine, leucine, and lysine (Zhang et al., 2011a,b).

D. officinale contains a variety of trace elements, such as Cu, Zn, Fe, Mn, Ca, Mg, K, Cr, Sr, and B, etc. (Li et al., 2008a).

However, due to over-exploitation and deterioration of its habitat, this valuable herbal plant has been on the verge of extinction since the 1980s, as listed in the *China Plant Red Data Book* in 1987 (Xing et al., 2013). The imbalance between the high demand and the short supply of *D. officinale* herb resulted in the prices soaring to around $3000 per kilogram (Ding et al., 2002; Li et al., 2008b). In order to meet market demands and to protect the wild sources of the herb, producers in China have developed plant tissue culture methods by inducing the explanted Dendrobium tissue segments to form protocorm-like bodies (PLBs). The PLBs are further developed into plantlets that can be grown into mature plants in farmlands or greenhouses. Recent statistics indicate that there are more than 130 companies now producing Dendrobium herbs for commercial purposes in China. There are only three Dendrobium herbs, namely, *D. officinale*, *D. nobile*, and *D. fimbriatum*, listed in the *Chinese Pharmacopoeia* (2005 edition) (Editorial Committee of the Chinese Pharmacopoeia, 2005), but many other Dendrobium herbs, such as *D. huoshanense*, are also consumed as folk medicines.

Dendrobium plants are rich in polysaccharides, and the amounts present in mature plants are controlled by many genetic and environmental factors, in particular species differences (Zhao et al., 2011; Xu et al., 2008), growth duration, and harvest time (Jin, 2012; Jin et al., 2011; Li and Zhu, 2010), blossoming (Zhu et al., 2010), plant geography (Xu et al., 2008), light intensities (Zheng et al., 2012), growing temperature, and humidity (Jiang et al., 2007; Zhu et al., 2010). The polysaccharide content of the Dendrobium plants cultured under favorable light, temperature, and humidity conditions in farmlands or greenhouses is equal to, or even higher than, those growing in the wild (Hua, 2005; Jiang et al., 2007). Thus, culturing Dendrobium plantlets in "indoor plant factories" could be a good alternative to harvesting mature Dendrobium plants in farmlands, in terms of the mass-production of polysaccharides (Wang et al., 2012).

Natural polysaccharides isolated from herbal plants have been shown to possess bioactivities. For example, the immune-stimulatory properties of aloe vera glucomannan have been confirmed and used for the treatment of immune-related diseases (Xing et al., 2013). Since the late 1980s, researchers in China have focused on the polysaccharides extracted from Dendrobium plants, with the intention of developing polysaccharide-based drugs or food ingredients with specific health-enhancing functions (Xing et al., 2013).

6.1.2 Introduction of Aloe

Aloe is named Monocotyledones Liliales Liliaceae in the genus Aloe. There are over 400 species within the genus Aloe. It is mainly distributed in tropical and subtropical regions in Africa. Among them, more than 270 kinds are produced from South Africa, and more than 40 kinds are produced from Madagascar. The Canary Islands, Seville, and the West Indies are also the main producing regions of aloe. In China, Guangdong, Guangxi, Yunnan, Fujian, and Zhejiang provinces are the main producing regions of both wild and cultivated aloe species. Recently, some places in northern China have also started to introduce and cultivate aloe.

The height of aloe can vary from less than 10 cm to over 20 m. Leaves are clustered at the top of the stem, with a round ball stem, thick and fleshy leaves. The base width and tip are thin and narrowly lanceolate, the leaf edges have thorns or teeth, with light green or green leaves The leaves of some varieties have white markings. Plants can have_different stem lengths, either very short, or up to several meters. Flowers have long stems, which are branched or unbranched with racemes of tubular flowers, and a six-lobed perianth; colors can be yellow, orange, or red, etc. Rhizomes at the base can occur, which can produce tips which sprout into shoots, and soon grow into plants.

Aloe prefers to grow in warm and high temperature areas, and is intolerant of cold weather. The optimum temperature for aloe to grow is around 25°C. When the temperature is around 5°C, it will stop growing, and when the temperature is below 10°C, it will become frozen. Aloe prefers bright and drought conditions, and dislikes shady areas. It prefers fertile land with good ventilation and drainage, while it dislikes heavy clay soil.

Aloe barbadensis Mill, *Aloe arborescens* Mill, *Aloe vera* var. chinesis Berg and *Aloe ferox* Mill are four commonly studied varieties of aloe. *A. barbadensis* Mill are largely planted and processed in America. *A. arborescens* Mill are commonly planted and processed in Japan. *A. ferox* Mill are mainly produced in South Africa. However, Chinese researchers are more interested in *A. vera* var. chinesis Berg. In recent years, the aloe planting industry has developed rapidly. In 2010, the planting area of aloe was about 3000 ha, and its industry output is worth over 9 billion yuan.

Aloe is reported to have numerous bioactivities, and can be of benefit to our health. The leaf gel contained in aloe is said to have antiinflammatory effects (Balan et al., 2014), antioxidant effects, wound healing effects, skin hydration effects, radioprotective effects (Saini and Saini, 2011), antidiabetic effects, immunomodulatory effects, anticancer effects, antimicrobial activities (Balan et al., 2014), hepatoprotective activities (Kim et al., 2009), and modulative effects on gastric acid secretion and ulcers (Hamman, 2008). Oral administration of aloe vera gel, an antimicrobial and antiinflammatory herbal remedy, stimulates cell-mediated immunity and antibody production in a mouse model (Balan et al., 2014).

A. vera (L.) Burm.f. (*A. barbadensis* Miller) is a perennial succulent xerophyte, which develops water storage tissue in the leaves to survive in dry areas of low or erratic rainfall. The aloe leaf can be divided into two major parts, namely the outer green rind, including the vascular bundles, and the inner colorless parenchyma containing the aloe gel. The innermost part of the leaf is a clear, soft, moist and slippery tissue that consists of large thin-walled parenchyma cells in which water is held in the form of a viscous mucilage, as is shown in Fig. 6.2. Therefore, the thick fleshy leaves of aloe plants contain not only cell-wall carbohydrates such as cellulose and hemicellulose, but also storage carbohydrates such as acetylated mannans (Hamman, 2008). Many of the health benefits associated with aloe vera have been attributed to the polysaccharides contained in the gel of the leaves (Radha and Laxmipriya, 2015).

Many effects could influence the polysaccharide content in aloe, such as plant species, plant age, and growth conditions. It is said that the polysaccharide content is higher in *A. vera* than in *A. arborescens*. Plants grown under reduced light intensities had significantly lower aloin and polysaccharide concentrations (Lucini et al., 2013).

FIGURE 6.2 (A) Aloe vera herbs; (B) The inner part of aloe vera.

Aloe is rich in nutrients. Clinical research at home and abroad has confirmed that aloe is efficient in a variety of chronic diseases, such as gastrointestinal problems, high blood pressure, heart disease, diabetes, liver disease, constipation, and stomatitis embolism. Some scientists in Germany have found that aloe can not only heal external wounds, but also has an effect on curing intestinal ulcers, and promoting the digestion and absorption of nutrients. Therefore, it has been applied in making functional foods. It is reported that the anthraquinone, cinnamate, and coumaric acid in aloe can be beneficial in resisting ultraviolet rays and diminishing pigmentation. The hydroxyl groups contained in aloe polysaccharides can form hydrogen groups with water, which contribute to the moisturizing effect of aloe. In addition, the catalase (CAT), vitamins, and carotenoids contained in aloe can help scavenging the free radicals in our body. Therefore, aloe is also widely used in numerous cosmetics.

6.2 PREPARATION PROCESS

6.2.1 Preparation Process for Dendronan

6.2.1.1 Pretreatment

After Dendrobium plants are harvested, the plants should be dried as soon as possible, because the physiological activities in the fresh plants continue as polysaccharides are hydrolyzed to supply energy (Chen, 2003). Traditionally, the fresh herb is dried by sunlight, but oven-drying substantially reduces the loss of polysaccharide, by deactivating the degrading enzymes and terminating physiological activities (Chen, 2003). Recently, freeze-drying has been increasingly favored by the Dendrobium producers in China, because they believe that this drying method results in products with a good appearance, and does not cause nutrient loss. Dry herbal materials should be powdered in order to generate a larger surface and thus increase extraction efficiency. Grinding, however, is the most important factor to ensure high yields of water-extractable polysaccharides from dry Dendrobium herbs (Qian and Lv, 2005).

Since the Dendrobium plant is also rich in small hydrophobic molecules, such as lipids and alkanes, organic solvents are used to remove these small hydrophobic molecules from the herbal material (Chen et al., 2002). Solvents thus used have been pure ethanol (Qin et al., 2011), ethanol solutions (80% or 85%, v/v) (Chen, 2010; Luo et al., 2010a), diethyl ether (Luo et al., 2010a), petroleum ether (Zheng et al., 2008), and acetone (Hua et al., 2004). A traditional method to process Dendrobium herbs is to soak fresh Dendrobium stems in a hot ethanol solution and then dry in the sun (Qu et al., 2011). This method can significantly increase the yield of the hot-water-extractable polysaccharide of the processed Dendrobium herb (Chen et al., 2002).

6.2.1.2 Extraction of Polysaccharides Using the Hot-Water Method

Hot-water extraction is the most commonly used method for the isolation of water-soluble polysaccharides from Dendrobium species. In order to obtain high yields, a long soaking time, high water temperature, high water-to-plant ratio, and multiple extraction steps are usually necessary. Wang et al. (2013b) optimized the extraction conditions of *D. officinale* polysaccharide. The results indicated that the optimum conditions for *D. officinale* polysaccharide were an extraction temperature of 70°C, solid/liquid ratio of 1:40, extraction time of 2 h, with two extraction times. Zhou et al. (2010) optimized the extraction conditions based on single factor tests and an orthogonal experiment. The results suggested that the highest yield would be 24.9%, with the extraction temperature being 85°C, extraction time of 3 h, and the solid/liquid ratio at 1:20. Huang et al. (2013b) increased the extraction temperature to 90°C, and found that the extraction rate could be increased to 30.56%. Wang et al. (2013a) increased the water/material ratio to 75, and found that the yield could be increased to 34.96%. Wu et al. (2013) increased the extraction temperature to 92°C, increased the extraction times to 3 times, and found that the extraction rate could be increased to 38.569%. Li et al. (2011a,b) increased the extraction temperature to 100°C, and found that the yield could be increased to 51.08%.

The extraction process can be further facilitated by physical methods such as ultrasonication, microwave treatment, and ultrahigh voltage (Fan et al., 2009b; Xia et al., 2012b; Liu and Xiong, 2009; Shang and Wang, 2010). Ultrasonication could destroy the plant tissue, and rupture the plant cells, and therefore release the polysaccharide contained in the herbs, and accelerate its dissolution in solvent. The ultrasonic extraction method has the advantage of a shorter time and higher yield compared with the conventional water extraction method, and without requiring a heating process. Ye (2009) compared the *D. officinale* polysaccharide extraction process with or without the assistance of ultrasonication, and found that the extraction rate would increase by 5% with the help of ultrasonication. With the assistance of a microwave electromagnetic field, the polysaccharide contained in herbs can easily dissolve and diffuse, and therefore the extraction rate is improved. Shang and Wang (2010) discussed the effects of microwave power, microwave time, solid/liquid ratio, extraction time, and extraction agent pH on polysaccharide yield, and found that with a microwave power of 400 W, pH of 8.0, microwave time of 6 min, and material/water ratio of 1:50, it could reach the best extraction rate of 17.48%. Ultra-high pressure (UHP) could break the plant cell wall, and therefore promote the dissolution of *D. officinale* polysaccharides. This method could shorten the extraction time and save energy consumption. Zong and Li (2012) compared the effects of UHP power, material/water ratio, UHP time, and grinding degree on extraction yield, and found that the effect of UHP power>solid/water ratio>UHP time > grinding degree. For example, ultrasonication treatment can reduce the extraction time to 0.5–1.5 h, and ensure relatively high polysaccharide yields (Qin et al., 2011; Jin et al., 2011; Chen et al., 2002; Fan et al., 2009b). The disadvantage of the two methods is that polysaccharide could be depolymerized during such treatments (Wang et al., 2012; Zheng et al., 2012).

Enzymatic methods have also been used to assist water extraction. Cellulase, papain, and pectin were thus used (Li and Zhu, 2010; Zhang et al., 2011a,b). The combined use of the two enzymes also facilitates hot-water extraction of *D. huoshanense* polysaccharides (Wei et al., 2009). It should be noted that cellulose breaks down *O*-acetylated β–D-1,4-glucomannan, a main polysaccharide component of *D. huoshanense* (Hsieh et al., 2008), *D. officinale* (Hua et al., 2004), and *Dendrobium moniliforme* (Lin et al., 2011). Although enzyme application can improve the extraction rate, there still remain some problems during the extraction process. Firstly, the enzymes are usually not cheap, therefore, the application of enzymes will raise the cost; secondly, the incomplete inactivation of enzymes can largely influence the purity and yield of the products; thirdly, the industrial application of the enzyme hydrolysis method is still immature, and needs further exploration. Following hot water extraction, polysaccharides are usually recovered by ethanol precipitation. A solution of the ethanol precipitate is subsequently dialyzed and freeze-dried to obtain a crude water-soluble polysaccharide.

6.2.1.3 *Isolation of Mucilaginous Polysaccharides and Cell-Wall Polysaccharides From Fresh Dendrobium Leaves and Stems*

Dendrobium plants contain mucilage cells, therefore a thick, viscous substance (mucilage) can be squeezed out of the fresh stems and leaves (Li et al., 2000; Zhu et al., 2011). Hsieh et al. (2008) extracted the mucilage by homogenizing fresh *D. huoshanense* stems and leaves with 70% ethanol, followed by treatments using sonication, dialysis, and freeze-drying. In the same study, cell-wall polysaccharides of *D. huoshanense* were extracted sequentially using 50 mM CDTA solution (pH6.5), 50 mM Na_2CO_3 solution (containing 25 mM $NaBH_4$), 1 M KOH solution (containing 25 mM $NaBH_4$), and 4 M KOH solution (containing 25 mM $NaBH_4$). The function of $NaBH_4$ is to prevent and reduce the β-elimination or peeling reactions of polysaccharides caused by alkaline treatment. Generally, cell-wall polysaccharides cannot be extracted by hot water, but these polysaccharides are water-soluble once they are extracted by CDTA or alkaline solutions. Qin et al. (2011) reported that *Dendrobium findlayanum* herb soaked in alkaline solution (pH8.0) at 95°C for 2.5 h would release polysaccharides in high yields (60.6%).

6.2.1.4 *Purification and Fractionation of Dendrobium Polysaccharides*

Based on the experiences of our research group, low-level protein contamination tends not to materially interfere with the detailed structural elucidation of the extracted polysaccharides using 2D NMR spectroscopy and methylation analysis. However, polysaccharides with little or no protein are better for conformational, functional, and bioactivity studies. The Sevag method is a favored method for the deproteinization treatment of Dendrobium polysaccharides. In this method, an aqueous solution of Dendrobium polysaccharides is mixed with *n*-butanol and chloroform, shaken vigorously to facilitate protein denaturation, and then the denatured (precipitated) proteins are removed (Zhao et al., 2010). Protein precipitation using trichloroacetic acid (TCA) or sulfosalicylic acid can also remove protein from polysaccharides. Compared with that of the TCA method, the experimental condition of the Sevag method is milder; and the time for deproteinization is shorter. However, only parts of the protein can be removed each time by the Sevag method, therefore, it usually needs 4–5 rounds of deproteinization. Thus, it causes loss of polysaccharide during this process. Meanwhile, the TCA method needs less reagent and fewer steps. Furthermore, parts of the pigment can dissolve in the TCA solution, and therefore can be removed. Alternatively, proteolytic enzymes, such as papain, have been used to hydrolyze proteins in Dendrobium polysaccharide extracts (Shang, 2010). We have shown that 20% (w/w) ethanol precipitation and repeated freeze – thaw – centrifugation treatments can remove considerable amounts of protein from the crude water-extractable *D. officinale* polysaccharide. Colorimetric or automatic nitrogen quantitative methods (based on the Kjeldahl method or Dumas method) can be used to monitor the total protein content in Dendrobium polysaccharide samples before and after purification (Hsieh et al., 2008; Hua et al., 2004).

Because starch can be gelatinized and solubilized in hot water, the crude polysaccharide extracted by hot water can be contaminated by starch. Indeed, histochemical studies have demonstrated the presence of starch granules in the parenchyma cells of Dendrobium plants (Davies and Turner, 2004; Guan et al., 2010; Kaminska and Stpiczynska, 2011). Starch granules have been shown by scanning electron microscopy to be present in the stem vascular bundles of eight Dendrobium species, namely, *D. officinale*, *Dendrobium tosaense*, *Dendrobium cumulatum*, *Dendrobium linawianum*, *D. moniliforme*, *Dendrobium aurantiacum*, *D. huoshanense*, and *Dendrobium nindii* (Wu et al., 2009). Using scanning electron microscopy, we also observed considerable numbers of starch granules in the cross-section of a *D. officinale* stem. Colorimetric methods have also shown the presence of starch in the plantlets (Ferreira et al., 2011), as well as in the mature plants (Pan, 2005) of *D. nobile* cultivars. The presence of starch in Dendrobium samples was also noticed by two research groups (Hsieh et al., 2008; Lertwiriyawong et al., 2011), when purifying Dendrobium DNA and non-starch polysaccharides. Thus, α-amylase treatment can be an effective method to remove starch from Dendrobium polysaccharides, provided that the α-amylase preparation is free from other contaminating glucanase activity (Hsieh et al., 2008).

The pigments remaining in Dendrobium polysaccharides might cause the polysaccharide to turn dark during the drying process. Therefore, decolorization is required during the purification procedure. Ethanol, H_2O_2, active carbon, and DEAE-52 are commonly used in the decolorization process. Some researchers have compared these four methods, and found that DEAE-52 performs best.

Many studies have used ion exchange chromatography (AEC) and/or size exclusion chromatography (SEC) methods to fractionate and purify Dendrobium polysaccharides. The AEC method using diethylaminoethyl (DEAE)-cellulose columns is useful for the separation of neutral polysaccharides from acidic ones and proteins; the SEC method using Sepharose columns is suitable for the separation of polysaccharides based on their sizes (Wang et al., 2010b).

Chromatographic techniques can definitely assist in the fractionation of extracted polysaccharide mixtures, but they are very time-consuming. The gradient ethanol precipitation method can be used to fractionate *D. officinale* polysaccharides, based on the solubility differences of these fractions (Bao et al., 2009). Using a sequential ethanol precipitation method, we have been able to separate a *D. officinale O*-acetyl glucomannan sample into five fractions with different molecular weights, mannose-to-glucose ratios, and levels and distributions of acetyl groups. In this way, large amounts of high-purity Dendrobium polysaccharide fractions (PFs) can be obtained in a relatively short period of time.

It is worth noting that some researchers consider a polysaccharide to be pure if it gives a single and symmetric peak during size exclusion or ion exchange chromatography. This could be rather misleading, since the purity of a polysaccharide sample is not determined only by molecular weight and/or charge; we have often found that two or more polysaccharides can have the same molecular weight and molecular weight distribution, thus coeluting in a given chromatographic system. Therefore, it is essential to have information about the starches, proteins, and other polymeric components present, as well as the monosaccharide composition, in order to assess the purity of a particular polysaccharide concentrate or isolate.

6.2.2 Preparation Process for Acemannan

Hot-water extraction followed by ethanol precipitation is the most commonly used method for aloe polysaccharide extraction. In addition, ultrasonication, microwave treatment, ultrahigh voltage, and enzymes have also been employed to promote the yield of aloe polysaccharide (Kang et al., 2014). In Kang's study (2014), ten digestive enzymes including five carbohydrases (Viscozyme, Celluclast, AMG, Termamyl, and Ultraflo) and five proteases (Kojizyme, Alcalase, Protamex, Flavourzyme, and Neutase) were employed for the preparation of aloe PFs. Among these enzymes, the highest yield was obtained with the application of Viscozyme, and this extract exhibited the highest radical-scavenging activity. The extraction process could be described as follows: an aloe gel sample was homogenized with water, and mixed with an adequate amount of different enzyme. Each reactant was adjusted to the optimum pH and temperature range of certain enzymes for the subsequent chemical reaction. The mixtures were maintained for 24 h for enzymatic reactions, followed by boiling at 100°C for 10 min to inactivate the enzyme. After that, samples were clarified by centrifugation (3000 rpm, for 20 min at 4°C) to remove the residue. These extracts were adjusted to pH 7.0 to achieve the enzymatic extract.

6.3 STRUCTURAL FEATURES

6.3.1 Structural Features of Dendronan

6.3.1.1 Methods Used for Structural Feature Analysis

Thin-layer chromatography, gas chromatography (GC), gas chromatography–mass spectrometry (GC–MS), and high performance anion-exchange chromatography coupled with pulsed amperometric detection (HPAEC-PAD) are commonly used methods for identification and quantification of the monosaccharides in Dendrobium polysaccharides (Ferreira et al., 2011; Hua et al., 2004; Luo et al., 2010a; Wang et al., 2010b). Recently, the monosaccharide composition of *D. officinale* polysaccharide has been found by chemically modifying acid-hydrolyzed polysaccharide samples with 1-phenyl-3-methyl-5-pyrazolone (PMP), and subsequently quantifying the monosaccharide derivatives by reverse-phase high performance liquid chromatography (RC-HPLC) (Yuan et al., 2011; Zhou and Lv, 2011).

Oligosaccharides (OS) produced by enzymatic hydrolysis of Dendrobium polysaccharide have also been characterized using HPSEC-PAD and MALDI-TOF-MS analysis (Hsieh et al., 2008). The combined application of classical chemical methods such as methylation, periodate oxidation–Smith degradation, and partial acid hydrolysis can provide information about the glycosidic linkages and branching in Dendrobium polysaccharides (Hua et al., 2004; Wang et al., 2010b). The weakness of chemical methods, i.e., incomplete chemical reactions and/or degradation events, in some instances, can result in misinterpretation of the structural features of a polysaccharide. For example, based on our experience, the full methylation of *D. officinale O*-acetyl glucomannans is fairly difficult to achieve, and under-methylated molecules can give peaks in the chromatogram of subsequent GC–MS analysis, leading to wrongly interpreted "branch-like" structures. One-dimensional ^{1}H and ^{13}C NMR spectroscopes have been widely used for structural characterizations of Dendrobium polysaccharides. Two-dimensional NMR techniques have been used for detailed structural analysis of Dendrobium polysaccharides in several studies (Hsieh et al., 2008; Li et al., 2012a; Wang et al., 2010b). Hsieh et al. (2008), e.g., using high-resolution 2D NMR spectra including COSY, TOCSY, HSQC, HMBC, and ROESY, elucidated the structure of a 2,3-*O*-acetyl glucomannan isolated from *D. huoshanense*. In contrast, the 2D NMR spectra in the other three studies were not informative, because of the low resolution of the spectra.

6.3.1.2 Structures of D. officinale *Polysaccharide*

D. officinale polysaccharides typically consist of glucose and mannose, sometimes with a small amount of arabinose and rhamnose (Xing et al., 2014; He et al., 2016; Liu et al., 2011). The molecular weight varied from 100 to 500 kDa (He et al., 2016; Xing et al., 2014; Liu et al., 2011).

Sequential ethanol precipitation and column chromatography (DEAE-Sephadex A-50 and Sephadex G-200) were used to fractionate a *D. officinale* polysaccharide sample into three fractions (designated I, II, and III) with molecular weights of 1000, 500, and 120 kDa, respectively (Wang et al., 1988). Detailed structural studies using periodate oxidation−Smith degradation analysis, methylation analysis, partial acid hydrolysis analysis, and 1D 1H and ^{13}C NMR analyses showed that the three fractions shared common structural features, notably, a backbone of -4)-β-D-Glcp-(1- and -3)-β-D-Manp-(1-. Some branches of -4)-β-D-Glcp-(1- or pentoses were found attached at the O-2, O-3, or O-6 positions of some Glcp residues in the backbone. All three fractions were found to be *O*-acetylated. Another research group (Huang et al., 1994) used hot water to extract a water-soluble polysaccharide from *D. officinale*, and the crude extract was purified using cetyl-trimethylammonium bromide treatment. The results of TLC and HPLC analysis indicated that the polysaccharide was composed of D-xylose (15.9%), L-arabinose (19.4%), and D-glucose (64.7%), which is not consistent with the findings of Wang et al. (1988).

Hua et al. (2004) concluded that a *D. officinale* glucomannan fraction (designated DOP-1-A1) had a backbone consisting of 1-4 linked β-D-Manp and β-D-Glcp, with acetyl groups attached to the O-2 position of both β-D-Manp and β-D-Glcp residues. Short branches, containing 1-3 linked β-D-Manp residues, were reported to be connected to some β-D-Manp residues in the backbone by 1-6 linkages. The Man/Glc ratio of DOP-1-A1 was approximately 4.8:1.

A PF (designated DCPP1a-1) isolated from the protocorms of *D. officinale* consisted of mannose and glucose in a ratio of around 7.0:1 (He et al., 2007a). The molecular weight of DCPP1a-1 was 189.0 kDa. Periodate oxidation analysis demonstrated that 4.0% of the residues in DCPP1a-1 were in the form of -6)-pyranose-(1- and terminal pyranose-(1-, with 52.1% in the form of -3)-pyranose-(1-, and 44.9% in the form of -2)- pyranose-(1- and -4)-pyranose-(1-. DCPP3c-1, another purified fraction from the same plant source contained mannose, rhamnose, galactouronic acid, glucose, galactose, and arabinose in a molar ratio of 1.23:1.00:1.05:23.35:3.83:1.05 (He et al., 2007a). Periodate oxidation analysis of DCPP3c-1 showed that 14.0% of the residues were in the form of -6)-pyranose-(1-, 40.7% of the residues were in the form of -2)-pyranose-(1- and -4)-pyranose-(1-, and 44.9% of the residues were in the form of -3)-pyranose- (1-.

A series of studies (Xing et al., 2014, 2015b) on the structural and viscous characteristics of Dendronan were undertaken by our group, as shown in Fig. 6.3. A crude polysaccharide designated as Dendronan was separated from *D. officinale*, with a purity of 90.4%. The Mw was 312 kDa, and the monosaccharide composition analysis showed that it predominantly consisted of mannose and glucose in a ratio of 6.9:1.

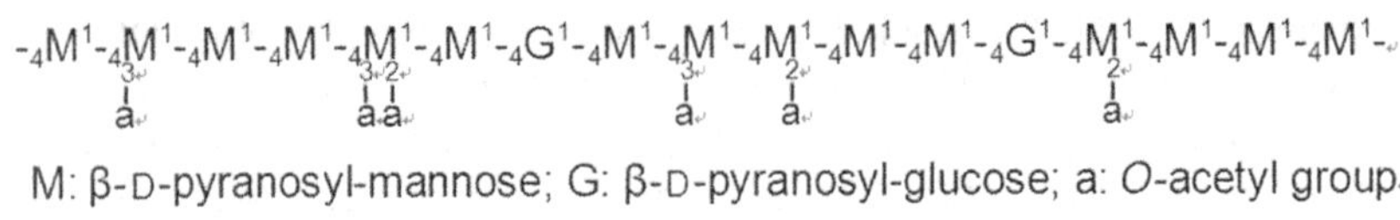

FIGURE 6.3 The structure of Dendronan (Xing et al., 2014, 2015b).

A gradient ethanol precipitation method was employed to fractionate Dendronan. Four major fractions were obtained with different Man/Glc ratios, molecular weights, and intrinsic viscosity. It was observed that as the ethanol concentration increased, the fractions tended to contain more *O*-acetyl groups, and the difference between O-2 and O-3 substitution tended to decrease. The experiments also indicated that the fragments rich in *O*-acetyl groups were highly resistant to cellulase and mannanase hydrolysis. However, even though the four fractions were different in detailed structural features, they shared the same major structure as 2,3-*O*-acetyl-glucomannan.

A more recent study (He et al., 2016) found a neutral heteropolysaccharide (named DOP-1-1) in *D. officinale* which consisted of mannose and glucose (5.9:1) with an average molecular weight of about 1.78×10^5 Da. The report suggested that DOP-1-1 was also an *O*-acetylated glucomannan with a β-D configuration in pyranose sugar forms. This is in accordance with the results of other reports.

There is, thus, agreed evidence that the neutral polysaccharides of *D. officinale* are composed dominantly of β-D-glucopyranoses and β-D-mannopyranoses, but there is disagreement about the ratio of mannose to glucose, the substitution position of *O*-acetyl groups, the existence of branches, and the building units of branches of *D. officinale* glucomannan. A combination of periodate oxidation − Smith degradation analysis, partial acid hydrolysis analysis, methylation analysis, and NMR spectroscopy has been used only by Hua et al. (2004). High-resolution 2D NMR spectra of *D. officinale* glucomannans have not been reported.

6.3.1.3 *Structure of* Dendrobium huoshanense *Polysaccharides*

D. huoshanense polysaccharides consist of glucose, mannose, arabinose, galactose, rhamnose, and xylose with different ratios. Sometimes, a trace of galacturonic acid is also found (Zha et al., 2007a; Li et al., 2014, 2015; Zha et al., 2013; Tian et al., 2013; Pan et al., 2013). The molecular weight varied from 3 to 1000 kDa (Zha et al., 2007a, 2013; Li et al., 2014; Tian et al., 2013; Pan et al., 2013). For example, a PF from *D. huoshanese* named DHPD1 had a Mw of 3.2 kDa (Zha et al., 2013); however, the polysaccharide extracted from PLBs of *D. huoshanese* had a Mw of 232 kDa (Li et al., 2015).

Hsieh et al. (2008) successfully elucidated the detailed structure of a destarched glucomannan (designated Fraction B) isolated from the mucilage of *D. huoshanense* stem, based on high quality 2D NMR spectra, including

TOCSY, HSQC, HMBC, and ROESY. The assignments of the 2D NMR spectra indicated the existence of -4)-β-D-Manp-(1-, -4)-β-D-Glcp-(1-, -4)-2-*O*-acetyl-β-D-Manp-(1-, and -4)-3-*O*-acetyl-β-D-Manp-(1-. However, evidence of branching was not found. This glucomannan had a mannose-to-glucose ratio of around 3.8:1. MALDI-TOF-MS analysis of β-D-mannanase hydrolysate of the glucomannan indicated that the hydrolysate was glucomannan oligosaccharide (3-9 DP) with partial acetylation. In another study (Zha et al., 2007a), a water-soluble *D. huoshanense* polysaccharide sample was obtained using hot-water extraction (at 50–60°C) and ethanol precipitation. The water-soluble polysaccharide was fractionated by DEAE-cellulose anion-exchange and gel filtration chromatography (Sephacryl S-200 and Sephadex G-75/G-100) to produce a PF (designated HPS-1B23) which was composed of glucose, mannose, and galactose in the ratio of 31:10:8. The results of oxidation – Smith degradation, partial acid hydrolysis, and methylation analysis, as well as 1D 1H and 13C NMR analyses, indicated that the fraction contained a backbone of -6)-α-D-Glcp-(1-, -4)-α-D-Glcp- (1-, and-6)-α-D-Manp-(1-, with α-D-Gal and an *O*-acetyl group attached to the O-3 positions of -6)-α-D-Manp-(1- and -6)- α-D-Glcp-(1-, respectively (Zha et al., 2007a). Also in their group, other two PFs, designed as DHPD1 (Zha et al., 2013) and DHPD2 (Li et al., 2014), were separated, and the detailed structural characteristics of DHPD2 were studied. DHPD2 contained (1→5)-linked α-L-Araf, (1→6)-linked α-D-Glcp, (1→6)-linked β-D-Glcp, (1→4)-linked β-D-Glcp, (1→3,6)-linked β-D-Galp, and (1→6)-linked β-D-Galp, with branches of terminal α-D-Xlyp and β-D-Manp.

Tian et al. (2013) isolated a PF (named DHP1A) from *D. huoshanense*. It predominantly consisted of mannose, glucose, and a trace of galactose, with a molecular weight of 6700 Da. Its backbone contained (1→4)-linked α-D-Glcp, (1→6)-linked α-D-Glcp, and (1→4)-linked β-D-Manp, with a branch of terminal β-D-Galp.

Based on methylation analysis and 2D NMR spectra including COSY, HSQC, and HMBC, it was found that a PF (designated DHP-W2) isolated from *D. huoshanense* had a backbone consisting of 1,6-linked β-D-Glc*p*, 1,4-linked β-D-Glc*p*, and 1,4,6-linked β-D-Glc*p* with branches at O-4 – 6. Sugar residues of branches were 1,2,4-linked α-D-Xyl*p*, 1,4-linked α-D-Xyl*p*, 1-linked α-D-Xyl*p*, 1-linked α-D-Gal*p*, and 1-linked α-D-Gal*p*A (Pan et al., 2013).

A more recent study by Li et al. (2015) obtained a homogeneous PF (named DHP-4A) from PLBs of *D. huoshanense* by anion-exchange chromatography and gel permeation chromatography. The backbone of DHP-4A consisted of (1→6)-linked glucose, (1→6)-linked mannose, and (1→3,6)-linked mannose. The β-L-Rha*p*-(1→2)-β-L-Rha*p*-(1→4)-β-D-Man*p*-(1→ and α-L-Ara*f*-(1→3)-α-L-Ara*f* -(1→3)-α-L-Ara*f*-(1→ were regarded as the branches attached to the C-3 position of (1→6)-linked mannose in the backbone.

6.3.1.4 *Structure of* Dendrobium nobile *Polysaccharides*

D. nobile polysaccharides consist of galactose, mannose, glucose, and small amounts of rhamnose, arabinose, and xylose. Sometimes, a trace of galacturonic acid is also found (Lucini et al., 2013; Luo et al., 2010b; Wang et al., 2010a). The molecular weight can vary from 11.4 to 1000 kDa (Lucini et al., 2013; Luo et al., 2010b; Wang et al., 2010a). For example, a PF from *D. nobile* named DNP4-2 had a Mw of 11.4 kDa (Luo et al., 2010b); however, a water-soluble polysaccharide extracted from *D. nobile* had a Mw of 1073 kDa (Lucini et al., 2013).

A neutral polysaccharide fraction (designated DNP-W2) was isolated from the stem of a *D. nobile* plant (Wang et al., 2010c). The DNP-W2 fraction contained glucose, mannose, and galactose in a molar ratio of 6.1:2.9:2.0. Detailed structural study using periodate oxidation–Smith degradation analysis, partial acid hydrolysis analysis, methylation analysis, and 1D ^{1}H and ^{13}C NMR spectroscopy indicated that the polysaccharide contained a glucomannan backbone consisting of -4)- β-D-Glcp-(1-, -6)-β-D-Glcp-(1-, and-4)-β-D-Manp-(1-residues, with acetyl groups substituted at the O-2 position of some -4)-β-D-Manp-(1- residues, and with α-Glap residues attached to the O-6 position of -4)-β-D-Glcp-(1- and -4)- β-D-Manp-(1- residues. A polysaccharide fraction (designated DNP-W3) from the same plant source contained galactose, rhamnose, and arabinose in a molar ratio of 3.1:1.1:1.0 (Wang et al., 2010a). The DNP-W3 had a backbone of 1-3 linked β-D-Galp residues, with branches at the O-4 position of some Galp residues. The branches consisted of 1-4 linked α-Rhap residues and terminated with β-L-Arap residues. A polysaccharide fraction (designated DNP-W5) isolated from the same *D. nobile* herbal material consisted of galactose (29.2%), glucose (28.8%), rhamnose (14.9%), galacturonic acid (13.9%), mannose (11.0%), and xylose (2.1%) (Wang et al., 2010b). About 21% of the galacturonic acids in DNP-W5 were in methyl-esterified form. The degree of substitution of *O*-acetyl groups was about 6.9%. The DNP-W5 had a backbone made of -4)-α-GalAp-(1- and -2)-α-Rhap-(1-, with branches at O-4 of the Rhap and O-3 of the GalAp. The side chains consisted of Galp, Manp, Glcp, and Xylp residues.

A water-soluble polysaccharide fraction was isolated from the stem of *D. nobile* by another research group (Luo et al., 2009). The polysaccharide demonstrated an average molecular weight of around 87.6 kDa, and the monosaccharide composition included glucose (63.3%), mannose (17.0%), galactose (16.5%), and trace amounts of xylose (1.5%), arabinose (1.2%), and rhamnose (0.5%). Using 1D ^{1}H and ^{13}C NMR spectroscopy, it was proposed that the polysaccharide had a backbone containing 1-6 linked α-D-Glcp and α-D-Glap. The O-4 position of each α-D-Glcp in the backbone was substituted by an α-D-Glcp residue, and the O-4 position of the α-D-Glap was substituted by an α-D-Manp. The structural information from this study remains speculative, since there was no explanation how the detailed structure was deduced based on the 1D ^{1}H and ^{13}C NMR spectra.

6.3.1.5 *Structure of* D. moniliforme *Polysaccharides*

A detailed structure of a purified *O*-acetyl glucomannan fraction (designated DMP1a-1) isolated from *D. moniliforme* has been reported (Chen, 2003). It is made up of a main chain consisting of 1-4 linked β-D-mannopyranosyl and β-D-gluco-pyranosyl residues. Terminal glucopyranosyl residues (in α or β configuration) were found attached to the β-D-mannopyranoses in the main chain through a 1-6 glycosidic bond. Acetyl groups were found substituted at the O-3 position of β-D-mannopyranosyl residues. The ratio of mannose to glucose was approximately 2.3:1, based on the results of methylation analysis.

The structure of a polysaccharide fraction (designated DMP2a-1) isolated from *D. moniliforme* was also studied using methylation analysis, periodate oxidation analysis, and 1D ^{1}H and ^{13}C NMR spectroscopy (Xu et al., 2004). The DMP2a-1 was composed of glucose and mannose in a molar ratio of 12.6: 1. It had a backbone consisting of -4)-α-D-Glcp- (1- residues, one sixth of which were found attached to β-D-Manp or β-D-Glcp residues through 1-6 linkages.

A fraction (designated DMP4a-1) of *D. moniliforme* polysaccharides was reported to have an average molecular weight of 3049 Da (Chen, 2003). The DMP4a-1 was composed mainly of glucose, mannose, rhamnose, arabinose, and galactose, in a molar ratio of 2.873:2.850:1.762:1.279:1. Periodate analysis demonstrated a molar ratio of 1.194:2.430:1 for 1-4, 1-3, and 1-6 glycosidic bonds, respectively. The authors hypothesized that the polysaccharide was made of β-D-pyranoses based on FT-IR spectroscopy; it is worth noting that FT-IR spectroscopy alone is not a reliable method to confirm the α or β configurations of monosaccharide residues in polysaccharides. In practice, NMR spectroscopy is now widely accepted for identifying the configurations of sugar units in polysaccharides.

6.3.1.6 *Structures of Polysaccharides From* Dendrobium densiflorum *and* Dendrobium denneanum

A novel neutral heteropolysaccharide (designated DDP-1-D) was extracted from the stems of a *D. densiflorum* plant (Li et al., 2012a). The DDP-1-D had an average molecular weight of 9.4 kDa. It was composed predominantly of glucose and mannose in a molar ratio of around 3:1. Detailed structural analysis, using 2D COSY, TOCSY, HSQC, and HMBC spectroscopy, coupled with periodate oxidation – Smith degradation analysis, indicated a linear polysaccharide chain consisting of -4)-α-D-Glcp-(1-, -6)-α-D-Glcp-(1-, -2)-α-D-Manp-(1-, and -4)-β-D-Manp-(1-. Three fractions (DDP1-1, DDP2-1, and DDP3-1) were purified from the crude polysaccharide of *D. denneanum* by DEAE-Cellulose and Sephadex G-200 column chromatography. The average molecular weights were 51.5, 26.1, and 6.95 kDa, respectively. The DDP1-1 and DDP2-1 were composed of arabinose, xylose, mannose, glucose, and galactose in different ratios, while DDP3-1 did not contain xylose (Fan et al., 2009b). A water-soluble polysaccharide (designated DDP) isolated from *D. denneanum* plants contained α-D-Galp-(1-, α-D-Manp-(1-, -4)-α-D-Glcp-(1-, and -4,6)-α-D-Glcp-(1-, based on 1D ^{1}H and ^{13}C NMR spectroscopy data (Luo, 2006). More detailed studies should be carried out to elucidate the complete structure of these polysaccharides.

6.3.1.7 Other Dendrobium Polysaccharides

Water-soluble polysaccharides (DCPP) were extracted from the stems of *Dendrobium chrysotoxum* by boiling water and ultrasound. Three fractions, termed DCPP-I DCPP-I-a, and DCPP-II, were isolated from DCPP by chromatography on DEAE-Cellulose-52 and Sephadex G-200 columns. Among the three fractions, DCPP-I-a was the most abundant at 79.5%. Its weight-average molecular weight was 122 kDa, and it was composed of xylose, glucose, and galactose in a molar ratio of 1.44:6.93:12.79. The structural analysis indicated that it was a β-D pyran ring with a β-configuration.

6.3.2 Structures of Acemannan

The methods mentioned in 6.3.1.1 are also employed in the structural feature analysis of aloe polysaccharide. Furthermore, a specific hydrolysis with an endo-β-(1→4)-D-mannanase is also applied in the investigation of the acetylation pattern of acemannans. After hydrolysis, the resulting OS were fractionated by SEC and characterized by ESI-MS, ESI-MS/MS, and MALDI-MS (Simões et al., 2012).

Early in 1989, Wozniewski et al. (1990) investigated the structural features of a neutral glucomannan and an acidic arabinogalactan from fresh leaf material of *A. arborescens*. The findings suggested that the polysaccharide was composed of mannose, glucose, arabinose, and galactose in the molar ratios of 60:17:11:11. In addition, glucuronic acid was also found in this polysaccharide, and the total hexose content of the crude polysaccharide fraction was 40%.

McConaughy et al. (2008) isolated highly purified galacturonate polysaccharide from the aloe vera plant (Avp), and studied its chemical composition, molecular weight, and rheological properties. This polysaccharide had a relatively higher molecular weight, with the unique chemical composition of a high galacturonic acid (GalA) content and a low degree of methyl ester substitution. These factors facilitate gel formation upon exposure to low concentrations of calcium ions, leading to potential applications in formulations designed for in situ nasal or subcutaneous protein delivery. Thorough examination of classic dilute solution properties, and the $[\eta]$-*M*w, and *R*g-*M*w relationships, showed an extended random coil Avp in salt solutions by identified persistence length (*L*p), and inherent chain stiffness (*B* parameter). The critical transition concentration of Ce from dilute to concentrated solution was measured by using the probe molecule 1,8-anilino-1-naphthalene sulfonic acid (1,8-ANS) as a function of polymer concentration to study the zero shear viscosity (ηo) and fluorescence emission. Determination of the zeta potential and use of *C*e as a function of ionic strength pointed out that the 0.60–0.30 wt% shift in Ce is correlated with an elevated appearance of intermolecular interactions at high salt concentration solutions. Furthermore, the presentation of dynamic rheology data indicated the gel forming ability of Avp at low polymer and calcium ion concentrations, demonstrating the possibility of the application of Avp to in situ drug delivery.

It has been reported that high hydrostatic pressure (HHP) treatment on aloe vera suspension will cause some effects on its rheological properties. Opazo-Navarrete et al. (2012) studied the rheological properties of aloe vera suspension during storage at 4°C. Aloe vera suspension was pressed at 300, 400, and 500 MPa for 1, 3, and 5 min, respectively. It has been shown that aloe vera suspension presented shear-thinning behavior, and the data fitted into the Herschel–Bulkley model perfectly ($R^2 > 0.99$). The samples pressurized at 300 MPa for 1, 3, and 5 min and 400 MPa for 1 and 3 min presented a decline in their yield stress (σ0) values ($P < 0.05$), while the samples treated at 300 MPa for 1, 3, and 5 min and 400 MPa for 1 min presented a growth in the consistency coefficient (*K*), and a decline in the flow index (*n*) values ($P < 0.05$). However, regardless of how much it was pressure treated, the aloe vera suspension exhibited thixotropy and behaved like a weak-gel ($G' > G''$) when kept at 4°C. The HHP treatment had no influence on the gel properties, it was the pressure – time treatments applied that had an effect on the rheological properties of aloe vera suspension.

In addition, Campestrini (Campestrini et al., 2013) studied the structural and rheological properties of aloe extract (AE) and the polysaccharide fraction (PF) obtained from the leaf pulp of *A. barbadensis* Miller. Structural analyses completed by composition, methylation analysis, and NMR spectroscopy indicated that PF mainly consisted of a partially acetylated 4-linked β-D-glucomannan. The acetyl groups were located at C-2, C-2, and C-3, C-3 and/or C-6. AE and PF aqueous solutions at 25°C showed non-Newtonian behavior (with pseudoplastic characteristics); however, PF showed higher apparent viscosity than AE. Dynamic oscillatory analyses showed that both samples, at the same concentration, behaved as a concentrated solution. The PF presented higher values of G' compared with those of AE, and this behavior could be a consequence of its higher content of partially acetylated glucomannan.

In recent years, the market for commercial aloe vera gel or juice products has increased in popularity. Thus, some researchers have compared the rheological characteristics of native aloe gel and juice under dynamic and steady shear (Lad and Murthy, 2013). Due to the fact that aloe gel has a weak, fibrous, and random polysaccharide structure, under oscillatory shear tests it causes the damping of the elastic moduli (G′) and viscous moduli (G″) at different temperatures. It has been shown that with an increase in temperature the moduli for gel increased, and that for juice decreased. Before reaching the plateau region after a certain shear rate, both aloe vera gel and juice demonstrated shear-thinning behavior. Within the shear-thinning region, the aloe gel samples obtained the flow behavior index of 0.1.

6.4 CHEMICAL MODIFICATIONS

6.4.1 Chemical Modifications of Dendronan

In Qian's research (Qian et al., 2014), a homogenous polysaccharide (DHPD1) was modified using the chlorosulfonic acid–pyridine (CSA–Pyr) method. The conditions for sulfated modification were optimized by the response surface method to get a high degree of substitution (DS). A maximum DS of 1.473 was obtained when the reaction condition was optimized to a reaction temperature of 60°C, reaction time of 160 min, and volume ratio of Pyr to CSA of 2:1. Under these conditions, sulfation occurred to the C-2 and C-6 glycosyl residues in DHPD1. Furthermore, the inhibition of sulfated DHPD1 (SDHPD1) in inhibiting Amadori product formation, dicarbonyl compound formation, or AGEs formation was higher than unsulfated DHPD1 at the same concentration. Also in their group, another polysaccharide fraction from *D. huoshanense* termed DHPD2 was further modified with the chlorosulfonic acid–pyridine method, two sulfated derivatives were generated with the substitution degree of 0.475 and 0.940. Furthermore, these derivatives exhibited stronger inhibitory abilities on protein glycation than that of DHPD2.

6.4.2 Chemical Modifications of Acemannan

Hydrolysis with cellulose is applied in the modification of acemannan. Qiu et al. (2000) incubated aloe gel extract with cellulose at a ratio of 12 mg enzyme per liter of extract for 2 h at room temperature. The mixture was heated at 90°C for 30 min to digest the aloe gel. Some small molecules, like anthraquinones, were removed by incubating the gel with charcoal for an hour followed by centrifugation. The average yield of crude modified aloe polysaccharide (MAP) in the supernatant was 4.5 g solid per liter of gel extract. Then MAP was purified by extensive dialysis against distilled water, followed by subsequent precipitation with 25%, 50%, and 80% ethanol. The precipitates between 50% and 80% ethanol were further purified with a Sepharose Cl-4B (Pharmacia) column (2.5 × 65 cm) and Sephadex G-100 (Pharmacia) column (2.5 × 68 cm) to obtain the purified modified fractions.

6.5 BIOACTIVITIES

6.5.1 Bioactivities of Dendronan

6.5.1.1 Immunostimulatory Properties

6.5.1.1.1 In Vitro Immunostimulatory Properties

Mouse macrophages (RAW264.7 and mouse peritoneal macrophages) and mouse plenocytes (T and B lymphocytes) are commonly used cell lines for the in vitro immune-stimulatory activities of Dendrobium polysaccharides. These cells are incubated with different doses of polysaccharides, followed by determination of a series of parameters, mainly including cell growth rates, nitric oxide (NO) level, and the phagocytic ability and cytokine production of macrophages. Growth rates of cells are calculated based on the cell numbers counted using a hemocytometer (Chen et al., 2011). The determination of NO includes reaction with Griess agent, followed by a colorimetric measurement (at 570 nm) with $NaNO_2$ as standard (Xia et al., 2012a). The phagocytic ability of macrophages can be studied using a neutral red phagocytosis assay system (Xia et al., 2012a). Many cytokines, such as tumor necrosis factor-α (TNF-α), interleukin-10 (IL-10), interleukin-6 (IL-6), interleukin-1α (IL-1α), interferon-γ (IFN-γ), granulocyte/macrophage colony stimulating factor (GM-CSF), and granulocyte-colony stimulating factor (G-CSF) can be tested using enzyme-linked immune sorbent assays (e.g., ELISA), and the levels of mRNA molecules for the expression of these cytokines can be determined using a real-time PCR method.

The in vitro immune-enhancing properties of three *D. officinale* polysaccharide fractions (designated DOP, DOP-1, and DOP-2) were investigated (Xia et al., 2012a). With proper polysaccharide dosages, DOP, DOP-1, and DOP-2 were all proved to increase the proliferation of splenocytes, activate the bioactivity of RAW264.7 macrophage cells, and promote natural killer cell-mediated cytotoxicity. Among the three fractions, DOP-1 exhibited the best pro-proliferative effect on splenocytes, and could stimulate the macrophages to produce the highest levels of IL-2 and IL-4. The DOP-2 induced the highest TNF-α and IL-1β secretion by macrophages, and DOP caused the macrophages to produce the highest levels of NO. In our series studies, RAW 264.7 was employed to evaluate the in vitro immunomodulatory activity of crude and purified *D. officinale* polysaccharide. The results showed that both the crude and the purified polysaccharide were noncytotoxic to cells, and could activate macrophages, resulting in enhanced phagocytic activity, modulated cytokine production, and up-regulated NO expression at both gene levels and protein levels (Cai et al., 2015) (see Fig. 6.4). In a more recent study, He et al. (2016) found that the expression of ERK1/2 and NF-κB were involved in the possible immunomodulatory mechanism of the *D. officinale* polysaccharide fraction.

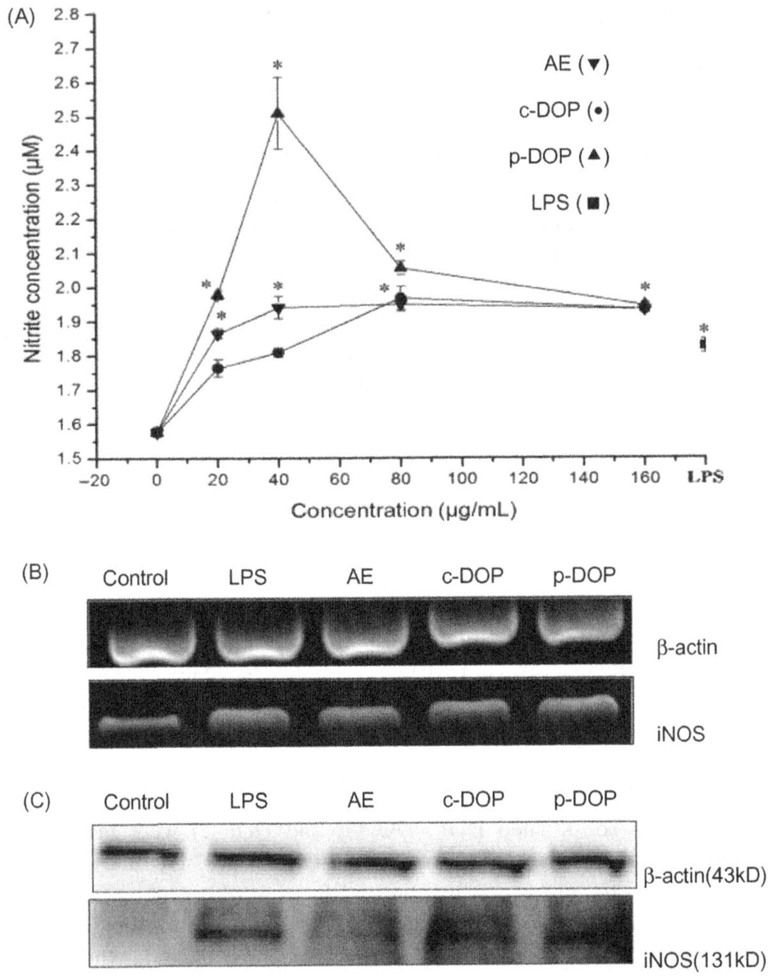

FIGURE 6.4 The *in vitro* immunomodulatory activity of Dendronan. (A) NO levels in macrophages; (B) The expression of iNOS in gene level; (C) The expression of iNOS in protein level.

The in vitro immune-enhancing properties of a crude polysaccharide from *D. huoshanense* (designated HPS) were examined (Zha et al., 2007a). Results showed that HPS, with a concentration of 200 μg/mL in the medium, stimulated peritoneal macrophages to produce 542.9 pg/mL TNF-α in the medium after 24 h incubation. Mice splenocytes in a medium containing 800 μg/mL HPS produced 2.5 times higher IFN-γ than a control group treated with 5 μg/mL concanavalin A. It was also found in this study that the purified fractions (designated HPS 1A, HPS 1B, and HPS 2B) could cause the immune cells to secrete higher levels of IFN-γ and TNF-α than the crude fraction (HPS). In another study (Huang et al., 1994), mouse peritoneal macrophages were incubated in a medium containing a purified *D. huoshanense* polysaccharide fraction (designated HPS-1B23) for 24 h. After the 24 h incubation, there was no difference in the cell numbers between the cells with and without polysaccharide treatment. However, the TNF-α level in the cell medium containing polysaccharide was significantly higher than that in the cell medium without polysaccharide. The medium with a HPS-1B23 at a concentration of 200 μg/mL resulted in TNF-α production of 1130.4 pg/mL in the medium. It was also found in the same study that HPS-1B23 stimulated mouse splenocytes to produce IFN-γ, and that the splenocytes in the medium containing 200 μg/mL HPS-1B23 and 5 μg/mL concanavalin A induced 3.6 times more IFN-γ production than the splenocytes with only 5 μg/mL concanavalin A treatment, suggesting a possible synergism between HPS-1B23 and concanavalin A. Hsieh et al. (2008) reported that a 2,3-*O*-acetyl glucomannan from *D. huoshanense* exhibited specific immune-enhancing properties in vitro. The polysaccharides induced macrophages to produce several cytokines, including IFN-γ, IL-10, IL-6, IL-1α, GM-CSF, and G-CSF. In sharp comparison, the deacetylated glucomannan obtained from alkaline treatment of the same polysaccharides did not induce cytokine production, indicating the significance of *O*-acetyl groups in the immune-biological properties of the glucomannan.

A crude water-soluble *D. nobile* polysaccharide (designated DNCP) and a deproteinized sample of the crude polysaccharide (designated DNPP) were reported to have positive effects on the proliferation of mouse splenocytes in vitro (Chen et al., 2011). The DNCP and DNPP (with dosages of 25 and 50 mg/L) could significantly increase ($P < 0.01$) the concanavalin A-induced proliferation of mice lymphocyte T cells and the LPS-induced proliferation of mice lymphocyte B cells. With the same dosage (25 or 50 mg/L), DNPP treatment tended to induce more lymphocyte B or T cell proliferation than DNCP treatment, though significant differences between the two treatments were not observed ($P < 0.05$). The DNP-W2, DNP-W3, and DNP-W5 fractions isolated from *D. nobile* plants showed in vitro immune-enhancing activities on T and B lymphocytes (Wang et al., 2010a,b,c).

The DMP4a-1 fraction isolated from *D. moniliforme* plants demonstrated the ability to induce RAW264.7 macrophage cells to produce TNF-α, and higher dosages of polysaccharide could result in higher TNF-α production (Chen, 2003). The DMP4a-1 fraction showed interesting dosage-dependent effects on the propagation of BALB/C mouse spleen lymphocyte cells in vitro. Moderate dosages of DMP4a-1 (3.2×10^{-5} or 4.9×10^{-5} mol/L) significantly increased the growth rate of lymphocyte cells, while neither a high dosage (6.4×10^{-5} mol/L) nor a low dosage of the fraction (1.6×10^{-5} mol/L) showed an effect (Chen, 2003).

A polysaccharide sample obtained from *Dendrobium chryseum* was reported to affect the growth rates of lymphocyte cells and macrophage cells isolated from the spleen and abdominal cavity of mice, respectively

(Zhong, 2011). The polysaccharide with a dosage of 100 or 500 mg/L significantly increased the growth rates of macrophage cells and B lymphocyte cells ($P < 0.01$). Compared with the T lymphocyte cells treated with only concanavalin A (5 mg/L), the T lymphocyte cells treated with both concanavalin A (5 mg/L) and polysaccharide (100 or 500 mg/L) exhibited significantly higher growth rates ($P < 0.01$). The structural features of the polysaccharide fraction, however, were not reported.

Meng et al. (2013) compared the effects on macrophage function of crude polysaccharides from five Dendrobium species, including *D. officinale*, *D. fimbriatum*, *D. huoshanense*, *D. nobile*, and *D. chrysotoxum*, and found that the polysaccharide from *D. officinale* had the best in vitro immunomodulating effects. The diversity in their chemical characteristics might contribute to the differences in bioactivity.

6.5.1.1.2 In Vivo Immunostimulatory Properties

In vivo bioactivity studies of Dendrobium polysaccharides include the determination of white blood cell counts (Huang et al., 1995), the growth rates of spleen lymphocytes (Song et al., 2006), and the phagocytic ability of macrophages using a gallinaceous serum globulin injection method (Cai et al., 1989) or Indian ink injection method (Song et al., 2006).

The first report concluding that *D. officinale* polysaccharides could increase the immune function of mice peritoneal macrophages in vivo date back to the late 1980s (Cai et al., 1989). In this study, the immune function of the mice in the control group was reduced by injecting each animal intraperitoneally with 0.2 mL of 0.5% (w/w) hydrocortisone saline solution each day for 5 days. The mice in the treatment group were injected intraperitoneally with the same dosage of hydrocortisone saline solution for the same period of time, with the exception that 0.2 mL of 0.5% (w/w) of *D. officinale* polysaccharide saline solution was injected intraperitoneally following each hydrocortisone injection. On the fourth and fifth days of the experiment, each mouse was subjected to the intraperitoneal administration of 1 mL of 10% (w/w) of starch saline solution and 1 mL of 2% (w/w) gallinaceous serum globulin saline solution, respectively. Thirty minutes after the injection of gallinaceous serum globulin, peritoneal macrophages were collected and stained, and the ability of the macrophages to engulf gallinaceous serum globulin was examined. The macrophage samples collected from the mice treated with the *D. officinale* polysaccharide demonstrated a significantly higher phagocytotic rate than those collected from the control animals (52.69% vs 30.54%, $P < 0.01$).

In another study (Huang et al., 1995), cyclophosphamide (CPA)-induced immunocompromised mice were fed with 0.5 mg of *D. officinale* polysaccharide each day for 9 days, then the white blood cell count of each mouse was tested. It was found that the white blood cell counts (37.33/mL) of the mice treated with polysaccharide were significantly higher ($P<0.01$) than either those of the immune-compromised mice without polysaccharide treatments (13.10/mL) or those of the normal mice without polysaccharide treatment (29.03/mL), indicating that the oral administration of *D. officinale* polysaccharide could increase the white blood cell counts of immunosuppressed mice.

Liu et al. (2011) orally administered *D. officinale* and its polysaccharide to healthy BALB/c mice, and found that the cellular immunity and nonspecific immunity could be remarkably stimulated by both the herb and its polysaccharide, while humoral immunity could only be improved by the administration of *D. officinale* plant.

In our series studies (Huang et al., 2015a), the immunomodulatory activities of *D. officinale* herbs, and crude and purified polysaccharides were tested on both cyclophosphamide-induced immunosuppressed mice and healthy mice. The results showed that all three materials could stimulate the function of spleen lymphocytes, promote the secretion of serum cytokines, immunoglobulin, and serum hemolysin, and improve the phagocytotic function of peritoneal macrophage, as well as protecting the organ from oxidative stress (Fig. 6.5). In addition, the results also proved that polysaccharide should be the main bioactive compound that contributes to the immunomodulating function.

The immunomodulatory effects of *D. tosaense* polysaccharide (DTP) were studied by oral administration to BALB/c mice for 3 weeks by Lin's group. The results demonstrated that DTP substantially increased the proliferation of natural killer (NK) cells, NK cytotoxicity, macrophage phagocytosis, and cytokine induction in splenocytes (Yang et al., 2014).

Zha's group (Zha et al., 2014) has investigated the immunomodulating responses in intestine, spleen, and liver in vivo. The results showed that *D. huoshanense* polysaccharide (DHP) significantly promoted the proliferation of splenic cells and hepatic cells, and remarkably stimulated the secretion of IFN-γ in spleen and liver. When marrow cells were cocultured with small intestinal lymphocytes isolated from mice treated with DHP, the proliferation of marrow cells was significantly increased. This indicated that DHP had the ability to increase immuno-regulatory activity both in the intestinal immune system and in the systemic immune system.

The in vivo immune-enhancing properties of a *D. thyrsiflorum* polysaccharide fraction were investigated by injecting mice with Indian ink and subsequently testing the effects of *D. thyrsiflorum* polysaccharide treatments on the ability of mice macrophages to engulf and clean the carbon particles (carbon clearance index) (Song et al., 2006). It was found that, compared with the mice without polysaccharide treatment, the mice orally given a daily dose of 600 mg/kg *D. thyrsiflorum* polysaccharide for 10 days demonstrated a significantly higher carbon clearance index (0.0126 vs 0.0052, $P<0.01$). This study also showed that the spleen B lymphocyte cells of the mice in the

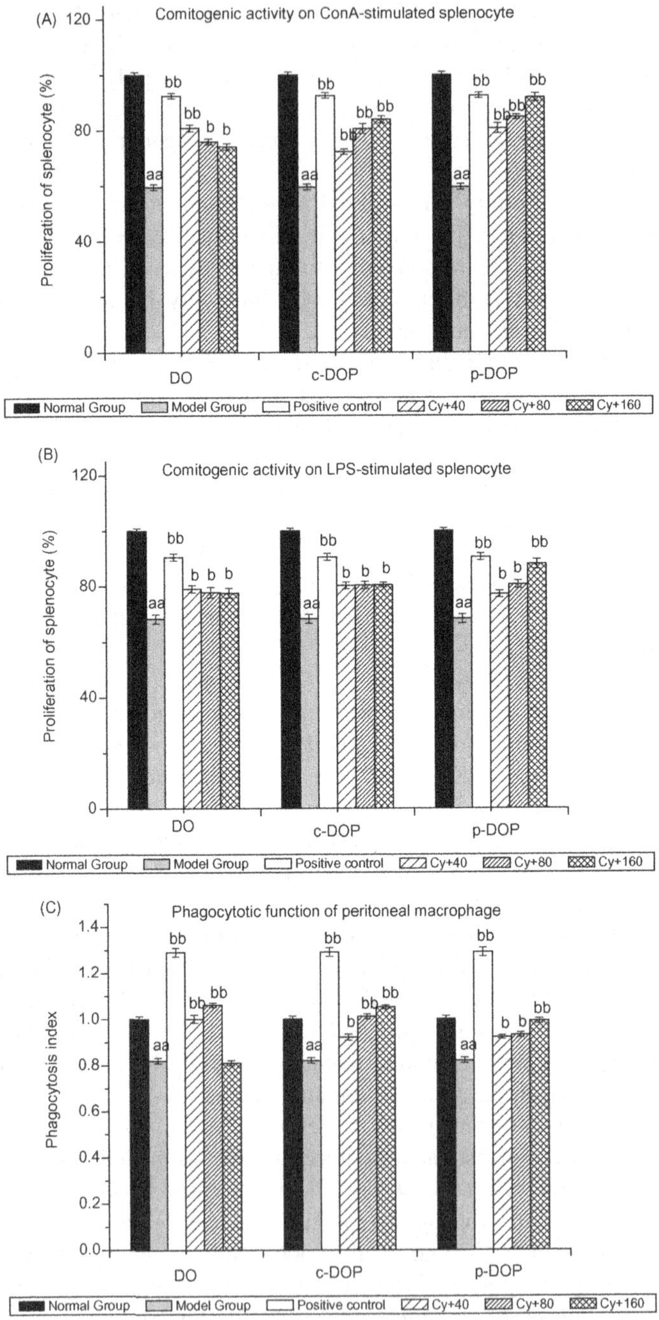

FIGURE 6.5 The *in vivo* immunomodulatory activity of Dendronan. (A) The effects of Dendronan on T lymphocyte subsets; (B) The effects of Dendronan on B lymphocyte subsets; (C) The effects of Dendronan on macrophages proliferation. a $P<.05$ and aa $P<.01$, as compared with Normal group, b $P<.05$ and bb $P<.01$, as compared with Model group.

high-dosage group (600 mg/kg) could grow significantly more quickly than those of the control animals ($P < 0.05$).

Lin investigated the effects of polysaccharides isolated from *D. huoshanense* (DH-PS) on inducing a panel of cytokines/chemokines in mice in vivo and humans in vitro (Lin et al., 2014). They discovered that DH-PS induced TH1, TH2, inflammatory cytokines, and chemokines in mouse in vivo and human cells in vitro. In addition, the list of in vivo DH-PS expanded mouse splenocytes are CD4 + T cells, CD8 + T cells, B cells, NK cells, NKT cells, monocytes/macrophages, granulocytes, and regulatory T cells. In addition, the levels of IL-1β, an antiinflammatory molecules produced by mouse or human immune cells, can also be increased by DH-PS. The injection of DH-PS-produced IL-1ra was over 10-fold that of IL-1β, indicating that DH-PS-induced antiinflammatory activities might supersede the IL-1β-mediated inflammatory activities. It has been shown that ERK/ELK, p38 MAPK, PI3K, and NFkB were involved in the signaling pathways of DH-PS-induced IL-1β production. Finally, the researchers observed that IL-1β levels induced by DH-PS were significantly higher than that by F3, which is a polysaccharide extract isolated from a Chinese herbal medicine called *Ganoderma lucidum*. Overall, this evidence indicated that DH-PS potentially could provide applications for ameliorating IL-1β-induced pathogenic situations.

6.5.1.2 Antioxidant Properties

6.5.1.2.1 In Vitro Antioxidant Properties

In order to test the in vitro antioxidant properties of Dendrobium polysaccharides, many systems have been used to generate free radicals and malondialdehyde (MDA). For example, hydroxyl radicals can be produced using an $Fe^{2+} - H_2O_2$ system (He et al., 2007b), or using an Fe^{2+} − EDTA system incubated with deoxyribose and ascorbic acid (Zhao et al., 2007). Superoxide radicals can be generated by a hypoxanthine (HX) − xanthine oxidase (XOD) system (Zhao et al., 2007), or by an ammonium persulfate − tetramethylethylen ediamine system (He et al., 2007b). The ABTS radicals can be formed in an ABTS − potassium persulfate system (Fan et al., 2009a). The MDA production can be induced in liver homogenate incubated with H_2O_2 or $FeSO_4$ or in mouse mitochondria incubated with $FeSO_4$ and Vitamin C (Vc) (He et al., 2007b). Using an MTT assay, the cell viability of Jurkat cells with glucose oxidase-mediated oxidative stress is also an in vitro antioxidant parameter (Zhao et al., 2007). The DNA nicking assay using super coiled pBR322 plasmid DNA can be performed to examine the in vitro antioxidant properties of Dendrobium polysaccharides (Luo, 2006).

A polysaccharide fraction isolated from the suspension-cultured protocorms of *D. officinale* plants, DCPP1a-1, could significantly inhibit ($P < 0.01$) hydroxyl radicals and superoxide anion radicals in vitro in a dose-dependent manner with half inhibitory concentration (IC_{50}) of polysaccharide of 1.181 mg/mL and 0.727 mg/mL, respectively (He et al., 2007b). The DCPP1a-1 could significantly inhibit ($P < 0.01$) MDA formation in mice liver homogenate caused by auto-oxidation, $FeSO_4$-induced oxidation, or H_2O_2-induced oxidation, with polysaccharide concentrations of 0.60, 1.05, or 1.50 mg/mL. The DCPP1a-1 also demonstrated significant ($P < 0.01$) inhibitory effects on the MDA production in mice liver mitochondria in vitro. In another study (He et al., 2007b), the in vitro antioxidant properties of a crude polysaccharide sample (designated DCPP) and a purified polysaccharide sample (designated DCPP3c-1) isolated from *D. officinale* plants were compared. Both DCPP and DCPP3c-1 can scavenge free radicals and inhibit lipoperoxidation. Both fractions showed very similar IC50 in scavenging hydroxyl radicals and superoxide radicals. Compared with DCPP, DCPP3c-1 demonstrated better inhibitory effects on the liver lipoperoxidation induced by Fe^{2+} and H_2O_2 in vitro. The in vitro free radical-scavenging effects of three purified *D. officinale* polysaccharide fractions designated DSP1, DSP2, and DSP3, with different molecular weight ranges of $15.7-157$, $35.3-74.4$, and $115.1-157$ kDa, respectively, were studied (Bao et al., 2009). It was found that, among the three fractions, DSP1 demonstrated the highest DPPH radical-scavenging capacity, the highest inhibitory ability against $Vc-Fe^{2-}$-induced liver lipoperoxidation, and the highest inhibitory ability against H_2O_2-induced erythrocyte hemolysis. The hydroxyl radical-scavenging ability of the three fractions was in the order of DSP34 >DSP14 >DSP2. All three fractions were found to be able to protect the pBR322 DNA from hydroxyl radical-induced oxidation. *Dendrobium officinale* polysaccharide also possessed the ability to scavenge superoxide anion radicals and hydroxyl radicals in vitro (Wang et al., 2009). The in vitro antioxidant activities of polysaccharides from *D. huoshanense* and *D. officinale* were evaluated and compared by some researchers (Zha et al., 2007b). Both polysaccharide samples exhibited excellent abilities to scavenge superoxide anion radicals and hydroxyl radicals, and to inhibit the oxidation of linoleic acid induced by alkyl radicals. The *D. huoshanense* polysaccharide extract tended to show higher in vitro antioxidant properties than the *D. officinale* polysaccharide, but the investigators did not mention whether the difference was significant or not.

The polysaccharide fraction (designated DCLP) extracted from *D. chrysotoxum* demonstrated a strong concentration-dependent scavenging effect on the hydroxyl radicals generated by Fe^{3+} using a cell-free hydroxyl radical generating system (Zhao et al., 2007). In this study, it was also observed that the dissolution of pBR322 plasmid DNA to the hydroxyl radical generating system resulted in a time-dependent increment in the formation of single-stranded nicked DNA, while the existence of DCLP in the system inhibited Fe^{3+}-dependent DNA nicking. The DCLP also had strong scavenging effects on superoxide anions and DPPH free radicals in a dose-dependent manner. Using DCLP with a dosage of 4 mg/mL caused a reduction of superoxide production by 50.5%. The DCLP showed antioxidant effects on the glucose/glucose oxidase-mediated cytotoxicity of Jurkat cells. The addition of DCLP with dosages of 60 and 120 μg/mL to the cell media enhanced the viability of Jurkat cells by 30.9% ($P < 0.05$) and 78.9% ($P < 0.001$), respectively. The antioxidant effect of DCLP on Jurkat cells was quite close to that of the control antioxidant, acteoside.

The abilities of *D. moniliforme* polysaccharide and *D. huoshanense* polysaccharide to scavenge hydroxyl radical and superoxide anion radicals in vitro were similar to each other in a concentration-dependent manner (Shu et al., 2009). Both polysaccharide samples were found to inhibit MDA formation in mouse liver homogenates in vitro induced by auto-oxidation or by a $Vc - Fe^{2+}$ system significantly ($P < 0.01$), but the inhibition effects were not concentration-dependent. Each polysaccharide had the highest inhibition ability for auto-oxidation-induced MDA formation at a concentration of 1 mg/mL, while the optimal concentration of each sample for the $Vc - Fe^{2-}$-induced MDA formation was 3 mg/mL.

The in vitro antioxidant properties of three purified polysaccharide fractions (designated DDP1-1, DDP2-1, and DDP3-1) isolated from *D. denneanum* plants were compared, with Vc as an antioxidant control (Fan et al., 2009a). Results showed that all three fractions had very weak 2,2′-azino-bis (3-ethylbenzothiazoline-6-sulfonic acid) (ABTS) radical-scavenging power, DDP1-1 and DDP3-1 had low 2,2-diphenyl-1-picrylhydrazyl (DPPH) radical-scavenging ability and moderate hydroxyl radical-scavenging ability, and DDP2-1 demonstrated a very strong ability to scavenge DPPH radical and hydroxyl radicals. At concentrations ranging from 0.5 to 2 mg/mL, the hydroxyl radical-scavenging power of DDP2-1 was close to that of Vc. Using the same methodology, the research team found that a water-soluble polysaccharide isolated from *D. fimbriatum* (designated DFHP) had strong hydroxyl radical and ABTS radical- scavenging power at polysaccharide concentrations ranging from 1 to 3 mg/mL, and demonstrated a moderate DPPH radical-inhibitory rate of 46.67% at a concentration of 3 mg/mL (Luo and Fan, 2011). The same research team further tested the in vitro antioxidant properties of a crude water-soluble *D. denneanum* polysaccharide (designated DDP) (Luo and Fan, 2011). Interestingly, they observed that DDP demonstrated a strong in vitro DPPH radical-scavenging power, strong in vitro hydroxyl radical-scavenging power, and weak in vitro ABTS radical-scavenging power, quite different from the previously reported radical-scavenging performance of purified *D. denneanum* polysaccharides (Fan et al., 2009a).

In He's study (He et al., 2016), a neutral heteropolysaccharide fraction (named DOP-1-1) showed no obvious scavenging capacity of DPPH in contrast with other studies of *D. officinale* polysaccharide. Furthermore, in our study on the in vitro antioxidant activity of *D. officinale* polysaccharide, a similar result was obtained. We found that DOP showed no remarkable effects on scavenging hydroxyl radical, superoxide anion radical, and DPPH free radical.

6.5.1.2.2 In Vivo Antioxidant Properties

The levels of superoxide dismutase (SOD), glutathione peroxidase (GSH-Px), and MDA in serum and organs are frequently used as molecular parameters for assessment of the in vivo antioxidant activity of polysaccharides (Luo et al., 2011; Wang et al., 2009). An observation has been made that mice orally taking *D. officinale* polysaccharide for 15 days demonstrated significantly higher ($P < 0.05$) levels of liver SOD, liver GSH-Px, serum SOD, and serum GSH-Px in vivo than control mice without polysaccharide feeding (Yuan et al., 2011). In our series study (Huang et al., 2015b), *D. officinale* polysaccharide was demonstrated to improve the SOD, CAT, and GSH levels, while it decreased the MDA contents in liver, thymus, and serum of both cyclophosphamide-induced mice and healthy mice. This indicated that Dendronan had effective activity in protecting the body from oxidative stress in vivo. A polysaccharide fraction isolated from *D. huoshanense*, termed DHP1A, proved to be active in decreasing the MDA levels and improving the SOD, CAT, and GPx, as well as GSH levels in livers of CCl_4-treated mice (Xu et al., 2013). It was observed in another study that mice injected intraperitoneally with *D. denneanum* polysaccharide for 20 days exhibited lower serum MDA and higher serum SOD than control mice without polysaccharide administration (Luo et al., 2011). However, the researchers did not report whether the differences were significant or not. The reported in vivo antioxidant effects followed a dose-dependent manner.

6.5.1.3 Antitumor Properties

6.5.1.3.1 In Vitro Antitumor Properties

Lymphokine activated killer (LAK) cells are widely applied to kill tumor cells in cancer patients (Rosenberg et al., 1989). The ability of Dendrobium polysaccharides to induce LAK cells in vitro can be an indicator of the antitumor properties of the polysaccharides. The E-rosette tests on lymphocytes isolated from blood samples of cancer patients are also useful for the evaluation of in vitro antitumor properties of Dendrobium polysaccharides (Cai et al., 1989). For example, Cai et al. (1989) took peripheral blood samples from 31 cancer patients (19 nasopharyngeal carcinoma patients and 12 malignant lymphoma patients), isolated lymphocytes from the blood samples, and investigated the effects of *D. officinale* polysaccharides on the production of E-rosette forming lymphocytes. The results showed that the level of E-rosette forming lymphocytes in the polysaccharide-treated group was significantly higher than that in the control group (69.87% vs 52.35%, $P < 0.01$), and was close to that in the thymosin treated group (69.60%). The results indicated that *D. officinale* polysaccharides had the same ability as thymosin to stimulate the E-rosette formation of peripheral blood lymphocytes in cancer patients. In another study, cord blood (CB) samples were collected from newborn baby umbilical cords, and peripheral blood (PB) samples from 10 cancer patients. The IL-2 (coupled with or without *D. officinale* polysaccharide) was used to induce the growth of LAK cells in those blood samples, followed by examination of the effects of polysaccharide treatment on the activities of LAK cells (the ability to kill Raji cells). It was found that the combined use of IL-2 and *D. officinale* polysaccharide significantly increased ($P < 0.01$) the activities of LAK cells compared with the treatment using only IL-2.

6.5.1.3.2 In Vivo Antitumor Properties

Hepatoma 22 (H22) tumor-bearing mice and sarcoma 180 (S180) tumor-bearing mice are useful animal models for in vivo antitumor bioactivity tests of Dendrobium polysaccharides. The in vivo parameters include tumor inhibitory rate, thymus index, T lymphocyte transformation rate, spleen index, activity of NK cells, phagocytic rate of peritoneal macrophages, and half value of hemolysin (HC_{50}) (Luo and Fan, 2011; Fan and Luo, 2011; Zhang et al., 2009). In addition, serum IL-2, TNF-α, IFN-γ, and MDA in tumor-bearing mice can be tested (Fan and Luo, 2011). The effects of the dosages of a polysaccharide fraction (designated DCPP1a-1) isolated from suspension-cultured protocorms on the tumor weight, tumor inhibitory rate, spleen index, and thymus index of H22 tumor-bearing mice were investigated (He et al., 2007a). Results showed that, as the polysaccharide dosage increased from 50 to 250 mg/kg, tumor weight increased from 0.741 to 0.876 g, tumor inhibitory rate decreased from 28.6% to 15.7%, spleen index decreased from 7.50 to 6.50, and thymus index decreased from 4.50 to 3.68. Compared with the tumor-bearing mice without polysaccharide treatments, the mice treated with DCPP1a-1 at a dosage of 50 mg/kg demonstrated significantly lower tumor weight ($P < 0.01$), significantly higher spleen index ($P < 0.01$), and significantly higher thymus index ($P < 0.01$). The results of this study indicated that low-dosage DCPP1a-1 polysaccharide treatment (50 mg/kg) had better antitumor effects than high-dosage polysaccharide treatment (150 and 250 mg/kg).

It was reported that the S180 tumor-bearing mice, gavage-fed with a daily dosage of *D. officinale* polysaccharides of 1600 mg/kg body weight for 10 days, demonstrated 72.2% of the tumor weight of the control tumor-bearing animals without polysaccharide treatments (Zhang et al., 2009). Compared with the control S180-bearing mice, the polysaccharide-treated ones showed significantly higher T lymphocyte transformation rates (7147 cpm vs 5072 cpm, $P < 0.01$), significantly higher NK cell activities (33.4CI% vs 26.5CI%, $P < 0.05$), significantly higher phagocytic rates of peritoneal macrophages (34.3% vs 27.9%, $P < 0.01$), and significantly higher half values of hemolysin (HC_{50}) (130.1 vs 72.4, $P < 0.01$).

There was another report that the S180 tumor-bearing mice orally taking *D. officinale* polysaccharides for 8 days exhibited significantly lower ($P<0.01$) tumor weight and significantly higher ($P>0.01$) levels of serum cytokines, including IL-2 and TNF-α, than the tumor-bearing animals without polysaccharide treatments (Jin et al., 2010). It was also shown that a 70-h *D. officinale* polysaccharide treatment could inhibit the growth of human hepatoma SMMC27721 cells in vitro by more than 30%. Aoxue Luo's group compared the antitumor activity of three water-soluble fractions from *D. denneanum* (designated as DDP1-1, DDP2-1, and DDP3-1) on S180-bearing mice. The results indicated that DDP1-1 had a better effect on inhibiting the increment of tumor at a low dose of 12.5 mg/kg (Fan and Luo, 2011).

In another study, mice inoculated with S180 were injected intraperitoneally with different dosages of each of the four polysaccharide fractions, designated DNP1-1, DNP2-1, DNP3-1, and DNP4-2, isolated from *D. nobile*, once a day for 10 days (Luo and Fan, 2011). Results showed that, among all the fractions, DNP4-2 demonstrated the strongest antitumor effects. The antitumor effects of DNP4-2 were not in a concentration-dependent manner. The DNP4-2 inhibited the tumor growth by 67.01% at a dosage of 2.5 mg/mL. It was also found that the tumor cell-bearing mice taking DNP4-2 at a dosage of 2.5 mg/mL showed significantly higher ($P<0.01$) levels of serum cytokines such as IL-2, TNF-α, and IFN-γ, and significantly lower ($P<0.01$) level of serum MDA than the control animals. The researchers proposed that the antitumor properties resulted from the excellent immune-enhancing and antioxidant properties of DNP4-2. Using the same animal model, the in vivo antitumor properties of three polysaccharide fractions (designated DDP1-1, DDP2-1, and DDP3-1) extracted from *D. denneanum* were examined in a separate study by the same group (Fan and Luo, 2011). The antitumor effect of DDP1-1 was found to be dose-dependent. The DDP1-1 with a dosage of 12.5 mg/kg had the highest tumor inhibition effect of 72.04%. However, significant differences in the levels of serum IL-2, TNF-α, IFN-γ, and MDA between each polysaccharide treatment group and the control group were not found ($P>0.05$).

A study has compared the antitumor activities of six water extract fractions (termed DNP-W1, DNP-W2, DNP-W3, DNP-W4, DNP-W5, and DNP-W6), a 5% NaOH extract fraction (DNP-OH), and a 5% HCl extract fraction (DNP-H), and found that DNP-W1 and DNP-W3 showed higher antitumor activities against Sarcoma 180 in vivo and HL-60 in vitro than any other fraction (Lucini et al., 2013). The inhibition ratio of the total water extract fraction (DNP-W) against S180 tumor was three times higher than that of DNP-OH, and six times higher than that of DNP-H. This suggested that the water-soluble polysaccharide was the main antitumor source in *D. nobile* herbs (Lucini et al., 2013).

6.5.1.4 Hypoglycemic Activity

A polysaccharide fraction (designated DMP1a-1) isolated from *D. moniliforme* was found capable of significantly lowering the blood glucose level of experimental diabetic mice, while the polysaccharide had no significant influence on the serum glucose level of normal mice (Chen, 2003). Some reported that *D. denneanum* polysaccharide could significantly reduce ($P<0.01$) the serum glucose level of alloxan-induced diabetic mice, and significantly increase ($P<0.01$) the glucose tolerance of diabetic rats (Luo, 2006). However, the serum glucose level of normal mice and the glucose tolerance of normal rats were not significantly affected by *D. denneanum* polysaccharide treatments. A polysaccharide fraction (designated DCLP) isolated from *D. chrysotoxum* was also reported to have antidiabetic effects on alloxan-induced hyperglycemic mice (Zhao et al., 2007). Mice with a 7-day treatment of DCLP at dosages of 200 and 500 mg/kg body weight showed blood glucose levels of 303.6 and 231.7 mg/d L, respectively, which were both significantly lower ($P<0.05$) than that of the diabetic mice without polysaccharide treatment (360 mg/dL). In another study (Chen and Liu, 2011), a DMEM high glucose medium was used to induce human vascular endothelial-like ECV 304 cells to express nuclear fact or kappa-light-chain-enhancer of activated B cells (designated NF-κB), a protein complex controlling the DNA transcription related to the progress of diabetic vascular diseases. It was found that the ECV304 cells growing in the DMEM high glucose medium containing *D. officinale* polysaccharides showed much lower NF-κB expression than the control cells growing in the media without the polysaccharides, indicating that the polysaccharide was capable of restraining the over-expression of NF-κB, and could thus be helpful in preventing diabetic vascular diseases. Some researchers (Li et al., 2012b) treated streptozotocin (STZ)-induced diabetic rats with different dosages of *D. huoshanense* polysaccharide for 45 days. After treatment, it was found that the rats taking high dosages of *D. huoshanense* polysaccharide demonstrated significantly higher ($P<0.01$) glutathione (GSH) level, GSH-Px activity, glutathione reductase (GR) activity, glutathione S-transferase (GST) activity, SOD activity, and CAT) activity in eye lens than the control rodents. There were significantly lower ($P<0.01$) levels of MDA and carbonyl group in the eye lens of polysaccharide-treated rats than those of control rats. The results indicated that *D. huoshanense* polysaccharide could help to prevent diabetic cataract. The advanced glycation end products (AGE) produced by the glycation reactions of proteins are one of the important causes of the complications of diabetes (Ulrich and Cerami, 2000). It was reported that the water-soluble polysaccharide (designated DHP-W2) isolated from *D. huoshanense* could inhibit the formation of AGE between bovine serum albumin and glucose (Pan, 2005). In a later study, DHP-W2 was added to a phosphate buffer solution (pH7.4) containing 2% bovine serum albumin (w/v) and 0.1 M glucose, and the solution was incubated at 37°C in the dark for 3 weeks. The solution containing 2% DHP-W2 (w/w) was found to inhibit protein glycation by 28%.

The antihypertensive and antistroke effects of *D. officinale* polysaccharide were studied using spontaneously hypertensive rats (stroke prone) (SHR-SP) as an animal model (Wu et al., 2011). It was found that the SHR-SP rats fed with *D. officinale* polysaccharides for 12 weeks demonstrated lower death rates, longer survival times, lower stroke rates, lower blood pressure, lower heart index, lower brain index, and lower kidney index than the control SHR-SP rats without polysaccharide treatments, implying that the *D. officinale* polysaccharide might be helpful in alleviating hypertension and stroke.

Pan et al. (2014) compared the hypoglycemic and antioxidative activities of four polysaccharides from different Dendrobium species, including *D. huoshanense* (DHP), *D. officinale* (DOP), *D. nobile* (DNP), and *D. chrysotoxum* (DCP). These four polysaccharides differed in intrinsic viscosities and monosaccharide compositions. The polysaccharides were orally administered to alloxan-induced diabetic mice. The fasting blood glucose, glycosylated serum protein, and serum insulin levels were determined. The results showed that DHP, DOP, and DNP, but not DCP, had a significant hypoglycemic effect with the decreasing order of DHP>DNP>DOP. Histopathological observation indicated that DHP, DOP, and DNP possessed the ability to intervene in the damage of pancreatic tissues. The analysis of SOD, CAT, malonaldehyde, and L-glutathione levels in the liver and kidney showed that DHP, DOP, and DNP could protect the organ from oxidative stress induced by alloxan, and DHP had the best protective effect. All these together suggested that there were significant differences in hypoglycemic and antioxidative activities between the four Dendrobium polysaccharides, which may contribute to their physicochemical properties.

6.5.1.5 Liver Protective Properties

The protective effects of *D. officinale* polysaccharides on cyclosporine A (CsA)-induced liver injury in Wistar rats were evaluated (Jin and Liu, 2009). The results of a hematoxylin and eosin (HE) staining study showed that the liver-injured rats taking *D. officinale* polysaccharides for 21 days had fewer histological liver lesions compared with the control rats with liver injury but without polysaccharide treatments. It was also found that the 21-day *D. officinale* polysaccharide treatment could reduce the levels of serum total bilirubin (TBIL), serum alanine transaminase (ALT), serum as part transaminose (AST), liver MDA, and liver triglyceride (TRIG), and could increase the levels of serum albumin (ALB), liver SOD, liver glutathione (GSH), Na^+-K^+ ATPase, and Ca^{2+}-Mg^{2+} ATPase of CsA-challenged Wistar rats. The results indicated that *D. officinale* polysaccharide could prevent and alleviate

CsA-induced liver injury. These effects could be related to the in vivo antioxidant bioactivities of *D. officinale* polysaccharide.

Pan et al. (2012) carried out a study to investigate the preventative effects of galactoglucomannan (GGM), a homogeneous polysaccharide from *D. huoshanense*, on liver injury and fibrosis induced by sodium selenite. Liver damaged Sprague-Dawley rats were orally administered with three different dosages of GGM. Biochemical and histological analysis discovered that GGM significantly ameliorated selenite-induced liver injury and fibrosis in rats. The decreased activities of the serum hepatic enzymes, including alanine aminotransferase (ALT), aspartate aminotransferase (AST), and lactate dehydrogenase (LDH) indicated that GGM could alleviate the toxicity of selenite to liver tissue. In addition, the administration of GGM could also down-regulate the levels of H_2O_2 and MDA and GSH, recover the fluidity of hepatic plasma membrane, and maintain the activities of endogenous enzymes including SOD, CAT, and GST. Meanwhile, the expression of TGF-β1 and type I collagen were also repressed. All these together indicated the protective effect of GGM on selenite-induced liver injury and fibrosis.

Wang et al. (2014) obtained a homogeneous polysaccharide (DHP) from *D. huoshanense* and studied its possible liver protection mechanisms in ethanol-induced subacute liver injury mice. The results showed that DHP could effectively restrain the increased ratio of liver weight to body weight, reduce the levels of serum AST, total cholesterol, total bilirubin, and low density lipoprotein, and down-regulate hepatic steatosis in liver-injured mice. Notably, hepatic proteomics analysis performed by two-dimensional difference gel electrophoresis (2D-DIGE) coupled with matrix-assisted laser desorption ionization time-of-flight mass spectrometry (MALDI-TOF/TOF-MS) discovered that cystathionine betasynthase (Cbs) and D-lactate dehydrogenase (Ldhd) were two key proteins regulated by daily DHP intervention, which may assist in correcting the abnormal hepatic methionine metabolism pathway and down-regulating the level of hepatic methylglyoxal generated from disordered metabolic pathways caused by ethanol. Therefore, it could be concluded that DHP could protect liver function from alcoholic injury by modulating the expression of Cbs and Ldhd.

In their later study, a UHPLC/LTQ Orbitrap XL MS-based metabolomics approach was used to investigate the metabolic profiles of serum and liver tissues from experimental mice. The results suggested that DHP ameliorated early steatosis and inflammation in liver histology. Furthermore, the metabolomic analysis of serum and hepatic tissue revealed that ethanol treatment mainly altered phosphatidylcholines (PCs), including PC (13:0) and phosphocholine, arachidonic acid metabolites including 20-ethyl PGF2α, and amino acids including L-proline; while DHP treatment modulated the altered metabolic levels particularly involved in phosphocholine and L-proline. These data indicated that DHP might restore the abnormal metabolism pathways caused by ethanol exposure to prevent the progression of alcoholic liver injury (Wang et al., 2015).

In Huang's study (Huang et al., 2013a), they also proved that *D. huoshanense* polysaccharide could restore the damage induced by CCl_4 in mice, which was proved by the alleviation in lipid peroxidation and TNF-α expression.

6.5.1.6 *The Effects on Sjögren's Syndrome*

The effects of *D. officinale* polysaccharide treatment on Sjögren's syndrome (SS), a chronic disease with the characteristics of xerostomia (dry mouth) and xerophthalmia (dry eyes), were studied (Lin et al., 2010). In this study, an in vivo auto allergic mouse model and an in vitro human salivary gland cell line A-253 were used for the bioactivity test. Results showed that immunization using SG autoantigen could decrease the body weight of mice, increase water intake, and decrease AQP-5 expression in organs related to body fluid. Sera from model mice could induce the apoptosis of A-253 cells by the activation of caspase-3. The treatment with *D. officinale* polysaccharide could alleviate a number of pathological changes in both in vivo and in vitro experimental models. The researchers concluded that the *D. officinale* polysaccharide was a promising candidate for the alleviation of Sjögren's syndrome by up-regulating AQP-5 expression and preventing cell apoptosis (Lin et al., 2010). In another study (Lin et al., 2011), the possible mechanisms of the preventative effects of *D. officinale* polysaccharide on Sjögren's syndrome were investigated. Experimental mice with Sjögren's syndrome were treated with or without the *D. officinale* polysaccharide with a daily dosage of 4 mg for 11 days. It was found that the polysaccharide had the ability to reduce progressive lymphocyte infiltration and apoptosis, and to alleviate the disorders of proinflammatory cytokines in the mice with Sjögren's syndrome. The polysaccharide could help the mice to normalize the level of AQP-5 and maintain proper saliva secretion. The protective effects of the polysaccharide on AQD-5 were confirmed by an in vitro experiment using A-253 cell line in the study. It was also reported that *D. officinale* polysaccharides had the ability to reduce the SRBC-induced production of serum complement and hemolysin of mice (Hua, 2005). Results of the study pointed to the fact that the croton oil-induced auricle tumefaction and the zymosan-induced foot edema of mice could be greatly alleviated by treatment with *D. officinale* polysaccharides.

6.5.1.7 *Antibacterial Activity*

The effects of polysaccharides from *D. officinale* and *D. hookerianum* on the growth of *Staphylococcus aureus*, *Escherichia coli*, and *Bacillus subtilis* in vitro were investigated using a filter paper disk method (Li et al., 2000). It was found that *D. hookerianum* had inhibitory effects on all three bacteria studied, while *D. officinale* polysaccharide could only inhibit the growth of *E. coli* and *B. subtilis* strains. The *D. officinale* polysaccharide had the most remarkable antibacterial effect on *E. coli*, by demonstrating a minimal inhibitory concentration (MIC) of 0.5% and an inhibiting zone diameter of 15.8 mm. The polysaccharide from *D. hookerianum* demonstrated the best antibacterial effect on *B. subtilis*, with an MIC of 0.5% and an inhibiting zone diameter of 12.8 mm.

What deserves to be noted is that, although polysaccharides are the predominant component of these aforementioned bioactive samples, and many of the samples were purified by chromatographic methods, it is still likely that some minor components, such as proteins and phenolics, could be responsible or partially responsible for the demonstrated bioactivities of the polysaccharide samples under test. For this reason, we suggest that the effects of these minor components should be considered in future bioactivity studies of Dendrobium polysaccharides by completely removing these impurities or carrying out parallel experiments to exclude the effects caused by these nonpolysaccharide components.

6.5.2 Bioactivities of Acemannan

A. vera (L.) Burm. f. (*A. barbadensis* Miller) is a perennial succulent xerophyte. They allow water storage tissue to grow inside the leaves, so aloe vera can survive in a dry climate with low or erratic rainfall. The majority of the inner leaf contains a clear, soft, moist and slippery tissue composed of large thin-walled parenchyma cells, therefore water stored inside is in a viscous mucilage form. Hence, the aloe plants' leaves contain not only cell-wall carbohydrates like cellulose and hemicellulose, but also carbohydrates like acetylated mannans (Liu et al., 2010). Many of the health benefits associated with aloe vera have a close relationship with the polysaccharides, especially the acetylated mannans contained in the gel of the leaves. These biological activities include immunomodulatory activity, antioxidant activity, antitumor activity, the promotion of wound healing, antimicrobial activity, hypoglycemic activity, antiinflammatory properties, gastroprotective properties, and so on (Hamman, 2008).

6.5.2.1 Immunomodulatory Activity of Acemannan

In Liu's study (Liu et al., 2006), a mannose-rich polysaccharide isolated from *A. vera* L. var. chinensis (Haw.) Berg. has been proved to have a potential stimulatory activity on B and T cells. Subsequently, they discussed the modulatory activity of this polysaccharide on murine peritoneal macrophages. The polysaccharide treatment of mice could promote the migration of macrophages to the peritoneal cavity. In vitro treatment of polysaccharide on peritoneal macrophage could increase expression of MHC-II and FcγR, and stimulate endocytosis, phagocytosis, NO production, TNF-α secretion, and tumor cell cytotoxicity. In addition, the administration of the polysaccharide into allogeneic ICR mice stimulated systemic TNF-α production dose-dependently, and prolonged the lifetime of tumor-bearing mice. Therefore, it can be speculated that acemannan can stimulate immunity by modulating the function of lymphocytes or macrophages.

In another study by Lee et al. (2001), the stimulatory property of acemannan on dendritic cells (DCs) was proved. Immature DCs were derived from mouse bone marrow (BM) cells by culturing in a medium with GM-CSF and IL-4 added, and then they were cocultured with acemannan, sulfated acemannan, and LPS, respectively. The resultant DCs were examined for phenotypic and functional properties. Phenotypic analysis confirmed that acemannan could induce the maturation of immature DCs, by modulating the expression of class II MHC molecules and major costimulatory molecules such as B7-1, B7-2, CD40, and CD54. Functional maturation of immature DCs was supported by increased allogeneic mixed lymphocyte reaction (MLR) and IL-12 production. The differentiation-inducing activity of acemannan was almost completely abolished by chemical sulfation. Based on these results, it could be proposed that the immunostimulatory activity of acemannan is at least in part due to its capacity to promote differentiation of immature DCs.

Yu et al. (2009) found that aloe vera polysaccharides could improve the immunity of oral ulcer mice. The supplementation of aloe vera polysaccharides could enhance blood CD8+ and reduce CD4+/CD8+ in a dose-dependent manner; up-regulate the blood IL-2 and IFN-γ levels; decrease oral mucosal expression of endothelin-1 (ET-1); and increase the plasma IgG, IgA, and IgM levels. All together indicated that aloe vera polysaccharides had the ability to improve body immunity and reduce oxidative stress.

Sun et al. (2011) discussed the immunomodulatory effects of aloe polysaccharides on chickens immunized with Bordetella avium inactivated vaccine. The experimental results found that aloe polysaccharide remarkably increased serum and bile antibody level, blood lymphocyte ratio, and the proliferation of splenic T lymphocytes. These indicated that aloe polysaccharide could greatly promote the immune effect of B. avium inactivated vaccine on chickens.

Farahnejad et al. (2011) found that the immunomodulatory effects of aloe vera and its fractions could contribute to its antibiotic effects. In their study, aloe vera extract was fractionated into six fractions based on their molecular weights; R100 (>100 kDa), R50 (100 kDa >R50 >50 kDa), R30 (50 kDa >R30 >30 kDa), R10 (30 kDa >R10 >10 kDa), R5 (10 kDa >R5 >5 kDa), and F5 (<5 kDa). These fractions were cocultured with macrophages infected with *C. albicans*. The results found that the extract, as well as R100, R50, R30, and R10 fractions of aloe vera, could significantly improve the cell viability of macrophages in most doses.

6.5.2.2 *Antioxidant Activity*

Kaithwas et al. (2014) studied the antioxidant activity of polysaccharides from aloe vera in vitro and in vivo. The in vitro study showed that the polysaccharides exhibited significant inhibitory activity against DPPH-, superoxide, NO, and hydrogen peroxide scavenging assay with significant reducing activity at all concentrations used. The in vivo study showed that the polysaccharide produced a marked protection against cardiotoxicity induced by doxorubicin (DOX), which was evidenced by significant reductions in serum LDH, serum creatine phosphokinase, cardiac TBARS, CAT, and SOD, along with increased levels of blood and tissue GSH in a dose-dependent manner.

Similar results were observed in Kang's experiment (Kang et al., 2014). Purified aloe vera polysaccharide (APS) strongly scavenged radicals including DPPH, hydroxyl, and alkyl radicals. Moreover, APS also showed a protective effect against AAPH-induced oxidative stress and cell death in Vero cells, as well as in the in vivo zebrafish model.

Liu investigated the in vitro antioxidant activity of polysaccharides isolated from the gel (GAPS-1) and the skin (SAPS-1) of *A. barbadensis* Miller respectively (Liu et al., 2007). The results indicated that both polysaccharides produced in vitro antioxidant activity. However, this antioxidant effect was higher in GAPS-1 than in SAPS-1, which might be due to the greater number of acetyl group in GAPS-1.

In addition, Wu pointed out that aloe polysaccharide also had good preventive and therapeutic significance in some free radical associated health problems such as coronary heart ailments, Parkinson's, and Alzheimer's diseases (Wu et al., 2006).

In Yu's study (Yu et al., 2009), the antioxidant effects of aloe vera polysaccharides in oral ulcer mice were also found. The plasma SOD activity and MDA levels of oral ulcer mice were modulated to a normal standard after the administration of aloe vera polysaccharides.

6.5.2.3 *Antitumor Activity*

Antitumor activity was also observed in polysaccharide fractions separated from aloe. This bioactivity was indirectly stimulated by its immunomodulatory activity, including the function in stimulating the peritoneal macrophages, splenic T and B cell proliferation, and activating these cells to secrete TNF-α, IL-1β, INF-γ, IL-2, and IL-6 (Leung et al., 2004).

6.5.2.4 *Wound Healing Activity*

Wound healing activity is widely observed in acemannans. One possible mechanism for acemannan to exhibit cell proliferation and skin wound healing effects is activating the AKT/mTOR signaling pathway. In Xing's study (Xing et al., 2015a), a mouse skin wound model and skin primary fibroblasts were used to evaluate the positive effects of acemannan on cutaneous wound healing. The results found that acemannan significantly accelerated skin wound closure and cell proliferation, and promoted the expression of cyclin D1 in cultured fibroblasts through the AKT/mTOR signal pathway. This lead to an improved activity of the eukaryotic translation initiation factor-4F (eIF4F), and increased translation of cyclin D1. In contrast, pharmaceutical blockade of AKT/mTOR signaling by mTOR inhibitor rapamycin or AKT inhibitor VIII abolished acemannan-induced cyclin D1 translation and cell proliferation. Therefore, it could be confirmed that the stimulation of the AKT/mTOR signal pathway played an important role in wound healing.

Tabandeh et al. (2014) found another mechanism for aloe polysaccharide to repair skin wounds. Rats with cutaneous wounds on the back were treated daily with aloe vera polysaccharide (AVP) for 30 days. The levels of MMP-3 and TIMP-2 gene expression were detected using real-time PCR. The levels of *n*-acetyl glucosamine (NAGA), *n*-acetyl galactosamine (NAGLA), and collagen content were also analyzed using standard biochemical methods. Faster wound closure was observed at day 15 postwounding after the treatment with aloe polysaccharide. With the administration of AVP, MMP-3 gene expression was inhibited at day 10 postwounding, but later was up-regulated. Furthermore, TIMP-2 gene expression, collagen, NAGLA, and NAGA synthesis were enhanced in relation to untreated wounds. These results suggested a new perspective for the wound repair activity of aloe vera polysaccharide at a molecular level.

6.5.2.5 Periodontal Health Care Ability

Chantarawaratit investigated the effect of acemannan on periodontal tissue regeneration (Chantarawaratit et al., 2014). Acemannan was added when culturing primary human periodontal ligament cells in vitro. New DNA synthesis, expression of growth/differentiation factor 5 and runt-related transcription factor 2, expression of vascular endothelial growth factor, bone morphogenetic protein-2 and type I collagen, alkaline phosphatase activity, and mineralized nodule formation were determined. The results showed that acemannan remarkably promoted the proliferation of periodontal ligament cell, up-regulated the expression of growth/differentiation factor 5, runt-related transcription factor 2, vascular endothelial growth factor, bone morphogenetic protein-2, type I collagen and alkaline phosphatase activity, and mineral deposition, as compared with the untreated control group. Furthermore, an animal trial was completed later. Premolar class II furcation defects were made in four mongrel dogs. Acemannan sponges were applied into the defects. The amount of new bone, cementum, and periodontal ligament formation was evaluated 30 and 60 days after the operation, and the results indicated that acemannan could remarkably accelerate new alveolar bone, cementum, and periodontal ligament formation in class II furcation defects.

The two major cells corresponding to new alveolar bone and dentin formation are periodontal fibroblasts and pulpal fibroblasts. Specifically, bone morphogenic protein-2 (BMP-2) has clinical potential for inducing bone and dentin repair and renewal. Jittapiromsak et al. (2007) studied the influence of acemannan on BMP-2 expression in periodontal fibroblasts and pulpal fibroblasts of humans. The levels of BMP-2 mRNA and protein expression were analyzed after the cells were treated with various concentrations of acemannan for 24 and 48 h. The results suggested that acemannan was able to induce BMP-2 mRNA expression and protein synthesis in both cells. This suggested that acemannan is a potential polysaccharide for promoting the expression of BMP-2 in periodontal fibroblasts and pulpal fibroblasts (Jittapiromsak et al., 2007).

6.5.2.6 *Gastrointestinal Protective Effect*

It is reported that acemannan can be regarded as a potential prebiotic in protecting gastrointestinal health. In Gullón's study (2015), stimulation of acemannan fermentation was carried out using artificial intestinal microbiota made from stool samples of six healthy donors. The prebiotic activity was assessed by the quantification of short chain fatty acids (SCFA) and the evaluation of dynamic bacterial population in mixed fecal cultures by fluorescence in situ hybridization (FISH) (Gullón et al., 2015). These findings support the idea that acemannan could be developed as a prebiotic to improve gastrointestinal health.

Furthermore, researchers also found that both the whole leaf and the inner fillet gel liquid preparations of aloe vera have the ability to enhance the intestinal absorption and bioavailability of coadministered compounds (Hamman, 2008).

6.5.2.7 *Other Bioactivities*

In spite of the previously mentioned activity, aloe polysaccharides also have many other biological activities, such as antiviral activity, liver protective ability, hematopoietic and hematologic activity, cell proliferation effects, and so on.

Farahnejad et al. (2011) indicated that aloe polysaccharide also had the ability to resist viral infections, and that this activity was induced by modulating macrophage function.

Cui did a study focused on the hepatoprotective potential of aloe vera polysaccharides (AVGP) against chronic alcohol-induced hepatotoxicity in mice (Cui et al., 2014). The findings pointed out that AVGP supplements significantly attenuated the levels of serum ALTs, lipids, and hepatic TG, and ameliorated histopathological modifications in liver damaged mice. Remarkably, AVGP markedly increased lipolytic genes (AMPK-α2 and PPAR-α) hepatic expression, but had no influence on lipogenic gene expression. Treatment with AVGP could also surpress alcohol-dependent oxidative stress, which was achieved by a decline in MDA and growth in GSH and SOD. In addition, AVGP treatment caused a significant decrease in LPS and TNF-α, down-regulation of TLR-4 and MyD88, as well up-regulation of IκB-α, thus these responses could mitigate alcohol-induced inflammation. Collectively, it could be supposed that AVGP had a potentially protective ability against chronic alcohol-induced liver injury. Also, its hepatoprotective effect appears to corresponded with its ability to accelerate lipolysis and inhibit inflammatory response, as well as its antioxidant capacity.

In addition, the recovery effects of aloe gel polysaccharide on oxidative DNA were also observed (Kim et al., 1999). Hikino et al. (1986) proved that the polysaccharide fractions of the leaves of a number of aloe species could significantly reduce blood sugar levels in normal and alloxan-induced hyperglycemic mice. Talmadge et al. (2004) revealed that a purified carbohydrate fraction from aloe vera extracts could increase hematopoietic and hematologic activity compared to the starting material. In addition, this fraction also modulated liver and lung cytokine mRNA levels, resulting in significant increases in messages for hematopoietic cytokines, including granulocyte-colony stimulating factor (G-CSF) and stem cell factor (SCF) (Talmadge et al., 2004).

The stimulating effect of aloe polysaccharide on human fibroblasts has also been proved. Yao et al. (2009) administered a polysaccharide fraction from *A. barbadensis* Miller (API) to the superficial II scald model mice. The results suggested that API could stimulate the proliferation of human fibroblasts in vitro, and this stimulation could result from the positive effect of API on the generation of glucogen, DNA synthesis, and the formation of collagen in human fibroblast.

6.6 DISCUSSION: STRUCTURE – BIOACTIVITY RELATIONSHIP

Wang et al. (2010a,b) extracted three polysaccharide fractions from the stems of *D. nobile*, including the water extract fraction (DNP-W), 5% NaOH extract fraction (DNP-OH), and 5% HCl extract fraction (DNP-H). Subsequently, anion-exchange chromatography was applied to isolate six subfractions (DNP-W1, DNP-W2, DNP-W3, DNP-W4, DNP-W5, and DNP-W6) from the water extract fraction (DNP-W) (Lucini et al., 2013). The monosaccharide composition, protein content, uronic acid content, total carbohydrate content, viscosity, and molecular weight, as well as the antitumor activity of these nine polysaccharide fractions were analyzed. Results showed that DNP-W1 and DNP-W3 exhibited higher antitumor activities against Sarcoma 180 in vivo and HL-60 in vitro, compared with that of the other polysaccharide fractions. The structural characteristics analysis showed that DNP-W1 had the smallest Mw of 9.2×10^4 g/mol and relatively high viscosity of 28.6 cm^3/g, which indicated that DNP-W1 had a more expanded flexible chain conformation. Meanwhile, DNP-W3 had the largest Mw of 107.3×10^4 g/mol and the highest viscosity of 41.3 cm^3/g (Lucini et al., 2013). All these together pointed out that the antitumor activities of these polysaccharide fractions were correlated positively with the amount of mannose, as shown in the case of DNP-W1. Additionally, a relatively higher molecular mass and better water solubility seemed to increase the antitumor activities, as shown in the case of DNP-W3.

Zha et al. (2013) obtained a homogeneous active polysaccharide, termed DHPD1, from *D. huoshanense*. An enzymatic degradation method was applied to achieve DHPD1-derived fragments. Under optimal conditions, three DHPD1-derived fractions with different molecular weights were obtained. The structural features and bioactivities of the three fragments were determined, and the results found that the inhibitory action of DHPD1 on protein nonenzymatic glycation reduced with a decrease in molecular weight.

Another example of aloe polysaccharide also found that the antitumor activity of glucomannan had a close relationship with the content of mannose and molecular weight (Leung et al., 2004). Three purified polysaccharide fractions designated as PAC-I, PAC-II, and PAC-III were prepared from *A. vera* L. var. chinensis (Haw.) Berg. Their antitumor activities were evaluated in a murine model. The results showed that PAC-I, which had the highest mannose content and molecular weight, exhibited the highest biological response. Hence, the bioactivity of polysaccharide was found to correlate with the mannose content and molecular weight (Li et al., 2008b).

The molecular size is another critical factor that could influence the immunomodulatory activity of MAP. Purified polysaccharide from aloe was achieved and later separated into three fractions, G2E1DS3 (MW > 400 kDa), G2E1DS2 (5 kDa < MW < 400 kDa), and G2E1DS1 (MW < 5 kDa) (Lei et al., 2011). The immunomodulatory activities of these polysaccharide fractions were examined on RAW 264.7 and in an ICR strain of mouse implanted with sarcoma 180 cells (Lei et al., 2011). The results suggested that polysaccharides between 400 and 5 KDa had the most potent macrophage-activating activity, as determined by increased cytokine production, NO release, expression of surface molecules, and phagocytic activity. Similar results were found in an in vivo antitumor study (Lei et al., 2011). Therefore, it might be speculated that molecular weight is correlated with the bioactivity of acemannan.

The existence of acetyl groups in glucomannan is an important factor that greatly affects the biological activity of polysaccharide. Chokboribal prepared a deacetylated acemannan (DeA-cAM) from aloe, and studied its physical properties and microscopic structure, as well as cell proliferation and gene expression activity (Zhang et al., 2003). The results suggested that increased deacetylation in acemannan could decrease its water solubility and hydrophilicity, and completed deacetylation could alter its conformation to a partial crystal structure (Zhang et al., 2003). Meanwhile, biological studies found that acemannan could induce cell proliferation, and VEGF and Collagen I expression; however, 100% DeAcAM could not (Zhang et al., 2003). This indicated that the bioactivity of acemannan was reduced corresponding to its deacetylation. Furthermore, it could be concluded that the acetyl groups affect acemannan's structure and physical/biological properties.

Liu et al. (2007) isolated two polysaccharides, GAPS-1 and SAPS-1, from the gel and the skin of *A. barbadensis* Miller, and studied their antioxidant activities. They found that GAPS-1 demonstrated more effective antioxidant activities than SAPS-1. And it was also observed that more acetyl groups were found in GAPS-1. Therefore, this could be further proof that acetyl groups play an important role in the bioactivity of polysaccharide.

Furthermore, Simões also pointed out that lower branching, shorter chains, and higher acetylation seem to promote the immunostimulatory activity attributed to acemannans (Simões et al., 2012).

REFERENCES

Balan, B.J., Niemcewicz, M., Kocik, J., Jung, L., Skopinska-Rozewska, E., Skopinski, P., 2014. Oral administration of *Aloe vera* gel, anti-microbial and anti-inflammatory herbal remedy, stimulates cell-mediated immunity and antibody production in a mouse model. Cent. Eur. J. Immunol. 39, 125–130.

Bao, S., Zha, X., Hao, J., Luo, J., 2009. In vitro antioxidant activity of polysaccharides with different molecular weights from *Dendrobium candidum*. Food Sci. 30, 123–127.

Cai, H.-L., Huang, X.-J., Nie, S.-P., Xie, M.-Y., Phillips, G.O., Cui, S.W., 2015. Study on *Dendrobium officinale* O-acetyl-glucomannan (Dendronan®): Part III–Immunomodulatory activity in vitro. Bioactive Carbohydrates and Dietary Fibre 5, 99–105.

Cai, T., Liu, Q., Li, D., Chen, A., Huang, B., Cheng, S., et al., 1989. Effects of *Dendrobium officinale* polysaccharides on the activities of T lymphocytes and macrophages. Acad. J. Sun Yat-Sen Univ. Med. Sci. 10, 66–67.

Campestrini, L., Silveira, J., Duarte, M., Koop, H., Noseda, M., 2013. NMR and rheological study of *Aloe barbadensis* partially acetylated glucomannan. Carbohyd. Polym. 94, 511–519.

Chantarawaratit, P., Sangvanich, P., Banlunara, W., Soontornvipart, K., Thunyakitpisal, P., 2014. Acemannan sponges stimulate alveolar bone, cementum and periodontal ligament regeneration in a canine class II furcation defect model. J. Periodontal. Res. 49, 164–178.

Chen, R., 2010. Extraction technology of polysaccharide in Dendrobium. China Med. Herald 7, 140–143.

Chen, Y., 2003. Structural characterization and biological activities of polysaccharides from *Dendrobium moniliforme* (L.) Sw. (Vol. Thesis for the degree of doctor of philosophy). *Zhejiang University, Hangzhou.*

Chen, Y., Liu, W., 2011. Optimal extraction procedure of polysaccharides from *Dendrobium candidum* and effects of polysaccharides on over expression of NF-κB agent in vascular endothelial cells induced by high glucose. J. Shanxi Coll. Tradit. Chin. Med. 12, 28–31.

Chen, Z., Lai, P., Lin, Q., 2002. Effects of different herbal processing methods on levels of water-extractable Dendrobium alkaloids and polysaccharides. J. Zhejiang Coll. Tradit. Chin. Med. 26, 79–81.

Chen, Z., Ye, S., Fan, Y., Hou, J., Bian, H., Peng, Y., 2011. Optimization of extraction technology of polysaccharide from *Dendrobium nobile* and effects on proliferation of mouse splenocytes. Chin. J. Exp. Tradit. Med. Form. 17, 27–32.

Cui, Y., Ye, Q., Wang, H., Li, Y., Yao, W., Qian, H., 2014. Hepatoprotective potential of *Aloe vera* polysaccharides against chronic alcohol-induced hepatotoxicity in mice. J. Sci. Food Agric. 94, 1764–1771.

Davies, K., Turner, M., 2004. Pseudopollen in *Dendrobium unicum* Seidenf. (Orchidaceae): reward or deception?. Ann. Bot. 94, 129–132.

Ding, X., Xu, L., Wang, Z., Zhou, K., Xu, H., Wang, Y., 2002. Authentication of stems of *Dendrobium officinale* by rDNA ITS region sequences. Planta Med. 68, 191–192.

Fan, Y., Luo, A., 2011. Evaluation of anti-tumor activity of water-soluble polysaccharides from *Dendrobium denneanum*. Afr. J. Pharm. Pharmaco. 5, 415–420.

Fan, Y., He, X., Zhou, S., Luo, A., He, T., Chun, Z., 2009a. Composition analysis and antioxidant activity of polysaccharide from *Dendrobium denneanum*. Int. J. Biol. Macromol. 45, 169–173.

Fan, Y., Luo, A., Luo, A., Ze, C., 2009b. Research on the ultrasonic-assisted extraction process for polysaccharide from D. denneanum. Chin. J. Hosp. Pharm. 22, 008.

Farahnejad, Z., Ghazanfari, T., Yaraee, R., 2011. Immunomodulatory effects of *Aloe vera* and its fractions on response of macrophages against *Candida albicans*. Immunopharmacol. Immunotoxicol. 33, 676–681.

Ferreira, W.M., Suzuki, R.M., Pescador, R., Figueiredo-Ribeiro, R.L., Kerbauy, G.B., 2011. Propagation, growth, and carbohydrates of Dendrobium Second Love (Orchidaceae) in vitro as affected by sucrose, light, and dark. In Vitro Cell Dev. Biol. 47, 420–427.

Guan, Y., Li, H., Wang, Y., Li, X., 2010. Microexamination of histological differences between *Dendrobium devonianum* and *Dendrobium officinale*. Chin. Med. Mat. 33, 1869–1871.

Gullón, B., Gullón, P., Tavaria, F., Alonso, J.L., Pintado, M., 2015. In vitro assessment of the prebiotic potential of *Aloe vera* mucilage and its impact on the human microbiota. Food Funct. 6, 525–531.

Hamman, J.H., 2008. Composition and applications of *Aloe vera* leaf gel. Molecules 13, 1599–1616.

He, T., Yang, L., Li, Y., Wang, C., Huang, D., 2007a. Physicochemical properties and antitumor activity of polysaccharide DCPP1a-1 from suspension-cultured protocorms of *Dendrobium candidum*. Nat. Prod. Res. Dev. 19, 578–583.

He, T., Yang, L., Li, Y., Wang, C., Hu, J., 2007b. Effects of the polysaccharides DCPP1a-1 from suspension-cultured protocorms of *Dendrobium candidum* on oxygen radical and lipid peroxidation. Nat. Prod. Res. Dev. 19, 410–414.

He, T.B., Huang, Y.P., Yang, L., Liu, T.T., Gong, W.Y., Wang, X.J., et al., 2016. Structural characterization and immunomodulating activity of polysaccharide from *Dendrobium officinale*. Int. J. Biol. Macromol. 83, 34–41.

Hikino, H., Takahashi, M., Murakami, M., Chohachikonno, Mirin, Y., Karikura, M., et al., 1986. Isolation and hypoglycemic activity of arborans A and B, glycans of *Aloe arborescens* var. natalensis leaves. Int. J. Crude Drug Res. 24, 183–186.

Hsieh, Y.S., Chien, C., Liao, S.K., Liao, S.F., Hung, W.T., Yang, W.B., et al., 2008. Structure and bioactivity of the polysaccharides in medicinal plant *Dendrobium huoshanense*. Bioorg. Med. Chem. 16, 6054–6068.

Hua, Y., 2005. Study on *Dendrobium officinale* polysaccharide. Thesis for the Doctor of Philosophy. Zhejiang University, Hangzhou.

Hua, Y.F., Zhang, M., Fu, C.X., Chen, Z.H., Chan, G.Y., 2004. Structural characterization of a 2-O-acetylglucomannan from *Dendrobium officinale* stem. Carbohyd. Res. 339, 2219–2224.

Huang, J., Li, S., Zhao, H., Pan, L., Sun, H., Luo, J., 2013a. Protective effects of polysaccharides from *Dendrobium huoshanense* on CCl4-induced acute liver injury in mice. Chin. J. Chin. Mater. Med. 38, 528–532.

Huang, X., Nie, S., Wang, Y., Wang, Y., Cai, H., Tian, F., et al., 2013b. Optimized Extraction and Compositional Analysis of Polysaccharides from Dried Stems of *Dendrobium officinale*. Chin. J. Food Sci. 34, 21–26.

Huang, M., Huang, B., Cai, T., Liu, Q., Li, D., Chen, I., 1994. Extraction, purification, and structural characterization of *Dendrobium officinale* polysaccharides. Chin. Tradit. Herbal Drugs 25, 128–129.

Huang, M., Cai, T., Liu, Y., 1995. Effects of polysaccharides from *Dendrobium candidum* on white blood cells and lymph cell moving inhibition factor of mice. Nat. Prod. Res. Dev. 8, 39–41.

Huang, X., Nie, S., Cai, H., Zhang, G., Cui, S.W., Xie, M., et al., 2015a. Study on *Dendrobium officinale* O-acetyl-glucomannan (Dendronan): Part IV. Immunomodulatory activity in vivo. J. Funct. Foods 15, 525–532.

Huang, X., Nie, S., Cai, H., Zhang, G., Cui, S.W., Xie, M., et al., 2015b. Study on *Dendrobium officinale* O-acetyl-glucomannan (Dendronan®): Part VI. Protective effects against oxidative stress in immunosuppressed mice. Food Res. Int. 72, 168–173.

Jiang, D., Liu, Z., Shen, H., Liao, C., 2007. Comparative study on *Dendrobium officinale* polysaccharide content at different habitats in Karstic environment. Carsologica Sinica 26, 226–229.

Jin, H., 2012. The content change of polysaccharides in 3 huoshan species of Dendrobium with the variation of time. J. Anhui Agric. Sci. 9, 024.

Jin, L., Liu, C., 2009. Preventive and therapeutic effect of Dendrobium polysaccharides on rat liver injury induced by cyclosporine A. Chin. Hosp. Pharm. J. 29, 1891–1894.

Jin, L., Liu, C., Tang, T., Shen, L., 2010. Experimental study on anti-tumor effect of *Dendrobium candidum* polysaccharides. Chin. Pharm. J. 22, 13.

Jin, X., Yuan, H., Si, J., Zhang, Z., Yu, Q., Wang, L., 2011. The effects of blossoming on the contents and monosaccharide compositions of polysaccharides in *Dendrobium officinale* plants. Chin. J. Chin. Mater. Med. 36, 2176–2178.

Jittapiromsak, N., Jettanacheawchankit, S., Lardungdee, P., Sangvanich, P., Thunyakitpisal, P.D., 2007. Effect of acemannan on BMP-2 expression in primary pulpal fibroblasts and periodontal fibroblasts, in vitro study. J. Oral Tissue Engin. 4, 149–154.

Kaithwas, G., Singh, P., Bhatia, D., 2014. Evaluation of in vitro and in vivo antioxidant potential of polysaccharides from *Aloe vera* (*Aloe barbadensis* Miller) gel. Drug Chem. Toxicol. 37, 135–143.

Kaminska, M., Stpiczynska, M., 2011. The structure of the spur nectary in *Dendrobium finisterrae* Schltr. (Dendrobiinae, Orchidaceae). Acta Agrobotanica 64, 19–26.

Kang, M.C., Kim, S.Y., Kim, Y.T., Kim, E.A., Lee, S.H., Ko, S.C., et al., 2014. In vitro and in vivo antioxidant activities of polysaccharide purified from aloe vera (*Aloe barbadensis*) gel. Carbohyd. Polym. 99, 365–371.

Kim, H.S., Kacew, S., Lee, B.M., 1999. In vitro chemopreventive effects of plant polysaccharides (*Aloe barbadensis* Miller, *Lentinus edodes*, *Ganoderma lucidum* and *Coriolus versicolor*). Carcinogenesis 20, 1637–1640.

Kim, K., Kim, H., Kwon, J., Lee, S., Kong, H., Im, S.A., et al., 2009. Hypoglycemic and hypolipidemic effects of processed *Aloe vera* gel in a mouse model of non-insulin-dependent diabetes mellitus. Phytomedicine 16, 856–863.

Kong, D., Zhong, Y., Shen, H., 2015. Effects of interaction between moisture and light on growth, photosynthetic characteristics and soluble sugar of *Dendrobium officinale*. J. Huazhong Agr. Univ. 34, 19–24.

Lad, V., Murthy, Z., 2013. Rheology of *Aloe barbadensis* miller: a naturally available material of high therapeutic and nutrient value for food applications. J. Food Eng. 115, 279–284.

Lee, J.K., Lee, M.K., Yun, Y.-P., Kim, Y., Kim, J.S., Kim, Y.S., et al., 2001. Acemannan purified from *Aloe vera* induces phenotypic and functional maturation of immature dendritic cells. Int. Immunopharmacol. 1, 1275–1284.

Lei, L., Changchun, D., Fuhui, L., 2011. Study on the antibacterial effects of two Dendrobium polysaccharides. Med. Plant 2, 21–22.

Lertwiriyawong, B., Phinija, K., Huehne, P.S., 2011. Sequence analysis of the 16SrRNA-rps12 inverted repeat region in chloroplast DNA of a Dendrobium orchid. Kasetsart J. Nat. Sci. 45, 461–472.

Leung, M., Liu, C., Zhu, L., Hui, Y., Yu, B., Fung, K., 2004. Chemical and biological characterization of a polysaccharide biological response modifier from *Aloe vera* L. var. chinensis (Haw.) Berg. Glycobiology 14, 501–510.

Li, C., Zhu, J., 2010. Comparison study of polysaccharides content in medicinal dendrobum herb of different collection period. Chin. J. Pharm. Anal. 6, 056.

Li, F., Cui, S.H., Zha, X.Q., Bansal, V., Jiang, Y.L., Asghar, M.N., et al., 2015. Structure and bioactivity of a polysaccharide extracted from protocorm-like bodies of *Dendrobium huoshanense*. Int. J. Biol. Macromol. 72, 664–672.

Li, G., Li, X., Sun, H., Dao, Z., 2011a. Optimization of extraction of *Dendrobium candidum* polysaccharides using response surface methodology. Mod. Chin. Med. 13, 29–33.

Li, J., Li, S., Huang, D., Zhao, X., Cai, G., 2011b. Advances in the of resources, constituents and pharmacological effects of *Dendrobium officinale*. Sci. Technol. Rev. 29, 74–79.

Li, M., Su, J., Wu, R., Yang, F., Zhao, G., Zhai, C., 2000. Anatomical study on the nutritive organs of *Dendrobium candidum*. Acta Agri. Univ. Henanensis 35, 125–129.

Li, Q., Xie, Y., Su, J., Ye, Q., Jia, Z., 2012a. Isolation and structural characterization of a neutral polysaccharide from the stems of *Dendrobium densiflorum*. Int. J. Biol. Macromol. 50, 1207–1211.

Li, X., Deng, Y., Pan, L., Huang, J., Luo, J., 2012b. Antioxidant effect of polysaccharide from *Dendrobium huoshanense* on lens tissue of diabetic cataract rats. Chin. Tradis. Patent Med. 34, 418–421.

Li, W., Hu, Z., Zhou, J., Zhu, J., Li, X., Lou, Y., et al., 2008a. Comparative analysis of amino acids, polysaccharides and trace elements in *Dendrobium officinale* from different sources. ACTA Univ. Tradit. Med. Sinen. Pharmacol. Shanghai 22, 80–83.

Li, X., Ding, X., Chu, B., Zhou, Q., Ding, G., Gu, S., 2008b. Genetic diversity analysis and conservation of the endangered Chinese endemic herb *Dendrobium officinale* Kimura et Migo (Orchidaceae) based on AFLP. Genetica 133, 159–166.

Li, X.L., Xiao, J.J., Zha, X.Q., Pan, L.H., Asghar, M.N., Luo, J.P., 2014. Structural identification and sulfated modification of an antiglycation *Dendrobium huoshanense* polysaccharide. Carbohyd. Polym. 106, 247–254.

Lin, J., Chang, Y.-J., Yang, W.-B., Alice, L.Y., Wong, C.-H., 2014. The multifaceted effects of polysaccharides isolated from *Dendrobium huoshanense* on immune functions with the induction of interleukin-1 receptor antagonist (IL-1ra) in monocytes. PLoS One 9, e94040.

Lin, X., Sze, S.C.-W., Tong, Y., Zhang, Z., Feng, Y., Ng, T.B., et al., 2010. Protective effect of *Dendrobium officinale* polysaccharides on experimental Sjogren's syndrome. J. Complement. Integr. Med.7.

Lin, X., Shaw, P.-C., Sze, S.C.-W., Tong, Y., Zhang, Y., 2011. *Dendrobium officinale* polysaccharides ameliorate the abnormality of aquaporin 5, pro-inflammatory cytokines and inhibit apoptosis in the experimental Sjögren's syndrome mice. Int. Immunopharmacol. 11, 2025–2032.

Ling, Z., Fang, Y., 2012. Chemical constituents and pharmacological effects of Dendrobium. Chin. Mod. Med. 19, 13–14.

Liu, C., Leung, M., Koon, J., Zhu, L., Hui, Y., Yu, B., et al., 2006. Macrophage activation by polysaccharide biological response modifier isolated from *Aloe vera* L. var. chinensis (Haw.) Berg. Int. Immunopharmacol. 6, 1634–1641.

Liu, C., Wang, C., Xu, Z., Wang, Y., 2007. Isolation, chemical characterization and antioxidant activities of two polysaccharides from the gel and the skin of *Aloe barbadensis* Miller irrigated with sea water. Process Biochem. 42, 961–970.

Liu, H., Yang, J., Huang, S., Huang, B., Lai, X., 2010. Determination of polysacharide in *Dendrobium officinale* by 3, 5-dinitrosalicylic acid colorimetry. Asia-Pacific Tradit. Med. 8, 011.

Liu, W., Xiong, Y., 2009. Process of extracting polysaccharides from *Dendrobium nobile* Lind. by microwave-assisted method. Chin. Arch. Tradit. Chin. Med. 27, 1315–1317.

Liu, X.-F., Zhu, J., Ge, S.-Y., Xia, L.-J., Yang, H.-Y., Qian, Y.-T., et al., 2011. Orally administered *Dendrobium officinale* and its polysaccharides enhance immune functions in BALB/c mice. Nat. Prod. Commun. 6, 867–870.

Lucini, L., Pellizzoni, M., Molinari, G.P., 2013. Anthraquinones and β-polysaccharides content and distribution in Aloe plants grown under different light intensities. Biochem. Syst. Ecol. 51, 264–268.

Luo, A., 2006. Effect of *Dendrobium denneanum* polysaccharide reducing blood glucose in vivo. Chin. J. Appl. Envir. Biol. 12, 334.

Luo, A., Fan, Y., 2011. Immune stimulating activity of water-soluble polysaccharide fractions from *Dendrobium nobile* Lindl. Afr. J. Pharm. Pharmacol. 5, 625–631.

Luo, A., He, X., Zhou, S., Fan, Y., He, T., Chun, Z., 2009. In vitro antioxidant activities of a water-soluble polysaccharide derived from *Dendrobium nobile* Lindl. extracts. Int. J. Biol. Macromol. 45, 359–363.

Luo, A., Fan, Y., Luo, A., He, T., Chun, Z., 2010a. Effect of different extracting methods on the content determination of polysaccharide from Caulis Derdrobii. China Pharm. 21, 601–602.

Luo, A., He, X., Zhou, S., Fan, Y., Luo, A., Chun, Z., 2010b. Purification, composition analysis and antioxidant activity of the polysaccharides from *Dendrobium nobile* Lindl. Carbohyd. Polym. 79, 1014–1019.

Luo, A., Ge, Z., Fan, Y., Luo, A., Chun, Z., He, X., 2011. In vitro and in vivo antioxidant activity of a water-soluble polysaccharide from *Dendrobium denneanum*. Molecules 16, 1579–1592.

Lv, G., Yan, M., Chen, S., 2013. Review of pharmacological activities of *Dendrobium officinale* based on traditional functions. Chin. J. Chin. Mater. Med. 38, 489.

Mcconaughy, S.D., Stroud, P.A., Boudreaux, B., Hester, R.D., Mccormick, C.L., 2008. Structural characterization and solution properties of a galacturonate polysaccharide derived from aloe vera capable of in situ gelation. Biomacromolecules 9, 472–480.

Meng, L.-Z., Lv, G.-P., Hu, D.-J., Cheong, K.-L., Xie, J., Zhao, J., et al., 2013. Effects of polysaccharides from different species of Dendrobium (Shihu) on macrophage function. Molecules 18, 5779–5791.

Nie, S.-P., Cai, H.-L., 2012. Research progress in bioactive components and functions of *Dendrobium officinale*. Chin. J. Food Sci. 23, 079.

Opazo-Navarrete, M., Tabilo-Munizaga, G., Vega-Gálvez, A., Miranda, M., Pérez-Won, M., 2012. Effects of high hydrostatic pressure (HHP) on the rheological properties of *Aloe vera* suspensions (*Aloe barbadensis* Miller). Innov. Food Sci. Emerg. Technol. 16, 243–250.

Pan, L.-H., Lu, J., Luo, J.-P., Zha, X.-Q., Wang, J.-H., 2012. Preventive effect of a galactoglucomannan (GGM) from *Dendrobium huoshanense* on selenium-induced liver injury and fibrosis in rats. Exp. Pathol. 64, 899–904.

Pan, L.-H., Feng, B.-J., Wang, J.-H., Zha, X.-Q., Luo, J.-P., 2013. Structural characterization and anti-glycation activityin vitroof a water-soluble polysaccharide fromdendrobium huoshanense. J. Food Biochem. 37, 313–321.

Pan, L.-H., Li, X.-F., Wang, M.-N., Zha, X.-Q., Yang, X.-F., Liu, Z.-J., et al., 2014. Comparison of hypoglycemic and antioxidative effects of polysaccharides from four different Dendrobium species. Int. J. Biol. Macromol. 64, 420–427.

Pan, S., 2005. Study on physiological, biochemical changes of Dendrobium during vernalization and cloning of MADS-box gene. Thesis for the degree of master of science. South China Normal University, Guangzhou.

Puchooa, D., 2004. Comparison of different culture media for the in vitro culture of Dendrobium (Orchidaceae). Int. J. Agric. Biol. 6, 884–888.

Qian, X.-P., Zha, X.-Q., Xiao, J.-J., Zhang, H.-L., Pan, L.-H., Luo, J.-P., 2014. Sulfated modification can enhance antiglycation abilities of polysaccharides from *Dendrobium huoshanense*. Carbohyd. Polym. 101, 982–989.

Qian, Y., Lv, B., 2005. Study on extraction process of polysaccharide from *Caulis Dendrobii*. Chin. J. Mod. Appl. Pharm. 22, 293–294.

Qin, X.-D., Ning, L., Yan, X.-Y., 2011. Distribution and extraction process of polysaccharides in *Dendrobium findlayanum*. J. Yunnan Agric. Univ. (Natural Science) 3, 029.

Qiu, Z., Jones, K., Wylie, M., Jia, Q., Orndorff, S., 2000. Modified *Aloe barbadensis* polysaccharide with immunoregulatory activity. Planta Med. 66, 152–156.

Qu, W., Liu, H., Shang, Z., 2011. Systematic philological research on the development of Dendrobium in traditional Chinese medicine. Asia-Pacific Tradit. Med. 7, 133–134.

Radha, M.H., Laxmipriya, N.P., 2015. Evaluation of biological properties and clinical effectiveness of *Aloe vera*: a systematic review. J. Tradit. Complement Med. 5, 21–26.

Rosenberg, S.A., Lotze, M.T., Yang, J.C., Aebersold, P.M., Linehan, W.M., Seipp, C.A., et al., 1989. Experience with the use of high-dose interleukin-2 in the treatment of 652 cancer patients. Ann. Sug. 210, 474.

Saini, D.K., Saini, M.R., 2011. Evaluation of radioprotective efficacy and possible mechanism of action of Aloe gel. Environ. Toxicol. Pharmacol. 31, 427–435.

Shang, X., 2010. The comparative study of the extraction of *Dendrobium officinale* polysaccharides using hot water treatment coupled with or without enzymatic treatment. J. Anhui Agric. Sci. 38, 9787–9788.

Shang, X., Wang, C., 2010. Optimization of the extraction of *Dendrobium officinale* polysaccharide using orthogonal experiment. Liaoning J. Tradi. Chin. Med. 37, 708–709.

Shu, W., Fengjuan, W., Yongping, C., 2009. Anti-oxidation activity in vitro of polysaccharides of *Dendrobium huoshanense* and *Dendrobium moniliforme*. Agric. Sci. Technol. 10, 121–124.

Simões, J., Nunes, F.M., Domingues, P., Coimbra, M.A., Domingues, M.R., 2012. Mass spectrometry characterization of an *Aloe vera* mannan presenting immunostimulatory activity. Carbohyd. Polym. 90, 229–236.

Song, N., Lu, Y., Qiu, M., 2006. Studies on immunomodulation of polysaccharide from *Dendrobium thrsiflorum* Rchb. f. Nat. Prod. Res. Dev. 18, 445.

Sun, Z., Wei, K., Yan, Z., Zhu, X., Wang, X., Wang, H., et al., 2011. Effect of immunological enhancement of aloe polysaccharide on chickens immunized with Bordetella avium inactivated vaccine. Carbohyd. Polym. 86, 684–690.

Tabandeh, M.R., Oryan, A., Mohammadalipour, A., 2014. Polysaccharides of *Aloe vera* induce MMP-3 and TIMP-2 gene expression during the skin wound repair of rat. Int. J. Biol. Macromol. 65, 424–430.

Talmadge, J., Chavez, J., Jacobs, L., Munger, C., Chinnah, T., Chow, J.T., et al., 2004. Fractionation of *Aloe vera* L. inner gel, purification and molecular profiling of activity. Int. Immunopharmacol. 4, 1757–1773.

Tian, C.C., Zha, X.Q., Pan, L.H., Luo, J.P., 2013. Structural characterization and antioxidant activity of a low-molecular polysaccharide from *Dendrobium huoshanense*. Fitoterapia 91, 247–255.

Ulrich, P., Cerami, A., 2000. Protein glycation, diabetes, and aging. Recent. Prog. Horm. Res. 56, 1–21.

Wang, J., Gu, L., Kang, Z., Chen, J., Xu, W., Wang, S., 2013a. Optimization of polysaccharides extraction from *Dendrobium officinale* Kimura et Migo and investigation on its antioxidant activity. J. Northwest A&F Univ. 41, 143–148.

Wang, P., Lu, Q., Li, J., Huang, S., Chen, J., 2013b. Optimization of ultrasonic extraction technology and content determination of total amino acids from *Dendrobium officinale*. Chin. Chin. J. Exp. Tradit. Med. Form. 19, 30–33.

Wang, J.-H., Luo, J.-P., Yang, X.-F., Zha, X.-Q., 2010a. Structural analysis of a rhamnoarabinogalactan from the stems of *Dendrobium nobile* Lindl. Food Chem. 122, 572–576.

Wang, J.-H., Luo, J.-P., Zha, X.-Q., 2010b. Structural features of a pectic polysaccharide from the stems of *Dendrobium nobile* Lindl. Carbohyd. Polym. 81, 1–7.

Wang, J.-H., Zha, X.-Q., Luo, J.-P., Yang, X.-F., 2010c. An acetylated galactomannoglucan from the stems of *Dendrobium nobile* Lindl. Carbohyd. Res. 345, 1023–1027.

Wang, S., Zheng, G., He, J., Yu, X., Wu, Y., 1988. Studies on polysaccharides of *Dendrobium candidum*. Acta Botanica Yunnanica 10, 389–395.

Wang, S., Gong, J., Zhang, H.-J., Xu, X., Yang, Y., 2009. Study on the internal antioxidant activity of polysaccharides from Dendrobium. China Pract. Med. 4, 15–16.

Wang, X.-Y., Luo, J.-P., Chen, R., Zha, X.-Q., Wang, H., 2014. The effects of daily supplementation of *Dendrobium huoshanense* polysaccharide on ethanol-induced subacute liver injury in mice by proteomic analysis. Food Funct. 5, 2020–2035.

Wang, X.-Y., Luo, J.-P., Chen, R., Zha, X.-Q., Pan, L.-H., 2015. *Dendrobium huoshanense* polysaccharide prevents ethanol-induced liver injury in mice by metabolomic analysis. Int. J. Biol. Macromol. 78, 354–362.

Wang, Z., Shi, H., Zhang, Z., 2012. Studies on suspension culture of protocorm-like bodies of *Dendrobium candidum* and their accumulation of polysaccharides. J. Henan Agric. Sci. 41, 129–131.

Wei, M., Shao, P., Yao, H., Yang, C., 2009. Study on the extraction of polysaccharides from *Dendrobium huoshanense* by ultrasonic wave and cellulase. Sci. Technol. Food Ind. 30, 199–201.

Wozniewski, T., Blaschek, W., Franz, G., 1990. Isolation and structure analysis of a glucomannan from the leaves of *Aloe arborescens* var. Miller. Carbohyd. Polym. 198, 5.

Wu, J.H., Xu, C., Shan, C.Y., Tan, R.X., 2006. Antioxidant properties and PC12 cell protective effects of APS-1, a polysaccharide from *Aloe vera* var. chinensis. Life Sciences 78, 622–630.

Wu, R., Yang, B., Li, Y., Zhang, X., Yang, F., Xia, K., et al., 2011. Antihypertensive and anti-stroke effects of *Dendrobium officinale* polysaccharides on SHR-SP rats. Chin. J. Tradit. Med. Sci. Technol. 18, 204–210.

Wu, S.-J., Liu, Y.-S., Chen, T.-W., Ng, C.-C., Tzeng, W.-S., Shyu, Y.-T., 2009. Differentiation of medicinal Dendrobium species (Orchidaceae) using molecular markers and scanning electron microscopy. J. Food Drug Anal. 17, 474–488.

Wu, Y., Deng, J., Yang, T., Wu, J., Yang, L., Wu, Y., 2013. Optimization of extraction of water-soluble polysaccharides from tissue culture seedling of *Dendrobium officinale*. Nat. Prod. Res. Dev. 25, 1709–1714.

Xia, L., Liu, X., Guo, H., Zhang, H., Zhu, J., Ren, F., 2012a. Partial characterization and immunomodulatory activity of polysaccharides from the stem of *Dendrobium officinale* (Tiepishihu) in vitro. J. Funct. Foods 4, 294–301.

Xia, Q., Haili, D., Hong, L., 2012b. Optimization of the ultrasonic assisted extraction of polysaccharides from *Dendrobium huoshanense* by response surface method. Med. Plant 3, 78–80.

Xing, X., Cui, S.W., Nie, S., Phillips, G.O., Douglas Goff, H., Wang, Q., 2013. A review of isolation process, structural characteristics, and bioactivities of water-soluble polysaccharides from Dendrobium plants. Bioact. Carbohyd. Dietary Fibre 1, 131–147.

Xing, X., Cui, S.W., Nie, S., Phillips, G.O., Goff, H.D., Wang, Q., 2014. Study on *Dendrobium officinale* O-acetyl-glucomannan (Dendronan®): Part I. Extraction, purification, and partial structural characterization. Bioact. Carbohyd. Dietary Fibre 4, 74–83.

Xing, W., Guo, W., Zou, C.-H., Fu, T.-T., Li, X.-Y., Zhu, M., et al., 2015a. Acemannan accelerates cell proliferation and skin wound healing through AKT/mTOR signaling pathway. J. Dermatol. Sci. 79, 101–109.

Xing, X., Cui, S.W., Nie, S., Phillips, G.O., Goff, H.D., Wang, Q., 2015b. Study on *Dendrobium officinale* O-acetyl-glucomannan (Dendronan®): Part V. Fractionation and structural heterogeneity of different fractions. Bioact. Carbohyd. Dietary Fibre 5, 106–115.

Xu, C., Chen, Y., Zhang, M., 2004. Structural characterization of the polysaccharide DMP2a-1 from *Dendrobium moniliforme*. Chin. Pharm. J. 39, 900–902.

Xu, C., Zhan, Z.-G., Liao, S.-M., 2008. Studies on agronomic characteristics, polysaccharide and cellulose contents of 8 kinds *Dendrobium officinale* from different distribution areas. J. Zhejiang Univ. (Science Edition) 5, 024.

Xu, J., Han, Q.-B., Li, S.-L., Chen, X.-J., Wang, X.-N., Zhao, Z.-Z., et al., 2013. Chemistry, bioactivity and quality control of Dendrobium, a commonly used tonic herb in traditional Chinese medicine. Phytochem. Rev. 12, 341–367.

Xu, Y., Wang, L., Huang, B., Xie, X., Wu, Q., Shi, J., 2014. Comparison of contents of polysaccharides and alkaloids in *Dendrobium* from different harvest time. West China J. Pharm. 29, 288–291.

Yang, L.-C., Lu, T.-J., Hsieh, C.-C., Lin, W.-C., 2014. Characterization and immunomodulatory activity of polysaccharides derived from *Dendrobium tosaense*. Carbohyd. Polym. 111, 856–863.

Yang, X., Wang, W., 2015. Determination of total alkaloid from caulis dendrobii. Shandong Chem. Ind. 44, 89–90.

Yao, H., Chen, Y., Li, S., Huang, L., Chen, W., Lin, X., 2009. Promotion proliferation effect of a polysaccharide from *Aloe barbadensis* Miller on human fibroblasts in vitro. Int. J. Biol. Macromol. 45, 152–156.

Ye, Y., 2009. Study on the extraction of *Dendrobium officinale* polysaccharides using ultrasonication method. Chin. Med. Mat. 32, 617–620.

Yu, Q., Guo, Y., Si, J., Wu, L., Wang, L., 2014. Variation of polysaccharides and alcohol-soluble extracts content of *Dendrobium officinale*. Chin. J. Chin. Mater. Med. 39, 4769–4772.

Yu, Z., Jin, C., Xin, M., Jianmin, H., 2009. Effect of *Aloe vera* polysaccharides on immunity and antioxidant activities in oral ulcer animal models. Carbohyd. Polym. 75, 307–311.

Yuan, H., Bai, Y., Si, J., Zhang, A., Jin, X., 2011. Variation of monosacchride composition of polysacchrides in *Dendrobium officinale* by pre-column derivatization HPLC method. Chin. J. Chin. Mater. Med. 36, 2465–2470.

Zha, X.-Q., Luo, J.-P., Luo, S.-Z., Jiang, S.-T., 2007a. Structure identification of a new immunostimulating polysaccharide from the stems of *Dendrobium huoshanense*. Carbohyd. Polym. 69, 86–93.

Zha, X.-Q., Wang, J.-H., Pan, L.-H., Luo, J.-P., Lu, M.-M., 2007b. Study on antioxidant activity of polysaccharides from dendrobium species. Chin. J. Food Sci. 10, 019.

Zha, X.Q., Li, X.L., Zhang, H.L., Cui, S.H., Liu, J., Wang, J.H., et al., 2013. Pectinase hydrolysis of *Dendrobium huoshanense* polysaccharide and its effect on protein nonenzymatic glycation. Int. J. Biol. Macromol. 61, 439–447.

Zha, X.-Q., Zhao, H.-W., Bansal, V., Pan, L.-H., Wang, Z.-M., Luo, J.-P., 2014. Immunoregulatory activities of *Dendrobium huoshanense* polysaccharides in mouse intestine, spleen and liver. Int. J. Biol. Macromol. 64, 377–382.

Zhang, A., Wei, T., Si, J., Jin, L., Mo, Y., 2011a. Study on basic amino acid contents in *Dendrobium officinale*. Chin. J. Chin. Mater. Med. 36, 2632–2635.

Zhang, L., Fan, M., Feng, X., Liu, X., Wei, R., 2011b. Study on the extraction of *Dendrobium candidum* polysaccharides and dendrobine by cellulase. Chem. Res. Application 23, 356–359.

Zhang, G.-N., Bi, Z.-M., Wang, Z.-T., Xu, L.-S., Xu, G.-J., 2003. Advances in studies on chemical constitutents from plants of Dendrobium Sw. Chin. Tradit. Herbal Drugs 6, 041.

Zhang, H., Dai, G., Ma, C., Yang, F., 2009. Effects of *Dendrobium officinale* polysaccharides on the immune functions of Sarcoma 180-bearing mice. Zhejiang J. Tradit. Chin. Med. 44, 380–381.

Zhao, J., Ma, J., Duan, L., 2010. Study on the extraction technology and deproteinization technology of polysaccharide from *Dendrobium devonianum*. Lishizhen Med. Mater. Med. Res. 21, 1884–1886.

Zhao, J., Wang, Y., Li, X., Tang, D., Gao, W., 2011. Comparison on polysaccharides content of several species of Dendrobium in different producing area. China J. Tradit. Chin. Med. Pharm. 26, 2001–2005.

Zhao, Y., Son, Y.-O., Kim, S.-S., Jang, Y.-S., Lee, J.-C., 2007. Antioxidant and antihyperglycemic activity of polysaccharide isolated from *Dendrobium chrysotoxum* Lindl. BMB Rep. 40, 670–677.

Zheng, S., Ge, X., Li, X., Jin, H., Tang, Y., 2008. The optimization of *Dendrobium nobile* polysaccharides determine condition and extraction technology. Chin. Arch. Tradit. Chin. Med. 26, 2676–2677.

Zheng, Y., Jiang, W., Silva, E.N., Mao, L., Hannaway, D.B., Lu, H., 2012. Optimization of shade condition and harvest time for *Dendrobium candidum* plants based on leaf gas exchange, alkaloids and polysaccharides contents. Plant Omics 5, 253.

Zhong, Y., 2011. Study on Effect of Polysaccharides from *Dendrobinm chryseum* Rolfe on Immunocytes of Mice. J. Anhui Agric. Sci. 22, 039.

Zhou, G., Lv, G., 2011. Analysis of monosaccharide compositions and contents in polysaccharide of *Dendrobium officinale* from different sources and growth years by pre-column derivatization-HPLC. Chin. Pharm. J. 46, 626–629.

Zhou, S., Lei, Z., Yu, Q., Ye, Z., 2010. Study on the preferred extraction of polysaccharides from *Dendrobium candium* with uniform design and orthogonal design. Sci. Technol. Food Ind. 296–297.

Zhu, H., Teng, J., Cai, Y., Liang, J., Zhu, Y., Wei, T., 2011. Relativity among starch quantity, polysaccharides content and total alkaloid content of *Dendrobium loddigesii*. Chin. J. Chin. Mater. Med. 36, 3262–3264.

Zhu, Y., Si, J., Guo, B., He, B., Zhang, A., 2010. Quantitive variation of polysaccharides content in cultivated *Dendrobium candidum*. Chin. J. Chin. Mater. Med. 35, 427–430.

Zong, W., Li, C., 2012. Technique condition optimization for ultra high pressure extraction of Dendrobium polysaccharide. J. Zhengzhou Univ. Light Ind. (Nat. Sci.) 27, 36–39.

Chapter 7

Tea Polysaccharide

7.1 INTRODUCTION

Tea, a product made up from the leaf and bud of the plant, is the most widely consumed beverage in the world, and has become an important agricultural product (Nie and Xie, 2011). It has a long history of consumption in Asian countries, such as China, Japan, India, and Thailand, as early as 50,000 years ago (Xiao and Jiang, 2015). Originating from China, tea was first cited in 100 BC in *Shen Nong's Herbal Classic* (widely considered as the oldest book on oriental herbal medicine and the foundation of traditional Chinese medicine) for its detoxification effects (Nie and Xie, 2011). Tea plants prefer to grow in many regions that have a high humidity, moderate temperature, and acidic soils, from sea level to high mountains (Dufresne and Farnworth, 2001). Tea is largely consumed worldwide, well ahead of coffee, beer, wine, and carbonated soft drinks (Cabrera et al., 2006). According to McKay and Blumberg's 2002 research, the per capita mean consumption of tea in the world is 120 mL/day (Cabrera et al., 2006). The consumption habits of tea are different between eastern and western countries. In western countries, tea time is usually set during afternoon, and it is commonly drunk with milk added, while in eastern countries, people prefer to drink it without other additives (Cabrera et al., 2006).

According to the manufacturing process, tea leaves can be divided into three main kinds. They are unfermented green tea (Fig. 7.1A), semifermented oolong tea (Fig. 7.1B), and fermented black tea (Fig. 7.1C) (Xiao and Jiang, 2015). The differences in fermentation result in oxidative and enzymatic changes within the tea leaves, which occur during processing (Cao, 2013). For instance, green tea is produced by drying and steaming the fresh leaves to inactivate the polyphenol oxidase and therefore, no oxidation occurs (Cabrera et al., 2006). Oolong tea is produced when the fresh leaves are subjected to a partial fermentation stage before drying (Cabrera et al., 2006; Xie et al., 1993). Black tea experiences a postharvest fermentation stage before drying and steaming (Cabrera et al., 2006; Xie et al., 1993). Approximately 76%–78% of the tea produced and consumed is black tea, 20%–22% is green tea, and less than 2% is oolong tea (Xiao and Jiang, 2015). Black tea is largely drunk in Europe, North America, and North Africa (except Morocco), while green tea is widely consumed in China,

Bioactive Polysaccharides. DOI: http://dx.doi.org/10.1016/B978-0-12-809418-1.00007-1

FIGURE 7.1 Three main kinds of tea: (A) green tea; (B) oolong tea; (C) black tea.

Japan, Korea, and Morocco, and oolong tea is popular in China and Taiwan (Xiao and Jiang, 2015).

Green tea, black tea, white tea, Pu-erh tea, *Cyclocarya paliurus*, and mulberry leaf have been commonly studied in recent years. Green tea, also termed as *Camellia sinensis*, is the second most consumed beverage in the world. It tastes excessively bitter compared with other types of tea, due to a very high content of polyphenols (Xiao and Jiang, 2015). The first green tea was exported from India to Japan during the 17th century (Chacko et al., 2010). Commercially sold green tea always has a black color. This results from the heating process during green tea production, which kills the enzyme polyphenol oxidase, an important substance responsible for the conversion of the flavanols in the leaf into the dark polyphenolic compounds (Cabrera et al., 2006). Black tea is quite popular in European countries. The average amount of black tea consumed per person per year in the United Kingdom was 3.2 kg in 2010 (Drynan et al., 2010).

Lower grade green tea has traditionally been used for the treatment of diabetes in East Asia, especially in Japan (Nie and Xie, 2011). White tea, designated as "Baicha" in Chinese, is very similar to green tea. However, it is exceptionally prepared only from the buds and fresh albino tea leaves of the *C. sinensis* plant, while green tea is prepared from the matured tea leaves. In the summer, the buds are all green. These green leaves are "old" and not used in the manufacture of white tea (Zhang et al., 2016b; Islam, 2011). Pu-erh tea is a popular beverage in southwestern China and South Asian countries. The well-known variety of this type of tea is Pu-erh from Yunnan Province in China. Pu-erh traditionally begins as a raw product known as "rough" tea and can be sold in this form, or pressed into a round shape and sold as "raw" tea. Both of two forms then undergo a complex process of further fermentation and maturation with time. This process involves an accelerated fermentation turning the raw materials into "ripe" tea, which is then stored loose or pressed into a number of shapes. Pu-erh tea is valued for providing nutrition, aiding digestion, and preventing obesity. *C. paliurus* (Batal.) Iljinskaja (*C. paliurus*), also named as sweet tea tree, is a well-known edible and medicinal plant, grown on cloudy and foggy highlands in Southern China (Xie and Nie, 2006). The leaves of *C. paliurus* have been commonly used as a traditional Chinese medicine in the treatment of hyperglycemia, hyperlipidemia, and diabetes mellitus for thousands of years (Xie et al., 2013b). Mulberry (*Morus alba* L., family of Moraceae) leaf, a multipurpose agro-forestry plant that belongs to the family Moraceae, is widely distributed in tropical, subtropical and temperate areas (Yuan et al., 2015). It has been used as an alternative medicine in China and Japan, and it is also drunk as a kind of tea in China (Zhang et al., 2014b). Furthermore, mulberry leaf tea is promoted for its health benefits and control of diabetes in Asian nations (Naowaboot et al., 2012).

During the past two decades, research has revealed that tea possesses beneficial effects, including reduction of cholesterol, depression of hypertension, antioxidation, antimicrobial, protection against cardiovascular disease, and anticancer (Béliveau and Gingras, 2004; Koo and Cho, 2004). These beneficial effects have been partly attributed to its variety of chemical ingredients.

The major constituents in tea are polyphenols (including catechins and flavanoids), alkaloids (including caffeine, theobromine, theophylline, etc.), volatile oils, polysaccharides, amino acids, lipids, vitamins (e.g., vitamin C), and inorganic elements (e.g., aluminum, fluorine, and manganese), etc. However, polyphenols, pigments, and catechins were the focus of these studies (Dell'aica et al., 2004; Nie and Xie, 2011).

Recently, water-soluble tea polysaccharides (TPSs) have attracted scientists' attention. The polysaccharide content in tea varies from approximately 1.5% to nearly 13% (Cabrera et al., 2006; Zhang et al., 2013, 2014a, 2016b). The phenol–sulfuric acid method and anthrone–sulfate method are two commonly used methods in detecting the polysaccharide content. A new method for the determination of polysaccharide in tea has also been developed by Wei et al. This method is based on the increase of resonant light scattering (RLS) of TPS in the presence of cetylpyridinium chloride (CPC)–NaOH system. With optimized conditions, the RLS intensity of CPC is largely improved by adding TPS. Compared with the other two traditional polysaccharide determination methods, this new method has the advantages of easy operation, rapidity, and practicability (Wei et al., 2011b). It has been demonstrated that TPSs have many bioactivities, such as reducing blood sugar levels, immunological, antiradiation, antiblood coagulation, anticancer, anti-HIV, and hypoglycemic activities (Xie and Nie, 2006; Wang et al., 2001a,b). Based on these bioactivities, TPS has been developed into a functional food recently. In addition, TPSs were found to be mostly glycoconjugates, in which a protein carries one or more carbohydrate chains covalently attached to a polypeptide backbone, usually via *N*- or *O*-linkages (Iso et al., 2006; Xie and Nie, 2006). However, the high order structure of this active component, as well as the relationship between the bioactivity and the chemical structure, is still not well established. In this chapter, we would like to summarize the different types of extraction and purification techniques, as well as its structure and bioactivities.

7.2 EXTRACTION METHODS

TPSs exist as a structural constituent of the cell wall of tea, so the extraction method depends on the cell wall structure. In general, TPSs are extracted by using a solvent extraction method, such as water, acidic solution, and diluted alkali solution after pretreatments (Nie and Xie, 2011; Wang and Xia, 2006).

Furthermore, some additional methods have also been used to improve this extraction process, such as freeze–thaw method, microwave, ultrasonic, enzymatic method, etc. (Nie and Xie, 2011; Thirugnanasambandham et al., 2015; Zhang et al., 2016a).

In Xie's study (2010b), microwave-assisted extraction was employed for the extraction of polysaccharide from *C. paliurus*. The optimal conditions for microwave-assisted extraction of polysaccharide were: a time for microwave radiation of 20.8 min, temperature of 100.9°C, and water/material ratio of 22.1:1. Under certain conditions, the polysaccharide yield was increased to

5.04% ± 0.09%. Sivakumar (Thirugnanasambandham et al., 2015) applied the microwave in polysaccharide extraction from mulberry leaves. The effects of three extraction factors were examined. A three factor-three level Box–Behnken response surface design coupled with response surface methodology (RSM) was introduced to optimize the extraction conditions. The results indicated that with a microwave power of 170 W, and extraction time of 10 min, the polysaccharide yield could be improved to 9.41%. In Xie's series of studies, an ultrasonic method was also used in the *C. paliurus* polysaccharide extraction (Xie et al., 2012). RSM was employed to optimize the extraction conditions. The results demonstrated that the best extraction conditions could be with a liquid/material ratio of 8, extraction time of 59 min, and extraction temperature of 58°C. Under optimal extraction conditions, the polysaccharide yield could be improved to 4.91% ± 0.11%. However, this yield is lower than that using the microwave-assisted method.

An ultrasonic extraction method is commonly applied in polysaccharide extraction from mulberry leaves. Zhang et al. (2016a) optimized the ultrasound-assisted extraction process of mulberry leaf polysaccharides (MLPs), and found that the optimum conditions were an extraction temperature of 57°C, an extraction time of 80 min, and a liquid/solid ratio of 53 mL/g. Under these circumstances, the polysaccharide yield could be improved to 6.92% ± 0.29%. The optimized extraction factors in Ying's research (2011) in MLP extraction were an extraction power of 60 W, extraction temperature of 60°C, extraction time of 20 min, and ratio of water to raw material of 15:1 (mL/g). In addition, they also compared the yields of three different extraction methods, including an ultrasonic extraction method, microwave extraction method, and a conventional solvent extraction method, and found that the ultrasonic extraction method gave the highest extraction rate.

In general, TPSs usually have a dark color due to the small molecules contained in tea. Therefore, decolorization is an important procedure in purifying the crude polysaccharides. A novel oxidation process, ultrasound/H_2O_2 oxidation, was employed in the decolorization of *C. paliurus* (Batal.) Iljinskaja polysaccharides (CPP) in Xie's study (2011). The initial concentration of polysaccharide solution, dosage of H_2O_2, temperature, pH, and ultrasonic irradiation designated for the decolorization process were investigated. The results indicated that the optimum operational conditions were set as follows: concentration of the polysaccharide solution, 0.5 mg/mL; H_2O_2, 0.623 mM; temperature, 40°C; and pH, 9.0. Under these circumstances, the ultrasound/H_2O_2 oxidation process provided good decolorizing ability, and the decolorization efficiency could reach at 84.1%. Additionally, this decolorization procedure would not cause any significant variation in polysaccharide structure.

After decolorization, the TPS solution was obtained after filtering and concentrating. By removing the free protein with Sevag reagent and then precipitating with ethanol, filtering, and drying, the crude tea polysaccharides (CTP) were obtained. Then the crude polysaccharides are redissolved in double-distilled water. This polysaccharide solution was later applied to different column chromatography, such as ion-exchange chromatography, gel filtration chromatography, or affinity chromatography, to get further purification. The polysaccharide concentration will elute with an appropriate running buffer, and later experience the process of concentration, dialyzation, and freeze-drying before getting the final purified fractions (Xie and Nie, 2006; Nie and Xie, 2011).

In Yuan's study (2015), a combination use of DEAE-Cellulose 52 and Sephadex G-100 was applied in purifying crude MLP. The MLP was dissolved in double-distilled water and first loaded onto a DEAE-52 cellulose column (2.6 × 50 cm). The column was eluted with 0, 0.1, 0.3, and 0.5 M sodium chloride (NaCl) solution at a flow rate of 60 mL/h. Then three separate polysaccharide fractions, MLP-1, MLP-2, and MLP-3, were obtained by evaluating the absorbance at 490 nm using the phenol–sulfuric acid method. The main fraction of MLP-3 was chosen for further purification using Sephadex G-100. The column was eluted with 0.2 M NaCl solution at a flow rate of 15 mL/h. The elution was evaluated as described above. Therefore, two purified polysaccharide fractions, designated as MLP-3a and MLP-3b, were obtained.

A reverse micelle extraction method is an efficient and economical alternative for extraction of bioactive molecules, due to its high selectivity, fast mass transfer, and relatively low cost. Sodium di-2-ethylhexyl sulfosuccinate (AOT) is a good candidate for extraction of biomolecules, because of its wide range of reverse micellar core sizes in the organic phase. Therefore, an AOT/heptane reverse micellar system was used in the purification of TPS in Li's research (Li and Cao, 2014). In addition, chaotropes (GuHCl and urea) were employed to promote the extraction yield of TPSs in the system. Several crucial factors in this purifying process were optimized, including pH, types and concentrations of salts, and types and concentrations of alcohols. In forward extraction, a low concentration of GuHCl in this AOT/heptane system may affect the TPS structure, inhibit its aggregation formation, and promote its solubilization into the water cores. With optimum conditions of pH 4.6, 0.05 mol/L NaCl, 7% [v/v] methanol, and 0.06 mol/L GuHCl, the forward extraction recovery could reach 34%. In backward extraction, 100% recovery was obtained by adding GuHCl or urea.

In conclusion, the general process of separation and fractionation of crude TPS from tea materials can be described as follows (Yuan et al., 2015): the tea materials are pretreated with ethanol to remove small molecules. The dry residue is then extracted with hot water under optimized conditions. After that, the extracts are filtered and evaporated under vacuum. Subsequently, the concentrated liquids are precipitated with ethanol and centrifuged. The final precipitate is redissolved in double-distilled water. Protein is removed with Sevag reagent. After removing the protein, the polysaccharide solution is put into the distilled water for dialysis. The dialysate is concentrated and freeze-dried to achieve crude TPS. Later, TPS can be fractionated with ion-exchange chromatography, gel filtration column chromatography, or affinity chromatography. The eluting solution and flow rate are two critical conditions that should be taken into consideration during this purifying procedure. The collected solution is then determined by the phenol–sulfuric acid method and UV spectroscopy to decide which tubes should be collected. The tubes are combined with solutions having maximum absorbance, and subsequently dialyzed, concentrated, and lyophilized to obtain purified tea polysaccharide fractions (TPFs). For example, a kind of Se-polysaccharide from Se-enriched tea leaves were purified using a DEAE–sepharose fast flow gel column (2.5×60 cm) in Wang's study (2015b). Three polysaccharide fractions, labeled Se-TPS1, Se-TPS2, and Se-TPS3, were obtained with yields of 6.5%, 37.14%, and 8.57%, respectively.

Apart from water, alkali is also used to isolate alkali-soluble TPS. In Chen's study (2010), hot water and sequential hot alkaline were used to separate the water-soluble and alkaline-soluble polysaccharides. Low-grade green tea material was suspended in double-distilled water, and incubated at 90°C for 2 h with continuous stirring. The mixture was filtered, and the supernatant was concentrated, vacuumed, and freeze-dried to achieve the water-soluble polysaccharide, while the residue was collected and subsequently incubated at 50°C for 2 h with 1.0% alkali aqueous solution. This alkaline extract was filtered, concentrated, deproteinated, precipitated, and freeze-dried to obtain alkali-soluble TPS conjugates.

It is worth noticing that different extraction methods might lead to a difference in the composition and bioactivity of polysaccharide from tea or tea flowers. Zhang did an extrusion treatment isolating the polysaccharide from coarse tea (Zhang et al., 2013). The results showed that the extrusion treatment could not only increase the polysaccharide yield, but also could vary the monosaccharide composition, molecular weight (Mw), thermal properties, the morphological properties, and even the bioactivity of this coarse TPS (Zhang et al., 2013). The molar ratios on monosaccharide composition changed from Rha:Ara:Man:Glc:Gal:Uronic acid = 0.88:1.19:0.34:1.00:1.00:1.93, to Rha:Ara:Man:Glc:Gal:Uronic acid = 0.20:1.11:0.18:1.00:0.63:1.23 after extrusion treatment (Zhang et al., 2013). The Mw distribution range varied from $33.0 \times 10^4 - 0.1 \times 10^4$ Da to $40.5 \times 10^4 - 0.4 \times 10^4$ Da after extrusion treatment (Zhang et al., 2013). As to the thermal properties, the enthalpy of gelatinization (ΔHgel) had increased more than 1.4 times after the extrusion treatment (Zhang et al., 2013). The polysaccharide from extruded tea had an even better scavenging ability against DPPH free radical. The traditional extraction method, microwave-assisted extraction method, and ultrasound-assisted extraction method were compared in the isolation of polysaccharide from tea flowers. The polysaccharide produced by the traditional water extraction method exhibited a strong inhibitory effect on α-glucosidase (Wei et al., 2010a).

It is also noteworthy that different drying patterns might also affect the chemical composition and biological properties of TPSs. In Wei's study (2010a), four polysaccharide fractions, labeled TPS-F, TPS-V, TPS-S, and TPS-M, were achieved from green tea leaves by freeze-drying, vacuum-drying, spray-drying, and microwave-vacuum drying, respectively. Their chemical composition and bioactivities were studied comparatively. Results showed that these four fractions were similar in IR, UV absorption, and Mw distribution. However, they differed significantly in polysaccharide yields, protein content, and total polyphenol content. Nevertheless, morphological study by scanning electron microscope (SEM) indicated that their surface had differences in size and shape. The surface of TPS-F was rough and porous, the surface of TPS-S was evenly particulate, while the surfaces of the other two were similar to anomalistic bricks. Moreover, the biological activities of these four polysaccharides were also investigated. It was found that TPS-F exhibited relatively higher activity in metal chelating and superoxide radical scavenging assays, and TPS-V exhibited a better ability in inhibiting the activities of α-glycosidase and α-amylase.

In addition, different tea fermentation degrees will also influence the yield and bioactivities of TPS. Tieguanyin, Fenghuangdancong, and Dahongpao are three typical kinds of oolong teas with the lightest, intermediate, and the highest degree of fermentation, respectively. Wang et al. (2012b) compared the physicochemical characteristics, in vitro antioxidant activity, and α-glucosidase inhibitory effect of polysaccharides from these three oolong teas. The measurements of chemical composition revealed that they were all acid heteropolysaccharides bound with protein. The contents of protein and uronic acid proportionally increased as the degree of fermentation increased. Similarly, their ability in antioxidant and α-glucosidase inhibition also correlated with the degree of fermentation. They contained the same monosaccharides, but these differed in molar ratio. An investigation of their Mw by high-performance gel-permeation chromatography (GPC) indicated that 92.9% of TTPS and 94.4% of DTPS had a peak Mw of 0.82×10^6 and 2.64×10^6 respectively, but that of FTPS had 34.2% with 0.93×10^6, and 68.8% with 0.01×10^6, respectively. Fourier Transform IR spectra proved that they had similar characteristic absorption bands of uronic acid, protein, polysaccharide, and pyran-glycosides. Moreover, the analysis of antioxidant activities by assays of DPPH, ABTS, and ferric-reducing antioxidant power (FRAP) and inhibitory abilities against α-glucosidase demonstrated that their antioxidant abilities decreased in the order of DTPS > FTPS > TTPS. In this study, it was also noticed that the potencies of the bioactivities of these three oolong teas had an apparent correlation with their content of protein and uronic acid. Therefore, it was concluded that a higher degree of fermentation could result in a higher content of protein-bound acid heteropolysaccharides, accompanied by higher antioxidant and α-glucosidase inhibitory activities. However, Zhao's research (2014) had the totally opposite result. They found that the polysaccharides from more deeply fermented tea had a lower antioxidant activity than polysaccharides from less intensively fermented tea. Therefore, how the degree of fermentation of the tea could affect the bioactivity of their polysaccharides remains to be seen. One possible reason accounting for certain phenomenon is that the biological activities of different TPSs are more likely to be affected by the functional groups in polysaccharides, rather than simply by fermentation degree.

7.3 PHYSICOCHEMICAL AND STRUCTURAL FEATURES

The physicochemical and structural features of a polysaccharide are defined by Mw, monosaccharide composition, configuration of glycosidic linkages, position of glycosidic linkages, sequence of monosaccharide, conformation, particle size, solubility, and rheological properties, etc. (Zhang et al., 2007). As for Mws and compositional components of TPSs, different results were given in various reports due to different tea raw material or purification processes. The TPS complex with the hypoglycemic effects obtained by Mori et al. (1988) was composed of L-arabinose, D-ribose, and D-glucose in the ratio of 5.1:4.7:1.7, with an average Mw of 40,000 Da. Takeo et al. (1992) reported that the monosaccharide composition of the polysaccharide with hypoglycemic effect extracted from green tea was galactose and glucose. Wang and Wang (1991) reported that the TPS extracted from Tunxi green tea was composed of L-fucose, D-mannose, D-glucose, D-galactose, and arabinose in the ratio of 0.23:1.04:0.62: 2.43:1.00, with a Mw of about 91,000 Da. Wang et al. (2001a,b) analyzed the monosaccharide composition of a TPS and found that it was a heteropolysaccharide consisting of arabinose, xylose, fucose, glucose, and galactose in the ratio of 5.52:2.21:6.08:44.20:41.99, with a Mw of about 107,000 Da. In addition, a small quantity of rhamnose was also found. Zhou et al. (2002) studied the TPSs purified from a green tea sampled from Wuyuan of Jiangxi, and found its monosaccharide composition was rhamnose, arabinose, xylose, mannose, glucose, and galactose. An acidic polysaccharide from green tea with selective antiadhesive effects against certain pathogenic bacteria was found that consisted mainly of galacturonic acid (49%) and glucuronic acid (26%), as well as small quantities of rhamnose, arabinose, fucose, galactose, mannose, and xylose, with a Mw of approximately 8.0×10^4 Da (Lee et al., 2006). An acidic polysaccharide without conjugation to protein, with glucokinase-stimulating activity from coarse green tea was obtained that consisted of rhamnose, arabinose, mannose, glucose, and galactose in the ratio of 12.57:22.95:4.4:39.34:20.77, and contained 8% galacturonic acid with a Mw of approximately 6×10^4 Da (Wang and Xia, 2006). A nonreductive acidic polysaccharide bound to protein from oolong tea was isolated that contained uronic acid, protein, and neutral sugar at 40.65%, 19.59%, and 26.66%, respectively, and its Mw was 1.28×10^6 Da, the neutral sugar was composed of D-rhamnose, L-arabinose, D-galactose, and D-glucose with a molar ratio of 1.37:1.89:1.00:1.30 (Chen et al., 2009). An acidic polysaccharide (ATPS-2) and a neutral polysaccharide (NTPS-1) were purified from green tea with Mws of 4430 Da and 21247 Da, respectively. The ATPS was composed of rhamnose, arabinose, galactose, and galacturonic acid. The NTPS-1 was found to consist of galactose (Wang et al., 2009a,b).

Gas chromatography (GC) was a method commonly used in monosaccharide composition studies in the early days. A water-soluble polysaccharide isolated from *C. paliurus* (Batal.) Iljinskaja by Xie's team (2010a) was named as CPP-1. The Mw of CPP-1 was approximately 1167 kDa, detected by high-performance GPC. Monosaccharide composition analysis completed by GC indicated that it was a heteropolysaccharide composed of D-xylose, L-arabinose, D-glucose, D-galactose, L-rhamnose, and D-mannose in a molar ratio of 1.00:9.67:9.65:4.96:3.29:2.70. In addition, CPP-1 contained 8.44% protein and 17 general amino acids, and it was rich in glutamic acid, asparaginic acid, leucine, glycine, arginine, tyrosine, and alanine.

Ion-exchange chromatography is another method frequently used recently in monosaccharide composition analysis. It is more convenient and accurate than the GC method. Wang obtained three Se-polysaccharide fractions from Se-enriched tea leaves by DEAE–Sepharose fast flow gel column (2.5 × 60 cm), designated as Se-TPS1, Se-TPS2, and Se-TPS3. Their yields were 6.5%, 37.14%, and 8.57%, respectively. The average sizes of Se-TPS1 and Se-TPS2 were analyzed by an HPGPC system, with Mws of 1.1×10^5 and 2.4×10^5 Da, respectively. Se-TPS3 was a polysaccharide polymer with two peaks with Mws of 9.2×10^5 and 2.5×10^5 Da. Monosaccharide components studied by ion chromatography indicated that the three Se-polysaccharide fractions were all acidic polysaccharoses with the same monosaccharide composition, but different in molar ratios. They were all composed of fucose, rhamnose, arabinose, galactose, glucose, glucuronic acid, and galacturonic acid with a molar ratio of 0.07:0.21:0.58:1.00:0.47:0.17:1.75, 0.07:0.28:0.59:1.00:0.49:1.24, and 0.07:0.38:0.72:1.00:0.30:0.19:0.88 in Se-TPS1, Se-TPS2, and Se-TPS3, respectively.

Two purified polysaccharide fractions, named MPS-1 and MPS-2, were obtained from mulberry leaves (Ying et al., 2011). The analysis of Mw showed that MPS-1 and MPS-2 were 24.898 and 61.131 kDa, respectively. The monosaccharide composition study indicated that MPS-1 consisted of Sor, Ara, Xyl, and Glc, and MPS-2 consisted of Rha, Ara, Xyl, Glc, Gal, and Man. Another study (Zhang et al., 2014a) in the monosaccharide composition of a high-purity polysaccharide from mulberry leaf by Zhang's group indicated a different result. It showed that this polysaccharide, with its purity of 99.8%, was composed of D-arabinose, D-xylose, D-glucose, D-rhamnose, and D-mannose in a molar ratio of 1:2.13:6.53:1.04:8.73.

In Lv's study (2009), a reverse-phase high-performance liquid chromatographic (HPLC) method was employed for simultaneous detection of aldoses and uronic acids. The standard monosaccharides, including mannose, ribose, rhamnose, glucuronic acid, galacturonic acid, glucose, xylose, galactose, arabinose, and fucose, were reacted with 1-phenyl-3-methyl-5-pyrazolone to form precolumn monosaccharide derivatives. These monosaccharide derivatives were separated using an RP-C18 column (4.6 mm i.d. × 250 mm, 5 um, Venusil, USA). Based on the retention time and peak area of the standard monosaccharide derivatives, the quantitative analysis of component monosaccharides could be completed. The results showed that the polysaccharide from *Gynostemma pentaphyllum* Makino tea was a typical heteropolysaccharide and consisted of mannose, ribose, rhamnose, glucuronic acid, galacturonic acid, glucose, xylose, galactose, and arabinose with molar contents of 16.3, 10.3, 47.1, 5.6, 24.0, 128.4, 25.0, 101.4, and 71.1. Furthermore, the quantitative recoveries using this method were in the range of 94.6%–108.0%, and the RSD values were lower than 4.9%, which indicated that the proposed method was practical for the monosaccharide composition analysis of *G. pentaphyllum* TPS.

In Jia's study (2014), the extraction, purification, and characterization of polysaccharides from Hawk mature leaf tea (HMP) were investigated. The crude HMP was purified using DEAE-52 chromatography, giving two purified fractions, named HMP-1 and HMP-2. The HMP-1 and HMP-2 mainly consisted of arabinose, galactose, glucose, and mannose, with Mws of 133 and 100 kDa, respectively.

As well as tea leaves, the polysaccharides isolated from tea seeds or tea flowers have also attracted scientists' attention in recent years. Wei et al. (2011a) investigated the composition of polysaccharides from tea seed (TSPS). Methods including HPGPC, IC, and IR were used in the analysis of the properties and chemical composition of TSPS. The results demonstrated that TSPS was composed of three kinds of polysaccharide with Mws of 500, 130, and 5 kDa. The TSPS consisted of rhamnose, xylose, arabinose, glucose, and galactose, GalA, and GulA, in a molar ratio of 4.9:1.7:11.1:27.2:14.0:3.4:1. The sugar backbone of TSPS might consist of glucose, while the branched chain may consist of rhamnose, xylose, arabinose, and galactose.

Two water-soluble polysaccharide fractions, coded as TFP-1 and TFP-2, were achieved from green tea flower. Their molecular characterization was investigated with the application of Ubbelohde viscometer, laser light scattering, a particle size analyzer, and thermo-gravimetric analyzer (Han et al., 2011b). The results of intrinsic viscosity studies, weight-average molar mass studies, gyration radius studies, and hydrodynamic radius studies for TFP-1 and TFP-2 were described as follows: $[\eta] = 0.767$ and 0.207 dL/g, Mw = 15.9 × 104 and 1.12 × 104 g/mol, Rg = 37.2 and 14.1 nm, Rh = 44.8 and 14.4 nm. Furthermore, the molecule morphology in NaCl solution appeared to be a branched spherical shape, and the typical thermal behaviors were presented in analysis curves for both fractions.

Olennikov et al. (2015) investigated the monosaccharide composition of polysaccharides from five different gentian teas, including *Gentiana algida*,

Gentiana decumbens, *Gentiana macrophylla*, and *Gentiana triflora* herb teas. Fructose, glucose, sucrose, gentiobiose, and gentianose were commonly identified in all five gentian teas studied. In spite of these, pectic polysaccharides with a high content of rhamnogalacturonans and arabinogalactans were also characterized for the first time in gentian decoctions. In addition, six organic acids, malic, citric, tartaric, oxalic, succinic, quinic, were also found to be distributed in the five gentian teas.

It has been found that the Mw and monosaccharide composition could vary among polysaccharides extracted from teas with different harvest years. Xu et al. (2014) studied the structural characteristics of polysaccharides isolated from Pu-erh tea (PTPS) with aging (mild fermentation) for 1 year (PTPS-1), 3 years (PTPS-3), and 5 years (PTPS-5). The results obtained indicated that all of these three kinds of PTPS were acid heteropolysaccharides bound with proteins. But their Mws and monosaccharide compositions were different: PTPS-1 had one major fraction (92%) with its Mw of 2.70×10^6; PTPS-3 had two major fractions (47% and 52%) with Mws of 1.93×10^6 and 6.31×10^5, respectively; PTPS-5 also had two major fractions (33% and 60%) with Mws of 3.9×10^6 and 1.16×10^6, respectively. The monosaccharide composition analysis showed that PTPS-1, PTPS-3, and PTPS-5 consisted of L-rhamnose, L-arabinose, D-xylose, D-mannose, D-galactose, D-glucose, and D-fucose in a molar ratio of 5.34: 21.86: 4.04: 21.59: 26.93: 16.52: 3.64, 6.82: 26.22: 0.35: 13.83: 39.34: 10.23: 3.21, and 15.98: 20.84: 0.15: 15.29: 40.33: 6.08: 1.68.

Xiao and Jiang (2015) summarized the characteristics of Mw, monosaccharide composition, and protein content of 32 kinds of currently investigated TPSs. They found that 29 of 32 kinds of TPSs (90.62%) contain arabinose, and 30 of 32 kinds of TPSs (93.75%) contain galactose. However, only 5 of 32 kinds of TPSs (15.62%) contain glucuronic acid and galacturonic acid.

The TPSs not only contain monosaccharide, but also combine with uronic acid, proteins, inorganic elements, and so on. For instance, despite repeated deproteination, the protein could still be detectable and the content remained basically unchanged (Buqing et al., 1996; Wang et al., 2000). Three different TPSs were found that contained uronic acid, with 30.0%, 47.6%, and 51.8% respectively determined by an improved HPLC method for the accurate quantification of uronic acid (Chen et al., 2004a). The study also confirmed the existence of rare earth elements in TPS complexes, and 8% of the hot water-soluble rare earth elements existed in the form of rare earth-polysaccharide. About 1 mg/g of rare earth elements were found in the rare earth-TPS with La, Ce, Nd, Pr, and Y as the five main elements which account for about 90% of its total (Wang et al., 2000). The experimental results of UV and IR spectrum showed that the purified TPS could be a protein-bound polysaccharide complex composed of glycan, peptide, and uronic acids, etc. (Nie et al., 2008).

As well as the monosaccharide composition, various compositions of amino acids in TPSs were also reported. Seventeen general amino acids were found in TPSs (Wang et al., 2000; Drynan et al., 2010). But Chen et al. (2005) found 15 general amino acids in TPSs from Hubei green tea. Nie et al. (2008) extracted and purified a TPS which contains 8.625% of protein, 65.3% of sugar, 25.14% of uronic acid, and 18 kinds of amino acids. The monosaccharide compositions were found to be ribose, rhamnose, arabinose, xylose, mannose, glucose, galactose, glucuronic acid, and galacturonic acid in a molar ratio of 1.71:5.88:13.70:1.99:1.00:1.84:33.75:17.24: 7.73. All the results showed that the monosaccharide composition, as well as the amino acid composition, of the TPS varied. These differences may be due to the fact that the materials of coarse tea and the preparation and purifying methods were different in each study.

Although many different TPSs were obtained by different research groups, little structural and conformational information was published. The structure of TPS extracted from coarse tea in Wuyuan, Jiangxi province was analyzed by periodate oxidation, Smith degradation, partial hydrolysis, ^{1}H NMR, and GCeMS (Zhou, 2004), and the conclusions were drawn as follows: the backbone of TPS was composed of rhamnose, glucose, and galactose by β1/3 glycosidic linkages, and the sugar residues chain was branched with arabinose chains which were connected probably by β1/2, β1/3, and β2/3 glycosidic linkages. Xyloses were represented as end groups by β1/glycosidic linkage in the side chain or backbone chain.

Based on the data obtained by Fourier transform infrared (IR) spectroscopy and ^{13}C nuclear magnetic resonance (NMR) spectroscopy, a protein-bound acidic TPS named TPC-3e1 was found to have an anomeric carbon sign of both a and b configurations and highly-branched chains (Chen et al., 2008b).

The structure of purified NTPS-1 was characterized by IR spectrum, GC, periodate oxidation, Smith degradation, 1D NMR, and 2D NMR spectroscopy. It was found that NTPS-1 was a galactan consisting of β-(1/4)-linked galactopyranosyl units (Wang et al., 2009a,b).

Structural features of the purified ATPS-2 were investigated by a combination of chemical and instrumental analysis, such as Smith degradation, ^{1}H NMR, ^{13}C NMR, and 2D NMR analysis. It was found that ATPS-2 was a pectic polysaccharide. The backbone was proved to consist of the smooth region of α-1,4-D-galactopyranosyluronan blocks interconnected by 1,2-linked rhamnose residues involved in the linear sugar chain. The ramified hairy regions were shown to contain the following backbone: →4)-α-D-GalpA-(1→2)-α-L-Rhap-(1→4)-α-D-GalpA-(1→, and the side chains attached to the residues of rhamnopyranose of these fragments. The side chains contain linear and branched α-L-Araf and β-D-Galp residues (Wang et al., 2009a,b).

A water-soluble polysaccharide, termed 7WA, was isolated from the leaves of green tea (Wang et al., 2015a). Its average Mw was 7.1×10^4 Da.

The monosaccharide composition analysis revealed that 7WA mainly consisted of arabinose and galactose in a molar ratio of 1.0:0.96. Methylation analysis, partial hydrolysis, and NMR were applied to characterize the structural features of 7WA. The results showed that 7WA had a backbone consisting of 1,3- and 1,6-linked galactopyranosyl residues, with branches attached to *O*-3 of 1,6-linked galactose residues, and *O*-4 and *O*-6 of 1,3-linked galactose residues.

By using methods of high-performance size-exclusion chromatography, multiangle laser light scattering, and viscosimetry, the molecular characteristics of TPS were investigated in Guo's research (2010). The results indicated that the molar masses of TPS ranged from 2.287×10^5 to 2.762×10^5 g/mol, the RMS radii distributed from 132.1 to 145.9 nm, and *Mw/Mn* was 1.028. The Mark–Houwink equation was established as $[\eta] = 0.5423\text{M}0.5379$ w, and the intrinsic viscosity and molecular chain parameters were described as follows: $[\eta] = 1.007$ dL g^{-1}, $k_H = 0.845$, $k_K = 0.387$, $\alpha = 0.5379$, and $K = 0.5423$. Furthermore, based on the slope of the RMS radius versus molar mass conformational plot being 0.24 ± 0.00, they inferred that the molecular morphology of this TPS was a homogeneous and spherical polymer with branches in solution.

A protein-bound acidic polysaccharide, designed as TPC3-1, was extracted and purified from low-grade green tea (*C. sinensis* L.) in Chen's research (2008a). The TPC3-1 was composed of L-arabinose, D-ribose, D-xylose, D-glucose, and D-galactose with a molar ratio of 4.9:2.2:3.1:1.8:1.0, with its average Mw around 120 kDa. Fifteen kinds of amino acids were identified in the polymer. In addition, TPC3-1 was determined to have an anomeric carbon sign of both α and β configurations and high-branched chains. The network structure of TPC3-1 was observed. Therefore, they concluded that this TPS TPC3-1 was an acid protein-bound polysaccharide with an image of network structure.

Scoparo et al. (2013) compared the structural features of water-soluble polysaccharides from green and black teas, and found that the two kinds of polysaccharides had some similarities in structural characteristics. Starch was present in both teas, and both polysaccharides were composed of a rhamnogalacturonan containing a long sequence of →4)-6-*O*-Me-α-D-GalpA-(1→, interrupted by α-L-Rhap residues, which were substituted by type II arabinogalactan chains. There were also some differences between these two polysaccharides. For instance, the black TPS contained less galacturonic acid residues (35%) than the green TPS (65%).

A novel white TPS (WTPS) was extracted from white tea (Jin et al., 2015). Methods including UV-vis, an amino acid analyzer, GC, Fourier transform infrared spectroscopy (FTIR), and NMR spectroscopy were employed to investigate the structure and ingredients of WTPS. The results showed that WTPS was an acidic polysaccharide mainly consisting of D-mannose and L-arabinose, with a small amount of D-glucose, L-rhamnose, and D-galactose with molar ratios of 4.5:4.2: 2.2:1.1:1. Seventeen kinds of amino acids were identified as components of WTPS. The polysaccharide appeared to be homogenous, with an average Mw of 2.9×10^4 Da. Furthermore, it was proved that WTPS was a hydrophilic polysaccharide composed of monosaccharide and acidic heteropolysaccharide chains based on the previously mentioned analysis.

There are very few reports concerning the chemical composition and structural features of polysaccharide conjugates isolated from tea flowers (Xiao and Jiang, 2015). Wang et al. (2010c) obtained a polysaccharide conjugate from tea flowers, and named it tea flower polysaccharide (TFPS). The TFPS mainly consisted of two kinds of polysaccharide (Wang et al., 2010c; Wei et al., 2010b). Therefore, they purified the crude TFPS using a DEAE-Sepharose FF gel filtration column to achieve a polysaccharide fraction named TFPS1, with a yield of 18%. The TFPS1 was a protein glycoconjugate consisting of rhamnose, arabinose, mannose, glucose, and galactose with a molar ratio of 1.0:2.9:0.5:1.3:3.3, with an average Mw of 450–500 kD. Glucose and galactose were the main components of the backbone of TFPS1, while arabinose and galactose were the main components of the side chain or edge backbone. The main glycosyl residues of TFPS1 were of α- and β-configuration, including signals of β-D-galactose, α-L-arabinose, α-D-mannose, α-rhamnose, and α-D-glucose. Wang et al. (2010c) also compared the composition of tea polysaccharide (TPF) and TFPS obtained by diverse extraction methods. The results indicated that TFPS obtained by different processes were composed of rhamnose, arabinose, glucose, galactose, xylose, mannose, and galacturonic acid, but glucuronic acid was not found in the TFPS. However, as to the TPS obtained from hot-water extraction, they mainly consisted of three kinds of polysaccharide with Mws of 413, 104, and 1165 D, respectively. The TFPS obtained from hot-water extraction was made up mainly of four kinds of polysaccharide with Mws of 483, 168, 120, and 1059 D, respectively. The average Mw of TFPS was higher than that of TPS. Han et al. (2011b) isolated a water-soluble crude polysaccharide from tea flower (TFP) by boiling-water extraction and ethanol precipitation. Two TPS fractions were separated from crude TFP, and they were termed TFP-1 and TFP-2, respectively. The structural features of TFP-1 and TFP-2 were identified by HPLC, GPC, rheometer, IR spectra, NMR spectroscopy, atomic force microscope, and SEM. Results showed that TFP-1 consisted of glucose, xylose, rhamnose, and galactose in a molar ratio of 1.0:1.2:0.81:0.98, with a Mw of 167.5 kDa. The TFP-2 was composed of glucose, xylose, rhamnose, and arabinose in a molar ratio of 1.0:0.76:2.3:2.3, with a Mw of 10.1 KDa (Han et al., 2011a,b).

The conformational transition of TPS was found by spectroscopy techniques including circular dichroism (CD) and ultraviolet (UV) spectroscopy (Chen et al., 2004b). The solution behaviors of TPSs in mediums at different temperatures, pH values, and ionic strengths were compared. Results showed that the native conformation of TPS was partially ordered. The CD and UV analysis showed that the TPS transferred from the ordered conformation (single or double helix) to the random coil varied with the polymer concentration, temperature, pH value, and ionic strength. The results indicated that external factors play important roles in the stability of the conformation of TPS (Chen et al., 2004b).

For polysaccharide (OTPS 2-1) extracted from oolong tea, the three-dimensional conformation and solution behavior were also studied using the techniques of AMF, QELS, CD, and LSCM (Ni et al., 2004). The results also showed that the molecules of OTPS 2-1 twisted each other, which formed a short stick of 0.2e 0.5 mm in diameter and 0.3e 0.6 mm in length with strong fluorescent properties. The conformation could be changed by the factors of pH, ion intensity, and complex compound. By using SEM, the folding and curling image under a low-powered microscope could be seen, while a smooth figure under a high-powered microscope was also observed. Thermal analysis of OTPS 2-1 also indicated that there appeared to be three endothermic peaks and three exothermic peaks during heating.

In conclusion, the research results indicated that the separation methods, tea source, and partial fermentation stage in tea production may have induced the alteration of chemical properties and structural features of polysaccharides, such as monosaccharide composition, Mw, and intermolecular interaction, thus causing transition of the configuration image, the chain conformation, and aggregation behavior, which increases its bioactivity beneficially.

The rheological properties of TPSs have seldom been reported. Lin and Lai (2009) extracted polysaccharides from mulberry leaves with water or sodium bicarbonate, and studied their rheological properties in a highly dilute to a semidilute regime. Mulberry leaf hydrocolloid contained a high level of uronic acid. The hydrodynamic behavior of mulberry leaf hydrocolloid in deionized water illustrated a typical polyelectrolyte behavior in the dilute domain, as indicated by a nonlinear Huggins' plot. Intrinsic viscosity values for alkaline-extracted (Alk) and water extracted (Hw) hydrocolloid were approximately 7.25 and 3.61 dL/g, respectively. The specific viscosity versus concentration plot for both Alk and Hw exhibited a power-law dependence on concentration, with an onset of molecular entanglement occurring at a concentration of about 0.6 and 1.8 g/dL, respectively. The coil-overlap parameter for Alk and Hw was in the range of 4.35–6.61. The relative chain stiffness parameter (B) for Alk and Hw was in the range of 0.004–0.013, which indicated a stiff backbone.

7.4 CHEMICAL MODIFICATION OF TEA POLYSACCHARIDES

Sulfation, carboxymethylation, acetylation, and enzyme treatment are commonly used methods in the chemical modification of TPSs.

Wang et al. (2013a) isolated two homogeneous water-soluble polysaccharides (TPSR4-2B and TPSR4-2C) from preinfused green tea, with average Mws of 41 and 28 kDa, respectively. A series of investigations indicated that both TPSR4-2B and TPSR4-2C were poly-(1→4)-α-D-galactopyranosyl uronic acid. Subsequently, the two water-soluble polysaccharides were suspended in dry formamide and kept at room temperature for 15 min. Then chlorosulfonic acid and pyridine were added in a ratio of 2:1. The mixture was reacted at 0°C for 2 h and kept at room temperature for 3 h. After that, the mixture was adjusted to pH 7 – 8 with NaOH solution. Then the solution was first dialyzed in saturated $NaHCO_3$ solution overnight, followed by extensive deionized water treatment. The residues were freeze-dried to achieve sulfated derivatives (Sul-R4-2B and Sul-R4-2C). A comparison of the anticomplementary activity studies of the sulfated and unsulfated polysaccharides showed that the sulfated derivatives of homogalacturonans in green tea might be used as drug candidates for therapeutic complement inhibition.

Wang et al. (2010b) isolated a NTPS and an ATPS from tea leaves with the application of D315 macroporous anion-exchange resin column chromatography. The two polysaccharides were sulfated by the pyridine–sulfonic acid method to achieve sulfated polysaccharides (NTPS-S and ATPS-S). It was found that NTPS was more easily sulfated than ATPS. The FTIR spectra showed that there were strong characteristic absorption peaks located at 1258, 1146, 832, and 617 cm^{-1} for sulfated polysaccharides. The DSC scan analysis indicated that the sulfated polysaccharides would affect the endothermic and exothermic peaks, and the appearance of exothermic peaks in both NTPS-S and ATPS-S indicated that a redox reaction might occur. Moreover, the biological activity study concluded that the sulfation of polysaccharides significantly improved their hypoglycemic activity.

In Xie's series study (2016), polysaccharide isolated from *C. paliurus* was chemically modified using the chlorosulfonic acid–pyridine method to achieve its sulfated derivatives. The sulfation process could be described as follows: firstly, chlorosulfonic acid was added to pyridine drop-by-drop at ratios of 1:1, 1:4, 1:6, and 1:8 (v/v) in an ice bath to make the sulfated product. The mixed reagents were vigorously stirred for 30 min at room temperature. Subsequently, polysaccharide was suspended in *N,N*-dimethyl formamide under agitation for 15 min at room temperature, then sulfated reagent was added drop-by-drop. The solution was stirred at 60°C for 4 h to allow for reaction. When the reaction finished, the mixture was cooled down to room temperature in an ice bath, and the solution was adjusted to pH 7 by adding NaOH. After that, the solution was precipitated with 95% (v/v) ethanol at 4°C for 12 h. The precipitate was collected and dialyzed in distilled water. The dialyzed solution was filtered, concentrated, and freeze-dried to achieve the sulfated polysaccharides. In addition, the morphological characterizations and antioxidant properties were further studied in these sulfated polysaccharide derivatives (Wang et al., 2015c). Scanning electron microscopy showed a remarkable change in the surface morphology of sulfated polysaccharides, but the degradation of the main chain of polysaccharides was inconspicuous, according to the results of monosaccharide composition. Hydroxyl radical assay, DPPH radical assay, superoxide radical assay and β-carotene–linoleic acid assay were applied in the in vitro studies (Xie et al., 2016). The results indicated that the physicochemical and antioxidant properties of polysaccharides were changed after sulfation. The sulfated derives with the lowest degree of substitution (DS), and Mw, and the highest protein content, exhibited excellent antioxidant activities in hydroxyl, superoxide, DPPH radical assay, and β-carotene–linoleic acid assay. Meanwhile, the antioxidant activity of sulfated and unsulfated polysaccharide were also tested on H_2O_2-induced oxidative mouse macrophage cell RAW 264.7 (Zhang et al., 2016b). The results showed that the sulfated polysaccharide derivatives could improve the viability of oxidative cells, increase the superoxide dismutase (SOD) activity in cells, and suppress the levels of malondialdehyde (MDA) in cells. These indicated that the sulfated polysaccharides had a better protective effect against H_2O_2-induced oxidative stress when compared with that of the unmodified polysaccharide. Based on the discussion above, it was concluded that the biological activity of sulfated polysaccharides from *C. paliurus* might have a close relationship with the combination of DS, Mw, sulfate content, protein, and uronic acid content.

In Xie's study, carboxymethylation was also applied to modify the *C. paliurus* polysaccharides (Wang et al., 2016). Three chemically carboxymethyl polysaccharides (CM-CPs) were derived from *C. paliurus* polysaccharides. The process for producing CM-CPs could be described as follows: the polysaccharide was stirred until it dissolved in 20% sodium hydroxide solution for 60 min at room temperature. After that, an appropriate amount of absolute ethyl alcohol mixed with chloroacetic acid of different quality was added to this polysaccharide solution. The mixture was stirred at 55°C for 5 h for reaction. When the reaction completed, the mixture was cooled down to room temperature. Then HCl was added to adjust the solution to a pH of 7, and the solution was dialyzed in tap water, distilled water, and ultra-pure water for 24 h, respectively. The carboxymethylated derivatives were obtained by ethanol precipitation and freeze-drying. The carboxymethylation of the polysaccharide was pronounced successful by the results of DS and FT-IR analysis.

The physicochemical properties and in vitro antioxidant activity of carboxymethyl derivatives were determined. Compared with native polysaccharides, the content of carbohydrate and protein decreased in carboxymethylated polysaccharide, while the uronic acid increased. The carboxymethyl derivatives mainly consisted of arabinose, galactose, glucose, mannose, and galacturonic acid, with a Mw of $1.03-1.08 \times 10^6$ Da. As to the antioxidant activity, the modified polysaccharides with the highest DS and Mw exhibited the highest antioxidant activity in a β-carotene–linoleic acid assay, but showed lower activity in scavenging the superoxide free radicals and hydroxyl free radicals.

Acetylation was another chemical modification method used in Xie's study (2015). Three acetylated derivatives, termed Ac-CP1, Ac-CP2, and Ac-CP3, were prepared. The physicochemical characteristics and antioxidant activities of acetylated derivatives were investigated. The acetylation was proved by the differences between acetylated derivatives and unmodified polysaccharide, which were shown in the results of chemical and FT-IR spectra. Compared with the native polysaccharide, the protein contents in acetylated derivatives decreased, while carbohydrate values increased. The acetylated derivatives mainly consisted of arabinose, galactose, mannose, and galacturonic acid, with Mws which varied from 1.05 to 1.09×10^6 Da. Furthermore, SEM investigations indicated that acetylation also changed the surface morphology of *C. paliurus* leaf polysaccharides. As to the in vitro antioxidant activity, Ac-CP1 with a relatively low DS (0.13 ± 0.01) exhibited excellent antioxidant activity in a DPPH radical assay and in a β-carotene–linoleic acid assay ($34.64\% \pm 2.07\%$) at a concentration of 0.5 mg/mL.

In Deng's study (2012), a polysaccharide from mulberry leaves (MPS) was isolated with hot water, and later purified with AB-8 macroporous adsorptive resins and Sephadex G-100 gel resins. After that, ethylenediamine was applied to chemically modify the polysaccharide chains to cationic mulberry leaf polysaccharide (CMPS). This cationic mulberry leaf polysaccharide could efficiently combine with plasmid transforming growth factor β1 (TGF-β1) to form nanoscaled particles. When the CMPS/pTGF-β1 weight ratio was increased to 30:1, a retarded plasmid migration could be observed by electrophoresis assay. In addition, in vitro bioactivity studies indicated that these nanoscaled particles were nontoxic to mesenchymal stem cells (MSCs), and the complex with a CMPS/pTGF-β1 weight ratio of 50:1 exhibited the highest cell transfection effect. Furthermore, the results of living cell imaging proved that more of the CMPS/plasmid TGF-β1 nanoparticles could be taken up by the MSCs than by the positive control, and a faster rate of uptake was observed. These results suggested that the CMPS/pTGF-β1 nanoparticle was efficient in gene delivery.

Enzymes were also used frequently in the chemical modification of polysaccharide. In Baik's work (Baik et al., 2015), pectinase treatment of TPS resulted in the change of monosaccharide composition. Furthermore, pectinase-driven hydrolysis in polysaccharides dramatically increased Interleukin 6 (IL-6) production in macrophages (Li and Cao, 2014).

7.5 BIOACTIVITY OF TEA POLYSACCHARIDES

Tea in general, and green tea in particular, have long been valued by human beings throughout the world for their pharmaceutical properties. A good number of animal and clinical studies suggest that the polysaccharides in tea play an important role in overall human health. The health benefits derived through the consumption of tea are summarized below.

7.5.1 Hypoglycemic Effect

Since ancient times in China and Japan, green tea has been used traditionally to control blood sugar in the body. Its hypoglycemic effects have been most widely studied, and animal studies suggest that TPS may help prevent the development of diabetes, and slow its progression once it has developed.

CTP and a TPF were found that could significantly decrease fasting blood glucose (FBG) and glucosylated serum protein (GSP) in alloxan-induced diabetic mice (Zhou et al., 2007). It has been shown that a soluble TPS is the major hypoglycemic factor in tea, and that this polysaccharide may be developed into a potential natural hypoglycemic functional ingredient. The findings of another group also suggested that TPSs were responsible for a reduction in blood glucose (Jie et al., 1997).

The effects of TPS on blood glucose and diabetic symptoms in mice were investigated, the results showed that TPS could significantly improve diabetic mice, alleviating the role of diabetic symptoms after oral administration with TPS at doses of 1.0, 2.0, and 4.0 g/kg body weight for 4 weeks. The symptoms of diabetic mice fed TPS were improved (Chen et al., 2003).

Two TPSs, NTPS and ATPS, were found that had remarkable hypoglycemic activity. After sulfuric esterification, the hypoglycemic activity of NTPS improved greatly, but that of ATPS did not change markedly (Wang and Jin, 2005).

The hypoglycemic activity of TPSs on normal mice and alloxan-induced diabetes mellitus mice's blood sugar was studied (Jiang and Zhen, 2004). The results showed that the LD50 of TPSs was 4.19 g/kg, and TPSs could reduce the rise in blood sugar in normal mice and alloxan-induced diabetes mellitus mice.

The hypoglycemic effects of the polysaccharides of green tea, oolong tea, and black tea were determined, and the results indicated the hypoglycemic effect of TPSs from Hubei tea were the best (Ni et al., 2003b). Moreover, with a low amount of injection, the hypoglycemic effects of TPSs extracted from semifermented oolong tea and fermented black tea were superior to that of nonfermented green tea, but at high dosages there was no obvious difference. It indicated that the habitats, species, and processing technology of TPS have a significant impact on its hypoglycemic effects.

Type 2 diabetes mellitus (T2DM) mice were treated with TPS in Li's study (2015), to investigate the hypoglycemic effects and the potential mechanism of TPS on diabetes. Results showed that TPS administration caused body weight increases and blood glucose decreases in diabetic mice. Moreover, the levels of total cholesterol (TC) and low-density lipoprotein cholesterol (LDL-c) were significantly down-regulated ($P < 0.05$), while the levels of triglyceride (TG) and high-density lipoprotein cholesterol (HDL-c) almost recovered to a normal level. These results suggested that TPS had the potential to alleviate insulin resistance. Additionally, the enzyme activities of SOD and glutathione peroxidase (GPX) were markedly improved in both liver and kidney tissues ($P < .05$) after TPS administration. Furthermore, the expressions of PI3Kp85/p-Akt/GLUT4 were also up-regulated with the treatment of TPS, which indicated that the PI3K/Akt signal pathway was involved in the hypoglycemic mechanism of TPS.

Crude polysaccharide was obtained by hot-water extraction from fresh tea leaves (Chen et al., 2014). Subsequently, DEAE-cellulose DE-52 column chromatography was used to yield a polysaccharide-conjugate fraction named TPC-L. Chemical components, Mw and its distribution, water vapor sorption properties, zeta potentials, and optical characteristics of TPC-L were investigated. The results showed that the two dosages of TPC-L (150 and 300 ug/mL) were discovered to have a significantly protective effect in a dose-dependent manner ($P < .05$, $P < .001$, respectively) on human umbilical vein endothelial cells against the injury caused by high glucose levels. Compared with a normal group, the administration of 40 mg/kg of TPC-L could remarkably reduce blood glucose levels of normal mice ingesting starch, and a significant difference of AUC (area under the curve of blood glucose) and ΔAUC ($P < .05$, $P < .01$) at the postprandial time point at 0.5 and 1.0 h were observed. In conclusion, TPC-L could promote starch tolerance to prevent impaired glucose tolerance (IGT) from developing into diabetes, and protect the HUVE cells from impairments caused by high glucose levels. Therefore, it was suggested that TPC-L could improve IGT through its ability in inhibiting the digestive enzymes.

The water-soluble polysaccharide (7WA) isolated from the leaves of green tea in Wang's research (Wang et al., 2015a), also proved to have the ability to augment insulin secretion at high glucose levels (25 mM). However, it did also work well at a low glucose level (5 mM). The possible mechanism for 7WA to exhibit the antidiabetic effect was to increase the glucose-stimulated insulin secretion through the cAMP-PKA pathway.

Chen et al. (2011) investigated the protective effects of green TPS-conjugate fractions, gTPC1, gTPC2, and gTPC3, on human umbilical vein endothelial cells against impairment triggered by high glucose levels. Compared with a cell injury group, three dosages of these fractions were found to exhibit remarkably protective effects on HUVE cells against impairments induced by high glucose dose-dependently. In order to understand the cell-based protective mechanism of the polysaccharide fractions, self-oxidation by a 1,2,3-phentriol assay was further applied. The scavenging effects of gTPC1, gTPC2, and gTPC3 were observed to be 55.1%, 47.6%, and 47.9%, respectively, at a concentration of 300 ug/mL.

It is accepted that high glucose-induced endothelial dysfunction is associated with the overproduction of reactive oxygen species (ROS), and contributes to vascular complications in patients with diabetes. Thus, the inhibitory effects of these polysaccharide fractions on high glucose-mediated HUVE cell loss are partly correlated with their potential scavenging abilities of ROS. Taken together, gTPC1−3 could be developed as noncytotoxic candidates of therapeutic agents for diabetic vascular complications.

Water-soluble polysaccharide conjugates (TPC-W) and alkali-soluble polysaccharide conjugates (TPC-A) were extracted from green tea by hot and alkaline water, respectively (Chen et al., 2010). The physicochemical properties of TPC-W and TPC-A were discussed. Nonobese diabetic (NOD) mice were employed to evaluate the antidiabetic activities of TPC-W and TPC-A. The administration of 150 mg/kg TPC-W could obviously decrease the blood glucose levels and the antiglutamic acid decarboxylase (GAD) antibody levels in NOD mice. Meanwhile, compared with a model control group, diabetic symptoms were also alleviated with treatment of TPC-W or TPC-A. In addition, the result of an organ index also showed that both TPC-W and TPC-A could protect the thymus from shriveling. In conclusion, the hypoglycemic effects of TPC-W and TPC-A were identified.

It has been reported that black tea is effective against T2DM, due to its modulation of postprandial hyperglycemia. However, the mechanism of its hypoglycemic effects remains unclear. Satoh et al. (2015) did some experiments to discuss the effects of black tea on the carbohydrate digestion and absorption process in the gastrointestinal tract. The hypoglycemic effects of water-extracts from black tea (JAT) were discussed from both in vitro and in vivo studies. Caco-2 cells were employed for the analysis of α-amylase activity, α-glucosidase activity, and glucose uptake. An ex vivo study of small intestinal α-glucosidase activity, as well as animal trails of nonobese T2DM, were further applied to reveal the possible mechanisms. Results suggested that JAT was able to inhibit the degradation of disaccharides into monosaccharides by α-glucosidase in the small intestine. In this way, it indirectly decreased the absorption of dietary glucose, mediated by SGLT1 and GLUT2 transporters localized at the apical side of enterocytes in the small intestine.

A carbose is a microbial carbohydrate and a α-glucosidase inhibitor, commonly used as an agent for alleviating type 2 diabetes (Satoh et al., 2015). However, the application of acarbose is usually accompanied with many side effects, such as abdominal distention, flatulence, diarrhea, and meteorism. These side effects might be caused by its strong inhibition of α-amylase, leading to the accumulation of several undigested carbohydrates. The bacteria colonized in the colon can further ferment the undigested carbohydrate and release gas. Therefore, it is necessary to find a new α-glucosidase inhibitor with a low inhibitory effect on α-amylase. Deng et al. (2015) found a group of polysaccharides isolated from Pu-erh tea could suppress the activity of α-glucosidase, but had little inhibitory effect on α-amylase. Their investigations on mice proved that Pu-erh TPSs might be better than acarbose at decreasing blood glucose levels after ingesting a carbohydrate diet.

Du et al. (2012) also studied the hypoglycemic effect of the water extract of Pu-erh tea (WEPT) on diabetes in vivo and in vitro. The in vitro investigation showed that WEPT significantly increased the glucose uptake by HepG2 cells in a dose-dependent manner. The in vivo test found that WEPS decreased rat intestinal sucrase, maltase, and porcine pancreatic amylase activity, and suppressed the increase in blood insulin and glucose levels. These results together indicated that WEPT was of benefit to glucose homeostasis in type 2 diabetes and in adjusting insulin resistance (Du et al., 2012).

In Xu's (Xu et al., 2014) work, they compared the hypoglycemic effects of polysaccharides from Pu-erh tea of different ages. The results indicated that the aging year had a relationship with the α-glycosidase inhibitory activities, and the longer the aging year, the higher the anti-α-glycosidase activity.

In Wang's study, both ethanol extracts (CPEE) and aqueous extracts (CPAE) were achieved from *C. paliurus*, and their antihyperglycemic, antihyperlipidemic, and antioxidant activities were evaluated with high fat diet (HFD) and streptozotocin (STZ)-induced diabetic rats (Wang et al., 2013b). After orally administering CPEE and CPAE at three different dosages to type 2 diabetic mice, blood glucose, serum insulin, oral glucose tolerance test (OGTT), insulin tolerance test (ITT), free fatty acid (FFA), TC, TG, low-density lipoprotein-cholesterol (LDL-c), high-density lipoprotein-cholesterol (HDL-c), MDA, SOD, glutathione peroxidase (GSH-Px), blood urea nitrogen (BUN), creatinine (CREA), and GSP were all measured. The results showed that both CPEE and CPAE could up-regulate the levels of OGTT, ITT, HDL-c, SOD, and GSH-Px, while they down-regulated the levels of FFA, TC, TG, LDL-c, MDA, BUN, CREA, and GSP. Therefore, it could be concluded that both the polysaccharides and flavonoids contained in tea leaves exhibited similar antihyperglycemic, antihyperlipidemic, and antioxidant effects in type 2 diabetic rats.

It is said that mulberry leaf is a traditional medicine applied in treating diabetes clinically. Ren et al. (2015) did a series of investigations to reveal the possible mechanisms of how MLPII improved hepatic glucose metabolism and insulin resistance in type 2 diabetic rats induced by high fat and STZ. MLP was given to type 2 diabetic Wistar rats for 6 weeks. At the end of the experiment, oral glucose tolerance, liver glycogen content, glucose synthase (GS) activity, and insulin resistance were assayed. Expression patterns of proteins and genes associated with insulin signaling, as well as biomarkers of oxidative stress and antioxidant enzyme activities, were determined. The results showed that MLPII administration significantly promoted oral glucose tolerance ($P < .01$) and restored the glycogen level ($P < .01$), as well as GS activity ($P < .05$) in diabetic rats, as compared with normal control rats. Insulin resistance was also improved in MLPII-treated diabetic rats

($P < .01$). Moreover, the expression levels of proteins involved in insulin signaling, like insulin receptor substrate 2 (IRS2), phosphatidylinositol 3-kinase (PI3K), and protein kinase B (PKB/AKT), were also notably increased ($P < .01$). However, protein-tyrosine phosphatase 1B (PTP 1B) expression was significantly reduced ($P < .01$). The levels of 8-hydroxy-2-deoxyguanosine (8-OHdG) and MDA in livers were also significantly decreased ($P < .01$) with the treatment by MLPII, while the activities of the antioxidant enzymes, such as SOD, catalase (CAT), and GPX, were significantly increased ($P < .01$, $P < .01$, $P < .01$, respectively). These results indicated that the administration of MLPII effectively normalized hepatic glucose metabolism and insulin signaling by inhibiting the expression of PTP1B, activating the PI3K-AKT pathway, and mitigating oxidative stress in the livers of type 2 diabetic rats.

Zhang et al. (2014b) investigated the possible mechanisms of antidiabetic activity of MLPs from both in vitro and in vivo aspects. In their studies, pancreatic islet cell and diabetic rats induced by a HFD and STZ were employed. The results showed that MLPs could inhibit the apoptosis of pancreatic islet cells, and normalized the insulin secretory capacity of pancreatic β-cells in diabetic rats. In addition, a chronic treatment of diabetic rats with MLPs could lead to up-regulation of antiapoptotic B-cell leukemia/lymphoma 2 (Bcl-2) protein, and down-regulation of pro-apoptotic Bcl2-associated X (Bax) and caspase-3 protein in pancreatic islet cells. Furthermore, MLPs remarkably recovered pancreatic duodenal homeobox-1 (PDX-1) protein nuclear localization, and promoted the mRNA and protein expression of PDX-1 and its downstream targets, glucose transporter 2 (GLUT2) and glucokinase (GCK), in pancreatic islet cells of diabetic rats. These results indicated that MLPs played an important role in preventing pancreatic islet cells from apoptosis via elevation of the Bcl-2/Bax ratio. Additionally, it also indicated that the polysaccharide could alleviate insulin secretory capacity of pancreatic β-cells via restoration of PDX-1 nuclear localization and expression levels in diabetic rats.

In Zhang's work (2014a), the antidiabetic effect of polysaccharide from mulberry leaf (MLP) was also studied in vivo. MLP was orally administered to STZ-induced diabetic rats for 5 weeks. The results found that MLP could significantly decrease the levels of FBG, GSP, serum TC, and serum TG, and up-regulate the body weight, fasting insulin (FINS), C-peptide (C-P), liver glycogen, liver glucokinase, and serum high-density lipoprotein cholesterol (HDL-c). Furthermore, it could stimulate the regeneration of marked pancreatic β-cell and insulin secretion, as well as down-regulate the liver fat accumulation in diabetic rats. It was also proved in their studies that the antidiabetic effect of MLP was similar to that of the antidiabetic drug glibenclamide.

It was mentioned above that tea flower polysaccharides isolated by Han's group were also proved to have inhibitory effects against α-glucosidase and α-amylase (Han et al., 2011b), and continuous treatment with these tea flower polysaccharides could result in a decrease in blood glucose levels in alloxan-induced diabetic mice.

In addition, Cai et al. (2013) also studied the hypoglycemic effects of polysaccharides from green tea flower (TFPS), and their possible mechanisms in a Sprague-Dawley (SD) rat model (Du et al., 2012). It was demonstrated that TFPS could decrease blood glucose in alloxan-treated S-D rats. Additionally, TFPS had a strong hydrogen donor ability and could protect cell membrane from over-oxidation. Moreover, it also inhibited α-amylase and α-glucosidase in vitro (Du et al., 2012).

7.5.2 Hypolipidemic and Antiatherogenic Effects

Several animal experimental studies have demonstrated that TPSs possess strong hypolipidemic effects.

The experimental results showed that the serum TC dropped by 18% and 24%, respectively, after intraperitoneal injection of 25 and 50 mg/kg of TPS to normal mice; when orally fed 50 and 100 mg/kg of TPS, the formation of hypercholesterolemia was significantly inhibited, and serum TC decreased by 34% and 43%. Hyperlipidemia rats orally fed TPSs at 22.5 10d and 45 mg/kg 10d will have a decrease in serum TC by 12% and 17%, reduction in TGs by 15% and 23%, decreased LDL-c by 6% and 29%, and increase in high-density lipoprotein cholesterol of 26% (Wang and Wang, 1991). It has been indicated that TPS can lower plasma TC, and prevent hypercholesterolemia, which in turn reduces the hyperlipidemia plasma TC, TG, LDL, and neutral fat, and increases high-density lipoprotein.

The antiatherogenic effect of TPS on laboratory rats was tested after treatment with TPS by intraperitoneal injection at a dosage of 50 mg/kg. The results showed that high-density lipoprotein cholesterol was increased, whereas no significant effect was found on serum cholesterol and serum TG. Over 12 h later, high-density lipoprotein cholesterol was boosted by 7.1%, and 24 h later it rose by 15%, which contributed to the inhibition of atherosclerosis (Jie et al., 1997).

The effects of TPS on lipoprotein lipase (LPL) activity were studied. The results showed that treatment with 160 mg/mL of TP could increase the activity of LPL by 7%, and also reduce the sensitivity of LPL to inhibitors such as NaCl. As heparin did, this enzyme could combine with LPL, which promotes arterial wall LPL flowing into the blood. Therefore, it might prove useful for the prevention of atherosclerosis (Lijun and Shuru, 1992).

7.5.3 Anticoagulant and Antithrombotic Effects

The study found that TPS could significantly prolong clotting time both in vivo and in vitro (Wang and Wang, 1992). After oral administration with

TPS of 50 mg/kg body weight, the clotting time in mice can be extended by 319%. After oral administration of 37 mg/kg, the thrombin time of rabbits can be extended by 40%. After the addition of different amounts of TPS into mixed human plasma, the recalcification time was determined, and the results showed that recalcification time could be extended with supplementation by TPS at 0.05 mg, and at 0.4 mg the plasma coagulation was completely inhibited. Oral administration of 37 mg/kg of TPS to rabbits could induce the inhibition of thrombosis. It was found that thrombus formation time was significantly prolonged, thrombosis length decreased, and a reduction of 20%, 43%, 16%, 11%, and 20% was found for platelet count, platelet adhesion, whole blood viscosity, plasma viscosity, and hematocrit, respectively, while an increase of 75% was found for erythrocyte sedimentation rate. After oral administration of 40 mg/kg, guinea fibrinolysis enzyme activity increased 77%. The results showed that TPS had high blood anticoagulation activity, significantly prolonged the time to thrombosis, and shortened the length of thrombosis, and was involved in anticoagulant and antithrombotic effects.

Four polysaccharide fractions, termed TPS-1, TPS-2, TPS-3, and TPS-4, were separated and purified from crude green TPS by anion exchange chromatograph in Cai's work (2013). In vitro investigations into the anticoagulant activities of TPS-4 indicated that TPS-4 could significantly prolong activated partial thromboplastin time (APTT) and thrombin time (TT), but not prothrombin time (PT).

7.5.4 Lowering Blood Pressure, Hypoxia Tolerance, and Increasing the Flow of Coronary Blood

A study found that after the administration of TPS at 22.5 mg/kg body weight to narcotic S-D rats, the rats could decrease their blood pressure by 9.60 kPa, and their heart rate also decreased. After the mice were treated with an intraperitoneal injection of TPS at a dose of 50 and 100 mg/kg body weight under normal pressure, the survival time of normal mice was prolonged by 59% and 66% ($P < .01$), respectively, compared with the control group. The results showed that TPS had hypoxia tolerance effects. In addition, by the injection of TPS at a dose of 1.0 mg/kg body weight, coronary blood flow rate increased by 37%. The difference is extremely significant ($P < .01$), indicating strong effects on the coronary blood flow (Wang and Wang, 1991).

7.5.5 Immunomodulatory Activity

According to a study performed by Xiao and Jiang (2015), experimental rats were treated with TPSs by subcutaneous injection, and then immunized with sheep red blood cells by intraperitoneal injection. After 6 days, a sample of venous blood was analyzed; it was found that the TPS at a concentration in the range from 3.0 to 10.0 mg/mL could enhance the role of humoral immunity, with serum agglutinin as the indicator. The best results were found at a concentration of 3.0 mg/mL. Compared with the control group, the significant difference was at the $P < .001$ level. Meanwhile, it was also found that TPS has the potential to promote the phagocytic activity of the monocyte–macrophage system, enhancing self-protection ability. It also exhibited recovering effects on the low transforming rate of spleen lymphocytes and the low secretion of interleukin-2 (LI-2) in AA rats. It also showed inhibitory effects on too high secretion of IL-1. The results indicated that the immune function of normal mice could be enhanced by this TPS.

Zhou et al. (1997) studied the immunological function of TPS. It was found that the immunological function was improved by oral administration at a daily dosage of 25 mg/kg BW for 14 days, with an increase of 5.0% and 5.2% for the spleen index and thymus index (Jie et al., 1997). According to the acute toxicity test, toxicity did not occur until a total dosage of 1000 mg/kg BW within 24 h was used.

The effects of TPS extracted by different methods on the immune function of rats with adjuvant arthritis were studied (Jiang et al., 1998). The results revealed that all of the TPS tended to increase the lower level of spleen lymphocyte proliferation and IL-2 in rats with adjuvant arthritis. The P2 of TPS could be enhanced markedly ($P < .05$). All of the TPS could decrease the higher level of IL-1 in rats with adjuvant arthritis, P2 of the TPS could also decrease markedly ($P < .05$), and TPS had marked immunoregulatory effects on the immune index of rats with adjuvant arthritis.

Yang et al. (1997) reported that TPS in the range of 3.0–10.0 per g/L could increase agglutinin and clear carbon in blood serum of rats. TPS was proved to have the function of increasing immunity.

Yang et al. (2004) discussed the effects of TPS on the immune response of mice and the underlying mechanism with the injection of sheep red blood cell (SRBC) and ovalbumin as antigens to mice. After administration with TPS, the levels of corresponding antibody (anti-SRBC and antiovalbumin) and cytokines (IL-2 and IFN-γ) in experimental groups were all significantly higher than those of the control groups, which suggests TPSs can intensify the immunity of mice.

The immunostimulating activity of TPS was also investigated (Monobe et al., 2008). They reported that the immunostimulating activity of TPS from immature leaf extract was higher than that of TPS from mature leaves, and its activities were dependent on the content of strictinin in the leaf extract. Furthermore, a mixture of catechin and TPS that removed polyphenols did not increase the immunostimulating activity. These results suggest that the catechine–polysaccharide complex is a very important molecule in the immunomodulating activity of tea extracts.

Wang et al. (2014b) obtained a purified polysaccharide fraction (TPS1-2) from green tea by hot-water extraction and subsequent precipitation with ethanol of different concentrations. It was later determined to be a homogalacturonan pectin. The immunomodulatory activity of TPS1-2 in vitro was studied by cell culture experiments. The results found that this homogalacturonan pectin exhibited a phagocytosis-enhancing activity in HL-60 cells.

Huang et al. (2012) extracted and purified a tea glycoprotein (TGP) from the leaves of green tea (*C. sinensis*) and evaluated its immunomodulatory effects on dendritic cells. Murine bone marrow cells were cultured with recombinant mouse (rm) GM-CSF and rmIL-4 for 6 days, followed by a subsequent 2-day treatment with TGP or lipopolysaccharide (LPS). The results showed that TGP did not have significant inhibition on cell proliferation and apoptosis. Dendritic cells cultured with 50 ug/mL of TGP exhibited higher levels of expression of MHC class II molecules and major costimulatory molecules, such as CD 86, CD 80, and CD 40. However, the endocytic activity was impaired markedly. The TGP improved the level of IL-12 p70, but suppressed the levels of IL-10 and NO. The activation of antigen-presenting ability and the lymphocyte proliferation of mixed lymphocyte reaction by dendritic cells were also enhanced with the treatment with TGP. In addition, the expression of the CCR7 mRNA of dendritic cells was also increased with the treatment by TGP. Therefore, it could be concluded that TGP was able to promote both phenotypic and functional maturation of murine bone marrow-derived dendritic cells in vitro, which proves the immunomodulatory activity of TGP.

Han et al. (2010) obtained TFP from tea flowers consisting of two kinds of polysaccharides with Mws of 31 and 4.4 kD. The effects of TFP on plasma IL-2 and IFN-γ levels were tested by ELISA kit. The results showed that TFP significantly increased the levels of IL-2 and IFN-γ, and the percentage of CD4+ T cells in plasma of S180-bearing mice in a dose-dependent manner. However, it had no effects on the proportion of CD8+ T cells. Furthermore, there was a remarkable trend toward an increased ratio of CD4+ T cells and CD8+ T cells in the polysaccharide-treated groups in a dose-dependent manner.

7.5.6 Anticancer and Antitumor Activities

The role of tea in protection against cancer has been supported by much evidence from studies in cell culture and animal models. Emerging animal and clinical studies are beginning to suggest that TPSs may play an important role in the prevention of cancer.

The anticarcinogenic effect of TPS was investigated using a panel of short-term cell biology assays: the cytokinesis block micronuclei test in V79 cells induced by mitomycin, the test of metabolic cooperation between V79 and M cells, and the test of growth ability of Hela cells in soft agar were used in the screening. The results showed that the TPSs tested were effective in the test involved in different stages of carcinogenesis, i.e., initiation, promotion, and progression (Chi and Jiang, 1995).

By assaying the activity of NAD(P)H:quinone reductase (QR) in Hep G2 cells exposed to inducing agents, the effects of TPS on inducing QR and preventing cancer were studied. The results showed that TPS improved QR activity in Hep G2 cells by 18.0% and 26.5% at the dose of 2 and 10 mg/L (Qi and Han, 1998). This shows that TPSs have a certain role in anticancer effects.

The anticancer activity of TGP at different concentrations was evaluated by MTT assay using two kinds of colon cancer cell lines (HCT-15, Caco-2). Dose-dependently, TPS exhibited good antiproliferation activity to HCT-15, whereas it exhibited very weak antiproliferation activity to Caco-2. Only at a very high concentration (409.6 mmol/L) did the TGP obviously inhibit the proliferation of Caco-2 (Nie et al., 2006).

Fan compared the effects of tea polysaccharides (TPS) and polyphenols (TPP) on the growth inhibition of hepatoma H22 cells in mice, and discussed their roles in immune stimulation (Fan et al., 2013). The results found that both TPS and TPP could remarkably suppress the growth of H22 transplantable tumor in mice, improve the proliferation of splenocytes induced by ConA or LPS, and significantly up-regulate macrophage phagocytosis towards neutral red. In addition, an increase in the thymus index, as well as a decrease in the spleen index, in the TPS or TPP-treated group was observed, when compared with that of the model group. Therefore, it could be concluded that both the polysaccharide component and the polyphenols in tea leaves had antitumor activities.

He et al. (2013) investigated the inhibitory effect of selenium-enriched tea polysaccharides (Se-GTPs) from a selenium-enriched Ziyang green tea against human MCF-7 breast cancer cells. The Se-GTPs produced effective inhibition of the growth of MCF-7 cells in a dose-dependent manner by inducing MCF-7 cancer cells to undergo G2/M phase arrest and apoptosis. The IC50 for the growth inhibition was 140.1 μg/mL. The blockade of cell cycle was associated with an up-regulation of p53 expression, an increase in Bax/Bcl-2 ratio and caspase-3, and caspase-9 activation. It was also found that the production of intracellular ROS was an important mediator in Se-GTPs-induced cell growth inhibition.

Wang et al. (2013c) also did a series of studies concerning the antitumor activity of Se-polysaccharide (Se-ZYTP) from Se-enriched Ziyang green tea. Human osteosarcoma U-2 OS cells and a U-2 OS cancer xenograft model were used for in vitro and in vivo studies, respectively. The results of the in vitro study indicated that Se-ZYTP could remarkably decrease the growth of human osteosarcoma U-2 OS cells in a dose-dependent manner. Furthermore, oral administration with Se-ZYTP of three different dosages for 28 days could lead to obvious tumor regression as compared to a model group. Additionally, there was no difference in mice body weight between the Se-ZYTP treated groups and control group, and no mice died during the experimental process. These two results suggested that Se-ZYTP was nontoxic to mice. In conclusion, it was inferred that Se-ZYTP might have cancer-preventive and cancer-therapeutic benefits for human osteosarcoma.

In Xie's series studies (2013a) on *C. paliurus* (Batal.) Iljinskaja polysaccharide (CPP), the anticancer activity of CPP in human gastric cancer HeLa cells was also evaluated by 3-(4,5-dimethylthiazol-2-yl)-2,5-diphenyltetrazolium bromide (MTT) assay. The results indicated that CPP exhibited a strong inhibitory effect on the proliferation of human gastric cancer HeLa cells.

As well as the tea leaves, the antitumor and anticancer activities of polysaccharides from tea flowers and tea seeds have also attracted researchers' attention. It is reported (Xu et al., 2012) that the polysaccharide isolated from tea flowers could inhibit the growth of human gastric cancer BGC-823 cells. Wei et al. (2011a) found that the polysaccharide from tea seeds could significantly inhibit the proliferation of K562 cells, especially at a concentration of 50 ug/mL. The highest inhibition ratio of tea seed polysaccharide was 38.44% ± 2.22% ($P < 0.01$). In addition, the higher the polysaccharide concentration, the higher was the proliferation effect on lymphocyte observed.

7.5.7 Antioxidation

The effect of TPSs on the activity of SOD in the red cells of mice was studied. The results showed that the activity of SOD in the red cells of mice was obviously increased by feeding TPSs (Deng and Xu, 1998).

Three fractions of water-soluble polysaccharide conjugates, coded as TPC-1, TPC-2, and TPC-3, were isolated and purified from low-grade green tea (*C. sinensis*) by absorbent chromatography and ion-exchange chromatography. Deoxyribose assay, photoreduction of nitro blue tetrazolium (NBT) assay, and lipid peroxidation inhibition assay were applied to test the antioxidant activities of TPS conjugates in vitro. The TPC-3 showed the highest antioxidant activity among the three fractions of polysaccharide conjugates, with an IC50 of 182 mg/mL for the deoxyribose assay, and 93 mg/mL for the photoreduction of NBT assay. The effects of the Mw and protein content of the polysaccharide conjugates on the improvement in antioxidant activities appeared to be significant (Chen et al., 2008b).

The impact of TPSs on the antioxidant function and the morphological changes in liver and kidney of diabetic rats were studied. It was found that after 4 weeks' administration of TPSs in diabetic rats, the liver and kidney SOD and GSH-PX activity were significantly improved, while the MDA content decreased greatly, which in turn enhanced the antioxidant activity (Ni et al., 2003a).

The antioxidative activities of tea glycoprotein (TGP) in green tea from Wuyuan of Jiangxi were evaluated by determining the change in value of heat-induced oxidation in a linoleic acid system with β-carotene, the decoloration of the 1,1-diphenyl-2-picrylhydrazyl (DPPH), and the superoxide generated from autoxidation of pyrogallol. The results show that the TGP possesses distinctive antioxidative activity (Nie et al., 2005a,b).

The antioxidant activities of crude TPSs from several kinds of out-of-date tea leaves, including Xihu Longjing (XTPS), Anxi Tieguanyin (TTPS), Chawentianxia (CTPS), and Huizhoulvcha (HTPS), in markets were investigated (Xiao et al., 2011). These TPSs showed similar DPPH scavenging activities. Higher concentrations of these TPS exhibited similar DPPH scavenging activity with Vc. In addition, HTPS showed a significantly higher superoxide anion scavenging activity than other TPSs and gallic acid. The XTPS exhibited a significantly higher inhibitory effect on α-glucosidase and α-amylase than other TPSs. The TTPS, XTPS, and HTPS showed a similar inhibitory ability on α-D-glucosidase and α-amylase.

In Chen's study (2005), the antioxidant activity of TPS was evaluated using a deoxyribose assay, a photoreduction of NBT assay, and a lipid peroxidation inhibition assay. The results showed that TPC exerted significant inhibitory effects on hydroxyl and superoxide radicals and lipid peroxidation with IC50 values of 101, 145, and 238 ug/mL, respectively. Furthermore, the antioxidant activity was also tested on alloxan-induced oxidative damage. The results indicated that TPC could also increase the activity of SOD ($P < .05$). Additionally, they also found that there might be a direct connection between antioxidant activity and hypoglycemic activity.

It was also reported that the antioxidant activity of TPS might be one reason accounting for its hepatoprotective activity. Wang et al. (2014a) investigated the hepatoprotective effects and antioxidant activity of tea polysaccharides (ZTPs) extracted from a selenium-enriched Ziyang green tea (*C. sinensis*). Administration of ZTPs to CCl_4-induced liver injury mice significantly prevented the increase in serum alanine aminotransferase, aspartate aminotransferase, and lactic dehydrogenase (LDH), as well as hepatic MDA levels. The ZTPs also prevented CCl_4-caused histological liver alteration. In addition, ZTP treatment normalized GPX and SOD activities, relative to the model group. These findings indicated that ZTPs could also protect the liver from oxidative stress induced by CCl_4.

In Yang's study (2012), the antioxidant activity of TPS was determined in gastric cancer mice. The results indicated that TPSs down-regulated the stomach MDA level, serum IL-6 and tumor necrosis factor-α (TNF-α) levels, increased serum immunoglobulin A (Ig A), immunoglobulin G (Ig G), immunoglobulin M (Ig M), interleukin-2 (IL-2), interleukin-4 (IL-4), interleukin-10 (IL-10) levels, and stomach antioxidant enzymes activities in gastric cancer mice.

Xie et al. (2010a) also found that the water-soluble polysaccharide from *C. paliurus* (Batal.) Iljinskaja also had the ability to scavenge DPPH free radicals. The scavenging ability of this polysaccharide was comparable to that of the reference controls of BHT or ascorbic acid.

The antioxidant effect of polysaccharides from tea of different harvest years differed. Xu et al. (2014) compared the antioxidant activities of polysaccharides from Pu-erh tea with different aging processes. The results indicated that a longer aging process induced a better antioxidant activity.

Zhang et al. (2016b) proved the excellent antioxidant activity of polysaccharide extracted from "Anji Baicha" (*C. sinensis* (L.) O. Kuntze). The study showed that "Anji Baicha" polysaccharide could scavenge the free radicals in vitro.

Hawk tea is a herbal tea, and has been a widely consumed traditional beverage in southwest China for hundreds of years. Jia et al. (2014) studied the extraction, purification, and characterization of polysaccharides from Hawk mature leaf tea (HMP). The results indicated that HMP had strong 2,2-diphenyl-1-picryl-hydrazyl (DPPH) radical scavenging activity and FRAP.

The existence of rare mineral elements could also affect the biological activity of polysaccharide. For example, in Wang's study (2015b), it was found that Se-polysaccharides exhibited higher antioxidant activities compared to ordinary polysaccharides.

Wang et al. (2012a) analyzed the antioxidant activities of polysaccharide from tea leaves, tea flowers, and tea seeds, and found that all three kinds of polysaccharide had strong antioxidant activities.

In the past decades, numerous kinds of TPSs have been widely proved, both in vitro and in vivo, to have potential antioxidant activities. However, there have been some different opinions recently. Some scientists argue that it is the polyphenols or flavonoids existing in TPS, not the carbohydrate itself, that actually cause the antioxidant effects.

Wang et al. (2013d) did an experiment to determine whether the antioxidant abilities of crude TPSs depend on tea polyphenols, or merely carbohydrate. They first compared the antioxidant effects of purified polysaccharide fractions and crude polysaccharide, and found that the crude fraction exhibited stronger antioxidant activity. Subsequently, they added EGCG to polysaccharide to see whether it could enhance the antioxidant activity of TPS. The results showed that EGCG remarkably increased DPPH radical scavenging potential and the reducing power of TPFs for each dose. These indicated that EGCG caused a synergistic increase in the antioxidant activities of the TPF. Thirdly, they also investigated the antioxidant activities of dextrans in the absence and presence of EGCG, and similar results were observed.

Three purified polysaccharide fractions were separated from mulberry leaves. Their antioxidant activities were compared, and the results showed that the antioxidant effects decreased with increasing polysaccharide purity (Zhang et al., 2016a). Highly concentrated polysaccharides were shown to have very little antioxidant activity. However, after adding 10 ug/mL of quercetin to the polysaccharide solution, the antioxidant activities of the polysaccharide were dramatically improved (Zhang et al., 2016a). This result indicated that polysaccharide and quercetin might have a synergistic effect on antioxidant activity.

7.5.8 Radiation Protection

As early as the 1970s, the Tea Research Institute of the Chinese Academy of Agricultural Sciences and Center for Disease Control and Prevention of Tianjin reported the radiation protection effects of TPS products with acute radiation sickness prevention experiments. According to this experiment, after subcutaneous administration of TPSs, mice were irradiated with ^{60}Co at an irradiation dose of 766–840-lun (R). The results showed that the TPS gave significant protection against radiation, with a 30% improvement in the survival rate of mice (Nie and Xie, 2011).

7.5.9 Skin Care

The protective effects on skin of three TPFs, termed TPS1 (92% TPS), crude TPS2 (20% TPS), and TPP (98%), were investigated by Wei et al. (2009). The protective effects on skin were assessed in four aspects: moisture absorption and retention, sunscreen, promoting the proliferation of fibroblast cells, and tyrosinase inhibitory effect. The TPS and TPP performed perfectly in moisture absorption and retention, and TPS with a higher purity had better moisture absorption and retention abilities. The TPS1 could hardly protect the skin from the sun UV, and had little promoting effect on the proliferation of fibroblasts. These results indicated that TPP and TPS had complementary advantages, and could be appropriately combined to achieve higher performance when applied as active components of cosmetics (Wei et al., 2009).

7.5.10 Toxicity

The toxicity of TPS is very crucial to the development of any product for medical treatments. An animal experiment was conducted to evaluate the toxicity of TPS, and the results found that no toxicity was exhibited to the liver, kidney, heart, thymus, or spleen of mice which were fed with the TPS conjugate, and none of the mice died throughout the period of the experiment. There was no significant difference between the thymus index, spleen index, and liver index of the mice from the test and control groups ($P > .05$). On the basis of this study, the TPS conjugate may be classified either as a very low toxicity substance, or as unclassified when orally administered to mice. It might be a candidate for dietary supplements beside the bioactivities as a polysaccharide (Chen et al., 2007).

7.5.11 Antifatigue Activity

The antifatigue activity of TPS has been reported in recent years. Chi et al. (2015) assessed the antifatigue activity of polysaccharides from Ziyang green tea. A 4-week weight-loaded swimming test of mice was established in their experiment. Polysaccharides were orally administered during the swimming exercise. The biochemical parameters relating to fatigue were evaluated, such as exhaustive time, BUN, blood lactate acid (BLA) levels, and LDH activity in serum; and SOD, GSH-Px activities, MDA, and glycogen levels in skeletal muscle. The results illustrated that polysaccharide from Ziyang green tea significantly prolonged exhaustive time, increased glycogen levels and GSH-Px activity in muscle, and markedly decreased BUN, BLA levels, and LDH activity in serum and MDA level in muscle. In conclusion, TPS has potential ability to improve exercise-induced fatigue and decrease oxidative stress induced by exhaustive exercise.

7.5.12 Hepatoprotective Effects

Hepatoprotective effects were also found in TPSs. Lu et al. (2013) examined the hepatoprotective effects of polysaccharides from green tea of HMTP against CCl_4-induced oxidative damage in vivo. The HMTP was orally administered to mice for 14 days before CCl_4 treatment. The results indicated that HMTP significantly reduced the impact of CCl_4 toxicity on the serum markers of liver damage, alanine aminotransferase, aspartate aminotransferase, TC, and TGs. This method of HMTP administration also markedly suppressed hepatic lipid peroxidation formation of MDA and 15-F_{2t} isoprostanes, and increased the antioxidant levels of hepatic glutathione and SOD. These results, together with liver histopathology, demonstrated that HMTP had excellent hepatoprotective effects against CCl_4-induced injury, and the effects were comparable to that of biphenyldicarboxylate. In addition, it was also concluded that the hepatoprotective effects of HMTP may probably be caused by both the inhibition of lipid peroxidation and the increase of antioxidant activity.

Sun et al. (2013) designed an experiment to prove the protective effects of Keemun black TPSs (KBTP) on acute CCl_4-induced oxidative hepatotoxicity in mice. Different dosages of KBTP solution were administered to mice before CCl_4 injection. The results showed that KBTP could decrease ALT, AST, TG, and TC in serum, and hepatic MDA, as well as 8-iso-PGF2a levels caused by CCl_4 injection. In addition, KBTP treatment displayed a better profile of hepatosomatic index, and promoted the activities of GSH and SOD, compared with the model group. These biochemical results, together with the liver histopathological assessment, supported the viewpoint that KBTP gave remarkable prevention against liver damage induced by CCl_4 in mice.

In addition, hepatoprotective activity is also found in tea flower polysaccharides. Xu et al. (2012) studied the in vivo hepatoprotective activity of crude polysaccharide from the TFPS. They suggested that TFPS could observably antagonize the increase in serum alanine aminotransferase and aspartate aminotransferase levels in CCl_4-induced liver injury mice.

7.5.13 Benefits on Gastrointestinal Function

Most of the naturally derived polysaccharides are nondigestible in the human intestine (Flint et al., 2008), but can be fermented and utilized by gut microbiota or interact with the immune cells in gut-associated lymphoid tissues. On the one hand, the end products of carbohydrate fermentation, such as various gases and short-chain fatty acids (SCFAs), can affect the physiological and metabolic functions of animals and humans. On the other hand, polysaccharide has the ability to balance the diversity of colon microbiota, adjust the colon pH value, and stimulate the intestinal immune system. Therefore, it is understandable that polysaccharides could be of benefit to our gastrointestinal tract. In Min's study (2014), the fermentation characteristics of *C. paliurus* leaves were investigated using human fecal inoculums in vitro. The variations in the fecal contents of neutral and reducing sugars and pH value, consumption of monosaccharide, and production of SCFAs were analyzed. Within the first half hour, the content of neutral sugar and reducing sugar increased, but it decreased in later hours. During fermentation, the pH value remarkably dropped from 7.2 to 6.04. Moreover, the greatest yields and the fastest consumption of galacturonic acid were observed, the yields of glucose and arabinose were relatively high, and the content of SCFAs were significantly increased. In conclusion, it indicated that the polysaccharide was partially fermented, the glycosidic bonds with galacturonic acid were more susceptible to be attacked by gut microbiota, and the galacturonic acid was regarded as the main producer of acetic acid.

Later in their study, the effects of this polysaccharide on mouse colon health were examined. Mice were orally administered with polysaccharide (150, 300, 600 mg/kg body weight) for 30 days. The parameters relating to colon health were investigated. The results showed that colon index and colon length were improved after the administration of polysaccharide for 30 days. In addition, the content of fecal moisture was significantly increased; and the pH values in cecum, colon, and feces were decreased in the polysaccharide supplemented group. Furthermore, the levels of acetate, propionate, and butyrate were found in the cecum and colon of mice fed with the highest dosage of polysaccharide. Therefore, it was demonstrated that *C. paliurus* polysaccharide was efficient in improving colon health.

It was also reported that polysaccharide could be used as a dietary supplement for early-weanling piglets. Zhao et al. (2015) determined the effects of MLPs on the growth performance, diarrhea, blood biochemical parameters, and gut microbiota of early-weanling piglets. Piglets were randomly divided into five groups. The dietary treatments can be described as follows: (1) control treatment (CT): basal diet (BD); (2) low-dose MLPs treatment (LT): 0.3 g/kg of MLPs + BD; (3) medium-dose MLPs treatment (MT): 0.6 g/kg of MLPs + BD; (4) high-dose MLPs treatment (HT): 0.9 g/kg of MLPs + BD; and (5) antibiotic treatment (AT): 0.15 g/kg chlortetracycline + BD. The fasting weight of each animal was measured and recorded in the morning at three time points (0, 10, and 21d), in addition to food intake per pigpen. These data were applied to calculate the average daily gain (ADG), average daily feed intake (ADFI), and feed-to-gain (F/G) ratio of each pig. The diarrheal status was recorded twice daily. After the dietary treatment, blood samples were collected for biochemical analysis, and the contents from the caecum, colon, and rectum were collected to identify and quantify relevant gut microbiota. The results indicated that there were no significant differences in the ADG, ADFI, or F/G ratio among MLPs-treated groups ($P < .05$), but the ADFI in LT, MT, and HT groups were all higher than those of both the CT and AT groups. Furthermore, the F/G ratios in LT, MT, and HT groups were lower than those in the CT and AT groups. The LT, MT, and HT groups exhibited a marked reduction in diarrheal incidence ($P < .05$) when compared to the CT and AT groups. Biochemical analysis showed that blood glucose (GLU) levels in all MLPs-treated groups were significantly ($P < .05$) lower than in either the CT or AT groups. The levels of blood urea nitrogen (BUN) in the MT and HT groups were significantly ($P < .05$) different from those in the CT and AT groups. The T3 levels in HT were obviously ($P < .05$) higher than those in either the CT or AT groups. Although the T4 levels in MLPs-treated groups were not remarkably different from those in the AT control group, both IGF-1 and GH levels in the MT group were considerably higher than those in either the CT or AT groups ($P < .05$). The results for the microbiota analysis indicated that the effects of LT, MT, and HT in both inhibiting gut *Escherichia coli* and in promoting gut lactobacilli and bifidobacteria were superior than those of both the CT and AT groups. Notably, the HT group (0.9 g/kg MLPs) showed a significantly better inhibitory effect on gut *E. coli*, as well as a significantly better promotional effect on gut-beneficial bacteria when compared to both the Ct and AT groups ($P < .05$). Collectively, these results suggest that dietary addition of MLPs can improve the ecology of gut microbiota, reduce diarrhea rate, and improve the overall growth performance in early-weanling pigs.

7.5.14 Other Bioactivity

The fatigue-alleviating effect of polysaccharides from *C. paliurus* (Batal) Iljinskaja (PCP) were found *in vivo* by a weight-loaded swimming test in Wang's study (Wang and Wang, 2012). The related biochemical parameters, including serum urea nitrogen (SUN), BLA, hemoglobin (Hb), and hepatic glycogen were measured. The results found that PCP could increase swimming time to exhaustion, and Hb and hepatic glycogen content, while reducing SUN and BLA content. These indicated that polysaccharide could alleviate the effects of exercise-induced fatigue in mice.

The antibacterial effects of polysaccharides have also been proved by scientists. Wang et al. (2010a) studied the antibacterial ability of polysaccharides from a mulberry leaf. The results found that mulberry polysaccharide exhibited a distinct antibacterial ability in a dose-dependent manner. In addition, similar antimicrobial activity of polysaccharide from *C. paliurus* (Batal.) Iljinskaja was proved in Xie's experiment (2012).

It has been reported that TPS also works well in polymicrobial sepsis. Scoparo et al. (2013) compared the effects on polymicrobial sepsis of polysaccharides from green tea and black tea, and found that their reducing mortality rates were 40% and 25%, respectively. Moreover, the influx of neutrophils was also affected, which resulted in a decreasing accumulation in lungs and tissue injury. Together with the chemical composition information on green and black TPSs, the different abilities to prevent mice deaths can be related to the different uronic acid contents.

A complex of TPS with ferric iron was prepared in Tang's study (2012), and its bioavailability in vitro and in vivo was analyzed. The in vitro digestion experiment found that the availability of TPIC was sufficient. The in vivo iron bioavailability of TPIC was determined by the rat Hb-repletion bioassay with ferrous sulfate ($FeSO_4$) as the positive control. Results indicated that at the end of a 21 day iron regeneration phase, the values of Hb concentration, free erythrocyte protoporphyrin (FEP), serum iron (SI) concentration, and mean cell hemoglobin (MCHC) of rats supplemented with $FeSO_4$ and TPIC increased quickly, when compared to those of normal ones, and the bioavailability of TPIC was even higher than that of the positive control.

7.6 STRUCTURE–FUNCTION RELATIONSHIP

Based on numerous studies, the biological activities of polysaccharides from different resources were closely correlated with their chemical composition information and structural characteristics, including the polysaccharide composition, protein contents in polysaccharide conjugates, Mw of polysaccharides, monosaccharide composition, and the DS in modified polysaccharides.

Pectinase treatment led to differences in polysaccharide composition in TPS by hydrolyzing polysaccharides. In addition, pectinase-driven hydrolysis in polysaccharides significantly increased TPS-induced IL-6 production in macrophages, which was mostly related to their differences in polysaccharide composition.

Three fractions of TPS conjugates, termed as TPC-1, TPC-2, and TPC-3, were obtained in Chen's study by absorbent chromatography and ion-exchange chromatography (2008b). The protein contents of TPC-1, TPC-2, and TPC-3 were 2.8%, 3.8%, and 4.0%, and the Mws were 26.8×10^4, 11.8×10^4, and 4.2×10^4, respectively. The antioxidant activity analysis showed that TPC-3 had the highest antioxidant activities among the three fractions of polysaccharide conjugates. The difference in their bioactivities was largely caused by their differences in Mw and protein content of polysaccharide conjugates.

Three TPFs, coded as TPS1, TPS2, and TPS3, with different Mws were obtained for study of their antioxidant activity (Wang et al., 2012c). Each fraction differed in their content of neutral sugar, uronic acid, protein, and total polyphenols. These differences were closely correlated with their differences in free radical scavenging activities. In general, TPS1 and CTPS had stronger antioxidant activity, and TPS2 and TPS3 had lower antioxidant activity. The TPS1 fraction had a higher activity for DPPH and lipid peroxidation inhibition, but lower ability in reducing power and metal chelating. This might be due to its higher content of hexuronic acid and larger Mw. The order of inhibition activity of lipid peroxidation of these fractions was similar to DPPH radical scavenging activity. The order of metal chelating activity of these polysaccharide fractions was similar to their hydroxyl radical scavenging activity. This indicated that the hydroxyl radical scavenging activity of polysaccharide relied heavily on the Fe^{2+} metal chelating capability.

As mentioned before, the chemical characterization and antioxidant activities of tea leaf polysaccharides (TLPS), tea seed polysaccharide (TSPS), and tea flower polysaccharide (TFPS) were investigated (Wang et al., 2012a). Results found that TLPS and TFPS had higher antioxidant activity than TSPS. Subsequently, they discussed the difference in structural features of these polysaccharide fractions and their bioactivities, and concluded that the nature of the monosaccharide and the higher Mw in TLPS and TFPS molecules probably contribute to their activities.

Studies on the anticomplementary activity of sulfated polysaccharide indicated that the DS in modified polysaccharides could also contribute to the variations in bioactivities (Wang et al., 2013a). The experimental results showed that the sulfated derivative of low DS exhibited similar anticomplementary activity with sulfated derivatives of high DS, and had even higher anticomplementary activity than the sulfated derivative of the highest DS. This might be because the high sulfation, where most of the hydroxyl groups of both the uronic acid and galactosamine residues were sulfated, might cause acute, rapid onset of significant side effects indicative of an allergic-type reaction.

7.7 SUMMARY

To date, tea has been reported to have a broad range of effects including lowering blood pressure, slowing heartbeat, antiblood coagulation, antitumor, anti-HIV, protecting blood phase, and enhancing human nonspecific immunity. TPSs are found to be one of the components responsible for those bioactivities. To clarify the chemical structures and chain conformations of TPSs it is important to understand their biological activities. However, polysaccharides belong to a structurally diverse class of macromolecules. Due to their inherent flexibility, the exact structures (high order structure) of most

TPSs in solution are still not readily characterized. Clearly, there is still a lot to learn about TPSs. Further structural analysis and evaluation of the bioactivities of the polysaccharide conjugates from tea will be important for their application in food and medicinal fields. The future challenge is to gain a better understanding of the high order structure of polysaccharides, and the structure–bioactivity relationship. This presents a good opportunity for scientists to elucidate the biological roles of polysaccharides, and design high potential antitumor drugs based on the high order structures.

REFERENCES

Baik, J.H., Shin, K.S., Park, Y., Yu, K.W., Suh, H.J., Choi, H.S., 2015. Biotransformation of catechin and extraction of active polysaccharide from green tea leaves via simultaneous treatment with tannase and pectinase. J. Sci. Food Agri. 95 (11), 2337–2344.

Béliveau, R., Gingras, D., 2004. Green tea: prevention and treatment of cancer by nutraceuticals. Lancet 364, 1021–1022.

Buqing, L., Huiling, Z., Qingling, S., Buchang, Z., Shengfang, G., 1996. Studies on the extraction of polysaccharide from middle and lower middle grade green tea and the effectiveness on blood-glucose depressing. J. Tea Sci. 1.

Cabrera, C., Artacho, R., Giménez, R., 2006. Beneficial effects of Green Tea—a review. J. Am. Coll. Nutr. 25, 79–99.

Cai, W., Xie, L., Chen, Y., Zhang, H., 2013. Purification, characterization and anticoagulant activity of the polysaccharides from green tea. Carbohyd. Polym. 92, 1086–1090.

Cao, H., 2013. Polysaccharides from Chinese tea: recent advance on bioactivity and function. Int. J. Biol. Macromol. 62, 76–79.

Chacko, S.M., Thambi, P.T., Kuttan, R., Nishigaki, I., 2010. Beneficial effects of green tea: a literature review. Chin. Med. 5, 1–9.

Chen, H., Zhang, M., Xie, B., 2004a. Quantification of uronic acids in tea polysaccharide conjugates and their antioxidant properties. J Agr. Food Chem. 52, 3333–3336.

Chen, H.X., Zhang, M., Xie, B.J., 2004b. Spectroscopy investigation on conformational transition of tea glycoconjugate from green tea. Chin. J. Chem. 22, 1387–1390.

Chen, H., Zhang, M., Xie, B., 2005. Components and antioxidant activity of polysaccharide conjugate from green tea. Food Chem. 90, 17–21.

Chen, H., Zhang, M., Qu, Z., Xie, B., 2007. Compositional analysis and preliminary toxicological evaluation of a tea polysaccharide conjugate. J. Agr. Food Chem. 55, 2256–2260.

Chen, H., Wang, Z., Lu, X., Xie, B., 2008a. Isolation and chemical characterisation of a polysaccharide from green tea (*Camellia sinensis* L.). J. Sci. Food Agri. 88, 2523–2528.

Chen, H., Zhang, M., Qu, Z., Xie, B., 2008b. Antioxidant activities of different fractions of polysaccharide conjugates from green tea (*Camellia sinensis*). Food Chem. 106, 559–563.

Chen, H., Wang, Z., Qu, Z., Fu, L., Dong, P., Zhang, X., 2009. Physicochemical characterization and antioxidant activity of a polysaccharide isolated from oolong tea. Europ. Food Res. Technol. 229, 629–635.

Chen, J.-G., Wang, Y., Mei, S., Lai, W.-Q., Fu, Y., Hu, X., 2003. The hypoglycemic effect and improvememt of diabetic symptoms by tea polysaccharide in mice. Acta Nutr. Sinica 3, 009.

Chen, X., Lin, Z., Ye, Y., Zhang, R., Yin, J., Jiang, Y., et al., 2010. Suppression of diabetes in non-obese diabetic (NOD) mice by oral administration of water-soluble and alkali-soluble polysaccharide conjugates prepared from green tea. Carbohyd. Polym. 82, 28–33.

Chen, X., Wang, Y., Wu, Y., Han, B., Zhu, Y., Tang, X., et al., 2011. Green tea polysaccharide-conjugates protect human umbilical vein endothelial cells against impairments triggered by high glucose. Int. J. Biol. Macromol. 49, 50–54.

Chen, X., Fang, Y., Nishinari, K., We, H., Sun, C., Li, J., et al., 2014. Physicochemical characteristics of polysaccharide conjugates prepared from fresh tea leaves and their improving impaired glucose tolerance. Carbohyd. Polym. 112, 77–84.

Chi, A., Li, H., Kang, C., Guo, H., Wang, Y., Guo, F., et al., 2015. Anti-fatigue activity of a novel polysaccharide conjugates from Ziyang green tea. Int. J. Biol. Macromol. 80, 566–572.

Chi, H., Jiang, T., 1995. Screening of active anticarcinogenic ingredients in tea-tea polyphonls. J. Hygiene Res. 4, 232–237.

Dell'aica, I., Dona, M., Tonello, F., Piris, A., Mock, M., Montecucco, C., et al., 2004. Potent inhibitors of anthrax lethal factor from green tea. EMBO Rep. 5, 418–422.

Deng, J., Xu, P., 1998. The effects of tea polysaccharide on the SOD activity in the red cell of mice. Nat. Sci. Ed. J. Yuzhou Univ. 15, 30–32.

Deng, W.W., Cao, X., Wang, M., Yang, Y., Su, W.Y., Wei, Y.W., et al., 2012. Efficient gene delivery to mesenchymal stem cells by an ethylenediamine-modified polysaccharide from mulberry leaves. Small 8, 441–451.

Deng, Y.T., Lin-Shiau, S.Y., Shyur, L.F., Lin, J.K., 2015. Pu-erh tea polysaccharides decrease blood sugar by inhibition of α-glucosidase activity in vitro and in mice. Food Func. 6 (5), 1539–1546.

Drynan, J.W., Clifford, M.N., Obuchowicz, J., Kuhnert, N., 2010. The chemistry of low molecular weight black tea polyphenols. Nat. Prod. Rep. 27, 417–462.

Du, W.-H., Peng, S.-M., Liu, Z.-H., Shi, L., Tan, L.-F., Zou, X.-Q., 2012. Hypoglycemic effect of the wáter extract of Pu-erh tea. J Agr. Food Chem. 60, 10126–10132.

Dufresne, C.J., Farnworth, E.R., 2001. A review of latest research findings on the health promotion properties of tea. J. Nutr. Biochem. 12, 404–421.

Fan, Y.L., Wang, W.H., Hu, Y.S., Rong, Q., Teng, A.G., Liu, A.J., 2013. Comparison of the roles of tea polysaccharides and polyphenols in growth inhibition of hepatoma H22 cells in mice. Adv. Mater. Res. 690–693, 1193–1197.

Flint, H.J., Bayer, E.A., Rincon, M.T., Lamed, R., White, B.A., 2008. Polysaccharide utilization by gut bacteria: potential for new insights from genomic analysis. Nat. Rev. Micro. 6, 121–131.

Guo, L., Du, X., Lan, J., Liang, Q., 2010. Study on molecular structural characteristics of tea polysaccharide. Int. J. Biol. Macromol. 47, 244–249.

Han, Q., Ling, Z.-J., He, P.-M., Xiong, C.-Y., 2010. Immunomodulatory and antitumor activity of polysaccharide isolated from tea plant flower. Prog. Biochem. Biophys. 37, 646–653.

Han, Q., Yu, Q.-Y., Shi, J., Xiong, C.-Y., Ling, Z.-J., He, P.-M., 2011a. Structural characterization and antioxidant activities of 2 water-soluble polysaccharide fractions purified from tea (*Camellia sinensis*) flower. J. Food Sci. 76, C462–471.

Han, Q., Yu, Q.-Y., Shi, J., Xiong, C.-Y., Ling, Z.-J., He, P.-M., 2011b. Molecular characterization and hypoglycemic activity of a novel water-soluble polysaccharide from tea (*Camellia sinensis*) flower. Carbohyd. Polym. 86, 797–805.

He, N., Shi, X., Zhao, Y., Tian, L., Wang, D., Yang, X., 2013. Inhibitory effects and molecular mechanisms of selenium-containing tea polysaccharides on human breast cancer MCF-7 cells. J. Agr. Food Chem. 61, 579–588.

Huang, D., Nie, S., Han, C., Chen, Y., Xie, M., Gong, D., 2012. Regulation of maturation and function of dendritic cells by tea glycoprotein. Europ. Food Res. Technol. 235, 1023–1032.

Islam, M.S., 2011. Effects of the aqueous extract of white tea (*Camellia sinensis*) in a streptozotocin-induced diabetes model of rats. Phytomedicine 19, 25–31.

Iso, H., Date, C., Wakai, K., Fukui, M., Tamakoshi, A., 2006. The relationship between green tea and total caffeine intake and risk for self-reported type 2 diabetes among Japanese adults. Ann. Intern. Med. 144, 554–562.

Jia, X., Ding, C., Yuan, S., Zhang, Z., Chen, Y., Du, L., et al., 2014. Extraction, purification and characterization of polysaccharides from Hawk tea. Carbohyd. Polym. 99, 319–324.

Jiang, H., Zhen, G., 2004. Studies of tea polysaccharides on lowering blood sugar of mice. Food Sci. 6, 42.

Jiang, X., Jing, Y., Li, J., 1998. Effects of TPS on immune index of rats with adjuvant arthritis. Acta Univ. Med. Anhui 2, 15–17.

Jie, Z., Jianping, D., Zenong, W., Xiaofeng, X., 1997. Effect of tea polysaccharides on blood glucose, blood lipid and immunological function of mice. J. Tea Sci. 1, 77–81.

Jin, F., Jia, L.-Y., Tu, Y.-Y., 2015. Structural analysis of an acidic polysaccharide isolated from white tea. Food Sci. Biotech. 24, 1623–1628.

Koo, M.W., Cho, C.H., 2004. Pharmacological effects of green tea on the gastrointestinal system. Europ. J. Pharm. 500, 177–185.

Lee, J.-H., Shim, J.S., Lee, J.S., Kim, J.K., Yang, I.S., Chung, M.-S., et al., 2006. Inhibition of pathogenic bacterial adhesion by acidic polysaccharide from green tea (*Camellia sinensis*). J. Agr. Food Chem. 54, 8717–8723.

Li, S., Cao, X., 2014. Extraction of tea polysaccharides (TPS) using anionic reverse micellar system. Separ. Purif. Technol. 122, 306–314.

Li, S., Chen, H., Wang, J., Wang, X., Hu, B., Lv, F., 2015. Involvement of the PI3K/Akt signal pathway in the hypoglycemic effects of tea polysaccharides on diabetic mice. Int. J. Biol. Macromol. 81, 967–974.

Lijun, Z., Shuru, W., 1992. Preparation of lipoprotein lipase and the effect of tea polysaccharide on lipoprotein lipase. J.-Chin. Pharm. Univ. 23, 287.

Lin, H.-Y., Lai, L.-S., 2009. Isolation and viscometric characterization of hydrocolloids from mulberry (*Morus alba* L.) leaves. Food Hydrocolloid. 23, 840–848.

Lu, X., Zhao, Y., Sun, Y., Yang, S., Yang, X., 2013. Characterisation of polysaccharides from green tea of Huangshan Maofeng with antioxidant and hepatoprotective effects. Food Chem. 141, 3415–3423.

Lv, Y., Yang, X., Zhao, Y., Ruan, Y., Yang, Y., Wang, Z., 2009. Separation and quantification of component monosaccharides of the tea polysaccharides from *Gynostemma pentaphyllum* by HPLC with indirect UV detection. Food Chem. 112, 742–746.

McKay, D.L., Blumberg, J.B., 2002. The role of tea in human health: an update. J. Am. Coll. Nutr. 21 (1), 1–13.

Min, F.F., Hu, J.L., Nie, S.P., Xie, J.H., Xie, M.Y., 2014. In vitro fermentation of the polysaccharides from *Cyclocarya paliurus* leaves by human fecal inoculums. Carbohyd. Polym. 112, 563–568.

Monobe, M., Ema, K., Kato, F., Maeda-Yamamoto, M., 2008. Immunostimulating activity of a crude polysaccharide derived from green tea (*Camellia sinensis*) extract. J. Agr. Food Chem. 56, 1423–1427.

Mori, M., Morita, N., Ikegaya, K., 1988. Polysaccharides from tea for manufacture of hypoglycemics, antidiabetics, and health foods. Patent-Japan Kokai Tokkyo Koho-63,308,001.

Naowaboot, J., Pannangpetch, P., Kukongviriyapan, V., Prawan, A., Kukongviriyapan, U., Itharat, A., 2012. Mulberry leaf extract stimulates glucose uptake and GLUT4 translocation in rat adipocytes. Am. J. Chin. Med. 40, 163–175.

Ni, D.-J., Chen, Y.-Q., Song, C.-H., Xie, B.-J., Zhou, S.-Q., 2003a. Effect of Oolong Tea Polysaccharide on Hepatic-nephritic Antioxidation and Histomorphology in the Diabetic Rats. J. Tea Sci. 1, 002.

Ni, D., Chen, Y., Xie, B., Song, C., Yuan, F., 2003b. Studies on composition, antioxidation and hypoglycemic effects of polysaccharides from green tea, oolong tea and black tea. [Ying yang xue bao]. Acta nutrimenta Sinica 26, 57–60.

Ni, D.-J., Chen, Y.-Q., Xie, B.-J., Zhang, Y., Zhou, J.-R., 2004. Spectrum, morphological and thermal characteristics of OTPS 2-1 in polysaccharides from oolong tea. Chem. Res. Chinese U. 12, 018.

Nie, S., Xie, M., Zhou, P., Cao, S., 2006. In vitro antioxidative and anticancer activities of tea glycoprotein in green tea. Europ. Food Res. Technol. 224, 437–442.

Nie, S., Xie, M., Fu, Z., Wan, Y., Yan, A., 2008. Study on the purification and chemical compositions of tea glycoprotein. Carbohyd. Polym. 71, 626–633.

Nie, S.-P., Xie, M.-Y., 2011. A review on the isolation and structure of tea polysaccharides and their bioactivities. Food Hydrocolloid. 25, 144–149.

Nie, S.-P., Xie, M.-Y., Luo, Z., 2005a. Extraction of tea polysaccharides by microwave technique. Food Sci. 11, 020.

Nie, S.-P., Xie, M.-Y., Luo, Z., 2005b. Studies on the antioxidative activity of tea polysaccharide. Nat. Product Res. Dev. 17, 549.

Olennikov, D.N., Kashchenko, N.I., Chirikova, N.K., Koryakina, L.P., Vladimirov, L.N., 2015. Bitter gentian teas: nutritional and phytochemical profiles, polysaccharide characterisation and bioactivity. Molecules 20, 20014–20030.

Qi, L., Han, C., 1998. Induction of NAD (P) H: quinone reductase by anticarcinogenic ingredients of tea. J. Hygiene Res. 27, 323–326.

Ren, C., Zhang, Y., Cui, W., Lu, G., Wang, Y., Gao, H., et al., 2015. A polysaccharide extract of mulberry leaf ameliorates hepatic glucose metabolism and insulin signaling in rats with type 2 diabetes induced by high fat-diet and streptozotocin. Int. J. Biol. Macromol. 72, 951–959.

Satoh, T., Igarashi, M., Yamada, S., Takahashi, N., Watanabe, K., 2015. Inhibitory effect of black tea and its combination with acarbose on small intestinal alpha-glucosidase activity. J. Ethnopharmacol. 161, 147–155.

Scoparo, C.T., De Souza, L.M., Rattmann, Y.D., Dartora, N., Paiva, S.M.M., Sassaki, G.L., et al., 2013. Polysaccharides from green and black teas and their protective effect against murine sepsis. Food Res. Int. 53, 780–785.

Sun, Y., Yang, X., Lu, X., Wang, D., Zhao, Y., 2013. Protective effects of Keemun black tea polysaccharides on acute carbon tetrachloride-caused oxidative hepatotoxicity in mice. Food Chem. Toxicol. 58, 184–192.

Takeo, C., Kinugasa, H., Oosu, H., Kawasaki, T., Takakuwa, N., Shimizu, M., et al., 1992. Extraction of hypoglycemics from tea. Japan, Kokai Tokkyo Koho JP 4.

Tang, M., Wang, D., Hou, Y., Buchili, P., Sun, L., 2012. Preparation, characterization, bioavailability in vitro and in vivo of tea polysaccharides–iron complex. Europ. Food Res. Technol. 236, 341–350.

Thirugnanasambandham, K., Sivakumar, V., Maran, J.P., 2015. Microwave-assisted extraction of polysaccharides from mulberry leaves. Int. J. Biol. Macromol. 72, 1–5.

Wang, D., Wang, S., 1991. The pharmaceutical effects of tea polysaccharides on cardiovascular diseases. Chin. Tradit. Herb Drugs 2, 4–5.

Wang, D., Li, J., Wang, C., Zhao, G., Jin, Y., Chen, D., et al., 2000. Study on the component and immune activity of polysaccharides from tea. J. Tea Sci. 20, 45–50.

Wang, D., Wang, C., Li, J., Zhao, G., 2001a. Components and activity of polysaccharides from coarse tea. J. Agr. Food Chem. 49, 507–510.

Wang, D., Wang, C., Zhao, G., Wei, Z., Tao, Y., Liang, X., 2001b. Composition, characteristic and activity of rare earth element-bound polysaccharide from tea. Biosci. Biotech. Biochem. 65, 1987–1992.

Wang, D., Zhao, Y., Sun, Y., Yang, X., 2014a. Protective effects of Ziyang tea polysaccharides on CCl4-induced oxidative liver damage in mice. Food Chem. 143, 371–378.

Wang, H., Wei, G., Liu, F., Banerjee, G., Joshi, M., Bligh, S.W., et al., 2014b. Characterization of two homogalacturonan pectins with immunomodulatory activity from green tea. Int. J. Mol. Sci. 15, 9963–9978.

Wang, F., Li, J., Jiang, Y., 2010a. Polysaccharides from mulberry leaf in relation to their antioxidant activity and antibacterial ability. J. Food Process Eng. 33, 39–50.

Wang, Y., Peng, Y., Wei, X., Yang, Z., Xiao, J., Jin, Z., 2010b. Sulfation of tea polysaccharides: synthesis, characterization and hypoglycemic activity. Int. J. Biol. Macromol. 46, 270–274.

Wang, Y., Yu, L., Zhang, J., Xiao, J., Wei, X., 2010c. Study on the purification and characterization of a polysaccharide conjugate from tea flowers. Int. J. Biol. Macromol. 47, 266–270.

Wang, H., Shi, S., Gu, X., Zhu, C., Wei, G., Bao, B., et al., 2013a. Homogalacturonans from preinfused green tea: structural characterization and anticomplementary activity of their sulfated derivatives. J. Agr. Food Chem. 61, 10971–10980.

Wang, Q., Jiang, C., Fang, S., Wang, J., Ji, Y., Shang, X., et al., 2013b. Antihyperglycemic, antihyperlipidemic and antioxidant effects of ethanol and aqueous extracts of *Cyclocarya paliurus* leaves in type 2 diabetic rats. J. Ethnopharmacol. 150, 1119–1127.

Wang, Y., Chen, J., Zhang, D., Zhang, Y., Wen, Y., Li, L., et al., 2013c. Tumoricidal effects of a selenium (Se)-polysaccharide from Ziyang green tea on human osteosarcoma U-2 OS cells. Carbohyd. Polym. 98, 1186–1190.

Wang, Y., Zhao, Y., Andrae-Marobela, K., Okatch, H., Xiao, J., 2013d. Tea polysaccharides as food antioxidants: an old woman's tale? Food Chem. 138, 1923–1927.

Wang, H., Shi, S., Bao, B., Li, X., Wang, S., 2015a. Structure characterization of an arabinogalactan from green tea and its anti-diabetic effect. Carbohyd. Polym. 124, 98–108.

Wang, Y., Li, Y., Liu, Y., Chen, X., Wei, X., 2015b. Extraction, characterization and antioxidant activities of Se-enriched tea polysaccharides. Int. J. Biol. Macromol. 77, 76–84.

Wang, Z.J., Xie, J.H., Kan, L.J., Wang, J.Q., Shen, M.Y., Li, W.J., et al., 2015c. Sulfated polysaccharides from *Cyclocarya paliurus* reduce H2O2-induced oxidative stress in RAW264.7 cells. Int. J. Biol. Macromol. 80, 410–417.

Wang, J., Wang, K., 2012. Fatigue-alleviating effect of polysaccharides from *Cyclocarya paliurus* (Batal) Iljinskaja in mice. Afr. J. Micro. Res. 6, 5243–5248.

Wang, L., Xia, W., 2006. Isolation and analysis of a novel acidic polysaccharide with glucokinase-stimulating activity from coarse green tea. J. Food Biochem. 30, 187–202.

Wang, S., Wang, D., 1992. The anticoagulant and antithrombotic effects of tea polysaccharide. Chin. Tradit. Herb Drugs 23, 254–256.

Wang, Y., Jin, Z., 2005. Isolation of polysaccharides from tea and their hypoglycemic activity. Chin. Tradit. Herb Drugs 36, 1453.

Wang, Y., Mao, F., Wei, X., 2012a. Characterization and antioxidant activities of polysaccharides from leaves, flowers and seeds of green tea. Carbohyd. Polym. 88, 146–153.

Wang, Y., Shao, S., Xu, P., Chen, H., Lin-Shiau, S.-Y., Deng, Y.-T., et al., 2012b. Fermentation process enhanced production and bioactivities of oolong tea polysaccharides. Food Res. Int. 46, 158–166.

Wang, Y., Yang, Z., Wei, X., 2012c. Antioxidant activities potential of tea polysaccharide fractions obtained by ultra filtration. Int. J. Biol. Macromol. 50, 558–564.

Wang, Y., Wei, X., Jin, Z., 2009a. Structure analysis of a neutral polysaccharide isolated from green tea. Food Res. Int. 42, 739–745.

Wang, Y., Wei, X., Jin, Z., 2009b. Structure analysis of an acidic polysaccharide isolated from green tea. Nat. Prod. Res. 23, 678–687.

Wang, Z.J., Xie, J.H., Shen, M.Y., Tang, W., Wang, H., Nie, S.P., et al., 2016. Carboxymethylation of polysaccharide from *Cyclocarya paliurus* and their characterization and antioxidant properties evaluation. Carbohyd. Polym. 136, 988–994.

Wei, X., Liu, Y., Xiao, J., Wang, Y., 2009. Protective effects of tea polysaccharides and polyphenols on skin. J. Agr. Food Chem. 57, 7757–7762.

Wei, X., Chen, M., Xiao, J., Liu, Y., Yu, L., Zhang, H., et al., 2010a. Composition and bioactivity of tea flower polysaccharides obtained by different methods. Carbohyd. Polym. 79, 418–422.

Wei, X., Yang, Z., Guo, Y., Xiao, J., Wang, Y., 2010b. Composition and biological activity of tea polysaccharides obtained by water extraction and enzymatic extraction. Lat. Am. J. Pharm. 29, 117–121.

Wei, X., Mao, F., Cai, X., Wang, Y., 2011a. Composition and bioactivity of polysaccharides from tea seeds obtained by water extraction. Int. J. Biol. Macromol. 49, 587–590.

Wei, X., Xi, X., Wu, M., Wang, Y., 2011b. A novel method for quantitative determination of tea polysaccharide by resonance light scattering. Spectrochim. Acta. A. Mol. Biomol. Spectrosc. 79, 928–933.

Xiao, J., Huo, J., Jiang, H., Yang, F., 2011. Chemical compositions and bioactivities of crude polysaccharides from tea leaves beyond their useful date. Int. J. Biol. Macromol. 49, 1143–1151.

Xiao, J.B., Jiang, H., 2015. A review on the structure-function relationship aspect of polysaccharides from tea materials. Crit. Rev. Food Sci. Nutr. 55, 930–938.

Xie, B., Shi, H., Chen, Q., Ho, C.-T., 1993. Antioxidant properties of fractions and polyphenol constituents from green, oolong and black teas. Proceedings of the National Sci. Council, Republic of China. Part B, Life Sci. 17, 77–84.

Xie, J.-H., Xie, M.-Y., Nie, S.-P., Shen, M.-Y., Wang, Y.-X., Li, C., 2010a. Isolation, chemical composition and antioxidant activities of a water-soluble polysaccharide from *Cyclocarya paliurus* (Batal.) Iljinskaja. Food Chem. 119, 1626–1632.

Xie, J.H., Xie, M.Y., Shen, M.Y., Nie, S.P., Li, C., Wang, Y.X., 2010b. Optimisation of microwave-assisted extraction of polysaccharides from *Cyclocarya paliurus* (Batal.) Iljinskaja using response surface methodology. J. Sci. Food Agr. 90, 1353–1360.

Xie, J.-H., Shen, M.-Y., Nie, S.-P., Li, C., Xie, M.-Y., 2011. Decolorization of polysaccharides solution from *Cyclocarya paliurus* (Batal.) Iljinskaja using ultrasound/H2O2 process. Carbohyd. Polym. 84, 255–261.

Xie, J.H., Shen, M.Y., Xie, M.Y., Nie, S.P., Chen, Y., Li, C., et al., 2012. Ultrasonic-assisted extraction, antimicrobial and antioxidant activities of *Cyclocarya paliurus* (Batal.) Iljinskaja polysaccharides. Carbohyd. Polym. 89, 177–184.

Xie, J.H., Liu, X., Shen, M.Y., Nie, S.P., Zhang, H., Li, C., et al., 2013a. Purification, physicochemical characterisation and anticancer activity of a polysaccharide from *Cyclocarya paliurus* leaves. Food Chem. 136, 1453–1460.

Xie, J.H., Shen, M.Y., Nie, S.P., Liu, X., Zhang, H., Xie, M.Y., 2013b. Analysis of monosaccharide composition of *Cyclocarya paliurus* polysaccharide with anion exchange chromatography. Carbohyd. Polym. 98, 976–981.

Xie, J.H., Zhang, F., Wang, Z.J., Shen, M.Y., Nie, S.P., Xie, M.Y., 2015. Preparation, characterization and antioxidant activities of acetylated polysaccharides from *Cyclocarya paliurus* leaves. Carbohyd. Polym. 133, 596–604.

Xie, J.-H., Wang, Z.-J., Shen, M.-Y., Nie, S.-P., Gong, B., Li, H.-S., et al., 2016. Sulfated modification, characterization and antioxidant activities of polysaccharide from *Cyclocarya paliurus*. Food Hydrocolloid. 53, 7–15.

Xie, M.-Y., Nie, S.-P., 2006. A review of the research progress on tea polysaccharide. J. Food Sci. Biotech. 2, 024.

Xu, P., Wu, J., Zhang, Y., Chen, H., Wang, Y., 2014. Physicochemical characterization of puerh tea polysaccharides and their antioxidant and α-glycosidase inhibition. J. Funct. Foods 6, 545–554.

Xu, R., Ye, H., Sun, Y., Tu, Y., Zeng, X., 2012. Preparation, preliminary characterization, antioxidant, hepatoprotective and antitumor activities of polysaccharides from the flower of tea plant (*Camellia sinensis*). Food Chem. Toxicol. 50, 2473–2480.

Yang, M., Zhao, W.H., Wang, S.F., 1997. The effects of TPS of coarse tea to the function of immunity. Shi Zhen J. Tradit. Chin. Med. Res. 8 (4), 310–311.

Yang, G., Li, F., Luo, H., 2004. Effects of tea polysaccharide on immune response of mice after antigen stimulation and their possible mechanisms. Chin. Arch. Tradit. Chin. Med. 22, 2294–2295.

Yang, J., Chen, B., Gu, Y., 2012. Pharmacological evaluation of tea polysaccharides with antioxidant activity in gastric cancer mice. Carbohyd. Polym. 90, 943–947.

Ying, Z., Han, X., Li, J., 2011. Ultrasound-assisted extraction of polysaccharides from mulberry leaves. Food Chem. 127, 1273–1279.

Yuan, Q., Xie, Y., Wang, W., Yan, Y., Ye, H., Jabbar, S., et al., 2015. Extraction optimization, characterization and antioxidant activity in vitro of polysaccharides from mulberry (*Morus alba* L.) leaves. Carbohyd. Polym. 128, 52–62.

Zhang, D.Y., Wan, Y., Xu, J.Y., Wu, G.H., Li, L., Yao, X.H., 2016a. Ultrasound extraction of polysaccharides from mulberry leaves and their effect on enhancing antioxidant activity. Carbohyd. Polym. 137, 473–479.

Zhang, Z., Wang, X., Li, J., Wang, G., Mao, G., 2016b. Extraction and free radical scavenging activity of polysaccharide from 'Anji Baicha' (*Camellia sinensis* (L.) O. Kuntze. Int. J. Biol. Macromol. 84, 161–165.

Zhang, M., Cui, S.W., Cheung, P.C.K., Wang, Q., 2007. Antitumor polysaccharides from mushrooms: a review on their isolation process, structural characteristics and antitumor activity. Trends Food Sci. Tech. 18, 4–19.

Zhang, X., Chen, H., Zhang, N., Chen, S., Tian, J., Zhang, Y., et al., 2013. Extrusion treatment for improved physicochemical and antioxidant properties of high-molecular weight polysaccharides isolated from coarse tea. Food Res. Int. 53, 726–731.

Zhang, Y., Ren, C., Lu, G., Cui, W., Mu, Z., Gao, H., et al., 2014a. Purification, characterization and anti-diabetic activity of a polysaccharide from mulberry leaf. Regul. Toxicol. Pharmacol. 70, 687–695.

Zhang, Y., Ren, C., Lu, G., Mu, Z., Cui, W., Gao, H., et al., 2014b. Anti-diabetic effect of mulberry leaf polysaccharide by inhibiting pancreatic islet cell apoptosis and ameliorating insulin secretory capacity in diabetic rats. Int. Immunopharmacol. 22, 248–257.

Zhao, X., Li, L., Luo, Q., Ye, M., Luo, G., Kuang, Z., 2015. Effects of mulberry (*Morus alba* L.) leaf polysaccharides on growth performance, diarrhea, blood parameters, and gut microbiota of early-weanling pigs. Livestock Sci. 177, 88–94.

Zhao, Z.-Y., Huangfu, L.-T., Dong, L.-L., Liu, S.-L., 2014. Functional groups and antioxidant activities of polysaccharides from five categories of tea. Ind. Crops Prod. 58, 31–35.

Zhou, P., 2004. Primary structure and configuration of tea polysaccharide. Sci. China Ser. C. 47, 416.

Zhou, J., Ding, J.P., Wang, Z.N., Xie, X.F., 1997. Effect of tea polysaccharides on blood-glucose, blood lipid and immunological function of mice. J. Tea Sci. 17 (1), 75–79.

Zhou, P., Shen, J., Xie, M., Zhuang, Z., Wang, X., 2002. Studies on monosaccharide composition of TGP extracted from tea and its ionization mechanism by GC-MS. J. Xiamen Univ. Nat. Sci. 42, 213–217.

Zhou, X., Wang, D., Sun, P., Bucheli, P., Li, L., Hou, Y., et al., 2007. Effects of soluble tea polysaccharides on hyperglycemia in alloxan-diabetic mice. J. Agr. Food Chem. 55, 5523–5528.

Chapter 8

Psyllium Polysaccharide

Plantago plants that belong to the plantaginaceae family are cosmopolitan. About 250 species are found around the world, and psyllium is a common name referring to these plants. Both leaves and seeds of the plant have been used in folk medicine for a long time (Fig. 8.1). These plants could also be used as a resource of food in many areas. The traditional medicinal uses of this plant include helping soften stools, using as a laxative, treatment for skin infections, insect bites, urinary infections yeast infections, and in weight loss (Chan and Wypyszyk, 1988). Psyllium polysaccharide is commonly prepared from the seed husk or the leaves of the plants of the Plantago genus.

There is no doubt that the consumption of a high-fiber diet provides benefits to human health, both in the prevention and in treatment of chronic diseases. Dietary fiber is also applicable for the treatment of various gastrointestinal and nongastrointestinal conditions. The National Academy of Sciences Institute of Medicine recommends that adults consume 20−35 g of dietary fiber per day (Eswaran et al., 2013). Dietary fiber from psyllium has been extensively used both as pharmacological supplements, and food ingredients to control weight, regulate glucose levels for diabetic patients, and reduce serum lipid levels in hypolipidemics. There are still other utilizations of this polysaccharide, due to its special physicochemical and rheological properties. In addition, several chemical modifications have been applied to psyllium to explore its utilization. For example, psyllium polysaccharide can be utilized for the absorption of toxic material from industrial waste (Hussain et al., 2015; Kumar et al., 2013), as an environmentally-friendly natural binder, or as drug delivery system (Singh, 2007).

This chapter discusses the preparation of psyllium polysaccharide, its therapeutic value for the treatment of constipation, diabetes, and hypercholesterolemia, and the relationship between its molecular characteristic and bioactivities.

8.1 PREPARATION PROCESS

The outermost cell layer of the seeds in the genus Plantago contains little but the mucilage that is the major source of psyllium polysaccharide. For instance, husk mucilage of *Plantago ovata* Forsk has about 85% polysaccharide content (Kennedy et al., 1979). According to published articles,

Bioactive Polysaccharides. DOI: http://dx.doi.org/10.1016/B978-0-12-809418-1.00008-3

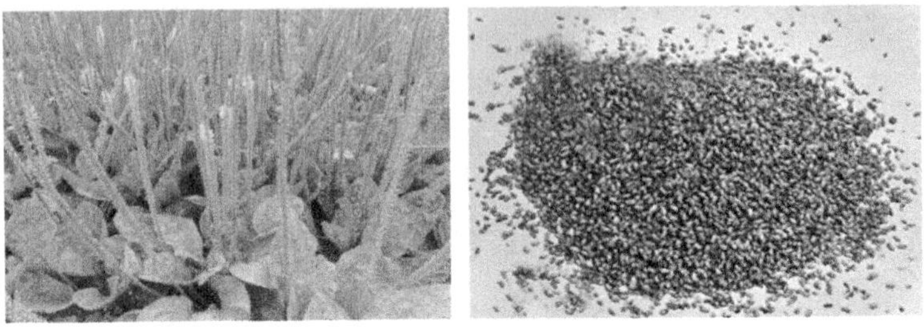

FIGURE 8.1 *Plantago asiatica* and the seeds of *Plantago asiatica*.

polysaccharides are obtained from many species of the Plantago family including *P. ovata*, *Plantago asiatica* L., *Plantago depressa* (Zhao et al., 2014), *Plantago palmata* (Biringanine et al., 2005), and *Plantago lanceolata* (Schmidgall et al., 2000), among which *P. asiatica* L. and *P. ovata* are the major source of the polysaccharide.

Aqueous extraction is the most common method used to prepare psyllium polysaccharide. Generally, dried seeds of Plantago plants are extracted with hot water, and then crude polysaccharide component is precipitated by adding ethanol, and separated by centrifuge. Guo et al. (2008) reported a typical sequential extraction method to extract polysaccharide from *P. ovata* (Guo et al., 2008). Briefly, 5 g of psyllium was dispersed in 1000 mL distilled water at 80°C for 2 h and stirred continuously. Finally, the dispersion became a homogenous gel. The dispersion was subjected to centrifugation at 21,000 g for 60 min to separate into two parts, the gel and the solution. The solution was concentrated to one-fourth of the original volume. Then, it was dialyzed against deionized water at room temperature for 48 h, and freeze-dried to obtain the water extracted fraction (WE). The gel was dissolved in 0.5 M NaOH solution at room temperature for 2 h. There were two parts, the solution and a small amount of residue. The residue was applied for further extraction with NaOH solution at a higher concentration. The extract was neutralized with 2 M HCl. A large amount of gel-like precipitate was observed, which could be separated by centrifugation. The soluble fraction, treated in a similar manner, was considered to be the alkaline extracted soluble fraction $AES_{0.5}$, while the gel fraction was homogenized for 1 h, washed three times with distilled water, and dialyzed at room temperature for 48 h, then freeze-dried to obtain the alkaline extracted gel fraction $AEG_{0.5}$. Finally, the residue was extracted with 1.2 M NaOH and then 2.0 M NaOH solutions at room temperature, respectively, to produce corresponding fractions of psyllium gum. Despite the differences between the species, factors in different protocols such as extraction temperature, extraction time, and ratio of raw material to water may also influence the preparation of psyllium polysaccharides. An optimized extraction condition for *P. asiatica* L. seeds was reported to be a ratio of seeds to water of 1:20, at 80°C for 3.5 h, with the extraction repeated twice (Ye et al., 2011). In addition, enzyme treatment and microwave-assisted methods were also developed to extract psyllium polysaccharide in some cases (Fu et al., 2005). Van Craeyveld et al. (2008) reported that a 48 h laboratory-scale ball mill treatment could reduce the average particle size of psyllium from 161 μm for the untreated sample to 6 μm. Concurrently, it increased the water-extractable arabinoxylan level from 13% to 90%, and increased the ratio of arabinose to xylose from 0.21 to 0.31.

Gel permeation chromatography is commonly used to separate and purify psyllium polysaccharide. In our group, a water-soluble polysaccharide from the seeds of *P. asiatica* L. was extracted by hot water. The crude polysaccharide was further dissolved in 15 mmol/L NaCl solution, and then applied for purification by a Sephacryl S-400 HR column. The column was eluted with 15 mmol/L NaCl solution at a flow rate of 1.5 mL/min. Finally, three purified fractions named PLP-1 (18.9%), PLP-2 (52.6%), and PLP-3 (28.5%) were obtained (Yin et al., 2012c). Ion-exchange chromatography was also used to purify the psyllium polysaccharide. In Samuelsen's group, crude polysaccharides from *Plantago major* L. were separated and purified by a DEAE-Sepharose fast flow column. The column was first eluted with water at 1 mL/min to obtain a neutral fraction, then gradient eluted by NaCl solution (0−1 M) to get acidic polysaccharide fractions (Samuelsen et al., 1995). Zhao et al. (2014) reported that the polysaccharide from seeds of *P. depressa* was purified using chromatography by DEAE-52 and Sephacryl S-400. The *P. depressa* seed polysaccharides were dissolved in distilled water and passed through two series connected resin columns (Amberlite FPA90-Cl and Amberlite IRC-84) eluting with water and NaCl solution to get two fractions (fraction A and fraction B), then the fractions were applied to a DEAE-52 cellulose column eluting with water and NaCl solution to get subfractions, respectively (Fraction A-1, Fraction A-2; Fraction B-1, Fraction B-2), and finally the subfractions were further purified by Sephacryl S-400 eluting with water (PDSP-1, PDSP-2, PDSP-3, PDSP-4).

Psyllium polysaccharide can also be isolated from the leaves of Plantago plants (Samuelsen, 2000; Deters et al., 2005). Polysaccharide fractions were isolated from *P. major* L. leaves by several processes in a sequence as 50°C water extraction, 100°C water extraction, and dimethylsulfoxide extraction; however, only the 50°C extraction is found to possess complement system activation and nitric oxide (NO) secretion stimulating activities (Samuelsen et al., 1995).

8.2 STRUCTURAL FEATURES

Lots of research confirms that the main monosaccharides in psyllium are arabinose and xylose, and it also contains other components such as rhamnose, mannose, and galacturonic acid. For instance, polysaccharide from the seed husk of *P. ovata* Frosk was reported to be an arabinosyl (galactosyluronic acid) rhamnosylxylan (Sandhu et al., 1981), and polysaccharide from *P. ovata* seeds contains 74.8% xylose and 23.2% arabinose (Saghir et al., 2008), while Van Craeyveld et al. (2009) reported that psyllium polysaccharide contained 62.5% arabinoxylan. The ratio of arabinose to xylose was 0.41. Although most psyllium polysaccharides are found to be heteroxylan, there are still some exceptions. A water-soluble polysaccharide from *P. lanceolata* was found to be composed of 20% arabinose, 28% galactose, 6% glucose, 2% rhamnose, 31% galacturonic acid, 7% glucuronic acid, and a minor amount of fucose and xylose (Bräutigam and Franz, 1985). Biringanine et al. (2012) reported that polysaccharide from *P. palmata* contained an unbranched galacturonan domain, and a domain bearing side chains which consisted of highly branched arabinan, galactan, and/or arabinogalactan.

The structural characteristics of psyllium polysaccharide have also been well studied. Methods including methylation and periodate oxidation, combined with gas chromatography, liquid chromatography, infrared spectrum, gas chromatography – mass spectrography are applied to characterize the primary structure of psyllium polysaccharide. In our group, the main fraction of polysaccharide from the seed of *P. asiatica* L. (PLP-2) was a highly branched heteroxylan which consisted of a β-(1→4)-linked xylopyranose (Xyl*p*) backbone with side chains attached to the O-2 or O-3 position. The side chains consisted of a β-terminal(T)-linked Xylp, α-T-linked arabinosefuranose (Ara*f*), α-T-linked-GlcA*p*, β-Xyl*p*-(1→3)-α-Ara*f*, and α-Ara*f*-(1→3)-β-Xyl*p* (Yin et al., 2012c). The structure of PLP-3 was also well studied. PLP-3 consists of a β-1,4-linked Xyl*p* backbone with short side chains attached to its O-2 (1,2,4-linked Xyl*p*, 17.87%) or O-3 (1,3,4-linked Xyl*p*, 24.24%) positions. The main terminal residues of PLP-3 were T-linked Ara*f* (8.13%), T-linked Xyl*p* (15.58%), and T-linked GlcA*p* (13.22%) (Fig. 8.2) (Yin et al.,

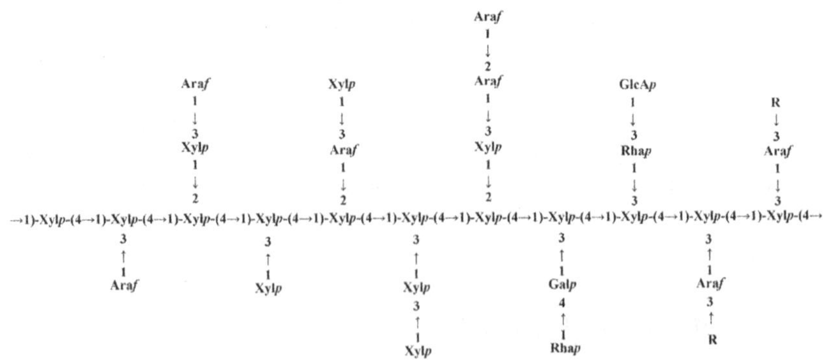

FIGURE 8.2 The possible structure of polysaccharide (PLP-2) from the seeds of *Plantago asiatica* L. (R, may be T-linked GlcAp, 1,6-linked Glcp, 1,4-linked Glcp, 1,3,6-linked Glcp or 1,3,4-linked Galp).

2012a). Similarly, Tomoda reported that the polysaccharide from *P. asiatica* L. is composed of L-arabinose, D-xylose, D-glucuronic acid, and D-galacturonic acid in a molar ratio of 4.0:10.8:3.3:0.7. It possesses a main chain composed of β-(1→4)-linked Xyl*p* with other Xyl*p* side chains at position 3, and branches composed of *O*-α-GlcA-(1→3)-α-Ara*f*, and *O*-α-GlcA*p*-(1→3)-α-Ara*f* at position 2 of the residual Xyl*p* unit (Tomoda et al., 1981). An acidic polysaccharide isolated from the seeds of *P. asiatica* L. by hot alkali solution was reported to consist of xylose, arabinose, glucuronic acid, and galactose. They were in a molar ratio of 18.8:7.2:6.1:1. The polysaccharide was a highly branched heteroxylan of a β-(1→4)-linked Xyl*p* backbone. Three α-GlcA*p*-(1→3)-Ara*f* were attached to the O-3 position of backbone, and one α-Ara*f*-(1→5)-Ara*f* and one α-T-linked-GlcA*p* were attached to the O-2 position every eight monosaccharide residues (Gong et al., 2015).

Samuelsen et al. (1999) reported polysaccharide extracted and isolated from the seeds of *P. major* L. consisted of a β-(1→4)-Xylp backbone. Side chains of β-(1→4)-linked residues were attached to the O-2 or O-3 positions, while polysaccharide from the leaves of *P. major* L. (PM II), which is composed of 71.7% galactose and 7.3% glucose, is partly *O*-acetylated with a region of (1→4) linked polygalacturonic acid (Samuelsen et al., 1996). The structure of the galactan side chain is complex and has at least two different hairy regions, in one of which the galactose side chains are linked to position 4 of rhamnose in the main chain, whereas in another hairy region arabinose is attached to position 3 or 6 of the falactose.

Kennedy et al. (1979) reported that psyllium polysaccharide of *P. ovata* is a highly branched and acidic arabinoxylan, and it consists of a xylan backbone with both (1→4) and (1→3) linkages. Saghir found the main chain of the psyllium polysaccharide consisted of β-(1→4)-linked Xyl*p*, some Xyl*p* residues in the backbone carried a single Xyl*p* moiety at its O-2, and other Xyl*p* residues bore trisaccharide branches at position 3 with a sequence of L-Ara*f*-α-(1→3)-D-Xyl*p*-β-(1→3)-L-Ara*f* (Saghir et al., 2008; Fischer et al., 2004). Guo et al. (2008) found that WE and $AEG_{0.5}$ from psyllium husk mainly contained (1→4) and (1→3) linked β-D-Xyl*p* residues in the backbone, and the side chains are primarily composed of T-linked Ara*f* and T-linked Xyl*p* connected to O-3 and/or O-2 of the xylose in the main chain. However, the backbone of $AES_{0.5}$ were found to be composed of (1→4) linked β-D-Xyl*p* residue, and the side chain only contained arabinosyl residues which were linked only through the O-3 position of the (1→4) linkedβ-D-Xyl*p* residues (Guo et al., 2008).

Structures of polysaccharides purified from the seeds of *P. depressa* in Zhao's team were also determined. The DPSP-1 may consist of a backbone of 1,3,6-linked Ara*f* and 1,3,4-linked Fuc*p*. Uronic acid in the fraction mainly appeared as linked-linked Gal*p*A. The PSDP-2 presented a linear chain, since almost 41.4% of 1,5-linked Ara*f* appeared in the main chain, while only 5.1% of 1,3,4-linked Gal*p*A and 4.6% of 1,2,4-linked Fuc*p* appeared in the branch. The PDSP-3 mainly contained 1,3-linked Ara*f* and 1.3-linked Man*p*. The DPSP-4 was found to be an acid polysaccharide. Theronic acid appeared mainly as 1.3.4-linked Gal*p*A, and smaller amounts of 1,4-linked Gal*p*A. All DPSPs were found to contain mannoses, DPSP-1 and DPSP-2 only contained 1,3-linked Man*p*, and DPSP-4 contained 1,3-linked Man*p* and 1,3,4-linked Man*p*. The mannose existed as 1,3-linked Man*p*, 1,3,4-linked Man*p*, and 1-linked Manp in PDSP-3. Besides, 1,3-linked Xyl*p* and 1,2-linked Xyl*p* were found in DPSP-2, DPSP-3, and DPSP-4, while 1,2-linked Xyl*p* was only found in DPSP-3 and DPSP-4, but xylose was found exist in DPSP-1 (Zhao et al., 2014).

Structural features of both water-extracted (IHWE) and alkaline-extracted (IHAE) polysaccharide from the seeds of *Plantago ispaghula* husk were reported (Edwards et al., 2003). The IHWE consists of ara, gal, gala, rha, and xyl with a molar ratio 201:17:44:19:719, while IHAE consists of the same monosaccharide in a ratio of 229:18:37:11:706. The polysaccharide was a highly branched arabinoxylan containing a linear β-(1→4)-Xylp backbone, and α-Araf units were attached as side residues through α-(1→2) and α-(1→3) linkages. Other more complex side groups were D-Glc*p*-(1→2)-Rha*p*-(1→4)-D-Xyl*p*-β-(1→4)-xylose, L-Ara*f*-(1→3)-D-Xyl*p*-β-(1→4)-Xyl*p*-β-(1→4)-D-xylose, and D-Glc*p*-(1→3)-L-Ara*f*-(1→4)-D-Xyl*p*-β-(1→4)-D-Xyl*p* (Edwards et al., 2003).

8.3 MOLECULAR CHARACTERISTICS

Psyllium polysaccharide could be dissolved in strong ammonia solution and dimethyl sulfoxide. However, it is insoluble in acetone, alcohol, ether, chloroform dichloromethane, dimethylamine, trimethylamine, diethylether, and ethylacetate.

Interestingly, psyllium polysaccharide could form a gel when it was dissolved in aqueous solutions. It was also found that the psyllium gel exhibits an extensive swelling property in water and 0.1 M HCl solution, whereas no swelling was observed in 0.5 M NaOH solution (Rao et al., 2011). Particularly, the gelling and strong water-absorbing capacities of psyllium polysaccharides are considered to be responsible for its therapeutic value for the treatment of constipation. Psyllium gum is an anionic polysaccharide which could bear a negative charge because of ionized carboxyl groups. Therefore, the intermolecular electrostatic repulsion due to homo-charges makes the polysaccharide chains fully extended and interpenetrated with each other to form intermolecular cross-linking which induces gelation (Farahnaky et al., 2010). The swelling could be a result of the entanglement of polysaccharide chains and development of inter- and intramolecular hydrogen bonds between the polysaccharide and water, which would cause more and more water to be trapped within the polysaccharide chains (Rao et al., 2011). Yu reported that solid-state xylanase treatment not only resulted in a smoother surface structure of psyllium polysaccharide, but also reduced the water uptake and gelling capacities of psyllium significantly. Scanning electron microscopy observation showed that the enzymatic modification could decrease the total surface area, which may contribute to the reduced water-absorbing ability of the modified psyllium polysaccharide gel (Yu and Perret, 2003; Yu et al., 2003). An acid treatment with 0.34%–1.44% HCl exhibited a similar effect on psyllium gel (Cheng et al., 2009).

Farahnaky investigated the effect of concentration, temperature, and pH on the dynamic rheology of psyllium gels. The results showed that psyllium gel at 2%–3% concentration was a weak gel, since the dynamic storage modulus (G′) was larger than the viscous modulus (G″), and there was no obvious melting point found for psyllium gel, indicating that the melting process was a continuous and long process. As for the pH value, the maximum functional properties of psyllium gels were observed at a pH of 4 and 7. It can be explained as net electrostatic repulsion decreases at low pH values, and the interaction of molecules also decreases, which leads to a reduction in gel elasticity. In addition, that the gel elasticity decreases in a higher pH value environment might be attributed to an alkaline depolymerization reaction, which results in less junction zones (Farahnaky et al., 2010).

Psyllium polysaccharide is rich in calcium, and that component has a significant influence on its appearance viscosity and gelling properties. Guo (2009) reported that when the psyllium polysaccharide solution was added to with 50 mM Ca^{2+}, the spectrum of G′ and G″ showed less frequency dependency, and lower tan δ, which meant the psyllium gel appeared more typical of a true gel in the presence of Ca^{2+}. Besides, by adding Ca^{2+}, the gel strength became much stronger and it was more tolerant of higher temperatures. As for the microstructure, the gel network nature of psyllium gel was filamentous. However, the filamentous gel strands were replaced by aggregates after adding Ca^{2+}, and it appeared that these small aggregates linked together to form a particulate network structure.

In our group, it was found that when treated with ethylenediaminetetraacetic acid (EDTA) to remove calcium ions, its intrinsic viscosity decreased significantly (Yin et al., 2015); however, adding Ca^{2+} resulted in a higher appearance viscosity of psyllium gel. The SEM micrographs showed EDTA chelation resulted in a more random flaky curly aggregated in psyllium gel compared with the linear style native one (Yin et al., 2012b). Besides, ferulic acid in arabinoxylans oxidized at the end of the arabinoxylan gelation process could probably yield higher ferulated oligomer cross-links, and physical entanglements were probably involved in the final gel structure (Carvajal-Millan et al., 2005). A strong interaction between Ca^{2+} and ferulic acid were found in our work. Psyllium gel was treated with sodium hydroxide to deplete ferulic acid, and the resultant material (PLCP-FAS) showed a lower viscosity, as well as psyllium gel treated with EDTA (PLCP-E). Although adding Ca^{2+} could increase the apparent viscosity of PLCP-FAS and PLCP-E, the viscosity of PLCP-FAS was still much lower than the natural gel, and only PLCP-E could keep a high viscosity when dialysis was carried out to remove free Ca^{2+} in the solution, which indicated that PLCP-FAS did not have a strong ability to bind free Ca^{2+} ions in solution. A possible interaction between Ca and psyllium polysaccharide could be proposed as follows: the chains of the polysaccharide were cross-linked to each other because of the ionic interaction of Ca^{2+} – ferulic acid and Ca^{2+} – glucuronic acid, which is responsible for its high apparent viscosity and weak gel properties (Yin et al., 2012b).

8.4 BIOACTIVITIES

Psyllium has long been used as a dietary fiber supplement, mostly as a gentle bulk-forming laxative with a good safety record. During the last decades, dietary supplementation with psyllium has also been considered to play an important role in the treatment of diabetes, hyperglycemia, and hyperlipidemia.

8.4.1 Laxative Effect

The term "constipation" is defined as difficult stool passage and infrequent bowel movements. It is characterized by discomfort, excessive straining, hard or lumpy stools, and unsatisfactory defecation (Rungsiprakarn et al., 2015). Constipation is a highly prevalent and difficult-to-cure health problem. It overlaps a large group of diseases, including functional dyspepsia, irritable bowel syndrome, and gastroesophageal reflux disease. Traditional methods to treat constipation include bulking agents, stool softeners, stimulant laxatives, osmotic laxatives, and polyethylene glycol. However, the treatment of constipation continues to evolve and remains challenging (Attaluri et al., 2011; Schiller, 2001). A lot of research has certainly demonstrated that psyllium seeds are very effective in treating constipation. For instance, Tomas-Ridocci reported that *P. ovata* increased the frequency of stools in patients with chronic constipation with or without irritable bowel syndrome (Tomas-Ridocci et al., 1992). Ashraf reported that stool frequency increased significantly in psyllium-treated subjects with diopathic constipation, as did stool weight. Psyllium treatment also improved stool consistency and pain on defecation (Ashraf et al., 1995). Odes (Odes and Madar, 1991) treated chronic constipation patients with capsules containing celandine–aloe vera–psyllium for 28 days, and found that bowel movements in the treated group became more frequent, the stools were softer, and laxative dependence was reduced. Psyllium was also reported superior to docusate sodium, a synthetic anionic detergent marketed as a stool softener laxative, for softening stools by increasing stool water content, and it has great overall laxative efficacy (McRorie et al., 1998).

Constipation is also a common symptom during pregnancy, and 25%–40% of pregnant women experience it (Cullen and O'Donoghue, 2007). A randomized double-blind clinical trial in pregnancy showed that psyllium powder consumption during the third trimester of pregnancy could significantly prevent anorectal complications, hemorrhoidal diseases, anal fissure, and constipation. A significant decrease in surgical complications, constipation, and fissure in women who had a normal delivery was also reported (Ghahramani et al., 2013).

Although many studies have demonstrated the clinical usefulness of psyllium in the treatment of constipation, only a few articles give details about the possible underlying mechanism. The laxative effects of dietary fiber are considered to depend on its physicochemical properties. Generally, it has a high water-holding capacity that can normalize stool consistency (Chutkan et al., 2012). A study conducted by Lewis (Lewis and Heaton, 1999) showed it is not the mass or number of particles that is important in altering intestinal function, but the size or shape, since volunteers who swallowed small granules found no change in whole-gut transit time and cecal transit time, while those who swallowed bran-like flakes did. Marlett et al. (2000) observed that psyllium consumption results in the passage of slick, gelatinous stools, and further study show an unfermented gel component of psyllium contributes to the laxative properties of psyllium by acting as a lubricating emollient to facilitate the propulsion of colon content. Interestingly, Marlett also reported that a poorly fermented gel-forming fraction (with ~30% of the constituent sugars disappearing) can also increase excreta moisture and bile acid excretion in rats, and this is a physiologically active component of psyllium (Marlett and Fischer, 2002; Marlett and Fischer, 2003). In healthy volunteers, psyllium consumption increased the neutral sugar output in feces, which were accounted for by the fecal excretion of arabinose and xylose in an intact highly polymerized form, since the apparent digestibility's of the polysaccharide were 24% and 53%, respectively (Marteau et al., 1994). This indicates that the bulking effect mainly results from intact material.

Recently, research has confirmed that there are chemical constituents responsible for the laxative effect of psyllium. Gilani et al. (1998) reported an aqueous-methanol extraction of psyllium caused a dose-dependent stimulatory effect in guinea pig ileum; at a dose range of 1–10 mg/mL, especially, this effect could be completely blocked by pretreatment with 1 μm atropine (a competitive blocker of acetylcholine at the muscarinic receptor) when the psyllium concentration was at 1–4 mg/mL, but it could be partially blocked at 6–10 mg/mL, indicating that the laxative effect at low doses is through an acetylcholine-like mechanism, while the effect of the high dose may partially be through another mechanism. A similarly atropine-sensitive effect was also observed by Mehmood et al. (2011). Mehmood also reported the relaxant effect of psyllium at 10 mg/mL was blocked in the presence of L-NAME (an NOS-inhibitor) or methylene blue (a guanylyl cyclase inhibitor). This suggests that psyllium has a gut-stimulatory component mediated through muscarinic or serotonin receptor activation.

8.4.2 Therapy for Intestinal Diseases

Irritable bowel syndrome (IBS) is a highly prevalent functional gastrointestinal disorder with characteristic symptoms of unexplained abdominal pain or discomfort and altered bowel habits (Chang and Talley, 2010). The treatment of IBS remains challenging. Hence, there is a need for complementary or alternative medicines, especially herbal therapies. A systematic review including 17 studies indicates that fiber was effective in the relief of global irritable bowel syndrome symptom, and soluble fiber such as psyllium showed a significant improvement, whereas insoluble fiber showed no significant difference compared with placebo, and in some cases the insoluble fiber caused a worse clinical outcome (Bijkerk et al., 2004). Hotz and Plein (1994) reported that psyllium seed treatment increased stool frequency and improved consistency in patients with IBS, and the effectiveness is superior to wheat brans. Another clinical trial showed that psyllium significantly improves overall well-being in IBS patients, and favorably affects bowel habit and transit time in those with constipation (Prior and Whorwell, 1987).

A comparative effectiveness trial evaluated the relative efficacy of psyllium in 275 patients aged from 18 to 65 years with IBS, who received a 12-week treatment of 10 g psyllium, or 10 g bran, or 10 g placebo. During the first month, the proportion of responders in the psyllium group was significantly greater than the placebo group. After 3 months' treatment, symptom severity in the psyllium group was significantly improved with a 90 points decrease in the IBS severity score, whereas the decrease was 40 in the placebo group (Bijkerk et al., 2009). The most recent clinical trial was reported by Nagafabadi et al. (2015), which compared the therapy effects of *P. major* and the drug colofac on IBS patients. *P. major* treatment was found to be as effective as colofac, particularly, the effectiveness of *P. major* in reducing the number of diarrhea defecations in a day was higher than that of colofac. Besides, in the treatment of patients with diarrhea predominant IBS, psyllium polysaccharide could effectively improve the symptoms including severity of abdominal pain, frequency of defecation, bloating, and urgency (Agah et al., 2010).

Psyllium treatment also shows effects to prevent diarrhea in clinical trials. Psyllium treatment could reduce diarrhea in a patient who is tube-fed (Heather et al., 1991), and increase the number of gelatinous stools and normal stools, but it could decrease the number of liquid stools in enterally fed patients (Belknap et al., 1997; Hart and Dobb, 1988). Treatment with normalization of diet and psyllium also showed therapeutic effects for chronic nonspecific diarrhea of childhood (Smalley et al., 1982). Diarrhea is a common medical symptom caused by chemotherapy or radiation in patients with cancer, it can cause dehydration, electrolyte imbalance, malnutrition, fluid depletion, and hospitalization, all of which can lead to compromise and death (Muehlbauer et al., 2009). Murphy et al. (2000) reported that psyllium significantly decreased the incidence and severity of radiation-induced diarrhea in cancer patients who were undergoing radiation therapy. In addition, Eherer et al. (1993) reported psyllium, but not wheat bran, was found to improve the fecal consistency and viscosity in subjects with experimentally-induced secretory diarrhea.

Wenzl et al. (1995) found that a normal intestine delivers stools differing widely in quantity, but maintains the percentage of fecal water within a narrow range, while stool looseness in diarrhea is determined by the ratio of fecal water to the water-holding capacity of insoluble solids. For patients with diarrhea with normal stool weights, loose stools are due to a low output of insoluble solids without the concomitant reduction in water output that occurs in normal subjects when insoluble solids are low. Hence, the ingestion of psyllium could increase the ratio of fecal water to water-holding capacity of insoluble solids to reduce diarrhea. Another study focused on the protective effect of psyllium against lactulose-induced diarrhea. Results showed that psyllium significantly delayed gastric emptying, but not small bowel transit. The progression through the colon was delayed, with an increase in the percentage of the dose at 24 h in the ascending and transverse colon, and with correspondingly less in the descending colon (Washington et al., 1998). In addition, Mehmood et al. (2011) found that blocking of Ca^{2+} channels and activation of the NO-cyclic guanosine monophosphate pathway could induce gut-inhibitory activity, which may be involved in the antidiarrhea effect of psyllium polysaccharide.

8.4.3 Intestinal Function Activity

Although it is nondigestible by human enzymes, many dietary fibers could be fermented by large-bowel microflora to produce short-chain fatty acids, which would result to a decreased luminal pH. Additionally, many studies indicate that fiber fermentation could be beneficial for human physiology, such as decreasing the proliferation of undesirable pathogens, affecting microbial enzyme activities, increasing mineral absorption, and serving as an energy source for the colon epithelium. In our research group, similar beneficial effects of psyllium polysaccharide from seeds of *P. asiatica* L. were investigated. Results showed levels of total short-chain fatty acid, acetic propionic, and *n*-butyric acid in mouse colonic content were all improved after polysaccharide intake, and a decreased pH in the mouse colon was also observed (Hu et al., 2012; Figs. 8.3–8.4). The polysaccharide was in vitro fermented by human fecal microbiota for 24 h. During fermentation, the pH in fecal culture decreased from 6.1 to 5.1, and the levels of total short-chain fatty acids, acetic, propionic, and *n*-butyric acid were significantly increased.

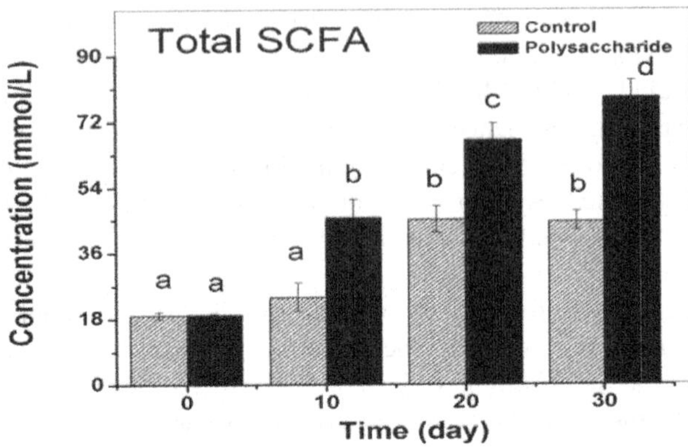

FIGURE 8.3 Total SCFA concentrations (mmol/L) in the mouse colon feces of polysaccharide treated groups compared to the control group. Values are expressed as the mean ± standard deviation of 12 mice. Different letters mean significantly different values ($P < .05$) from each other.

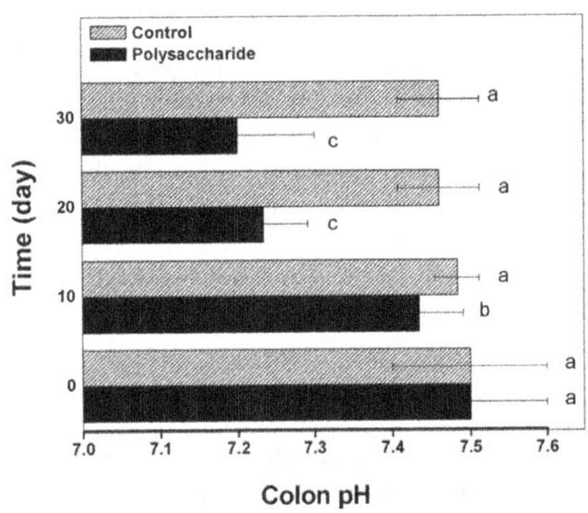

FIGURE 8.4 The pH change in mouse colon feces of the polysaccharide treated group compared to the control group. Values are expressed as the mean ± standard deviation of 12 mice. Different letters mean significantly different values ($P < .05$) from each other.

The increase in acetic and *n*-butyric acid production was mainly due to the fermentation of glucuronic acid and xylose in polysaccharide, while the increase in propionic acid primarily resulted from the fermentation of arabinose and xylose (Hu et al., 2013a; Fig. 8.5). Consistent with this result, in vitro study also found that a steady decrease in molecular weight of the polysaccharide from 1902 to 4.7 kDA was observed as the digestion time increased, meanwhile the reducing ends were increased from 0.157 to 0.622, which indicated that the decrease of Mw may be due to the breakdown

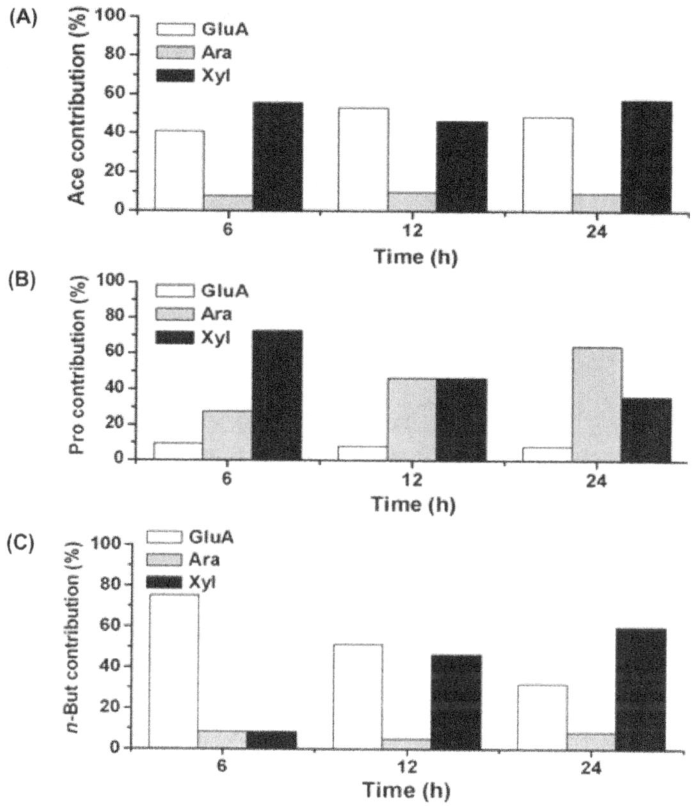

FIGURE 8.5 Contribution of xylose, arabinose, and glucuronic acid in the polysaccharide from *P. asiatica* L. for the production of acetic (A), propionic (B), and *n*-butyric acid (C) at different times during fermentation. Data are mean of three independent experiments. *Ace*, acetic acid; *Pro*, propionic acid; n-*But*, *n*-butyric acid; *Xyl*, xylose; *Ara*, arabinose; *GluA*, glucuronic acid.

of glycosidic bonds. However, there was no free monosaccharide released during the whole digestion period (Hu et al., 2013b; Figs. 8.6–8.7). *Bacteriodes* spp., *Eubacterium* spp., *butyrivibrio* spp., *Bifidobacterium bifidum*, *Lactobacillus fermentum*, and *Lactobacillus reuteri* in mouse colon microbiota were all enhanced due to psyllium polysaccharide intake (Hu et al., 2013c; Fig. 8.8). High pressure homogenization treatment was found to increase short-chain fatty acid production of psyllium polysaccharide (Hu et al., 2013d; Table 8.1).

The beneficial effects of *P. ovata* on intestinal function are also reported. *P. ovata* seed treatment could increase the length and weight of large intestine, fecal bacterial mass, and the total fecal bile acid excretion, but decrease β-glucuronidase activity in rats (Leng-Peschlow, 1991). Oral intake of the seeds also increased butyrate and acetate concentration in feces; however,

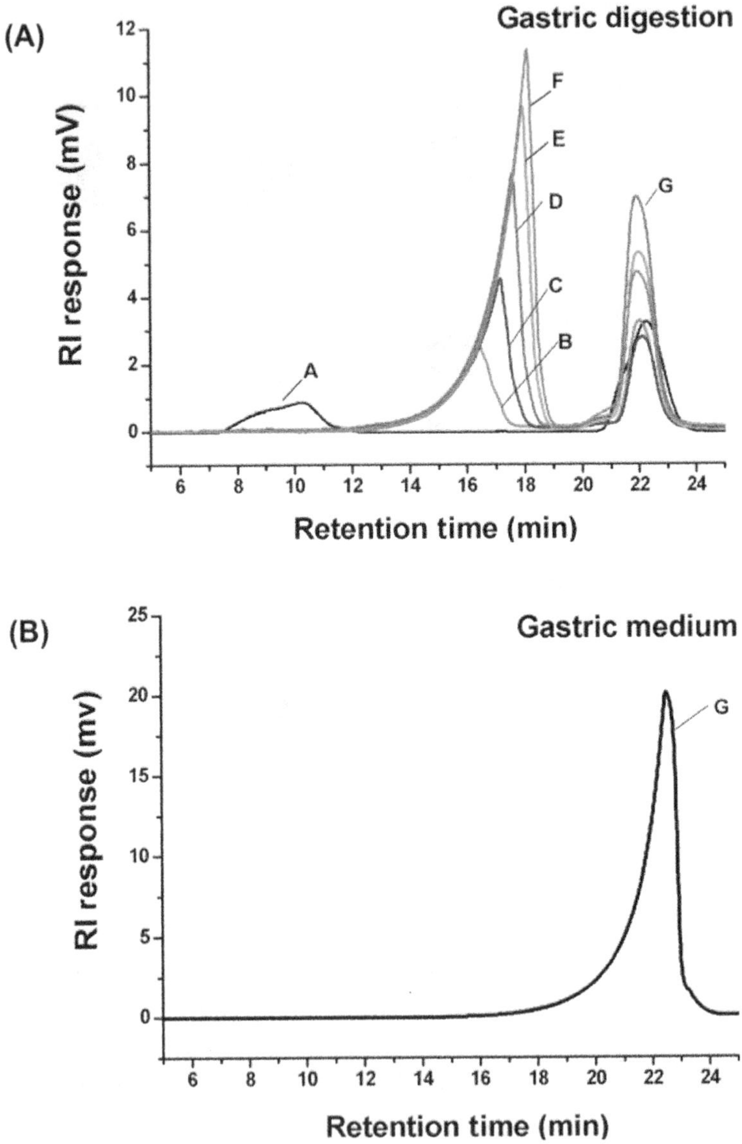

FIGURE 8.6 HPGPC chromatogram of the polysaccharide from *P. asiatica* L. in vitro gastric digestion at different times within 6 h (A), and the chromatogram of gastric medium before digestion (B), with RI detection on an Ultrahydrogel linear column (7.8 × 300 mm) with distilled water at 0.5 mL/min. Peaks (A−F) represent the polysaccharide at different digestion time points: A−0 h, B−0.5 h, C−1 h, D−2 h, E−4 h, F−6 h, while peak G represents the substances in gastric medium.

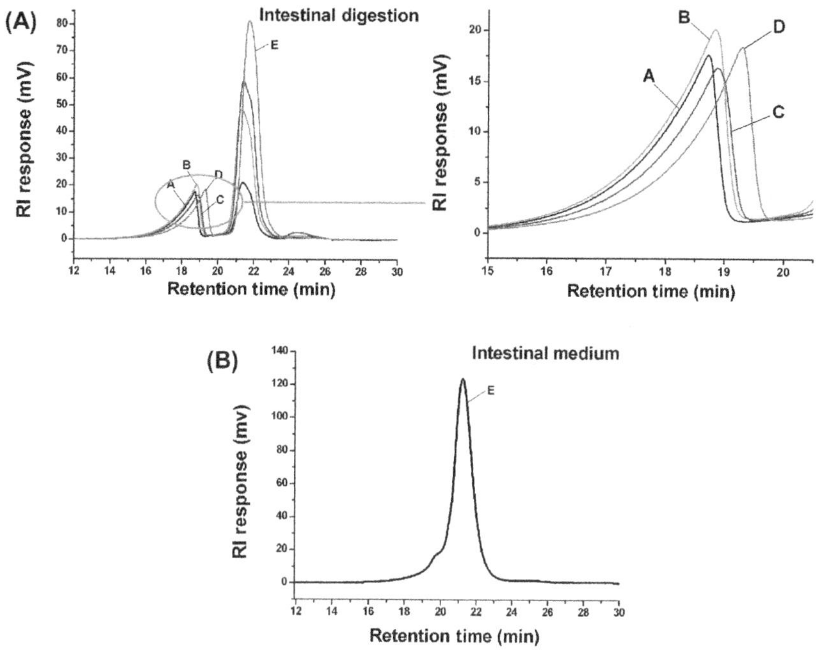

FIGURE 8.7 HPGPC chromatogram of the polysaccharide from *P. asiatica* L. in vitro small intestinal digestion at different times within 6 h (A), and the chromatogram of small intestinal medium before digestion (B), with RI detection on an Ultrahydrogel linear column (7.8 × 300 mm) with distilled water at 0.5 mL/min. Peaks (A–D) represent the polysaccharide at different digestion time points: A–0.5 h, B–2 h, C–4 h, D–6 h, while peak E represents the substances in intestinal medium.

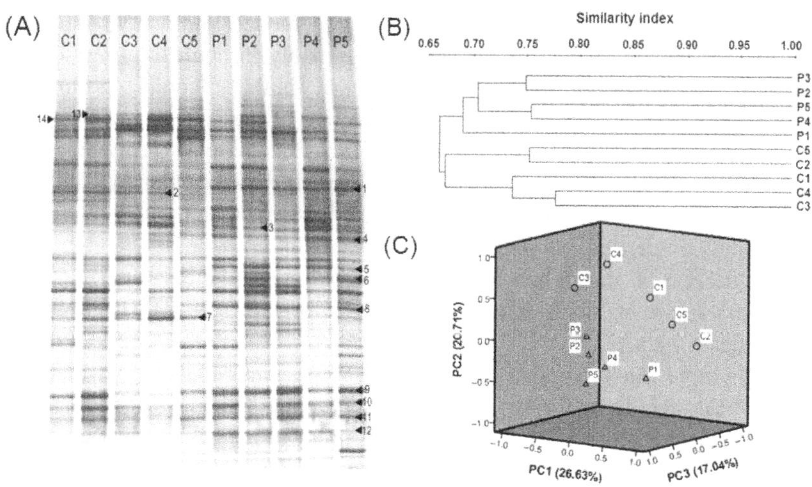

FIGURE 8.8 Microbiota analysis for the colon content in mice from the polysaccharide group and control group. (A) Denaturing gradient gel electrophoresis (DGGE) of V3 regions of bacterial 16S rDNA amplifications; (B) UPGMA cluster analysis of DGGE bacterial profiles; (C) PCA scores plot of DGGE bacterial profiles. PC1, PC2, and PC3 represent principle factors and their probability (%) of influence on the bacterial profiles. C1, C2, … C5: colon content samples individually from five mice randomly chosen from the control group after 30 days' administration; P1, P2, … P5: colon content samples individually from five mice randomly chosen from the polysaccharide group after 30 days' administration.

TABLE 8.1 Effect of Nontreated and Homogenization-Treated Polysaccharide From Seeds of *P. asiatica* L. on Short-Chain Fatty Acid Production in Mouse Colon

	Control Group	NTP Group[a]	HTP Group[b]
SCFA (μg/g of Colonic Content)			
Acetic acid	$2332 \pm 57^{a,c}$	3074 ± 61^{b}	3090 ± 49^{b}
Propionic acid	949 ± 39^{a}	1064 ± 27^{b}	1468 ± 40^{c}
Butyric acid	1167 ± 21^{a}	1509 ± 36^{b}	2206 ± 30^{c}
i-Butyric acid	171 ± 14^{a}	164 ± 15^{a}	160 ± 20^{a}
Valeric acid	62 ± 8^{a}	58 ± 6^{a}	60 ± 10^{a}
i-Valeric acid	146 ± 11^{a}	144 ± 13^{a}	146 ± 18^{a}
Total SCFA	4827 ± 147^{a}	6013 ± 142^{b}	7130 ± 156^{c}

[a]*NTP group, nontreated polysaccharide group.*
[b]*HTP group, homogenization-treated polysaccharide group.*
[c]*Each value is the mean ± standard deviation (n = 12); means in the same line not sharing a common letter are significantly different (P .05).*

the effect depended on continuity of treatment, since the concentration reversed to pretreatment within 1–2 months after cessation of fiber supplementation (Nordgaard et al., 1996). Sahagún et al. (2015) reported an antiulcerogenic effect on intestinal mucosa in a rabbit model with acetyl salicylic acid-induced lesion, and the mechanism is probably through limiting acetyl salicylic acid penetration into epithelial cells. In another randomized clinical trial, *P. ovata* seeds were found to be as effective as mesalamine to maintain remission in ulcerative colitis, and a significant increase in fecal butyrate levels may account for this beneficial effect (Fernandez-Banares et al., 1999). A fiber-supplemented diet containing 5% *P. ovata* seeds for 13 weeks significantly increased butyrate and propionate in a rat colitis model, and it also decreased the production of proinflammatory mediators such as NO, leukotriene B_4, and tumor necrosis factor α (TNF-α) (Rodríguez-Cabezas et al., 2003).

Psyllium, methylcellulose, and pectin were fermented in vitro with human fecal inoculums for 4, 12, and 24 h. Data indicate pectin fermentation results in the highest short-chain fatty acid production, gas production, and organic matter disappearance, followed by psyllium (Campbell and Fahey, 1997). Another comparative experiment showed yeast cell produced the highest short-chain fatty acid, followed by psyllium polysaccharide and soy fiber (McBurney, 1990). The effects of dietary cellulose and psyllium husk on monkey colonic microbial metabolism were compared. African green monkeys were fed with psyllium or dietary cellulose for more than 3 years, then the fecal or colonic samples were inoculated into a continuous culture system and fed with psyllium husk or cellulose, respectively. Psyllium-fed cultures showed a lower pH, higher volatile fatty acid concentration, and higher β-glucuronidase activity than cellulose-fed cultures. Besides, the ratio of anaerobes to aerobes was found to be lower in psyllium-fed cultures (Costa et al., 1989).

The main fermentation sites of dietary fiber are the cecum and proximal colon. Interestingly, a special fermentation site shifting the effect of psyllium was found. Rats were fed low amylose (LAS) or high amylose cornstarch diets (HAS), with or without psyllium. Compared to rats fed with LAS and LAS/psyllium diets, the concentration of cecal butyrate in rats fed with HAS and HAS/psyllium diets were significantly higher. However, butyrate and total SCFA concentrations in rats fed with the HAS diet decreased, along with the length of the colon, and fecal butyrate concentration was reduced to one-third of that in the cecum. In contrast, the HAS/psyllium diet maintained higher butyrate concentrations throughout the large bowel. The concentration of fecal butyrate in the HAS/psyllium diet-fed group significantly exceeded the sum of the concentrations in rats fed the LAS/psyllium and HAS diets. In a further experiment, ileorectostomized rats were fed with HAS and HAS/psyllium diets. Excluding the difference in fecal starch excretion between normal and ileorectostomized rats, starch degradation by large-bowel microflora in rats fed the HAS and HAS/psyllium diets was deduced to be 96% and 63%, respectively, indicating that PS may delay the fermentation rate of HAS in the cecum and shift the fermentation site of HAS toward the distal colon, leading to the higher butyrate concentration in the distal colon and feces (Morita et al., 1999).

However, long-term dietary supplementation with psyllium on monkeys showed that it may cause intestinal epithelial cell loss, jejunoileal muscle layer hypertrophy, and thinning of the colonic wall after prolonged intake. Although all animals in this study were healthy and had no diarrhea or abnormal feces, it still indicated that care should be taken in recommending long-term ingestion of gel-forming fibers (Paulini et al., 1987).

8.4.4 Hypoglycemic Effect

Diabetes mellitus (DM), commonly referred to as diabetes, is a serious disease that is becoming increasingly common in many places around the world. There are several forms of diabetes, including insulin-dependent diabetes (type 1 diabetes), noninsulin-dependent diabetes (type 2 diabetes), and gestational diabetes. Type 2 diabetes is the most common form, and the patient population with diabetes rises every year. Despite the tremendous developments in medicinal chemistry, therapeutic treatment is still not quite satisfactory. In many countries, traditional medicines are used as alternative medicine in the treatment of diabetes. Plantago plants are considered to be an effective antidiabetic medicinal herb, because of their hypoglycemic effect.

The hypoglycemic effect of psyllium was investigated using several animal models. Type 1 diabetes in rats was induced with a single intraperitoneal injection in anaesthetized, fasted rats at 65 mg streptozotocin/kg body weight, which was dissolved in 0.5 M $_M$-citrate buffer immediately before use. Meanwhile, type 2 diabetes was induced by a single intraperitoneal injection of 48 h old rats using 90 mg streptozotocin/kg body weight (Hannan et al., 2006). Administration of psyllium significantly improved glucose tolerance in the normal, type 1 and type 2 diabetic rat model. Especially, when psyllium polysaccharide was administered orally with sucrose solution on the rat model, it suppressed postprandial blood glucose and retarded small intestinal absorption, without inducing the influx of sucrose into the large intestine. Ahmed et al. (2010) reported that a 7-week treatment of psyllium husk powder decreased blood glucose and cholesterol level in rats with diabetes, as well as in rats with diabetes and hypercholesterolemia. A rabbit model was established by intravenous administration of alloxan. A diet treatment containing psyllium polysaccharide significantly decreased glucose levels and increased insulin concentration in the diabetic rabbits (Díez et al., 2013).

Some clinical trials also confirm the hypoglycemic effect of psyllium. One-hundred and twenty-five patients with type 2 diabetes were included in a study that consisted of a 6-week period of diet counseling followed by 6 weeks of treatment with psyllium product (containing 79% psyllium mucilloid and 21% inactive excipients). It showed that no significant changes in weight between the groups, but fasting plasma glucose, total cholesterol, LDL cholesterol, and triglyceride levels were significantly reduced, whereas the HDL cholesterol level significantly increased (Rodríguez-Morán et al., 1998). In another study, 37 type 2 diabetic patients were randomly assigned to three treatment groups, including a placebo group, a 3.4 g psyllium group, and a 6.8 g psyllium group, using a 20-week intent-to-treat analysis (8 weeks baseline, 12 weeks treatment). The results showed that the 6.8 g psyllium treatment significantly decreased the glycated hemoglobin at week 8, while both doses of psyllium at 3.4 g and 6.8 g could significantly lower fasting blood glucose at treatment weeks 8 and 12 (Feinglos et al., 2013). A similar observation was found by Gibb (Gibb et al., 2015). Siavash et al. (2007) also reported that psyllium could effectively reduce postprandial plasma glucose in type 2 diabetic patients.

A relationship between the dose of psyllium mucilage and its attenuating effect on hyperglycemia was investigated by Frati-Munari et al. (1988). An dose of 75 g of glucose was mixed with 0, 10, 20, and 30 g of mucilage, and consumed by healthy volunteers. The maximum peak of serum glucose was determined after 30 min. The areas under the curve of glucose were significantly lower in the test with 20 and 30 g of mucilage than in the tests with

0 and 10 g. The mechanism of hypoglycemic effect of the herb could be that inhibiting glucose absorption from the intestine improved insulin secretion from the pancreas, decreased glucose production from hepatocytes, and enhanced glucose uptake by adipose and muscle tissues (El-Abhar and Schaalan, 2014). It was shown that glucose absorption decreased significantly in the psyllium group without a sharp change in insulin, and consumption of the fiber also reduced total and LDL cholesterol and uric acid levels of the patients (Sierra et al., 2002). However, The effect of psyllium on insulin levels in type 2 diabetes was controversial. Watters reported that psyllium polysaccharide significantly enhanced insulin levels in mice with diabetes (Watters and Blaisdell, 1989). A cross-over experiment was designed by Pastors where type 2 patients were treated with psyllium polysaccharide or a placebo before breakfast and dinner, but not lunch. Then the postprandial glucose and insulin level was evaluated. It was found that those treated with psyllium polysaccharide before breakfast and dinner could reduce postprandial glucose by 14% at breakfast and 20% at dinner compared to the placebo, and a reduction of postprandial glucose at lunch was also observed in 31%. Meanwhile, the postprandial serum insulin concentration after breakfast was reduced by 12% (Pastors et al., 1991). In addition, the effect of the manner of administration of psyllium on its hypoglycemic effect was also investigated by Wolever et al. (1991). Three different administration methods of psyllium were designed that included just taking in water before a flaked bran cereal test meal, sprinkling on top of the cereal, or actually incorporating into the flake, to investigate the effect of treatment method on its hypoglycemic activity. The results showed that psyllium reduced the glycemic response when it was sprinkled onto or incorporated into the cereal, but not when it was taken before the cereal, indicating that reducing the rate of digestion may also be one mechanism to explain its hypoglycemic activity.

In healthy individuals, psyllium significantly delayed gastric emptying from the third hour after a meal, and it increased the sensation of satiety (Bergmann et al., 1992). Psyllium reduces feelings of hunger and therefore energy intake in normal volunteers without requiring mixing with the meal; however, it does not act by slowing down the gastric emptying of hydrosoluble nutrients, but by increasing the time allowed for intestinal absorption (Rigaud et al., 1998).

8.4.5 Hypocholesterolemic Effect

Coronary heart disease is the most common disease in many countries, where the incidence has been conclusively linked to elevated blood cholesterol levels. The second report of the National Cholesterol Education Program Adult Treatment Panel points out that dietary therapy remains the first line of treatment for high blood cholesterol, and drug therapy is reserved for patients who are considered to be at high risk of coronary heart disease (Grundy et al., 1993).

In the last decades, much research has confirmed the hypocholesterolemic effect of psyllium polysaccharide. Anderson reported that long-term use of a standard dose of psyllium (5.1 g twice daily) resulted in a significant decrease in serum total cholesterol and LDL-cholesterol concentration in patients with primary hypercholesterolemia (Anderson et al., 2000). In patients with mild-to-moderate hypercholesterolemia, an 8-week treatment with 3.4 g psyllium achieved a significant reduction in total cholesterol level, low-density lipoprotein cholesterol level, and apolipoprotein B level without affecting body weight, blood pressure, serum level of HDL cholesterol, or serum lipid level (Anderson et al., 1988). Psyllium treatment was also found to significantly decrease body mass index, fasting plasma glucose and insulin, homeostasis model assessment index, glycated hemoglobin, low-density lipoprotein cholesterol, and apolipoprotein B in metabolic syndrome patients. Particularly, after a 6-month diet treatment containing psyllium, a notable decrease in plasma triglyceride concentration (−13.3%), systolic (−3.9%), and diastolic blood pressure (−2.6%) was observed (Cicero et al., 2010). A meta-analysis conducted by Olson showed that consumption of a psyllium cereal-enriched diet resulted in lower total blood cholesterol and LDL cholesterol concentrations, and there was no effect of sex, age, or menopausal status on blood lipids (Olson et al., 1997). Wei et al. (2009) reported that such a serum cholesterol-lowering effect in mild and moderate hypercholesterolemic patients was in a dose- and time-dependent manner. In our research group, we also demonstrated that a treatment with psyllium polysaccharide from *P. asiatica* L. seeds could lower the apparent absorption of lipid, as well as total triglyceride and cholesterol in blood serum, in a mice model (Hu et al., 2013c) (Table 8.2).

Colestipol hydrochloride is a drug that reduces the concentration of cholesterol in the blood. A combination therapy consisting of 2.5 g psyllium mucilloid with 2.5 g colestipol reduced the ratio of total cholesterol to HDL by 18.2%, which was significantly lower than colestipol (10.6%) or psyllium (6.1%) alone (Spence et al., 1995). This finding suggests that adding psyllium to half the usual dose of bile acid sequestrant resins maintains the efficacy and improves the tolerability of these resins. A long-term effect of consuming foods containing psyllium seed husk on serum lipid in subjects with hypercholesterolemia was evaluated (Davidson et al., 1998). A group of 286 individuals with LDL-cholesterol concentrations between 3.36 and 5.68 mmol/L were randomly assigned to one of four treatment groups after following a low-fat diet for more than 8 weeks. At week 24, no significant effect of psyllium treatment on HDL cholesterol or triacylglycerol was observed. However, LDL cholesterol concentration in the high-dose psyllium group remained below baseline throughout the 24-week treatment period, indicating the potential for long-term benefit (Davidson et al., 1998).

TABLE 8.2 Effect of Polysaccharide From *P. asiatica* L. on Characteristic Blood Serum Chemistry of Mice

	Control Group	Polysaccharide Group
Glucose (mmol/L)	6.72 ± 0.51[a]	6.56 ± 0.39[a]
Total protein (g/L)	55.3 ± 1.1[a]	54.0 ± 1.1[a]
Albumin (g/L)	31.3 ± 0.9[a]	31.6 ± 0.5[a]
Albumin/globulin	1.3 ± 0.1[a]	1.4 ± 0.1[a]
Urea nitrogen (mmol/L)	9.5 ± 0.6[a]	9.4 ± 0.5[a]
Creatinine (μmol/L)	33 ± 2[a]	35 ± 3[a]
Triglyceride (mmol/L)	1.9 ± 0.11[a]	1.25 ± 0.12[b]
Total cholesterol (mmol/L)	3.27 ± 0.26[a]	2.65 ± 0.28[b]
HDL-cholesterol (mmol/L)	2.62 ± 0.11[a]	2.34 ± 0.14[a]
LDL-cholesterol (mmol/L)	0.58 ± 0.14[a]	0.32 ± 0.09[a]
Atherogenic index[b]	0.22 ± 0.01[a]	0.17 ± 0.01[b]
GOT (U/L)[c]	193 ± 21[a]	192 ± 29[a]
GPT (U/L)[d]	132 ± 19[a]	134 ± 21[a]

[a]*Each value is the mean ± standard deviation (n = 12); means in the same row not sharing a common letter are significantly different (P < .05).*
[b]*Atherogenic index = (LDL-cholesterol)/(HDL-cholesterol).*
[c]*GOT, glutamic oxalacetic pyruvic transaminase activity.*
[d]*GPT, glutamic.*

Seventy-two patients with triglyceride levels between 160 and 200 mg/dL and LDL cholesterol between 130 and 190 mg/dL were treated with psyllium or cellulose for 60 days, the results showed triglyceride level and serum LDL decreases in both groups, while HDL cholesterol levels decreased in the cellulose group but not in the psyllium group (Rahbar et al., 2014). Similar to its hypoglycemic effect, Wolever et al. (1994) reported that psyllium must be mixed with food to have the maximum effect on serum cholesterol. In patients either on high-fat diets or low-fat diets, the hypocholesterol effect of psyllium consumption could be observed without significant differences (Sprecher et al., 1993). Another study including 24 men, 23 premenopausal women, and 21 postmenopausal women evaluated the influence of sex and hormonal status on the effect of psyllium supplementation on parameters of plasma lipoprotein metabolism. It found that psyllium intake decreased plasma LDL cholesterol in men, pre-, and postmenopausal women. However, the results also showed the important influence of sex and hormonal status on the effects of psyllium on lipoprotein (Vega-López et al., 2002).

Many efforts focus on the mechanism of the hypocholesterolemic effects of psyllium. Everson found that psyllium consumption had no effect on cholesterol synthesis, and only decreased cholesterol absorption in subjects with high baseline absorption of cholesterol who did not lower LDL cholesterol in response to psyllium, but it increased the turnover rate of bile acids by impairing intestinal absorption of bile acid, and increased total bile acid synthesis. Especially, the decrease in LDL cholesterol was correlated directly with the increase in total bile acid synthesis (Everson et al., 1992). Hepatic 7a-hydroxylase is the initial and rate-limiting enzyme in the pathway that converts cholesterol to bile acids. Vergara-Jimenez et al. (1998) found that psyllium plasma significantly lowered LDL cholesterol, triacylglycerol, apolipoprotein B, and hepatic cholesteryl ester in guinea pigs fed with high fat or high sucrose diets. They also found psyllium up-regulated cholesterol 7α-hydroxylase and 3-hydroxy-3-methylglutaryl coenzyme A (HMG-CoA) reductase activities in the guinea pigs, while psyllium could also stimulate LDL fractional catabolic rate by as much as 100%. Similarly, Buhman et al. (1998) found that psyllium treatment increased fecal bile acid and total steroid excretion, as well as 7α-hydroxylase activity and 7α-hydroxylase mRNA levels in rats. Chan and Heng (2008) reported 3 weeks of psyllium fiber-enriched feeding down-regulated genes involved in lipogenesis in C57BL/6J mice, but genes involved in cholesterol and bile acid synthesis were up-regulated. HMG-CoA is a regulatory enzyme for cholesterol biosynthesis, it couples with LDL receptor to regulate cholesterol synthesis, once it is inhibited it would decrease cholesterol synthesis (Goldstein and Brown, 1984). Although psyllium increases the activity, it could be described as maintaining cholesterol homeostasis (Horton et al., 1994). So, it is more likely that the lower LDL-cholesterol resulted from increased production of bile acids through the up-regulating activity of cholesterol 7α-hydroxylase. Besides, one suggested mechanism of the LDL-lowering effect of the soluble fiber is by inhibition of cholesterol synthesis through the production of short-chain fatty acid from fermentation. Some research shows that psyllium polysaccharide also possesses a bile acid-binding capacity, which could contribute to the cholesterol-lowering activity (Gong et al., 2015; Buhman et al., 1998). Impairment of intestinal absorption of fats and sterols is another pathway for dietary fiber to lower the LDL cholesterol. Gastrointestinal mucin is believed to contribute to the intestinal diffusion barrier for nutrients, levels of colonic luminal and total gastric mucin were significantly higher in rats fed 10% psyllium than in rats fed the control diet, which may be one way to reduce cholesterol and fatty acid absorption (Satchithanandam et al., 1996).

In addition, failure of lipid-lowering by psyllium in some cases is also reported. A modest 7% reduction in LDL cholesterol concentration compared with the control cereal was observed when a psyllium-enriched diet was used as an adjunct to a low-fat diet in children (Davidson et al., 1996). Dennison and Levine (1993) found that psyllium polysaccharide had no additional lowering effect on total cholesterol or LDL-c levels in children with hypercholesterolemia who were already following low fat, low saturated fat, and low cholesterol diets. Sixteen men and 47 women with cholesterol levels that were normal or slightly elevated received an average daily dose of 12.7 g of psyllium hydrophilic mucilloid, and no effect on any lipid parameter was finally observed, despite it being hypothesized that psyllium may be more effective when taken with breakfast than at bedtime (Van Rosendaal et al., 2004). It was also found that in aging animals fiber supplementation could not prevent an age-related increase in lipid concentration (Schneeman and Richter, 1993). In elderly veterans, an intake of 10.4 g of psyllium per day only reduced LDL cholesterol by 2%, and had no effect on the LDL/HDL ratio (Schectman et al., 1993). Similarly, Stewart reported that there was no significant change in serum cholesterol concentration in elderly participants after psyllium consumption (Stewart et al., 1991).

8.4.6 Antitumor Effect

Although the underlying mechanisms of such cancer prevention effects are still unclear, many studies have shown that psyllium fiber possess anticarcinogenic properties, especially for colon cancer and breast cancer.

Adult male rats were fed semipurified diets containing 20% fat, no fiber, and 10% psyllium or cellulose for 22 weeks, and colon tumors were induced in one-half of the rats by the gastric intubation of 1,2-di-methylhydrazine (DMH) during weeks 3–11. Results indicated that psyllium strongly reduced the tumorigenicity of DMH (Roberts-Andersen et al., 1987). Colon tumor was induced in Fischer-344 rats by injecting azoxymethane. Having a diet containing wheat bran or psyllium both significantly reduced the morbidity of colon tumors. However, a combination of wheat bran and psyllium fiber at a ratio of 1:1 showed a greater protective effect than either wheat bran or psyllium fiber alone (Alabaster et al., 1993).

Friedman reported a protective effect of psyllium polysaccharide on isolated colonic epithelial cells primary-cultured from patients at high-risk of colon cancer. They showed that the polysaccharide provided colonocyte protection from deoxycholic acid-induced lysis, and propionic acid, which is a potent colonocyte mitogen, may indirectly protect the colon by providing colonocyte nutrients (Friedman et al., 1988). Sohn et al. (2012) reported a protective effect of partially fermentable *P. ovata* polysaccharide on series colorectal cancer cells. *P. ovata* polysaccharides were anaerobically fermented with colonic bacteria, and the fermentation products were found to induce apoptosis in colorectal cancer cells, including different phenotypes of Caco-2, HCT116, LoVo, HT-29, and SW480. This induction is mediated by both the intrinsic and extrinsic apoptosis pathways, and is the result of modulating important regulators of these pathways.

The short-chain fatty acids, especially butyrate, arising in the large bowel during bacterial fermentation of dietary fiber are considered to play an important role in this antitumor effect in the colon. The short-chain fatty acids are found to promote proliferation of normal epithelial cells, but induce G1 or G2-M arrest in transformed cells, which is coupled to differentiation and apoptosis (Scheppach et al., 1992). Gamet et al. (1992) investigated the effect of butyrate, acetate, and propionate on the growth of HT-29 cell lines. The results showed both propionate and butyrate inhibit the growth of HT29 cells, whereas acetate has no significant effect, and the antiproliferative effect of propionate or butyrate is associated with an inhibition of FCS-induced activation of ornithine decarboxylase, a key enzyme of polyamine metabolism. Whitehead et al. (1986) reported that adding butyrate to the growth medium of a colon carcinoma cell line, LIM1215, could affect the growth rate and the production of alkaline phosphatase, dipeptidyl peptidase IV, and carcinoembryonic antigen. Butyrate administered at 1 mmol/l increased the doubling time, decreased the cloning efficiency, and rapidly increased alkaline phosphatase concentrations more than sixfold. Levels of dipeptidyl peptidase IV and carcinoembryonic antigen were also increased after culture in butyrate-containing medium. Miyauchi found SLC5A8 is a Na^+-coupled transporter for short-chain fatty acids (Miyauchi et al., 2004). Reexpression of SLC5A8 in colon cancer cells leads to apoptosis, but only in the presence of butyrate, and this effect is associated with the inhibition of histone deacetylation. While Hinnebusch et al. (2002) reported that the anti-proliferative, apoptotic, and differentiating properties of various short-chain fatty acids on human colon cancer cells are linked to the degree of induced histone hyperacetylation, short-chain fatty acid-mediated growth arrest in colon carcinoma cells requires the p21 gene. Besides, GPR43, a G-protein-coupled receptor, was also found to be a tumor suppressor by mediating SCFA-induced cell proliferation and apoptotic cell death in colon cancer (Tang et al., 2011). The cancer prevention effect of psyllium may also be due to its effect on enzymes related to cancer promotion. Sphingomyelin metabolism may have implications for the pathogenesis of colonic tumors and ulcerative colitis. A psyllium-containing diet was found to increase the activities of alkaline sphingomyelinase and caspase-3, and decrease the activity of acid sphingomyelinase and neutral ceramidase, whereas a high-fat diet showed opposite effects. The change in caspase-3 activity was positively correlated with the activity of alkaline sphingomyelinase, but negatively with that of acid sphingomyelinase (Cheng et al., 2004). It has also been reported that short-chain fatty acid could inhibit the invasion of colon cancer cells through stimulating tissue inhibitor matrix metalloproteinases and inhibiting the activity of urokinase plasminogen activator (Emenaker and Basson, 1998).

An antitumor effect of psyllium polysaccharide from the seeds of *P. ovata* Frosk against mammary carcinogenesis under a condition of hypercholesterolemia was reported. Female rats were given an intragastric dose of 200 mg/kg body weight of dimethylbenz[*a*]-anthracene to induce a mammary tumor, meanwhile the rats were also given a high cholesterol diet with or without psyllium polysaccharide. In the psyllium-treated group, a significant decrease in the number of palpable mammary masses was observed at weeks 18 and 19 compared with the cholesterol group, and both total serum cholesterol level and LDL cholesterol level were also significantly decreased (Takagi et al., 1999). The synergistic effect between wheat bran and psyllium polysaccharide was also found in another rat mammary cancer model reported by Cohen. A mammary tumor was induced by *N*-methylnitrosourea in F344 rats; as the level of psyllium relative to that of wheat bran increased, the total tumor number and multiplicity of mammary adenocarcinomas in rats decreased at a statistically significant linear trend. Compared with the group given wheat bran alone, the group given the 1:1 (wheat bran:psyllium) combination had maximum protection against mammary tumorigenesis (Cohen et al., 1996). Many studies have also been conducted to reveal the possible cancer preventive effect of psyllium in breast cancer. Estrogen has been found to be closely related to the development and growth of breast cancer, and most free estrogens that enter the circulation are products of bacterial hydrolysis of estrogen conjugates in the colon. The mechanism by which dietary fiber inhibits breast cancer includes reduction of circulating tumor-promoting estrogens via suppression of bacterial β-D-glucuronidase activity, and direct binding of estrogens in the colon (Cohen et al., 1991). Cohen et al. (1996) found that fecal estrogen excretion was significantly decreased in a mouse group treated with a wheat−psyllium diet than an only wheat group, in a dose-dependent manner; further study confirmed that psyllium is a very effective suppressor of β-D-glucuronidase-producing cecal bacteria.

Besides its direct or indirect antitumor effects, psyllium polysaccharide also exhibits a protective effect in cancer therapy. Cyclophosphamide is widely used in cancer chemotherapy, in spite of its toxic effect. Mice were pretreated with psyllium polysaccharide obtained from *Plantago albicans* L. areal parts for 7 days, and were then treated with cyclophosphamide. Groups of mice pretreated with polysaccharide showed a statistically significant reduction in the percentage of chromosomal aberrations and micronuclei in bone marrow cells in a dose-dependent manner. It demonstrated an antimutagenic effect of psyllium against genotoxicity of the anticancer drug cyclophosphamide (El Souda et al., 2014).

8.4.7 Immune Enhancement Effect

An immune enhancement effect is also an important activity of psyllium polysaccharide. Kim et al. (1996) reported that treatment with polysaccharide from the seeds of *P. asiatica* at a dose of 63 mg/kg on ICR mice which were immunized and challenged with sheep red blood cells (SRBC) could significantly increase the body weight gain, and the relative weights of spleen and thymus. Not only plaque forming cells and hemagglutination titers to SRBC, but also delayed-type hypersensitivity reaction to SRBC, phagocyte activity, and circulating leukocyte were significantly increased in mice. These results demonstrated that the psyllium polysaccharide markedly enhanced both humoral immune and allergic reaction to SRBC. Türel et al. (2009) reported that the psyllium polysaccharide exhibited an antiinflammatory effect in a rat inflammation model induced by Lambda-carrageenan, as well as a protective effect in a CCl_4-induced hepatic injury rats model. In an in vitro study, polysaccharides from seeds of *P. depressa* were found to stimulate splenocyte proliferation, secretion of NO, and TNF-α of macrophages, with the PDSP-3 component presenting the highest effect (Zhao et al., 2014). Biringanine et al. (2005) reported that polysaccharide isolated from *Plantago palmate* leaves could induce NO and TNF-α production in interferon-γ-activated J774 macrophages, in particular, the fraction separated at 50°C showed the highest activity.

The complement system consists of over 20 serum proteins, including 9 complement components and their regulators, and is normally present in blood serum in an inactive form. The complement proteins can be activated through three cascade pathways including the classical pathway, alternative pathway, and antibody-independent lectin pathway (Yamada and Kiyohara, 1999). Yamada et al. (1985) reported that polysaccharide from the seeds of *P. asiatica* could activate the complement system through both the classical pathway and the alternative pathway. The PM II isolated from the leaves of *P. major* was also reported to activate the complement system via both classical and alternative pathways, since ETGA could inhibit C-3 activation, but not block PM II at the level of terminal complement activation (Michaelsen et al., 2000).

In our research group, we found that polysaccharide from seeds of *P. asiatica* could stimulate NO secretion of RAW 264.7 macrophage (Chen et al., 2009). Dendritic cells (DCs) are the most potent professional antigen presenting cells with distinct abilities to stimulate naïve T lymphocytes and to initiate primary immune responses (Steinman, 1991). DCs in the periphery capture and process antigens, express lymphocyte costimulatory molecules, migrate to lymphoid organs, and secrete cytokines to initiate immune responses. They not only activate lymphocytes, they also tolerize T cells to antigens that are innate to the body (self-antigens), thereby minimizing autoimmune reactions (Banchereau and Steinman, 1998). Although they exist in almost all tissues, most dendritic cells are found to stay in an immature state, and the development of maturation is considered as a key point to initiate the immune response. We cultured bone-marrow-derived dendritic cells by using a combination of granulocyte-macrophage colony-stimulating factor (GM-CSF) and interleukin-4 (IL-4). The polysaccharide increased the expression of maturation markers of major histocompatibility complex II, CD86, CD80, and CD40 on DCs. Consistent with the changes in the phenotypic markers, a functional assay for DC maturation showed that the polysaccharide decreased DCs endocytosis, but increased its section of IL-12 cytokine and allostimulatory activity. Furthermore, using a syngeneic T cell activation model, we found that polysaccharide-treated DCs presented ovalbumin antigen to T cells more efficiently, as demonstrated by increased T cell proliferation. Besides, the polysaccharide also has a promotional effect on dendritic immigration, since the mRNA of the chemokine receptor CCR7 expression in DCs was enhanced by the polysaccharide treatment. All of these data indicate that polysaccharide from the seeds of *P. asiatica* L. is an effective immune adjuvant as a stimulator of dendritic cell maturation (Huang et al., 2009a,b). In addition, we also found the effects of the polysaccharide on DCs were significantly impaired by treating the cells with antitoll-like receptor 4 (TLR4) antibody prior to polysaccharide treatment, implying that the polysaccharide may induce DC maturation through TLR4 (Huang et al., 2014).

8.4.8 Other Activities

Psyllium polysaccharides were also reported have beneficial effects on wound cleansing, healing, and limiting scar formation. Westerhof et al. (2000) found that psyllium polysaccharide has a strong adherence to wound bacteria which started after 2 h, and was more pronounced after 3 h.

Inbred NIH/OlaHsd and Fox Chase SCID mice were experimentally infected with *Streptococcus pneumoniae* serotype 6B, and treated with polysaccharide from the leaves of *P. major*, either once 3 days before challenge, or once to three times from 3 to 48 h after challenge. The results showed prechallenge administration of the polysaccharide gave a dose-dependent protective effect against *S. pneumoniae* type 6B infection, but it had no curative effect against *S. pneumoniae* 6B infection when the polysaccharide was given postchallenge (Hetland et al., 2000). The polysaccharide also exhibited an antiinfection effect against intracellular pathogen *Mycobacteria* spp. in vivo (Hetland, 2003). The mechanism for the protective effect is more likely dependent on stimulating the immune system.

Psyllium polysaccharides are found to possess antioxidant activity. In our group, the antioxidant activity of polysaccharide from the seeds of *P. asiatica* L. was investigated. The scavenging effect of the polysaccharide for DPPH radicals was 71.2% at a concentration of 1 mg/mL, the scavenging effect of the polysaccharide for superoxide radicals generated by 1,2,3-phentriol autoxidant was 23.0% at a concentration of 2 mg/mL, and the scavenging effect of the polysaccharide for hydroxyl radicals was 63.2% at a concentration of 4 mg/mL. The polysaccharide also exhibited a high inhibitory activity on lipid peroxidation, which was found to be 58.3% at 2 mg/mL. Interestingly, such an antioxidant effect could be increased by treatment with high pressure homogenization (Hu et al., 2013d). Similarly,

Ye et al. (2011) reported that the scavenging rates of the polysaccharides (0.75 mg/mL) on superoxide and 1,1-diphenyl-2-picrylhydrazyl radicals were 79.7% and 81.4%, respectively. It also exhibited a moderate concentration-dependent ABTS radical scavenging activity, ferrous ion chelating potency, and H_2O_2 scavenging activity. Gong et al. (2015) also confirmed the antioxidant activities of the polysaccharide.

In addition, an intestinal antiinflammatory activity of psyllium was observed in HLA-B27 transgenic rats. This effect was associated with an increased production of short-chain fatty acids, which can act synergistically in inhibiting the production of proinflammatory mediators such as NO, leukotriene B_4, and TNF-α (Rodríguez-Cabezas et al., 2003).

8.4.9 Possible Adverse Effects: Anaphylaxis, Rhinitis, and Asthma

Although psyllium use has increased significantly due to its pharmacological and food properties, there is also a potential allergy from eating psyllium-containing cereals (James et al., 1991; Lantner et al., 1990). Allergic symptoms may be severe, and consist of any of the following: rhinoconjunctivitis, skin reactions, asthma, gastrointestinal symptoms, and anaphylaxis (Gillespie and Rathbun, 1992).

Because the occupational allergy has been mainly described in pharmaceutical and health care workers, a number of epidemiologic and cross-sectional studies have been conducted to determine the prevalence rate of sensitization and occupational allergy (Khalili et al., 2003). An epidemiologic investigation conducted by Nelson Allergic showed that 136 (18%) of 743 surveyed health care workers who prepared psyllium laxative for patients reported allergic events, while 34 (5%) of the 743 surveyed workers reported shortness of breath, wheezing, or hives within 30 min after preparing psyllium laxatives. Bardy et al. (1987) assessed the prevalence of occupational asthma (OA) and IgE sensitization to psyllium in a pharmaceutical company producing psyllium hydrophilic mucilloid. The data show that the prevalence of OA and IgE sensitization to psyllium were at least 3.7% and 27.9%, respectively. It also noted that psyllium polysaccharides are associated with a spectrum of allergic events, ranging from mild to disabling (Nelson, 1987).

Although individuals sensitized by occupational exposure to psyllium dust are at high risk for allergic reactions to ingested psyllium-containing products, most cases of anaphylaxis usually occur after ingestion. For instance, Drake et al. (1991) reported the case of a 38-year-old female nurse who ingested a bowl of psyllium-containing Heartwise Cereal, and 25 min later developed severe systemic anaphylaxis manifested by hypotension, a feeling of constriction in the throat, hoarseness, dyspnea, wheezing, generalized pruritus, urticaria, and vomiting.

Rhinitis and rhinitis from psyllium hypersensitivity have also been reported. Vaswani et al. (1996) reported a case of a 69-year-old who was diagnosed with recurrent rhinitis and asthma related to psyllium exposure. Schwartz et al. (1989) reported a specific rhinitic reaction on a nurse that developed as a result of occupational exposure to a psyllium-based powdered laxative. Cartier et al. (1987) also reported psyllium-induced asthma in a nurse who had a history of asthmatic symptoms after being exposed to psyllium. A prevalence analysis showed that the prevalence of OA to psyllium among health personnel in chronic care hospitals was 4%, indicating a significant risk to respiratory health (Malo et al., 1990).

8.5 DISCUSSION: STRUCTURE–BIOACTIVITY RELATIONSHIP

8.5.1 Viscosity

Viscous dietary fibers induce thickening when mixed with liquids. An increase in the viscosity of food may have several effects, including slowing down the small intestinal transit time, reducing the interaction between food nutrients and digestive enzymes, and increasing the time for nutrient diffusion, by increasing the thickness of the unstirred water layer (Rigaud et al., 1998).

Dietary fibers with high viscosity have been shown to delay the gastric emptying of liquids and solids, and these effects were found to be closely related to the increase in meal viscosity. Russell and Bass (1985) reported that low-viscosity fiber meals emptied from the stomach much more rapidly compared with the high-viscosity meals in dogs. Vuksan et al. (2011) found that the well-documented lipid-lowering effects of dietary fiber may also be related to its viscosity. The lipid-lowering effect of low-viscosity wheat bran, medium-viscosity psyllium, and a high-viscosity viscous fiber blend (VFB) were compared, and the results showed that the reduction in LDL cholesterol level was greater with the VFB compared with the medium-viscosity psyllium and low-viscosity wheat bran. The magnitude of LDL cholesterol reduction showed a positive association with the fiber's apparent viscosity. Despite the smaller quantity consumed, the high-viscosity fiber lowered LDL cholesterol levels to a greater extent than lower-viscosity fibers. Psyllium polysaccharide showed an inhibitory effect on mineral absorption in a rat model; however, this reduced absorption was increased with a lower viscous treatment of partial hydrolysis on psyllium (Asvarujanon et al., 2004).

A mechanism for viscous dietary fibers in appetite regulation has been reported (Kristensen and Jensen, 2011). Viscous dietary fibers can absorb large quantities of water, and increase stomach distension, which may trigger afferent vagal signals of fullness. Besides, more time and effort are required to masticate foods rich in dietary fibers, and this allows more signals mediating satiety sensations to the brain. While gastric satiety originates mechanically, intestinal satiety is typically nutrient-dependent. Satiety signals are released upon close interaction between the intestinal wall and nutrients. Exposure of the intestinal mucosa to nutrients induces release of appetite regulating peptides, which either function as hormones or activate neural pathways. Viscous dietary fibers increase the viscosity of the digesta in the small intestine, and therefore prolong small intestine transit time and absorption rate of nutrients, which finally influence the release of peptide. These, in turn, affect gastric emptying and signaling to the central nervous system.

8.5.2 Structural Modifications

So far, some structural modifications aiming to improve the gel property or bioactivity have been applied to psyllium polysaccharide.

Treating psyllium husks with different concentrations of HCl in ethanol at ambient temperature can alter their gelling and water uptake properties. The results showed that 0.36%−1.44% HCl treatment in ethanol may significantly decrease the gelling, water uptake, and swelling capacities of psyllium, without reducing their bile acid-binding abilities (Cheng et al., 2009).

The reticuloendothelial system-potentiating and alkaline phosphatase-inducing activity of psyllium polysaccharide, and its derivatives obtained from deacetylation, periodate oxidation, controlled Smith degradation, and partial acid hydrolysis were also investigated (Tomoda et al., 1991). The study found that only deacetylation markedly increased both of the activities. Structural analysis showed the deacetylated polysaccharide lost all *O*-acetyl groups, all xylose branches, and many hexuronosyl arabinose side chains.

Sulfation of psyllium polysaccharide was carried out with sulfur trioxide−pyridine in dimethyl formamide. Sulfation reduced the gelling capacity and viscosity of psyllium polysaccharide, but significantly increased its bile acid-binding capacity in vitro. The bile acid-binding capacity of sulfated psyllium polysaccharide with a substitution degree of 0.58 was about 8.4-fold of that observed for the original psyllium (Liu et al., 2009).

Succinyl psyllium derivatives (S-Psys) were successfully prepared by reacting psyllium with succinic anhydride, 4-dimethylaminopyridine, and tributylamine in *N*,*N*-dimethylformamide (Niu et al., 2012). It was reported that psyllium had filamentous structures, which contained regular strands forming a network-like folding structure on the surface. Compared to the surface characteristics of psyllium, the surfaces of succinyl derivatives were still filamentous, but had been significantly eroded and also were found to be less fibrillary. S-Psys with three different succinyl DSs were studied. It was shown that S-Psy-L, with a low succinyl DS of 1.9, showed a reduction in gel hardness and gel adhesiveness as compared to those of the original material. Meanwhile, S-Psy-M and S-Psy-H, with DS values of 3.3 and 3.9, respectively, formed weaker gels with a dramatic reduction in gel hardness, while having a reduction in the forces to hold the compressing probe, respectively. S-Psy-L, S-Psy-M, and S-Psy-H could absorb water at an average rate of 2.36, 1.99, and 1.81 mg/g/min under the experimental conditions in the first 30 min, respectively, which was 1.96-fold, 1.65-fold, and 1.51-fold of that for psyllium, but the swelling volumes of S-Psys were 67.25%, 45.52%, and4 0.42% of that of the original psyllium, respectively. The bile acid-binding capacities of succinyl psyllium were higher than that of psyllium against both cholic and chenodeoxycholic acid. S-Psy-M had the greatest effect, it had a 2.80-fold greater capacity of binding CA, and a 2.75-fold stronger ability of binding CDCA compared to that of the original psyllium.

Hydroxypropylated psyllium derivatives were prepared with propylene oxide under alkaline conditions. It was found that hydroxypropylation significantly reduced the gelling properties of psyllium. Psyllium derivatives with a relatively lower hydroxypropyl substitution degree (0.15–0.26) had greater in vitro binding capacities against cholic and chenodeoxycholic acids, and higher swelling ability (Liu et al., 2010).

After sulfation, hydroxypropylation, and succinylation, the gelling, water uptake, swelling, and bile acid-binding capacities of psyllium were compared at the same molar substitution degree. Sulfated, hydroxypropylated, and succinylated psyllium were prepared with substitution levels of 1.02, 0.88, and 0.79, respectively. The results showed that sulfated polysaccharide had the greatest bile acid-binding capacities among the three derivatives, which was 6.00-fold greater in binding cholic acid and 6.23-fold stronger in binding chenodeoxycholic acid compared to those of the original psyllium polysaccharide. The bile acid-binding capacities of SFP were comparable to that of cholestyramine resin, which is commercially used as an adjunctive therapy for reducing plasma cholesterol level. The order of bile acid-binding capacity was sulfated psyllium polysaccharide > succinylated psyllium polysaccharide > hydroxypropylated psyllium polysaccharide > psyllium polysaccharide (Niu et al., 2013).

In our group, we found carboxymethylation modification could enhance the maturation-inducing activity in dendritic cells of polysaccharide from the seeds of *P. asiatica* L. (Jiang et al., 2014). Compared with the original polysaccharide, carboxymethylated polysaccharide with DS 0.50, 0.55, and 0.62 was found to be more effective in stimulating the maturation of dendritic cells.

Besides the chemical modification, enzyme treatments were also used to improve the bioactivity of psyllium polysaccharide. The hypolipidemic effects of two solid-state enzymatically modified psyllium derivatives were investigated in hamster. Although it did not influence the hypolipidemic effect of psyllium (Allen et al., 2004), enzyme treatment could reduce the water-absorbing ability, gel hardness, and gel adhesiveness of psyllium gum to make it more feasible for incorporation into food product (Yu et al., 2003)

8.6 APPLICATIONS

8.6.1 Food Industry and Dietary Fiber Supplements

Generally, there are two purposes in adding polysaccharides to food products. One is as modifiers of food texture, and the other is as dietary fibers with potential health benefits. Some products of fiber-enriched food have been developed or designed, such as ready-to-eat cereals containing extruded prewetted psyllium (Kincaid and Talbot, 1995).

Trautwein et al. (2000) designed a psyllium-enriched bread product, in which 15% of the flour was replaced by psyllium, so that 3–4 slices of the bread contained 10–15 g psyllium. The psyllium-enriched breads were evaluated as good tasting, and as palatable as conventional bread in sensory tests. Levels of psyllium husk on the basis of bread making and organoleptic properties of bread were evaluated by Kamaljit et al. (2011). They showed that baking absorption increases with level of incorporation of psyllium husk, and the increase in baking absorption with the addition of psyllium husk could be attributed to its high hydration capacity. Up to 3.0% incorporation of psyllium husk, dough handling was smooth. Loaf volume, loaf height, and specific volume decreased with an increase in the level of incorporation of psyllium husk. Further, analysis of organoleptic quality showed breads prepared with the incorporation of 3% psyllium husk had good overall acceptability. Besides, the shelf life of the psyllium bread was found to be 5 days at room temperature conditions. In the team of Raymundo, the maximum level of psyllium incorporation in a biscuit was found to be 9% (w/w), and dough with higher fiber incorporation failed to mold (Raymundo et al., 2014).

Celiac disease is an antibody-mediated enteropathy that presents permanent intolerance to ingestion of gluten. The only treatment available is to have a gluten-free diet. However, commercially prepared gluten-free products are expensive and difficult to find. To meet this demand, psyllium can be used as a substitute for gluten in bread. Sensory analysis of psyllium dough indicated that the gluten-free bread had good acceptance of more than 93% by individuals with and without celiac disease. Although its texture was rejected by 3.3% of individuals, its taste had 100% acceptance by the celiac disease subjects. As for product color, acceptance was observed by up to 96% of individuals (Zandonadi et al., 2009). The influence of psyllium on gluten-free dough properties and bread quality was determined by Cappa et al. (2013). Data shows that psyllium can improve the workability of doughs, especially playing a central role on gluten-free bread development both during proving and baking. The psyllium ingredient also provides an effective antistaleing effect, presumably due to its higher water-binding capacity. However, psyllium supplementation should be carefully modulated, since it can produce an excessive increase in bread hardness if a proper amount of water is not added to the recipe. A similar result was reported by

Mariotti et al. (2009), that 2% of psyllium enhanced the farinographic water absorption percentage, so that resulted in a higher dough moisture, as well as good structure and workability of the dough, but this also occurred in a corn starch-based gluten-free bread. Ultrastructural analysis of psyllium dough showed that psyllium fiber not only formed many particles with a highly organized network and elevated specific surface, which were presumed to enable high water absorption through capillarity, but also could provide more stability to the dough matrix. Psyllium inclusion reduced pasting temperature, compliance values, and bread specific volume, and increased G′ and G″ moduli vales and bread hardness, however, when dough hydration level was increased, there was a decrease in the influence of hydrocolloid on dough rheology and specific volume and hardness of breads (Mancebo et al., 2015).

Psyllium gum is also an ideal fat alternative that could be used to improve the production of an industrial low-fat yogurt gel (Ladjevardi et al., 2015). The combined effects of critical structural components on the textural, rheological, and chemical attributes of developed set yogurts were studied. The results showed that 0.12% PHG and 0.63% fat was the optimal formulation for achieving optimal pH (4.39) and titratable acidity (81% lactic acid), maximizing firmness (0.172 N) and viscosity (6.40 Pa s), and minimizing whey separation (36.21 mL/100 g). The yogurts manufactured at optimal points had more aroma, texture, and overall acceptability than the control yogurts by sensory characterization.

Aghdaei et al. (2014) reported that psyllium mucilage could be used as a suitable fat replacer in mayonnaise formulation. It was found that a mayonnaise sample, in which fat was partially substituted by mucilage gels at a level of 30%, showed similar textural characteristic as the control one. Sensory evaluation demonstrated that the modified mayonnaise could be more acceptable.

In addition, a postharvest application of psyllium mucilage as an edible coating on postharvest life and quality was reported. An immersion treatment of 12.5 mL/L of psyllium mucilage on strawberries showed a significant effect on retention of water (Yarahmadi et al., 2014). To obtain an antifungal edible coating for mandarin, a methanolic extract of garlic was incorporated into psyllium mucilage, and an effective antifungal effect was observed when garlic extracts were incorporated at a rate of 6%–8% (ur Rehman et al., 2015).

8.6.2 Drug Delivery Systems

Recently, there has been a marked drive in research activities related to the use of natural polysaccharide for the development of hydrogels as drug delivery devices.

Excipients are a set of pharmaceutical dosage forms, and comprise the greatest proportion in dosage units. Psyllium polysaccharide was used to prepare granules containing paracetamol as a model drug. The granules were found to have good flowability and compressibility, as well as sufficient mechanical strength. Especially, psyllium polysaccharide was found to have similar satisfactory binder properties compared to a conventional binding agent, hydroxyl propyl methyl cellulose (HPMC) (Rao et al., 2013).

Sustained release gastroretentive drug delivery systems enable prolonged, continuous release of drug to the upper parts of the gastrointestinal tract, and ensure optimal bioavailability. They can help in optimizing the oral controlled delivery of drugs having an "absorption window" continually releasing the drug prior to the absorption window for a prolonged period of time, thus ensuring optimal bioavailability A daily SR floating matrix tablet for theophylline using psyllium husk as a release-controlling polymer was designed (Yasir et al., 2010). Comparative analysis showed that the floating duration of the formulation containing psyllium husk alone was less than that containing a similar concentration of HPMC K100M. The floating duration of the formulation containing psyllium husk in combination with HPMC K15M was comparable with the formulation containing a similar amount of HPMC K100M. The release rate of the formulation containing psyllium husk in combination with HPMC K15M is higher than the formulation containing a similar amount of HPMC K100M. Dimensional stability increased with psyllium husk concentration, while floating lag time decreased. It was also found that the in vitro release data and drug release mechanism of the optimized formulation followed Higuchi kinetics and nonfickain type, respectively.

Siahi-Shadbad et al. (2011) investigated the release behavior of propranolol hydrochloride from psyllium matrices in the presence of different hydrophilic polymers such as HPMC K4M, sodium alginate (Na alginate), and sodium carboxy methyl cellulose. The HPMC and Na alginate were able to remarkably retard drug release from the psyllium matrices. The results also showed that the ternary mixture of the polymers produced different drug release profile manipulation and combination of polymers in different ratios and viscosity grades.

Another sustained release gastroretentive drug delivery system for ofloxacin was designed by Chavanpatil et al. (2005). It was also found that the dimensional stability of the formulation increases with increasing psyllium husk concentration. The optimized formulation in 24 healthy human volunteers was investigated, and the pharmacokinetic parameters of developed formulations were compared with the marketed once daily (Zanocin) formulation. Based on a parallel in vivo study design in healthy subjects, the percentage of relative bioavailability of the developed formulation was found to be 97.55%, which shows promise of being bioequivalent to the marketed product (Zanocin).

Thiolated psyllium was prepared by esterification with thioglycolic acid. Thiolation was observed to change the surface morphology of psyllium from fibrous to granular, and resulted in a slight increase in the crystallinity and swelling, but it significantly enhanced its mucoadhesive strength, and prolonged the release of drug metronidazole (Bhatia and Ahuja, 2013).

Phosphorylated psyllium polysaccharide has a high swelling index, and the percentage of increase in swelling of the phosphorylated polysaccharide as compared to the original polysaccharide was found to be 90.26%. The phosphorylated polysaccharide has a mucoadhesive property which was comparable with carbopol (Rao et al., 2016). Further study showed that tablets containing phosphorylated psyllium polysaccharide had swelling and dissolution profiles comparable to that of tablets containing hydroxypropyl methylcellulose. Phosphorylated psyllium polysaccharide enhances matrix rigidity, thereby sustaining the drug release rate when compared to psyllium polysaccharide (Rao et al., 2015).

Some modified psyllium polysaccharide hydrogels were also designed as potential drug delivery systems. It has been reported that an optimum condition for the synthesis of psyllium-poly(vinyl alcohol)(PVA)-poly(acrylic acid) blended hydrogels, was found to be 1% (v/v) acrylic acid, 2% (v/v) PVA, and 1 g psyllium. The release of the model antibiotic drug tetracycline HCl from this psyllium-poly(PVA)-poly(acrylic acid) hydrogel was observed to be more in a pH 2.2 buffer, indicating an effective drug delivery agent in treatment for peptic ulcer caused by *Helicobacter pylori* (Singh and Sharma, 2010). Besides, other psyllium-based hydrogels such as psyllium-*N*-NVP hydrogels (Singh and Kumar, 2008), Psy-*cl*-poly(AAm) hydrogels (Singh et al., 2007), Psy-*g*-PMA (Kumar and Sharma, 2013), and psy-*cl*-poly (HEMA-*co*-AAm) (Singh et al., 2008) are also designed as effective drug delivery systems.

8.6.3 Wastewater Treatment

Flocculation is an essential phenomenon in industrial wastewater treatment. Inorganic coagulants (salts of multivalent metals) are always used, due to their low cost and ease of use. However, their application is constrained by low flocculating efficiency and the presence of residue metal concentrations in the treated water. Mishra et al. (2002a) reported that psyllium polysaccharide is a very effective flocculant capable of removing suspended solid from sewage and tannery wastewater samples. A very low flocculant concentration (1.2 mg/L) was capable of removing an appreciable amount of suspended solid, and the most suitable pH was found to be neutral. Polyacrylamide-grafted psyllium (Sen et al., 2012) and polymethylmethacrylate-grafted psyllium (Mishra et al., 2014) were synthesized using a microwave as initiator, and both of them exhibited a high flocculation efficacy in kaolin suspension.

Landfill leachate is a heavily polluted and a likely hazardous liquid that is produced as a result of water percolation through solid wastes. As primary coagulant, psyllium husk was not effective because of its low surface charge and there was no effect on pH. However, psyllium husk could be used as a coagulant aid for the treatment of landfill leachate (Al-Hamadani et al., 2011). A graft copolymer of psyllium mucilage and polyacrylamide (Psy-*g*-PAM) was synthesized in the presence of nitrogen using a ceric ion−nitric acid redox initiator. Its flocculation efficiency in tannery and domestic wastewater were tested. The optimal dose was found to be 60 ppm, at which maximum solid removal from both the effluents took place. The maximum solid removal was seen after 1 h at an acidic pH in the case of the domestic effluent, and at an alkaline pH in the case of tannery effluent. Psy-*g*-PAM has been proved to be a better flocculant than pure psyllium mucilage (Agarwal et al., 2002). Psy-*g*-PAM was also reported to be an effective flocculant for treatment of textile wastewater, 1.6 mg l−1 was the optimum dose, at which maximum solids removal was seen. The most suitable pH for total dissolved solids and color removal was neutral (7.0), and for suspended solid removal an alkaline pH (9.2) was found to be most suitable. The optimum treatment duration for solid waste removal was 5 h (Mishra et al., 2004). Another study showed that psyllium polysaccharide could also reduce the dye concentration in textile wastewater by flocculation and settling, and it is more effective for removal of solubilized vat dye than for reactive black (Mishra et al., 2002b; Mishra and Bajpai, 2005). Besides, psyllium-grafted-polyacrylonitrile also was reported to produce a good flocculation effect on textile effluent (Mishra et al., 2002b).

8.6.4 Other Applications

In the landscape industry, a psyllium-based binding product was innovatively used to replace chemical binders for pavement in environmentally sensitive areas. By using response surface methodology (RSM), Guo et al. (2010) obtained an optimum formula for a psyllium-based binding product. It consists of 1.82% (w/w) psyllium, 1.05% (w/w) lime, and 0.55% (w/w) soda ash.

A wound dressing film using psyllium polysaccharide complexed with povidone iodine (SHP) was fabricated (Patil et al., 2011). The films showed good elasticity (80.27%−116.24%) and acceptable tensile strengths (8.33−22.13 N/mm^2). The films were able to swell in water up to 167%−191% w/w, and water vapor transmission rates were in the range between 8.23×10^{-3} and 1.12×10^{-4} g · cm^2/day. Selected formulation of SHP films with optimum characteristics were loaded with Povidone iodine (PI) solution by a soaking method, and were evaluated for antimicrobial and wound healing activity on an excision wound model. Results indicated that PI-loaded films showed significant antibacterial activity against both Gram-positive and Gram-negative selected bacteria. PI-loaded SHP films showed comparatively better wound healing properties compared to a control group, with faster epithelialization and greater rates of wound contraction.

Gharibzahedi et al. (2013) reported that psyllium gum is an attractive carbohydrate biopolymer for the production of stable canthaxanthin (CX) emulsions. The physical stability of the ultrasonically prepared emulsions containing CX produced by Dietzia natronolimnaea HS-1 strain was maximized using a face central composite design of RSM. The linear and interactive effects of whey protein isolate (WPI), psyllium husk gum, and coconut oil on stability were studied. Maximum stability of 98.8% was predicted at the optimum levels of formulation consisting of a WPI concentration of 1.20 wt%, PHG content of 3.30 wt%, and CO concentration of 5.43 wt%. The results also demonstrated that coconut oil and WPI concentration had a greater effect on the droplet size and density values, whereas the PHG:WPI ratio of psyllium husk gum to WPI had a rather greater effect on the turbidity values.

The application of psyllium-based hydrogels in the selective absorption of water from different oil/water emulsions was investigated. Mucilage from *P. ovata* has been modified through graft copolymerization and network formation using acrylic acid (AA) as the monomer, potassium persulfate as an initiator, and hexamethylenetetramine as a cross-linker to obtain a cross-linked Psy-*g*-poly(AA) hydrogel. It has been found that the functionalized polymer can absorb water up to 6710%, and it is effective in selective removal of water from petrol/water, diesel/water, kerosene/water, and petroleum ether/water emulsions (Kaith and Kumar, 2007a). A series of synthesis methods to obtain psyllium-based hydrogels were also reported (Kaith and Kumar, 2008; Kaith and Kumar, 2007a,b; Kumar et al., 2012).

REFERENCES

Agah, S., Fatali, S., Ashayeri, N., 2010. Treatment of diarrhea predominant Irritable Bowel Syndrome (IBS-D) using plantagel: a random double-blind clinical trial. Razi J. Med. Sci. 16, 17–23.

Agarwal, M., Srinivasan, R., Mishra, A., 2002. Synthesis of Plantago psyllium mucilage grafted polyacrylamide and its flocculation efficiency in tannery and domestic wastewater. J. Polym. Res. 9, 69–73.

Aghdaei, S.A., Aalami, M., Geefan, S.B., Ranjbar, A., 2014. Application of Isfarzeh seed (*Plantago ovate* L.) mucilage as a fat mimetic in mayonnaise. J. Food Sci. Technol. 51, 2748–2754.

Ahmed, I., Naeem, M., Shakoor, A., Ahmed, Z., Iqbal, H.M.N., 2010. Investigation of anti-diabetic and hypocholesterolemic potential of psyllium husk fiber (*Plantago psyllium*) in diabetic and hypercholesterolemic albino rats. Int. J. Bio. Life Sci. 6, 185–189.

Alabaster, O., Tang, Z., Frost, A., Shivapurkar, N., 1993. Potential synergism between wheat bran and psyllium: enhanced inhibition of colon cancer. Cancer Lett. 75, 53–58.

Al-Hamadani, Y.A., Yusoff, M.S., Umar, M., Bashir, M.J., Adlan, M.N., 2011. Application of psyllium husk as coagulant and coagulant aid in semi-aerobic landfill leachate treatment. J. Hazard. Mater. 190, 582–587.

Allen, K.G., Bristow, S.J., Yu, L., 2004. Hypolipidemic effects of modified psyllium preparations. J. Agric. Food Chem. 52, 4998–5003.

Anderson, J.W., Zettwoch, N., Feldman, T., Tietyen-Clark, J., Oeltgen, P., Bishop, C.W., 1988. Cholesterol-lowering effects of psyllium hydrophilic mucilloid for hypercholesterolemic men. Arch. Intern. Med. 148, 292–296.

Anderson, J.W., Davidson, M.H., Blonde, L., Brown, W.V., Howard, W.J., Ginsberg, H., et al., 2000. Long-term cholesterol-lowering effects of psyllium as an adjunct to diet therapy in the treatment of hypercholesterolemia. Am. J. Clin. Nutr. 71, 1433–1438.

Ashraf, W., Park, F., Lof, J., Quigley, E., 1995. Effects of psyllium therapy on stool characteristics, colon transit and anorectal function in chronic idiopathic constipation. Aliment. Pharmacol. Ther. 9, 639–647.

Asvarujanon, P., Ishizuka, S., Hara, H., 2004. Inhibitory effects of psyllium on rat mineral absorption were abolished by reduction of viscosity with partial hydrolysis. Biosci. Biotechnol. Biochem. 68, 1737–1742.

Attaluri, A., Donahoe, R., Valestin, J., Brown, K., Rao, S.S.C., 2011. Randomised clinical trial: dried plums (prunes) vs. psyllium for constipation. Aliment. Pharmacol. Ther. 33, 822–828.

Banchereau, J., Steinman, R.M., 1998. Dendritic cells and the control of immunity. Nature 392, 245.

Bardy, J.-D., Malo, J.-L., Séguin, P., Ghezzo, H., Desjardins, J., Dolovich, J., et al., 1987. Occupational asthma and IgE sensitization in a pharmaceutical company processing psyllium 1, 2. Am. Rev. Respir. Dis. 135, 1033–1038.

Belknap, D., Davidson, L.J., Smith, C.R., 1997. The effects of psyllium hydrophilic mucilloid on diarrhea in enterally fed patients. Heart Lung J. Acute Crit. Care 26, 229–237.

Bergmann, J., Chassany, O., Petit, A., Triki, R., Caulin, C., Segrestaa, J., 1992. Correlation between echographic gastric emptying and appetite: influence of psyllium. Gut 33, 1042–1043.

Bhatia, M., Ahuja, M., 2013. Thiol modification of psyllium husk mucilage and evaluation of its mucoadhesive applications. Sci. World J. 2013, Article ID 284182.

Bijkerk, C., Muris, J., Knottnerus, J., Hoes, A., De Wit, N., 2004. Systematic review: the role of different types of fibre in the treatment of irritable bowel syndrome. Aliment. Pharmacol. Ther. 19, 245–251.

Bijkerk, C., De Wit, N., Muris, J., Whorwell, P., Knottnerus, J., Hoes, A., 2009. Soluble or insoluble fibre in irritable bowel syndrome in primary care? Randomised placebo controlled trial. BMJ 339, b3154.

Biringanine, G., Vray, B., Vercruysse, V., Vanhaelen-Fastré, R., Vanhaelen, M., Duez, P., 2005. Polysaccharides extracted from the leaves of *Plantago palmata* Hook. f. induce nitric oxide and tumor necrosis factor-α production by interferon-γ-activated macrophages. Nitric Oxide 12, 1–8.

Biringanine, G., Ouedraogo, M., Vray, B., Samuelsen, A.B., Duez, P., 2012. Partial chemical characterization of immunomodulatory polysaccharides from *Plantago palmata* Hook. fs leaves. Int. J. Carbohyd. Chem. 2012, Article ID 458456, 7 pages.

Bräutigam, M., Franz, G., 1985. Structural features of *Plantago lanceolata* mucilage. Planta Med. 51, 293–297.

Buhman, K.K., Furumoto, E.J., Donkin, S.S., Story, J.A., 1998. Dietary psyllium increases fecal bile acid excretion, total steroid excretion and bile acid biosynthesis in rats. J. Nutr. 128, 1199–1203.

Campbell, J.M., Fahey, G.C., 1997. Psyllium and methylcellulose fermentation properties in relation to insoluble and soluble fiber standards. Nutr. Res. 17, 619–629.

Cappa, C., Lucisano, M., Mariotti, M., 2013. Influence of Psyllium, sugar beet fibre and water on gluten-free dough properties and bread quality. Carbohydr. Polym. 98, 1657–1666.

Cartier, A., Malo, J.L., Dolovich, J., 1987. Occupational asthma in nurses handling psyllium. Clin. Exp. Allergy 17, 1–6.

Carvajal-Millan, E., Landillon, V., Morel, M.-H., Rouau, X., Doublier, J.-L., Micard, V., 2005. Arabinoxylan gels: impact of the feruloylation degree on their structure and properties. Biomacromolecules 6, 309–317.

Chan, J., Wypyszyk, V., 1988. A forgotten natural dietary fiber: psyllium mucilloid. Cereal Food. World (USA) 33 (11), 919–922.

Chan, M.-Y., Heng, C.-K., 2008. Sequential effects of a high-fiber diet with psyllium husks on the expression levels of hepatic genes and plasma lipids. Nutrition 24, 57–66.

Chang, J.Y., Talley, N.J., 2010. Current and emerging therapies in irritable bowel syndrome: from pathophysiology to treatment. Trends Pharmacol. Sci. 31, 326–334.

Chavanpatil, M., Jain, P., Chaudhari, S., Shear, R., Vavia, P., 2005. Development of sustained release gastroretentive drug delivery system for ofloxacin: in vitro and in vivo evaluation. Int. J. Pharm. 304, 178–184.

Chen, Y.-Q., Nie, S.-P., Huang, D.-F., Han, C., Xie, M.-Y., 2009. Effect of polysaccharide isolated from the seed of *Plantago asiatica* L. on the nitric oxide production by RAW264. 7 cells. Chin. Pharmacol. Bull. 8, 035.

Cheng, Y., Ohlsson, L., Duan, R.-D., 2004. Psyllium and fat in diets differentially affect the activities and expressions of colonic sphingomyelinases and caspase in mice. Br. J. Nutr. 91, 715–723.

Cheng, Z., Blackford, J., Wang, Q., Yu, L., 2009. Acid treatment to improve psyllium functionality. J. Funct. Foods 1, 44–49.

Chutkan, R., Fahey, G., Wright, W.L., Mcrorie, J., 2012. Viscous versus nonviscous soluble fiber supplements: mechanisms and evidence for fiber-specific health benefits. J. Am. Acad. Nurse Pract. 24, 476–487.

Cicero, A.F., Derosa, G., Bove, M., Imola, F., Borghi, C., Gaddi, A.V., 2010. Psyllium improves dyslipidaemia, hyperglycaemia and hypertension, while guar gum reduces body weight more rapidly in patients affected by metabolic syndrome following an AHA Step 2 diet. Med. J. Nutr. Metab. 3, 47–54.

Cohen, L., Kendall, M., Zang, E., Meschter, C., Rose, D., 1991. Modulation of N-nitrosomethylurea-induced mammary tumor promotion by dietary fiber and fat. J. Natl. Cancer Inst. 83, 496–501.

Cohen, L.A., Zhao, Z., Zang, E.A., Wynn, T.T., Simi, B., Rivenson, A., 1996. Wheat bran and psylium diets: effects on N-methylnitrosourea-induced mammary tumorigenesis in F344 rats. J. Natl. Cancer Inst. 88, 899–907.

Costa, M.A., Mehta, T., Males, J.R., 1989. Effects of dietary cellulose and psyllium husk on monkey colonic microbial metabolism in continuous culture. J. Nutr. 119, 979–985.

Cullen, G., O'donoghue, D., 2007. Constipation and pregnancy. Best Pract. Res. Cl. Ga. 21, 807–818.

Davidson, M.H., Dugan, L.D., Burns, J.H., Sugimoto, D., Story, K., Drennan, K., 1996. A psyllium-enriched cereal for the treatment of hypercholesterolemia in children: a controlled, double-blind, crossover study. Am. J. Clin. Nutr. 63, 96–102.

Davidson, M.H., Maki, K.C., Kong, J.C., Dugan, L.D., Torri, S.A., Hall, H.A., et al., 1998. Long-term effects of consuming foods containing psyllium seed husk on serum lipids in subjects with hypercholesterolemia. Am. J. Clin. Nutr. 67, 367–376.

Dennison, B.A., Levine, D.M., 1993. Randomized, double-blind, placebo-controlled, two-period crossover clinical trial of psyllium fiber in children with hypercholesterolemia. J. Pediatr. 123, 24–29.

Deters, A., Schröder, K., Smiatek, T., Hensel, A., 2005. Ispaghula (*Plantago ovata*) seed husk polysaccharides promote proliferation of human epithelial cells (skin keratinocytes and fibroblasts) via enhanced growth factor receptors and energy production. Planta Med. 71, 33–39.

Díez, R., García, J.J., Diez, M.J., Sierra, M., Sahagún, A.M., Calle, P., et al., 2013. Hypoglycemic and hypolipidemic potential of a high fiber diet in healthy versus diabetic rabbits. BioMed Res. Int. 2013, 960568.

Drake, C.L., Moses, E.S., Tandberg, D., 1991. Systemic anaphylaxis after ingestion of a psyllium-containing breakfast cereal. Am. J. Emerg. Med. 9, 449–451.

Edwards, S., Chaplin, M.F., Blackwood, A.D., Dettmar, P.W., 2003. Primary structure of arabinoxylans of ispaghula husk and wheat bran. Proc. Nutr. Soc. 62, 217–222.

Eherer, A., Santa Ana, C., Porter, J., Fordtran, J., 1993. Effect of psyllium, calcium polycarbophil, and wheat bran on secretory diarrhea induced by phenolphthalein. Gastroenterology 104, 1007.

El-Abhar, H.S., Schaalan, M.F., 2014. Phytotherapy in diabetes: review on potential mechanistic perspectives. World J. Diabetes 5, 176–197.

El Souda, S.S.E.D., Mohammed, R.S., Marzouk, M.M., Fahmy, M.A., Hassan, Z.M., Farghaly, A.A., 2014. Antimutagenicity and phytoconstituents of Egyptian *Plantago albicans* L. Asian Pac. J. Trop. Dis. 4, S946–S951.

Emenaker, N.J., Basson, M.D., 1998. Short chain fatty acids inhibit human (SW1116) colon cancer cell invasion by reducing urokinase plasminogen activator activity and stimulating TIMP-1 and TIMP-2 activities, rather than via MMP modulation. J. Surg. Res. 76, 41–46.

Eswaran, S., Muir, J., Chey, W.D., 2013. Fiber and functional gastrointestinal disorders. Am. J. Gastroenterol. 108, 718–727.

Everson, G., Daggy, B., Mckinley, C., Story, J., 1992. Effects of psyllium hydrophilic mucilloid on LDL-cholesterol and bile acid synthesis in hypercholesterolemic men. J. Lipid. Res. 33, 1183–1192.

Farahnaky, A., Askari, H., Majzoobi, M., Mesbahi, G., 2010. The impact of concentration, temperature and pH on dynamic rheology of psyllium gels. J. Food Eng. 100, 294–301.

Feinglos, M.N., Gibb, R.D., Ramsey, D.L., Surwit, R.S., Mcrorie, J.W., 2013. Psyllium improves glycemic control in patients with type-2 diabetes mellitus. Bioact. Carbohydr. Diet. Fibre 1, 156–161.

Fernandez-Banares, F., Hinojosa, J., Sanchez-Lombrana, J., Navarro, E., Martinez-Salmeron, J., Garcia-Puges, A., et al., 1999. Randomized clinical trial of *Plantago ovata* seeds (dietary fiber) as compared with mesalamine in maintaining remission in ulcerative colitis. Am. J. Gastroenterol. 94, 427–433.

Fischer, M.H., Yu, N., Gray, G.R., Ralph, J., Anderson, L., Marlett, J.A., 2004. The gel-forming polysaccharide of psyllium husk (*Plantago ovata* Forsk). Carbohydr. Res. 339, 2009–2017.

Frati-Munari, A., Flores-Garduno, M., Ariza-Andraca, R., Islas-Andrade, S., Chávez, N.A., 1988. Effect of different doses of Plantago psyllium mucilage on the glucose tolerance test. Arch. Investigat. Med. 20, 147–152.

Friedman, E., Lightdale, C., Winawer, S., 1988. Effects of psyllium fiber and short-chain organic acids derived from fiber breakdown on colonic epithelial cells from high-risk patients. Cancer Lett. 43, 121–124.

Fu, Z.-H., Xie, M.-Y., Nie, S.-P., 2005. Polysaccharides extraction in seeds of *Plantago asiatica* L. by Microwave Technique. Food Sci. 3, 151–153.

Gamet, L., Daviaud, D., Denis-Pouxviel, C., Remesy, C., Murat, J.C., 1992. Effects of short-chain fatty acids on growth and differentiation of the human colon-cancer cell line HT29. Int. J. Cancer 52, 286–289.

Ghahramani, L., Hosseini, S.V., Rahimikazerooni, S., Bananzadeh, A.M., Jahromi, B.N., Samsam, A., et al., 2013. The effect of oral psyllium herbal laxative powder in prevention of hemorrhoids and anal fissure during pregnancy, a randomized double blind clinical trial. Ann. Colorectal Res. 1, 23–27.

Gharibzahedi, S.M.T., Razavi, S.H., Mousavi, S.M., 2013. Psyllium husk gum: an attractive carbohydrate biopolymer for the production of stable canthaxanthin emulsions. Carbohydr. Polym. 92, 2002–2011.

Gibb, R.D., Mcrorie, J.W., Russell, D.A., Hasselblad, V., D'alessio, D.A., 2015. Psyllium fiber improves glycemic control proportional to loss of glycemic control: a meta-analysis of data in euglycemic subjects, patients at risk of type 2 diabetes mellitus, and patients being treated for type 2 diabetes mellitus. Am. J. Clin. Nutr. 102, 1604–1614.

Gilani, A.U.H., Aziz, N., Khan, M., Khan, S., Zaman, V., 1998. Laxative effect of ispaghula: physical or chemical effect? Phytother. Res. 12, S63–S65.

Gillespie, B., Rathbun, F., 1992. Adverse effects of psyllium. Can. Med. Assoc. J. 146, 16.

Goldstein, J.L., Brown, M.S., 1984. Progress in understanding the LDL receptor and HMG-CoA reductase, two membrane proteins that regulate the plasma cholesterol. J. Lipid Res. 25, 1450–1461.

Gong, L., Zhang, H., Niu, Y., Chen, L., Liu, J., Alaxi, S., et al., 2015. A novel alkali extractable polysaccharide from *Plantago asiatic* L. seeds and its radical-scavenging and bile acid-binding activities. J. Agric. Food Chem. 63, 569–577.

Grundy, S.M., Bilheimer, D., Chait, A., et al., 1993. Summary of the second report of the national cholesterol education program (ncep) expert panel on detection, evaluation, and treatment of high blood cholesterol in adults (adult treatment panel ii). J. Am. Med. Assoc. 269, 3015–3023.

Guo, Q., Cui, S.W., Wang, Q., Young, J.C., 2008. Fractionation and physicochemical characterization of psyllium gum. Carbohydr. Polym. 73, 35–43.

Guo, Q., Riehm, M., Defelice, C., Cui, S.W., 2010. Formulation optimization of psyllium-based binding product by response surface methodology. J. Food Agric. Enviro. 8, 882–889.

Hannan, J., Ali, L., Khaleque, J., Akhter, M., Flatt, P., Abdel-Wahab, Y., 2006. Aqueous extracts of husks of *Plantago ovata* reduce hyperglycaemia in type 1 and type 2 diabetes by inhibition of intestinal glucose absorption. Br. J. Nutr. 96, 131–137.

Hart, G.K., Dobb, G.J., 1988. Effect of a fecal bulking agent on diarrhea during enteral feeding in the critically ill. J. Parenter. Enter. Nutr. 12, 465–468.

Heather, D., Howell, L., Montana, M., Howell, M., Hill, R., 1991. Effect of a bulk-forming cathartic on diarrhea in tube-fed patients. Heart Lung J. Crit. Care 20, 409–413.

Hetland, G., 2003. Anti-infective action of immuno-modulating polysaccharides (β-glucan and *Plantago major* L. Pectin) against intracellular (*Mycobacteria* sp.) and extracellular (*Streptococcus Pneumoniae* sp.) respiratory pathogens. Curr. Med. Chem. Antiinfect. Agents 2, 135–146.

Hetland, G., Samuelsen, A., Loslash, V., Paulsen, B., Aaberge, I., Groeng, E.C., et al., 2000. Protective effect of *Plantago major* L. pectin polysaccharide against systemic *Streptococcus pneumoniae* infection in mice. Scand. J. Immunol. 52, 348–355.

Hinnebusch, B.F., Meng, S., Wu, J.T., Archer, S.Y., Hodin, R.A., 2002. The effects of short-chain fatty acids on human colon cancer cell phenotype are associated with histone hyperacetylation. J. Nutr. 132, 1012–1017.

Horton, J.D., Cuthbert, J.A., Spady, D.K., 1994. Regulation of hepatic 7 alpha-hydroxylase expression by dietary psyllium in the hamster. J. Clin. Invest. 93, 2084.

Hotz, J., Plein, K., 1994. Effectiveness of plantago seed husks in comparison with wheat brain on stool frequency and manifestations of irritable colon syndrome with constipation. Med. Klin. (Munich) 89, 645–651.

Hu, J.-L., Nie, S.-P., Min, F.-F., Xie, M.-Y., 2012. Polysaccharide from seeds of *Plantago asiatica* L. increases short-chain fatty acid production and fecal moisture along with lowering pH in mouse colon. J. Agric. Food Chem. 60, 11525–11532.

Hu, J.-L., Nie, S.-P., Li, C., Xie, M.-Y., 2013a. In vitro fermentation of polysaccharide from the seeds of *Plantago asiatica* L. by human fecal microbiota. Food Hydrocolloid. 33, 384–392.

Hu, J.-L., Nie, S.-P., Min, F.-F., Xie, M.-Y., 2013b. Artificial simulated saliva, gastric and intestinal digestion of polysaccharide from the seeds of *Plantago asiatica* L. Carbohydr. Polym. 92, 1143–1150.

Hu, J.-L., Nie, S.-P., Wu, Q.-M., Li, C., Fu, Z.-H., Gong, J., et al., 2013c. Polysaccharide from seeds of *Plantago asiatica* L. affects lipid metabolism and colon microbiota of mouse. J. Agric. Food Chem. 62, 229–234.

Hu, J.-L., Nie, S.-P., Xie, M.-Y., 2013d. High pressure homogenization increases antioxidant capacity and short-chain fatty acid yield of polysaccharide from seeds of *Plantago asiatica* L. Food Chem. 138, 2338–2345.

Huang, D., Nie, S., Jiang, L., Xie, M., 2014. A novel polysaccharide from the seeds of *Plantago asiatica* L. induces dendritic cells maturation through toll-like receptor 4. Int. Immunopharmacol. 18, 236–243.

Huang, D.-F., Tang, Y.-F., Nie, S.-P., Wan, Y., Xie, M.-Y., Xie, X.-M., 2009a. Effect of phenylethanoid glycosides and polysaccharides from the seed of *Plantago asiatica* L. on the maturation of murine bone marrow-derived dendritic cells. Eur. J. Pharmacol. 620, 105–111.

Huang, D.-F., Xie, M.-Y., Yin, J.-Y., Nie, S.-P., Tang, Y.-F., Xie, X.-M., et al., 2009b. Immunomodulatory activity of the seeds of *Plantago asiatica* L. J. Ethnopharmacol. 124, 493–498.

Hussain, M.A., Muhammad, G., Jantan, I., Bukhari, S.N.A., 2015. Psyllium arabinoxylan: a versatile biomaterial for potential medicinal and pharmaceutical applications. Polym. Rev. 3, 1–30.

James, J.M., Cooke, S.K., Barnett, A., Sampson, H.A., 1991. Anaphylactic reactions to a psyllium-containing cereal. J. Allergy Clin. Immunol. 88, 402–408.

Jiang, L.-M., Nie, S.-P., Zhou, H.-L., Huang, D.-F., Xie, M.-Y., 2014. Carboxymethylation enhances the maturation-inducing activity in dendritic cells of polysaccharide from the seeds of *Plantago asiatica* L. Int. Immunopharmacol. 22, 324–331.

Kaith, B., Kumar, K., 2007a. In air synthesis of Psy-cl-poly (AAm) network and its application in water-absorption from oil-water emulsions. Express Polym. Lett. 1, 474–480.

Kaith, B., Kumar, K., 2007b. Selective absorption of water from different oil-water emulsions with Psy-cl-poly (AAm) synthesized using irradiation copolymerization method. Bull. Mater. Sci. 30, 387–391.

Kaith, B., Kumar, K., 2008. In vacuum synthesis of psyllium and acrylic acid based hydrogels for selective water absorption from different oil–water emulsions. Desalination 229, 331–341.

Kamaljit, K., Amarjeet, K., Pal, S.T., 2011. Analysis of ingredients, functionality, formulation optimization and shelf life evaluation of high fiber bread. Am. J. Food Technol. 6, 306–313.

Kennedy, J.F., Sandhu, J.S., Southgate, D.A., 1979. Structural data for the carbohydrate of ispaghula husk ex *Plantago ovata* Forsk. Carbohydr. Res. 75, 265–274.

Khalili, B., Bardana, E.J., Yunginger, J.W., 2003. Psyllium-associated anaphylaxis and death: a case report and review of the literature. Ann. Allergy Asthma Immunol. 91, 579–584.

Kim, J.-H., Kang, T.-W., Ahn, Y.-K., 1996. The effects of plantago-mucilage A from the seeds of *Plantago asiatica* on the immune responses in ICR mice. Arch. Pharm. Res. 19, 137–142.

Kincaid, J.G., Talbot, M.W., 1995. Ready-to-eat cereals containing extruded pre-wetted psyllium. U.S. Patents No. 5,382,443. 17 Jan. 1995.

Kristensen, M., Jensen, M.G., 2011. Dietary fibres in the regulation of appetite and food intake. Importance of viscosity. Appetite 56, 65–70.

Kumar, K., Kaith, B., Jindal, R., Mittal, H., 2012. Gamma-radiation initiated synthesis of Psyllium and acrylic acid-based polymeric networks for selective absorption of water from different oil-water emulsions. J. Appl. Polym. Sci. 124, 4969–4977.

Kumar, R., Sharma, K., 2013. Biodegradable polymethacrylic acid grafted psyllium for controlled drug delivery systems. Front. Chem. Sci. Eng. 7, 116–122.

Kumar, R., Sharma, K., Tiwary, K., Sen, G., 2013. Polymethacrylic acid grafted psyllium (Psy-g-PMA): a novel material for waste water treatment. Appl. Water Sci. 3, 285–291.

Ladjevardi, Z.S., Gharibzahedi, S.M.T., Mousavi, M., 2015. Development of a stable low-fat yogurt gel using functionality of psyllium (*Plantago ovata* Forsk) husk gum. Carbohydr. Polym. 125, 272–280.

Lantner, R.R., Espiritu, B.R., Zumerchik, P., Tobin, M.C., 1990. Anaphylaxis following ingestion of a psyllium-containing cereal. J. Am. Med. Assoc. 264, 2534–2536.

Leng-Peschlow, E., 1991. Plantago ovata seeds as dietary fibre supplement: physiological and metabolic effects in rats. Br. J. Nutr. 66, 331–349.

Lewis, S.J., Heaton, K.W., 1999. Roughage revisited (the effect on intestinal function of inert plastic particles of different sizes and shape). Dig. Dis. Sci. 44, 744–748.

Liu, W., Zhang, B., Wang, Q., Xie, Z., Yao, W., Gao, X., et al., 2009. Effects of sulfation on the physicochemical and functional properties of psyllium. J. Agric. Food Chem. 58, 172–179.

Liu, W., Xie, Z., Zhang, B., Wang, Q., Yao, W., Gao, X., et al., 2010. Effects of hydroxypropylation on the functional properties of psyllium. J. Agric. Food Chem. 58, 1615–1621.

Malo, J.-L., Cartier, A., L'archevêque, J., Ghezzo, H., Lagier, F., Trudeau, C., et al., 1990. Prevalence of occupational asthma and immunologic sensitization to psyllium among health personnel in chronic care hospitals. Am. Rev. Respir. Dis. 142, 1359–1366.

Mancebo, C.M., San Miguel, M., Martínez, M.M., Gómez, M., 2015. Optimisation of rheological properties of gluten-free doughs with HPMC, psyllium and different levels of water. J. Cereal Sci. 61, 8–15.

Mariotti, M., Lucisano, M., Pagani, M.A., Ng, P.K., 2009. The role of corn starch, amaranth flour, pea isolate, and *Psyllium flour* on the rheological properties and the ultrastructure of gluten-free doughs. Food Res. Int. 42, 963–975.

Marlett, J.A., Fischer, M.H., 2002. A poorly fermented gel from psyllium seed husk increases excreta moisture and bile acid excretion in rats. J. Nutr. 132, 2638–2643.

Marlett, J.A., Fischer, M.H., 2003. The active fraction of psyllium seed husk. Proc. Nutr. Soc. 62, 207–209.

Marlett, J.A., Kajs, T.M., Fischer, M.H., 2000. An unfermented gel component of psyllium seed husk promotes laxation as a lubricant in humans. Am. J. Clin. Nutr. 72, 784–789.

Marteau, P., Flourie, B., Cherbut, C., Correze, J., Pellier, P., Seylaz, J., et al., 1994. Digestibility and bulking effect of ispaghula husks in healthy humans. Gut 35, 1747–1752.

Mcburney, M., 1990. Potential water-holding capacity and short-chain fatty acid production from purified fiber sources in a fecal incubation system. Nutrition (Burbank, Los Angeles County, Calif.) 7, 421–424.

Mcrorie, J., Daggy, B., Morel, J., Diersing, P., Miner, P., Robinson, M., 1998. Psyllium is superior to docusate sodium for treatment of chronic constipation. Aliment. Pharmacol. Ther. 12, 491.

Mehmood, M.H., Aziz, N., Ghayur, M.N., Gilani, A.-H., 2011. Pharmacological basis for the medicinal use of psyllium husk (Ispaghula) in constipation and diarrhea. Dig. Dis. Sci. 56, 1460–1471.

Michaelsen, T., Gilje, A., Samuelsen, A., Høgåsen, K., Paulsen, B., 2000. Interaction between human complement and a pectin type polysaccharide fraction, PMII, from the leaves of *Plantago major* L. Scand. J. Immunol. 52, 483–490.

Mishra, A., Bajpai, M., 2005. Flocculation behaviour of model textile wastewater treated with a food grade polysaccharide. J. Hazard. Mater. 118, 213–217.

Mishra, A., Agarwal, M., Bajpai, M., Rajani, S., Mishra, R., 2002a. Plantago psyllium mucilage for sewage and tannery effluent treatment. Iranian Polymer J. 11, 381–386.

Mishra, A., Srinivasan, R., Dubey, R., 2002b. Flocculation of textile wastewater by *Plantago psyllium* mucilage. Macromol. Mater. Eng. 287, 592–596.

Mishra, A., Srinivasan, R., Bajpai, M., Dubey, R., 2004. Use of polyacrylamide-grafted Plantago psyllium mucilage as a flocculant for treatment of textile wastewater. Colloid. Polym. Sci. 282, 722–727.

Mishra, S., Sinha, S., Dey, K.P., Sen, G., 2014. Synthesis, characterization and applications of polymethylmethacrylate grafted psyllium as flocculant. Carbohydr. Polym. 99, 462–468.

Miyauchi, S., Gopal, E., Fei, Y.-J., Ganapathy, V., 2004. Functional identification of SLC5A8, a tumor suppressor down-regulated in colon cancer, as a Na^+-coupled transporter for short-chain fatty acids. J. Biol. Chem. 279, 13293–13296.

Morita, T., Kasaoka, S., Hase, K., Kiriyama, S., 1999. Psyllium shifts the fermentation site of high-amylose cornstarch toward the distal colon and increases fecal butyrate concentration in rats. J. Nutr. 129, 2081–2087.

Muehlbauer, P.M., Arlene Davis, R., Rachael Drabot, M., Cnsd, R., Rawlings, B.L., Elizabeth Kiker, R., 2009. Putting evidence into practice: evidence-based interventions to prevent, manage, and treat chemotherapy-and radiotherapy-induced diarrhea. Clin. J. Oncol. Nurs. 13, 336.

Murphy, J., Stacey, D., Crook, J., Thompson, B., Panetta, D., 2000. Testing control of radiationinduced diarrhea with a psyllium bulking agent: a pilot study. Can. Oncol. Nurs. J. 10, 96–100.

Nagafabadi, M.S., Aghah, S., Yousefi, S.S., Kamalinejad, M., Mahdaviyan, F., Jafarian, A.A., 2015. Comparison of the plant *Plantago major* and the drug colofac on clinical symptoms in ibs patients. Int. J. Life Sci. 9, 108–112.

Nelson, W.L., 1987. Allergic events among health care workers exposed to psyllium laxatives in the workplace. J. Occup. Environ. Med. 29, 497–499.

Niu, Y., Xie, Z., Hao, J., Yao, W., Yue, J., Yu, L.L., 2012. Preparation of succinylated derivatives of psyllium and their physicochemical and bile acid-binding properties. Food Chem. 132, 1025–1032.

Niu, Y., Xie, Z., Zhang, H., Sheng, Y., Yu, L., 2013. Effects of structural modifications on physicochemical and bile acid-binding properties of psyllium. J. Agric. Food Chem. 61, 596–601.

Nordgaard, I., Hove, H., Clausen, M., Mortensen, P., 1996. Colonic production of butyrate in patients with previous colonic cancer during long-term treatment with dietary fibre (*Plantago ovata* seeds). Scand. J. Gastroenterol. 31, 1011–1020.

Odes, H., Madar, Z., 1991. A double-blind trial of a celandin, aloevera and psyllium laxative preparation in adult patients with constipation. Digestion 49, 65–71.

Olson, B.H., Anderson, S.M., Becker, M.P., Anderson, J.W., Hunninghake, D.B., Jenkins, D.J., et al., 1997. Psyllium-enriched cereals lower blood total cholesterol and LDL cholesterol, but not HDL cholesterol, in hypercholesterolemic adults: results of a meta-analysis. J. Nutr. 127, 1973–1980.

Pastors, J.G., Blaisdell, P.W., Balm, T.K., Asplin, C.M., Pohl, S.L., 1991. Psyllium fiber reduces rise in postprandial glucose and insulin concentrations in patients with non-insulin-dependent diabetes. Am. J. Clin. Nutr. 53, 1431–1435.

Patil, B.S., Mastiholimath, V.S., Kulkarni, A.R., 2011. Development and evaluation of psyllium seed husk polysaccharide based wound dressing films. Orient. Pharm. Exp. Med. 11, 123–129.

Paulini, I., Mehta, T., Hargis, A., 1987. Intestinal structural changes in African green monkeys after long term psyllium or cellulose feeding. J. Nutr. 117, 253–266.

Prior, A., Whorwell, P.J., 1987. Double blind study of ispaghula in irritable bowel syndrome. Gut 28, 1510–1513.

Rahbar, A., Poolasi, S., Rahbar, A., 2014. The effect of psyllium on serum lipid profile in patients with hyperlipidemia. Nutr. Food Sci. Res. 1, 217.

Rao, M., Khambete, M., Lunavat, H., 2011. Study of rheological properties of Psyllium polysaccharide and its evaluation as suspending agent. Int. J. Pharm. Tech. Res. 3, 1191–1197.

Rao, M., Sadaphule, P., Khembete, M., Lunawat, H., Thanki, K., Gabhe, N., 2013. Characterization of psyllium (*Plantago ovata*) polysaccharide and its use as a binder in tablets. Ind. J. Pharm. Edu. Res. 47, 154–159.

Rao, M.R., Warrier, D.U., Rao, S.H., 2015. Evaluation of phosphorylated psyllium seed polysaccharide as a release retardant. Indian J. Pharm. Sci. 77, 605.

Rao, M.R., Warrier, D.U., Gaikwad, S.R., Shevate, P.M., 2016. Phosphorylation of psyllium seed polysaccharide and its characterization. Int. J. Biol. Macromol. 85, 317–326.

Raymundo, A., Fradinho, P., Nunes, M.C., 2014. Effect of Psyllium fibre content on the textural and rheological characteristics of biscuit and biscuit dough. Bioact. Carbohydr. Diet. Fibre 3, 96–105.

Rigaud, D., Paycha, F., Meulemans, A., Merrouche, M., Mignon, M., 1998. Effect of psyllium on gastric emptying, hunger feeling and food intake in normal volunteers: a double blind study. Eur. J. Clin. Nutr. 52, 239–245.

Roberts-Andersen, J., Mehta, T., Wilson, R.B., 1987. Reduction of DMH-induced colon tumors in rats fed psyllium husk or cellulose. Nut. Cancer 10, 129–136.

RodríGuez-Cabezas, M., Galvez, J., Camuesco, D., Lorente, M., Concha, A., Martinez-Augustin, O., et al., 0030. Intestinal anti-inflammatory activity of dietary fiber (*Plantago ovata* seeds) in HLA-B27 transgenic rats. Clin. Nutr. 22, 463–471.

RodríGuez-Morán, M., Guerrero-Romero, F., Lazcano-Burciaga, G., 0030. Lipid-and glucose-lowering efficacy of plantago psyllium in type II diabetes. J. Diabetes Complica. 12, 273–278.

Rungsiprakarn, P., Laopaiboon, M., Sangkomkamhang, U.S., Lumbiganon, P., Pratt, J.J., 2015. Interventions for treating constipation in pregnancy. Cochrane Library. Available from: http://dx.doi.org/10.1002/14651858.CD001142.

Russell, J., Bass, P., 1985. Canine gastric emptying of fiber meals: influence of meal viscosity and antroduodenal motility. Am. J. Physiol. 249, G662–7.

Saghir, S., Iqbal, M.S., Hussain, M.A., Koschella, A., Heinze, T., 2008. Structure characterization and carboxymethylation of arabinoxylan isolated from Ispaghula (*Plantago ovata*) seed husk. Carbohydr. Polym. 74, 309–317.

Sahagún, A.M., Vaquera, J., García, J.J., Calle, P., Diez, M.-J., Fernández, N., et al., 2015. Study of the protective effect on intestinal mucosa of the hydrosoluble fiber Plantago ovata husk. BMC Complement. Altern. Med. 15, 1.

Samuelsen, A., Lund, I., Djahromi, J., Paulsen, B., Wold, J., Knutsen, S., 1999. Structural features and anti-complementary activity of some heteroxylan polysaccharide fractions from the seeds of *Plantago major* L. Carbohydr. Polym. 38, 133–143.

Samuelsen, A.B., 2000. The traditional uses, chemical constituents and biological activities of *Plantago major* L. A review. J. Ethnopharmacol. 71, 1–21.

Samuelsen, A.B., Paulsen, B.S., Wold, J.K., Otsuka, H., Yamada, H., Espevik, T., 1995. Isolation and partial characterization of biologically active polysaccharides from *Plantago major* L. Phytother. Res. 9, 211–218.

Samuelsen, A.B., Paulsen, B.S., Wold, J.K., Otsuka, H., Kiyohara, H., Yamada, H., et al., 1996. Characterization of a biologically active pectin from *Plantago major* L. Carbohydr. Polym. 30, 37–44.

Sandhu, J.S., Hudson, G.J., Kennedy, J.F., 1981. The gel nature and structure of the carbohydrate of ispaghula husk ex *Plantago ovata* Forsk. Carbohydr. Res. 93, 247–259.

Satchithanandam, S., Klurfeld, D.M., Calvert, R.J., Cassidy, M.M., 1996. Effects of dietary fibers on gastrointestinal mucin in rats. Nutr. Res. 16, 1163–1177.

Schectman, G., Hiatt, J., Hartz, A., 1993. Evaluation of the effectiveness of lipid-lowering therapy (bile acid sequestrants, niacin, psyllium and lovastatin) for treating hypercholesterolemia in veterans. Am. J. Cardiol. 71, 759–765.

Scheppach, W., Bartram, P., Richter, A., Richter, F., Liepold, H., Dusel, G., et al., 1992. Effect of short-chain fatty acids on the human colonic mucosa in vitro. J. Parenter. Enter.Nutr. 16, 43–48.

Schiller, L., 2001. The therapy of constipation. Aliment. Pharmacol. Ther. 15, 749–763.

Schmidgall, J., Schnetz, E., Hensel, A., 2000. Evidence for bioadhesive effects of polysaccharides and polysaccharide-containing herbs in an ex vivo bioadhesion assay on buccal membranes. Planta Med. 66, 48–53.

Schneeman, B.O., Richter, D., 1993. Changes in plasma and hepatic lipids, small intestinal histology and pancreatic enzyme activity due to aging and dietary fiber in rats. J. Nutr. 123, 1328–1337.

Schwartz, H.J., Arnold, J.L., Strohl, K.P., 1989. Occupational allergic rhinitis reaction to psyllium. J. Occup. Environ. Med. 31, 624–626.

Sen, G., Mishra, S., Rani, G.U., Rani, P., Prasad, R., 2012. Microwave initiated synthesis of polyacrylamide grafted Psyllium and its application as a flocculant. Int. J. Biol. Macromol. 50, 369–375.

Siahi-Shadbad, M.R., Asare-Addo, K., Azizian, K., Hassanzadeh, D., Nokhodchi, A., 2011. Release behaviour of propranolol HCl from hydrophilic matrix tablets containing psyllium powder in combination with hydrophilic polymers. AAPS PharmSciTech. 12, 1176–1182.

Siavash, M., Salehioun, M., Najafian, A., Amini, M., 2007. A randomized controlled study for evaluation of psyllium effects on kinetics of carbohydrate absorption. J. Res. Med. Sci. 12, 125–130.

Sierra, M., Garcia, J., Fernandez, N., Diez, M., Calle, A., 2002. Therapeutic effects of psyllium in type 2 diabetic patients. Eur. J. Clin. Nutr. 56, 830–842.

Singh, B., 2007. Psyllium as therapeutic and drug delivery agent. Int. J. Pharm. 334, 1–14.

Singh, B., Kumar, S., 2008. Synthesis and characterization of psyllium-NVP based drug delivery system through radiation crosslinking polymerization. Nucl. Instrum. Meth. B 266, 3417–3430.

Singh, B., Sharma, V., 2010. Design of psyllium–PVA–acrylic acid based novel hydrogels for use in antibiotic drug delivery. Int. J. Pharm. 389, 94–106.

Singh, B., Chauhan, G., Kumar, S., Chauhan, N., 2007. Synthesis, characterization and swelling responses of pH sensitive psyllium and polyacrylamide based hydrogels for the use in drug delivery (I). Carbohydr. Polym. 67, 190–200.

Singh, B., Chauhan, N., Kumar, S., Bala, R., 2008. Psyllium and copolymers of 2-hydroxylethylmethacrylate and acrylamide-based novel devices for the use in colon specific antibiotic drug delivery. Int. J. Pharm. 352, 74–80.

Smalley, J.R., Klish, W.J., Campbell, M.A., Brown, M.R., 1982. Use of psyllium in the management of chronic nonspecific diarrhea of childhood. J. Pediatr. Gastroenterol. Nutr. 1, 361–364.

Sohn, V.R., Giros, A., Xicola, R.M., Fluvia, L., Grzybowski, M., Anguera, A., et al., 2012. Stool-fermented Plantago ovata husk induces apoptosis in colorectal cancer cells independently of molecular phenotype. Br. J. Nutr. 107, 1591–1602.

Spence, J.D., Huff, M.W., Heidenheim, P., Viswanatha, A., Munoz, C., Lindsay, R., et al., 1995. Combination therapy with colestipol and psyllium mucilloid in patients with hyperlipidemia. Ann. Intern. Med. 123, 493–499.

Sprecher, D.L., Harris, B.V., Goldberg, A.C., Anderson, E.C., Bayuk, L.M., Russell, B.S., et al., 1993. Efficacy of psyllium in reducing serum cholesterol levels in hypercholesterolemic patients on high-or low-fat diets. Ann. Intern. Med. 119, 545–554.

Steinman, R.M., 1991. The dendritic cell system and its role in immunogenicity. Annu. Rev. Immunol. 9, 271–296.

Stewart, R.B., Hale, W.E., Moore, M.T., May, F.E., Marks, R.G., 1991. Effect of psyllium hydrophilic mucilloid on serum cholesterol in the elderly. Dig. Dis. Sci. 36, 329–334.

Takagi, H., Mitsumori, K., Onodera, H., Takegawa, K., Shimo, T., Koujitani, T., et al., 1999. A Preliminary study of the effect of plantago ovata forsk on the development of 7, 12-dimethylbenz [a] anthracene-initiated rat mammary tumors under the influence of hypercholesterolemia. J. Toxicol. Pathol. 12, 141–145.

Tang, Y., Chen, Y., Jiang, H., Robbins, G.T., Nie, D., 2011. G-protein-coupled receptor for short-chain fatty acids suppresses colon cancer. Int. J. Cancer 128, 847–856.

Tomas-Ridocci, M., Anon, R., Minguez, M., Zaragoza, A., Ballester, J., Benages, A., 1992. The efficacy of Plantago ovata as a regulator of intestinal transit. A double-blind study compared to placebo. Revista española de enfermedades digestivas: organo oficial de la Sociedad Española de Patología Digestiva 82, 17–22.

Tomoda, M., Yokoi, M., Ishikawa, K., 1981. Plant Mucilages. XXIX. Isolation and Characterization of a Mucous Polysaccharide, "Plantago-mucilage A," from the Seeds of *Plantago major* var. asiatica. Chem. Pharm. Bull. (Tokyo) 29, 2877–2884.

Tomoda, M., Takada, K., Shimizu, N., Gonda, R., Ohara, N., 1991. Reticuloendothelial system-potentiating and alkaline phosphatase-inducing activities of plantago-mucilage A, the main mucilage from the seed of Plantago asiatica, and its five modification products. Chem. Pharm. Bull. (Tokyo) 39, 2068–2071.

Trautwein, E., Carls, C., Erbersdobler, H., Hisserich, D., 2000. Development of types of psyllium-enriched bread as part of a cholesterol lowering diet. Deutsche Lebensmittel-Rundschau 96, 58–64.

Türel, I., zbek, H., Erten, R., ner, A.C., Cengiz, N., Yilmaz, O., 2009. Hepatoprotective and anti-inflammatory activities of *Plantago major* L. Indian J. Pharmacol. 41, 120.

Ur Rehman, H., Farooq, U., Akram, K., Sidhu, A.I., Shafi, A., Sarfraz, F., 2015. Incorporation of garlic extract as antifungal agent in psyllium based edible coating for mandarin. Int. J. Food Allied Sci. 1, 11–17.

Van Craeyveld, V., Delcour, J.A., Courtin, C.M., 2008. Ball milling improves extractability and affects molecular properties of psyllium (*Plantago ovata* Forsk) seed husk arabinoxylan. J. Agri. Food Chem. 56, 11306–11311.

Van Craeyveld, V., Delcour, J.A., Courtin, C.M., 2009. Extractability and chemical and enzymic degradation of psyllium (*Plantago ovata* Forsk) seed husk arabinoxylans. Food Chem. 112, 812–819.

Van Rosendaal, G.M., Shaffer, E.A., Edwards, A.L., Brant, R., 2004. Effect of time of administration on cholesterol-lowering by psyllium: a randomized cross-over study in normocholesterolemic or slightly hypercholesterolemic subjects. Nutr. J. 3, 1.

Vaswani, S., Hamilton, R., Valentine, M., Adkinson, N., 1996. Psyllium laxative-induced anaphylaxis, asthma, and rhinitis. Allergy 51, 266–268.

Vega-López, S., Conde-Knape, K., Vidal-Quintanar, R.-L., Shachter, N.S., Fernandez, M.L., 2002. Sex and hormonal status influence the effects of psyllium on lipoprotein remodeling and composition. Metabolism 51, 500–507.

Vergara-Jimenez, M., Conde, K., Erickson, S.K., Fernandez, M.L., 1998. Hypolipidemic mechanisms of pectin and psyllium in guinea pigs fed high fat–sucrose diets: alterations on hepatic cholesterol metabolism. J. Lipid Res. 39, 1455–1465.

Vuksan, V., Jenkins, A.L., Rogovik, A.L., Fairgrieve, C.D., Jovanovski, E., Leiter, L.A., 2011. Viscosity rather than quantity of dietary fibre predicts cholesterol-lowering effect in healthy individuals. Br. J. Nutr. 106, 1349–1352.

Washington, N., Harris, M., Mussellwhite, A., Spiller, R.C., 1998. Moderation of lactulose-induced diarrhea by psyllium: effects on motility and fermentation. Am. J. Clin. Nutr. 67, 317–321.

Watters, K., Blaisdell, P., 1989. Reduction of glycemic and lipid levels in db/db diabetic mice by psyllium plant fiber. Diabetes 38, 1528–1533.

Wei, Z., Wang, H., Chen, X., Wang, B., Rong, Z., Su, B., et al., 2009. Time-and dose-dependent effect of psyllium on serum lipids in mild-to-moderate hypercholesterolemia: a meta-analysis of controlled clinical trials. Eur. J. Clin. Nutr. 63, 821–827.

Wenzl, H.H., Fine, K.D., Schiller, L.R., Fordtran, J.S., 1995. Determinants of decreased fecal consistency in patients with diarrhea. Gastroenterology 108, 1729–1738.

Westerhof, W., Das, P., Middelkoop, E., Verschoor, J., Storey, L., Regnier, C., 2000. Mucopolysaccharides from psyllium involved in wound healing. Drugs Exp. Clin. Res. 27, 165–175.

Whitehead, R., Young, G., Bhathal, P., 1986. Effects of short chain fatty acids on a new human colon carcinoma cell line (LIM1215). Gut 27, 1457–1463.

Wolever, T., Vuksan, V., Eshuis, H., Spadafora, P., Peterson, R., Chao, E., et al., 1991. Effect of method of administration of psyllium on glycemic response and carbohydrate digestibility. J. Am. Coll. Nutr. 10, 364–371.

Wolever, T., Jenkins, D., Mueller, S., Boctor, D.L., Ransom, T.P., Patten, R., et al., 1994. Method of administration influences the serum cholesterol-lowering effect of psyllium. Am. J. Clin. Nutr. 59, 1055–1059.

Yamada, H., Kiyohara, H., 1999. Complement-activating polysaccharides from medicinal herbs. In: Basel, W.H. (Ed.), Immunomodulatory Agents From Plants. Birkhauser Verlag, Basel, pp. 161–202.

Yamada, H., Nagai, T., Cyong, J.-C., Otsuka, Y., Tomoda, M., Shimizu, N., et al., 1985. Relationship between chemical structure and anti-complementary activity of plant polysaccharides. Carbohydr. Res. 144, 101–111.

Yarahmadi, M., Azizi, M., Morid, B., Kalatejari, S., 2014. Postharvest application of gum and mucilage as edible coating on postharvest life and quality of strawberry fruit. Int. J. Ad. Biol. Biomed. Res. 2, 1279–1286.

Yasir, M., Asif, M., Bhattacharya, A., Bajpai, M., 2010. Development and evaluation of gastroretentive drug delivery system for theopylline using psyllium husk. Int. J. Chem. Tech. Res. 2, 792–799.

Ye, C.-L., Hu, W.-L., Dai, D.-H., 2011. Extraction of polysaccharides and the antioxidant activity from the seeds of *Plantago asiatica* L. Int. J. Biol. Macromol. 49, 466–470.

Yin, J.-Y., Lin, H.-X., Nie, S.-P., Cui, S.W., Xie, M.-Y., 2012a. Methylation and 2D NMR analysis of arabinoxylan from the seeds of *Plantago asiatica* L. Carbohydr. Polym. 88, 1395–1401.

Yin, J.-Y., Nie, S.-P., Li, J., Li, C., Cui, S.W., Xie, M.-Y., 2012b. Mechanism of interactions between calcium and viscous polysaccharide from the seeds of *Plantago asiatica* L. J. Agri. Food Chem. 60, 7981–7987.

Yin, J., Lin, H., Li, J., Wang, Y., Cui, S.W., Nie, S., et al., 2012c. Structural characterization of a highly branched polysaccharide from the seeds of *Plantago asiatica* L. Carbohydr. Polym. 87, 2416–2424.

Yin, J.-Y., Nie, S.-P., Guo, Q.-B., Wang, Q., Cui, S.W., Xie, M.-Y., 2015. Effect of calcium on solution and conformational characteristics of polysaccharide from seeds of *Plantago asiatica* L. Carbohydr. Polym. 124, 331–336.

Yu, L., Perret, J., 2003. Effects of xylanase treatments on gelling and water-uptaking properties of psyllium. J. Agri. Food Chem. 51, 492–495.

Yu, L., Perret, J., Parker, T., Allen, K.G., 2003. Enzymatic modification to improve the water-absorbing and gelling properties of psyllium. Food Chem. 82, 243–248.

Zandonadi, R.P., Botelho, R.B.A., Araújo, W.M.C., 2009. Psyllium as a substitute for gluten in bread. J. Am. Diet. Assoc. 109, 1781–1784.

Zhao, H., Wang, Q., Sun, Y., Yang, B., Wang, Z., Chai, G., et al., 2014. Purification, characterization and immunomodulatory effects of *Plantago depressa* polysaccharides. Carbohydr. Polym. 112, 63–72.

Chapter 9

Cereal Beta-Glucan

9.1 INTRODUCTION

Cereal β-glucan is a cell wall polysaccharide in the grains of cereals, including oats, barley, rye, and wheat. Beta-β-D-glucans present in cereal endosperm and aleurone cells, as shown in Fig. 9.1.

Beta-D-glucan is evenly distributed across the grain of barley, oats, and rye, although some enrichment was observed in the subaleurone layer of oats (Autio, 2006; Fulcher and Miller, 1993). In contrast, wheat β-D-glucan is primarily in the subaleurone layer, with little found in the endosperm cell walls (Cui et al., 2000).

Our knowledge about the health benefits of oats can be traced back a long time. Just like all the other polysaccharides described in this book which have been well-documented in historic use in Asian countries, the use of oats and barley is deeply rooted in western culture. For example, there is a saying in English "Oats, a grain that in England is fed to horses, and in Scotland to people ... this is why England has such fine horses, and Scotland has such fine people." The coincidence is that the Food and Drug Administration of the US approved the following health claims in 1997: "diets low in saturated fat and cholesterol that include soluble fiber from whole oats and barley 'may' or 'might' reduce the risk of heart disease" (FDA, 1997, 2005). Such a claim recognized β-glucan as the primary bioactive component, and specified that an effective daily intake of β-glucan was 3 g. As a result, cereal β-glucans have been accepted as functional and bioactive ingredients which are widely used by food companies in many industrialized countries. There is increasing global acceptance that viscous soluble fibers like cereal β-D-glucan can lower serum LDL cholesterol (LDL-C) in both Caucasians and non-Caucasians (Wolever et al., 2011). Clinical trials showed the relationship between LDL-C levels and viscosity (Wood, 2007). It was concluded that the efficacy of various soluble fiber preparations could not simply be related to the amount of β-D-glucan, as there are other parameters which need to be considered, including molecular weight (MW) and MW distribution, and coexistence of other food ingredients such as fat, sugar, and/or starch. Wood's group established the relationships between the solution properties of cereal β-glucans and observed physiological effects; the LDL-C lowering and the ability to control blood glucose of oat β-glucan

Bioactive Polysaccharides. DOI: http://dx.doi.org/10.1016/B978-0-12-809418-1.00009-5

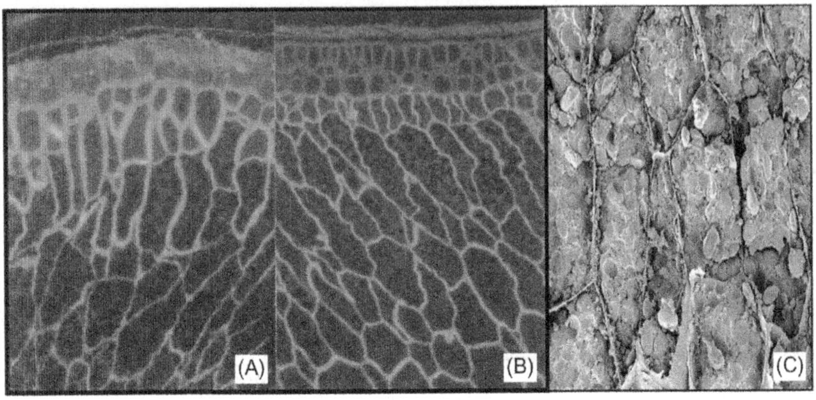

FIGURE 9.1 Distribution of β-D-glucan in oats (A) and barley (B); (C) is the scanning microscope image of a portion of endosperm of barley showing plant cell walls (Wood, 2002; Vasanthan, 2015).

may depend on its viscosity in solution, which is controlled by the MW, structure, and concentration in the intestine. Significant efforts have been made to establish the structure – function relationship of cereal β-D-glucans (Doublier and Wood, 1995; Lazaridou and Biliaderis, 2004; Li et al., 2006, 2012) with substantial success. Most recently, significant in vivo studies testing the efficacy of cereal β-glucans using animal and human subjects has led to health claims for this material in many industrialized countries (AbuMweis et al., 2016; Wolever et al., 2010; Panahi et al., 2014; FDA, 2017; Health Canada, 2016; EFSA, 2011). This chapter will review some basic understanding of cereal β-glucans, including their structure – functional property relationship, health attributes, and available commercial products on the market.

9.2 EXTRACTION AND STRUCTURAL CHARACTERIZATION

9.2.1 Solubility and Extractability

The content of β-D-glucans in cereals follows the order of barley, 3%–11%, oat, 3.2%–6.8%, rye 1%–2%, and wheat, <1%, as shown in Table 9.1.

The solubility of β-D-glucan is an important parameter that will not only affect the extractability from raw material, but will also affect the physico – chemical and physiological functions in the intestine after consumption. The solubility of β-D-glucan depends on the linkage pattern and MW. Cereal β-D-glucans contain approximately 30% (1–3)- and 70% (1–4)-linkage glucosyl residues which are organized into blocks of two, three, and sometimes more consecutive (1–4)-linked blocks. The (1–3)-linkages act as kinks to disrupt the regularity of the (1–4)-linkages (Cui and Wood, 2000) (Fig. 9.2). The higher solubility of mixed-linked β-D-glucans

TABLE 9.1 β-D-Glucan Content and Compositions of Cereals

	Starch (%)	Protein (%)	β-D-Glucan (%)	Pentosans (%)
Wheat	70	14	0.5–1	2–6
Barley	64	11	3–10	Smaller amount
Oat	59	16	2–6	Smaller amount
Rye	65	12	2	10

A B C A B C D A B C

$$\ldots{}^4G_1{}^3G_1[{}^4G_1{}^4G_1{}^3G_1]_m\ [{}^4G_1{}^4G_1{}^4G_1{}^3G_1]_n{}^4G_1\ldots$$

Ratio of m/n = 2.2 Oat
3.3 Barley
4.5 Wheat

FIGURE 9.2 Structural features of cereal β-D-glucans.

compared to cellulose is attributed to the presence of the irregular (1−3)-β-bonds.

The solubility of β-D-glucan in water is defined as the relative maximum amount of material that can be introduced into water. However, it is not easy to obtain a true solution for polysaccharides, due to aggregations caused by hydrogen bonding between two or more polymer chains. Therefore, solubility is frequently referred to as the extractability of β-D-glucan under specified conditions. Researchers should be aware that to compare data from different studies regarding the solubility and/or extractability of β-glucans from cereals are not recommended unless the extraction conditions and analytical methods are exactly the same (Beer, et al., 1996; Wood et al., 1991a,b; McCleary and Codd, 1991; Åman et al., 1989; Anderson et al., 1978) (Table 9.2).

Solubilization of β-D-glucan in aqueous solvents involves heating the polysaccharides at different temperatures for a period of time. Extraction of cereal β-glucan with mild reagents and conditions frequently results in incomplete extraction. For example, under mild extraction conditions (neutral pH and room temperature) only 30%−70% of the total β-D-glucan can be extracted (Beer et al., 1997a,b). Vigorous conditions such as in alkaline conditions could achieve a total extraction; however, it might depolymerize the chain of polysaccharides to give smaller MW β-D-glucan (Table 9.3). Due to the differences in solubility and stability of β-glucans in various solvents, decisions need to be made whether to use mild conditions and get a

TABLE 9.2 Comparison of Extraction Rate and Molecular Weight of β-D-Glucans From Oat Bran Samples Using Hot Water and Physiological Conditions

		β-Glucan Extracted (% of Total)		Molecular Weight ($\times 10^{-3}$)	
Sample	β-Glucan (g/100 g)	Hot Water	Physiological	Hot Water	Physiogical
Bran A	13.4	51.3	12.9	1400	1100
Bran B	8.9	56.7	25.1	1600	1800
Bran C	7.6	64.1	28.7	1800	1900
Bran C[a]	7.6	30.2	–[b]	1800	–[b]
Rolled oats[a]	4.2	69.5	33.4	1500	1500

[a]*Samples cooked before extraction.*
[b]*Not determined.*
Source: Adapted from Beer et al. (1997a,b).

lower extraction rate, or use more vigorous conditions and risk the degradation of polysaccharides. Using hot water or carbonate buffer (pH 10) as an extraction media (McCleary, 1988; Beer et al., 1997a,b) only 50%–70% of total β-D-glucan was solubilized, leaving over 30% of the β-glucan left unextractable (unknown MW). Sodium hydroxide appeared to give total extraction, but work by Beer et al. showed that the MW of β-D-glucan was significantly decreased by NaOH (Beer et al., 1997a,b; Carr et al., 1990).

McCleary (1988) reported that increasing the temperature of extraction leads to an increase in the MW of barley β-glucans, as determined by intrinsic viscosity of extracts. Similarly, Beer and coworkers found that the MW of oat β-D-glucan extracted at 90°C was up to 30% higher than that of β-D-glucan extracted at 37°C (Beer et al., 1997a,b). These observations suggest that high MW species are more soluble and therefore, extractable at higher temperatures.

Processing such as cooking could also affect the extractability of β-D-glucan. Cooked and uncooked samples gave substantially different results; e.g., only 13%–29% of the total β-D-glucans were extracted from an uncooked oat bran sample, compared to 30%–85% of the total β-D-glucans from a cooked sample (Table 9.3).

9.2.2 Solubility of Freeze-Dried β-D-Glucan Samples

Drying methods have a significant effect on the solubility of cereal β-D-glucans. Past experience in our laboratory also showed that freeze-dried oat

TABLE 9.3 Effect on Extraction Conditions on the Yield and Molecular Weight of Cereal β-D-Glucans

Source	Pt[a]	Extraction Solvent and Condition			Detection Method[b]	M_W ($\times 10^{-6}$)
		Solvent	T (°C)	Duration (min)		
Wheat bran preprocessed	2	1M NaOH	25	120	B,C	0.3–0.6
Wheat bran AACC soft white	2	1M NaOH	25	120	B,C	0.4–0.7
Wheat bran AACC red hard	2	1M NaOH	25	120	B,C	0.27–0.6
Barley flour	2	Na_2CO_3	45	90	B,C	0.8
Barley flour	2	H_2O	65	120	A,C	0.6
Barley cell wall (endosperm)	2	H_2O	40–65	–	D	~40
Oat bran	1	H_2O	70	60	A	1.5
Oat flour	2	H_2O	90	120	B,C	2.0–2.5
Oat bran	1	Na_2HPO_4 (20 mM)	37	135	B,C	1.1–1.9
Oat bran	1	Na_2CO_3	70	120	A	0.4–1.5
Oat bran	2	Na_2CO_3	60	120	B,C	2.7
Flour	2	Na_2CO_3	60	120	B,C	3.0
Oat bran	2	Na_2CO_3	60	120	B,C	3.1

[a]*Sample pretreat: (1) no pretreatment; (2) sample treated with hot aqueous or ethanol prior to extraction.*
[b]*Detection method: (A) multiangle laser light scattering detection; (B) calcofluor postcolumn detection with β-glucan standards; (C) refractive index detection with pullulan standards; (D) sugar concentration.*
Source: Reprinted from Cui and Wood (2000) with permission.

β-D-glucan is very difficult to dissolve in water. Freeze-dried wheat β-D-glucan is essentially insoluble in water, regardless of heating temperature and time (Cui and Wood, 2000). To solve this problem, Wood developed a solvent exchange method in our laboratory which is briefly described below: β-D-glucan solutions at 1%–2% can be precipitated in three volumes of ethanol. The precipitate is gently homogenized into particles in 100% ethanol. The top ethanol is decanted; the precipitate is merged in 100% isopropanol at 4°C overnight. This process can be repeated twice. The final precipitate-free of solvent is gently heated on a hot plate to give a porous white powder of β-D-glucan. Beta-D-glucan samples prepared using this solvent exchange method proved to be soluble in water; but sometimes may need gentle heating.

9.2.3 Structural Characterizations of Cereal Beta-Glucans

9.2.3.1 Monosaccharide Composition and Linkage Pattern

Monosaccharide analysis of cereal β-D-glucan revealed that it only contains one sugar unit: glucose. However, arabinose and xylose could be detected in the extract from wheat bran beta-glucan which was derived from arabinoxylan. Use of xylanase could remove the arabinoxylans to give a purified β-D-glucan (Cui et al., 2000). The monosaccharide composition of the extracted nonstarch polysaccharides (NSP) and purified β-D-glucan from wheat bran is illustrated in Fig. 9.3.

The linkage patterns of cereal β-D-glucan are determined by methylation analysis. This is a classic method which includes two steps: the first step is chemical derivatization of the polysaccharide into partially methylated and acetylated alditols (PMAA); PMAA is subsequently analyzed by GC-MS (Cui, 2005). The gas chromatogram and mass spectra of the two types of substitutions are presented in Fig. 9.4: the peak at retention time 20.27 min is confirmed to be (1−3)-linkage, and the peak at retention time 23.63 min is (1−4) linkage. The peak area or peak height is used to calculate the relative amount of each linkage. In this case, it was estimated at about 30% (1−3)-linkage and 70% (1−4)-linkage (β-D-glucan from wheat bran, Cui et al., 2000).

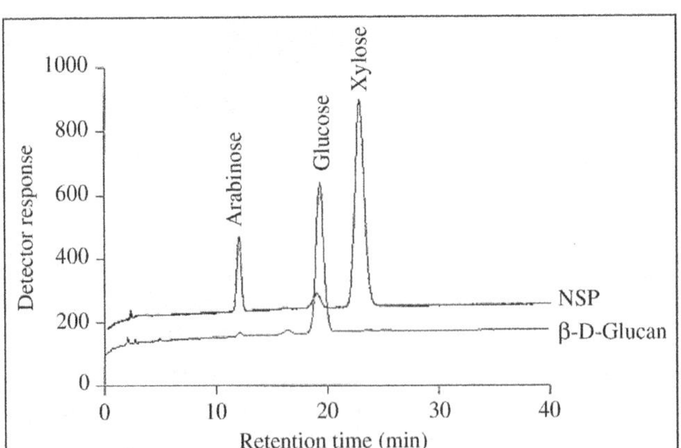

FIGURE 9.3 Monosaccharide composition analysis of non-starch polysaccharides (NSP) and β-D-glucan isolated from wheat bran (Cui et al., 2000).

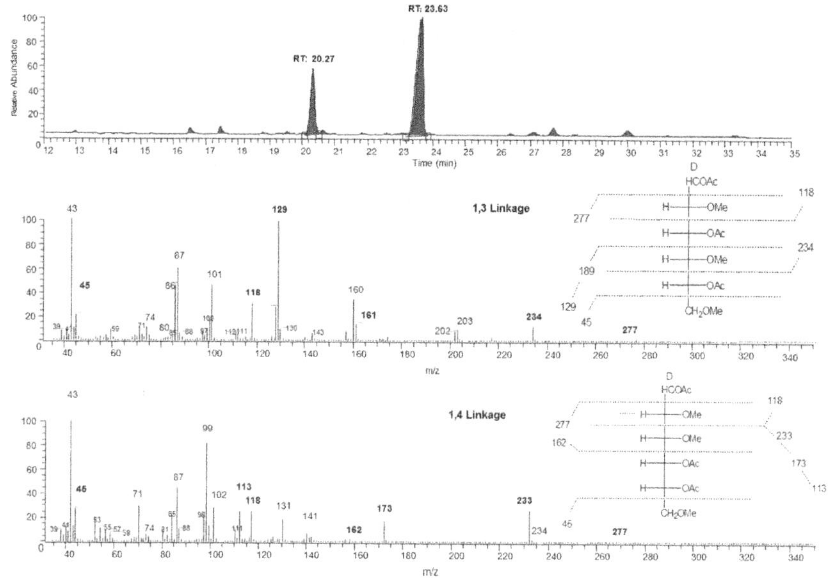

FIGURE 9.4 Chromatogram of partially permethylated acetal alditols of cereal β-glucan and their mass spectra (Cui, 2005).

9.2.3.2 Enzyme Hydrolysis

Lichenase is a (1→3)(1→4)-β-D-glucan-4-glucanohydrolase (E.C. 3.2.1.73) which specifically cleaves the (1→4)-linkage of the 3-*O*-substituted glucose unit in cereal β-D-glucans. Purified β-D-glucan, raw materials, and food products containing β-D-glucan can be dissolved in 0.05M phosphate buffer (pH 6.9), then digested by this enzyme (McCleary and Codd, 1991). The oligosaccharide release by lichenase is analyzed by high performance anion exchange chromatography (e.g., a Dionex system with a gold electrode equipped with a PAD) (Wood et al., 1989).

The relative amounts of oligosaccharides produced by lichenase treatment constitute a fingerprint of the structure of a β-D-glucan. The total of tri- and tetrasaccharide of cereal β-D-glucan is similar (92%–93%) among cereal β-D-glucans; however, the amount of trisaccharide follows the order of wheat (~72%), barley and rye (~66%), and oat (~58%) (Table 9.4). This is also reflected by the ratio of tri- to tetrasaccharides, which also follows the order of wheat (4.6), barley (3.3), rye (2.7), and oat (2.2). The ratio of tri- to tetrasaccharides can be used as a parameter which defines the identity (fingerprint) of the structure of a particular cereal β-D-glucan.

It is worth noting that there is some precipitate observed in the lichenase digested product which was formed by hydrogen bonding of longer chains of the consecutive β-(1–4)-linked glucosyl residues (greater than 5). Because the enzyme action is specific, these oligosaccharides may contain greater

TABLE 9.4 Comparison of Oilgosaccharide Profiles of Cereal β-Glucans

β-D-Glucan Source	Peak Area (%)		Total (%)		Molar Ratio
	Tri[a]	Tetra[a]	Tri + tetra[b]	Penta ~ nona[b]	Tri/tetra[c]
Wheat bran (soft white)	70.6	22.9	93.5	6.5	4.2
Wheat bran (red hard)	72.0	21.0	93.0	7.0	4.5
Barley	63.7	28.5	92.2	7.8	3.0
Oat	58.3	33.5	91.9	8.1	2.3

[a]*HPLC peak area percentage of tri and tetrasaccharides after lichenase hydrolysis.*
[b]*Total percentage of tri and tetrasaccharides and penta to nonasaccharides, respectively.*
[c]*The molar ratio of trisaccharide over tetrasaccharide from lichenase hydrolysis.*
Source: Reprinted from Cui and Wood (2000) with permission.

than five consecutive β-(1→4)-linked glucopyranosyl units with one terminal reducing end of β-(1→3)-linked glucose (Wood and Beer, 1998). The higher DP oligosaccharides simulate cellulose-like structures in β-D-glucan, which may partially contribute to the stiff conformation of this polysaccharide in solution.

9.2.3.3 NMR Spectroscopy

The detailed structural properties of cereal β-D-glucans have been further confirmed by 1D and 2D NMR spectroscopies. The ^{13}C NMR spectrum of wheat β-D-glucan at 90°C appeared similar to that of oat and barley, as demonstrated in Fig. 9.5.

Three distinct signals were observed for anomeric carbons of β-D-glucan in chemical shifts above 100 ppm (Fig. 9.5), suggesting three types of linkage environment. The ^{13}C-^{1}H heteronuclear correlation (Fig. 9.6A) allowed a one-to-one coupling of the ^{13}C and ^{1}H resonances. The ^{1}H/^{1}H correlation (COSY) spectrum (Fig. 9.6B), in combination with TOCSY (Fig. 9.6C), are used to assign each proton signal on one sugar residue; this assignment is the foundation for establishing the interresidue connectivity; any misassignment could lead to a wrong conclusion of the linkage, and therefore, an incorrect structure (Cui et al., 2000). The long range heteronuclear multiband correlation spectrum (HMBC) allows the establishment of interresidue connectivity of the two neighboring sugar residues (Fig. 9.6D). The combination of the 2D NMR spectra allowed the complete assignment of all the major signals which are summarized in Table 9.4.

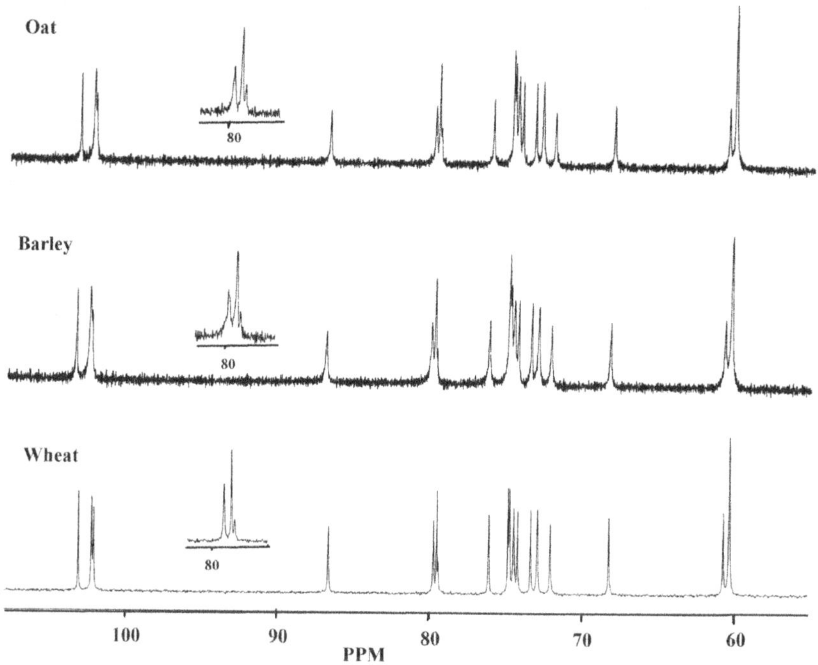

FIGURE 9.5 C-13 NMR spectra of three β-glucans.

The (1→3)-linked β-D-glucose residue (B in Scheme 9.1) is confirmed by carbon peaks at 101.9, 86.4, 75.9, 71.8, 68.0, and 60.5 ppm, which can be assigned to C-1, C-3, C-5, C-2, C-4, and C-6, respectively. The presence of only a single resonance (86.4 ppm) for C-3 confirmed that there are no consecutive (1→3)-linkages in cereal β-D-glucan. These assignments, along with the remaining correlations defining the two (1→4) linked β-D-glucan residues, are summarized in Table 9.5.

9.2.4 Structural Features

From the above information, the structural features of cereal β-D-glucans are summarized below: cereal β-glucan is typically of a linear homopolysaccharide containing a single type of sugar unit: β-D-glucopyranose (β-D-Glc*p*). Most of the β-D-Glc*p* residues in cereal β-glucans are grouped as blocks of tri- and tetrasaccharide units: two or three consecutive (1→4)-linked units separated by a single (1→3)-linkage. These building blocks can also be described by a cellotriosyl unit and a cellotetraosyl unit, as shown in Fig. 9.2. About 10% of the polymer chains are composed of longer cellulosic sequences (from 5 up to 14 β-D-Glc*p* residue units). The structural features of cereal β-glucans are represented by the oligosaccharide profiles released

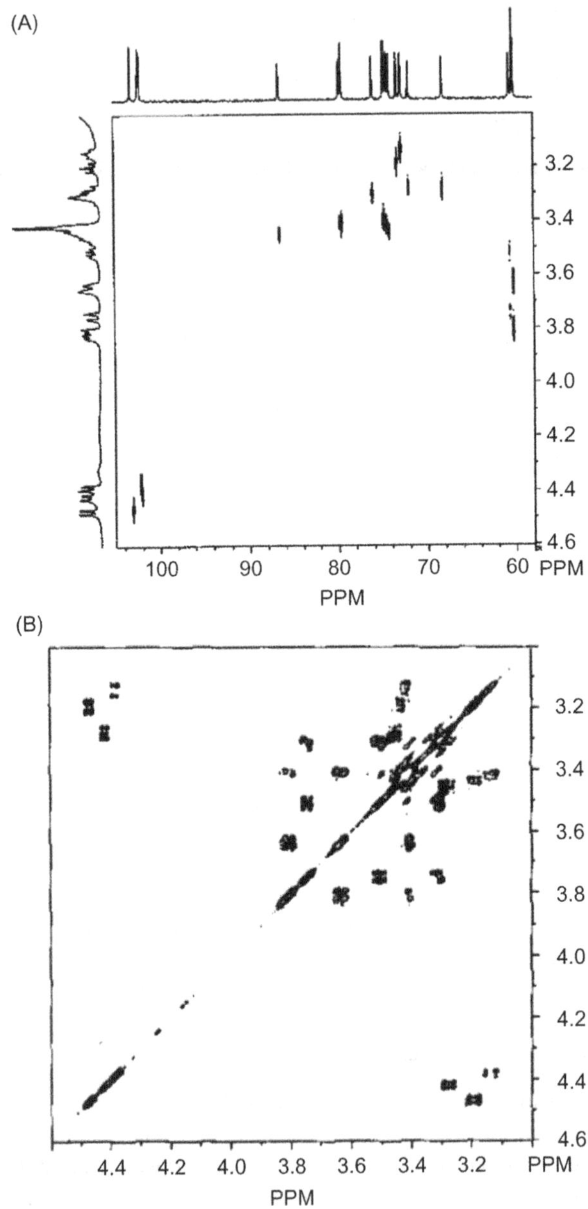

FIGURE 9.6 Two dimensional NMR spectroscopy of wheat β-glucan. (A) The ^{13}C-^{1}H heteronuclear correlation; (B) The 1H/1H correlation (COSY) spectrum; (C) Total correlation spectrum (TOCSY); and (D) Heteron nuclear multiband correlation (HMBC).

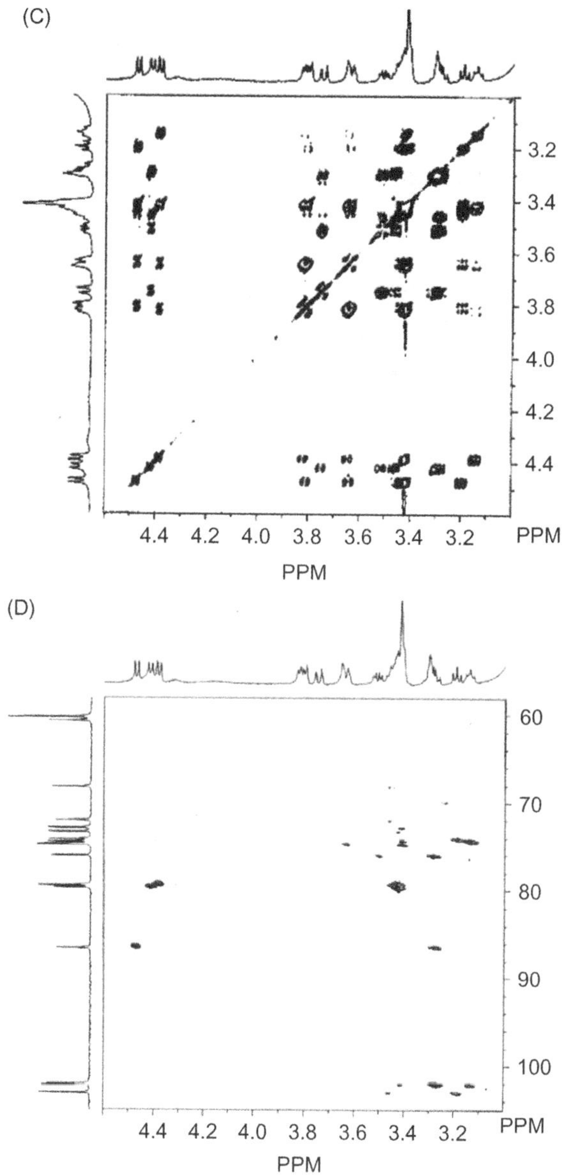

FIGURE 9.6 (Continued)

by the specific enzymatic hydrolysis (lichenase: EC 3.2.1.73, *endo*-1,3(4)-β-D-glucanase) (Cui et al., 2000). Of the oligomers released, the trisaccharide and tetrasaccharide units are particularly important, because their ratios are used as the fingerprint of the β-glucan from a specific cereal source. The tri/

$$\begin{matrix} & \text{A} & \text{B} & & \text{C} & \text{A} & \text{B} & & \text{C} & \text{D} & \text{A} & \text{B} & \text{C} \\ \ldots & {}^{4}G_{1} & {}^{3}G_{1}[& & {}^{4}G_{1} & {}^{4}G_{1} & {}^{3}G_{1}]_{m}[& & {}^{4}G_{1} & {}^{4}G_{1} & {}^{4}G_{1} & {}^{3}G_{1}]_{n} & {}^{4}G_{1}\ldots \end{matrix}$$

SCHEME 9.1

tetrasaccharide ratio for β-glucan from wheat, barley, and oats are 4.2, 3.2, and 2.2, respectively, as shown in Table 9.4.

9.3 MOLECULAR WEIGHT, CONFORMATION, AND RHEOLOGICAL PROPERTIES

9.3.1 Molecular Weight Determination

Parallel to the primary structure, molecular weight and molecular weight distribution are other fundamental structural properties of polysaccharides that govern the overall functionality. Molecular weight and molecular weight distribution affect the functionality of polysaccharides by affecting their solubility, hydration, and conformational properties in aqueous systems, which are reflected in food systems as viscosity and rheological behavior in the intestine after consumption. Because of the complexity and polydispersity of natural polysaccharides, there is no simple single way to accurately express their molecular weight.

Number − average molecular weight, Mn, is defined as the total weight of substance divided by the total number of moles present:

$$Mn = \Sigma MiNi/\Sigma Ni \tag{9.1}$$

Mn can be determined by the colligative properties of solutions (e.g., osmometric method).

Weight average MW is defined as:

$$Mw = \Sigma NiMi^2/\Sigma NiMi = \Sigma CiMi^2/\Sigma CiMi \tag{9.2}$$

A number of methods can be used to determine the weight average MW, including light scattering, sedimentation, diffusion, and flow birefringence. However, the most commonly used method is high performance size exclusion chromatography (Cui, 2005). Recently, size exclusion chromatography systems have been equipped with a light scattering detector in combination with refractive index, viscometric and UV detectors. The combination of size exclusion chromatographic profile (retention time or volume) and the parameters calculated from multi detectors allows simultaneous characterization of both MW/shape and MW distribution. Fig. 9.7 demonstrates the responses of four different detectors to one β-D-glucan sample. The signals from the light scattering detector often come out earlier (smaller retention volumes) than the other three detectors. This is because the light scattering detector is sensitive to large molecules which appeared earlier, even if the amount is very small, but the detected signals can be very high.

TABLE 9.5 Complete Assignment of 13C and 1H NMR Spectra of Wheat β-D-Glucan Based on Heteronuclear Correlation and Shift-Correlated Spectroscopy (COSY)

Glucose Residue	Assigned C, H Position	^{13}C Resonance	^{1}H Resonance
(1–4)-Linked b-D-glucose (residue A in Scheme 9.1)	1	103	4.47
	2	73.2	3.19
	3	74	3.43
	4	79.3	3.42
	5	74.7	3.42
	6	60.1	3.64
			3.81
(1–3)-Linked b-D-glucose (residue B in Scheme 9.1)	1	101.9	4.42
	2	71.9	3.28
	3	86.4	3.46
	4	68	3.3
	5	75.9	3.3
	6	60.5	3.51
			3.75
(1–4)-Linked b-D-glucose (residue C in Scheme 9.1)	1	102.1	4.38
	2	72.7	3.14
	3	74.3	3.42
	4	79.5	3.42
	5	74.6	3.42
	6	60.1	3.64
			3.82
(1–4)-Linked b-D-glucose (residue D in Scheme 9.1)	1	102.1	4.38
	2	72.7	3.14
	3	74.3	3.42
	4	79.2	3.42
	5	74.7	3.42
	6	60.1	3.64
			3.81

Source: Reprinted from Cui et al. (2000) with permission.

TABLE 9.6 Linkage Sequence and Connectivities of Wheat β-D-Glucan Derived From Long Range Correlation (HMBC) Spectra

Resonance	Correlated Resonance
4.47 ppm (H-1 of A)	86.4 ppm (C-3 of B)
4.42 ppm (H-1 of B)	79.5 ppm (C-4 of C)
4.38 ppm (H-1 of C)	79.3 and 79.2 ppm (C-4 of A and D, respectively)
103.3 ppm (C-1 of A)	3.46 ppm (H-3 of B)
101.9, 102.1 ppm (C-1 of B and C)	3.42 ppm (H-4 of A and C)
3.19 ppm (H-2 of A)	74.0 ppm (C-3 of A) and 103.0 ppm (C-1 of A)
3.14 ppm (H-2 of C)	74.3 ppm (C-3 of C) and 102.1 ppm (C-1 of C)
3.28 ppm (H-2 of B)	86.4 ppm (C-3 of B) and 101.9 ppm (C-1 of B)
68.0 ppm (C-4 of B)	3.46 ppm (H-3 of B)
75.9 ppm (C-5 of B)	3.30 ppm (H-4 of B)

A, B, C, and D are the sugar residues as described in Scheme 9.1.
(A) is the (1→4)-β-D-glucose residue glycosidically linked to the 3 position of the (1→3)-β-D-glucose (B) and attached through C-4 to a (1→4)-β-D-glucose.
(B) is the (1→3)-β-D-glucose residue.
(C) is the (1→4)-β-D-glucose residue which glycosidically linked to the 4 position of the (1→4)-β-D-glucose (A or D) and attached by a (1→3)-β-D-glucose at the C-4 position.
(D) is the (1→4)-β-D-glucose residue which is franked on either side by a (1→4)-β-D-glucose residues.
Source: Reprinted from Cui et al. (2000) with permission.

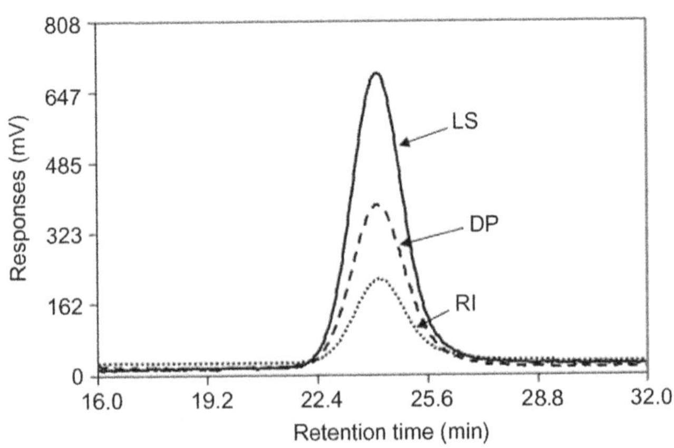

FIGURE 9.7 HPSEC profiles of polysaccharides (Cui, 2005).

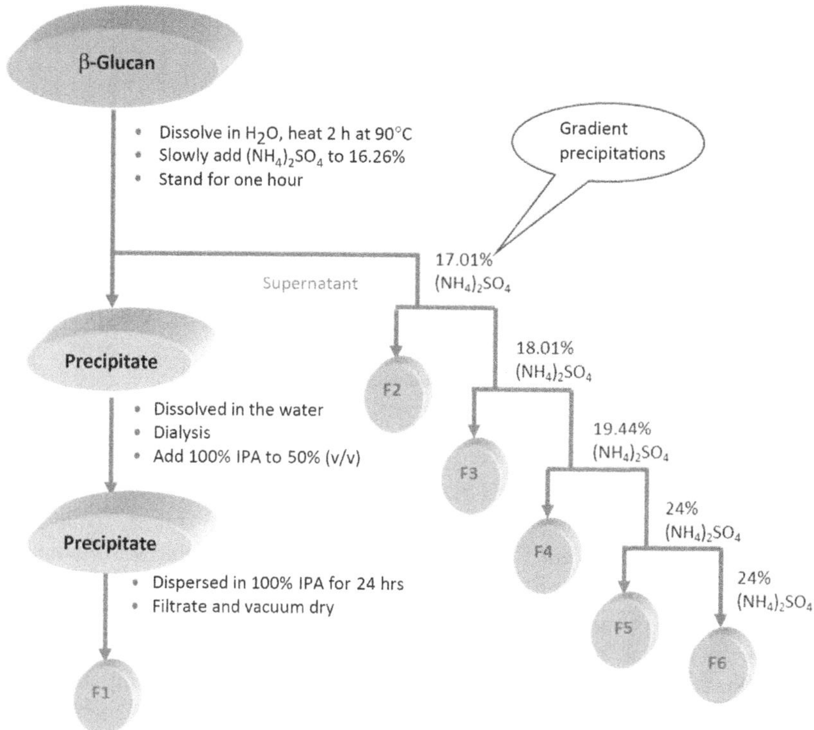

FIGURE 9.8 Gradient precipitation of wheat β-glucan by step-wise addition of ammonium sulfate.

Fig. 9.9 illustrates the MW and MW distribution of six fractions of cereal β-D-glucans obtained by ammonium sulfate step-wise precipitation (Fig. 9.8) (Li et al., 2011).

The MW of β-D-glucan in crude extracts from oat, barley, and rye and food products could be determined without a prior purification process. This method used a specific dye binding (e.g., calcofluor) which can selectively bind with β-D-glucan (Wood and Fulcher, 1978). In this method, purified β-glucan standards are used to calibrate the column (Wood et al., 1991a,b). A set of high purity barley β-D-glucan MW standards was prepared by our laboratory which has been used widely in the measurement of MW and MW distribution (Fig. 9.9) (Wang et al., 2003). It is unlikely that the "true" MW or MW distribution of β-glucan from cereals can be obtained as its original location is in the cell wall (Fig. 9.1). As soon as it is extracted by solvent, depolymerization can occur. However, it is worth the effort to aim to determine the maximum possible MW that can be extracted.

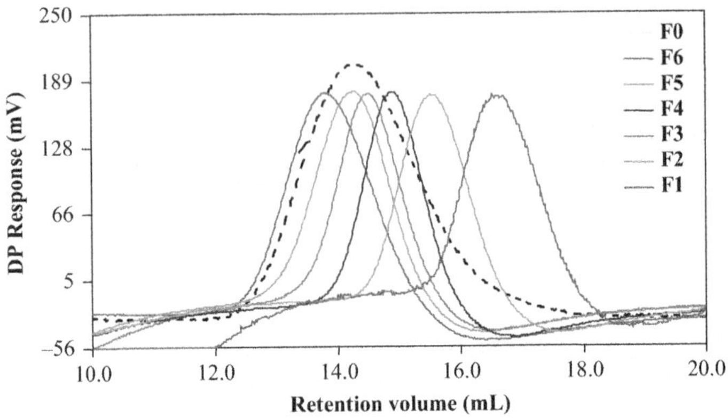

FIGURE 9.9 Molecular weight distribution of six fractions.

The isolated MW of wheat β-glucan is in the range of 300,000–600,000. Isolated and purified wheat β-glucan is in the lower MW range, while the direct measurement of crude extract gives a higher MW (Table 9.3), indicating that the purification process might cause some degradation of the polysaccharides or loss of the large molecular species during the purification process.

Measurement of MW of polysaccharides requires that the polymers are properly dispersed, ideally molecularly. Vårum et al. (1992) reported an aggregation phenomenon for depolymerized oat β-glucan. It has also been observed that wheat and barley β-glucans can form gels under storage conditions (4–25°C) (Cui et al., 1999; Böhm and Kulicke, 1999). Both lower temperature and repeated cycles of short-term steady shear and rest periods accelerated the sol – gel transition. Therefore, heating of the sample before HPSEC analysis is necessary. A column temperature of 77.5°C was used by Böhm and Kulicke (1999) in order to avoid aggregations during measurement. This aggregation behavior of cereal β-D-glucan will be discussed in more detail in the following section, due to its importance in affecting the functionality.

9.3.2 Solution and Conformational Properties

9.3.2.1 *Solution and Aggregation Behavior*

Similar to many other polysaccharides, β-D-glucans exist in solution as a conformationally-disordered "random coil" whose shape fluctuates continually under Brownian motion. At low concentrations, the individual coils are separated from each other and are free to move independently. When concentration is increased to a stage those independent random coils start to touch, overlap and interpenetrate with one another. In company with the relative status of the coils are their solution properties, such as solution viscosity. The concentration dependence of limiting specific viscosity of oat β-D-glucans is shown in Fig. 9.15B.

There is a clear transition from a dilute solution of independently-moving coils to an entangled network, evidenced by the dramatic change of the slope from 1.08 to 3.90 (Doublier and Wood, 1995).

The solution properties of β-D-glucan chains also varied significantly in different solvent systems. These variations in conformation will affect the hydrodynamic volume, and hence, the MW determined based on it. As shown in Table 9.6, the measured MW of a β-D-glucan sample isolated from beer has the highest MW, radius of gyration, and intrinsic viscosity in water, followed by 90% in DMSO, 2M GHCl, Cadoxen, and Cuoxam I and II (Grimm et al., 1995).

The metal complexing solvents, cuoxam and cadoxen, are known to be good solvents for cellulose, which can minimize the interactions between β-D-glucan molecules. These solvents dissociated the aggregates, and thus allowed determination of the MW of the unimer. In a double logarithmic plot of Rg versus Mwapp (Fig. 9.10), a low exponent $a_{Rg} = 0.22$ was found (Grimm et al., 1995):

$$Rg = KMw, app^{0.22} \tag{9.3}$$

To clearly demonstrate the aggregation behavior and develop a method of eliminating the aggregate in order to study the conformational properties of cereal β-D-glucans, our laboratory has carried out a systematic study examining the particle size distributions of β-D-glucans in dilute solution dynamic light scattering measurements, as shown in Fig 9.11A (Li et al., 2008).

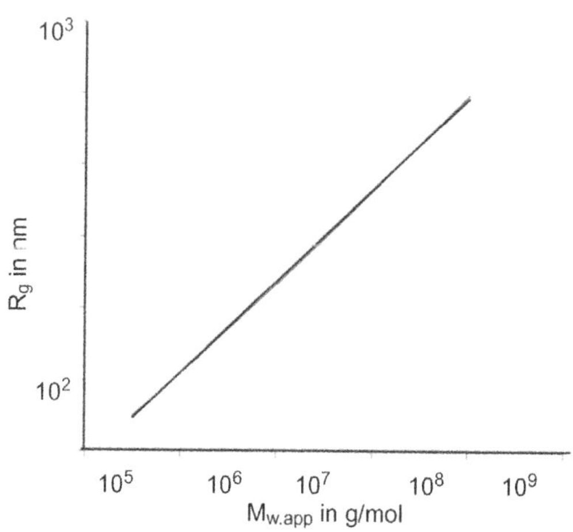

FIGURE 9.10 Double logarithmic plot of R_g against Mw of 0.1% cereal β-glucans in various aqueous and non-aqueous solvents (redrawn from Grimm et al., 1995).

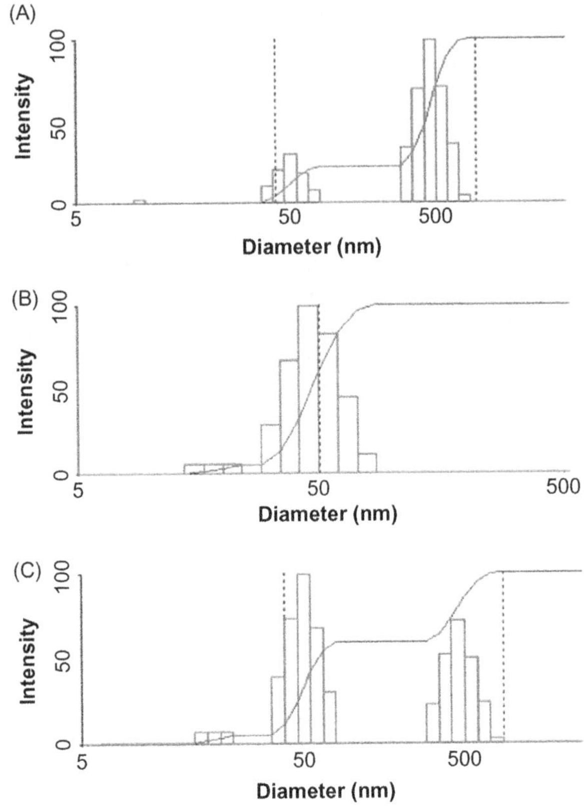

FIGURE 9.11 The molecular size distribution of wheat β-D-glucan. (A) In pure water; (B) In 0.5M NaOH solution; (C) In neutralized solution.

Subsequently, such aggregation behavior was also confirmed by atomic force microscope and NMR relaxation techniques (Agbenorhevi et al., 2011; Wu et al., 2017). This aggregation behavior will prevent the accurate measurement and characterization of the β-D-glucans. Therefore, these aggregates must be eliminated before conformational studies. Aggregation phenomena of cereal β-D-glucans are also important for their applications in foods and human health, which will be described in detail in later sections.

9.3.2.2 Conformational Properties: Light Scattering and Computer Modeling

Fig. 9.11 shows the size distribution of wheat β-D-D-glucan in dilute water solution (0.075 mg/mL), with two populations with apparent mean diameter of 67.2 and 278.2 nm, respectively. The second population were considered to be aggregates. Although the samples were prepared by heating for 3 h at 90°C and filtering five times with a 0.45 μm membrane, the aggregates were still observed. The aggregates accounted for 3.89% of the total amount of wheat β-D-D-glucan molecules by weight, but accounted for 81.5% by intensity. The presence of larger particles contributed more to the scattering intensity than the smaller ones.

The formation of aggregates of neutral polysaccharides are usually caused by intermolecular hydrogen bonding. Addition of 0.5M NaOH in the solution effectively eliminated the aggregates by decreasing the strength of hydrogen bonding between β-D-glucan molecules (Fig. 9.11B). By neutralizing the NaOH, the aggregates reappeared (Fig. 9.11C), confirming that the aggregates were formed via hydrogen bonding between two β-D-glucan molecules, which dissociated in the presence of hydrogen bond breaking agent NaOH, and reformed after NaOH was neutralized (Li et al., 2008).

With the removing the aggregates from β-D-glucan solutions, Li et al. (2008) successfully measured the absolute MW of cereal β-D-glucans based on the Zimm plot, as shown in Fig 9.12. The molecular characteristics of six fractions from each of the three cereal β-D-glucans, i.e., wheat, barley, and oats are summarized in Table 9.7 (Fig. 9.13).

Based on these important molecular parameters, a plot of the Mark – Houwink – Sakurada equation ($[\eta] = KM_w^{\alpha}$) was derived. The conformation parameter α was calculated from the slope, which gives a value of 0.5–0.8, indicating a random coil conformation for all the β-D-glucans. Li et al. (2006, 2012) further use this technique to study the conformational properties, and dynamic and kinetics, of the aggregation behavior of cereal β-D-glucans. Table 9.8. summarized the conformational characteristics of the three cereal β-D-glucans based on the worm-like cylinder model (Li et al.,

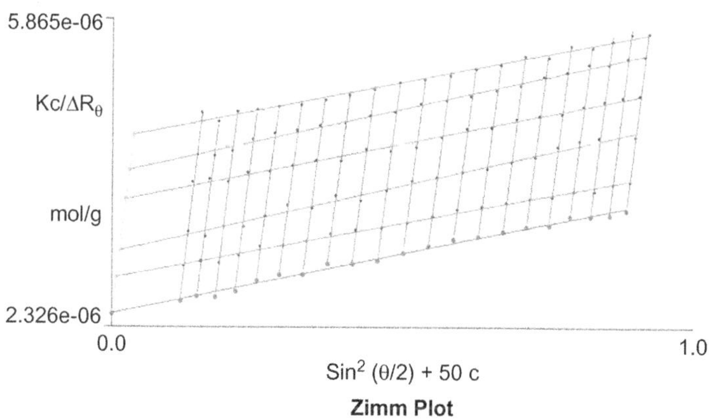

FIGURE 9.12 The Zimm plot of wheat β-D-glucan in 0.5 M NaOH solution.

TABLE 9.7 Molecular Characteristics of Cereal β-Glucan Samples in Aqueous Solution

	M_w (g/mol)	A_2 ($\times 10^4$ mol mL/g^2)	R_g (nm)	R_h (nm)	[η] (dL/g)
WF1	919,000	−2.18	80.5	99.0	6.5
WF2	575,000	4.49	62.2	79.9	6.1
WF3	367,100	2.01	53.2	73.4	5.2
WF4	319,500	2.46	51.9	72.5	3.7
WF5	171,500	2.71	47.3	69.8	2.8
WF6	103,500	−1.44	44.6	53.9	1.5
BF1	581,100	−2.70	60.4	109.5	5.9
BF2	492,800	−1.69	55.2	105.7	4.4
BF3	290,300	−1.20	45.4	96.2	4.0
BF4	242,000	−0.82	49.8	86.3	3.5
BF5	150,100	−3.30	39.2	73.8	2.3
OF1	2,025,000	−4.98	75.8	191.4	4.2
OF2	2,076,000	−2.09	73.9	183.6	4.2
OF3	811,000	−2.11	52.0	165.3	3.2
OF4	438,300	−2.71	48.8	128.2	1.8
OF5	364,300	−1.71	50.7	135.9	1.4
OF6	128,100	−3.50	44.3	70.7	0.9

Note: Molecular weight (M_w), the second virial coefficient (A_2), and radius of gyration (R_g) were measured by SLS; hydrodynamic radius (R_h) was measured by DLS.

2012). The structural features of cereal β-glucans played an important role in their aggregation behaviors. As MW increased, the degree of aggregation decreased, due to the lower diffusion rate of large molecules. For the more rigid conformation (lower diffusion rate) of β-D-glucans with a higher tri/tetra ratio, their degrees of aggregation were lower. These results suggested that the aggregation process was diffusion-limited.

Li et al. (2012) further used computer modeling techniques to calculate those molecular parameters and visualize the shape of the random coil polysaccharide in aqueous solution. Fig. 9.14A illustrates the side and top views of the single helix structure of six consecutive cellotriosyl units, which was considered as the active site for the formation of junction zones between two β-D-glucan molecules (Li et al., 2012).

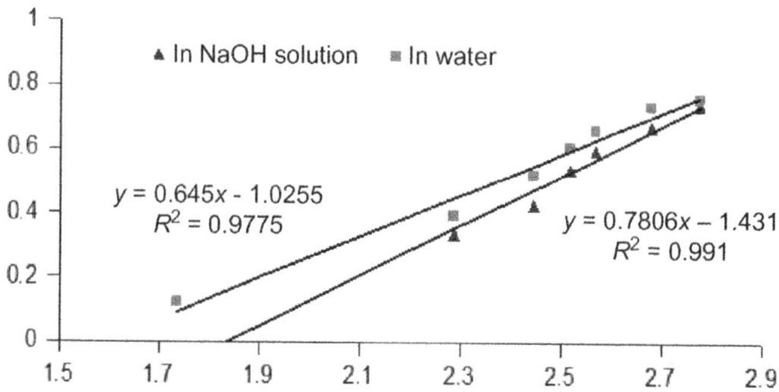

FIGURE 9.13 Plot of the Mark – Houwink – Sakurada equation: $[\eta] = KM_w^{\alpha}$. The double logarithmic plot of $[\eta]$ against MW of wheat β-D-glucan in pure water and in 0.5 m NaOH solution. Slope value: $\alpha = 0.5–0.8$, suggesting a random coil conformation in both solvents (Li et al., 2011).

TABLE 9.8 Worm-Like Chain Parameter of Cereal β-Glucans

	α	ν	ν'	q (nm)	M_L (g/mol nm)	n	q_0 (nm)
Wheat β-glucan	0.68	0.24	0.31	25.35	1459.4	4.7	4.42
Barley β-glucan	0.61	0.29	0.29	21.05	1332.8	4.3	3.47
Oat β-glucan	0.58	0.32	0.25	8.41	1217	3.9	2.3

Note: α, ν, and ν' are exponents from plot log $[\eta]$, R_g, R_h versus log M_w, respectively. q and q_0 are the persistent length with and without aggregates, respectively. M_L is linear mass density, and $n = M_L/M_{L0}$, multiplicity of aggregate structure (Li et al., 2011).

Molecular modeling was used to study the conformational properties of cereal β-glucans and was compared with experimental results. Following the steps of exploring global minima of β 1-4- and β 1-3-linked disaccharides and building up the repeating units, the consecutive cellotriosyl units, which are considered as reaction sites of β-glucan chains, were created and then characterized as a threefold helix with a pitch of 41.35 Å (Fig. 9.14A; Li et al., 2012). Most importantly, the 3D molecular model of cereal β-glucans was created, and the moderately extended sinuous chain conformation was first visualized, as shown in Fig. 9.14B. The conformational parameters of cereal β-glucans were calculated by RMMC simulation, which are in good agreement with experimental results. The calculated parameters also revealed that the chain stiffness of β-glucans increased with the tri/tetra ratio. Explicit

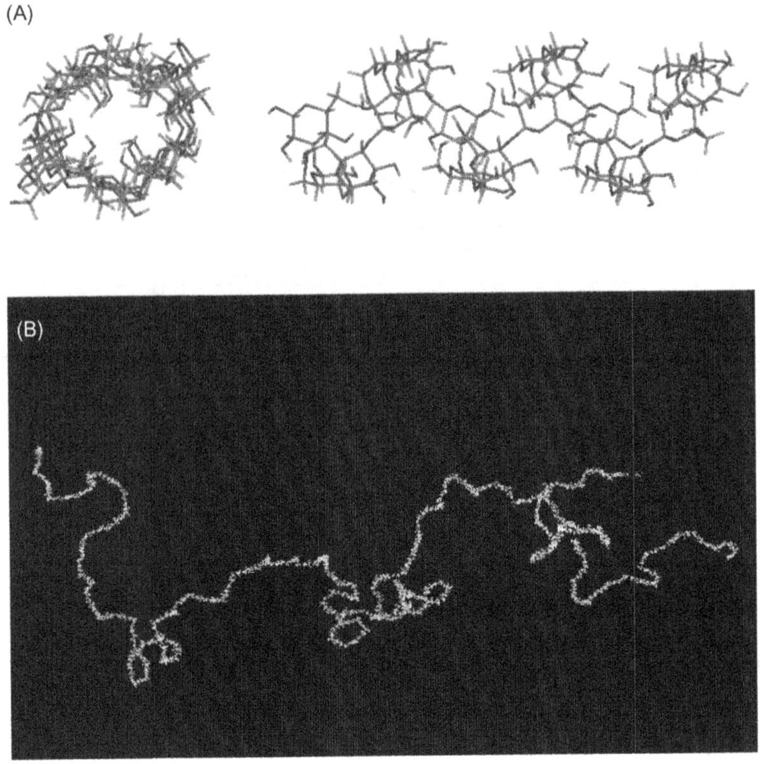

FIGURE 9.14 (A) Side and top view of single helix structure of six consecutive cellotriosyl units. (B) Snapshot view of a β-D-glucan molecule in random coil shape (Li et al., 2012).

conformations obtained from the new approach of computer modeling provided an insight into the molecular information favorable for understanding the mechanisms of molecular interactions (Fig. 9.14B; Li et al., 2012). These studies provide fundamental information for understanding the physiological effects and functional properties of cereal β-glucans.

9.3.3 Rheology: Viscoelastic and Gelling Properties

9.3.3.1 *Steady Flow*

Cereal β-glucan is a linear, high MW polysaccharide. The shear-rate ($\dot{\gamma}$) dependence of viscosity of β-D-glucans is typical of a "random coil" solution: there is a Newtonian plateau at low shear rate region and a shear thinning region at a higher shear-rate region, as shown in Fig. 9.15A (Autio et al., 1987; Doublier and Wood, 1995). This typical rheological behavior of the "random coil" type of polysaccharides is well understood,

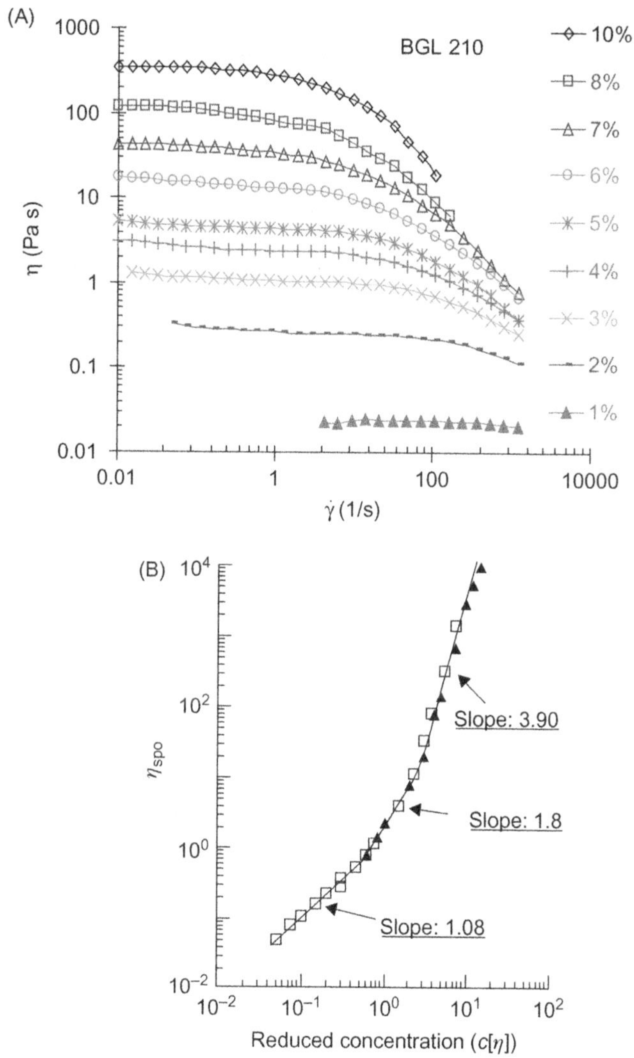

FIGURE 9.15 Effect of concentration on the apparent viscosity of barley beta-glucan (A) (Biliaderis, 2015) and specific viscosity (B) of oat beta-glucan (Doublier and Wood, 1995).

and can be described as: at very low shear-rates ($\dot{\gamma}$), the viscosity (η) remain unchanged with the increase of shear rate; this is also called "zero shear rate viscosity" (η_0) at which the β-D-glucan chain entangled at its maximum value (η_0). The increase of shear rate will cause the disentanglement of the junction zone; however, since the shear rate is very slow, there is sufficient time for new entanglements to form to compensate the disentanglement caused by shear flow, therefore, the viscosity remains unchanged. As the shear-rate (γ) increased to a certain level, the rate of reentanglements falls behind the rate of disruption of the existing entanglements, and the overall junction zone of the network decreases progressively, and the viscosity falls. This is the shear thinning flow behavior. The onset shear rate value at which the shear thinning flow behavior begins is both concentration- and molecular-weight dependent (Lazaridou and Biliaderis, 2007).

The viscosity of a polysaccharide in solution is related to the volume occupied by the polysaccharide. From Fig. 9.15B it could be observed that higher MW β-glucans have occupied larger volumes, and have a much lower critical concentration (c^*) at which the onset of inter molecular chain entanglements begins. It is obvious that at concentrations above c^*, the viscosity increases much more rapidly with increase of concentration, compared to the concentrations below c^* (Doublier and Wood, 1995).

9.3.3.2 *Viscoelastic Properties and Gelation*

As discussed in the previous section, cereal β-D-glucans exhibit flow behavior of "random-coil" polysaccharides. The frequency sweep rheological test of a β-D-glucan solution has a typical "random coil" type of polysaccharide, as shown in Fig. 9.16 (Ren et al., 2003), where the storage modulus G' is lower than the loss modulus G'' at low frequencies; as the frequency increased, G' becomes greater than that of G'' (Cui, 2005; Ren et al., 2003). However, an unusual behavior was observed for partially hydrolyzed oat β-D-glucan by Doublier and Wood (1995), who found that the storage modulus G' was higher than the loss modulus G'' at a lower frequency range, and G' become weaker than G'' at a higher frequency region, as shown in Fig. 9.17. Similar observation was reported on barley β-glucan by (Gómez et al., 1997). This phenomenon was later confirmed by Cui and Wood (2000), who observed that the entanglement of β-D-glucans in solution were time-dependent, which essentially lead to the formation of a gel. It has been noticed in the brewing industry that barley β-D-glucans form gels under storage conditions (Bamforth, 1982; Böhm and Kulicke, 1997).

During our studies on cereal β-glucans, we observed that commercial β-D-glucans (Magezyme Intl. MW ~ 200,000–300,000) and wheat β-D-glucans (MW 300,000–600,000) at 2.0% (w/w) formed gels when stored at 4°C over a period of 2–5 days. The gel forming properties of β-D-glucans were examined at 2.0% (w/w) and 4°C. After 24 h, wheat β-glucan solution formed a firm gel with a small but visible amount of water separated from the gel (Fig. 9.18). The mechanical spectrum of the wheat β-D-glucan gels at different MW is shown in Fig. 9.19. This gel started to melt at 55–60°C (Fig. 9.20) (Cui and Wood, 2000). It is interesting to observe that the gelation process is structural and MW-dependent: the higher ratio of the tri- to

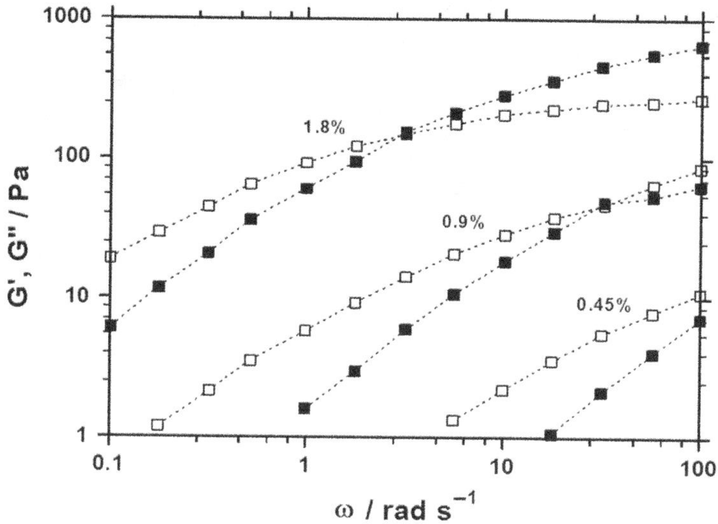

FIGURE 9.16 Frequency dependence of storage (G′ ■) and loss (G″ □) moduli of oat β-glucan solutions (Ren et al., 2011).

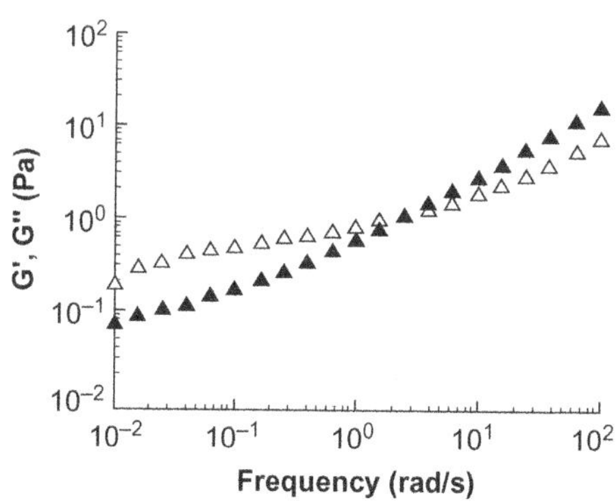

FIGURE 9.17 Unusual rheological behaviors of partially hydrolyzed oat β-glucan (Doublier and Wood, 1995): Storage modulus (G′) and loss modulus (G″) as a function of angular frequency of acid-hydrolyzed oat gum (OG60) solutions. Concentration: 2.75% (Δ = G′, ▲ = G″).

tetrasaccharide, the faster and stronger the gelation (Fig. 9.18). The effect of MW on the gelation of β-glucan was unexpected, as the lower MW fraction of wheat β-glucan (100K) exhibited the fastest and strongest gelation property compared to the larger MWs (200K and 300K, respectively). This unusual gelation behavior was further investigated at molecular level in dilute solutions using dynamic and static light scattering and computer modeling methods, which helped to explain the underlying mechanisms of

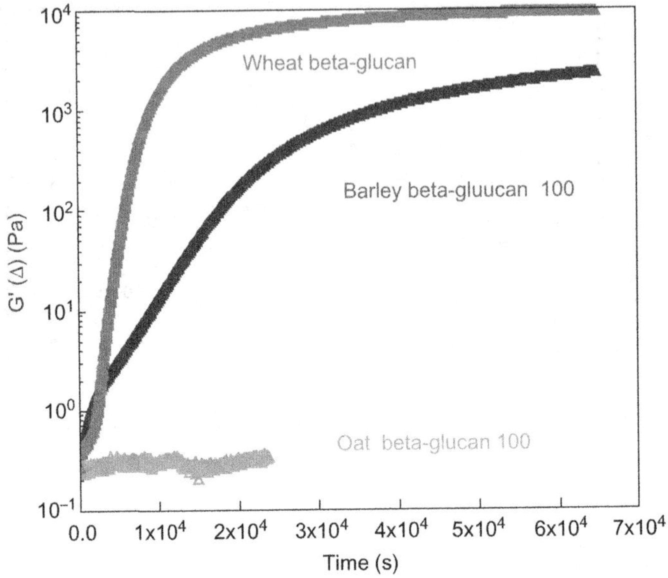

FIGURE 9.18 Time-dependent of storage modulus G′ of three cereal β-glucans at MW100k Dalton at 5°C.

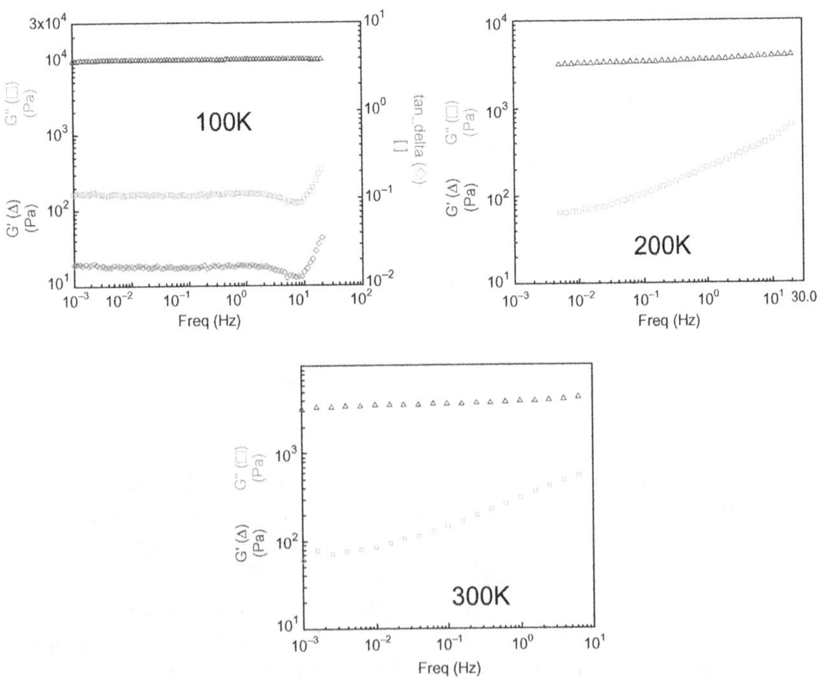

FIGURE 9.19 Mechanical spectra of wheat β-glucan at different molecular weight.

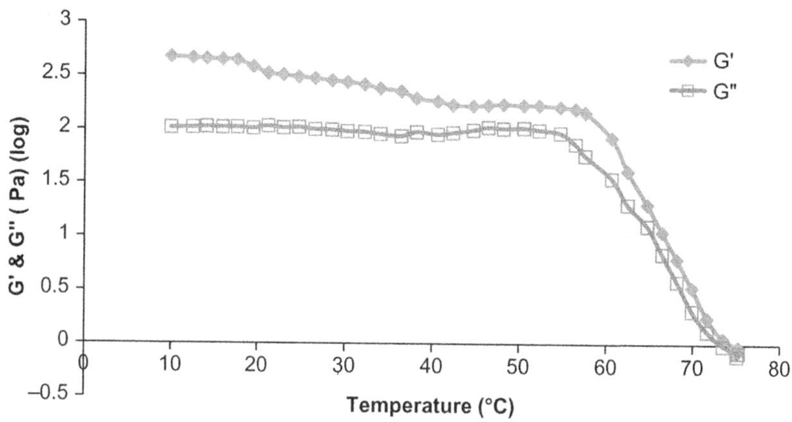

FIGURE 9.20 Melting curve of a β-glucan gel (Cui and Wood, 2000).

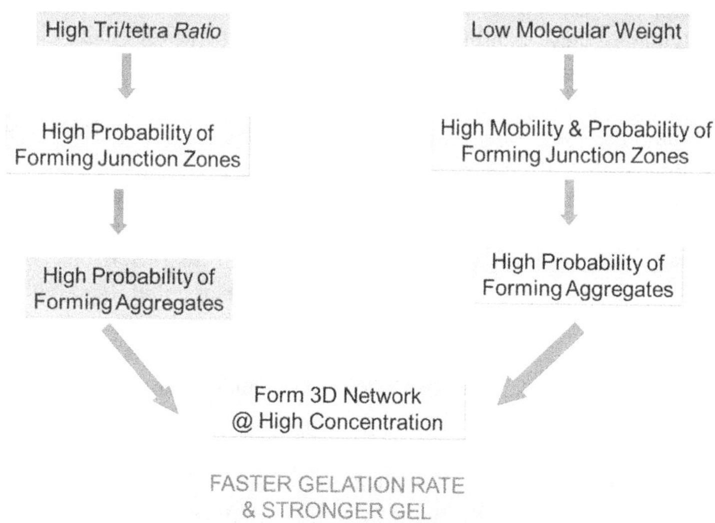

FIGURE 9.21 Mechanism of molecular association and gelation of cereal β-D-glucans.

the gelation process (Li et al., 2007, 2012). Following is the hypothesis derived from those studies, as illustrated in Fig. 9.21.

The gelation rate and strength of cereal β-D-glucans are determined by a combination of the mobility of the polymer chains (Mw) and the entanglement density of the "junction zone" of the polymer chain in solutions (structural features and concentrations). An increase in the tri/tetra ratio in the β-D-glucans leads to a more rigid conformation in dilute solutions, and easier formation of gels at higher concentrations; smaller MW β-D-glucans have higher mobility, which have more opportunity to interact with neighboring chains to form entanglements; this results in a higher degree of aggregation. The association of three dimensional networks (entanglements) or the aggregates at higher concentrations leads to the formation of gel.

9.4 BIOACTIVITIES AND HEALTH BENEFITS

This section starts by honoring one of our outstanding scientists in Canada (Guelph Food Research Centre, Agriculture and Agri-Food Canada) who laid the foundation in the science of oat β-D-glucan research in human health. "Dr. Peter Wood was an exceptional scientist who helped lay the foundation for our understanding of the health benefits of soluble fiber. As a carbohydrate chemist, he developed methods to characterize polysaccharides and used them to better understand the structure and behavior of cereal $(1\rightarrow3)(1\rightarrow4)$- β-D-glucans. With collaborators, he demonstrated the distribution of these mixed linkage β-glucans in cereal grains and food products. He studied the behavior of β-glucans in solution and demonstrated that, like other high MW polymers, they exhibited shear thinning behavior and viscosity increased exponentially with increasing concentration. By using size exclusion chromatography, with postcolumn calcofluor detection, Peter was able to determine the MW distributions of β-glucans in unpurified extracts. He investigated the differences in fine structure among oat, barley, rye, and wheat β-glucans and showed that the differences influenced the gelation properties of these polymers. He drew together multidisciplinary collaborations to investigate how the physico – chemical characteristics of oat β-glucans influenced their ability to attenuate postprandial blood glucose levels and lower serum cholesterol concentrations. Using oat gum extracts which Peter and coworkers had produced, they demonstrated that β-glucan was the major bioactive component of oat bran which affected blood glucose and cholesterol levels in humans. A series of clinical trials showed that solubility and MW of oat β-glucan were modified by food processing, which in turn affected bioactivity. Peter's dedication to excellent science and drive to establish the underlying mechanisms of health benefits of oat fiber have provided us with a wealth of knowledge for future research." (Cited from Tosh, 2012, with permission).

9.4.1 Beta-D-Glucan and Blood Glucose

Oat and barley based products are accepted by several industrialized countries as a health food which can bring benefits to humans, including lowering cholesterol levels and attenuating postprandial glycemic response (Health Canada, 2016; FDA, 2017; EFSA, 2011). A well accepted study was by Wood and coworkers, who demonstrated that oat gum (mostly β-D-glucans) attenuated blood glucose on a drink model in a dose-dependent manner, as shown in Fig. 9.22 (Braaten et al., 1991).

In order to elucidate the possible mechanisms as to how β-glucan could attenuate blood glucose levels, Wood et al. further demonstrated an inverse linear relationship between peak postprandial blood glucose and log (viscosity) in healthy subjects using the drink model which used both extracted and purified β-glucan samples at various doses and MWs (Wood, 2011) (Fig. 9.23).

Healthy subjects consuming oat or barley β-glucan containing foods, such as muffins, bread, and granola bars were investigated (Östman et al., 2006; Tosh et al., 2008; Regand et al., 2009). The inverse viscosity and blood glucose reduction relationship was confirmed in a clinical trial using foods under specific storage/processing conditions to give various viscosities in the intestine.

In a recent study, Tosh et al. (2010) demonstrated that the solubility of β-glucan in fresh baked oat muffins and muffins went through two and four

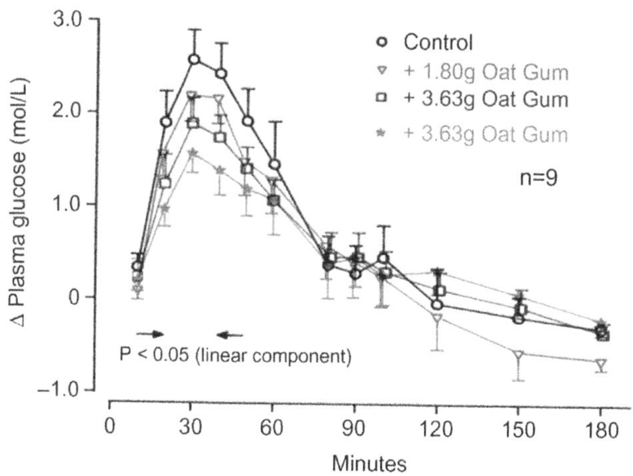

FIGURE 9.22 Effect of β-D-glucan in beverages on blood glucose (Wood, 2002).

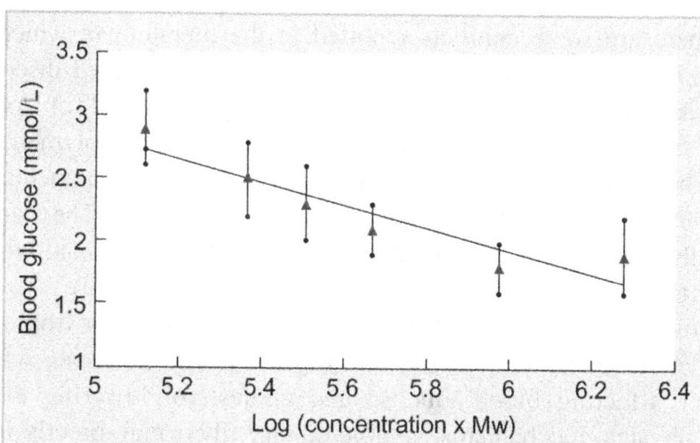

FIGURE 9.23 Relationship of drink viscosity blood glucose levels (Wood, 2002)

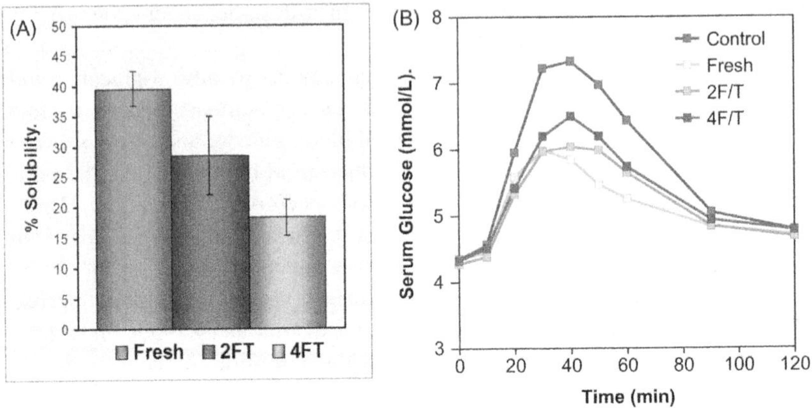

FIGURE 9.24 Effect of freeze – thaw process on solubility and blood glucose control in subjects fed with oat bran muffins containing 8 g or 12 g β-D-glucan/serving (130 g) (Tosh, 2008).

freeze – thaw cycles was very different, which could affect its function. The solubility of β-glucan reduced by more than 50% after four freeze – thaw (FT) cycles, which means a lower concentration of β-glucan in the intestine, and a reduced ability to attenuate serum glucose (Lan-Pidhainy et al., 2007) (Fig. 9.24).

9.4.2 Beta-D-Glucan and Cholesterol Levels

In Wood's earlier study in clinical trials in collaboration with hospitals, he demonstrated that oat β-glucan, the main component of the soluble fiber of oats, significantly reduced the total and LDL-C levels of hypercholesterolemic adults without changing HDL cholesterol (Fig. 9.25), (Braaten et al., 1991). In a recent human intervention trial using extruded breakfast cereal containing oat β-glucan of various MWs, Wolever and colleagues (Wolever et al., 2010) demonstrated an inverse relationship between serum LDL-C and log viscosity. However, there are some conflicts reported in the literature in which not all studies give positive results (Keogh et al., 2003). The observed discrepancies can be attributed to factors affecting the amount and MW of solubilized β-glucan in the intestine. Regand and coworkers observed depolymerization of β-glucan in bread and pasta; this resulted in reduced efficacy in attenuating the peak blood glucose in healthy subjects (Regand et al., 2009). The same group also found that extrusion could break down the MW of β-glucan, which ultimately affected the efficacy by up to 50% (Wolever et al., 2010).

These studies provided convincing evidence that β-glucan from oats and barley can increase the viscosity of the digesta in the intestine, which is a major factor affecting blood glucose and cholesterol lowering ability. In addition to physical mechanisms, soluble dietary fibers can directly influence

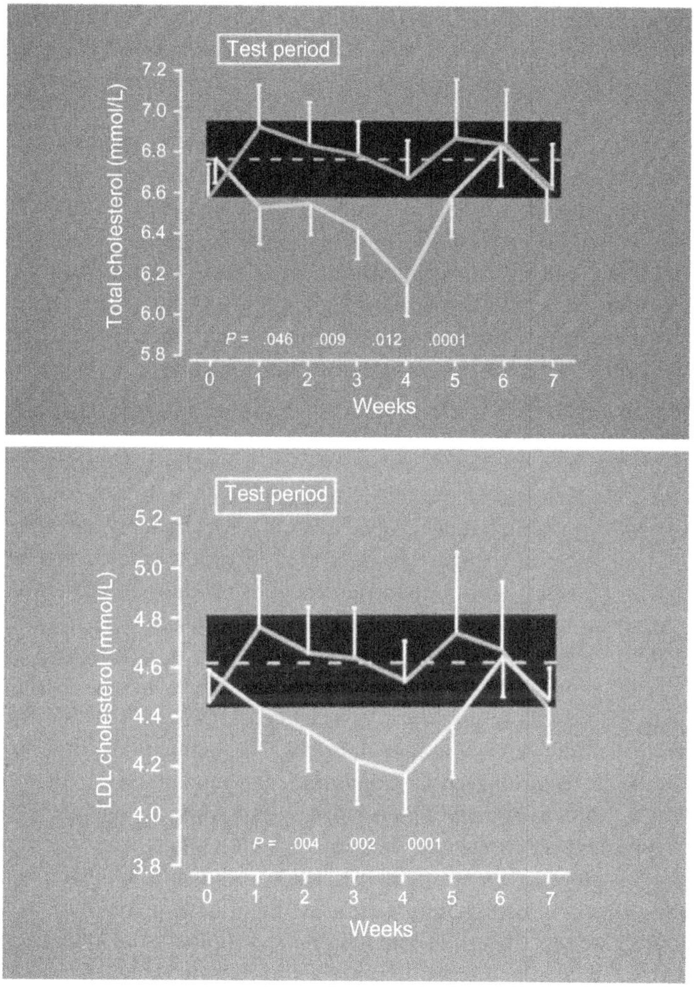

FIGURE 9.25 Effect of oat beta-glucan on total and LDL cholesterols (Wood, 2002).

the degradation of starch by slowing down the actions of enzymes and acting as prebiotics, which have shown great potential in human health (please see next section).

9.4.3 Fermentation Properties and Gut Health and Other Bioactivities

One of the major biological functions of dietary fiber is to be utilized by gut microorganisms, preferable the beneficial ones, and the production of short chain fatty acids. Most of the soluble fibers, especially the oligosaccharides, can modulate colonic microbiota composition by promoting beneficial species, such as Bifidobacterium, and Lactobacilli. This has been reported to promote gut health and immune responses of the host during diseases and health (Round and Mazmanian, 2009; Mittal, 2012; Hooper et al., 2012; Clemente et al., 2012; Lam and Cheung, 2013). Although cereal β-D-glucans have been considered as one of the most studied dietary fibers in the literature, it is surprising to note that the number of studies devoted to the prebiotic or fermentation properties of cereal β-D-glucans is rather limited. Hughes et al. studied the fermentation properties of β-glucan fractions from barley using human fecal microbiota (Hughes et al., 2008). A few other studies also used human or animal microbiota to evaluate the fermentation efficiency of cereal β-glucans in comparison with other soluble dietary fibers (Crittenden et al., 2002; Kedia et al., 2008). Dietary fibers differ in their chemical structures and types of linkages; hence, their ability to be utilized by gut microorganisms varies with the type and genetics of the bacteria, as well as their own structural characteristics and MW. It is generally agreed that lower MW polysaccharides, especially oligosaccharides, are more susceptible to gut bacteria (Hughes et al., 2008). A recent study compared oat β-glucan with flaxseed gum and fenugreek gum on their fermentation efficiencies using bacteria from different gut locations of pig digesta, as shown in Fig. 9.26.

The results showed that cultures with β-D-glucan as the sole carbon and energy source appeared to produce more lactate (a maximum 43.58% of total SCFA with cecal digesta as the inoculum, 58.57% with colonic digesta) than the cultures with flaxseed gum or fenugreek gum. In contrast, cultures grown on flaxseed or fenugreek gum produced more principal SCFAs (acetate − propionate − butyrate), butyrate in particular, than the cultures containing β-D-glucan (Lin et al., 2011). These results are in agreement with the observation on barley β-glucan reported by Hughes using human feacal bacteria. Both studies suggested that the prebiotic potential of cereal β-D-glucans are less effective compared to other soluble fibers (Hughes et al., 2008; Lin et al., 2011). However, recent studies suggest cereal β-D-glucans may activate nuclear factor-kappa B in intestinal leukocytes and enterocytes from mice (Volman et al., 2010). A more recent study indicated that the specific structures of oat and barley β(1,3)(1,4)-glucans induced different *in vitro* immunomodulatory effects in bone marrow-derived dendritic cells (BMDC) fromTLR2/4 knock out mice (Rösch et al., 2016). In other studies, cereal β-glucans and probiotics can enhance the antiinflammatory activity on host cells by a synergistic effect (Arena et al., 2016). However, most current studies are based on cell cultures, which is by no means conclusive. In-depth study including *in vivo* tests should be carried out on promising results obtained from cell culture studies.

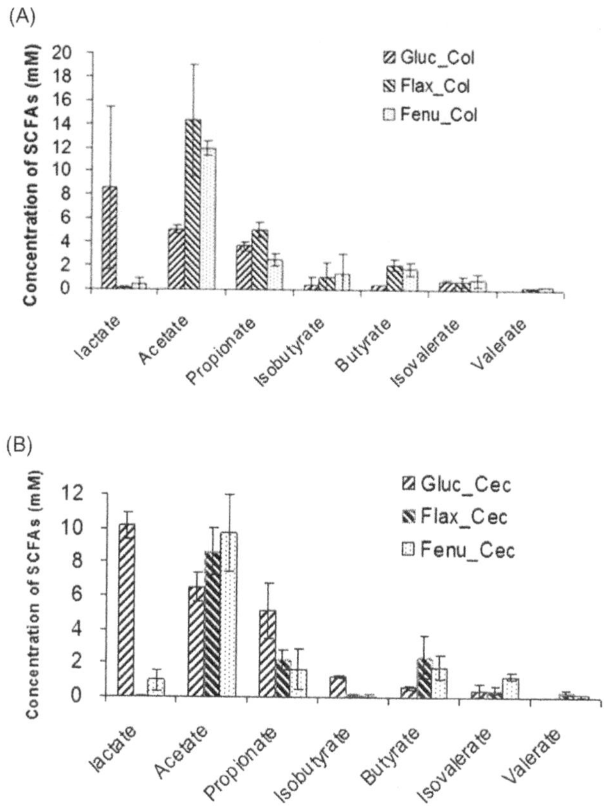

FIGURE 9.26 Short fatty acid production in batch cultures containing different types of dietary fiber (β-D glucan, flaxseed gum, or fenugreek gum) and inoculated with digesta from the cecum (A) or the colon (B). The graph was prepared on basis of the second and third 11 subculture measurements. The error 12 bars are standard deviation of two treatments (Lin et al., 2011).

9.5 SUMMARY

In summary, β-D-glucan is a linear polymer which exhibits a wide range of rheological properties, including viscoelastic fluid, weak-gel, and real gel. These rheological properties and other physical properties (e.g., solubility) depend on two important structural features: MW and structural regularity (expressed by tri/tetra ratio). The higher tri/tetra ratio will increase the opportunity to form ordered junction zones in solution. These structural characteristics of cereal β-D-glucans, together with their MW and MW distribution, have significant effects on their physiological efficiency, i.e., controlling and modulating blood cholesterols and glucose response.

Beta-D-glucans have not been applied in foods as ingredients or stabilizers, largely due to their commercial unavailability. However, as active components in oat and barley-based products, such as whole oats, oat bran, oat meal, barley flour, etc., β-D-glucans exist broadly in breakfast cereals, snack foods, and more recently, made into flat bread (Ames et al., 2006). Purified oat and β-D-glucan solution has been used in the cosmetic industry as a moisturizing agent and thickener; the lack of a consistent supply of the product has limited its wider application in the personal care industry. Extensive research is needed for commercial production and application in order to further expand the application of β-D-glucans for both food and nonfood uses. The other specific biactivities of cereal β-glucans are still limited at cell culture levels; this might be due to the much lower potency exhibited by cereal β-glucans compared to many other bioactive polysaccharides described in other chapters of this book.

REFERENCES

AbuMweis, S., Thandapilly, S.J., Storsley, J., Ames, N., 2016. Effect of barley β-glucan on postprandial glycaemic response in the healthy human population: a meta-analysis of randomized controlled trials. J. Funct. Foods 27, 329–342.

Agbenorhevi, J.K., Kontogiorgos, V., Kirby, A.R., Morris, V.J., Tosh, S.M., 2011. Rheological and microstructural investigation of oat β-glucan isolates varying in molecular weight. Int. J. Biol. Macromol. 49, 369–377.

Åman, P., Graham, H., Tilly, A.-C., 1989. Content and solubility of mixed-linked (1→ 3), (1→ 4)-β-d-glucan in barley and oats during kernel development and storage. J. Cereal Sci. 10, 45–50.

Ames, N., Rhymer, C., Rossnagel, B., Therrien, M., Ryland, D., Dua, S., et al., 2006. Utilization of diverse hulless barley properties to maximize food product quality. Cereal Foods World 51, 23–28.

Anderson, M., Cook, J., Stone, B., 1978. Enzymatic determination of 1, 3: 1, 4-β-glucans in barley grain and other cereals. J. Inst. Brewing 84, 233–239.

Arena, M.P., Russo, P., Capozzi, V., Rascón, A., Felis, G.E., Spano, G., et al., 2016. Combinations of cereal β-glucans and probiotics can enhance the anti-inflammatory activity on host cells by a synergistic effect. J. Funct. Foods 23, 12–23.

Autio, K., 2006. Functional aspects of cereal cell-wall polysaccharides. Food Sci. Technol. NY Marcel Dekker 159, 167.

Autio, K., Myllymäki, O., Mälkki, Y., 1987. Flow properties of solutions of oat β-glucans. J. Food Sci. 52, 1364–1366.

Bamforth, C., 1982. Barley β-glucans: their role in malting and brewing. Brewers Digest 57, 22–35.

Beer, M.U., Arrigoni, E., Amado, R., 1996. Extraction of oat gum from oat bran: effects of process on yield, molecular weight distribution, viscosity and (1--> 3)(1--> 4)-b-D-glucan content of the gum. Cereal Chem. 73, 58–62.

Beer, M.U., Wood, P.J., Weisz, J., 1997a. Molecular weight distribution and (1→ 3)(1→ 4)-β-D-glucan content of consecutive extracts of various oat and barley cultivars. Cereal Chem. 74, 476–480.

Beer, M.U., Wood, P.J., Weisz, J., Fillion, N., 1997b. Effect of cooking and storage on the amount and molecular weight of (1→ 3)(1→ 4)-β-D-glucan extracted from oat products by an in vitro digestion system. Cereal Chem. 74, 705–709.

Biliaderis, C. 2015. Personal communication.

Böhm, N., Kulicke, W.-M., 1997. Optimization of the use of polyelectrolytes for dewatering industrial sludges of various origins. Colloid Polymer Sci. 275, 73.

Böhm, N., Kulicke, W.-M., 1999. Rheological studies of barley (1→ 3)(1→ 4)-β-glucan in concentrated solution: mechanistic and kinetic investigation of the gel formation. Carbohydr. Res. 315, 302–311.

Braaten, J.T., Wood, P.J., Scott, F.W., Riedel, K.D., Poste, L.M., Collins, M.W., 1991. Oat gum lowers glucose and insulin after an oral glucose load. Am. J. Clin. Nutr. 53, 1425–1430.

Carr, J., Glatter, S., Jeraci, J., Lewis, B., 1990. Enzymic determination of β-glucan in cereal-based food products. Cereal Chem. 67, 226–229.

Clemente, J.C., Ursell, L.K., Parfrey, L.W., Knight, R., 2012. The impact of the gut microbiota on human health: an integrative view. Cell 148, 1258–1270.

Crittenden, R., Karppinen, S., Ojanen, S., Tenkanen, M., Fagerström, R., Mättö, J., et al., 2002. In vitro fermentation of cereal dietary fibre carbohydrates by probiotic and intestinal bacteria. J. Sci. Food. Agric. 82, 781–789.

Cui, S.W., 2005. Structural analysis of polysaccharides. In: Cui, S.W. (Ed.), Food Carbohydrates. CRC Press, Boca Raton, FL, pp. 105–160.

Cui, W., Wood, P., 2000. Relationships between structural features, molecular weight and rheological properties of cereal β-D-glucans. In: Nishinari, K. (Ed.), Hydrocolloids—Part 1. Elsevier Science BV, Amsterdam, pp. 159–168.

Cui, W., Wood, P.J., Weisz, J., Beer, M.U., 1999. Nonstarch polysaccharides from preprocessed wheat bran: carbohydrate analysis and novel rheological properties. Cereal Chem. 76, 129–133.

Cui, W., Wood, P., Blackwell, B., Nikiforuk, J., 2000. Physicochemical properties and structural characterization by two-dimensional NMR spectroscopy of wheat β-D-glucan—comparison with other cereal β-D-glucans. Carbohydr. Pol. 41, 249–258.

Doublier, J.-L., Wood, P.J., 1995. Rheological properties of aqueous solutions of (1--> 3) (1--> 4)-bD-glucan from oats (Avena sativa L.). Cereal Chem. 72, 335–340.

European Food Safety Authority (EFSA), 2011. https://www.efsa.europa.eu/en/efsajournal/pub/2207.

FDA (1997, 2005) <https://www.fda.gov/food/guidanceregulation/guidancedocumentsregulatoryinformation/labelingnutrition/ucm064919.htm>.

FDA, 2017. https://www.ecfr.gov/cgi-bin/text-idx?SID = c7e427855f12554dbc292b4c8a7545a0&mc = true&node = pt21.2.101&rgn = div5#se21.2.101_176.

Fulcher, R., Miller, S., 1993. Structure of oat bran and distribution of dietary fiber components. Oat Bran1–24.

Gómez, C., Navarro, A., Gamier, C., Horta, A., Carbonell, J., 1997. Physical and structural properties of barley (1→ 3),(1→ 4)-β-D-glucan—III. Formation of aggregates analysed through its viscoelastic and flow behaviour. Carbohydr. Pol. 34, 141–148.

Grimm, A., Krüger, E., Burchard, W., 1995. Solution properties of β-D-(1, 3)(1, 4)-glucan isolated from beer. Carbohydr. Pol. 27, 205–214.

Health Canada, 2016. http://www.hc-sc.gc.ca/fn-an/label-etiquet/claims-reclam/assess-evalu/oat-avoine-eng.php.

Hooper, L.V., Littman, D.R., Macpherson, A.J., 2012. Interactions between the microbiota and the immune system. Science 336, 1268–1273.

Hughes, S.A., Shewry, P.R., Gibson, G.R., McCleary, B.V., Rastall, R.A., 2008. Invitro fermentation ofoat and barleyderivedb-glucans by human faecal microbiota. FEMS. Microbiol. Ecol. 64 (3), 482–493.

Kedia, G., Vázquez, J.A., Pandiella, S.S., 2008. Evaluation of the fermentability of oat fractions obtained by debranning using lactic acid bacteria. J. Appl. Microbiol. 105, 1227–1237.

Keogh, G.F., Cooper, G.J., Mulvey, T.B., McArdle, B.H., Coles, G.D., Monro, J.A., et al., 2003. Randomized controlled crossover study of the effect of a highly β-glucan–enriched barley on cardiovascular disease risk factors in mildly hypercholesterolemic men. Am. J. Clin. Nutr. 78, 711–718.

Lam, K.L., Cheung, C.K., 2013. Non-digestible long chain beta-glucans as novel prebiotics. Bioact. Carbohydr. Diet. Fibers 2, 45–64.

Lan-Pidhainy, X., Brummer, Y., Tosh, S.M., Wolever, T.M., Wood, P.J., 2007. Reducing beta-glucan solubility in oat bran muffins by freeze-thaw treatment attenuates its hypoglycemic effect. Cereal Chem. 84, 512–517.

Lazaridou, A., Biliaderis, C., 2004. Cryogelation of cereal β-glucans: structure and molecular size effects. Food Hydr. 18, 933–947.

Lazaridou, A., Biliaderis, C., 2007. Molecular aspects of cereal β-glucan functionality: physical properties, technological applications and physiological effects. J. Cereal Sci. 46, 101–118.

Li, W., Cui, S.W., Wang, Q., Yada, R.Y., 2011. Studies of aggregation behaviours of cereal β-glucans in dilute aqueous solutions by light scattering: Part I. Structure effects. Food Hydrocoll 25, 189–195.

Li, W., Wang, Q., Cui, S., Huang, X., Kakuda, Y., 2006. Elimination of aggregates of (1→ 3) (1→ 4)-β-D-glucan in dilute solutions for light scattering and size exclusion chromatography study. Food Hydr. 20, 361–368.

Li, W., Wang, Q., Cui, S.W., Burchard, W., Yada, R., 2007. Carbanilation of cereal β-glucans for molecular weight determination and conformational studies. Carbohydr. Res. 342, 1434–1441.

Li, W., Cui, S.W., Wang, Q., Yada, R.Y., 2012. Study of conformational properties of cereal β-glucans by computer modeling. Food hydrocolloids 26, 377-382.McCleary, B.V., 1988. Purification of (1→ 3),(1→ 4)-β-D-glucan from barley flour. Methods Enzymol. 160, 511–514.

Li, X., Deng, F., Hua, Y., Qiu, A., Yang, C., Cui, S., 2008. Effect of molecular weight of dextran on the phase behavior and microstructure of preheated soy protein/dextran mixtures. Carbohydr. Pol. 72, 160–168.

Lin, B., Gong, J., Wang, Q., Cui, S.W., Yu, H., Huang, B., 2011. In-vitro assessment of the effects of dietary fibres on microbial fermentation and communities from large intestinal digesta of pigs. Food Hydr. 25, 180–188.

McCleary, B.V., 1988. Purification of (1→ 3),(1→ 4)-β-D-glucan from barley flour. Methods Enzymol. 160, 511–514.

McCleary, B.V., Codd, R., 1991. Measurement of (1→ 3),(1→ 4)-β-D-glucan in barley and oats: a streamlined enzymic procedure. J. Sci. Food. Agric. 55, 303–312.

Mittal, R., 2012. Polysaccharide A: an immunomodulatory molecule having therapeutic potential against inflammatory gut diseases (Editorial). J. Microb. Biochem. Technol. 4, 1.

Östman, E., Rossi, E., Larsson, H., Brighenti, F., Björck, I., 2006. Glucose and insulin responses in healthy men to barley bread with different levels of (1→ 3; 1→ 4)-β-glucans; predictions using fluidity measurements of in vitro enzyme digests. J. Cereal Sci. 43, 230–235.

Panahi, S., Ezatagha, A., Jovanovski, E., Jenkins, A., Temelli, F., Vasanthan, T., et al., 2014. Glycemic effect of oat and barley beta-glucan when incorporated into a snack bar: a dose escalation study. J. Am. Coll. Nutr. 33, 442–449.

Regand, A., Tosh, S.M., Wolever, T.M., Wood, P.J., 2009. Physicochemical properties of β-glucan in differently processed oat foods influence glycemic response. J. Agric. Food. Chem. 57, 8831–8838.

Ren, Y., Ellis, P., Ross-Murphy, S., Wang, Q., Wood, P., 2003. Dilute and semi-dilute solution properties of (1→3),(1→4)-β-D-glucan, the endosperm cell wall polysaccharide of oats (Avena sativa L.). Carbohydr. Pol. 53, 401–408.

Rösch, C., Meijerink, M., Delahaije, R.J.B.M., Taverne, N., Gruppen, H., Wells, J.M., et al., 2016. Immunomodulatory properties of oat and barley β-glucan populations on bone marrow derived dendritic cells. J. Funct. Foods 26, 279–289.

Round, J.L., Mazmanian, S.K., 2009. The gut microbiota shapes intestinal immune responses during health and disease. Nat. Rev. Immunol. 9, 313–323.

Tosh, S.M., 2008. Personal communication.

Tosh, S.M., 2012. Glycemic response to extruded oat bran cereals processed to vary in molecular weight. Cereal Chem. 89, 255–261.

Tosh, S.M., Brummer, Y., Miller, S.S., Regand, A., Defelice, C., Duss, R., et al., 2010. Processing affects the physicochemical properties of β-glucan in oat bran cereal. J. Agric. Food. Chem. 58, 7723–7730.

Tosh, S.M., Brummer, Y., Wolever, T.M., Wood, P.J., 2008. Glycemic response to oat bran muffins treated to vary molecular weight of beta-glucan. Cereal Chem. 85, 211.

Vasanthan, 2015. Personal communication.

Vårum, K.M., Smidsrød, O., Brant, D.A., 1992. Light scattering reveals micelle-like aggregation in the (1→ 3),(1→ 4)-β-d-glucans from oat aleurone. Food Hydr. 5, 497–511.

Volman, J. J., Mensink, R. P., Ramakers, J. D., de Winther, M. P., Carlsen, H., Blomhoff, R., Buurman, W.A. and Plat, J. (2010). Dietary (1→3), (1←4)-β-d-glucans from oat activate nuclear factor-B in intestinal leukocytes and enterocytes from mice. Nutrition Research, 30 (1), 40–48.

Wang, Q., Wood, P.J., Huang, X., Cui, S.W., 2003. Preparation and characterization of molecular weight standards of low polydispersity from oat and barley (1→3)(1→4)- β-D-glucan. Food Hydr. 17, 824–853.

Wolever, T.M., Tosh, S.M., Gibbs, A.L., Brand-Miller, J., Duncan, A.M., Hart, V., et al., 2010. Physicochemical properties of oat β-glucan influence its ability to reduce serum LDL cholesterol in humans: a randomized clinical trial. Am. J. Clin. Nutr. 92, 723–732.

Wolever, T.M., Gibbs, A.L., Brand-Miller, J., Duncan, A.M., Hart, V., Lamarche, B., et al., 2011. Bioactive oat β-glucan reduces LDL cholesterol in Caucasians and non-Caucasians. Nutr. J. 10, 130.

Wood, P., Beer, M., 1998. Functional oat products. Funct. Foods Biochem. Proces. Aspects1–37.

Wood, P., Fulcher, R., 1978. Interaction of some dyes with cereal beta-glucans. Cereal Chem. 55, 952–966.

Wood, P., Anderson, J., Braaten, J., Cave, N., Scott, F., Vachon, C., 1989. Physiological effects of beta-D-glucan rich fractions from oats. Cereal Foods World (USA).

Wood, P., Weisz, J., Blackwell, B., 1991a. Molecular characterization of cereal β-D-glucans. Structural analysis of oat β-D-glucan and rapid structural evaluation of β-D-glucans from different sources by high-performance liquid chromatography of oligosaccharides released by lichenase. Cereal Chem. 68, 31–39.

Wood, P., Weisz, J., Fedec, P., 1991b. Potential for β-glucan enrichment in brans derived from oat (Avena stativa L.) cultivars of different (1→ 3),(1→ 4)-BD-glucan concentrations. Cereal Chem. 68, 48–51.

Wood, P.J., 2002. Personal communication.

Wood, P.J., 2007. Cereal β-glucans in diet and health. J. Cereal Sci. 46, 230–238.

Wood, P.J., 2011. Oat beta-glucan: properties and function. In: Webster, F.H., Wood, P.J. (Eds.), Oats: Chemistry and Technology. AACC International, Inc, St. Paul, MN, pp. 219–254.

Wu, J., Zhao, L., Li, J., Jin, S., Wu, S., 2017. Aggregation and gelation of oat beta-glucan in aqueous solution probed by NMR relaxometry. Carbohydr. Pol. 163, 170–180.

FURTHER READING

Barsanti, L., Passarelli, V., Evangelista, V., Frassanito, A.M., Gualtieri, P., 2011. Chemistry, physico-chemistry and applications linked to biological activities of β-glucans. Nat. Prod. Rep. 28, 457–466.

Chapter 10

Other Herbal Polysaccharides

The classical Chinese pharmacopeia describes a large number of herbal formulations that are used for the treatment of various diseases. Traditional Chinese medicine (TCM) has thousands of years of history. It consists of a unique system of practice including theory, methodology, prescription formulation, and drugs. It is priceless as a precious cultural heritage, since the majority of Chinese traditional drugs are of plant origin. Many of the TCMs have been found to possess immunomodulatory bioactivity without side-effects. In the past decades, fungal and plant-derived polysaccharides with a broad range of human beneficial bioactivities have been isolated and purified from TCM. Research, including animal tests and clinical trials, has demonstrated that the polysaccharide is one of the major pharmacological components. Some such polysaccharides have been developed into adjuvants in cancer therapy. In this chapter, some polysaccharides isolated from TCM including *Panax ginseng*, *Angelica sinensis*, *Lycium barbarum*, and *Astragalus* are discussed.

10.1 SOURCES AND PREPARATION

Most herb polysaccharides exist as a structural constituent of cells, and how to extract the polysaccharide depends on the cell wall structure. The basic extraction principle is to break the cell wall from the outer layer to the inner layer using mild-strong extraction conditions to avoid modifying the polysaccharide content. Because of the dominant advantages such as simplicity and economy, the hot-water extraction method is used as a classical way to isolate bioactive polysaccharides from medicinal herbs. Sometimes, a combination method of hot-water extraction with some supplementary methods such as microwave, ultrasonic, or freeze-thaw methods are applied to prepare the herb polysaccharides.

10.1.1 *Panax ginseng* Polysaccharide

Panax is one of the most medicinally important genera in the orient. Panax species are found in the northern hemisphere from the central Himalayas onward through China, Korea, and Japan to North America. *P. ginseng* C.A. Meyer is one the most popular natural tonics in TCM. Since the first research

Bioactive Polysaccharides. DOI: http://dx.doi.org/10.1016/B978-0-12-809418-1.00010-1

on *P. ginseng* polysaccharide was reported by Ovodov and Solov'eva (1966), there have been a series of polysaccharide fractions isolated from the roots and leaves of this material reported (Sun, 2011).

Many polysaccharide fractions were isolated from the roots of ginseng using a water extraction method. For example, the roots of ginseng were extracted with distilled water at 100°C for 4 h. The solid residues were collected by filtering through four sheets of gauze, and then extracted twice again. Water-insoluble materials in the filtrates were further removed through centrifugation. The supernatant was precipitated by the addition of 95% ethanol (4 volumes). The precipitate was collected by centrifugation and dried by solvent exchange, first using 95% ethanol, then absolute ethanol, and finally ether to obtain crude polysaccharide. To remove the protein content, the crude polysaccharide was redissolved in distilled water and treated with Sevag reagent (1:4 *n*-butanol:chloroform, v/v). The deproteinated polysaccharide fraction was obtained by ethanol precipitation and drying by solvent exchange. Further, the polysaccharide was dissolved in distilled water and loaded on a DEAE-Cellulose column, then the column was eluted first with distilled water at a flow rate of 25 mL/min to obtain the unbound neutral fraction, and then with 0.5 mol/L NaCl to obtain the bound acidic fraction (Zhang et al., 2009). Liu reported that the content of ginseng polysaccharide was enhanced by an ultrasonic process at a fixed 200 W, 30 min, ratio of material to distilled water of 1.5, grain size 80 mesh, and an enhancement rate of 2.3% (Zhong et al., 2006).

Yamada isolated a series of polysaccharides from the leaves of ginseng using the hot-water extraction method (Gao et al., 1988). The leaves of ginseng were extracted with water for 3 h at 100°C, and the extracts were precipitated with ethanol to give GL-1. An aqueous solution of GL-1 was dialyzed against distilled water for 8 days, then centrifuged, and precipitated with ethanol to get GL-2. The GL-2 was added to aqueous 8% cetyltrimethylammonium bromide (Cetavlon). After storage for 6 h at 20°C, the precipitate was bulleted through centrifugation, redissolved in aqueous 10% NaCl, and then precipitated with ethanol to give GL-3. A 1% boric acid solution was added to the supernatant, and the pH of the mixed solution was adjusted to 8.8 using a 2 mol/L NaOH solution. After stirring for another 2 h, the mixed solution was centrifuged and the precipitate was treated as for GL-3, to obtain GL-4. The remaining supernatant was acidified using acetic acid and diluted with ethanol together with potassium acetate. Then the precipitate obtained was dialyzed and freeze-dried to give polysaccharide fraction GL-5. Further, GL-3 was further fractionated into five fractions using ion-exchange chromatography, ethanol fractionation, and gel-filtration chromatography. The GL-4 was applied to an anion-exchange chromatography on a DEAE-Sepharose CL-6B column to get a purified fraction, GL-4IIb2 (Sun et al., 1994). In order to remove the colored materials in the polysaccharide, GL-4IIb2 was further purified on a Q-Sepharose column (Cl^- form), and the column was washed with water and eluted sequentially with NaCl aqueous solution of gradient concentration. The major fraction which was eluted with 0.3 mol/L NaCl solution was fractionated by gel filtration on a Bio-Gel P-30 column to obtain the purified GL-4IIb2 (Shin et al., 1997).

10.1.2 *Angelica sinensis* Polysaccharide

The root of *A. sinensis* has a record in traditional Chinese medicinal prescriptions going back for thousands of years. It has been used as a tonic, hematopoietic, and antiinflammatory medicine in therapy for gynecological diseases such as menstrual disorders, amenorrhea, and dysmenorrhea. In the Asian area, the root of *A. sinensis* was widely marketed as a health food for women's care. In Europe and the American area, it was used as a dietary supplement.

Before the extraction of *A. sinensis* polysaccharide, the powdered roots of *A. sinensis* were preextracted three times with ethanol at 80°C to remove coloring matter. The residue was extracted three times with water at 80°C for 2 h. Then the polysaccharide content in the supernatant was precipitated by adding three times volume of ethanol. The gel-like precipitate was redissolved in water and dialyzed against distilled water. The nondialyzable portion was frozen at −20°C, then thawed and centrifuged to remove insoluble materials. Such a freeze-thaw process was repeated six times, and then the supernatant was lyophilized to give crude polysaccharide. To purify the polysaccharide, the crude polysaccharide was dissolved in distilled water, filtered through a 0.65 Lm membrane filter, and loaded onto a DEAE-Sephadex A-25 column. The column was first eluted with distilled water, and the largest water-eluted fraction was pooled, dialyzed, lyophilized, and designated as APS-1. Further, APS-1 was fractionated on a column of Sephadex G-100, eluted with 0.1 mol/L NaCl, and separated into four subfractions (Cao et al., 2006b). Yang et al. (2007b) extracted *A. sinensis* polysaccharide using hot-water extraction, and deproteinized the polysaccharide using the freeze-thaw method, and further purified the polysaccharide by Sephacryl S-400 gel filtration column chromatography eluted with 0.1 mol/L NaCl solution. Wang et al. (2012) optimized the extraction conditions of *A. sinensis* polysaccharide using response surface methodology. The results showed the optimal conditions were as follows: a time of 2.15 h, the ratio of material to water at 1:8.27, and three extraction times. It has also been reported that ultrasonic-assistant extraction is a quick and efficient method to prepare *A. sinensis* polysaccharide (He et al., 2010).

In addition, Fan et al. (2009) purified *A. sinensis* polysaccharide using ultrafiltration technology. In this study, *A. sinensis* polysaccharide was ultrafiltrated with different aperture membranes, and the separation parameters on the membrane separation efficiency of polysaccharide of *A. sinensis* were optimized by an orthogonal test. The results showed that the main component of *A. sinensis* polysaccharide was the fraction with 200 kD (65.42%), and the optimum condition of ultrafiltration was obtained as follows: one volume extraction, at 35°C, and 0.3 MPa (Yan et al., 2009).

10.1.3 *Lycium barbarum* Polysaccharide

The fruits of *L. barbarum*, small red berries, have been used for thousands of years in TCM and cuisine, since they are considered to be beneficial to the eyes, liver, and kidney. Generally, the *L. barbarum* polysaccharide (LBP) was extracted using the common water extraction method. Briefly, the dry fruit of *L. barbarum* were first extracted three times in 80°C hot water, and then the polysaccharide component was precipitated using 80% ethanol. The precipitate was collected and washed successively with ethanol and acetone, and then dried to obtain crude polysaccharide (Zou et al., 2010). Jing et al. (2009) optimized the hot-water extraction conditions of LBPs using an orthogonal experiment. The results showed that the highest yield was obtained under extraction conditions at extraction temperature 93°C, extraction time 4.3 h, and ratio of water to raw material 43. Yin and Dang (2008) also reported optimal extraction conditions for the polysaccharide as extraction temperature 100°C, extraction time 5.5 h, ratio of water to raw material 31.2, and number of extraction times five.

Wang et al. (2010a) reported another method for LBP extraction. A 10 g sample of fruit of *L. barbarum* was mixed with 100 mL of deionized water and homogenized for 1 min, followed by heating in boiling water for 30 min, and centrifuging at 6000 rpm for 25 min. After filtering, the crude extract was precipitated with 80% ethanol overnight to obtain the crude polysaccharide. To further purify the polysaccharide, crude polysaccharide was dissolved in phosphoric acid-buffered solution (pH 8), after which the mixture was heated in a 60°C water bath for 5 min, and proteinase was added to hydrolyze the protein. Then, 5% trichloroacetic acid was added to terminate the reaction. The supernatant was collected, and poured into a column of DEAE-Sepharose CL-6B. The neutral fraction of polysaccharide was eluted with deionized water, and the acidic fraction of the polysaccharides were eluted with NaCl solution.

An enzyme-assistant isolation method for LBP was reported. The concentrations of papain and cellulose were firstly optimized by an orthogonal experimental design. The optimal concentrations of the enzymes were found to be a cellulose concentration of 2.0% and a papain concentration of 1.0% of *L. barbarum* powder. The optimal enzyme-assistant extraction conditions were an extraction time of 91 min, extraction temperature of 59.7°C, at pH 5.0. The optimal experimental yield was 6.81% (Zhang et al., 2011).

In addition, an efficient ultrasound-assisted enzymatic extraction procedure for LBP was investigated and optimized. The optimized condition obtained was an extraction time of 20.29 min, ultrasonic output power of

78.6 W, cellulose concentration of 2.15%, and extraction temperature of 55.79°C. The result showed that the optimal yield of the polysaccharides was 6.31% (Qiu et al., 2014).

10.1.4 *Astragalus* Polysaccharides

The dried root of *Astragalus* is one health-promoting herbal medicine that has been used as an immunomodulating agent for the treatment of common cold, diarrhea, fatigue, and anorexia for more than 2000 years. So far, the root of *Astragalus* is also used as an adjuvant extensively in the treatment of various cancers including lung, digestive tract, urinary system, and so on.

The root of *Astragalus* is the major source for obtaining *Astragalus* polysaccharide. To isolate the *Astragalus* polysaccharide, the fresh plant material was extracted twice with water for 2.5 h at 100°C. After deproteinization with Sevag reagent, the crude polysaccharide was precipitated using ethanol overnight at 4°C. The crude polysaccharide was applied to a column of Sephadex G-100, which was eluted with a pH 6.0 phosphate buffer containing 20 mM Na/K and 0.15 mol/LNaCl. The peak with the highest polysaccharide content was collected and then freeze-dried to give the purified polysaccharide (Li et al., 2009a). Rui et al. (2009) reported an optimum condition as extraction time 3 h, ratio of water to raw material 4, and particle size 33.8 mesh, under which the maximum polysaccharide extraction rate was 16%. Mu et al. (2009) reported another optimum extraction procedure as 8-times water extracting for three times at 100°C, at 90 min per time. They also found deproteinization using pepsase combined with Sevag agent and depigmentation with DEAE fibrin could obtain a high purification rate.

Gong and Yang (2004) investigated microwave-assisted extraction technology for preparing *Astragalus* polysaccharide. The optimal extraction conditions were obtained as follows: the ratio of water to material of 12, pH value 9 (regulated by saturated limewater), and microwave radiation at a dose of 300 W for 10 min. The yield of crude *Astragalus* polysaccharide was 14.6%, and a purity of 88.1% was then obtained after the precipitate was washed with ethanol several times and dried in a vacuum. Compared with the direct-heating extraction, microwave-assisted extraction can shorten the extraction duration, reduce the consumption of the solvent, and improve the yield of *Astragalus* polysaccharide.

A comparison of different extraction methods including reflux, potass, microwave assistance, ultrasound, and an enzymic method for *Astragalus* polysaccharide has been reported (Yang et al., 2010). The extraction yield and purity of *Astragalus* polysaccharides with microwave assistance were both obviously higher than those of the other four methods, and second was the refluxing method. In addition, the microwave assistance extraction time was 6–7 times shorter than that of refluxing, so microwave assistance extraction was found to be the optimal extraction method for practical significance. As a traditional method, refluxing is extensively applied in extracting the polysaccharide.

10.2 STRUCTURAL FEATURES

10.2.1 *Panax ginseng* Polysaccharide

A water-soluble polysaccharide isolated from the roots of *P. ginseng* C.A. Meyer consists of 77.9% glucose, 6.8% galactose, 4.6% arabinose, 1% mannose, 1.1% rhamnose, and 8.7% galacturonic acid. Further, the crude polysaccharide was fractionated and purified using a combination of ethanol precipitation, ion-exchange, and gel permeation chromatographies. Two neutral fractions (WGPN and WGPA-N), and six acidic fractions (WGPA-1-RG, WGPA-2-RG, WGPA-1-HG, WGPA-2-HG, WGPA-3-HG, and WGPA-4-HG) were obtained. The WGPN fraction was found to be a starch-like glucan, while WGPA-N was a mixture of starch-like glucan and arabinogalactan. The acidic fractions WGPA-1-RG and WGPA-2-RG consist of major neutral sugars and minor acidic sugars that belong to type-I rhamnogalacturonan (RG-I)-rich pectins. The acidic fractions WGPA-1-HG to WGPA-4-HG mainly consist of galacturonic acid (GalA, 62.4%–92.1%), and these two fraction were found to be homogalacturonan (HG)-rich pectins with different degrees of methyl-esterification, ranging from 0% to 30% (Zhang et al., 2009).

The molecular weight of two homogeneous acidic polysaccharides (ginsenan PA and ginsenan PB) isolated from ginseng root were estimated to be 1.6×10^5 and 5.5×10^4, respectively. It also found that PA was composed of L-Ara, D-Gal, L-Rha, D-GalA, and D-GlcA in molar ratios of 11:22:1:6:1, and that of PB was 3:7:2:8:1. It was also found that almost all of the hexuronic acid residues of these acidic polysaccharides exist as methyl esters. Their structural features include mainly both α-arabino-β-3,6-galactan type and rhamnogalacturonan type structural units (Tomoda et al., 1993). Two neutral polysaccharide fractions from ginseng root were isolated using a DEAE-Sepharose CL-6B column chromatography (GPII and GPIII). Both GPII and GPIII mainly consisted of glucose. The molecular weights of GPII and GPIII were 3×10^5 and 4×10^5, respectively. Further analysis showed that GPII contains 60.06% (1→)- or (1→6)-glycosidic linkages and 39.94% (1→3)-glycosidic linkages, while GPIII contains 16.23% (1→)- or (1 → 6)-glycosidic linkages, 25.98% (1→2)-glycosidic linkages, and 57.79% (1→3)-glycosidic linkages (Luo and Fang, 2008).

Solov'eva et al. (1969) reported that one pectin polysaccharide from ginseng roots mainly contains residues of galacturonic acid, galactose, and arabinose. After treatment by partial acidic hydrolysis, a branched galactan and an acidic heteropolysaccharide were obtained. The galactan contains chains of (1→3)- and (1→6)-linked D-galactopyranose residues, while the heteropolysaccharide contains D-galactose, D-galacturonic acid, D-xylose, and L-rhamnose. In addition, Tomoda et al. (1985) reported that a peptidoglycan isolated from the roots of *P. ginseng* mainly consists of α-(1→6)-linked D-glycopyranose residues with branching at the C-3 position, and the ratio of terminals, branching positions, and intermediate units was about 1:1:1.8.

Structural features of polysaccharides from the leaves of *P. ginseng* were also well studied. Shin et al. (1997) reported a complex pectic polysaccharide (GL-4IIb2) isolated from the leaves of *P. ginseng* C.A., which was found to consist of 15 different monosaccharides and with 34 different glycosyl linkages. It was found that BL-4IIb2 comprised an α-(1→4)-linked oligogalacturonide backbone, and mainly four different side chains including Ara*f*-(1→5)-Dha, α-Rha*p*-(1→5)-Kdo, AceA-containing oligosaccharides, and a uronic acid-rich octasaccharide. Another polysaccharide (GL-BIII) from the leaves was found to mainly consist of terminal Ara*p*, 4- or 5-substituted Ara, 2,4-disubstituted Rha, 4- and 6-substituted Gal, and 3,6-disubstituted Gal. The GL-BIII was also found to contain terminal, 4-substituted, and 3,4-disubstituted GalA, and terminal and 4-substituted GlcA (Kiyohara et al., 1994).

10.2.2 *Angelica sinensis* Polysaccharide

Two water-soluble glucans named as APS-1cI and APS-1cII were extracted from the roots of *A. sinensis*, and further purified using anion-exchange and gel-filtration chromatography. The molecular weights of the two polysaccharides were determined to be 1.7×10^5 and 3.9×10^4 Da, respectively. The APS-1cI was a linear α-glucan only composed of (1→6)-α-D-Glc*p*, and APS-1cII contained a repeating unit consisting of (1→4)-α-D-Glc*p* and (1→6)-α-D-Glc*p* in a molar ratio of 4:1 (Cao et al., 2006a). An arabinoglucan with a molecular weight of 5.1 kDa was extracted from the roots of *A. sinensis*. The monosaccharides of the arabinoglucan consist of Glc and Ara in a molar ratio of 13.8:1. The backbone of the arabinoglucan comprised a 1,4-α-D-glucopyranosyl residue with some branches attached to the O-6 position. Such branches were found to be composed of 1,6-α-D-Glc*p* residues, and terminated with β-L-arabinofuranose residues (Cao et al., 2006b).

Crude water-soluble *A. sinensis* polysaccharides were fractionated into neutral and acidic fractions using anion-exchange chromatography. The neutral polysaccharide (ASP1) was found to be a mixture of glucan and arabinogalactan rich in glucose, galactose, and arabinose. The acidic polysaccharide consisted mainly of galacturonic acid, rhamnose, arabinose, and galactose (Sun et al., 2005). Two novel homogeneous polysaccharides (APS-1a and APS-3a) were obtained from the root of *A. sinensis*. Both of these homogeneous polysaccharides mainly consist of 1,4-linked galactose, 1,3,6-linked galactose, T-galactose, and T-arabinose. The molecular weights of APS-1a APS-3a were 49.0 and 65.4 kDa, respectively. The APS-1a was composed of galactose, arabinose, and glucose in a relatively molar percentage of 57.34%, 27.67%, and 14.98%, while APS-3a was composed of galactose, arabinose, and glucose in a relatively molar percentage of 84.54%, 6.50%, and 8.96% (Zhao et al., 2012). Yamada et al. (1990) purified an anticomplementary polysaccharide fraction AR-4E-2 from *A. sinensis*. The AR-4E-2 contained 14.5% galacturonic acid and 3.2% protein, and it was composed of arabinose, galactose, and rhamnose in molar ratios of 3.3:1.0:0.7. Methylation analysis and base-catalyzed beta-elimination analysis found that AR-4E-2 contained a rhamnogalacturonan moiety in which 2,4-di-substituted rhamnose residues were attached to 4-substituted galacturonic acid through position 2 of rhamnose. The AR-4E-2 also contained highly branched 3,5-arabinan and (1→4)-galactan. Zhang et al. (2010b) isolated one polysaccharide from *A. sinensis* root, and found the major sugar content of the polysaccharide was saccharose (18.55%), while it also contained glucose and fructose as minor components.

Sun et al. (2010) reported that a pectic polysaccharide isolated from roots of *A. sinensis* consisted of a backbone of linear homogalacturonan fragments as "smooth regions" and rhamnogalacturonan fragments as "hairy regions," with repeating units of (→ 4)-α-D-Gal*p*A-(1 → 2)-α-L-Rha*p*-(1 →). There were 58.8% of the rhamnopyranose residues in the backbone substituted at the O-4 position by the side chains. The side chains contained mainly β-1,6- and β-1,4-galactopyranan having 3,6- and 4,6-substituted β-D-galactopyranose residues as branched points and short α-1,5-arabinofuranan possessing 3,5-substituted α-L-arabinofuranose residues as branching points. The β-1,6-galactopyranan side chains were found to be highly branched, with α-1,5-arabinofuranan bearing 3-*O*-substituents (1,3,6-Gal) terminated by the α-arabinofuranose residues which form arabinogalactan.

The molecular weight and the proposed structure of the *A. sinensis* polysaccharide – iron complex (APIC) were also reported. The APIC was found to be a polynuclear ferrihydrite core chelated firmly by an encircling framework of ASP chains, forming a core molecule, which was surrounded by a removable outer protective sheath of colloidal ASP. The molecular weight of ASIC was 270 kDa, and the molecular formula of APIC was proposed to be $[(Fe_2O_3 \cdot 2.2H_2O)_{1043}(ASP)_{32}](ASP)_{12}$ (Wang et al., 2008).

10.2.3 *Lycium barbarum* Polysaccharide

Tian (Tian and Wang, 2006) prepared LBP using hot-water extraction, and further purified the polysaccharide by DEAE ion-exchange cellulose and gel chromatography. It was found that LBP has four water-soluble fractions with a mean molecular weight of 152.4 kDa. The polysaccharide was composed of six kinds of monosaccharides (Ara, Rha, Xyl, Man, Gal, and Glc), galacturonic acid, and 18 kinds of amino acids. It consists of acidic heteropolysaccharides and polypeptide or protein, and has a Glycan-*O*-Ser glycopeptide structure.

An acidic polysaccharide, LBP-1, isolated from the fruit of *L. barbarum* L. was purified by ion-exchange column. It was composed of rhamnose, arabinose, xylose, galactose, mannose, and galacturonic acid at a ratio of 1.00:7.85:0.37:0.65:3.01:8.16. The polysaccharide has an average molecular weight of 2250 kDa. The backbone of LBP-1 was mainly composed of (1→5)-linkage arabinose, (1,4)-linkage galacturonic acid, -(1)-mannose-(3→6)-linkage, and terminated with -(1)-mannose (Zou et al., 2010). Zhang and Zhang (2007) reported that a LBP fraction named LBP-4 consists of a xylose, galactose, mannose, rhamnose, glucose, and arabinose with a molar ratio of 4.45:1.20:1.0:0.98:0.90:0.85. The content of metal elements calcium, magnesium, zinc, and iron in LBP-4 were found to be 127.907, 32.395, 0.384, and 0.293 mg/g, respectively. The content of galacturonic acid in LBP-4 was 55.45%, while the content of protein was 3.80%. The indican bonds in the backbone of sugar residues chain were composed mainly of 1→3 bonds. Another acidic LBP, LBP-s-1, was reported. The LBP-s-1 was found to be a heteropolysaccharide consisting of rhamnose, arabinose, xylose, mannose, glucose, galactose, and galacturonic acid with molar ratio of 1.00:8.34:1.25:1.26:1.91:7.05:15.28 (Zhu et al., 2013a).

A water-soluble polysaccharide, LbGp1, with an average molecular weight of 49.1 kDa was reported. The LbGp1 was composed of Ara and Gal in a molar ratio of 5.6:1. It is a highly branched polysaccharide with a backbone of →6)-Gal*p*-(1 → linked galactose substituted at O-3 by galactosyl or arabinosyl groups. The branches contain (1 → 3)-linked-Gal*p*, (1 → 4)-linked-Gal*p*, (1 → 2)-linked-Ara*f* and (1 → 3)-linked-Ara*f*, and arabinose was located at the terminal of the branches. The existence of an *O*-glycopeptide bond in LbGp1 was demonstrated by β-elimination reaction (Wang et al., 2014). Peng and Tian (2001) reported a structural feature of another glycoconjugate, LbGp2. The molecular weight of LbGp2 was 68.2 kDa, and the glycoconjugate comprised Ara, Gal, and amino acids. The LbGp2 possesses a backbone consisting of (1→6)-β-galactosyl residues, about 50% of which are substituted at C-3 by galactosyl or arabinosyl groups, and the major nonreducing end is made of Ara (1→). In addition, Huang et al. (2000) reported that both LbGp3 and LbGp4 contained a backbone consist of (1 → 4)-linked-Gal*p*, but the side chains of LbGp3 consist of (1 → 3)-linked-Gal, while that of LbGp4 consist of (1 → 3)-linked-Gal, (1 → 3)-linked-Gal-(1 → 3)-Gal, and (1 → 3)-linked-Gal-(1 → 6)-Gal.

10.2.4 *Astragalus* Polysaccharide

Zhu et al. (2011) isolated two polysaccharides from the root of *Astragalus*. The molecular weights of the two polysaccharides were 4.77×10^6 and 8.68×10^3 Da, respectively. These two polysaccharides were mainly composed of glucose, and a small amount of arabinose and xylose. Further study showed that the structures of the two polysaccharides were similar. The main chains of the polysaccharides were mainly composed of major α-(1 → 3) glucose and a few 1 → 4, 1 → 6 glucoses, while the side chain contained arabinoses and xyloses. Xu et al. (2008) investigated the structure of another two polysaccharides (APS-I and APS-II). The APS-I consists of arabinose and glucose in a molar ratio of 1:3.45, while APS-II consists of rhamnose, arabinose, and glucose in a molar ratio of 1:6.25:17.86. The molecular weights of APS-I and APS-II were 1699.1 and 1197.600 kDa, respectively.

Two glucans AG-1, and AG-2, and two heteropolysaccharides AH-1, and AH-2 were isolated and purified from *Astragalus*. The AG-1 was identified as an α-glucan which had a structure of α-(1→4)- and (1→6)-glucan. The ratio of α-(1→4) and α-(1→6) linkages was about 5:2. While AG-2 was elucidated as an α-(1→4) glucan, AH-1 was an acidic polysaccharide, and consisted of hexuronic acid (galacturonic acid and glucuronic acid), glucose, rhamnose, and arabinose with a molecular ratio of about 1:0.04:0.02:0.01. The AH-2 was composed of glucose and arabinose with a molecular ratio of 1:0.15 (Huang et al., 1982). Similarly, Niu et al. (2011) reported an α-glucan with a molecular weight of 2.07×10^4 Da which has a repeating (1 → 4)-linked backbone with a (1 → 6)-linked branch every 10 residues. A α-(1→4)-D-glucan from *Astragalus* was also reported, which contains a single α-D-glucose at the C-6 position every nine residues, on average, along the main chain. The molecular weight of the α-(1→4)-D-glucan was found to be 3.6×10^4 Da (Li and Zhang, 2009).

A water-soluble heteropolysaccharide was obtained from the roots of *Astragalus*. The molecular weight of the polysaccharide was estimated to be 5.79×10^5 Da. The minimal repeat unit of the polysaccharide was composed of one terminal arabinose, one 1,5-linked arabinose, one 1,3-linked rhamnose, one 1,3,4-linked rhamnose, five 1,4-linked methyl galacturonates, and six 1,4-linked methyl glucuronates (Wang et al., 2006). Yin et al. (2012) reported one *Astragalus* heteropolysaccharide that has a molecular weight of 1334 kDa. Further study showed the polysaccharide was composed of Rha, Ara, Glc, Gal, and GalA in a molar ratio of 0.03:1.00:0.27:0.36:0.30. The backbone of the heteropolysaccharide consisted of 1,2,4-linked Rha*p*, α-1,4-linked Glc*p*, α-1,4-linked GalA*p*, β-1,3,6-linked Gal*p*, with a branch at the O-4 of the 1,2,4-linked Rha*p* and O-3 or O-4 of β-1,3,6-linked Gal*p*, while the side chains mainly consisted of α-T-Ara*f* and α-1,5-linked Ara*f*, with O-3 as branching points. The terminal residues of the heteropolysaccharide were T-linked Ara*f*, T-linked Glc*p*, and T-linked Gal*p*.

10.3 BIOACTIVITIES

10.3.1 *Panax ginseng* Polysaccharide

10.3.1.1 Antitumor Effect

It has been confirmed that ginseng consumption not only reduces the risk of cancer, but it also improves the quality of life (QOL) in patients with cancer. A human case-control study with 905 human pairs conducted by the Korea

Cancer Center Hospital showed that the odds ratio (OR) of cancer in relation to ginseng intake was 0.56. Another human case-control study on 1987 pairs showed that ginseng intake resulted in a decreased risk of cancer, and the ORs were 0.47 for cancer of the lip, oral cavity, and pharynx; 0.20 for esophageal cancer; 0.36 for stomach cancer; 0.42 for colorectal cancer; 0.48 for liver cancer; 0.22 for pancreatic cancer; 0.18 for laryngeal cancer; 0.55 for lung cancer; 0.15 for ovarian cancer; and 0.48 for other cancers (Yun, 1996). Cui et al. (2006) evaluated the association of ginseng use as a complementary therapy with survival and QOL in a cohort of 1455 breast cancer patients. Compared with patients who never used ginseng, the patients who took ginseng regularly had a significantly reduced risk of death. The adjusted hazard ratios associated with ginseng use were 0.71 for total mortality, and 0.70 for disease-specific mortality/recurrence. Ginseng consumption after cancer diagnosis, especially current use, was positively associated with scores of QOL, with the strongest effect in the psychological and social well-being domains. QOL also improved as cumulative ginseng use increased.

In vivo studies showed that polysaccharide of ginseng is a potential antitumor components. Kim et al. (1990) reported antitumor activity against a solid form of sarcoma 180 tumor in ICR mice. Tumor cells were inoculated subcutaneously into the left groin of ICR mice. After 1 day of tumor inoculation, polysaccharide samples including crude ginseng polysaccharide, neutral polysaccharide fraction, and acidic polysaccharide fraction were intraperitoneally administered once daily for 10 days. The tumor weights were measured on the 25th day. The results showed that the crude polysaccharide suppressed tumor growth by 42%, while the neutral polysaccharide and acidic polysaccharide suppressed tumor growth by 36% and 47%, respectively. The neutral and acidic polysaccharide also significantly increased the ratio of spleen-to-body weight, as well as activating B cell and T cells. It was also found that the neural polysaccharide possessed a synergistic effect with 5-fluorouracil in treatment for sarcoma 180 tumor-bearing mice. The neural polysaccharide also stimulated the proliferation of lymphocytes, improved natural killer cell cytotoxicity, increased the phagocytosis and nitric oxide production by macrophages, as well as increasing the level of tumor necrosis factor-α (TNF-α) in serum in mice (Ni et al., 2010).

A synergistic antitumor effect of ginseng polysaccharide and paclitaxel were reported. It was found that a combined treatment with paclitaxel and ginseng polysaccharide could increase the life span of ICR mice bearing sarcoma 180 tumor cells by 28%−43%, while no significant effect was observed on the sole paclitaxel group. When a combination of 10 mg/kg paclitaxel and 100 mg/kg ginseng polysaccharide was administered to C57BL/6 mice implanted with B16 melanoma, the tumor weight also decreased by 76.3%. Such a synergistic antitumor effect is supposed to be the result of the immunomodulating antitumor effect of the polysaccharide, since it could restore the proliferation of splenocytes and natural killer cell activity which were suppressed by paclitaxel, as well as increase the tumoricidal activity of macrophages (Shin et al., 2004).

The antitumor effect of ginseng polysaccharide was also evaluated in vitro. The acidic ginseng polysaccharide was found to significantly inhibit the growth of HT-29 colon cancer cells, the WGPA-3-HG fraction of the acidic polysaccharide was especially found to be responsible for this antiproliferative effect. Further study showed that WGPA-3-HG exerts an antiproliferative effect by cell cycle arrest at the G2/M phase, and inducing apoptosis of tumor cells via the caspase-3 pathway (Cheng et al., 2011). It has been reported that ginseng polysaccharide dose-dependently inhibited migration and invasiveness of HGC-27 cells by regulation of Twist, AKR1C2, NF1, E-cadherin, vimentin, and N-cadherin expression (Cai et al., 2013; Li et al., 2014). Fan et al. (2010) also found that WGPA-3-HG could suppress the migration of human fibrosarcoma HT-1080 cells, but not human adenocarcinoma of cervical HeLa cells. Further study showed the polysaccharide could inhibit cell migration via decreasing adhesion of the tumor cells.

Ginseng polysaccharide also exhibits a protective effect in radiotherapy for cancer. A clinical trial on nasopharyngeal carcinoma (NPC) treated with combined radiotherapy and ginseng polysaccharide injection was reported by Xie et al. (2001). A group of 131 patients with NPC were randomly divided into two, and treated with conventional radiotherapy and combined therapy, respectively. After a 3-month clinical examination, the complete remission rate in the combined therapy group was 96.6%, and in the control group 93.3%, the complete remission rate of cervical lymph node metastasis in the two groups was 85.7% and 78.0%, and the NPC remission rate in the two groups was 60.3% and 51.7%, respectively. Reexamination carried out 1 year after treatment showed that the total survival rates in the combined therapy group and conventional radiotherapy group were 100% and 96. 5%, tumor free survival rates were 84.4% and 74.6%, and no remote metastasis survival rates were 93.8% and 88.1%, respectively. The activity of NK cells and LAK cells, as well as T3 and T4 values in peripheral blood were also significantly increased in the combined therapy group. Ivanova et al. (2006) reported that ginseng polysaccharide intake could significantly reduce the frequency of micronucleated polychromatic erythrocytes, and decrease polychromatic erythrocyte/normochromatic erythrocytes ratios in mice treated with gammairradiation. It also found a pretreatment of ginseng polysaccharide significantly increased the viability of bone marrow cells against gammairradiation, as well as stimulating the secretion of interleukin-12 (IL-12), and expression of major histocompatibility complex class II (MHC II) molecules. Further in vivo study showed ginseng polysaccharide-treated mice had a larger number of bone marrow cells after gammairradiation (Kim et al., 2007).

10.3.1.2 Immune Enhancement Effect

Many efforts also focus on the immune enhancement effect of ginseng polysaccharide, and the antitumor activity is considered to be closely related to its immunoregulatory activity. It has been confirmed that ginseng polysaccharide could activate macrophages both in vitro and in vivo. A high level of nitric oxide synthase (iNOS) was shown in female BALB/c mice administered intraperitoneally with the acidic polysaccharide from ginseng (Park et al., 2001). It was also found that ginseng polysaccharide treatment significantly increased the lysosomal phosphatase activity and the phagocytic index of the peritoneal macrophage (Lim et al., 2004). Song et al. (2002) reported that treatment by ginseng polysaccharide in vitro could induce the mRNA of cytokines such as TNF-α, IL-1β, IL-6, and IL-12 in murine peritoneal macrophages. The tumoricidal activity of polysaccharide-treated macrophage on Yac-1 cells was enhanced in a dose-dependent manner. Similarly, the cytotoxic activity of the polysaccharide-treated macrophage against B16 melanoma cells was significantly induced (Shin et al., 2002). Byeon et al. (2012) reported that ginseng acidic polysaccharide treatment strongly induced NO production in RAW264.7 cells, and the polysaccharide-induced NO production was accompanied with enhanced mRNA levels of iNOS. Lim et al. (2002) also reported a macrophage-mediated antisepticemic effect of ginseng polysaccharide in *Staphylococcus aureus*-infected C57BL/6J. Treatment of the polysaccharide at a 0.025 mg/kg concentration showed a potent antisepticemic activity, since a low mortality of polysaccharide-treated mice was observed. However, it was found that the polysaccharide did not directly inhibit *S. aureus* in a minimum inhibitory concentration test in vitro, while the polysaccharide significantly stimulated the secretion of TNF-α, IL-1, and IL-6, as well as enhancing the phagocytic activity of the macrophages. The NO production of macrophages in polysaccharide-treated mice was increased approximately four times more than in the untreated group at 24 and 48 h incubation. Compared with the untreated group, the intracellular concentration of *S. aureus* in macrophages decreased by approximately 50% in the polysaccharide-treated group.

Acidic ginseng polysaccharide could induce the proliferation of T cells and B cells. Spleen cells became cytotoxic to a wide range of tumor cells without major histocompatibility complex-restriction after 4 or 5 days culture in vitro with the polysaccharide (Lee et al., 1996). A synergistic effect of a combined treatment with ginseng acidic polysaccharide and pidotimod in cyclophosphamide-treated mice was reported. The combination of pidotimod and ginseng polysaccharide was found to significantly restore concanavalin A-induced splenic T-cell proliferation and lipopolysaccharide-induced B cell proliferation. The production of NO from peritoneal macrophages was also increased by the combination treatment. A synergistic increase in the level of serum IL-12 and IFN-γ, as well as the level of serum C-reactive protein was observed (Du et al., 2008). Gamma radiation causes suppression of the immune function. However, ginseng polysaccharide could induce the mRNA expressions of Th1 and Th2-type cytokines, and also restore the mRNA expression of IFN-γ, and Th1 cytokine in mice with whole-body gamma irradiation. It was also found that the polysaccharide could restore T lymphocyte function which had been suppressed by gamma irradiation in allogenic mixed lymphocyte reactions (Han et al., 2005). Hwang et al. demonstrated a therapeutic efficacy of acidic ginseng polysaccharide that induces the suppression of an encephalitogenic response during experimental autoimmune encephalomyelitis (EAE). The polysaccharide also significantly ameliorates the progression of EAE through inhibiting the proliferation of autoreactive T cells and the production of inflammatory cytokines IFN-γ, IL-1β, and IL-17. In particular, it promotes the generation of immunosuppressive regulatory T cells (Tregs) through activating Foxp3, a transcription factor (Hwang et al., 2011).

The effect of ginseng polysaccharide on dendritic cells (DCs) was also investigated. The polysaccharide markedly increased the levels of IL-12 and TNF-α production in DCs. It also profoundly enhanced the expression of CD86 on DC surfaces. In ^{3}H-thymidine incorporation assays, polysaccharide-treated DCs induced a significantly higher proliferation of allogeneic CD4$^+$ T lymphocytes. The data indicates that ginseng polysaccharide could promote the maturation of DCs (Kim et al., 2009).

In addition, ginseng polysaccharide also showed a protective effect against radiation in vivo. It was found that the polysaccharide significantly increased the number of bone marrow cells, spleen cells, granulocyte-macrophage colony-forming cells (GM-CFC), and circulating neutrophils, lymphocytes, and platelets in irradiated mice. Pretreatment of the polysaccharide protected mice from the lethal effects of ionizing radiation more effectively than when it was given immediately after or at various times after irradiation (Song et al., 2003). Wang et al. (2003) reported that ginseng polysaccharide could stimulate thymocytes, splenocytes, or bone marrow stromal cells in a hematopoietic microenvironment, and it could also induce the secretion of GM-CSF and other hematopoietic regulation factors.

10.3.1.3 Other Effects

Some literature has demonstrated an antibacterial effect of ginseng polysaccharide. *Helicobacter pylori* infection increases the risk of gastric cancer development (Uemura et al., 2001). Lee et al. (2006) reported that ginseng polysaccharide could inhibit the attachment of *H. pylori* to the human gastric adenocarcinoma epithelial cell line, and the minimum inhibitory concentration was as low as 0.2 mg/mL(Lee et al., 2004). The antiadhesive effects against *Actinobacillus actinomycetemcomitans*, *Propionibacterium acnes*, and *S. aureus* were also evaluated by Lee's group. The result showed that the minimum inhibitory concentrations were found to be in the range of 0.25−0.5 mg/mL. In particular, it also showed that the polysaccharide has no inhibitory effects against the beneficial and commensal bacteria, such as *Lactobacillus acidophilus*, *Escherichia coli*, and *Staphylococcus epidermidis*.

Xie et al. (2004) evaluated the antihyperlipidemic effect of ginseng polysaccharide using a hyperlipidemic rats model which was established by Triton WR1339 or corn oil intravenous injection. Oral administration of the polysaccharide reduced the serum levels of triglyceride (TG) up-regulated by Triton WR1339 in a dose-dependent manner. Moreover, the polysaccharide treatment was shown to significantly decrease the levels of nonesterified fatty acid (NEFA) concomitant with TG reduction. Similar suppressive effects were also observed in hepatic total lipids and TG. Such antihyperlipidemic activities of the polysaccharide in serum and hepatic parameters of TG and NEFA were also observed in the exogenous hyperlipidemic rat model induced by corn oil. Konno (Konno and Hikino, 1987) also reported that ginseng polysaccharide could significantly lower blood sugar levels in both normal mice and alloxan-induced hyperglycemic mice models.

An acidic *P. ginseng* polysaccharide fraction GRA-4 was administered to mice from 50 to 200 mg/kg orally. The polysaccharide treatment inhibited gastric lesions induced by HCl-ethanol or absolute ethanol, in a dose-dependent manner. The cytoprotective activity of GRA-4 decreased after its treatment with periodate, but not after protein digestion, indicating that the carbohydrate moiety of this acidic polysaccharide contributes to this activity (Sun et al., 1991). Baek et al. (2010) evaluated the antidiarrheal effect of ginseng in an in vitro model of rotavirus infection, which is the leading cause of severe diarrhea. Two pectic ginseng polysaccharides were found that could rescue cell viability from rotavirus infection dose-dependently. The antirotavirus effect of the polysaccharide resulted from inhibiting the rotaviral attachment to cells.

Li et al. (2009b) investigated the effects of ginseng polysaccharide on energy metabolism and mitochondrial protection. It was found that ginseng polysaccharide could inhibit mitochondrial injury and swelling induced by Fe^{2+}-L-Cys. A 10-day oral administration of the polysaccharide inhibited the formation of malondialdehyde (MDA) in mice brain, increased levels of adenosine triphosphate (ATP), adenosine diphosphate (ADP), total adenylate pool (TAP) and adenylate energy charge (AEC), and ratio of ATP/ADP and ATP/AMP in liver cells. The polysaccharide treatment could also increase creatine kinase (CK) activities in mice skeletal muscle under chronic hypoxia condition. This indicates that ginseng polysaccharide not only protects mitochondria by inhibiting mitochondrial swelling, and improving energy metabolism, but also functions as a potential antioxidant by increasing CK activities.

In addition, an antifatigue activity of ginseng polysaccharide was also reported. The polysaccharide was found to reduce immobility in the forced-swim test on mice. Moreover, the polysaccharide treatment could inhibit the reduction in glucose and glutathione peroxidase (GPx), and increase in creatine phosphokinase, lactic dehydrogenase, and malondialdehyde levels induced in the forced-swim test on the mice model (Wang et al., 2010d).

10.3.2 *Angelica sinensis* Polysaccharide

10.3.2.1 Antitumor Effect

Several kinds of murine tumor model in vivo, including sarcoma 180, leukemia L1210, and Ehrlich ascetic cancer (EAC) were established to evaluate the antitumor effect of *A. sinensis* polysaccharide (Shang et al., 2003). For each kind of tumor model, three experimental groups were respectively given the polysaccharide at doses of 30, 100, and 300 mg/kg a day for 10 days. The results showed that *A. sinensis* polysaccharide had no obvious inhibitory effect on the growth of sarcoma 180, but it could significantly decrease the thymus weights of mice bearing sarcoma 180. The polysaccharide could significantly reduce the production of ascitic liquids and prolong the life of EAC-bearing mice, and it could also increase the survival time of L1210-bearing mice. An in vitro study showed that both *A. sinensis* and its subfractions could significantly inhibit the invasion of human hepatocellular carcinoma cell line (HHCC) into matrigel reconstituted basement membrane, as well as block the adhesion of HHCC to extracellular matrix proteins.

Cao et al. (2010a) investigated the proapoptotic effects of *A. sinensis* polysaccharide in human cervical cancer HeLa cells both in vitro and in vivo. *A. sinensis* polysaccharide decreased HeLa cell proliferation in a concentration- and time-dependent manner in vitro. An in vivo study showed the polysaccharide could significantly inhibit tumor growth in athymic nude mice. Characteristic manifestations of apoptosis, including apoptotic morphological features and sub- G0/G1 peaks, were observed when HeLa cells were treated with the polysaccharide. Further analysis showed that the polysaccharide-induced apoptosis was associated with the regulation of Bcl-2 family protein expression, a decrease in the mitochondrial membrane potential, and an increase in the cytosolic cytochrome *c* levels. Sequentially, the polysaccharide increased the activities of caspase-9, -3, and poly (ADP-ribose) polymerase in a concentration-dependent manner. Particularly, pretreatment of Z-LEHD-FMK, a specific inhibitor of caspase-9, significantly attenuated polysaccharide-induced cell apoptosis, and activation of caspase-3, indicating that *A. sinensis* polysaccharide induces apoptosis of Hela cells primarily through the activation of the intrinsic mitochondrial pathway. Two acidic polysaccharide subfractions were also found to significantly inhibit the growth of sarcoma 180 tumors and increase the life spans of sarcoma 180-bearing mice (Cao et al., 2010b).

Leukemia stem cells (LSCs) play important roles in leukemia initiation, progression, and relapse, and thus represent a critical target for therapeutic intervention. *A. sinensis* polysaccharide was found to inhibit human acute myelogenous leukemia (AML) $CD34^{+}CD38^{-}$ cell proliferation in in vitro culture, in a dose-dependent manner, while sparing normal hematopoietic stem and progenitor cells at physiologically achievable concentrations. The polysaccharide also exhibited cytotoxic effects on AML K562 cells, especially LSC-enriched $CD34^{+}CD38^{-}$ cells. Colony formation assays confirmed that *A. sinensis* polysaccharide significantly suppressed the formation of colonies derived from AML $CD34^{+}CD38^{-}$ cells, but not those from normal $CD34^{+}CD38^{-}$ cells. Furthermore, it was found that *A. sinensis* polysaccharide-induced AML $CD34^{+}CD38^{-}$ cell senescence was closely associated with a series of characteristic events, including up-regulation of p53, p16, p21, and Rb genes, and changes in related cell cycle regulation proteins P16, P21, cyclin E, cyclin-dependent kinase-4, and telomere end attrition, as well as repression of telomerase activity (Liu et al., 2013). Zheng (Zheng and Wang, 2002) also reported that *A. sinensis* polysaccharide not only inhibited the proliferation of K562 cells in vitro and inhibited the cell from entering the active proliferative phase, but it also stimulated the differentiation of K562 cells to erythrocyte series and granulo-monocyte series.

10.3.2.2 Immune Enhancement Effect

A. sinensis polysaccharide is also a potential natural immune enhancement agent. The immunomodulatory activities of *A. sinensis* polysaccharide, oligosaccharides, and total amino acids were compared on murine peritoneal macrophages. The polysaccharide exhibited the most significant growth facilitative activity, while the oligosaccharide did better in activation of H_2O_2 release, and the total amino acid exerted the best activation of phagocytic activity, production of NO, and intracellular adhesion molecule-1 (ICAM-1) expression. The immune regulatory activity of all these compounds may be related to the up-regulation of TLR4 mRNA (Chen et al., 2010).

A study involving 90 female middle-aged participants showed that *A. sinensis* polysaccharide not only up-regulated immunity activities such as serum vascular cell adhesion molecule-1 (s-VCAM-1), IL-1β, IL-6, and TNF-α levels, but also enhanced antioxidant activities such as superoxide dismutase (SOD), catalase (CAT), GPx, and reduced glutathione (GSH) and lipid peroxidation levels in the middle-aged female subjects (Juan et al., 2009).

Three *A. sinensis* polysaccharide fractions were isolated and identified as the immunomodulators of murine peritoneal macrophages. It was found that the fractions induced a significant increase in cellular lysosomal enzyme activity, NO formation, reactive oxygen species (ROS) production, and TNF-α secretion in macrophages in vitro (Yang et al., 2008). Similarly, an acid *A. sinensis* polysaccharide was also reported to significantly enhance NO production, TNF-α secretion, and cellular lysosomal enzyme activity in murine peritoneal macrophages, both in vivo and in vitro. Further study also showed the polysaccharide activated the macrophage through triggering Toll-like receptor-4 (Yang et al., 2007a).

Yang et al. (2006) reported that *A. sinensis* polysaccharide could promote the proliferation of total spleen cells, macrophages, and T cells. The polysaccharide could also increase the production of IL-2 and IFN-γ on T cells, but decreased the synthesis of IL-4. Accordingly, the IL-2 and IFN-γ gene expression were enhanced in the polysaccharide-treated T cells, but the IL-4 gene expression was decreased. At the same time, the percentage of $CD4^+$ T cells in total spleen cells was remarkably increased by the polysaccharide, while that of $CD8^+$ T cell was slightly decreased.

Ma et al. (2009) reported a hepatoprotective effect against hepatocyte apoptosis induced by exhaustive exercise on rats. It found that the expression of Bax in liver tissue was decreased, while expression of Bcl-2 was increased, and the ratio of Bax/Bcl-2 was decreased in the rat when exposed to exhaustive exercise after being force-fed with *A. sinensis* polysaccharide by gastrolavage. In particular, *A. sinensis* polysaccharide could inhibit the hepatocytes apoptosis in exhaustively-exercised rats' liver tissue, with better results obtained at high doses.

A. sinensis polysaccharide supplemented in diet could also enhance the innate cellular immune response and disease resistance in the *Epinephelus malabaricus* group. After feeding with diets containing *A. sinensis* polysaccharide for 12 weeks, the respiratory burst activities of the fish in polysaccharide-treated groups were increased significantly. The phagocytic activities in the polysaccharide-treated groups were significantly higher than that of the control group. *A. sinensis* polysaccharide also stimulated the head kidney leukocyte proliferation significantly. The cumulative mortalities of fish fed with a diet containing the polysaccharide at 3000 mg/kg were significantly lower than those fed with a diet containing the polysaccharide at 500 per kg after 96 h of challenge (Wang et al., 2011b).

In addition, the angelica polysaccharide was selenizingly modified by the nitric acid–sodium selenite method. Then the effect of the derivatives on chicken peripheral lymphocytes proliferation in vitro was investigated using the natural polysaccharide as control. The results showed that selenylated polysaccharide possessed an enhanced activity in stimulating the proliferation of lymphocytes, suggesting that selenylation modification could improve the immune-enhancing activity of the angelica polysaccharide (Qin et al., 2013a,b).

10.3.2.3 Hematopoietic Effect

The hematopoietic effect is another important beneficial effect of *A. sinensis* polysaccharide. Bradley et al. (1999) reported a case of the hematopoietic effect of *A. sinensis* on an anemia patient who experienced marked improvement in the anemia after self-initiating regular consumption of *A. sinensis*. A blood-deficient chicken model was induced by cyclophosphamide. The model chickens were dull, shrinking into themselves, broken winded, loose feathered, waxy eyelided, and pale tongued, which were all symptoms of a blood-deficient syndrome. Meanwhile, red blood cell count and hemoglobin content decreased significantly. *A. sinensis* polysaccharide could enrich the blood directly by increasing the number of RBCs and hemoglobin, and it could also regulate the hematopoietic factors to enrich the blood indirectly (Hou et al., 2012). It was also reported that administration of a low dose of *A. sinensis* polysaccharide at 2.3 mg/kg body weight per day on blood-loss mice could significantly accelerate the recovery of hemoglobin level to its original value. Moreover, the colony-forming ability of bone marrow cells that were isolated from the polysaccharide-fed mice was also significantly increased in an ex vivo test. The hematopoietic activity of the polysaccharide was through the stimulation of secretion of IL-6 and GM-CSF, and the amounts of these hematopoietic growth factors secreted, in general, agreed with the number of GM colony formations (Liu et al., 2010b). Besides, Lee fractionated the *A. sinensis* polysaccharide into four fractions (F1, F2, F3, and F4) using a DEAE-Sepharose CL-6B column, and found the F2 fraction had the highest hematopoietic activity. The F2 fraction not only stimulated the human peripheral blood MNCs to secret GM-CSF and IL-3, but also protected the hematopoietic function of $CD34^+$ cells from Adriblastina (Lee et al., 2012).

A haematocytopenia with thrombocytopenia model was established using 4-Gy-irradiated mice. The mice were treated with *A. sinensis* polysaccharide by injection daily for 21 days. The polysaccharide treatment not only enhanced the recovery of platelets, white blood cells, red blood cells, and their progenitor cells significantly, but also stimulated the formation of the hematopoietic colony-forming unit (CFU) of the myeloid (CFU-GM), erythroid (BFU/CFU-E), mixed (CFU-GEMM), and megakaryocytic (CFU-MK) lineages, and bone marrow stromal cells (CFU-F). It was also found that *A. sinensis* polysaccharide possessed an antiapoptosis effect on megakaryoblastic cell line M-07e cells induced by cytokine and serum depletion. Interestingly, treatment of Ly294002, a phosphatidylinositol 3-kinase inhibitor, could attenuate the antiapoptosis effect of the polysaccharide (Liu et al., 2010a). *A. sinensis* polysaccharide could also markedly increase the expression of GM-CSF in protein and mRNA level in bone marrow stromal cells, thymocytes, and splenocytes. Plenocytes, thymocytes or bone marrow stromal cell-conditioned media prepared with *A. sinensis* polysaccharide could significantly enhance the proliferation and differentiation of CFU-GM (Wang et al., 2000).

Huang et al. (2009) reported that *A. sinensis* polysaccharide cooperated with erythropoietin has a significant impact on the expression of STAT5 of mononuclear cells. The expression of STAT5 in the cells has significant differences between the polysaccharide group and the control group at four time points. The JAK2 and STAT5 expressed in the cytoplasm and nucleus of mononuclear cells in the polysaccharide group was significantly increased by five times those of the control group, indicating that JAK2, STAT5 signal transduction plays an important effect in *A. sinensis* polysaccharide cooperated with erythropoietin in promoting hematopoiesis. Hepcidin is also thought to be a principal iron-regulatory hormone that inhibits the entry of iron into the plasma compartment, and *A. sinensis* polysaccharide significantly reduced the expression of hepcidin expression through inhibiting the expression of signal transducer and activator of transcription 3/5 (STAT3/5) and sons of mothers against decapentaplegic protein 4 (SMAD4) in liver, and stimulating the secretion of erythropoietin, which further down-regulated hepcidin by repressing CCAAT/enhancer-binding protein α (C/EBPα), SMAD4, and the phosphorylation process of STAT3/5(Wang et al., 2011a). Further study demonstrated that the polysaccharide can interrupt the BMP–SMAD and JAK–STAT pathways of hepcidin regulation, which would lead to a decreasing expression of hepcidin in a dose-related manner in vivo (Liu et al., 2012; Zhang et al., 2012).

X-ray irradiation showed an increased percentage of senescence-associated beta-galactosidase (SA-beta-Gal) positive cells, an increased ratio of G1 stages and the production of ROS, and an up-regulated expression of p16 in mRNA, while the capacity of colony-forming and total-antioxidant (T-AOC) were decreased. *A. sinensis* polysaccharide could significantly decrease the percentage of SA-beta-Gal positive cells, the ratio of G1 stages, and the production of ROS in HSCs, and down-regulate the expression of p16 in mRNA levels in HSCs. The polysaccharide could remarkably increase T-AOC and the capacity for colony forming in HSCs (Zhang et al., 2013b). Further study indicated that such an effect of the polysaccharide to antagonize X-ray-induced aging of HSCs may be related to the increase in the length of telomere and the activity of telomerase, as well as the down-regulation of the expression of P53 protein (Zhang et al., 2013a).

In addition, Ling et al. (2001) reported that *A. sinensis* polysaccharide could promote the proliferation and differentiation of BFU-E, CFU-E, CFU-GM, and CFU-MK in healthy and anemic mice. The culture media of the polysaccharide-treated splenocyte, macrophage, fibroblast, and skeletal muscle had an effective stimulating effect on hematopoietic progenitor cells (Wang and Zhu, 1996). It also reported that *A. sinensis* polysaccharide could increase the index of thymus, index of spleen, IgG, IgM, and C5 in rats with blood stasis, as well as increasing the activity of SOD, but the amount of MDA from heart, brain, kidney, and pancreas decreased.

10.3.2.4 Other Effects

An immunological colitis rat model was induced by using a combined intracolon enema with 2,4,6-trinitrobenzene sulfonic acid (TNBS) and ethanol. In the model rats, enhanced colonic mucosal injury, inflammatory response, and oxidative stress were observed. In colonic tissues, the CMDI, HS, OBT, and MPO activity, and the level of MDA and NO contents, as well as the levels of TNF-α and IL-2, were significantly increased, while the SOD activity and IL-10 content were significantly decreased. However, these parameters were found to be significantly ameliorated in colitis rats treated intracolicly with *A. sinensis* polysaccharide at doses of 400 and 800 mg/kg. In the meantime, colonic EGF protein expression in colitis rats was remarkably up-regulated (Liu et al., 2003).

A. sinensis polysaccharide also has potent anticoagulant and hemostatic effects. It has been found that the polysaccharide prominently enhanced platelet aggregation at 5 min, while it showed less effects on maximum platelet aggregation. The polysaccharide markedly prolonged coagulation time, but shortened bleeding time. It also significantly prolonged thrombin time and active partial thromboplastin time in a dosage of 3 mg/kg and 10 mg/kg, while it showed no obvious effect on prothrombin time (Yang et al., 2002).

A. sinensis polysaccharide exhibits a protective effect against nerve cell injury and impairment caused by oxidative stress. The polysaccharides not only protected PC12 neuronal cells from H_2O_2-induced cytotoxicity, but also reduced apoptosis and intracellular ROS levels. It was also found that *A. sinensis* polysaccharide could increase the number of microvessels, and improve blood flow after ischemia in a rat model of local cerebral ischemia (Lei et al., 2014).

Qin et al. (2013a,b) investigated the effect of *A. sinensis* polysaccharide on a rat osteoarthritis (OA) model in vivo and rat IL-1β-stimulated chondrocytes in vitro. The polysaccharide was administered into rat OA knee joints, and it showed protective effects on rat OA cartilage in vivo. Primary rat articular chondrocytes were cotreated with *A. sinensis* polysaccharide and IL-1β in vitro. Further study showed that the significant changes were concentrated in the genes which were involved in proteoglycan synthesis. It also found that polysaccharide treatment increased the mRNA expression of aggrecan and glycosyltransferases which were inhibited by IL-1β, but it did not affect the mRNA expression of matrix-degrading enzymes (Kim et al., 2013). Further study demonstrated that the polysaccharide presented anti-OA activity through stimulating insulin-like growth factor 1 (IGF-1) and IGF1 receptor (IGF1R) gene expression, but it did not directly activate the IGF1R signaling pathway, which consequently promoted uridine diphosphate sugars (UDP-sugars) and glycosaminoglycan synthesis due to up-regulating gene expression of UDP-sugar synthases (Wen et al., 2014).

In addition, intragastric administration of an *A. sinensis* polysaccharide-enriched fraction was found to prevent liver toxicity induced by acetaminophen in mice, without affecting the concentration of serum acetaminophen. The polysaccharide treatment normalized the rise in serum alanine transferase (ALT) and hepatic nitric oxide synthase (NOS) activity, and the decrease of glutathione level in the liver, as well as reducing the hepatic MDA concentration (Ye et al., 2001).

10.3.3 *Lycium barbarum* Polysaccharide

10.3.3.1 Antitumor Effect

Seventy-nine patients with various advanced cancers including malignant melanoma, renal cell carcinoma, colorectal carcinoma, lung cancer, NPC, and malignant hydrothorax were treated with LBP in combination with lymphokine activated killer cell (LAK) and IL-2 in a clinical trial. The data showed that regression of cancer was achieved in 75 of the 79 polysaccharide-treated patients. The effectiveness rate of patients in the combination therapy group was 40.9%, while that of patients treated with LAK/IL-2 only was 16.1%. The mean length of time patients treated with the combination therapy remained in remission was significantly longer than for patients in the LAK/IL-2 only treatment group. The combination treatment significantly increased the activity of natural killer (NK) and LAK cells (Gau et al., 1994).

LBP was given to H22-bearing mice orally for 2 weeks to evaluate the effects of LBP on tumor microenvironment T-lymphocyte subsets and DCs in tumor-bearing mice. LBP significantly increased the numbers of $CD4^+$ and $CD8^+$ T cells in TIL. The polysaccharide also increased the number of DCs in the tumor microenvironment, which was decreased markedly in the tumor-bearing mice (He et al., 2005).

The influence of LBP on human prostate cancer cells was reported. The in vitro experiments showed that LBP inhibited the growth of both PC-3 and DU-145 cells in both dose- and time-dependent manners. The polysaccharide caused the breakage of DNA strands of PC-3 and DU-145 cells, and induced the apoptosis of these cells. The highest apoptosis rates of PC-3 and DU-145 cells were 41.5% and 35.5%, respectively. The ratio of Bcl-2/Bax protein expression was significantly decreased by the polysaccharide treatments. The in vivo effect of the polysaccharide on PC-3 cells was also assessed in the nude mouse xenograft tumor model, and the results indicated that LBP could significantly inhibit PC-3 tumor growth in nude mice, since both the tumor volume and weight in mice of the polysaccharide-treated group were significantly decreased (Luo et al., 2009).

The antitumor effect was demonstrated on various cancer cell lines in vitro. Mao et al. (2011) reported that LBP treatment inhibited the growth of human gastric cancer MGC-803 and SGC-7901 cells, with cell-cycle arrest at the G0/G1 and S phase, respectively (Miao et al., 2010). The polysaccharide could also inhibit the growth of human colon cancer SW480 and Caco-2 cells lines in a dose-dependent manner through G0/G1 phase arresting. Changes in cell-cycle-associated protein, cyclins, and cyclin-dependent kinase were consistent with the changes in cell-cycle distribution. Gan et al. (2001) found that LBP inhibited the growth of HL-60 cells in a dose-dependent manner, and decreased the membrane fluidity of the cell. Further study showed that inducing apoptosis of HL-60 cells maybe an important mechanism for the antitumor effect of the polysaccharide. LBP was also reported to have anticancer activity against the human hepatoma QGY7703 cell line, and the polysaccharide treatment caused inhibition of QGY7703 cell growth with cycle arrest in the S phase and apoptosis induction (Zhang et al., 2005). LBP was also reported to arrest the cell cycle of breast carcinoma cell line MCF-7 cell in the S phase (Shen and Du, 2012), and inhibit the growth of human bladder carcinoma cell line BIU87 by inducing apoptosis (Ke et al., 2011).

The antitumor effect of *L. barbarum* was also considered to be closely related to its immune enhancement effect. The modulation of a polysaccharide–protein complex from *L. barbarum* on the immune system in sarcoma 180-bearing mice was also investigated. The data showed that a 10-day treatment of the polysaccharide – protein complex could significantly inhibit the growth of transplantable sarcoma 180, but could increase macrophage phagocytosis, the form of antibody secreted by spleen cells, spleen lymphocyte proliferation, CTL activity, IL-2 mRNA expression level, and could also reduce lipid peroxidation in the tumor-bearing mice. The most effective concentration was found to be 10 mg/kg (Gan et al., 2004).

In addition, LBP possesses a protective effect in irradiation or chemotherapy, since it can promote the recovery of peripheral blood in irradiation or chemotherapy-induced myelo-suppressive mice. The polysaccharide at a concentration of 50 mg/kg significantly ameliorated the decrease in peripheral white blood cells of irradiated myelo-suppressive mice on day 13, while the polysaccharide at a concentration of 100 mg/kg did the same on days 17 and 21. LBP significantly ameliorated the decrease in peripheral red blood cells of irradiated myelo-suppressive mice as well. A significant increase in peripheral platelet counts of chemotherapy-induced myelo-suppressive mice was also observed. The polysaccharide could obviously stimulate human PBMCs to produce G-CSF (Gong et al., 2005).

10.3.3.2 Immune Enhancement Effect

Clinical trials showed that the CD3, CD4, and CD4/CD8 ratio of T cells, lymphocyte blastogenesis rate, phagocytosis rate of macrophages, and leukocyte and lymphocyte counts of peripheral blood in patients decreased markedly after radiotherapy. However, LBP treatment could increase the above parameters significantly, except that leukocyte and lymphocyte counts of peripheral blood maintained unchanged, indicating that LBP can raise the immune responses of cancer patients during radiotherapy (Liu et al., 1995).

Subhealth has been described as a chronic condition of unexplained deteriorated physiological function, which falls between health and illness. Zhao et al. (2015) reported that a 4-week treatment with LBP could improve the subhealth state in a mice model caused through increasing antioxidant ability, promoting T lymphocyte proliferation, inhibiting thymus lymphocyte apoptosis, and alleviating fatigue.

LBP could markedly up-regulate the expression of CD40, CD80, CD86, and MHC class II molecules on peritoneal macrophage, and it significantly enhanced endocytic and phagocytic capacities of the macrophage. The polysaccharide could also activate transcription factors NF-κB and AP-1, as well as secretion of TNF-α in RAW 264.7 macrophages (Chen et al., 2009).

The effects of LBP on the phenotypic and functional maturation of DCs were investigated. Zhu et al. (2007) reported the polysaccharide was capable of promoting both the phenotypic and functional maturation of murine bone marrow DCs, since not only the coexpression of MHC class II and CD11c, and the secretion of IL-12 p40 by DCs were increased by the polysaccharide treatment, but also the activation of proliferation of allogenic lymphocytes by DCs was enhanced, while the endocytosis of the cells was impaired. Further study found the maturation of DCs induced by the polysaccharide was able to directly activate the nuclear transcription factor NF-κB p65, and the activity of the polysaccharide on DCs was significantly inhibited by treating the cells with pretreatment by anti-TLR2 or anti-TLR4 antibody, indicating that LBP induces the maturation of DCs via TLR2- and/or TLR4-mediated NF-κB signaling pathways (Zhu et al., 2013b). The polysaccharide was also reported to promote the expression of the molecular phenotype, capacity to stimulate allogeneic lymphocyte proliferation, and the levels of IL-12p70 and IFN-γ secretion of DCs via the NF-κB signaling pathway in a tumor microenviroment, which were established using supernatant from hepatocellular carcinoma cells hepG2.2.15 and hepG2 cells (Chen et al., 2012).

A polysaccharide–protein complex from *L. barbarum* significantly stimulated the proliferation of T cells. Cell cycle profile analysis indicated that it markedly reduced sub-G1 cells, and also found that the polysaccharide–protein complex could activate transcription factors NFAT and AP-1, prompt CD25 expression, and induce IL-2 and IFN-γ gene transcription and protein secretion (Chen et al., 2008). Another study of the LBP – protein complex showed that the polysaccharide could increase the expression of IL-2 and TNF-α at both mRNA and protein levels in a dose-dependent manner (Gan et al., 2003).

Four sulfated LBPs (sLBPSs) with different degrees of sulfation (DS), $sLBPS_{0.7}$, $sLBPS_{1.1}$, $sLBPS_{1.5}$, and $sLBPS_{1.9}$, were prepared, and their effect on the cellular infectivity of Newcastle disease virus (NDV) were evaluated. NDV was added into the cultivating system of chick embryo fibroblast (CEF) respectively in three ways, pre- and postadding polysaccharide, and simultaneously adding polysaccharide and virus after being mixed. The results showed that $sLBPS_{1.5}$, $sLBPS_{1.9}$, and $sLBPS_{1.1}$ could significantly inhibit the infectivity of NDV to CEF. The viral inhibitory rate of $sLBPS_{1.5}$ in a pre- and simultaneous adding manner, and $sLBPS_{1.9}$ in postadding was the highest (Wang et al., 2010b,c).

10.3.3.3 Antioxidant Activity

Oxidative damage of biomolecules increases with age and is postulated to be a major causal factor of various physiological functional disorders. The antioxidant effect of LBPs on age-induced oxidative stress in different organs of aged mice was evaluated. It was found that increased endogenous lipid peroxidation, and decreased antioxidant activities, as assessed by SOD, CAT, glutathione peroxidase (GSH-Px), and T-AOC, and immune function were observed in aged mice. However, oral administration of LBPs restored such parameters to normal levels (Li et al., 2007b). LBP was also reported have an antioxidant effects against membrane damage induced by the free radicals generated during γ-irradiation in rat liver mitochondria. The polysaccharide exhibited a significant protective effect against irradiation-induced loss of protein thiols and inactivation of SOD, CAT, and GSH-Px in a dose-dependent manner. Moreover, the polysaccharide was more effective than α-tocopherol (VE) in inhibiting irradiation-induced oxidative injury (Li et al., 2007c).

Male Kunming mice were placed on either a high-fat diet or a normal laboratory diet for 30 consecutive days. It was found that the high-fat diet not only increased low density lipoprotein (LDL), total cholesterol (TC), triacylglycerols (TAG), and blood glucose, but also decreased the activities of antioxidant enzymes and levels of nonenzymic antioxidants. However, administration of LBPs significantly decreased the levels of LDL, TC, TAG, blood glucose, and thiobarbituric acid reactive substances, and increased the activities of antioxidant enzymes, as well as nonenzymic antioxidant levels, suggesting that LBPs showed a noticeable inhibition against lipid oxidation induced by free radicals caused by a high-fat diet intake on the basis of their antioxidant activities (Ming et al., 2009).

Niu et al. (2008) investigated the modulatory effect of LBPs on the oxidative stress induced by exhaustive exercise. Thirty-two male Wistar rats were involved in a 30-day exhaustive exercise program, and the lipid peroxidation, glycogen levels, and antioxidant enzyme activities in skeletal muscle were determined. It was found that the polysaccharide treatment significantly increased glycogen level and antioxidant enzyme activities, and decreased MDA level and CK activities in the rats.

LBPs could also provide a protective effect against acute doxorubicin (DOX)-induced cardiotoxicity through suppressing oxidative stress. Doxorubicin treatment lead to conduction abnormalities, and it decreased heart-to-body weight ratio, increased serum CK, and myofibrillar disarrangement in the rat model. The DOX treatment also increased MDA and decreased SOD and GSH-Px activity in cardiac tissues. However, pretreatment with LBPs significantly attenuated cardiac myofibrillar disarrangement, and improved conduction abnormalities induced by DOX. The polysaccharide treatment also increased SOD and GSH-Px activity, and decreased the MDA level of heart tissues damaged by DOX exposure in rats, as well as decreasing the levels of serum CK (Xin et al., 2011).

The antioxidant activity of the LBPs was also evaluated using various systems in vitro, including superoxide radical (O_2^-) scavenging activity, reducing power, β-carotene–linoleate model, inhibition of mice erythrocyte hemolysis mediated by peroxyl free radicals, 1,1-diphenyl-2 picrylhydrazyl ($DPPH^-$) radical-scavenging, and metal chelating activity. The polysaccharides showed a notable inhibitory activity in the β-carotene–linoleate model system, and a moderate inhibition in the $DPPH^-$ radical system. Meanwhile, LBPs also showed significant reducing power, superoxide scavenging ability, and ferrous ion chelating potency. The polysaccharide could also inhibit mice erythrocyte hemolysis induced by peroxyl free radicals (Li et al., 2007a). Similarly, Ke et al. (2011) reported that the scavenging effect of LBP at a concentration of 1.6 mg/mL on the superoxide radical, hydroxyl radical, and DPPH radical were 35.6%, 57.8%, and 78.73%, respectively. The polysaccharide could also suppressed oxidative hemolysis of erythrocytes induced by 2,2′-azobis-(amidinopropane) dihydrochloride in a concentration- and time-dependent manner.

In addition, LBP was modified by the HNO_3–Na_2SeO_3 method to obtain nine selenizing LBPs (sLBPs). In vivo tests showed that three sLBPs could significantly enhance GSH-Px and SOD activities, and decrease MDA content, in 14-day-old chickens. The antioxidant activity of sLBPs were significantly stronger than that of unmodified polysaccharide (Qiu et al., 2014).

10.3.3.4 Other Effects

Neuronal cell death, glial cell activation, retinal swelling, and oxidative injury are complications in retinal ischemia/reperfusion (I/R) injuries. A retinal I/R injury mice model was established by surgical occlusion of the internal carotid artery. Apoptosis and decreased viable cell count were found in the ganglion cell layer (GCL) and the inner nuclear layer (INL) of the vehicle-treated I/R retina. Additionally, increased retinal thickness, glial fibrillary acidic protein activation, aquaporin-4 up-regulation, IgG extravasations, and poly(ADP-ribose) expression levels were observed in the vehicle-treated I/R retina. A pretreatment with LBP at a dose of 1 mg/kg once a day for 1 week effectively diminished or abolished the change in the parameters. This protected the retina from neuronal death, apoptosis, glial cell activation, aquaporin water channel up-regulation, disruption of BRB, and oxidative stress, indicating that LBP may have a neuroprotective role to play in ocular diseases (Li et al., 2011). Further study demonstrated that the protective effect of the polysaccharide after I/R damage was mediated by activating the Nrf2/HO-1-antioxidant pathway in the retina. It was found that treatment with LBP significantly increased Nrf2 nuclear accumulation and HO-1 expression in the retina after I/R injury. Meanwhile, the increased apoptosis and decrease in the number of viable cells which were observed in the ganglion cell layer and INL in the I/R retina were reversed by polysaccharide treatment. The HO-1 inhibitor could also diminish the polysaccharide treatment-induced protective effects in the retina after I/R injuries (He et al., 2014).

The antiathletic fatigue activities of LBP were evaluated using a mice weight-loaded swimming model. It was found that LBP has a noticeable anti-athletic fatigue effect, since it prolonged the weight-loaded swimming time, prevented the increase of blood lactic acid of mice after swimming, and increased the content of hepatic glycogen of mice after swimming. The strongest effect on most biomarkers was seen with a 60 mg/kg dose (Jing et al., 2009).

In addition, the effect of LBP on heat-induced rat testis damage, on H_2O_2-induced DNA damage of mouse testicular cells, and on sexual behavior and reproductive function of hemicastrated male rats were investigated. It was found that the polysaccharide significantly increased testis and epididymis weights, improved SOD activity, and raised sexual hormone levels in the damaged rat testes in the heat-induced group. The polysaccharide also showed a dose-dependent protective effect against the H_2O_2-induced DNA oxidative damage of mouse testicular cells. Polysaccharide treatment improved the copulatory performance and reproductive function of hemicastrated male rats, such as shortened penis erection latency and mount latency, regulated secretion of sexual hormones and increased hormone levels, raised accessory sexual organ weights, and improved sperm quantity and quality (Luo et al., 2006).

10.3.4 *Astragalus* Polysaccharide

10.3.4.1 Antitumor Effect

A platinum-based two-drug regimen is currently the standard of care for patients with advanced nonsmall-cell lung cancer (NSCLC). Guo et al. (2012a) evaluated the effect of a combined treatment of *Astragalus* polysaccharide injection integrated with vinorelbine and cisplatin (VC) on improving QOL for advanced NSCLC patients. A group of 136 NSCLC patients were randomized to receive either VC (VC group) or VC combined with *Astragalus* polysaccharide (VC-APS group). The objective response rate was 42.64% in the VC-APS group, and 36.76% in the VC group. The median survival time was 10.7 and 10.2 months in the VC-APS group and VC group, and 1-year survival rates of the VC-APS group and VC group were 35.3% and 32.4%, respectively. The QOL, physical function, fatigue, nausea and vomiting, pain, and loss of appetite of the NSCLC patients significantly improved after three cycles of combined therapy.

Li et al. (2008) found that a combination of polysaccharopeptide and *Astragalus* polysaccharides could be used as an adjuvant for aiding adriamycin (AMD) chemotherapy. EAC tumor-bearing mice were given a treatment with the combination formula for 30 days, then a significant increase in the percentage of $CD3^+$ and $CD4^+$ T lymphocytes, the ratio of $CD4^+/CD8^+$, and the expression of IL-2/IL-2R in spleen and Bax in tumor tissue were observed, while the levels of Bcl-2 and CDK4 in tumor tissue decreased. In addition, the combination prescription could also restore the immunological effects against AMD-induced immunosuppression, such as the subset of leukomonocyte, the expression of IL-2/IL-2R in the spleen, and the thymus index.

The antitumor activity of *Astragalus* polysaccharide was investigated on a H22 hepatocarcinoma mice model. It was found that the solid tumor growth of H22 hepatocarcinoma transplanted in BALB/c mice was significantly suppressed by the polysaccharide treatment. The polysaccharide could also improve the body weight, spleen/thymus indexes, and phagocytotic function of macrophage of H22 tumor-bearing mice. Furthermore, it also found that the polysaccharide treatment could promote the secretion of IL-2, IL-12, and TNF-α, and decrease the IL-10 level in serum (Yang et al., 2013). Tian et al. (2012) reported a synergistic antitumor effect of *Astragalus* polysaccharide with adriamycin on the H22-bearing mice model. Not only inhibition rates, but also the spleen indexes of the Adriamycin − *Astragalus* polysaccharide-treated groups were significantly higher than that of the ADM alone group. There was also a synergistic effect on up-regulating serum IL-1α, IL-2, IL-6, and TNF-α levels. The polysaccharide was also found to alleviate the decrease in the sizes of the spleen and thymus induced by ADM. The *Astragalus* polysaccharide − adriamycin combined treatment could down-regulate the expression of P-glycoprotein (P-GP) and multidrug resistance (MDR) 1 mRNA in tumor tissues, which was related to drug resistance in tumor cells. Wu and Wang (Wu and Wang, 2010) also found *Astragalus* polysaccharide possessed a strong antitumor effect against liver cancer cell HepG2. The polysaccharides at doses of 25 mg/mL decreased more than 40.5% (24 h) and 67.3% (48 h) of liver cancer HepG2 cell viability. Particularly, 23.9% and 38.2% of HepG2 cells experienced apoptosis when treated with *Astragalus* polysaccharide for 24 and 48 h, respectively.

The effects of *Astragalus* polysaccharide on the proliferation of basal-like breast cancer cell line MDA-MB-468 cells was reported. The polysaccharide at 1 and 0.5 mg/mL concentrations effectively inhibited the proliferation of MDA-MB-468 cells. Further study suggested that the antiproliferation mechanisms may be related to its effects of up-regulating the expressions of p53, phosphatase, and tensin homolog deleted on chromosome ten (Ye et al., 2011).

Astragalus polysaccharide also possesses a radio-protective effect. Mice were pretreated with *Astragalus* polysaccharides, and then given a 5 Gy ^{60}Co γ-irradiation. At 24 h postirradiation, the polysaccharide-pretreated mice showed significantly decreased alanine aminotransferase (ALT), aspartate aminotransferase (AST), lactate dehydrogenase levels, and NF-κB expression. A high dose of polysaccharide pretreatment could lead to remarkably less morphologic features of IR-induced hepatic and pulmonary injury. The polysaccharide-pretreated mice also showed attenuation of the IR-induced increase in thiobarbituric acid reactive substance, and resolution of IR-induced decreases in SOD, CAT, and glutathione activities, suggesting that *Astragalus* polysaccharide exerts its radio-protective effect through suppressing the radiation-induced oxidative stress reaction (Liu et al., 2014).

10.3.4.2 Immune Enhancement Effect

Many efforts have focused on studying the immune enhancement effect of *Astragalus* polysaccharide. The polysaccharide has been found to be a potential macrophage activator both in vivo and in vitro. Shao et al. (2004) reported that *Astragalus* polysaccharide could activate mouse macrophages and B cells by stimulating proliferation or cytokine production. It found that fluorescence-labeled *Astragalus* polysaccharide was able to selectively stain murine macrophages and B cells. Interestingly, the polysaccharide could effectively stimulate the proliferation of splenic B cells from C3H/HeJ mice which have a mutated TLR4 molecule incapable of signal transduction. However, macrophages from C3H/HeJ mice were unable to respond to the polysaccharide treatment, indicating that *Astragalus* polysaccharide could activate B cells via a TLR4-independent manner, while it activates the macrophage through a TLR4-mediated pathway. Zhao et al. (2011) also found that fluorescence material 2-aminoacridone (2-AMAC) labeled *Astragalus* polysaccharide binds to macrophage in a time- dependent manner. *Astragalus* polysaccharide could increase the level of cytokines including TNF-α, GM-CSF, and the production of NO on RAW264.7 macrophage. The NF-κB protein level was also increased in response to the polysaccharide stimulation. However, blocking NF-κB resulted in a decreased level of NO and TNF-α, suggesting that NF-κB/Rel activation was involved in stimulating the cytokine secretion. Similarly, Lee (Lee and Jeon, 2005) also reported that treatment of RAW 264.7 cells with *Astragalus* polysaccharide produced strong induction of NF-κB/Rel-dependent reporter gene expression. Nuclear translocation and DNA binding activity of NF-κB/Rel was also significantly induced by the polysaccharide. It was also found that an administration of intraperitoneal injection of 500 mg/kg of *Astragalus* polysaccharide at a dose of 500 mg/kg for 2 h could significantly increase in the deposition of C3 on peritoneal macrophages in ICR mouse. The proportion of C3 positive macrophages in mice injected with a five-time concentration of the polysaccharide was much more than that of a one-time injection, indicating *Astragalus* polysaccharide could enhance the activity of the third component of complement (C3) of macrophages (Wang et al., 1989).

Astragalus polysaccharide also showed an immunomodulatory effect on other immune cells. For instance, *Astragalus* polysaccharide could suppress negative immunoregulation of $CD4^+CD25^+$ regulatory T cells, and stimulate a shift of Th2 to Th1 with activation of $CD4^+$ T cells in burned mice with *P. aeruginosa* infection (Liu et al., 2011a). It was also found that *Astragalus* polysaccharide has a synergistic effect with sulfated epimedium polysaccharide to resist the immunosuppression induced by cyclophosphamide in chickens (Guo et al., 2012b). Du et al. (2011) evaluated the adjuvant effect of *Astragalus* polysaccharide on immune responses to hepatitis B subunit vaccine on a vaccine-injected mice model. The data showed that coadministration of *Astragalus* polysaccharide with recombinant hepatitis B surface antigen significantly increased antigen-specific antibody production, T-cell proliferation, and cytotoxic T cell activity. The secretion of IFN-γ, IL-2, and IL-4 in $CD4^+$ T cells, and of IFN-γ in $CD8^+$ T cells were significantly enhanced. Meanwhile, the frequency of $CD4^+CD25^+$ regulatory T cells was down-regulated. Further study showed *Astragalus* polysaccharide exhibited such an immune enhancement effect by activating the TLR4 signaling pathway. Dang et al. (2009) also reported administration of *Astragalus* polysaccharide could decrease HBV DNA levels, as well as the contents of HBsAg, HBeAg, and HBcAg in HBV transgenic mice. In addition, Yuan et al. (2006) found that dietary supplementation of *Astragalus* polysaccharide increased average daily gain, the numbers of white blood cells and lymphocytes, the proportion of $CD4^+$T cells, and the secretion of IL-2 and IFN-γ in weaned pigs, while the contents of specific ovalbumin antibodies, Ig G, IL-4, and IL-10 were not affected.

Shao et al. (2006) reported that *Astragalus* polysaccharide could increase the coexpression of CD11c and MHC class II surface molecules, and enhance IL-12 secretion on DCs. Meanwhile, DCs treated with *Astragalus* polysaccharide displayed a more mature morphology, with long protrusions, and exhibited a down-regulated endocytotic activity. Liu et al. (2011b) found that the polysaccharide could induce the differentiation of splenic DCs to $CD11c^{high}CD45RB^{low}$ DCs, which was further demonstrated to be closely related to IL-12 synthesis. The $CD11c^{high}CD45RB^{low}$ DCs further enhanced T cell immune function by stimulating T cell differentiation to Th1. Zhang et al. (2010a) also reported that pretreatment of *Astragalus* polysaccharide could significantly enhance the efficacy of foot-and-mouth disease virus vaccination, since it not only increased the expression of CD40, CD86, and MHC II on DCs, but also enhanced the phagocytic capacity of peritoneal macrophage, proliferation of splenic lymphocyte, serum antibody titer, and the level of IL-4 and IL-10 in mice.

Thirteen polysaccharides isolated from an extract of the aerial portions of *Astragalus* were demonstrated to possess immunomodulatory activity against Peyer's patch immunocompetent cells. Nine of these polysaccharides were found to be arabinogalactans, pectic arabinogalactans, or pectins. The activities of the arabinogalactans and pectic arabinogalactans were closely associated with β-D-(1 → 3)-galactan moieties branched with β-D-(1 → 6)-galactooligosaccharide side chains, since their immunomodulating activities were dramatically decreased with the degradation of the β-D-(1 → 3)-galactan or β-D-(1 → 6)-galactosyl side chains in the arabinogalactans. As for the pectin-enriched polysaccharide, rhamnogalacturonan I with β-D-(1 → 3,6)-galactosyl side chains and a terminal β-D-GlcA is the active component (Kiyohara et al., 2010).

In addition, the immunoenhancement effect of sulfated angelica polysaccharides was also reported. Yang et al. (2012) reported that sulfated angelica polysaccharide not only inhibited the replication of murine leukemia virus, but also increased the thymus/body weight index. Both the percentage of $CD4^+$ cells and $CD4^+/CD8^+$ ratio in peripheral blood cells were significantly enhanced by the sulfated polysaccharide. The sulfated *Astragalus* polysaccharides were also reported have a beneficial effect on inhibiting bursal disease virus to infect CEF (Huang et al., 2008), stimulating proliferation of mouse selenocytes (Yang et al., 2000).

10.3.4.3 Other Effects

A nonobese diabetic (NOD) mice model was applied to investigate the effects of *Astragalus* polysaccharide in preventing type 1 diabetes mellitus (DM) and Th1/Th2-subtype cytokines. The results showed that NOD mice in the *Astragalus* polysaccharide-administration group had a lower incidence rate of type 1 DM, lower serum C-peptide levels, better histologic findings of pancreatic islets, and a lower $CD4^+/CD8^+$ ratio of T lymphocytes from the spleen and the infiltrated islets. Gene expression of IL-1β, IL-2, IL-6, IL-12, TNF-α, INF-γ, Fas, and iNOS in the pancreatic tissue was decreased in the polysaccharide-administration group, while the gene expression of IL-4, IL-5, IL-10, TGF-β, Bcl-2, and SOD was increased. It indicates that the effects of *Astragalus* polysaccharides on the prevention of type 1 DM in NOD mice works by correcting the imbalance between the Th1/Th2 cytokines (Chen, 2008). Mao et al. (2007) reported a hypoglycemic effect of *Astragalus* polysaccharide on an obese and type 2 diabetic KKAy mouse model. After an oral treatment by the polysaccharide at 700 mg/kg for 8 weeks, the hyperglycemic status, systemic insulin sensitivity, fatty liver disease, and insulin action in the liver of diabetic mice were partly normalized in the mice. Such a hypoglycemic effect partly depended on alleviating endoplasmic reticulum stress and enhancing the adaptive capacity of the endoplasmic reticulum. Wu et al. (2005) also found that *Astragalus* polysaccharide could decrease body weight, plasma glucose, and improve insulin sensitivity in a streptozotocin-induced type 2 diabetic rats model. The polysaccharide, in particular, enables insulin-sensitizing and hypoglycemic activity by reducing both expression and activity of protein tyrosine phosphatase 1B (a negative regulator of insulin-receptor signal transduction) in the muscle, which would result in an increase in insulin-induced tyrosine phosphorylation of the insulin receptor β-subunit and insulin receptor substrate-1. It was also found that *Astragalus* polysaccharide can alleviate endoplasmic reticulum stress in a high glucose-induced HepG2 cell model in vitro. The polysaccharide could improve hyperglycemic status, systemic insulin sensitivity, fatty liver disease, and insulin action in the liver of diabetic mice in vivo, suggesting that the polysaccharide enables insulin-sensitizing and hypoglycemic activity partly through enhancing the adaptive capacity of the endoplasmic reticulum (Mao et al., 2007).

A synergistic hepatoprotective effect of *Astragalus* polysaccharide with lignan from Fructus Schisandrae chinensis on chronic liver injury in male Sprague-Dawley rats was reported. After treatment with a subcutaneous injection of 10% CCl_4 for 3 months, significantly elevated serum ALT, AST, and alkaline phosphatase (ALP) activities were observed. This lead to significantly increased levels of MD and lowered levels of reduced GSH and CAT, and SOD in the liver of rats. However, a combination treatment of Fructus Schisandrae chinensis lignan and *Astragalus* polysaccharide displayed hepatoprotective and antioxidative effects, as evidenced by lower levels of serum ALT, AST, ALP, and hepatic MDA concentration, as well as higher SOD CAT activities, and GSH concentration. Meanwhile, histopathological examinations revealed severe fatty degeneration in the toxin group, while mild damage in the combined treatment group was observed (Yan et al., 2009). Similarly, Dang et al. (2008) also reported a synergistic effect of emodin and *Astragalus* polysaccharide in reducing ALT and restoring ALB in the serum on a CCl_4-induced chronic hepatic injury mice model. The combination treatment of emodin and *Astragalus* polysaccharide could also ameliorate the CCl_4-induced hepatic injury in rats by elevating antioxidant–enzyme activities and reducing lipid peroxidation.

Pathological cardiac hypertrophy induced by increased sympathetic drive can subsequently lead to congestive heart failure, which represents the major cause of morbidity and mortality worldwide. *Astragalus* polysaccharide is confirmed to be a potential antihypertrophic agent on a cardiac hypertrophic cells model. Cardiomyocytes was treated with isoprenaline to establish a cardiac hypertrophic model, in which a significant increase in cell surface area, total protein content, and protein synthesis, as well as the expression of hypertrophic markers, including atrial natriuretic peptide and B-type natriuretic peptide, were observed. However, such parameters were effectively inhibited by *Astragalus* polysaccharide treatment in a dose-dependent manner. Moreover, the polysaccharide treatment also alleviated the augmentation of intracellular free calcium during cardiac hypertrophy in the model cells. Further study revealed that the polysaccharide exerted its antihypertrophic action via inhibiting Ca^{2+}-mediated calcineurin/cytoplasmic-3 and CaMKII signaling cascades (Dai et al., 2014).

In addition, *Astragalus* polysaccharide was also reported to have an anti-inflammatory effect against bronchopulmonary dysplasis (BDP) in an EA. hy926 BDP cells model. The inflammatory response plays a crucial role in the development of BPD. A variety of cytokines, including IL-6, IL-8, IL-10, and ICAM-1, are considered to participate in lung inflammation, which is closely related to the development of BPD. *Astragalus* polysaccharide not only exhibited an antiinflammatory effect by inhibiting the activation of NF-κB p65, which further suppressed the expression IL-8, ICAM-1, and the inflammatory response, but also showed an antioxidant effect by decreasing the ROS and MDA levels and increasing SOD production in the model cells (Huang et al., 2013).

10.4 SUMMARY

To date, a great deal of research in the polysaccharide field have been reported over the past half century. Although the knowledge of structural analysis and biological activity evaluation has increased explosively, the understanding of polysaccharides is still far behind that of the nucleic acids and proteins. Research needs in the field of polysaccharides is huge, but it is balanced by the potential health benefits and the enormous size of the market. Research into the quality, safety, mechanism of action, and structure – activity relationship are needed.

With the concept of "Pattern Recognition Receptors" and "Pathogen Associated Molecular Pattern" being put forward, bioactive polysaccharides are now considered as a mainstay of biological response modifiers that could interact with hosts by triggering a series of receptors, such as toll-like receptors and Dectin-1 (Leung et al., 2006; Gordon, 2002). Some intracellular signal transduction pathways are also found to be involved in polysaccharide-induced beneficial activities. However, how the polysaccharide interacts with pattern recognition, and what the relationship is between the structure and bioactivity is a challenge to further elucidate.

Recently, a concept of "glycoinformatics" was proposed as a new branch of bioinformatics, which refers to informatics tools available for assessing primary data (the covalent and three-dimensional structure of glycans and glycoconjugates), and organizing this primary data into a database that can be used for speeding up the production of primary data, predicting new features, and characterizing the structure – activity relationship is necessary (Pérez and Mulloy, 2005). Polysaccharides are dramatically different from proteins and nucleic acids, not only due to their extremely complex structure, but also to their nontemplate-driven biosynthesis. Classical bioinformatics algorithms developed for proteomics or genomics are not suitable for glycoinformatics. However, high-throughput and automated techniques, which allow increasingly rapid structural analysis of polysaccharides, combined with utilization of carbohydrate databases and computer molecular simulated technology would strongly promote the vigorous development of glycoinformatics.

REFERENCES

Baek, S.-H., Lee, J.G., Park, S.Y., Bae, O.N., Kim, D.-H., Park, J.H., 2010. Pectic polysaccharides from *Panax ginseng* as the antirotavirus principals in ginseng. Biomacromolecules 11, 2044–2052.

Bradley, R.R., Cunniff, P.J., Pereira, B.J., Jaber, B.L., 1999. Hematopoietic effect of Radix Angelicae sinensis in a hemodialysis patient. Am. J. Kidney Dis. 34, 349–354.

Byeon, S.E., Lee, J., Kim, J.H., Yang, W.S., Kwak, Y.S., Kim, S.Y., et al., 2012. Molecular mechanism of macrophage activation by red ginseng acidic polysaccharide from Korean red ginseng. Mediators Inflamm. 2012, 732860.

Cai, J.-P., Wu, Y.-J., Li, C., Feng, M.-Y., Shi, Q.-T., Li, R., et al., 2013. *Panax ginseng* polysaccharide suppresses metastasis via modulating Twist expression in gastric cancer. Int. J. Biol. Macromol. 57, 22–25.

Cao, W., Li, X.-Q., Liu, L., Wang, M., Fan, H.-T., Li, C., et al., 2006a. Structural analysis of water-soluble glucans from the root of *Angelica sinensis* (Oliv.) Diels. Carbohydr. Res. 341, 1870–1877.

Cao, W., Li, X.-Q., Liu, L., Yang, T.-H., Li, C., Fan, H.-T., et al., 2006b. Structure of an antitumor polysaccharide from *Angelica sinensis* (Oliv.) Diels. Carbohydr. Polym. 66, 149–159.

Cao, W., Li, X.-Q., Wang, X., Fan, H.-T., Zhang, X.-N., Hou, Y., et al., 2010a. A novel polysaccharide, isolated from *Angelica sinensis* (Oliv.) Diels induces the apoptosis of cervical cancer HeLa cells through an intrinsic apoptotic pathway. Phytomedicine 17, 598–605.

Cao, W., Li, X.-Q., Wang, X., Li, T., Chen, X., Liu, S.-B., et al., 2010b. Characterizations and anti-tumor activities of three acidic polysaccharides from *Angelica sinensis* (Oliv.) Diels. Int. J. Biol. Macromol. 46, 115–122.

Chen, J.-R., Li, E.-Q., Dai, C.-Q., Yu, B., Wu, X.-L., Huang, C.-R., et al., 2012. The inducible effect of LBP on maturation of dendritic cells and the related immune signaling pathways in hepatocellular carcinoma (HCC). Curr. Drug Del. 9, 414–420.

Chen, W., 2008. Astragalus polysaccharides: an effective treatment for diabetes prevention in NOD mice. Exp. Clin. Endocr. Dia. 116, 468–474.

Chen, Y., Duan, J.-A., Qian, D., Guo, J., Song, B., Yang, M., 2010. Assessment and comparison of immunoregulatory activity of four hydrosoluble fractions of *Angelica sinensis*in vitro on the peritoneal macrophages in ICR mice. Int. Immunopharmacol. 10, 422–430.

Chen, Z., Tan, B.K.H., Chan, S.H., 2008. Activation of T lymphocytes by polysaccharide–protein complex from *Lycium barbarum* L. Int. Immunopharmacol. 8, 1663–1671.

Chen, Z., Soo, M.Y., Srinivasan, N., Tan, B.K.H., Chan, S.H., 2009. Activation of macrophages by polysaccharide–protein complex from *Lycium barbarum* L. Phytother. Res. 23, 1116–1122.

Cheng, H., Li, S., Fan, Y., Gao, X., Hao, M., Wang, J., et al., 2011. Comparative studies of the antiproliferative effects of ginseng polysaccharides on HT-29 human colon cancer cells. Med. Oncol. 28, 175–181.

Cui, Y., Shu, X.-O., Gao, Y.-T., Cai, H., Tao, M.-H., Zheng, W., 2006. Association of ginseng use with survival and quality of life among breast cancer patients. Am. J. Epidemiol. 163, 645–653.

Dai, H., Jia, G., Liu, X., Liu, Z., Wang, H., 2014. Astragalus polysaccharide inhibits isoprenaline-induced cardiac hypertrophy via suppressing Ca 2+-mediated calcineurin/NFATc3 and CaMKII signaling cascades. Environ. Toxicol. Pharmacol. 38, 263–271.

Dang, S.-S., Zhang, X., Jia, X.-L., Cheng, Y.-A., Song, P., Liu, E.-Q., et al., 2008. Protective effects of emodin and astragalus polysaccharides on chronic hepatic injury in rats. Chin. Med. J. 121, 1010–1014.

Dang, S.-S., Jia, X.-L., Song, P., Cheng, Y.-A., Zhang, X., Sun, M.-Z., et al., 2009. Inhibitory effect of emodin and Astragalus polysaccharide on the replication of HBV. World J. Gastroenterol. 15, 5669–5673.

Du, X., Chen, X., Zhao, B., Lv, Y., Zhang, H., Liu, H., et al., 2011. Astragalus polysaccharides enhance the humoral and cellular immune responses of hepatitis B surface antigen vaccination through inhibiting the expression of transforming growth factor β and the frequency of regulatory T cells. FEMS Immunol. Med. Microbiol. 63, 228–235.

Du, X.F., Jiang, C.Z., Wu, C.F., Won, E.K., Choung, S.Y., 2008. Synergistic immunostimulating activity of pidotimod and red ginseng acidic polysaccharide against cyclophosphamide-induced immunosuppression. Arch. Pharm. Res. 31, 1153–1159.

Fan, Q., Li, Y., Zhao, W., et al., 2009. Purification of polysaccharide from *Angelica sinensis* by ultrafiltration membrane technology. Chin. J. Inf. Tradit. Chin. Med. 16, 54–55.

Fan, Y., Cheng, H., Li, S., Wang, J., Liu, D., Hao, M., et al., 2010. Relationship of the inhibition of cell migration with the structure of ginseng pectic polysaccharides. Carbohydr. Polym. 81, 340–347.

Gan, L., Wang, J., Zhang, S., 2001. Inhibition the growth of human leukemia cells by *Lycium barbarum* polysaccharide. J. Hygiene Res. 30, 333–335.

Gan, L., Zhang, S.-H., Liu, Q., Xu, H.-B., 2003. A polysaccharide-protein complex from *Lycium barbarum* upregulates cytokine expression in human peripheral blood mononuclear cells. Eur. J. Pharmacol. 471, 217–222.

Gan, L., Zhang, S.H., Yang, X.L., Xu, H.B., 2004. Immunomodulation and antitumor activity by a polysaccharide–protein complex from *Lycium barbarum*. Int. Immunopharmacol. 4, 563–569.

Gao, Q.-P., Kiyohara, H., Cyong, J.-C., Yamada, H., 1988. Characterisation of anti-complementary acidic heteroglycans from the leaves of *Panax ginseng* CA Meyer. Carbohydr. Res. 181, 175–187.

Gau, G., Yang, W., Du, P., 1994. Observation of the effects of *Lycium barbarum* polysaccharides (LBP) in combination with LAK/IL-2 therapy in the treatment of 75 cancer patients. Chin. J. Oncol. 16, 1190–1197.

Gong, H., Shen, P., Jin, L., Xing, C., Tang, F., 2005. Therapeutic effects of *Lycium barbarum* polysaccharide (LBP) on irradiation or chemotherapy-induced myelosuppressive mice. Cancer Biother. Radiopharm. 20, 155–162.

Gong, S.Z., Yang, Z.R., 2004. Investigation into the Microwave-assisted Extraction Technology for Astragalus Polysaccharide. J. South China Univ. Technol. (Natural Science) 8, 21.

Gordon, S., 2002. Pattern recognition receptors: doubling up for the innate immune response. Cell 111, 927–930.

Guo, L., Bai, S.-P., Zhao, L., Wang, X.-H., 2012a. Astragalus polysaccharide injection integrated with vinorelbine and cisplatin for patients with advanced non-small cell lung cancer: effects on quality of life and survival. Med. Oncol. 29, 1656–1662.

Guo, L., Liu, J., Hu, Y., Wang, D., Li, Z., Zhang, J., et al., 2012b. Astragalus polysaccharide and sulfated epimedium polysaccharide synergistically resist the immunosuppression. Carbohydr. Polym. 90, 1055–1060.

Han, S.-K., Song, J.-Y., Yun, Y.-S., Yi, S.-Y., 2005. Ginsan improved Th1 immune response inhibited by gamma radiation. Arch. Pharm. Res. 28, 343–350.

He, M., Pan, H., Chang, R.C.-C., So, K.-F., Brecha, N.C., Pu, M., 2014. Activation of the Nrf2/HO-1 antioxidant pathway contributes to the protective effects of *Lycium barbarum* polysaccharides in the rodent retina after ischemia-reperfusion-induced damage. PLoS One 9, e84800.

He, X.Y., Xu, J.F., Wang, Q.S., Chen, Y.Y., 2010. Ultrasonic extraction and determination of polysaccharide in radix angelica sinensis. China Pharm. 19, 024.

He, Y.-L., Ying, Y., Xu, Y.-L., Su, J.-F., Luo, H., Wang, H.-F., 2005. Effects of *Lycium barbarum* polysaccharide on tumor microenvironment T-lymphocyte subsets and dendritic cells in H22-bearing mice. J. Chin. Integrat. Med. 3, 374–377.

Hou, H., Bao, Y., Li, Q., Shi, W., 2012. Preparation of blood-deficient model and research of *Angelica polysaccharide* on enriching blood in chickens. Evid. Based Compl. Alternat. Med. 2012, 965947.

Huang, D.-F., Xie, M.-Y., Yin, J.-Y., Nie, S.-P., Tang, Y.-F., Xie, X.-M., et al., 2009. Immunomodulatory activity of the seeds of *Plantago asiatica* L. J. Ethnopharmacol. 124, 493–498.

Huang, L., Tian, G., Qi, C., Zhang, Y., 2000. Structure elucidation and immunoactivity studies of glycan of glycoconjugate LbGp4 isolated from the fruit of *Lycium barbarum* L. Chem. J. Chin. Univ. 22, 407–411.

Huang, Q., Lu, G., Li, Y., Guo, J.-H., Wang, R.-X., 1982. Studies on the polysaccharides of "Huang Qi"(*Astragalus mongholicus* Bunge). Acta Pharma. Sinica 3, 007.

Huang, W.M., Liang, Y.Q., Tang, L.J., Ding, Y., Wang, X.H., 2013. Antioxidant and anti-inflammatory effects of Astragalus polysaccharide on EA. hy926 cells. Exp. Ther. Med. 6, 199–203.

Huang, X., Wang, D., Hu, Y., Lu, Y., Guo, Z., Kong, X., et al., 2008. Effect of sulfated astragalus polysaccharide on cellular infectivity of infectious bursal disease virus. Int. J. Biol. Macromol. 42, 166–171.

Hwang, I., Ahn, G., Park, E., Ha, D., Song, J.-Y., Jee, Y., 2011. An acidic polysaccharide of *Panax ginseng* ameliorates experimental autoimmune encephalomyelitis and induces regulatory T cells. Immunol. Lett. 138, 169–178.

Ivanova, T., Han, Y., Son, H.-J., Yun, Y.-S., Song, J.-Y., 2006. Antimutagenic effect of polysaccharide ginsan extracted from *Panax ginseng*. Food Chem. Toxicol. 44, 517–521.

Jing, L., Cui, G., Feng, Q., Xiao, Y., 2009. Orthogonal test design for optimization of the extraction of polysaccharides from *Lycium barbarum* and evaluation of its anti-athletic fatigue activity. J. Medicinal Plant Res. 3, 433–437.

Juan, J., Yingjie, G., Aijun, N., 2009. Extraction, characterization of *Angelica sinensis* polysaccharides and modulatory effect of the polysaccharides and Tai Chi exercise on oxidative injury in middle-aged women subjects. Carbohydr. Polym. 77, 384–388.

Ke, M., Zhang, X.-J., Han, Z.-H., Yu, H.-Y., Lin, Y., Zhang, W.-G., et al., 2011. Extraction, purification of *Lycium barbarum* polysaccharides and bioactivity of purified fraction. Carbohydr. Polym. 86, 136–141.

Kim, H.-J., Kim, M.-H., Byon, Y.-Y., Park, J.W., Jee, Y., Joo, H.-G., 2007. Radioprotective effects of an acidic polysaccharide of *Panax ginseng* on bone marrow cells. J. Vet. Sci. 8, 39–44.

Kim, H.S., Shin, B.R., Lee, H.K., Park, Y.S., Liu, Q., Kim, S.Y., et al., 2013. Dendritic cell activation by polysaccharide isolated from *Angelica dahurica*. Food Chem. Toxicol. 55, 241–247.

Kim, M.-H., Byon, Y.-Y., Ko, E.-J., Song, J.-Y., Yun, Y.-S., Shin, T., et al., 2009. Immunomodulatory activity of ginsan, a polysaccharide of *Panax ginseng*, on dendritic cells. Korean J. Physiol. Pha. 13, 169–173.

Kim, Y.S., Kang, K.S., Kim, S.I., 1990. Study on antitumor and immunomodulating activities of polysaccharide fractions from *Panax ginseng*: comparison of effects of neutral and acidic polysaccharide fraction. Arch. Pharm. Res. 13, 330–337.

Kiyohara, H., Hirano, M., Wen, X.-G., Matsumoto, T., Sun, X.-B., Yamada, H., 1994. Characterisation of an anti-ulcer pectic polysaccharide from leaves of *Panax ginseng* CA Meyer. Carbohydr. Res. 263, 89–101.

Kiyohara, H., Uchida, T., Takakiwa, M., Matsuzaki, T., Hada, N., Takeda, T., et al., 2010. Different contributions of side-chains in β-d-(1→3, 6)-galactans on intestinal Peyer's patch-immunomodulation by polysaccharides from Astragalus mongholics Bunge. Phytochemistry 71, 280–293.

Konno, C., Hikino, H., 1987. Isolation and hypoglycemic activity of panaxans M, N, O and P, glycans of *Panax ginseng* roots. Int. J. Crude Drug Res. 25, 53–56.

Lee, J.-G., Hsieh, W.-T., Chen, S.-U., Chiang, B.-H., 2012. Hematopoietic and myeloprotective activities of an acidic *Angelica sinensis* polysaccharide on human CD34+ stem cells. J. Ethnopharmacol. 139, 739–745.

Lee, J.-H., Lee, J.S., Chung, M.-S., Kim, K.H., 2004. In vitro anti-adhesive activity of an acidic polysaccharide from *Panax ginseng* on *Porphyromonas gingivalis* binding to erythrocytes. Planta Med. 70, 566–568.

Lee, J.-H., Shim, J.S., Lee, J.S., Kim, M.-K., Chung, M.-S., Kim, K.H., 2006. Pectin-like acidic polysaccharide from *Panax ginseng* with selective antiadhesive activity against pathogenic bacteria. Carbohydr. Res. 341, 1154–1163.

Lee, K.Y., Jeon, Y.J., 2005. Macrophage activation by polysaccharide isolated from *Astragalus membranaceus*. Int. Immunopharmacol. 5, 1225–1233.

Lee, Y., Chung, I., Lee, I., Kim, K., Hong, W., Yun, Y., 1996. Activation of multiple effector pathways of immune system by the antineoplastic immunostimulator acidic polysaccharide ginsan isolated from *Panax ginseng*. Anticancer Res. 17, 323–331.

Lei, T., Li, H., Fang, Z., Lin, J., Wang, S., Xiao, L., et al., 2014. Polysaccharides from *Angelica sinensis* alleviate neuronal cell injury caused by oxidative stress. NRR 9, 260.

Leung, M., Liu, C., Koon, J., Fung, K., 2006. Polysaccharide biological response modifiers. Immunol. Lett. 105, 101–114.

Li, C., Tian, Z.-N., Cai, J.-P., Chen, K.-X., Zhang, B., Feng, M.-Y., et al., 2014. *Panax ginseng* polysaccharide induces apoptosis by targeting Twist/AKR1C2/NF-1 pathway in human gastric cancer. Carbohydr. Polym. 102, 103–109.

Li, J., Bao, Y., Lam, W., Li, W., Lu, F., Zhu, X., et al., 2008. Immunoregulatory and anti-tumor effects of polysaccharopeptide and Astragalus polysaccharides on tumor-bearing mice. Immunopharmacol. Immunotoxicol. 30, 771–782.

Li, R., Chen, W.-C., Wang, W.-P., Tian, W.-Y., Zhang, X.-G., 2009a. Extraction, characterization of Astragalus polysaccharides and its immune modulating activities in rats with gastric cancer. Carbohydr. Polym. 78, 738–742.

Li, X.-T., Chen, R., Jin, L.-M., Chen, H.-Y., 2009b. Regulation on energy metabolism and protection on mitochondria of *Panax ginseng* polysaccharide. Am. J. Chin. Med. 37, 1139–1152.

Li, S.-G., Zhang, Y.-Q., 2009. Characterization and renal protective effect of a polysaccharide from *Astragalus membranaceus*. Carbohydr. Polym. 78, 343–348.

Li, S.-Y., Yang, D., Yeung, C.-M., Yu, W.-Y., Chang, R.C.-C., So, K.-F., et al., 2011. *Lycium barbarum* polysaccharides reduce neuronal damage, blood-retinal barrier disruption and oxidative stress in retinal ischemia/reperfusion injury. PLoS One 6, e16380.

Li, X., Li, X., Zhou, A., 2007a. Evaluation of antioxidant activity of the polysaccharides extracted from *Lycium barbarum* fruits in vitro. Eur. Polym. J. 43, 488–497.

Li, X., Ma, Y., Liu, X., 2007b. Effect of the *Lycium barbarum* polysaccharides on age-related oxidative stress in aged mice. J. Ethnopharmacol. 111, 504–511.

Li, X., Zhou, A., Li, X., 2007c. Inhibition of *Lycium barbarum* polysaccharides and *Ganoderma lucidum* polysaccharides against oxidative injury induced by γ-irradiation in rat liver mitochondria. Carbohydr. Polym. 69, 172–178.

Lim, D., Bae, K., Jung, I., Kim, C., Yun, Y., Song, J., 2002. Anti-septicaemic effect of polysaccharide from *Panax ginseng* by macrophage activation. J. Infect. 45, 32–38.

Lim, T.-S., Na, K., Choi, E.-M., Chung, J.-Y., Hwang, J.-K., 2004. Immunomodulating activities of polysaccharides isolated from *Panax ginseng*. J. Med. Food 7, 1–6.

Ling, X., Ding, H., Luo, S., Zhang, X., Zhang, L., 2001. Study on the pharmacological mechanisms of *Angelica polysaccharide* on the immunocompetence and the effects of its antioxidation. Chin. J. Hosp. Pharm. 22, 584–586.

Liu, C., Li, J., Meng, F.Y., Liang, S.X., Deng, R., Li, C.K., et al., 2010a. Polysaccharides from the root of *Angelica sinensis* promotes hematopoiesis and thrombopoiesis through the PI3K/AKT pathway. BMC Complement. Altern. Med. 10, 1.

Liu, P.-J., Hsieh, W.-T., Huang, S.-H., Liao, H.-F., Chiang, B.-H., 2010b. Hematopoietic effect of water-soluble polysaccharides from *Angelica sinensis* on mice with acute blood loss. Exp. Hematol. 38, 437–445.

Liu, J., Cheng, B., Zhang, J., Tang, X., Ji, Y., 1995. Effect of lycium polysaccharide on immune responses of cancer patients following radiotherapy. Chin. J. Radiol. Med. Protect. 16, 18–20.

Liu, J., Xu, C.-Y., Cai, S.-Z., Zhou, Y., Li, J., Jiang, R., et al., 2013. Senescence effects of *Angelica sinensis* polysaccharides on human acute myelogenous leukemia stem and progenitor cells. Asian Pac. J. Cancer Prev. 14, 6549–6556.

Liu, J.-Y., Zhang, Y., You, R.-X., Zeng, F., Guo, D., Wang, K.-P., 2012. Polysaccharide isolated from *Angelica sinensis* inhibits hepcidin expression in rats with iron deficiency anemia. J. Med. Food 15, 923–929.

Liu, Q.-Y., Yao, Y.-M., Yu, Y., Dong, N., Sheng, Z.-Y., 2011a. Astragalus polysaccharides attenuate postburn sepsis via inhibiting negative immunoregulation of CD4+ CD25 high T cells. PLoS One 6, e19811.

Liu, Q.-Y., Yao, Y.-M., Zhang, S.-W., Sheng, Z.-Y., 2011b. Astragalus polysaccharides regulate T cell-mediated immunity via CD11c high CD45RB low DCs in vitro. J. Ethnopharmacol. 136, 457–464.

Liu, S.-P., Dong, W.-G., Wu, D.-F., Luo, H.-S., Yu, J.-P., 2003. Protective effect of *Angelica sinensis* polysaccharide on experimental immunological colon injury in rats. World J. Gastroenterol. 9, 2786–2790.

Liu, Y., Liu, F., Yang, Y., Li, D., Lv, J., Ou, Y., et al., 2014. Astragalus polysaccharide ameliorates ionizing radiation-induced oxidative stress in mice. Int. J. Biol. Macromol. 68, 209–214.

Luo, D., Fang, B., 2008. Structural identification of ginseng polysaccharides and testing of their antioxidant activities. Carbohydr. Polym. 72, 376–381.

Luo, Q., Li, Z., Huang, X., Yan, J., Zhang, S., Cai, Y.-Z., 2006. *Lycium barbarum* polysaccharides: protective effects against heat-induced damage of rat testes and H 2 O 2-induced DNA damage in mouse testicular cells and beneficial effect on sexual behavior and reproductive function of hemicastrated rats. Life Sci. 79, 613–621.

Luo, Q., Li, Z., Yan, J., Zhu, F., Xu, R.-J., Cai, Y.-Z., 2009. *Lycium barbarum* polysaccharides induce apoptosis in human prostate cancer cells and inhibits prostate cancer growth in a xenograft mouse model of human prostate cancer. J. Med. Food 12, 695–703.

Ma, H., Lin, H., Feng, H., Putheti, R., 2009. Effects of *Angelica polysaccharide* on hepatocytes apoptosis induced by exhaustive exercise. Afr. J. Microbiol. Res. 3, 774–777.

Mao, F., Xiao, B., Jiang, Z., Zhao, J., Huang, X., Guo, J., 2011. Anticancer effect of *Lycium barbarum* polysaccharides on colon cancer cells involves G0/G1 phase arrest. Med. Oncol. 28, 121–126.

Mao, X.-Q., Wu, Y., Wu, K., Liu, M., Zhang, J.-F., Zou, F., et al., 2007. Astragalus polysaccharide reduces hepatic endoplasmic reticulum stress and restores glucose homeostasis in a diabetic KKAy mouse model. Acta. Pharmacol. Sin. 28, 1947–1956.

Miao, Y., Xiao, B., Jiang, Z., Guo, Y., Mao, F., Zhao, J., et al., 2010. Growth inhibition and cell-cycle arrest of human gastric cancer cells by *Lycium barbarum* polysaccharide. Med. Oncol. 27, 785–790.

Ming, M., Guanhua, L., Zhanhai, Y., Guang, C., Xuan, Z., 2009. Effect of the *Lycium barbarum* polysaccharides administration on blood lipid metabolism and oxidative stress of mice fed high-fat diet in vivo. Food Chem. 113, 872–877.

Mu, L., Zhu, L., Zhao, A., Zhou, M., Jia, W., 2009. Study on extraction and purification of polysaccharides from Astragalus membranaceus. J. Chin. Med. Mater. 32, 1741–1745.

Ni, W., Zhang, X., Wang, B., Chen, Y., Han, H., Fan, Y., et al., 2010. Antitumor activities and immunomodulatory effects of ginseng neutral polysaccharides in combination with 5-fluorouracil. J. Med. Food 13, 270–277.

Niu, A.-J., Wu, J.-M., Yu, D.-H., Wang, R., 2008. Protective effect of *Lycium barbarum* polysaccharides on oxidative damage in skeletal muscle of exhaustive exercise rats. Int. J. Biol. Macromol. 42, 447–449.

Niu, Y., Wang, H., Xie, Z., Whent, M., Gao, X., Zhang, X., et al., 2011. Structural analysis and bioactivity of a polysaccharide from the roots of *Astragalus membranaceus* (Fisch) Bge. var. mongolicus (Bge.) Hsiao. Food Chem. 128, 620–626.

Ovodov, Y.S., Solov'eva, T., 1966. Polysaccharides of *Panax ginseng*. Chem. Nat. Compd. 2, 243–245.

Park, K.M., Kim, Y.S., Jeong, T.C., Joe, C.O., Shin, H.J., Lee, Y.H., et al., 2001 Nitric oxide is involved in the immunomodulating activities of acidic polysaccharide from *Panax ginseng*. Planta Med. 67, 122–126.

Peng, X., Tian, G., 2001. Structural characterization of the glycan part of glycoconjugate LbGp2 from *Lycium barbarum* L. Carbohydr. Res. 331, 95–99.

Pérez, S., Mulloy, B., 2005. Prospects for glycoinformatics. Curr. Opin. Struct. Biol. 15, 517–524.

Qin, J., Liu, Y.S., Liu, J., Li, J., Tan, Y., Li, X.J., et al., 2013a. Effect of *Angelica sinensis* polysaccharides on osteoarthritis in vivo and in vitro: a possible mechanism to promote proteoglycans synthesis. Evid. Based Compl. Alt. Med. 2013, 15.

Qin, T., Chen, J., Wang, D., Hu, Y., Wang, M., Zhang, J., et al., 2013b. Optimization of selenylation conditions for Chinese *Angelica polysaccharide* based on immune-enhancing activity. Carbohydr. Polym. 92, 645–650.

Qiu, S., Chen, J., Chen, X., Fan, Q., Zhang, C., Wang, D., et al., 2014. Optimization of selenylation conditions for *lycium barbarum* polysaccharide based on antioxidant activity. Carbohydr. Polym. 103, 148–153.

Rui, L., Chen, W.-C., Wang, W.-P., Tian, W.-Y., Zhang, X.-G., 2009. Optimization of extraction technology of *Astragalus polysaccharides* by response surface methodology and its effect on CD40. Carbohydr. Polym. 78, 784–788.

Shang, P., Qian, A.R., Yang, T.H., Jia, M., Mei, Q.B., Cho, C.H., et al., 2003. Experimental study of anti-tumor effects of polysaccharides from *Angelica sinensis*. World J. Gastroenterol. 9, 1963–1967.

Shao, B.-M., Xu, W., Dai, H., Tu, P., Li, Z., Gao, X.-M., 2004. A study on the immune receptors for polysaccharides from the roots of *Astragalus membranaceus*, a Chinese medicinal herb. Biochem. Biophys. Res. Commun. 320, 1103–1111.

Shao, P., Zhao, L.-H., Pan, J.-P., 2006. Regulation on maturation and function of dendritic cells by *Astragalus mongholicus* polysaccharides. Int. Immunopharmacol. 6, 1161–1166.

Shen, L., Du, G., 2012. *Lycium barbarum* polysaccharide stimulates proliferation of MCF-7 cells by the ERK pathway. Life Sci. 91, 353–357.

Shin, H.J., Kim, Y.S., Kwak, Y.S., Song, Y.B., Park, J., 2004. Enhancement of antitumor effects of paclitaxel (taxol) in combination with red ginseng acidic polysaccharide (RGAP). Planta Med. 70, 1033–1038.

Shin, J.-Y., Song, J.-Y., Yun, Y.-S., Yang, H.-O., Rhee, D.-K., Pyo, S., 2002. Immunostimulating effects of acidic polysaccharides extract of *Panax ginseng* on macrophage function. Immunopharmacol. Immunotoxicol. 24, 469–482.

Shin, K.-S., Kiyohara, H., Matsumoto, T., Yamada, H., 1997. Rhamnogalacturonan II from the leaves of *Panax ginseng* CA Meyer as a macrophage Fc receptor expression-enhancing polysaccharide. Carbohydr. Res. 300, 239–249.

Solov'eva, T., Arsenyuk, L., Ovodov, Y.S., 1969. Some structural features of *Panax ginseng* CA Mey pectin. Carbohydr. Res. 10, 13–18.

Song, J.-Y., Han, S.-K., Son, E.-H., Pyo, S.-N., Yun, Y.-S., Yi, S.-Y., 2002. Induction of secretory and tumoricidal activities in peritoneal macrophages by ginsan. Int. Immunopharmacol. 2, 857–865.

Song, J.-Y., Han, S.-K., Bae, K.-G., Lim, D.-S., Son, S.-J., Jung, I.-S., et al., 2003. Radioprotective effects of ginsan, an immunomodulator. Radiat. Res. 159, 768–774.

Sun, X., Matsumoto, T., Yamada, H., 1994. Purification of immune complexes clearance enhancing polysaccharide from the leaves of *Panax ginseng*, and its biological activities. Phytomedicine 1, 225–231.

Sun, X.-B., Matsumoto, T., Kiyohara, H., Hirano, M., Yamada, H., 1991. Cytoprotective activity of pectic polysaccharides from the root of *Panax ginseng*. J. Ethnopharmacol. 31, 101–107.

Sun, Y., 2011. Structure and biological activities of the polysaccharides from the leaves, roots and fruits of *Panax ginseng* CA Meyer: an overview. Carbohydr. Polym. 85, 490–499.

Sun, Y., Tang, J., Gu, X., Li, D., 2005. Water-soluble polysaccharides from *Angelica sinensis* (Oliv.) Diels: preparation, characterization and bioactivity. Int. J. Biol. Macromol. 36, 283–289.

Sun, Y., Cui, S.W., Tang, J., Gu, X., 2010. Structural features of pectic polysaccharide from *Angelica sinensis* (Oliv.) Diels. Carbohydr. Polym. 80, 544–550.

Tian, M., Wang, M., 2006. Studies on extraction, isolation and composition of *Lycium barbarum* polysaccharides. China J. Chin. Mater. Med. 31, 1603–1607.

Tian, Q.-E., Li, H.-D., Yan, M., Cai, H.-L., Tan, Q.-Y., Zhang, W.-Y., 2012. Astragalus polysaccharides can regulate cytokine and P-glycoprotein expression in H22 tumor-bearing mice. World J. Gastroenterol. 18, 7079–7086.

Tomoda, M., Shimada, K., Konno, C., Hikino, H., 1985. Structure of panaxan B, a hypoglycaemic glycan of *Panax ginseng* roots. Phytochemistry 24, 2431–2433.

Tomoda, M., Takeda, K., Shimizu, N., Gonda, R., Ohara, N., Takada, K., et al., 1993. Characterization of two acidic polysaccharides having immunological activities from the root of *Panax ginseng*. Biol. Pharm. Bull. 16, 22–25.

Uemura, N., Okamoto, S., Yamamoto, S., Matsumura, N., Yamaguchi, S., Yamakido, M., et al., 2001. *Helicobacter pylori* infection and the development of gastric cancer. N. Engl. J. Med. 345, 784–789.

Wang, C., Chang, S., Inbaraj, B.S., Chen, B., 2010a. Isolation of carotenoids, flavonoids and polysaccharides from *Lycium barbarum* L. and evaluation of antioxidant activity. Food Chem. 120, 184–192.

Wang, J., Hu, Y., Wang, D., Liu, J., Zhang, J., Abula, S., et al., 2010b. Sulfated modification can enhance the immune-enhancing activity of *Lycium barbarum* polysaccharides. Cell Immunol. 263, 219–223.

Wang, J., Hu, Y., Wang, D., Zhang, F., Zhao, X., Abula, S., et al., 2010c. *Lycium barbarum* polysaccharide inhibits the infectivity of Newcastle disease virus to chicken embryo fibroblast. Int. J. Biol. Macromol. 46, 212–216.

Wang, J., Li, S., Fan, Y., Chen, Y., Liu, D., Cheng, H., et al., 2010d. Anti-fatigue activity of the water-soluble polysaccharides isolated from *Panax ginseng* CA Meyer. J. Ethnopharmacol. 130, 421–423.

Wang, J., Ito, H., Shimura, K., 1989. Enhancing effect of antitumor polysaccharide from Astragalus or Radix hedysarum on C3 cleavage production of macrophages in mice. Jpn. J. Pharmacol. 51, 432–434.

Wang, K.P., Chen, Z.X., Zhang, Y., Wang, P.P., Wang, J.H., Dai, L.Q., 2008. Molecular weight and proposed structure of the *Angelica sinensis* polysaccharide-iron complex. Chin. J. Chem. 26, 1068–1074.

Wang, K.-P., Zeng, F., Liu, J.-Y., Guo, D., Zhang, Y., 2011a. Inhibitory effect of polysaccharides isolated from *Angelica sinensis* on hepcidin expression. J. Ethnopharmacol. 134, 944–948.

Wang, Q.-K., Chen, C.-X., Guo, Y.-J., Zhao, H.-Y., Sun, J.-F., Ma, S., et al., 2011b. Dietary polysaccharide from *Angelica sinensis* enhanced cellular defence responses and disease resistance of grouper *Epinephelus malabaricus*. Aquac. Int. 19, 945–956.

Wang, S., Wang, Y., Zheng, M., Jiang, R., 2000. Study on effect of *Angelica polysaccharide* on modulation of human CFU-GM. Acta Anatomica Sinica 32, 241–245.

Wang, S.C., Shan, J.J., Wang, Z.T., Hu, Z.B., 2006. Isolation and structural analysis of an acidic polysaccharide from *Astragalus membranaceus* (Fisch.) Bunge. J. Integr. Plant Biol. 48, 1379–1384.

Wang, Y., Wang, T., Ding, L., 2012. Optimization of polysaccharide extraction from *Angelica sinensis* using response surface methodology. Food Sci. 33, 146–149.

Wang, Y., Zhou, K., Jiang, R., 2003. Effect of *Ginseng polysaccharide* on the proliferation and differentiation of granulocyte-monocyte progenitor cells. Basic Med. Sci. Clin. 24, 52–55.

Wang, Y., Zhu, B., 1996. The effect of *Angelica polysaccharide* on proliferation and differentiation of hematopoietic progenitor cell. Zhonghua yi xue za zhi 76, 363–366.

Wang, Z., Liu, Y., Sun, Y., Mou, Q., Wang, B., Zhang, Y., et al., 2014. Structural characterization of LbGp1 from the fruits of *Lycium barbarum* L. Food Chem. 159, 137–142.

Wen, Y., Li, J., Tan, Y., Qin, J., Xie, X., Wang, L., et al., 2014. *Angelica sinensis* polysaccharides stimulated UDP-sugar synthase genes through promoting gene expression of IGF-1 and IGF1R in chondrocytes: promoting anti-osteoarthritic activity. PLoS One 9.

Wu, Y., Wu, K., Wang, Y., 2005. Hypoglycemic effect of *Astragalus polysaccharide* and its effect on PTP1B1. Acta Pharmacol. Sin. 26, 345–352.

Wu, Y.J., Wang, Y.Q., 2010. Inhibition of *Astragalus membranaceus* polysaccharides against liver cancer cell HepG2. Afr. J. Microbiol. Res. 4, 2181–2183.

Xie, F.-Y., Zeng, Z.-F., Huang, H.-Y., Zhao, C., Lu, T.-X., 2001. Clinical observation on nasopharyngeal carcinoma treated with combined therapy of radiotherapy and ginseng polysaccharide injection. Chin. J. Integ. Tradit. Western Med. 7, 273–276.

Xie, J., Wu, J., Mehendale, S., Aung, H., Yuan, C.-S., 2004. Anti-hyperglycemic effect of the polysaccharides fraction from American ginseng berry extract in ob/ob mice. Phytomedicine 11, 182–187.

Xin, Y.F., Wan, L.L., Peng, J.L., Guo, C., 2011. Alleviation of the acute doxorubicin-induced cardiotoxicity by *Lycium barbarum* polysaccharides through the suppression of oxidative stress. Food Chem. Toxicol. 49, 259–264.

Xu, D.-J., Xia, Q., Wang, J.-J., Wang, P.-P., 2008. Molecular weight and monosaccharide composition of *Astragalus polysaccharides*. Molecules 13, 2408–2415.

Yamada, H., Komiyama, K., Kiyohara, H., Cyong, J.-C., Hirakawa, Y., Otsuka, Y., 1990. Structural characterization and antitumor activity of a pectic polysaccharide from the roots of *Angelica acutiloba*. Planta Med. 56, 182–186.

Yan, F., Zhang, Q.-Y., Jiao, L., Han, T., Zhang, H., Qin, L.-P., et al., 2009. Synergistic hepatoprotective effect of Schisandrae lignans with Astragalus polysaccharides on chronic liver injury in rats. Phytomedicine 16, 805–813.

Yang, B., Xiao, B., Sun, T., 2013. Antitumor and immunomodulatory activity of Astragalus membranaceus polysaccharides in H22 tumor-bearing mice. Int. J. Biol. Macromol. 62, 287–290.

Yang, T., Jia, M., Shang, P., Mei, Q., Zhao, D., 2000. Synthesis of *Angelica sinensis* polysaccharide sulfate and their effects on splenocyte proliferation in vitro. J. Fourth Milit. Med. Univ. 22, 432–434.

Yang, T., Jia, M., Mei, Q., Shang, P., 2002. Effects of Angelica polysaccharide on blood coagulation and platelet aggregation. J. Chin. Med. Mater. 25, 344–345.

Yang, T., Jia, M., Meng, J., Wu, H., Mei, Q., 2006. Immunomodulatory activity of polysaccharide isolated from *Angelica sinensis*. Int. J. Biol. Macromol. 39, 179–184.

Yang, T., Jia, M., Zhou, S., Pan, F., Mei, Q., 2012. Antivirus and immune enhancement activities of sulfated polysaccharide from *Angelica sinensis*. Int. J. Biol. Macromol. 50, 768–772.

Yang, X., Zhao, Y., Wang, H., Mei, Q., 2007a. Macrophage activation by an acidic polysaccharide isolated from *Angelica sinensis* (Oliv.) Diels. J. Biochem. Mol. Biol. 40, 636–643.

Yang, X., Zhao, Y., Zhou, Y., Lv, Y., Mao, J., Zhao, P., 2007b. Component and antioxidant properties of polysaccharide fractions isolated from *Angelica sinensis* (OLIV.) DIELS. Biol. Pharm. Bull. 30, 1884–1890.

Yang, X., Zhao, Y., Li, G., Wang, Z., Lv, Y., 2008. Chemical composition and immunostimulating properties of polysaccharide biological response modifier isolated from Radix *Angelica sinensis*. Food Chem. 106, 269–276.

Yang, Y., Feng, J., Xu, H., Peng, D.-M., Hu, J.-P., 2010. Influence of different extraction and purification methods on astragalus polysaccharides and pharmacological evaluation. Chin. Herb. Med. 2, 54–61.

Ye, M., Chen, H., Zhou, R., Liao, M., 2011. Effects of *Astragalus polysaccharide* on proliferation and Akt phosphorylation of the basal-like breast cancer cell line. J. Chin. Integ. Med. 9, 1339–1346.

Ye, Y., Liu, E., Li, Y., So, H., Cho, C., Sheng, H., et al., 2001. Protective effect of polysaccharides-enriched fraction from *Angelica sinensis* on hepatic injury. Life Sci. 69, 637–646.

Yin, G., Dang, Y., 2008. Optimization of extraction technology of the *Lycium barbarum* polysaccharides by Box-Behnken statistical design. Carbohydr. Polym. 74, 603–610.

Yin, J.-Y., Chan, B.C.-L., Yu, H., Lau, I.Y.-K., Han, X.-Q., Cheng, S.-W., et al., 2012. Separation, structure characterization, conformation and immunomodulating effect of a hyperbranched heteroglycan from Radix Astragali. Carbohydr. Polym. 87, 667–675.

Yuan, S., Piao, X., Li, D., Kim, S., Lee, H., Guo, P., 2006. Effects of dietary *Astragalus polysaccharide* on growth performance and immune function in weaned pigs. Anim. Sci. 82, 501–507.

Yun, T.-K., 1996. Experimental and epidemiological evidence of the cancer-preventive effects of *Panax ginseng* CA Meyer. Nutr. Rev 54, S71.

Zhang, J., Jia, S., Liu, Y., Wu, S., Ran, J., 2011. Optimization of enzyme-assisted extraction of the *Lycium barbarum* polysaccharides using response surface methodology. Carbohydr. Polym. 86, 1089–1092.

Zhang, M., Chen, H., Huang, J., Li, Z., Zhu, C., Zhang, S., 2005. Effect of *Lycium barbarum* polysaccharide on human hepatoma QGY7703 cells: inhibition of proliferation and induction of apoptosis. Life Sci. 76, 2115–2124.

Zhang, M., Zhang, S., 2007. Study on structure of *Lycium barbarum* L. polysaccharide. Food Res. Devel. 28, 74–77.

Zhang, N., Li, J., Hu, Y., Cheng, G., Zhu, X., Liu, F., et al., 2010a. Effects of astragalus polysaccharide on the immune response to foot-and-mouth disease vaccine in mice. Carbohydr. Polym. 82, 680–686.

Zhang, S., He, B., Ge, J., Li, H., Luo, X., Zhang, H., et al., 2010b. Extraction, chemical analysis of *Angelica sinensis* polysaccharides and antioxidant activity of the polysaccharides in ischemia–reperfusion rats. Int. J. Biol. Macromol. 47, 546–550.

Zhang, X., Liu, J., Xu, C., Wei, Q., Li, J., Wang, L., et al., 2013a. Effect of *Angelica sinensis* polysaccharide on expression of telomere, telomerase and P53 in mice aging hematopoietic stem cells. China J. Chin. Mater. Med. 38, 2354–2358.

Zhang, X., Wang, Q., Chen, B., Wei, Q., Xu, C., Jiang, R., et al., 2013b. *Angelica sinensis* polysaccharides delay aging of hematopoietic stem cells through inhibitting oxidative damge. China J. Chin. Mater. Med. 38, 407–412.

Zhang, X., Yu, L., Bi, H., Li, X., Ni, W., Han, H., et al., 2009. Total fractionation and characterization of the water-soluble polysaccharides isolated from *Panax ginseng* CA Meyer. Carbohydr. Polym. 77, 544–552.

Zhang, Y., Li, M.-M., Zeng, F., Yao, C., Wang, K.-P., 2012. Study to establish the role of JAK2 and SMAD1/5/8 pathways in the inhibition of hepcidin by polysaccharides from *Angelica sinensis*. J. Ethnopharmacol. 144, 433–440.

Zhao, L., Wang, Y., Shen, H.-L., Shen, X.-D., Nie, Y., Wang, Y., et al., 2012. Structural characterization and radioprotection of bone marrow hematopoiesis of two novel polysaccharides from the root of *Angelica sinensis* (Oliv.) Diels. Fitoterapia 83, 1712–1720.

Zhao, L.-H., Ma, Z.-X., Zhu, J., Yu, X.-H., Weng, D.-P., 2011. Characterization of polysaccharide from *Astragalus radix* as the macrophage stimulator. Cell Immunol. 271, 329–334.

Zhao, R., Hao, W., Ma, B., Chen, Z., 2015. Improvement effect of *Lycium barbarum* polysaccharide on sub-health mice. Iran. J. Basic Med. Sci. 18, 1245–1252.

Zheng, M., Wang, Y., 2002. Experimental study on effect of *Angelica polysaccharide* in inhibitory proliferation and inducing differentiation of K562 cells. Chin. J. Integ. Tradit. West. Med. 22, 54–57.

Zhong, Z.-S., Sun, L.-J., Feng, Y., Mu, X.-M., 2006. Study on the extraction of polysaccharides from ginseng with ultrasonic wave. Chem. Bioeng. 6, 013.

Zhu, J., Zhao, L.-H., Zhao, X.-P., Chen, Z., 2007. *Lycium barbarum* polysaccharides regulate phenotypic and functional maturation of murine dendritic cells. Cell Biol. Int. 31, 615–619.

Zhu, J., Liu, W., Yu, J., Zou, S., Wang, J., Yao, W., et al., 2013a. Characterization and hypoglycemic effect of a polysaccharide extracted from the fruit of *Lycium barbarum* L. Carbohydr. Polym. 98, 8–16.

Zhu, J., Zhang, Y., Shen, Y., Zhou, H., Yu, X., 2013b. *Lycium barbarum* polysaccharides induce toll-like receptor 2-and 4-mediated phenotypic and functional maturation of murine dendritic cells via activation of NF-κB. Mol. Med. Report 8, 1216–1220.

Zhu, Z.-Y., Liu, R.-Q., Si, C.-L., Zhou, F., Wang, Y.-X., Ding, L.-N., et al., 2011. Structural analysis and anti-tumor activity comparison of polysaccharides from *Astragalus*. Carbohydr. Polym. 85, 895–902.

Zou, S., Zhang, X., Yao, W., Niu, Y., Gao, X., 2010. Structure characterization and hypoglycemic activity of a polysaccharide isolated from the fruit of *Lycium barbarum* L. Carbohydr. Polym. 80, 1161–1167.

Chapter 11

Practical Applications of Bioactive Polysaccharides

11.1 BIOACTIVE POLYSACCHARIDES AS CLINICAL DRUGS AND MEDICINES

It is widely accepted that naturally derived polysaccharides have many bioactivities, such as antitumor activity, immunomodulatory activity, gastrointestinal benefit activity, liver protection activity, hypoglycemic function, hypolipidemic function, antivirus activity, and so on (Zong et al., 2012). The unique structural features and biological properties of polysaccharides can be utilized successfully in various pharmacological, nutraceutical, and functional applications. Therefore, some of the bioactive polysaccharides have been made into clinical drugs or medicines for the treatment of certain diseases. One advantage for the clinical application of polysaccharides is that these carbohydrates are noncytotoxic to body cells. These bioactive polysaccharides have also been approved for different commercial products in the form of injections, granules, tablets, capsules, etc.

11.1.1 Anticancer Drugs

Most of the bioactive polysaccharides have shown promising potential as anticancer agents. It has been found that the β-D-glucan obtained from the cell wall of fungi is more likely to have anticancer activity (Dong and Fang, 2001). Certain polysaccharides, like lentinan, *Schizophyllum commune* polysaccharide, and polysaccharide krestin have already been used in cancer immunotherapy. The polysaccharides' anticancer activity can be mainly mediated through three approaches: (1) direct cytotoxicity, (2) immune-regulatory ability, and (3) synergistic effects in combination treatment with anticancer drugs (Zong et al., 2012).

Direct cytotoxicity includes the interference of polysaccharides with cancer induction, growth, and progression, such as inducing cellular apoptosis and cell cycle arrest, and inhibiting tumor invasion, adhesion, and metastasis (Zong et al., 2012). An example is fucoidans. These are known to have antitumor effects, but the active mechanism is not fully understood. Alekseyenko et al. (2007) administered 10 mg/kg of fucoidan to mice with

Bioactive Polysaccharides. DOI: http://dx.doi.org/10.1016/B978-0-12-809418-1.00011-3

transplanted Lewis lung adenocarcinoma, and found that it had antitumor and antimetastatic effects (Li et al., 2008b). These polyanionic polysaccharides exhibited antiangiogenesis, and antiproliferation for tumor cells. They could inhibit tumor growth and decrease tumor size (Ellouali et al., 1993; Li et al., 2008a), restrain tumor cell adhesion to various substrata (Liu et al., 2005), and had direct anticancer effects on human HS-Sultan cells through caspase and ERK pathways (Aisa et al., 2005).

Immunoenhancement ability is considered as another main mechanism of some types of polysaccharides, such as β-glucans from fungi, or fucoidans (Zong et al., 2012). Take fucoidans as an example. They can also inhibit the development and diffusion of tumor cells by improving the body's immunomodulatory activities. Fucoidan induces tumor destruction by activating type 1 T helper (Th1) cell and NK cell responses (Maruyama et al., 2007). What is more, fucoidan has the potential to alleviate the toxic effect of cyclophosphamide at a dose of 25 mg/kg (Alekseyenko et al., 2007). Bioactive polysaccharides can stimulate the maturation, differentiation, and reproduction of various immune cells; balance the body's immune system; and engulf the tumor cells. Even if the immunoregulatory activity of polysaccharide has a close relationship with its antitumor activity, there are still some differences between the two activities (Dong and Fang, 2001). For example, the more complex plant polysaccharides, like arabinogalactan and rhamnogalacturonan, usually appear to have better immunostimulatory activities; however, their antitumor activity usually appears to be lower than that of fungal polysaccharides with a simpler structure. This indicates that the polysaccharides with immunomodulatory abilities might not have definite antitumor abilities as well, the antitumor ability may have some other requirements.

Synergistic studies of polysaccharides with other chemotherapeutic drugs have shown distinct improvements in the anticancer potential toward cancer treatment compared with single agents (Zong et al., 2012). This process is mediated by enhancing the sensitivity of tumors and enhancing immune response to the treatments. For example, the combined effects of ling-zhi polysaccharide fraction 3 (LZP-F3) and anticancer drugs indicated that LZP-F3 enhanced the cytotoxicity of chemotherapeutic agents in N/P(14) and N/As(0.5) cells, and perhaps involved Fas/FasL-mediated apoptosis through the signal pathway of cytochrome c release and caspase-8 and -3 activation. The potential ability of LZP-F3 was associated with p21 expression and subsequent induction of cell cycle G1 arrest in cancer cells. Thus, induction of p21 appeared to play a key role in the ability to inhibit cell cycle progression and suppress tumor cell growth (Huang et al., 2010a,b).

11.1.2 Immunomodulatory Drugs

As was mentioned before, a majority of polysaccharides have immunomodulatory activity. Therefore, they have been used in the pharmaceutical area.

Fucoidan, for instance, stimulates the immune system in several ways, most of which are largely related to their ability to modify cell surface properties (Usov et al., 2001). It has been reported that consumption of fucoidans could directly inhibit viral replication, and stimulate the innate and adaptive immune systems (Hayashi et al., 2008). In addition, fucoidan has been found to have the ability to recover the immunomodulatory activity of immune-suppressed mice and irradiated rats. It could directly stimulate macrophage, T lymphocyte, B cell, and natural killer cells (NK cell; Wang et al., 1994), and induce the generation of interleukin (IL-1) and interferon-γ (IFN-γ) in vitro, as well as promoting the primary antibody response in sheep red blood cells in vivo (Yang et al., 1995).

11.1.3 Antivirus Drugs

Early in the 1970s, it was found that some of the polysaccharides and their derivatives have antiviral effects (Dong and Fang, 2001). Later in the 1980s, it was found that those polysaccharides with sulfuric acid groups even have inhibitory effects on the AIDS virus (human immunodeficiency virus (HIV)). The sulfated polysaccharides can express antiviral effects from two aspects: (1) the sulfated polysaccharides can bind to the virus binding site (CD4 receptor) on the surface of the T cells, which interferes with the invasion of the virus, and this combination may be carried out by electrostatic interaction; (2) reverse transcriptase (RT) is essential for the replication of the HIV virus, and some sulfated polysaccharides can inhibit the activity of RT. Therefore, it has been speculated that the inhibition effects of polysaccharides on RT should be one possible mechanism for its anti-HIV effects. However, the replication of HIV happens in the cell internally, but the polysaccharide cannot enter the cell. Therefore, this presumption remains to be discussed. Fucoidan also has anti-HIV effect. The mechanism of the antiviral activities of fucoidan is to inhibit viral absorption, so as to inhibit viral-induced syncytium formation (Mandal et al., 2007). Some polysaccharides have no anti-HIV activity before sulfation, however, they can express this bioactivity after sulfation. This indicates that sulfate is necessary for the antiviral activity. Sulfate located at C-4 of (1-3)-linked fucopyranosyl units in fucoidan appears to be quite important for its antiherpetic activity (Mandal et al., 2007). It is reported that fucoidan could inhibit the infection of HIV, herpes simplex virus (Iqbal et al., 2000; Mandal et al., 2007; Witvrouw and DeClercq, 1997), poliovirus III, adenovirus III, ECH06 virus, coxsackie B3 virus, coxsackie A16 (Li et al., 1995), cytomegalovirus, and bovine viral diarrhea virus (Iqbal et al., 2000).

11.1.4 Hypoglycemic Drugs

Since the 1980s, with extensive research on the biological activity of polysaccharides, people have found that many polysaccharides in pharmacological experiments have a hypoglycemic effect on normal mice or drug-induced hyperglycemic mice (Dong and Fang, 2001). Zhu et al. (2014) designed an experiment to investigate the hypoglycemic and hypolipidemic effects of *Ganoderma atrum* polysaccharide (PSG-1) on type II diabetic rats, as well as the protective effects and probable mechanisms of PSG-1 on diabetic nephropathy, diabetes-induced liver injury, and endothelial dysfunction in diabetic rat aorta. The results indicated that PSG-1 has potential for the treatment of hyperglycemia, hyperlipidemia, hyperinsulinemia, and insulin resistance, as well as a protective effect on kidney injury in type II diabetes.

11.1.5 Anticoagulant Drugs

Some polysaccharides from animals, like heparin, have been applied in clinical practice for thromboembolic disease, myocardial infarction, cardiac surgery, and cardiac catheterization (Bai and Ahsan, 2009). Heparin is an endogenous anticoagulant substance, belongs to the sulfated glycosaminoglycan containing group, has a lot of negative charge, and can inhibit the activity of thrombin (Dong and Fang, 2001). The sulfated polysaccharides extracted from kelp also showed a similar anticoagulant activity. It has been proved that most of the sulfated polysaccharides are obtained from kelp. Sulfated polysaccharides commonly exist in seaweed, while they are rarely found in herbs. Brown seaweeds are rich in sulfated polysaccharides that could potentially be exploited as functional ingredients for human health. The sulfate fucosan fractions isolated from brown seaweeds have different structural features, and their anticoagulant activity is closely related to the structural features, molecular weight, and the number of sulfuric acid groups (Church et al., 1989).

11.1.6 Ulcer Healing Drugs

It has been supported by much research that polysaccharides extracted from the cell wall of lactic acid bacteria (LAB) have antiulcer effects (Nagaoka et al., 1994). These polysaccharides can improve the levels of epidermal growth factor (EGF) and basic fibroblast growth factor (bFGF), which help to restore ulcers in the stomach. Moreover, it has been found that those polysaccharides with more rhamnose have higher bioactivity (Hokputsa et al., 2004). In addition, some mucopolysaccharides can form a layer of protective film on gastric epithelial tissue, which helps to relieve the symptoms of ulcers.

11.1.7 Bioactive Polysaccharides as Vaccinum

The bacterial cell wall or capsules often contain a variety of species-specific polysaccharides, which are usually bacterial epitopes (Dong and

Fang, 2001). Therefore, they can be made into different kinds of vaccinum, like Meningococcal AC polysaccharide vaccinum with encephalitis prevention effects, ACY-W135 polysaccharide vaccinum, *Streptococcus pneumonia* polysaccharide vaccinum, *Haemophilus influenza* Hib polysaccharide vaccinum, *Streptococcus agalactiae* polysaccharide vaccinum, and *Salmonella typhi* polysaccharide vaccinum. Capsular polysaccharide is a T cell-independent antigen. It can stimulate the differentiation and proliferation of mature B cells, and thus improve antibody production. Since newborn babies lack certain cells, children or patients with a low immune state often appear have no response to capsular polysaccharide vaccine.

In order to overcome the disadvantages of capsular antigen vaccine, in recent decades, the complex formed by coupling of polysaccharides and protein has aroused people's attention. The proteins used are often bacteriotoxin or outer membrane proteins, like diphtheria toxoid and tetanus toxoid, which are T cell-dependent antigens, these complexes can work well in newborn babies.

11.2 PRACTICAL APPLICATIONS IN THE FOOD INDUSTRY

Hydrocolloids are finding increasing applications as food additives in several food products as thickening and gelling agents (Saha and Bhattacharya, 2010). The thickening effects are mainly provided by carboxymethyl cellulose (CMC), methylcellulose and hydroxypropylmethyl cellulose, guar gum, locust bean gum, tara gum, konjac maanan, gum tragacanth, gum ghatti, and gum Arabic. Modified starch, agar, carrageenans, pectins, gellan gum, and alginates, and methyl and hydroxypropylmethyl celluloses are several frequently applied gelling agents (Saha and Bhattacharya, 2010).

Due to its viscosity inducing and stabilizing properties, xanthan is probably the most commonly used bacterial polysaccharide as a food additive (Garcia-Ochoa et al., 2000). It is thermally stable, even under acidic conditions, and has good freeze-thaw stability, as well as excellent suspending properties (Soares et al., 2005). Therefore, xanthan has been applied as a viscosifier and stabilizer in sirups, sauces, dressings, bakery products, soft cheese, restructured meat, and so on.

Mannitol is another popular polysaccharide food additive. The United States, the United Kingdom, France, and Japan are the main centers of its production. It has been used in a variety of foods, candies, and chocolate-flavored compound coatings, because it can replace sucrose to make sugar-free compound coatings (Zumbe et al., 2001). For example, sugar-free chocolate with mannitol added instead of sugar is becoming especially popular for people with diabetes. In addition, it can also be used in chewable tablets because of its favorable feel in the mouth.

11.2.1 Applications in Food Preservation

One of the most popular applications of polysaccharide is to keep fruits and vegetables fresh. On the one hand, these polysaccharides have good film-forming properties, so that they can form a film on the surface of foods; on the other hand, their nonhygroscopic and chemically inert quality helps to keep moisture from being lost. For instance, mannitol can be used as food preservative to maintain proper moisture levels in foods so as to increase their shelf-life and stability (Zumbe et al., 2001).

Another good example is pullulan, which has already been applied in the fresh preservation of fruit, vegetables, and eggs. *Aureobasidium* pullulans can inhibit the accumulation of volatile basic nitrogen, which effectively protects the moisture in seafood products from being lost. In addition, pullulan–polyethylene films can be used to offer high water- and oxygen-resistance, and better rigidity and strength compared to expanded polystyrene films.

Chitosan-based films have attracted serious attention in food preservation and packaging technology (Aider, 2010). Chitosan, in particular, could produce chitosan – protein based films and chitosan – polysaccharide based films. Among the currently applied biopolymers, chitosan seems to be highly promising as an additive for bio-based active film preparation. This is mainly due to its particular physicochemical properties, short time biodegradability, biocompatibility with human tissues, antimicrobial activity, antifungal activity, and nontoxicity. A new food package with antimicrobial properties is urgently required in today's food industry. Thus, chitosan-based active packaging materials can be useful for improving food quality and preservation during processing and storage of different food materials.

Protein–polysaccharide complexes could also be used in the food preservation industry (Schmitt and Turgeon, 2011). Its interesting viscosifying and gelling properties are helpful in producing gel-like food products without a heating process. The interfacial properties of complexes and coacervates give us a new approach to stabilize dispersed food systems. The application of protein – polysaccharide complexes could lead to very promising structures, textures, and shelf-life stability. In addition, its application in protecting bioactive or sensitive molecules in food is also very attractive.

11.2.2 Edible Films

Most of the naturally derived hydrocolloids are nontoxic, and thus can be consumed together with preserved food. Therefore, some gelling polysaccharides can be applied to form edible coatings in the food industry. Novel galactomannans are commonly used as edible coatings for tropical fruits to extend their shelf-life. Different formulations of galactomannan and glycerol reflect diverse ranges of wettability, and the best values relate to the dispersive component of the fruits' surface (Cerqueira et al., 2009). Konjac glucomannan aqueous dispersion could be used as a food ingredient to overcome the negative effect of poor protein functionality of low-quality squid surimi, and achieve better gels from them (Iglesias-Otero et al., 2010). Furthermore, blended films formed by methycellulose or hydroxypropyl methylcellulose (HMC) with whey protein isolates show higher tensile strength and lower oxygen permeability than individual pure polysaccharide films (Yoo and Krochta, 2011).

11.2.3 Application in Beverages

The carbohydrate active enzymes (CAZymes) are not encoded in the human genome, but exit in the genes of bacteria. Therefore, the bacteria can utilize polysaccharides; meanwhile, the polysaccharides can balance the proliferation of probiotics and harmful bacteria. Ginseng polysaccharide can stimulate the proliferation and acid producing rate of *Lactobacillus bulgaricus* and *Streptococcus thermophilus* (Liang et al., 2013). Liang et al. (2013) developed a novel ginseng polysaccharide yogurt, which had a better texture and longer shelf-life compared with ordinary yogurt.

The application of bioactive polysaccharides can not only provide certain health benefit functions to the beverage, but can also improve the taste, due to its thickening properties (Paquin 1999). Soybean soluble polysaccharide and pectin have similar structural features, and can stabilize the suspension of protein particles. Thus, they have been commonly used as a stabilizer in protein drinks (Nakamura et al., 2006). Exopolysaccharides (EPSs) produced by LAB play an important role in the improvement of the physical properties of fermented dairy products (Yang et al., 2010). The use of EPSs in the dairy industry can help improve the rheological properties of fermented milk products by increasing their water-holding capacity and viscosity. Carrageenan is also frequently added in beverages, as it can function as an emulsifier, gelling agent, stabilizer, or thickener, and therefore modify the flowability and appearance of the food (Prajapati et al., 2014).

Beta-lactoglobulin polysaccharide complexes can be used as nanovehicles for hydrophobic nutraceuticals in nonfat foods and clear beverages, which allows the stabilization, protection, and delivery of hydrophobic bioactive compounds in clear nonfat aqueous systems (Ron et al., 2010). The bioactive compounds will bind within a carrier protein, and be covered with a secondary protective polyionic hydrocolloid of the opposite charge. Ron et al. (2010) have successfully entrapped VD_2, as a model HN, in electrostatically stable nanoparticles of about 50–70 nm in size, comprised of beta-Lg (to which the HN is bound) and low molecular mass, low-methoxyl pectin. In this case, the vitamin is significantly protected from degradation compared to the unprotected vitamin dispersion. The nanocomplex systems formed were transparent, and were more suitable to be applied in clear acid drinks and other nonfat acid beverages, or foods containing health promoting hydrophobic nutrients, like vitamin D.

11.2.4 Application in Fried Foods

Hydrocolloids are the principal category of food ingredients and additives that have been used in the past 40 years to improve fried foods (Varela and Fiszman, 2011). Hydrocolloids play two main roles in fried food processes. One is to form a fine "invisible" coating on the surface of the food ingredients, so as to avoid excessive oil absorption during the prefrying or frying processes. When they are added to the batter among its other ingredients, they are also used to avoid oil absorption (Varela and Fiszman, 2011). Moreover, they also act as viscosity control agents, improve adhesion, pick-up control, and freeze-thaw stability, or help to retain the crispness of the battered/breaded fried foods (Varela and Fiszman, 2011).

Gellan gum can be used as a water and oil barrier in a fried pastry. However, it is reported that gellan could increase the water loss by 100% as compared to the control, while the cellulose derivatives methylcellulose (MC) and HPMC decreased it by 30% (Williams and Mittal, 1999). The thickness of the gellan gum film is associated with the fat and moisture barrier. As the film thickness increased, both diffusivities increased. When comparing the application of tragacanth gum, guar gum, xanthan, and CMC films as oil barriers in potato chips, it was found that CMC had the best fat barrier performance (Garmakhany et al., 2008). Another comparison of pectin, alginate, and CMC as coatings in banana chips, discovered that CMC was the best oil barrier among the three. However, pectin had a very good performance, with no remarkable differences in fat content compared to CMC (Singthong and Thongkaew, 2009).

11.2.5 Application in Noodles, Bread, or Cake Making

Bioactive polysaccharides could also be used in producing novel noodles. On the one hand, the various biological activities of polysaccharides can improve the functionality of the noodles. On the other hand, its ability in thickening, stabilization, and gelling helps to improve the chewiness, elasticity, and texture of the noodles (Tan et al., 2008).

Red ginseng polysaccharide (RGP), as a food additive, has been used in bread making. The results described that 1.2% RGP added to the original bread recipe could significantly improve the sensory quality and water-binding capacity of the bread (Yu et al., 2012).

Xanthan gum can be used in bakery products to provide various functions, such as improving the volume and texture of gluten-free breads, increasing water-binding ability during baking, keeping freeze-thaw stability of refrigerated doughs, extending the products' shelf-life, replacing egg white in low calorie cakes, increasing flavor release, and reducing syneresis in creams and fruit fillings (Sharma et al., 2006). Dextrans from *Leuconostoc mesenteroides* or other lactic bacteria (e.g., *Lactobacillus curvatus*) are reported to be proposed as texturizers in bread, especially gluten-free bread, in which they could improve water-retaining capacity, elasticity, and the specific volume of bread (Ruhmkorf et al., 2012).

Pullulan can be used as a replacement for starch in pastas and baked products, for the reason that it can strengthen food consistency, moisture and gas retention, and dispersibility (Leathers, 2003). In frozen or heat-treated food like steamed foods and bakery products, scleroglucan can perform as a stabilizer and thickener (Viñarta et al., 2013).

11.2.6 Application in Meat Products

The addition of bioactive polysaccharide can improve the nutrient value of meat products, like selenium-rich meat products with selenic acid polysaccharide added, or they can produce certain functional meat products for special populations, like hyperlipidemia patients or diabetics. In addition, the application of polysaccharide in meat products can enhance the water retentiveness of meat, and at the same time suppress the deformation of protein at high temperatures (Kanatt et al., 2008).

Alginate can form coatings by evaporation of the solvent, electrolyte cross-linking (calcium), or injection of a water-miscible nonsolvent for alginate (Varela and Fiszman, 2011). It is commonly used in meat products to protect the meat from lipid oxidation, or to sacrifice itself for retarding dehydration. It has also been found to decrease shrinkage losses and retard off-odor formation. However, because of the high moisture of gelatinous alginate films, it is not necessary for controlling interphase migration in multilayered food systems. Low-methoxyl pectin solutions can be gelled by cross-linking with calcium ions, and the solvent can be evaporated to yield a gelatinous film. Gelatinous edible pectinate coatings have similar characteristics to alginate ones, and behave as sacrificing agents to retard moisture loss. Similarly, carrageenan coatings can also act as sacrificing agents. Enrobed meat products are more likely to be subject to microbial spoilage because of extra handling compared to noncoated products. Therefore, these gel materials can be applied in meat and precooked meat products to retard microbial growth in gel matrices containing antimicrobial agents, or used as oxygen barriers to delay lipid oxidation.

11.2.7 Application in Desserts

Apart from the previously mentioned areas, polysaccharides are also commonly applied in making desserts like ice cream and puddings.

Some of the principal uses of galactomannans are in thickening desserts, sorbets, ice creams, and low-energy fat substitutes, fruit-based water gels, hot milk puddings, coffee whiteners, baby milk formulations, seasonings, sauces, and soups. This broad range of applications reflects a number of different functional characteristics including high solution viscosity (like guar), stabilization of frozen systems (like guar or locust bean gum), and mixed gel formation with other polysaccharides and proteins (like locust bean gum).

Carrageenan can improve texture, thickness, and solubility of dairy products (McHugh, 2003). It successfully prevents separation and maintains texture in dairy products when added in small amounts, around 0.3% in milk gels, whipped cream, yogurt, and milkshakes, and 0.03% in frozen desserts and liquid milk products (Prajapati et al., 2014).

11.3 APPLICATIONS IN DRUG DELIVERY

11.3.1 Drug Carrier in Micro Particle Systems

Chitosans and pectins are natural polysaccharides which show great potential in drug delivery systems (Morris et al., 2010). Chitosan is a cationic natural polysaccharide obtained by deacetylation of chitin, the second most abundant polysaccharide in nature (Kalyan et al., 2010). Chitosan and its nanoparticles have gained more attention as drug delivery carriers because of their better stability, low toxicity, and simple and mild preparation methods. Recent applications of chitosan are in ophthalmic, nasal, sublingual, buccal, periodontal, gastrointestinal, colon-specific, vaginal, and transdermal drug delivery, and as mucosal-vaccine and gene carriers. Chitosans are a family of strongly polycationic derivatives of poly-*N*-acetyl-D glucosamine. This positive charge is very important in chitosan drug delivery systems, as it plays a very important role in muco-adhesion (adhesion to the mucosal surface). Other chitosan-based drug delivery systems involve complexation with ligands to form chitosan nanoparticles which can be used to encapsulate active compounds. For example, chitosan/chondroitin sulfate complex microcapsules were prepared to encapsulate the anticancer drug 5-fluorouracil (5-Fu) by an emulsion-chemical cross-linking method (Huang et al., 2010a,b). Infrared spectra indicated that there were electrostatic interactions between chitosan and chondroitin sulfate. The microcapsules were in a spherical form with a size range of 20–60 Lm. The 5-Fu was inside the microcapsules. The internal structure of the blank microcapsules is compact, while the internal structure of the microcapsules loaded with the drug was porous. The release performance of the microcapsules can be controlled by the degree of cross-linking, drug loading, and pH of the release medium.

Pectins are made of several structural elements, the most important of which are the homogalacturonan (HG) and type I rhamnogalacturonan (RG-I) regions. Pectin HG regions consist of poly-glacturonic acid residues which can be partially methyl esterified. Pectins with a degree of methyl esterification (DM) >50% are known as high methoxyl pectins, and consequently low methoxyl (LM) pectins have a DM <50%. LM pectins are of particular interest in drug delivery, as they can form gels with calcium ions (Ca^{2+}), which has potential applications especially in nasal formulations (Morris et al., 2010).

In the last decade, the use of polysaccharides, particularly chitosan and pectin, in drug delivery systems has become quite popular. It is obvious that both the polysaccharides either individually or in combination can perform very well. However, there are still many problems that remain to be solved. As to chitosans, there are mainly two questions. First, chitosans are soluble only at pH <6, therefore, it is hard to maintain their stability. Second, their construction into microparticles capable of surviving the large environmental variation between mouth and intestine for oral drug delivery is difficult. In order to solve these two issues, the degree of acetylation and molecular weight of the chitosan should be optimized. The low methoxy-pectin systems issues include: (1) optimal molecular weight and degree of esterification; (2) drug diffusivity; (3) interactions with mucosal tissues; and (4) stability (Morris et al., 2010).

There has been increasing interest in developing colon-targeted drug delivery systems during the recent decade, but only enteric-coated colonic tablets have been able to hit the market so far (Shukla and Tiwari, 2012). However, specific values of pH are required to activate this drug delivery system. Therefore, microflora-activated systems appear to be more promising, because the sudden increase in the microbial population and related enzyme activity in the colon reflect a noncontinuous issue independent of gastrointestinal transit time (Shukla and Tiwari, 2012).

Chemical modifications for polysaccharides are one possible way to postpone the release of drugs in the gut (Shukla and Tiwari, 2012). However, the kinetics of degradation and of solute release from hydrogels depend on various parameters and the physicochemical characteristics of drugs (Shukla and Tiwari, 2012). For these reasons, the formulation of a hydrogel designed for drug delivery in the colon depends on the properties of the drugs incorporated into the dosage forms. Combinations of polysaccharides and polymers that are either insoluble or soluble should be tested at colonic pH. The erosion and swelling of coatings all along the gastrointestinal tract, and the degradation of polysaccharides in the colon, are quite important in producing these combinations. Some in vitro studies seem to be very promising, but further investigation is still required. Moreover, in vivo studies must be carried out to confirm the promising results showed in vitro concerning these combined polymers. These special drugs appear to be promising therapeutic agents in treating inflammatory bowel disease, due to their ability to reduce the release of drugs at the target site of the colon. In addition, they can also reduce side-effects compared to parent drugs. These combinations can be formulated in a much easier manner, and many technical difficulties that occur in the preparation of other types of colon-specific delivery systems, like coated, multiple coated, systems, etc., can be avoided.

11.3.2 Tissue Repair

Polysaccharides form a class of materials which have generally been underutilized in the biomaterials field. In general, the ability for hydrogel formation is one of the most important properties of polysaccharides. Hydrogels can be formed using a number of mechanisms, and are strongly influenced by the types of monosaccharide involved, as well as the presence and properties of substituent groups (Suh and Matthew, 2000). Polysaccharide gel is formed generally in two ways: hydrogen bonded and ionic. Hydrogen-bonded gels are typical for agarose (thermal gellation) and chitosan (pH-dependent gellation), whereas ionic gels usually occur in alginates and carrageenans. However, the distinction is limited, since some charged polysaccharides exhibit hydrogen-bonded gel formation under neutral conditions. Proteoglycans are one of the major macromolecules found in articular cartilage. These molecules consist of a core protein and covalently attached GAG chains. The GAGs are long, unbranched heteropolysaccharides, composed of repeated disaccharide units, with the general structure described as $\{\text{uronic acid-amino sugar}\}_n$. These molecules can form hydrogel complexes with oppositely charged ionic polymers, particularly the cationic polysaccharide chitosan.

Chitosan is biocompatible and has bioactivities like antimicrobial and antifungal activities. This makes it popularly used in the biomedical area, especially in tissue growth, tissue repair, wound-healing, and bone regeneration aspects. Chitosan polymers are incorporated into hydrogels and microspheres which demonstrate good potential in delivery systems for drug proteins and genes. In dental medicine, chitosan is also used as a dressing for oral mucous wounds, and a tampon following radical treatment of maxillary sinusitis. In addition, it is being investigated as an absorbing membrane for periodontal surgery. Chitosan has various bioactivities and is advertised as a healthy food that is effective in the treatment of various disorders, arthritis, cancer, diabetes, hepatitis, etc. Chitosan and cellulose were used as examples of fibers with high, intermediate, and low bile acid-binding capacities, respectively (Kalyan et al., 2010).

11.4 APPLICATIONS IN AGRICULTURE

11.4.1 Feed Supplements

In recent decades, people's pursuit of high quality animal products with no hormone and pesticide residue contamination has become more and more anxious. Naturally derived polysaccharides are usually nontoxic, and have a great number of biological benefits like antiinflammation, antivirus, antioxidant, and so on. Therefore, they can be used as safe and functional feed additives (Shen et al., 2011).

In addition, it is accepted the many polysaccharides can work like "prebiotics," as they can modulate the gut microbiota community, by decreasing the proportion of harmful bacteria, and increasing the proportion of probiotics. For instance, the combined application of astragalus polysaccharides and probiotics in feed additives can improve the immunity of chicks. Another use of astragalus polysaccharides in pig feed shows that it can enhance the piglets' resistance to disease. In addition, epimedium polysaccharide is added into chicks' feed, while Krestin is added into fish feed to improve the immunity, or inhibit the lipid oxidant of the animals (Shen et al., 2011).

11.4.2 Application in Pesticides

Xanthan can stabilize the suspension of solid components in water solutions, emulsions, and multiphase liquid system stability, and therefore can modulate the rheological features of pesticides. In this way, it could improve the atomization performance and reduce insecticidal powder (Huang and Zhao, 2006).

REFERENCES

Aider, M., 2010. Review: chitosan application for active bio-based films production and potential in the food industry. Food Sci. Technol.-LEB. 43, 837–842.

Aisa, Y., Miyakawa, Y., Nakazato, T., Shibata, H., Saito, K., Ikeda, Y., et al., 2005. Fucoidan induces apoptosis of human HS-Sultan cells accompanied by activation of caspase-3 and down-regulation of ERK pathways. Am. J. Hematol. 78, 7–14.

Alekseyenko, T.V., Zhanayeva, S.Y., Venediktova, A.A., Zvyagintseva, T.N., Kuznetsova, T.A., Besednova, N.N., et al., 2007. Antitumor and antimetastatic activity of fucoidan, a sulfated polysaccharide isolated from the Okhotsk sea Fucus evanescens brown alga. Bull. Exp. Biol. Med. 143, 730–732.

Bai, S., Ahsan, F., 2009. Synthesis and evaluation of pegylated dendrimeric nanocarrier for pulmonary delivery of low molecular weight heparin. Pharm. Res. 26, 539–548.

Cerqueira, M.A., Lima, Á.M., Teixeira, J.A., et al., 2009. Suitability of novel galactomannans as edible coatings for tropical fruits. J. Food Eng. 94, 372–378.

Church, F.C., Meade, J.B., Treanor, R.E., Whinna, H.C., 1989. Antithrombin activity of fucoidan:the interaction of fucoidan with heparin cofactor II, antithrombin III and thrombin. J. Biol. Chem. 264, 3618–3623.

Dong, Q., Fang, J., 2001. Structural elucidation of a new arabinogalactan from the leaves of *Nerium indicum*. Carbohyd. Res. 332, 109–114.

Ellouali, M., Boissonvidal, C., Durand, P., Jozefonvicz, J., 1993. Antitumor-activity of low molecular-weight fucans extracted from brown seaweed *Ascophyllum nodosum*. Anticancer Res. 13, 2011–2019.

Garcia-Ochoa, F., Santos, V.E., Casas, J.A., Gomez, E., 2000. Xanthan gum: production, recovery, and properties. Biotechnol. Adv. 18, 549–579.

Garmakhany, A.D., Mirzaei, H.O., Nejad, M.K., Maghsudlo, Y., 2008. Study of oil uptake and some quality attributes of potato chips affected by hydrocolloids. Eur. J. Lipid Sci. Tech. 110, 1045–1049.

Hayashi, L., Yokoya, N.S., Ostini, S., Pereira, R.T.L., Braga, E.S., Oliveira, E.C., 2008. Nutrients removed by *Kappaphycus alvarezii* (Rhodophyta, Solieriaceae) in integrated cultivation with fishes in re-circulating water. Aquaculture 277, 185–191.

Hokputsa, S., Gerddit, W., Pongsamart, S., Inngjerdingen, K., Heinze, T., Koschella, A., et al., 2004. Water-soluble polysaccharides with pharmaceutical importance from Durian rinds (Durio zibethinus Murr.): isolation, fractionation, characterisation and bioactivity. Carbohyd. Polym. 56, 471–481.

Huang, C.Y., Chen, J.Y.F., Wu, J.E., et al., 2010a. Ling-Zhi polysaccharides potentiate cytotoxic effects of anticancer drugs against drug-resistant urothelial carcinoma cells. J. Agr. food Chem. 58, 8798–8805.

Huang, L., Sui, W., Wang, Y., et al., 2010b. Preparation of chitosan/chondroitin sulfate complex microcapsules and application in controlled release of 5-fluorouracil. Carbohyd. Polym. 80, 168–173.

Huang, X., Zhao, L., 2006. The research and application of bacterial extracellular polysaccharide. Shanxi Chem. Ind. 26, 10–13.

Iglesias-Otero, M.A., Borderías, J., Tovar, C.A., 2010. Use of Konjac glucomannan as additive to reinforce the gels from low-quality squid surimi. J. Food Eng. 101, 281–288.

Iqbal, M., Flick-Smith, H., McCauley, J.W., 2000. Interactions of bovine viral diarrhea virus glycoprotein E-rns with cell surface glycosaminoglycans. J. Gen. Virol. 81, 451–459.

Kalyan, S., Sharma, P.K., Garg, V.K., et al., 2010. Recent advancement in Chitosan based formulations and its pharmaceutical application. Der. Pharmacia. Sinica. 1, 195–210.

Kanatt, S.R., Chander, R., Sharma, A., 2008. Chitosan and mint mixture: a new preservative for meat and meat products. Food Chem. 107, 845–852.

Leathers, T.D., 2003. Biotechnological production and applications of pullulan. Appl. Microbiol. Biot. 62, 468–473.

Li, B., Lu, F., Wei, X.J., Zhao, R.X., 2008a. Fucoidan: structure and bioactivity. Molecules 13, 1671–1695.

Li, Y., Qian, Z.J., Le, Q.T., Kim, M.M., Kim, S.K., 2008b. Bioactive phloroglucinol derivatives isolated from an edible marine brown alga, *Ecklonia cava*. J. Biotechnol. 136, S578-S578.

Li, F., Tian, T.C., Shi, Y.C., 1995. Study on anti virus effect of fucoidan in vitro. J. Norman Bethune Univ. Med. Sci. 21, 255–257.

Liang, H., Zhang, T., Zhen, J., Guo, M., Wang, S., Liu, D., et al., 2013. Effect of ginseng polysaccharide on the fermentation characteristics of *lactic acid bacteria* and the texture of yogurt. China Dairy Ind. 41, 21–24.

Liu, R.M., Bignon, J., Haroun-Bouhedja, F., Bittoun, P., Vassy, J., Fermandjian, S., et al., 2005. Inhibitory effect of fucoidan on the adhesion of adenocarcinoma cells to fibronectin. Anticancer Res. 25, 2129–2133.

Mandal, P., Mateu, C.G., Chattopadhyay, K., Pujol, C.A., Damonte, E.B., Ray, B., 2007. Structural features and antiviral activity of sulphated fucans from the brown seaweed *Cystoseira indica*. Antivir. Chem. Chemoth. 18, 153–162.

Maruyama, H., Tanaka, M., Hashimoto, M., Inoue, M., Sasahara, T., 2007. The suppressive effect of Mekabu fucoidan on an attachment of *Cryptosporidium parvum* oocysts to the intestinal epithelial cells in neonatal mice. Life Sci. 80, 775–781.

McHugh, D.J., 2003. A Guide to the Seaweed Industry. Food and Agriculture Organization of the United Nations, Rome.

Morris, G.A., Kök, S.M., Harding, S.E., et al., 2010. Polysaccharide drug delivery systems based on pectin and chitosan. Biotechnol. Genet. Eng. 27, 257–284.

Nagaoka, M., Hasimoto, S., Watanabe, T., Yokokura, T., Mori, Y., 1994. Anti-ulcer effects of lactic acid bacteria and their cell wall polysaccharides. Biol. Pharm. Bull. 17, 1012–1017.

Nakamura, A., Yoshida, R., Maeda, H., Corredig, M., 2006. The stabilizing behaviour of soybean soluble polysaccharide and pectin in acidified milk beverages. Int. Dairy J. 16, 361–369.

Paquin, P., 1999. Technological properties of high pressure homogenizers: the effect of fat globules, milk proteins, and polysaccharides. Int. Dairy J. 9, 329–335.

Prajapati, V.D., Maheriya, P.M., Jani, G.K., et al., 2014. Carrageenan: a natural seaweed polysaccharide and its applications. Carbohyd. Polym. 105, 97–112.

Ron, N., Zimet, P., Bargarum, J., et al., 2010. Beta-lactoglobulin–polysaccharide complexes as nanovehicles for hydrophobic nutraceuticals in non-fat foods and clear beverages. Int. Dairy J. 20, 686–693.

Ruhmkorf, C., Rubsam, H., Becker, T., Bork, C., Voiges, K., Mischnick, P., et al., 2012. 'Effect of structurally different microbial homoexopolysaccharides on the quality of gluten-free bread. Europ. Food Res. Technol. 235, 139–146.

Saha, D., Bhattacharya, S., 2010. Hydrocolloids as thickening and gelling agents in food: a critical review. J. Food Sci. Technol. 47, 587–597.

Schmitt, C., Turgeon, S.L., 2011. Protein/polysaccharide complexes and coacervates in food systems. Adv. Colloid Interfac. 167, 63–70.

Sharma, B.R., Naresh, L., Dhuldhoya, N.C., Merchant, S.U., Merchant, U.C., 2006. Xanthan gum – a boon to food industry. Food Promot. Chronic. 1, 27–30.

Shen, L.H., Wang, J.S., Li, Y., 2011. Research and application of plant polysaccharide. Chin. Agri. Sci. Bull. 27, 349–352.

Shukla, R.K., Tiwari, A., 2012. Applications and recent advances in delivering drugs to the colon. Carbohyd. Polym. 88, 399–416.

Singthong, J., Thongkaew, C., 2009. Using hydrocolloids to decrease absorption in banana chips. Food Sci. Technol.-LEB. 42, 1199–1203.

Soares, R.M.D., Lima, A.M.F., Oliveira, R.V.B., Pires, A.T.N., Soldi, V., 2005. Thermal degradation of biodegradable edible films based on xanthan and starches from different sources. Polym. Degrad. Stabil. 90, 449–454.

Suh, J.K.F., Matthew, H.W., 2000. Application of chitosan-based polysaccharide biomaterials in cartilage tissue engineering: a review. Biomaterials 21, 2589–2598.

Tan, H.Z., Tan, B., Liu, M., Tian, X.H., Gu, W.Y., 2008. Mechanism of interaction between polysaccharide gums and sweet potato starch in production of its noodles. Chin. J. Food Sci. 5, 004.

Usov, A.I., Smirnova, G.P., Klochkova, N.G., 2001. Polysaccharides of algae: 55. Polysaccharide composition of several brown algae from Kamchatka. Russ. J Bioorganic Chem. 27, 395–399.

Varela, P., Fiszman, S.M., 2011. Hydrocolloids in fried foods. A review. Food Hydrocolloid 25, 1801–1812.

Viñarta, S.C., Yossen, M.M., Vega, J.R., Figueroa, L.I., Fariña, J.I., 2013. Scleroglucan compatibility with thickeners, alcohols and polyalcohols and downstream processing implications. Carbohyd. Polym. 92, 1107–1115.

Wang, W.T., Zhou, J.H., Xing, S.T., 1994. Immunomodulating action of marine algae sulfated polysaccharides on normal and immunosuppressed mice. Chinese J. Pharm. Toxicol. 8, 199–202.

Williams, R., Mittal, G.S., 1999. Water and fat transfer properties of polysaccharide films on fried pastry mix. Food Sci. Technol.-LEB. 32, 440–445.

Witvrouw, M., DeClercq, E., 1997. Sulfated polysaccharides extracted from sea algae as potential antiviral drugs. Gen. Pharmacol.-Vasc. S. 29, 497–511.

Yang, X.L., Sun, J.Y., Xu, H.N., 1995. An experimental study on immunoregulatory effect of fucoidan. Chin. J. Mar. Drugs 14, 9–13.

Yang, Z., Li, S., Zhang, X., et al., 2010. Capsular and slime-polysaccharide production by *Lactobacillus rhamnosus* JAAS8 isolated from Chinese sauerkraut: potential application in fermented milk products. J. Biosci. Bioeng. 110, 53–57.

Yoo, S.R., Krochta, J.M., 2011. Whey protein–polysaccharide blended edible film formation and barrier, tensile, thermal and transparency properties. J. Sci. Food Agr. 91, 2628–2636.

Yu, L., Yue, H., Liu, Y., et al., 2012. Effect of red ginseng polysaccharide on bread quality. Sci. Technol. Food Ind. 24, 078.

Zhu, K., Nie, S., Li, C., Gong, D., 2014. A newly identified polysaccharide from *Ganoderma atrum* attenuates hyperglycemia and hyperlipidemia. Carbohyd. Polym. 103, 520–527.

Zong, A., Cao, H., Wang, F., 2012. Anticancer polysaccharides from natural resources: a review of recent research. Carbohyd. Polym. 90, 1395–1410.

Zumbe, A., Lee, A., Storey, D., 2001. Polyols in confectionery: the route to sugar-free, reduced sugar and reduced calorie confectionery. Brit. J. Nutr. 85, S31.

Index

Note: Page numbers followed by "*f*" and "*t*" refer to figures and tables, respectively.

D

E

F

G

H

I

N

O

P

Q

R

S

X

Y

Z

Made in the USA
Monee, IL
13 September 2026

62977010R00312